GUIDELINES FOR
Evaluating Water in Pit Slope Stability

GUIDELINES FOR
Evaluating Water in Pit Slope Stability

EDITORS: GEOFF BEALE AND JOHN READ

 **CRC Press**
Taylor & Francis Group
Boca Raton London New York

CRC Press is an imprint of the
Taylor & Francis Group, an informa business
A BALKEMA BOOK

First published exclusively in Australia and New Zealand by
CSIRO Publishing

Published exclusively throughout the world (excluding Australia and New Zealand) by CRC Press/Balkema, with ISBN 978-0-367-57621-9
CRCPress/Balkema
P.O. Box 447, 2300 AK Leiden, The Netherlands
e-mail: Pub.NL@taylorandfrancis.com
www.crcpress.com – www.taylorandfrancis.com

First issued in paperback 2020

ISBN 13: 978-0-367-57621-9 (pbk)
ISBN 13: 978-1-138-00134-3 (hbk)

Visit the Taylor & Francis Web site at
http://www.taylorandfrancis.com

and the CRC Press Web site at
http://www.crcpress.com

National Library of Australia Cataloguing-in-Publication entry

Guidelines for evaluating water in pit slope stability/edited by John Read and Geoff Beale.

Includes bibliographical references and index.

Strip mining – Planning.
Strip mining – Design and construction.
Slopes (Soil mechanics)
Landslides.

Read, John Russell Lee, 1939– editor.
Beale, Geoff, 1954– editor.

622.292

Front cover: Diavik Diamond Mine, Northwest Territories, Canada (courtesy Diavik Diamond Mines Inc, Yellowknife, Canada)

Set in 10/12 Adobe Minion Pro and Optima
Cover and text design by James Kelly
Typeset by Desktop Concepts Pty Ltd, Melbourne
Index by Indexicana

Contents

Preface and acknowledgements — xiii

Introduction — 1
John Read, Geoff Beale, Marc Ruest and Martyn Robotham

1   Scope of LOP project hydrogeological studies — 1

2   General impact of water on mining — 4
    2.1   Water management issues — 4
    2.2   Consequences of mining below the water table — 6
    2.3   General goals for the water-control program — 6

3   Cost of managing water in slope stability — 8
    3.1   Introduction — 8
    3.2   Cost–benefit analysis — 8
    3.3   An example of managing early dewatering costs — 10
    3.4   An example of large-scale cost–benefit analysis for pit slope depressurisation — 13

4   Goals of managing water in slope stability — 14
    4.1   Opportunities — 14
    4.2   Passive pore pressure control — 14
    4.3   Active pore pressure control — 15
    4.4   Making the decision to implement an active program — 15

5   General planning for mine water management — 16

1   Framework: assessing water in slope stability — 19
Geoff Beale, Michael Price and John Waterhouse

1.1   Fundamental parameters — 19
    1.1.1   Porosity and storage properties — 19
    1.1.2   Permeability and transport properties — 24
    1.1.3   Pore pressure — 34
    1.1.4   Head and pressure conditions — 35
    1.1.5   Controls on pore pressure — 38
    1.1.6   The role of water pressure in slope stability — 41

1.2   The hydrogeological model — 45
    1.2.1   Basic regimes — 45
    1.2.2   Geology — 45
    1.2.3   Hydrology — 48
    1.2.4   Hydraulic controls — 49

1.3   Managing water in open pit mines — 49
    1.3.1   Key factors affecting the water-management program — 49
    1.3.2   General mine dewatering — 51
    1.3.3   Pit slope depressurisation and general mine dewatering — 52
    1.3.4   Steps required for implementing a slope depressurisation program — 56
    1.3.5   Mine water balance — 57
    1.3.6   Mine closure considerations — 58

2   Site characterisation   **65**
Greg Doubek, Ashley Creighton, Jeremy Dowling, Michael Price and Mark Hawley

2.1   **Planning field programs**   **65**
    2.1.1   Introduction   65
    2.1.2   Scale of the investigation   68
    2.1.3   Early-stage investigation   69
    2.1.4   Integrating the design process   69
    2.1.5   Required effort based on project level   73
    2.1.6   Planning for a Greenfield mine development   78
    2.1.7   Planning for a Brownfield site development   79
    2.1.8   Environmental baseline studies   80
    2.1.9   Water management practices during the field investigation program   81

2.2   **Implementing field programs**   **82**
    2.2.1   Background   82
    2.2.2   Drilling methods   83
    2.2.3   'Piggy-backing' of data collection   83
    2.2.4   Dedicated hydrogeological drilling programs   83
    2.2.5   Single-hole testing methods   88
    2.2.6   Monitoring installations   96
    2.2.7   Downhole geophysical logging   107
    2.2.8   Cross-hole and multi-hole testing   119
    2.2.9   Water quality testing   126
    2.2.10   Pilot drainage trials   128

2.3   **Presentation, analysis and storage of data**   **129**
    2.3.1   Types of data   129
    2.3.2   Display of time-series monitoring data   130
    2.3.3   Analysis of one-off data   140
    2.3.4   Levels of data analysis for a typical development program 145
    2.3.5   Databases   149

3   Preparing a conceptual hydrogeological model   **153**
Geoff Beale, Pete Milmo, Mark Raynor, Michael Price and Frederic Donzé

3.1   **Introduction**   **153**
    3.1.1   Background   153
    3.1.2   What is a conceptual model?   153
    3.1.3   Development of a sector-scale model   153
    3.1.4   Available data   155

3.2   **Components of the conceptual model**   **155**
    3.2.1   Components of a larger scale conceptual model   155
    3.2.2   The 'A-B-C-D' concept of fracture flow   156
    3.2.3   Components of the sector-scale conceptual model   157

3.3   **Research outcomes from Diavik**   **158**
    3.3.1   Background   158
    3.3.2   Diavik site setting   159
    3.3.3   Effects of blasting   166
    3.3.4   Influence of freeze-back   169
    3.3.5   Responses to changes in hydraulic stress   172
    3.3.6   Overall interpretation of the Diavik results   175

3.4   **Discrete Fracture Network (DFN) modelling**   **179**
    3.4.1   DFN development   179

3.4.2 Stochastic realisations of the DFN 180
3.4.3 The DFN as the basis for a groundwater flow model 180

3.5 **Summary of case studies** 181
3.5.1 Introduction 181
3.5.2 Diavik North-west wall: an interconnected rock mass that is strongly influenced by recharge and discharge boundaries 182
3.5.3 Escondida East wall: alteration in the fracture network and groundwater recharge from outside the pit crest 182
3.5.4 Chuquicamata, a very low-permeability system with little recharge or discharge 183
3.5.5 Antamina West wall: drainage of the slopes inhibited by structural barriers 183
3.5.6 Jwaneng East wall: poorly permeable but highly interconnected shale sequence 184
3.5.7 Cowal 184
3.5.8 Layered limestone sequence in Nevada, USA 184
3.5.9 Whaleback South wall 184

3.6 **Factors contributing to a slope-scale conceptual model** 185
3.6.1 Regimes 185
3.6.2 The influence of geology on the conceptual model 185
3.6.3 Hydrological input: recharge to the slope domain 189
3.6.4 Hydrological output: the role of discharge in slope depressurisation 192
3.6.5 Hydraulics 194
3.6.6 Deformation 201
3.6.7 Transient pore pressures 209

3.7 **Conclusions** 211
3.7.1 Key factors 211
3.7.2 Hydrogeological setting 211
3.7.3 Nature of the conceptual model 213

4   Numerical model 215
Loren Lorig, Jeremy Dowling, Geoff Beale and Michael Royle

4.1 **Planning a numerical model** 215
4.1.1 Background 215
4.1.2 Scale-specific application of the model 217
4.1.3 Focussing the model on the slope design process 220
4.1.4 General planning considerations 220
4.1.5 Timeframe and budget considerations 223
4.1.6 Modelling workflow 225
4.1.7 Data requirements and sources 225

4.2 **Development of numerical groundwater flow models** 229
4.2.1 Steps required for model development 229
4.2.2 Determining model geometry 231
4.2.3 Setting the model domain and boundaries 239
4.2.4 Defining the mesh or grid size 244
4.2.5 Determining whether to run steady-state, transient or undrained simulations 246
4.2.6 Determining whether the use of an equivalent porous medium (EPM) code is adequate 249
4.2.7 Selecting the appropriate time steps (stress periods) 251
4.2.8 Deciding whether a coupled modelling approach is required 252

4.2.9	Incorporating active drainage measures into the model	254
4.2.10	Calibrating the model	254
4.2.11	Interpreting model results	258
4.2.12	Validating model results	261
4.2.13	Using the model for operational planning	262

4.3 Use of pore pressures in numerical stability analyses — **262**

4.3.1	Background	262
4.3.2	How pore pressure modelling differs from stability analysis	264
4.3.3	Methods for inputting pore water pressure	264
4.3.4	Pore pressure profiles versus phreatic surface (water table) assumptions	265
4.3.5	Integration of the hydrogeology and geotechnical models	269
4.3.6	Model codes	271
4.3.7	Requirements for groundwater input to the slope design	272
4.3.8	Transferring output from the hydrogeological model to the geotechnical model	273
4.3.9	Input of transient pore pressures to the slope design model	275
4.3.10	Introducing *Slope Model*	277

5 Implementation of slope depressurisation systems — **279**

Geoff Beale, John De Souza, Rod Smith and Bob St Louis

5.1 Planning slope depressurisation systems — **279**

5.1.1	General factors for planning	279
5.1.2	Integration with mine planning	283
5.1.3	Development of targets	286

5.2 Implementing a groundwater-control program — **291**

5.2.1	Types of control systems	291
5.2.2	Passive drainage into the pit	295
5.2.3	Horizontal drain holes	296
5.2.4	Vertical and steep-angled drains	312
5.2.5	Design and installation of pumping wells	316
5.2.6	Drainage tunnels	333
5.2.7	Opening up drainage pathways by blasting	337
5.2.8	Protection of in-pit dewatering installations	338
5.2.9	Organisational structure	343

5.3 Control of surface water — **345**

5.3.1	Goals of the surface water management program	345
5.3.2	Sources of surface water	346
5.3.3	Control of surface water	346
5.3.4	Estimating flow rates	352
5.3.5	Control of recharge	353
5.3.6	In-pit stormwater management and maintenance	353
5.3.7	Maintenance of surface water control systems	355
5.3.8	Integration of in-pit groundwater and surface water management	357
5.3.9	Protection of the slope from erosion	360

6 Monitoring and design reconciliation — **363**

Chris Lomberg, Ian Ream, Rory O'Rourke and John Read

6.1 Monitoring — **363**

| 6.1.1 | Overview | 363 |
| 6.1.2 | Components of the monitoring system | 363 |

	6.1.3	Setting up monitoring programs	366
	6.1.4	Water level monitoring	370
	6.1.5	Telemetry	372
	6.1.6	Display of monitoring results	373
6.2	**Performance assessment**		**376**
	6.2.1	Overview	376
	6.2.2	Operational groundwater flow model	377
	6.2.3	Process for ongoing assessment	378
6.3	**Water risk management**		**379**
	6.3.1	Overview	379
	6.3.2	Process of risk analysis	379
	6.3.3	Risk assessment methodology	380
	6.3.4	Identifying the risks	383
	6.3.5	Defining the consequences	386
	6.3.6	Implementing a water risk management program	387
	6.3.7	Value of water risk management	389

Appendix 1   Hydrogeological background to pit slope depressurisation   391

1	Darcy's law	391
2	Head and pressure	391
3	Darcy's law in field situations	392
4	Flow in three dimensions	394

Appendix 2   Guidelines for field data collection and interpretation   398

1	Summary of drilling methods commonly used in mine hydrogeology investigations		398
	1.1	Direct push method	398
	1.2	Auger drilling	398
	1.3	Sonic drilling	398
	1.4	Cable tool drilling	398
	1.5	Rotary core drilling	398
	1.6	Conventional mud rotary drilling	399
	1.7	Conventional air/foam drilling	399
	1.8	Flooded reverse-circulation drilling	399
	1.9	Dual-tube reverse-circulation (RC) drilling with air	400
	1.10	Horizontal, angled and directional drilling	401
2	Standardised hydrogeological logging form for use with RC drilling		401
3	Interpretation of data collected while RC drilling		401
	3.1	Airlift pumping	401
	3.2	Submergence	403
	3.3	Examples of pilot hole comparison and data interpretation	405
4	Guidelines for drill-stem injection tests		408
5	Guidelines for running and interpreting hydraulic tests		410
	5.1	Single-hole variable-head tests	410
	5.2	Packer tests	412
	5.3	Pumping tests	418
6	Guidelines for the installation of grouted-in vibrating wire piezometer strings		424
	6.1	Drilling methods	424

	6.2	Depth setting for vibrating wire sensors	424
	6.3	Installation of multi-level VWPs using the guide-tremie pipe method	425
	6.4	Installation of multi-level VWPs using the wireline method	426
	6.5	Installation of VWP sensors in horizontal or positive inclined drill holes	429
	6.6	Installation of multi-level VWPs in underground boreholes	430
	6.7	Prefabricated multi-level VWP installations	430
	6.8	Commonly used grout mix	431
7	**Westbay multi-level system**		**432**
	7.1	Installation	432
	7.2	Operation	432

Appendix 3   Case study: Diavik North-west wall   **436**

1	**Background**		**436**
	1.1	Location	436
	1.2	The North-west wall, A154 Pit	436
	1.3	Climate	436
	1.4	Hydrogeological setting	436
	1.5	Depressurisation of the North-west wall	437
	1.6	Piezometer installations in the North-west wall	439
2	**Hydrograph analysis**		**440**
	2.1	Overview	440
	2.2	Analysis of specific events	443
3	**DFN modelling**		**456**
	3.1	DFN-based data analysis	456
	3.2	DFN model building	471
	3.3	DFN model validation	474
	3.4	Summary	482

Appendix 4   Case studies: Escondida East wall; Chuquicamata; Radomiro Tomic; Antamina West wall; Jwaneng; Cowal; Whaleback South wall; La Quinua (Yanacocha)   **486**

1	**Escondida East wall**		**486**
	1.1	Background	486
	1.2	Geology and hydrostratigraphy	486
	1.3	Dewatering and depressurisation	487
	1.4	Conceptual hydrogeological model	488
	1.5	Discussion	492
2	**Chuquicamata**		**493**
	2.1	Background	493
	2.2	Geology and hydrostratigraphy	493
	2.3	Drain hole results	496
	2.4	Conceptual hydrogeological model	496
	2.5	Discussion	500
3	**Radomiro Tomic**		**500**
	3.1	Background	500
	3.2	Response to mining pushbacks in the south-east area of the pit	500
	3.3	Piezometric responses in the West wall	501
	3.4	Responses in the northern sector of the West wall	504
4	**Antamina West wall**		**507**
	4.1	Background	507

4.2	Geology and hydrostratigraphy	507
4.3	Dewatering and depressurisation	509
4.4	Conceptual hydrogeological model	511
4.5	Discussion	512
5	**Jwaneng Diamond Mine South-east wall**	**513**
5.1	Background	513
5.2	Geology and hydrostratigraphy	513
5.3	Dewatering program	513
5.4	Slope depressurisation	516
5.5	Discussion	519
6	**Cowal Gold Mine (CGM)**	**520**
6.1	Introduction	520
6.2	Initial instability of the East high wall	520
6.3	Geology and geotechnical domains	520
6.4	Geotechnical considerations for the soft oxide zone	520
6.5	Hydrogeological results	522
6.6	Prediction of pore pressures	522
6.7	Depressurisation response	523
6.8	Dewatering well pumping and recovery test	524
6.9	Horizontal drainage response example for ultimate slope depressurisation	525
7	**Whaleback South wall**	**525**
7.1	Dewatering of the main ore body	525
7.2	Conditions in the South wall	526
7.3	Interpretation	528
8	**La Quinua, Peru**	**528**
8.1	General site setting	528
8.2	General mine dewatering	528
8.3	Phase 2C area	530
8.4	East high wall	530
8.5	Remediation plan	531

Appendix 5	**Cases studies for numerical modelling**	**536**
	Case study 1: Numerical modelling of the North-west wall of the Diavik A154 pit	536
1.1	Analysis of the DFN	536
1.2	Use of the EPM Model	548
1.3	Conclusions	555
	Case study 2: Numerical modelling of the marl sequence at the Cobre Las Cruces Mine, Andalucía, Spain	556
2.1	Background	556
2.2	Model development	558
2.3	Conclusions	567

Appendix 6	**The lattice formulation and the *Slope Model* code**	**568**
1	**The lattice formulation**	**568**
2	**Features of the lattice approach**	**569**
3	**Example application**	**570**
4	**Validation of *Slope Model* using experimental data from Coaraze test site**	**573**

4.1	Geometry	573
4.2	Boundary conditions	573
4.3	Building the model	574
4.4	Simulation results	575
4.5	Conclusions	575

Appendix 7 Lessons learnt and basic guidelines to monitoring for general dewatering **578**

1	**Introduction**	**578**
2	**Water level monitoring**	**578**
2.1	Reasons for monitoring	578
2.2	Water level measurement – practical guidelines	578
2.3	Monitoring frequency	580
2.4	Location of piezometers	580
2.5	Monitoring of pumping rates	581
2.6	Monitoring for hydrochemical fingerprinting	582
3	**Summary and conclusions**	**582**
	Symbols	583
	Abbreviations	584
	Glossary	585
	References	588
	Index	594

Preface and acknowledgements

Guidelines for Evaluating Water in Pit Slope Stability (the 'Water Book') is an outcome of the Large Open Pit (LOP) project, an international research and technology transfer project on the stability of rock slopes in open pit mines. It is a follow-up to Chapter 6 in the LOP project's previous publication *Guidelines for Open Pit Slope Design* (Read & Stacey, 2009), which outlined how the hydrogeological model is developed and applied in the design of rock slopes. It describes the outcomes of hydrogeological research performed by the LOP project since 2009 and has the objective of providing slope design practitioners with a road map that that will help them decide how to investigate and manage water pressures in pit slopes.

The LOP project was initiated in April 2005 and is managed on behalf of CSIRO Australia by John Read, CSIRO Earth Science and Resource Engineering, Brisbane, Australia. Since 2005 the project has been funded by the following mining companies: Anglo American plc; AngloGold Ashanti Limited; Barrick Gold Corporation; BHP Chile; BHPBilliton Innovation Pty Limited; Corporacion Naciónal de Cobre Del Chile ('Codelco'); Compañía Minera Doña Inés de Collahuasi SCM ('Collahuasi'); De Beers Group Services (Pty); Debswana Diamond Company; Newcrest Mining Limited; Newmont Australia Limited; Ok Tedi Mining Limited; Technological Resources Pty Ltd (the RioTinto Group); Teck Resources Limited; Vale; and Xstrata Copper Queensland. Debswana and Collahuasi ceased to be sponsors in 2010 and 2011 respectively, and Xstrata Copper Queensland ceased to be a sponsor in 2012; AngloGold Ashanti and Ok Tedi Mining joined the project in 2010, and Teck joined in 2011.

In addition to the research performed by the LOP project since 2009, principally by Itasca International Inc., Schlumberger Water Services, and CSIRO Earth Science and Resource Engineering, the book draws on the experience of the sponsors and a number of industry consultants and practitioners who have shared their knowledge and experience by contributing to several of the sections in the book. In particular, the efforts of the following people are gratefully acknowledged.

- Enrique Buschiazzo, Schlumberger Water Services, Lima, Peru
- Ned Clayton, Schlumberger Water Services, Tucson, Arizona, USA
- Sergio Cosio, Kumtor Operating Company, Kyrgystan
- Branko Damjanac, Itasca Consulting Group, Inc., Minneapolis, USA
- Christine Detournay, Itasca Consulting Group, Inc., Minneapolis, USA
- Frederic Donzé, Formerly CSIRO Earth Science and Resource Engineering, Brisbane, Australia
- Greg Doubek, Doubek Hydrologic Inc., Denver, USA
- Blair Douglas, BHPBilliton Iron Ore, Perth, Australia
- Jeremy Dowling, Schlumberger Water Services, Denver, USA
- John De Souza, AngloGold Ashanti, Johannesburg, RSA
- Greg Ghidotti, RioTinto Group, Superior, Arizona, USA
- Bart Harthong, Formerly CSIRO Earth Science and Resource Engineering, Brisbane, Australia
- Mark Hawley, Piteau Associates, Vancouver, Canada
- Jim Hazard, Itasca Consulting Group, Inc., Minneapolis, USA
- David Holmes, Schlumberger Water Services, Shrewsbury, UK
- Peter Knight, Newcrest Mining Limited, Lihir, Papua New Guinea
- Richard LeBreton, Diavik Diamond Mines Inc, Yellowknife, Canada
- Chris Lomberg, Aquastrat, Perth, Australia
- Paul Mansfield, VAI Groundwater Solutions, Santiago, Chile
- Gordon MacLear, AngloGold Ashanti, Dakar, Senegal
- Rowan McKittrick, Schlumberger Water Services, Shrewsbury, UK
- Peter Milmo, RioTinto Iron Ore, Simandou, Guinea
- Rory O'Rourke, Datum Monitoring, Bury, UK
- Brian Peck, Schlumberger Water Services, Reno, Nevada, USA
- Manuel Rapiman, Minera Escondida Ltd., Antofagasta, Chile
- Mark Raynor, SRK, Shrewsbury, UK
- Ian Ream, Schlumberger Water Services, Tucson, USA
- Martyn Robotham, the RioTinto Group, Brisbane, Australia
- Jorge Rodriguez, Schlumberger Water Services, Perth, Australia
- Steve Rogers, Golder Associates, Vancouver, Canada
- Michael Royle, SRK, Vancouver, Canada
- Patrick Rummel, De Beers, Canada
- Bob Sharon, Newmont Mining Corporation, Tucson, Arizona, USA
- Rod Smith, Smith Water Services, Vancouver, Canada
- Peter Stacey, Stacey Mining Geotechnical Limited, Vancouver, Canada
- Craig Stevens, RioTinto Technology and Innovation, Salt Lake City, Utah, USA
- Bob St. Louis, Newmont Mining, Elko, Nevada, USA

- Rick Tunney, Iam Gold, Suriname
- John Waterhouse, Golder Associates, Perth, Australia

The book has been edited by Geoff Beale (Schlumberger Water Services, Shrewsbury, UK) and John Read (CSIRO Earth Science and Resource Engineering, Brisbane, Australia), with the assistance of an editorial subcommittee comprising Ashley Creighton (the RioTinto Group, Brisbane, Australia), Loren Lorig (Itasca International, Inc., Minneapolis, USA), Marc Ruest (De Beers Global Mining, Johannesburg, RSA) and Michael Price (Independent Consultant, Church Stretton, UK).

Geoff Beale and John Read
August 2013

INTRODUCTION

John Read, Geoff Beale, Marc Ruest and Martyn Robotham

1 Scope of LOP project hydrogeological studies

The presence of water has a detrimental effect on slope stability. Water pressure acting in the pore spaces, fractures or other discontinuities in the materials that make up the pit slope will reduce the strength of those materials, and may therefore have a large influence on the performance, safety and economics of a mining operation.

The past 10 to 15 years has seen a significant improvement in the understanding of mining hydrogeology. Most mines now incorporate groundwater and surface water programs at all stages of the mining cycle. The programs are driven by both engineering and mine economics, and by increasing regulatory constraints, environmental laws and safety awareness. However, there remains a tendency to apply over-simplified approaches for assessing the distribution of pore water pressures in closely jointed rock masses and other pit slope materials. There is increasing awareness that an inadequate understanding of pore water pressures leads to mine production losses, either from conservatively over-designed slopes (i.e. slopes that are flatter than necessary), or under-designed slopes that fail.

In response to these issues, in 2009, the LOP project commenced a number of detailed hydrogeological studies to assess the effect of pore water pressures on the stability of jointed rock. The studies were designed to develop an understanding of the flow process in rock masses at different scales, particularly those with poor permeability and/or limited hydraulic connectivity between fractures, joints and other discontinuities. The objective was to develop a methodology that would help geotechnical design engineers decide how to treat water pressures in their pit walls.

The primary considerations for the studies have been as follows.

- The scale at which the water pressures are important for reducing rock strength, and how this relates to the grid and mesh sizes used in most geotechnical analyses.
- Whether water pressures are important throughout the rock mass or only in the major discontinuities.
- Whether analysis of the water pressure distribution can consider the rock mass as an equivalent continuum, so the flow analyses can employ some form of equivalent porous-medium approach.
- How slope deformation and dilation of the materials may change the water pressures within the rock mass and what are the resulting changes in the permeability of the fracture network.

For the geotechnical design engineer, a fundamental decision is whether to compute the water pressures only in the individual structures that cut through the rock mass; whether to compute the water pressures only in the intact rock within the intervening rock bridges (including the mélange of lesser joints, microfabric and intact rock substance included within the rock bridges); or whether to compute them in both.

To address these considerations, the hydrogeological studies focused on three core questions.

1. *How does a fractured rock slope depressurise*, which sought to define the role and the response of the various orders of fractures within the rock mass by:
   a. reviewing the literature from other industries with interests in the same or similar questions (e.g. nuclear, hydrocarbon and construction);
   b. analysing the structural and hydrogeological data from a number of large open pit mines to develop a conceptual model of how depressurisation occurred in fractured rock pit slopes for different rock types, alteration zones and structural settings, and in a range of different mine site environments.
2. *How should depressurisation be simulated numerically*, which sought to formulate a standardised

modelling approach for water pressure simulation for a range of environments and settings, with guidelines for determining:

a. the approach that should be used to construct, calibrate and run different types of numerical models;
b. the selection of appropriate model codes and procedures for each type of hydrogeological setting;
c. when it is appropriate to use an equivalent porous-medium flow code;
d. when it is necessary to use a fracture flow code;
e. whether to use a transient or steady-state modelling approach;
f. when it is necessary to use a coupled hydromechanical modelling approach;
g. how active drainage measures (e.g. wells, drains, tunnels) should be simulated within the model.

3. ***How should pore water pressures be incorporated into the slope design model***, which sought to outline recommended procedures and formats for providing basic water pressure distributions and responses into limit equilibrium and numerical slope stability programs, and assessing the impact of water pressures on rock mass strength.

Initially, it was intended that the information obtained from these three questions would form the basis of a handbook detailing the hydrogeological procedures that should be followed when performing open pit slope stability design studies. However, it was realised that such a handbook would not be complete without supporting sections that outlined how water pressures are characterised and reported, how slope depressurisation systems are implemented and maintained, and how predicted versus actual conditions and behaviour are monitored and reconciled.

Following the initial research using the data from Diavik and the other case study sites, it also quickly became apparent that the wider site-scale hydrogeology (and, in particular, groundwater recharge) is the most important factor for the assessment of water in the pit slope design. The scope of these guidelines was therefore further widened to include the characterisation and assessment of the wider site-scale hydrogeology. Furthermore, for a large number of pits, it is not possible to achieve the required pit slope pore pressure goals without the operation of an efficient general mine dewatering system. The general principles for modelling and design of mine dewatering systems are therefore also included.

The book was specifically designed to follow up on Chapter 6, (the Hydrogeological Model) in the LOP project's book *Guidelines for Open Pit Slope Design* (Read & Stacey, 2009). The contents are divided into the six sections shown in Figure 1 (The hydrogeological assessment process), starting with a section that outlines

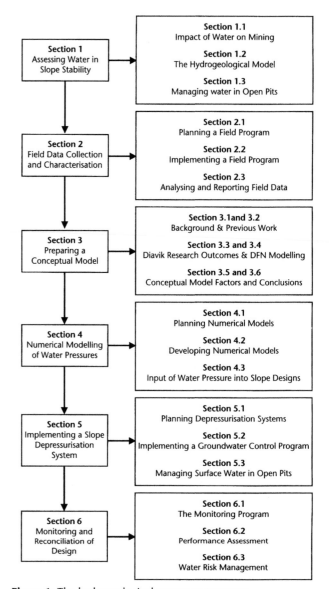

Figure 1: The hydrogeological assessment process

the framework used to assess the effect of water in slope stability (Section 1), followed by sections outlining how the water pressures are measured and tested in the field (Section 2), how a conceptual hydrogeological model is prepared (Section 3), how water pressures are modelled numerically (Section 4), how slope depressurisation systems are implemented (Section 5), and how the performance of a slope depressurisation program is monitored and reconciled with the design (Section 6).

The hydrogeological pathway followed through the slope design process (Figure 2) is illustrated in Figure 3. Sections 1, 2 and 3 are part of the Hydrogeology and Geotechnical Model sections of the process, Section 4 reports to the Groundwater Pressures and Stability Analyses sections, and Sections 5 and 6 relate to the Implementation section (general mine dewatering, slope depressurisation and monitoring).

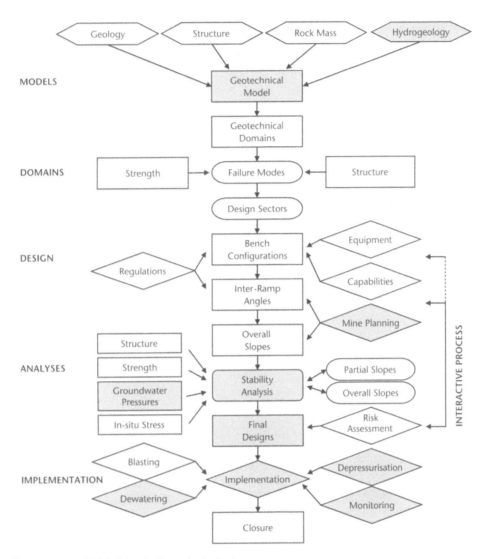

Figure 2: The slope design process, highlighting hydrogeologically important steps

When implementing the actions outlined in Figures 1 and 2, it is important to recognise that they are not completed just once but are required at every stage of the evaluation of a mineral deposit, from the initial conceptual and scoping-level designs through to the pre-feasibility, feasibility, final designs and expansion phases for an operating pit. They may also need to be considered for maintaining stability during mine closure and for long-term post-closure of the mine.

The level of effort required at each stage of project development is shown for each component of the geotechnical model in Table 1. As the developing project progresses from level to level, so there should be a corresponding increase in the level of confidence in the data. Target levels of data confidence are shown at the bottom of Table 1.

The target levels of data confidence shown in Table 1 are subjective, but were developed by the sponsors of the LOP project to provide guidelines to the level of certainty

required at each stage of project development. They can also be used to indicate the level of expenditure that may be required. For example, for the geological model, the greatest level of effort and expenditure will occur early in project development, with quite high levels of confidence being obtained by Level 2. In contrast, for the hydrogeological model, the greatest level of effort and expenditure will be at Levels 3 and 4, with confirmatory work being required to ensure there is the required level of confidence for the detailed slope design studies and the planning of any required depressurisation systems.

When implementing the agreed final slope design, it is vital that the geotechnical, mine planning and mine operations engineers understand the need to integrate the water monitoring system with the mine operations activities on a day-to-day basis from Day 1 of the project. Poorly organised communications between the designers and planners and mine operations will interfere with the organised collection of data, diminishing the value and

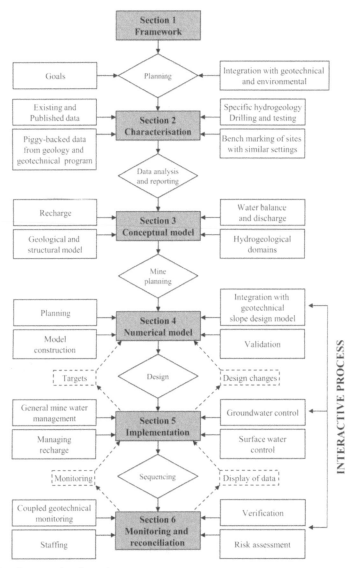

Figure 3: The hydrogeological pathway in the slope design process

level of confidence in the results, and increasing the risk that the design is not aligned with the performance criteria required by corporate mine management and the investment community. Such lack of integration will also probably lead to increased project costs.

2 General impact of water on mining

2.1 Water management issues

As the scale of open pit mines gets larger and regulatory constraints in most countries get tighter, the evaluation and management of water is becoming increasingly important for all mine operators and corporate mining houses. A 'typical' mine site will have to consider the following eight water management issues.

1. The control and management of groundwater and surface water within and around the open pit itself.
2. The provision of secure water supplies for mineral processing and other on-site activities.
3. The management of precipitation events, runoff and surface water throughout the mine site, including stormwater management.
4. The water balance of the individual processing facilities and how these fit into the site-wide water balance.
5. The need to discharge excess water from the site, and the associated engineering and environmental implications.
6. The protection of groundwater and surface water features and existing water users from the mine operations, including process facilities, dewatering and slope depressurisation activities.

Table 1: Levels of effort and suggested target levels of data confidence by project stage (Read & Stacey, 2009)

Project level status	Conceptual	Pre-Feasibility	Feasibility	Design and Construction	Operations
			PROJECT STAGE		
Geotechnical level status	Level 1	Level 2	Level 3	Level 4	Level 5
Geological model	Regional literature; advanced exploration mapping and core logging; database established; initial country rock model	Mine scale outcrop mapping and core logging, enhancement of geological database; initial 3D geological model	Infill drilling and mapping, further enhancement of geological database and 3D model	Targeted drilling and mapping; refinement of geological database and 3D model	Ongoing pit mapping and drilling; further refinement of geological database and 3D model
Structural model (major features)	Aerial photos and initial ground proofing	Mine scale outcrop mapping; targeted oriented drilling; initial structural model	Trench mapping; infill oriented drilling; 3D structural model	Refined interpretation of 3D structural model	Structural mapping on all pit benches; further refinement of 3D model
Structural model (fabric)	Regional outcrop mapping	Mine scale outcrop mapping; targeted oriented drilling; database established; initial stereographic assessment of fabric data; initial structural domains established	Infill trench mapping and oriented drilling; enhancement of database; advanced stereographic assessment of fabric data; confirmation of structural domains	Refined interpretation of fabric data and structural domains	Structural mapping on all pit benches; further refinement of fabric data and structural domains
Hydrogeological model	Regional groundwater survey	Mine scale airlift, pumping and packer testing to establish initial hydrogeological parameters; initial hydrogeological database and model established	Targeted pumping and airlift testing; piezometer installation; enhancement of hydrogeological database and 3D model; initial assessment of depressurisation and dewatering requirements	Installation of piezometers and dewatering wells; refinement of hydrogeological database, 3D model, depressurisation and dewatering requirements	Ongoing management of piezometer and dewatering well network; continued refinement of hydrogeological database and 3D model
Intact rock strength	Literature values supplemented by index tests on core from geological drilling	Index and laboratory testing on samples selected from targeted mine scale drilling; database established; initial assessment of lithological domains	Targeted drilling and detailed sampling and laboratory testing; enhancement of database; detailed assessment and establishment of geotechnical units for 3D geotechnical model	Infill drilling, sampling and laboratory testing; refinement of database and 3D geotechnical model	Ongoing maintenance of database and 3D geotechnical model
Strength of structural defects	Literature values supplemented by index tests on core from geological drilling	Laboratory direct shear tests of saw cut and defect samples selected from targeted mine scale drill holes and outcrops; database established; assessment of defect strength within initial structural domains	Targeted sampling and laboratory testing; enhancement of database; detailed assessment and establishment of defect strengths within structural domains	Selected sampling and laboratory testing and refinement of database	Ongoing maintenance of database
Geotechnical characterisation	Pertinent regional information; geotechnical assessment of advanced exploration data	Assessment and compilation of initial mine scale geotechnical data; preparation of initial geotechnical database and 3D model	Ongoing assessment and compilation of all new mine scale geotechnical data; enhancement of geotechnical database and 3D model	Refinement of geotechnical database and 3D model	Ongoing maintenance of geotechnical database and 3D model
Target levels of data confidence					
Geology	>50%	50–70%	65–85%	80–90%	>90%
Structural	>20%	40–50%	45–70%	60–75%	>75%
Hydrogeological	>20%	30–50%	40–65%	60–75%	>75%
Rock mass	>30%	40–65%	60–75%	70–80%	>80%
Geotechnical	>30%	40–60%	50–75%	65–85%	>80%

7. Water management and environmental protection during the period of active mine closure.
8. The creation of stable and sustainable groundwater and surface water conditions for long-term post-closure, ideally with long-term added value to local communities.

The guidelines presented in this book relate mostly to the first of these factors, the control and management of groundwater and surface water within and around the open pit itself, with particular reference to creating stable pit slopes. However, at many mine sites the water system is inter-related and the water produced from the open pit needs to be integrated into a site wide water management plan, as outlined in Section 5 of this introduction.

An increasing number of both new and mature mining operations are designing larger and deeper pits in response to rising commodity prices and deeper mineral exploration targets. As a result, many future pits will be excavated deeper below the water table than at present, leading to an increase in the required scale of the water-management program.

2.2 Consequences of mining below the water table

The consequences of mining below the water table can be divided into two general categories.

1. *Water will flow into the workings*. Inflows to an open pit usually lead to reduced operational efficiency and increased mining costs because of:
   - loss of access to parts of the pit area due to inundation and, in some cases, an ultimate loss of the ability to mine.
   - wet and inefficient drill-and-blast process, which can cause poor blast performance and, in some cases, detonation failure. Typically, a greater use of explosives or the use of slurry or emulsion is required.
   - increased equipment wear, including damage to tyres, electrical components or even equipment corrosion, leading to higher maintenance requirements.
   - a general loss of efficiency, which may lead to higher loading costs, reduced trafficability and increased haulage costs, increased roadway maintenance and, for some ore types, elevated product moisture contents that must be managed prior to shipping and delivery to the customer.

Figures 4a and 4b provide examples of how pit floor inundation may impact directly on mining activities.

2. *There will be a reduction in slope stability*. The presence of water invariably causes a loss of performance of the pit slopes. The water pressure acting within any discontinuities and pore spaces in the rock mass reduces the effective stress, with a consequent reduction in shear strength of the rock mass.

Either the slope must be depressurised, designed with a lower factor of safety, or flattened to compensate for the reduced rock mass strength. Where excess water pressures occur below the pit floor, heave may result.

Figure 5 shows an example of the adverse effects of groundwater on slope stability. Here, the ingress of water from the river to the left of the photograph into the perimeter wall of the coal mine to the right of the photograph resulted in the failure of the wall. The obvious consequences of an external high wall failure are reduced safety, production losses and rising mining costs, in addition to environmental damage. In other cases, concerns for slope stability may also lead to a deferment or sterilisation of ore and potentially a reputational loss for the operation or the corporate image.

2.3 General goals for the water-control program

Based on the two categories defined above, the general goals for an open pit water-control program are as follows.

1. To divert or remove water from the excavation in order to improve operational conditions, reduce mining costs and improve safety performance.
   This is the usual focus of a *general mine dewatering program*. The program often involves pumping from vertical wells to reduce groundwater levels and groundwater inflow to operating areas, and/or the general management and removal of surface water runoff using control structures beyond the open pit perimeter, together with collection ditches and sumps inside the pit.
2. To reduce the water pressure acting on the materials that form the pit slopes.
   This may require a dedicated *pit slope depressurisation program*. The program requires an integration of the hydrogeological and geotechnical models, and normally involves some form of drain hole installation that is geared toward reducing pore water pressures, thereby increasing the effective stress and consequently increasing the strength of the materials in the slope.

General mine dewatering and pit slope depressurisation are closely linked. A mine dewatering program will typically benefit the stability of the pit slopes. However, it is normal that, even if the pit is successfully dewatered, residual water pressures may remain in certain sectors of the slopes, so that a specific pit slope depressurisation program is required. For any given mine site setting, the

Figure 4: Adverse effect of pit floor inundation on operations and safety (Source: Schlumberger Water Services)

Figure 5: Example of an instability resulting from saturated slope conditions (Courtesy Tim Sullivan, Pells Sullivan Meynink, Sydney)

correct characterisation of the hydrogeological system and definition of the targets for dewatering and slope depressurisation are essential to ensure that a 'fit-for-purpose' system is designed and implemented. The successful planning and implementation of an integrated program can have a significant impact on the economic performance of the mine.

3 Cost of managing water in slope stability

3.1 Introduction

Dewatering of pits and depressurisation of pit slopes reduces mining costs and improves safety. However, the implementation of water control measures has become increasingly expensive, both in terms of the direct costs to construct and operate the system, and the requirement for additional infrastructure for subsequent management of the produced water. For larger mine projects, the management of water can often be a significant part of the annual capital and operating budget. The initial cash flow of a new project will often be affected by the short-term, high capital cost of implementing an initial groundwater-control program, because this is not adding to mine production. In the longer term, water management costs can be in the region of 20 per cent to 35 per cent of the total annual operating costs for mines that require a large water-control program.

3.2 Cost–benefit analysis

The costs associated with a dewatering or slope depressurisation program are always obvious and apparent but the consequential benefits may not be. Therefore, a cost–benefit analysis early in the mine development or expansion program can allow mine management to consider the cost outlay associated with implementing such a program against the long-term cost savings to be gained from the program.

In many cases, a relatively simple limit equilibrium analysis can be used to help quantify the potential cost–benefit of reducing pore pressure. The effective stress concept can be used to help quantify the balance between (i) the cost of reducing the pore pressures and (ii) the consequential benefit of increased slope performance, which may be presented in terms of an improved factor of safety for a given slope design or a higher slope angle for a given factor of safety.

Typical steps for preparing a quantitative cost–benefit analysis for a slope depressurisation program are as follows. The procedure is often an iterative process and Steps 3, 4 and 5 are often carried out simultaneously.

1. Calculate the slope angles and the associated safety factors assuming no depressurisation apart from passive drainage to the slope as mining progresses ('Base Case' – Case 1).
2. Analyse available data to determine the practicability and potential cost of implementing an active slope depressurisation program.
3. Calculate slope angles and the relevant safety factors assuming reduced pore water pressures as a result of the active depressurisation program. This is often done for a 'normal' active depressurisation program (Case 2) and an 'aggressive' active depressurisation program (Case 3).
4. Evaluate the difference in slope design and stripping costs for the passive program (Case 1), the 'normal' active program (Case 2), and the 'aggressive' active program (Case 3).
5. Prepare a cost estimate for achieving the depressurisation for the cases simulated (Case 2 and Case 3), and compare with the total reduction in mining costs (cost–benefit analysis).
6. Evaluate the risk and contingency costs if the slope depressurisation measures do not perform as expected.

The payback of a slope depressurisation program in terms of an increased slope angle and/or improved slope performance is dependent on many factors, including the geomechanical properties of the slope materials and the height of the slope. As a general rule, the payback of the program may be six-fold or more for a typical open pit, but may be significantly greater than this for large slopes cut in low-strength materials.

The cost–benefit achievable may be illustrated by looking at an open pit gold mine with dimensions of approximately 2000 m along strike by 1000 m width by 400 m depth designed at an average slope angle of 38 degrees and an average stripping cost of US$2.50 per tonne. For this particular example, the cost comparison was as follows.

▪ Short-term cost outlay. The slope depressurisation program consisted of seven production dewatering wells and a three phase program of horizontal drain holes, totalling about 4000 m of horizontal drain drilling. The entire cost of the program, including all infrastructure and monitoring, was about US$9.5 million.
▪ Long-term cost saving. As a consequence of the depressurisation program, the slope angle was increased by two degrees around 50 per cent of the pit perimeter. The associated reduction in stripping requirements was about 90 million tonnes, with an associated saving in stripping costs of about US$225 million.

Figure 6 shows an approximate relationship between the cost of implementing a slope depressurisation program and the savings associated with that program, prepared using a cost–benefit analysis carried out for a single slope sector at a number of larger mines.

Two specific case studies are presented below. The Victor Diamond Mine provides an example of managing the early cost outlay of a dewatering program and gaining the secondary benefit of depressurising the pit slopes. The

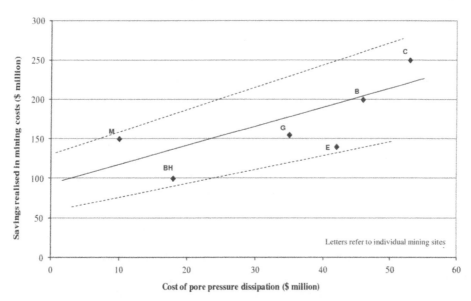

Figure 6: Approximate relationship between the cost of a slope depressurisation program and the consequent savings in mining costs (Source: Schlumberger Water Services)

Figure 7: (Left) Victor Mine at construction phase; (Right) Victor Mine two years into production (Source: De Beers Group Services)

Bingham Canyon Copper Mine provides a large-scale example illustrating the benefits of implementing a pro-active depressurisation approach in a low-permeability environment to achieve a substantial increase in the ore recovered.

3.3 An example of managing early dewatering costs

The Victor Diamond Mine, located in Ontario, Canada, is an example of an operating mine where the water table is near the surface and the permeability of the country rock is sufficiently high that mining could not proceed without an aggressive dewatering strategy commencing early in the life of the mine.

The photograph in Figure 7 (Left) shows the Victor Mine at construction phase. The muskeg, the numerous ponds on the property and the Attawapiskat River nearby all point to the high, near-surface water table in the area.

The mine area contains seven main hydrostratigraphic units.

1. The locally karstic and fossiliferous limestone comprising the upper part of the Attawapiskat Formation.
2. The mottled limestone/limey claystone of the lower part of the Attawapiskat Formation.
3. The limestone and interbedded, thin shales of the Ekwan River and Severn River Formations.
4. The mudstone of the combined Red Head Rapids Formation and Churchill River Group.
5. A relatively thin, poorly consolidated sandstone within the Red Head Rapids Formation.
6. The dolomite of the Bad Cache Rapids Formation.
7. The basement granite.

The three other local hydrogeological units identified in the Victor area include the heterogeneous peat, marine clays, and glacial overburden (primarily clay and silt with minor sand) which generally vary in thickness from 5 m to about 30 m over the general Victor site. The

conceptual model for Victor is illustrated in Figure 8. The associated hydrogeological properties are summarised in Table 2.

The photograph in Figure 7 (Right) shows the mine after two years of mining. The up-front dewatering costs were relatively high. The mine utilises peripheral dewatering wells to lower the water table below the pit floor which, in turn, provides adequate drainage and depressurisation of the pit slopes. The peripheral wells have 356 mm diameter casing installed within a 508 mm diameter drill hole, and the annulus is backfilled with one-half to three-quarter inch (12–20 mm) washed gravel. The wells discharge into 4 × 559 mm feeder lines around the perimeter that are connected to a single 914 mm main line that discharges 6 km away into the Attawapiskat River. The clear cut for the discharge lines can be observed as the white line to the upper right of the photograph.

At the end of 2011, the dewatering system at the Victor Mine comprised eleven perimeter wells equipped with variable speed submersible pumps. In 2011, the combined average discharge rate was approximately 78 560 m³/day. The average pumping cost was US$0.086 per cubic metre of water per day, or an average total cost of US$6735 per day. The average operating cost for the system is of the order of US$5M/yr, with a capital outlay of approximately US$1.6M for each equipped well. Given the relatively high cost of pumping, the operation has adopted a 'just-in-time' approach, where pumps are commissioned only to maintain approximately 15 m of freeboard below the current pit bottom. Figure 9 is a plot of the water table relative to the pit depth over time.

For mining operations like Victor, the general dewatering is on the critical path for developing the mine and the benefits to mine management are clear, but the financial benefit of depressurised slopes, optimised slope angles and reduced stripping was not as apparent.

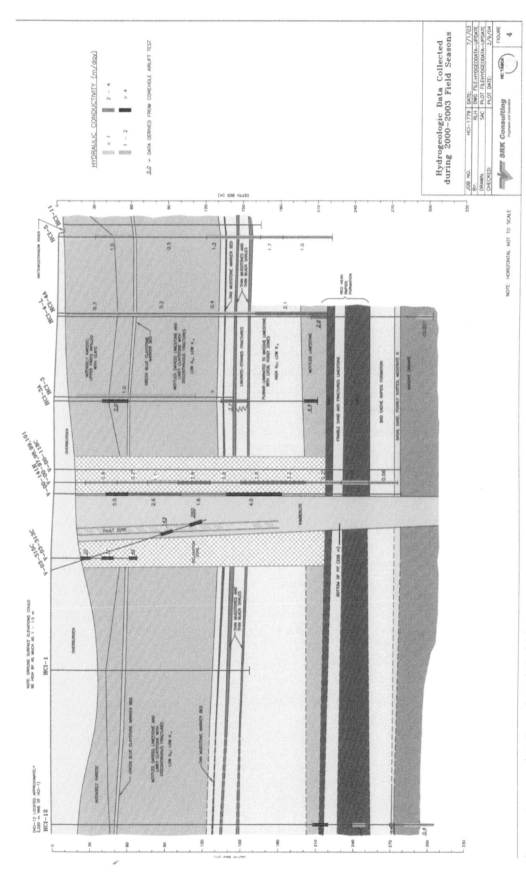

Figure 8: Conceptual Hydrogeological Model for Victor Mine (Source: HCI, 2004)

Table 2: Properties of Hydrogeological Stratigraphy at Victor Mine (Source: HCI, 2004)

Model zone	Hydrogeologic unit	Hydraulic conductivity (m/day)			Specific storage (m⁻¹)	Specific yield
		Kx	Ky	Kz		
1	Kimberlite (above Mudstone)	10	10	10	10^{-5}	0.01
2	Kimberlite (with Mudstone and Dolomite)	0.03	0.03	0.03	10^{-6}	0.005
3	Kimberlite (within Granite)	0.0001	0.0001	0.0001	10^{-6}	0.005
4	Kenogami River Formation (Dolomite)	0.5	0.5	0.005[3]	10^{-6}	0.005
5	Upper Attawapiskat Formation (Limestone)	1	1	0.01[3]	10^{-6}	0.005
6	Lower Attawapiskat Formation (Limestone)	0.5	0.5	0.005[3]	10^{-6}	0.005
7	Ekwan River Formation and Severn River Formation (Limestone)	1.5	1.5	0.015[3]	10^{-6}	0.005
8	Red Head Rapids Formation, Churchill River Group (Mudstone)	0.1	0.1	0.0001	10^{-6}	0.005
9	Bad Cache Rapids Formation (Dolomite)	5	5	5	10^{-6}	0.005
10	Granite	0.1	0.1	0.001	10^{-6}	0.005
11	Contact Zone in Lst between Kimberlite and Zones 4 through 6	0.0001	0.0001	0.0001	10^{-6}	0.005
12	Overburden Clay	5	5	5	10^{-6}	0.005
13	Sand within Overburden Trench[1]	0.001	0.001	0.001	10^{-5}	0.005
14	Fault Zone	3	3	3	10^{-6}	0.2
15	Silt within Overburden Trench	500	500	500	10^{-6}	0.005
16	Fine PK on bottom/sides of Polishing Pond[2]	0.01	0.01	0.01	10^{-5}	0.005
		0.04	0.04	0.04	10^{-5}	0.1

Notes:
1) A single layer 20 m thick that is not in direct connection with limestone
2) Thickness of 7 m
3) Hydraulic conductivity near river-bed is assumed to be isotropic (Kx = Ky = Kz)

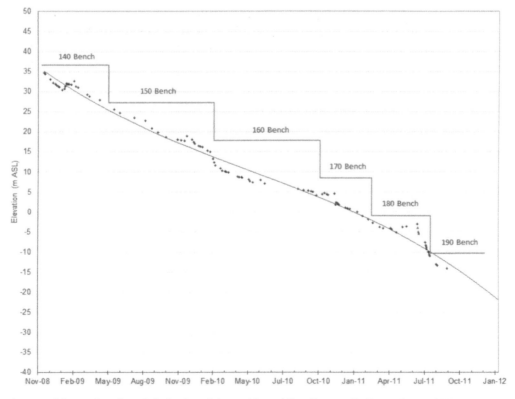

Figure 9: Plot of water table as a function of pit depth and time at Victor Mine (Source: De Beers Group, 2012)

Table 3: Bingham Canyon Mine, Highland Boy corner: geometric mean values of hydraulic conductivity from field RC drill-stem testing (Source: KUCC)

Geological unit	Sector				
	East	Q-Cut	North	South	Entire pit
Quartz Monzonite Porphyry	n/d	5.5×10^{-7}	n/d	1.4×10^{-7}	4.0×10^{-7}
Monzonite	4.6×10^{-6}	2.3×10^{-8}	n/a	1.3×10^{-6}	2.5×10^{-7}
Latite Porphyry	1.2×10^{-7}	2.3×10^{-7}	n/a	n/a	2.0×10^{-7}
Quartzite	6.6×10^{-7}	6.0×10^{-7}	2.6×10^{-7}	3.8×10^{-7}	4.0×10^{-7}
Limestone	1.4×10^{-7}	n/a	n/a	1.4×10^{-6}	1.4×10^{-6}

Note: All units in m/s: n/d = not determined: n/a = not applicable

3.4 An example of large-scale cost–benefit analysis for pit slope depressurisation

An example of a cost–benefit analysis of slope depressurisation for a large mining operation is provided by the Bingham Canyon Copper Mine, Utah, USA. Although general mine dewatering had already been achieved, a cost–benefit analysis identified the potential for localised depressurisation to allow associated slope steepening and increased ore recovery in the Highland Boy corner, located in the Q-cut sector of the pit.

The groundwater system at Bingham Canyon is generally characterised by low permeability, with strong anisotropy and compartmentalisation imparted by (i) the relative permeability contrasts between bedrock units and (ii) the variable shape, dip and orientation of geological units controlled by folding, faulting and intrusive emplacement. The hydrogeological conditions are thus strongly variable within the pit on a sector by sector basis. The hydraulic properties for the relevant units in the Q-cut sector are summarised in Table 3, along with the properties around the remainder of the pit for comparison purposes.

A range of depressurisation methods were considered, with the development of an underground drainage gallery identified as the most viable option. This decision was based on the general low permeability of the intrusive rocks in the immediate area of the pushback and the poor logistics associated with installing multiple horizontal drains on each bench for a specific pushback with a rapid mining rate.

Figure 10 shows the original slope design with no drainage gallery (cross-section 8I), the slope design including the drainage gallery (cross-section 8H), and the wall geometry that was actually achieved (as at June 2010). The total additional ore tonnage converted to reserves in the optimised design was estimated at 18M tonnes, with minimal additional waste stripping.

The drainage gallery was developed from June 2006 to July 2009 and comprised 2225 m of tunnel, with 262 dewatering holes, for a total drilled length of 50 760 m. The layout of the gallery and dewatering holes is shown

relative to the mine surface topography in Figure 11. The drainage program was designed to achieve the dual purpose of (i) reducing groundwater levels in the more permeable limestone units further behind the slopes and (ii) depressurising the low-permeability intrusive rocks in the immediate area of the pushback. Peak flow gained from the gallery was some 63 l/s (1000 USgpm), with most of the water derived from a limited number of high flow drain holes targeting the limestone. The sustained flow was around 22 l/s (350 USgpm), which continued throughout the mining of the Highland Boy corner. As a result of the depressurisation program and associated wall steepening an additional US$1.2B worth of ore was recovered leading to a net revenue increase of US$840M.

The plot in Figure 12 shows a comparison of the water levels obtained from the predictive model used for initial planning and the actual water levels following construction of the drainage gallery. The water levels achieved were lower than predicted by the model, indicating that depressurisation was more effective than anticipated. The monitoring data from the Highland Boy drainage gallery were subsequently used to calibrate later modelling efforts to assess the potential benefits of

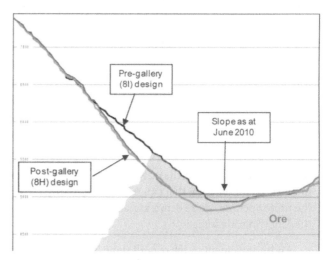

Figure 10: Cross-section showing increased ore recovery as a function of wall steepening (Source: KUCC)

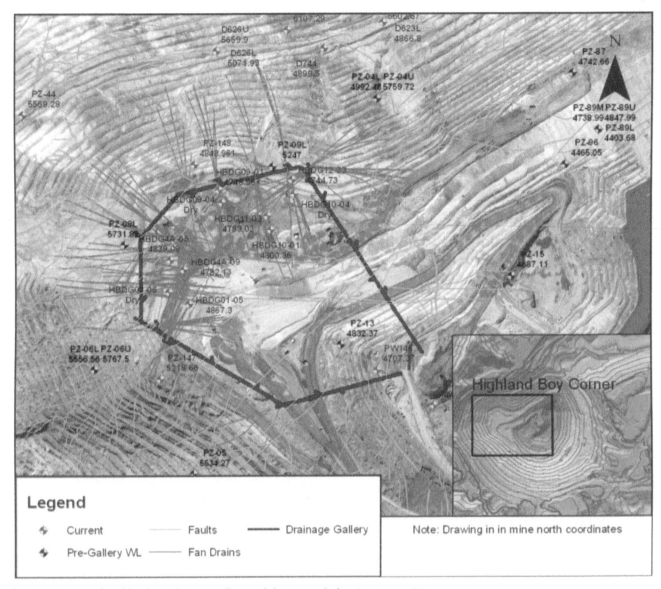

Figure 11: Layout of Highland Boy drainage gallery and dewatering holes (Source: KUCC)

drainage galleries on slope pushbacks being considered elsewhere in the mine.

4   Goals of managing water in slope stability

4.1   Opportunities

Groundwater pressure is the only geotechnical parameter in pit slope engineering that can readily be modified. For the mine operator, slope depressurisation often presents the following advantages.

- An improvement in the overall slope performance.
- The possibility of increasing the slope angle with an associated reduction in the cost of stripping.

- A reduced risk of an instability that could result in corrective remediation, potential sterilisation of ore, reduced mine safety and loss of mine access and infrastructure.

Therefore, at an early stage of the mine evaluation process (most often Pre-Feasibility), it is necessary to implement a hydrogeological characterisation program specifically to help quantify the probable pore pressure conditions in the pit slopes. Usually, the initial part of the hydrogeological characterisation of the pit slope can be tied to the geotechnical investigation.

4.2   Passive pore pressure control

As soon as the pit floor is excavated below the water table and the pit is dewatered, natural drainage of the wall

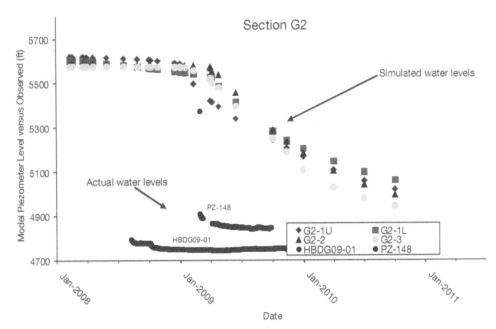

Figure 12: Sample comparison of modelled versus actual depressurisation (Source: KUCC)

rocks will usually occur in response to seepage and potentially in response to mining-induced relaxation of the rock mass. This drainage causes a reduction in the pore pressures in the wall rocks and can be thought of as *passive pore pressure control* (Figure 13). In some situations, the passive pore pressure dissipation achieved by natural drainage is adequate for achieving the desired pore pressure goals.

4.3 Active pore pressure control

In many cases, the rate of pore pressure dissipation achieved by passive drainage is too low to keep up with the mining advance rate or to achieve the target depressurisation levels associated with a new excavation or for a pushback. In these cases, the implementation of an

active pore pressure control program (e.g. wells, horizontal drains) can be used to enhance the rate of pore pressure dissipation in the wall rocks (Figure 14).

4.4 Making the decision to implement an active program

The following are key considerations for deciding whether to implement an active slope depressurisation program.

1. What are the hydrogeological conditions?
   - Will drainage occur passively, or would there be economic benefit of an active program?

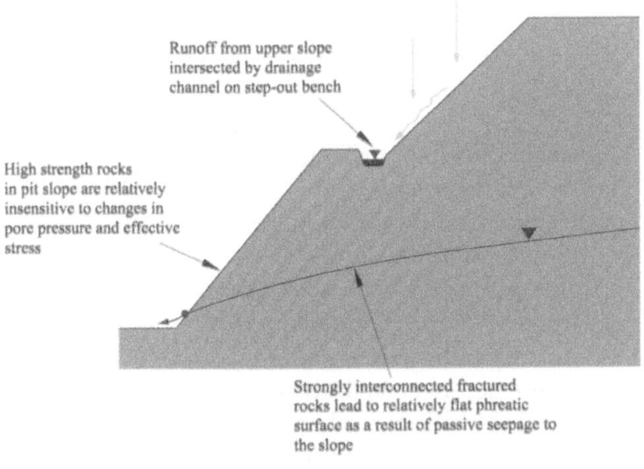

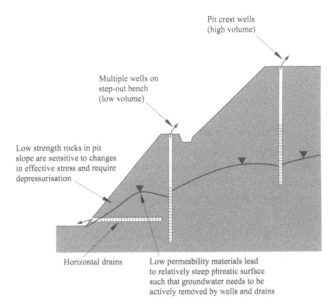

Figure 13: Schematic illustration of passive pore pressure control

Figure 14: Schematic illustration of active pore pressure control

Table 4: Guidelines for determining when an active slope depressurisation program may be necessary

An active slope depressurisation system is likely to have an increased economic benefit under the following conditions	Active slope depressurisation methods may be either unnecessary or not beneficial under these conditions
Depressurisation would lead to an increase in the effective stress of the materials in all or part of the slope	Depressurisation would create little change in the strength of the slope materials, such as in certain granites, limestones or other strong rocks
There are identified structural zones that would have a reduced tendency to shear or topple if the pore pressure were reduced	Passive drainage of pore spaces and fracture zones occurs rapidly throughout the entire slope sector in response to the normal mine excavation and/or the general mine dewatering system
There are weak, soft or erodible materials present in part or all of the slope	The overall design of the slope is dictated by factors other than pore pressure, for example where it is necessary to cut the slope along shallow dipping bedding or foliation
There are poorly permeable materials that will not drain freely in response to normal mining and/or general mine dewatering	It is impracticable to achieve the required slope depressurisation, for example in high rainfall areas where pore pressures result from shallow water in the blast damaged zone or in dense, poorly fractured rocks with insufficient permeability to yield the required water to wells or drain holes
There are structural compartments that will not drain freely in response to normal mining and/or general mine dewatering	It is not practicable to gain access for installation of passive slope depressurisation measures, given the pit geometry and the nature of the mine operations
The slope height is large	The rocks are of such low permeability that an active depressurisation program may not create sufficient pressure dissipation in the required time
There are increased mining risks such as single-ramp access or in-pit (and/or pit crest) infrastructure that may become affected by slope instability	The slopes are located adjacent to permeable aquifers or surface water bodies such that any active slope depressurisation measures would become 'swamped' by recharge that cannot be interrupted by the general dewatering system
Other forms of slope remediation are not viable	The slope height is small and/or the risks are short term and can be managed by other methods

■ Is the permeability of the wall rocks such that an active depressurisation program can achieve the desired results in the time available?
2. What are the geotechnical properties of the wall rocks?
   ■ Would an active slope depressurisation program increase the effective strength of the materials sufficiently to permit an increase in the slope angle?
   ■ Will the control of water reduce seepage-induced erosion or mitigate piping or deformation of the softer materials within the slope?
3. What is the nature of the geological structures?
   ■ Would depressurisation of the identified structural zones lead to reduced risk of failures associated with shearing or toppling?
   ■ Will the structural zones provide conduits to enhance the recharge to the wall rocks?
4. Is it necessary to reduce the amount of recharge to the slope materials to allow them to depressurise more efficiently?
   ■ Is the rate of recharge sufficiently high that the slopes will not depressurise unless the recharge source is cut off?
   ■ Is it necessary to protect the weaker, more deformable materials exposed in the slope from continuing infiltration of surface water?

As described in Section 3.6.3, recharge is identified as the single-most important factor influencing a slope depressurisation program. Recharge to the specific pit slope sector may be derived from precipitation, the infiltration of surface water, surrounding permeable formations, or leaky mine facilities.

Table 4 provides guidelines for deciding when it may be necessary to consider an active slope depressurisation program. Judgement and past experience are important in making a practical decision. The first step is usually to carry out a basic characterisation of field conditions.

5 General planning for mine water management

It is usually necessary to integrate the specific program for slope depressurisation into the wider water management plan for the mine site. Table 5 provides a list of the typical considerations for the overall open pit mine water management plan, based on the eight water management categories described in Section 2.1 of this introductory section.

It is also necessary (and advisable) for the mining operation to provide as good a groundwater and surface water baseline as possible, focused on providing an adequate characterisation of naturally elevated chemical parameters and any pre-existing disturbance to the natural system. A good baseline characterisation is important for placing potential future changes to the hydrological system into proper context as mining progresses. Increasingly, mining operators are finding it necessary to demonstrate to the public, regulators and NGOs that some changes to local hydrological systems may result from natural variance, or from other external developments, rather than from the mining operation itself.

Table 5: Typical considerations for an open pit mine water management plan

Requirement	Response
1 The control and management of groundwater and surface water within and around the pit.	• Design and operate the dewatering system to ensure that mining can be carried out in a dry environment and in a safe, timely and economic manner. • The system design may need to consider both groundwater and surface water. In some instances, it may be advantageous to segregate the 'dirty' in-pit water from the 'clean' groundwater, which typically shows less seasonal variation of flow and quality. • The system design needs to consider the sequencing of the mine plan and the availability of in-pit access. • If partial or sequential backfilling of the pit is planned, waste rock sequencing also should be considered. • The system design should minimise potential impacts to the surrounding surface and groundwater systems, including local groundwater users and other sensitive receptor areas. • If there are periods of the year when the availability of surface water is low, consideration can be given to designing the dewatering system in a manner that can provide a reliable and sustained source of water of suitable quality. • Pore pressure dissipation in the pit slopes often adds minimal additional water to the total dewatering flow rates, but this needs to be evaluated on a case-by-case basis.
2 The provision of secure water supplies for mineral processing and other on-site activities.	• Ideally, the water should be sourced from close to the mine site and the primary source would normally be the dewatering system. • However, water supplies from remote areas are becoming increasingly important as the scale of mining expands and the demand for water increases. • In seasonal climate zones, integrating the dewatering system into the water management plan can be important as groundwater typically exhibits less variability than surface water. • The use of the mine dewatering system as a source of water minimises the need to discharge water and may therefore minimise requirements for downstream water treatment. • In arid areas (e.g. northern Chile), there is an increasing need to utilise seawater or water from other saline sources. • The re-use of industrial water is becoming increasingly cost effective.
3 The management of precipitation events, runoff and surface water throughout the mine site, including stormwater management.	• Precipitation, evaporation, and other climatic controls must be evaluated at a level of detail sufficient for the stage of the project. • Predicted surface water flows need to be accounted for in the mine dewatering and surface water management plan, inside and outside the pit, and also need to be accounted for in the water balance, impact assessment and mitigation studies. • Diversion of non-contact surface water around the pit, waste rock and other key facilities is normally a key part of the management plan. • The management plan must include the capture, containment and management of contact waters. • Surface water diversion and control structures must be supported by an adequate cleaning and maintenance program as blocking and failure of a surface water drain has the potential to create slope failures and significant financial and environmental damage. • Surface water diversion and control structures must be designed to minimise the potential for recharge to the pit slope materials.
4 The water balance of individual processing facilities and how these fit into the site-wide water balance.	• The site-wide water balance must include potential runoff and seepage from the waste rock and stockpile areas. • The water balance must also include decant water and potential seepage from tailing ponds. The tailing pond water balance is often a large component of the site water balance. • It is often necessary to include upstream diversion and management of non-contact surface water for individual facilities, including the pit. • It is sometimes necessary to include upgradient diversion and management of non-contact groundwater for individual facilities. • Detailed hydrological and hydrogeological models are required to support the water balance for the complete mine area and the surrounding district. These models form the basis for optimising the dewatering and water management plan, impact assessment, and mine closure approach. • The integrated model may need to consider surface and groundwater interaction. • Preparation of the engineering design parameters for the open pit dewatering, surface water diversions and mine water management systems needs to be based on the water balance results, particularly the extreme high flow and low flow analyses.

(Continued)

Table 5: (Continued)

Requirement	Response
5 The need to discharge excess water from the site and the associated engineering environmental implications.	• The mine water balance should consider the need for mitigation water and for off-site discharge. • Any offsite discharge must meet regulatory requirements and controls, and must protect the downstream groundwater and surface water system, together with local users and other receptors. • The seasonality of the mine water balance often drives the need for off-site discharge. • Typical discharge options may include deep groundwater injection, shallow groundwater infiltration, irrigation, seasonal storage and enhanced evaporation. Direct discharge to a surface water body may carry more stringent discharge standards for water chemistry, temperature and flow, with less tolerance for short-term variations, and a higher regulatory risk. • The results of the water balance studies will usually feed into environmental assessment and permitting work.
6 The protection of groundwater and surface water features and existing water users from the mine operations, including process facilities, dewatering and slope depressurisation activities.	• The water management plan must include adequate protection of downstream receptors, habitat areas and water users. • Downstream conditions must be considered for the pit, tailing facilities, waste rock areas, process areas, and for the mine site as a whole. • It is often necessary to provide a source of good-quality water for impact mitigation.
7 Water management and environmental protection during the period of active mine closure.	• It is necessary to model the post-closure water balance, pit inflows and recovery of water levels, and also the water chemistry during short-term active closure period, prior to stabilisation of the hydrologic system (pre-stabilisation). • During the period that the pit actively fills with water, all hydraulic gradients are often toward the pit, so the pit is therefore a hydrological sink with no potential for outflow or discharge to the surrounding groundwater or surface water system. • The potential for, and consequences of, any post-closure changes in the pit slope pore pressures is often a major consideration for maintaining post-closure stability of the mine site. • Any instability of the pit walls may lead to changes in the void space of the pit and may therefore affect the filling rate. It may also lead to additional oxidation of the pit wall rocks which, in turn, may create a short- or long-term increase to the influx of chemical parameters to the pit lake. • During active mine closure, the groundwater system in the pit wall rocks often recovers faster than the rising water levels within the pit itself. This may cause a period where pore pressures in the wall rocks are elevated prior to them equalising with the water pressure in the pit lake. This is often a key factor for slope design, particularly if there is a need to protect facilities and infrastructure close to the crest of the pit.
8 The creation of stable and sustainable groundwater and surface water conditions for long-term post-closure, ideally with long-term added value to local communities.	• The closure water balance needs to extend from the active recovery (pre-stabilisation) phase to the long-term post-recovery (post-stabilisation) phase of the project. • In some areas, it may take several decades (or even centuries) for water levels to recover fully in the pit and for the seepage rates of other facilities to stabilise. In these cases, it may be beneficial for the mine operator to consider enhancing the pit filling rate using other external water sources to shorten the closure time and therefore shorten the period of post-closure monitoring and risk. • Where there is a negative water balance of the pit in arid climatic zones (an excess of evaporation over incident precipitation, pit wall runoff, and any runoff into the pit from the surrounding areas), the final water level in the pit may be permanently depressed below the pre-mining groundwater table. In this case the pit may not discharge to the surrounding groundwater or surface water system in the long term, and may therefore be a hydrological sink. This may have advantages for protecting downstream water chemistry and receptors. • Where there is a positive water balance of the pit in cold, temperate or tropical climatic zones (an excess of incident precipitation, pit wall runoff and any runoff into the pit from the surrounding areas over evaporation), the final water level in the pit may be permanently higher than the pre-mining groundwater table. In this case the pit may discharge to the surrounding groundwater system and may also potentially over-spill into the surface water system from the lowest point on the pit rim. • There may be considerable seasonal fluctuation of post-closure conditions and there may be also longer term cyclic fluctuations because of climatic trends. • The consequence of any post-closure changes in the pit slope pore pressure distribution is often a major consideration for maintaining post-closure stability of the mine site. For a pit lake with a surface water spill point the consequences of a slope failure and the potential for creation of a flood wave (seiche) may need to be considered.

1

FRAMEWORK: ASSESSING WATER IN SLOPE STABILITY

Geoff Beale, Michael Price and John Waterhouse

This section of the book outlines the framework used to assess the effect of water in slope stability. It describes the fundamental parameters used in the assessment (Section 1.1), the key elements that make up the hydrogeological model (Section 1.2), and general water management in open pit mines (Section 1.3), each of which is considered in the Hydrogeological and Geotechnical Modelling phases of the slope design process (Figure 1.1).

1.1 Fundamental parameters

1.1.1 Porosity and storage properties

1.1.1.1 Porosity and drainable porosity

The porosity of a rock mass (n) is defined as the fraction of the total volume of that rock mass that is occupied by void space. It is usually expressed as a percentage (though in calculations it must be expressed as a fraction). For saturated rock, the void space is occupied by water (Figure 1.2a). For unsaturated rock, the void space is occupied by a mixture of water and air (Figure 1.2b). Interconnected porosity is the fraction of the total volume of rock occupied by interconnected void space. The interconnected porosity can be subdivided into drainable porosity (or specific yield) and non-drainable porosity (or specific retention).

The sequences in Figure 1.2 illustrate that, for any given saturated rock mass, only part of the interconnected void space will drain under gravity. This space, expressed as a fraction or percentage of the bulk volume of the rock, is termed the drainable porosity or specific yield (S_y). Even after a long period of drainage time, some water will remain within the interconnected pore spaces due to surface tension, adhering as films to mineral grains and fracture surfaces, and sometimes completely filling the finer pores or fractures; this is termed pellicular water (Figure 1.2b). There will also be some pores that cannot drain because they are not interconnected; it is therefore

useful to distinguish between total porosity (which may include isolated, unconnected voids) and interconnected porosity, through which fluid movement is possible. Many igneous and metamorphic rocks, for example, will contain some primary porosity around or within crystals, but these voids are rarely interconnected to a significant extent and so this porosity is seldom significant for water storage. A sample of Ailsa Craig microgranite was found to possess a primary porosity of 0.9 per cent but this was mostly within the altered feldspar microphenocrysts (Odling *et al.*, 2007); it could not therefore contribute to either specific yield or to permeability, which were extremely low (Section 1.1.2).

Generally in this book the term porosity, without a qualifier, can be assumed to refer to interconnected porosity. The void space that does not drain under gravity is termed the non-drainable porosity. The fraction or percentage of retained water to the bulk volume of rock is termed the specific retention (S_r). The sum of the specific yield (S_y) and the specific retention (S_r) equals the interconnected porosity (n). The use of the once-popular term effective porosity is discouraged (Lohman *et al.*, 1972) because it has been used to mean different properties by different people.

In terms of gaining a practical understanding of porosity and pore pressure, most materials that are encountered within and around mine sites can be classified into three groups, as follows.

1. Unconsolidated or semi-consolidated materials that exhibit porous-medium conditions.
2. Hard rock materials that exhibit fracture flow conditions.
3. Dual-porosity materials that exhibit a combination of porous-medium and fracture flow conditions.

1.1.1.2 Porous-medium conditions

Porous-medium conditions usually occur within most unconsolidated or poorly consolidated strata, such as

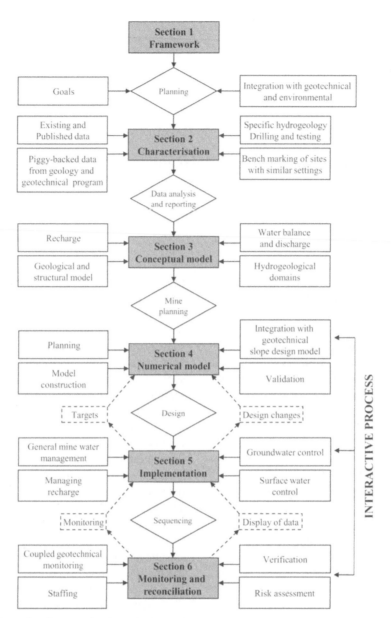

Figure 1.1: The hydrogeological pathway in the slope design process

sands, silts, clays, evaporite deposits and some poorly-consolidated clastic rock types. Within these rock and soil types, virtually all of the groundwater is contained within the primary interstitial pore spaces of the formation itself (Figure 1.2a).

The porosity of these materials is mostly controlled by the interstitial spaces between grains, which typically range from about 10 per cent to about 30 per cent of the total volume of the formation ($n = 0.1$–0.3), but may be more than 50 per cent ($n = 0.5$) in some poorly-consolidated fine-grained materials. A cubic metre of the rock mass may therefore typically contain 100 to 300 litres of groundwater. Clay materials may have a very high total porosity, in some cases in excess of 50 per cent ($n > 0.5$).

However, for many clays and other fine-grained sediments, the drainable porosity typically represents only a small proportion of the total porosity, in many cases less than five per cent of the total volume ($S_y < 0.05$) and in some cases less than one per cent ($S_y < 0.01$).

The porosity is typically higher in well-sorted materials where the grain size of the formation falls within a relatively narrow band. The drainable porosity will also be higher in well-sorted materials provided that the grain size is relatively large (e.g. sands and sandstones). Some well-sorted materials (e.g. chalks and some silts) may have very uniform grain sizes but the grains and hence the pores and pore connections are so small that very little water will drain from them under gravity. The total and

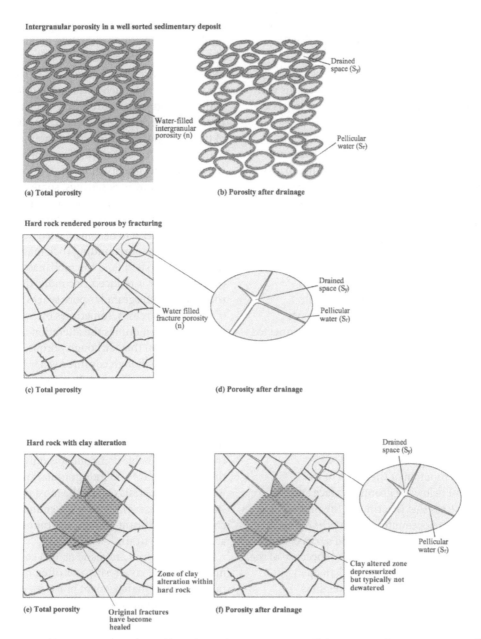

Figure 1.2: Porosity, drainable porosity (specific yield) and specific retention. (a) and (b) Intergranular porosity in a well-sorted sand before (a) and after (b) drainage. (c) and (d) Fracture porosity in a hard rock before (c) and after (d) drainage. (e) and (f) Hard rock with fracture porosity and clay alteration before (e) and after (f) drainage.

drainable porosity tend to decrease in poorly sorted materials where the void spaces between the coarser grains can be filled with finer particles. The total and drainable porosity also tend to decrease in materials where the void spaces have become filled with secondary chemical deposits (cements).

1.1.1.3 Fracture flow conditions

Fracture flow conditions usually occur within most competent (consolidated) rock types, such as igneous, metamorphic, cemented clastic and carbonate rocks, and in

consolidated coal formations. Although the unfractured rock may contain some pore space, this is mostly unconnected and much of the groundwater within these rocks occurs within fractures in the rock mass (Figure 1.2c). Groundwater movement in these materials therefore occurs mostly by fracture flow and drainage from the unfractured blocks will be effectively non-existent.

The interconnected porosity of competent (consolidated) rock types is dependent on the frequency of open fractures and joints, and may typically range from less than 0.1 per cent to about three per cent of the total volume of the formation ($n \leq 0.001$–0.03). A cubic metre of the fractured rock mass may therefore contain from less than 1 to around 30 litres of groundwater. As with intergranular pore spaces, of the total groundwater contained within the fractures, only a portion of this will drain out if the rock pore pressures are reduced by passive drainage or pumping (Figure 1.2d).

Both the total and drainable porosity are typically higher in settings where the rock mass is well fractured and jointed, and the joints are open. Porosity may also be significantly higher where the rock mass itself contains intergranular porosity or other forms of secondary porosity such as vugs or solution voids. The total and drainable porosity tend to decrease where fractures or voids have become filled during the mineralisation process or by secondary chemical deposits (cements).

A good rule of thumb is that the magnitude of groundwater flow to mine dewatering systems in fractured-rock settings tends to be higher and more sustained when the flow system is hydraulically connected to saturated alluvial deposits or other unconsolidated materials that have a high drainable porosity. Unconsolidated alluvium typically yields between 10 and 100 times more water per unit volume than many fractured rock types. An exception to this rule can occur when working in regional carbonate settings where there is sufficient regional-scale flow within the fractured rock system itself to sustain high dewatering flows.

Porosity is also a function of the depth of the hydrogeological unit below the ground surface and therefore the total stress that is acting on the unit. Where fractured rock units are encountered at depths of several hundred metres or more below the ground surface, the fracture porosity may be very low ($n \leq 0.001$) because of the weight of the overlying materials. Around an open pit mine, the stress caused by the overlying weight is continually changing as a result of the removal of material from the open pit and the placement of material within waste rock facilities. As a result, there is a tendency for both the total and drainable porosity to change with time. (This is described in more detail in Section 3.6.6.) Clearly, the mining of an open pit typically causes a reduction in total stress and an increase in porosity whereas waste rock facilities typically cause an increase in total stress and a decrease in porosity. However, the effect of changing porosity is often less of an issue for waste rock facilities because their height is often (but not always) limited to tens of metres so the proportionate change of total stress is lower than for the open pit.

1.1.1.4 Dual porosity or transition materials

Materials falling into this category include (i) altered or weathered rock, where the primary rock mass has been broken down into a clay material (or, in some cases, into a sandy material), (ii) hard rocks that behave like true dual-porosity media, such as vuggy limestones or fractured sandstones, (iii) many of the iron ore deposits, and (iv) some poorly consolidated coal formations. In these settings, much of the wider-scale groundwater movement occurs by fracture flow, but much of the groundwater storage occurs within intergranular pore space. In many cases, the actual release of water from these materials can be limited because of the fine-grained nature of the intergranular pore space, with much of the total void space still occupied by retained water following gravity drainage.

In many hard rock settings, some portion of the rock has been broken down by weathering, alteration and/or mineralisation. There are two main results of this. First, the fractures tend to become healed because of the softening of the surrounding rock mass, so the ability of the fracture system (and the rock mass as a whole) to transmit water is reduced. Secondly, the softening of the rock creates intergranular pore spaces, so the total porosity of the rock increases, sometimes by over an order of magnitude. Sometimes the alteration of the inter-fracture rock may mean that, although porous, it does not drain under gravity, though confined water may be released by depressurisation (Figure 1.2e–f and Section 1.1.1.5).

Figure 1.3 illustrates the behaviour of a dual-porosity medium. Most of the groundwater movement still occurs by fracture flow. Drainage into the fractures occurs from the surrounding interstitial pore spaces. However, in many cases, the drainable component is only a very small part of the total (fracture and intergranular) porosity. These weathered materials can be the most problematic for slope stability because (a) they are softer and weaker, (b) they contain a significant amount of groundwater, and (c) they are difficult to drain.

Most pit slopes are made up of a combination of material types. As a result, it is often necessary to employ a range of water management practices within a given mining operation. Virtually all hard rock settings contain zones where the rock mass has been broken down to some degree. For example, the slopes of a porphyry copper pit may exhibit fracture-controlled porosity in the unaltered primary rock, and porous-medium or dual porosity conditions in the oxidised zone or in zones of argillic or sericitic alteration. Many hard rock settings and most sedimentary settings include parts of the pit slopes that encompass unconsolidated porous-medium deposits or are hydraulically connected to such deposits.

1.1.1.5 Confined (elastic) storage

In an unconfined formation, the upper limit of fully-saturated conditions is the water table. Below the water table, pore pressure is greater than atmospheric, above the water table it is less than atmospheric and at the water table (by definition) it is exactly at atmospheric pressure. When water is removed from the formation by pumping or by drainage to natural discharge points, the water table

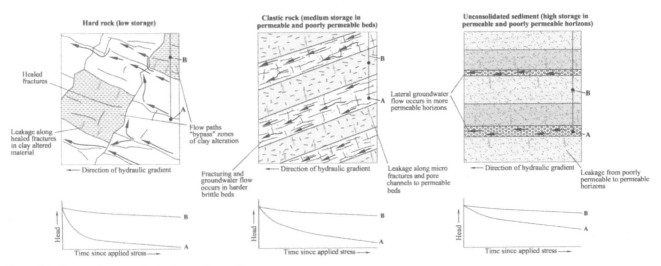

Figure 1.3: Flow and storage in dual-porosity media

falls and the unsaturated zone increases in thickness; some of the pore space that was completely full is now occupied only by pellicular water (Figure 1.2b).

In confined portions of formations, the pore and fracture space in the full thickness of the rock is completely saturated and the pore pressure throughout is greater than atmospheric pressure, so the potentiometric surface is above the top of the formation and the water level in a borehole drilled into it will rise above the formation top. When water is pumped from such a unit the pore space will stay full but the pore pressure will be reduced and the potentiometric surface will fall: as a result the water will expand slightly and, in response to the resulting increase in effective stress, the formation (especially any compressible layers or zones) will compact. The water that is withdrawn is therefore taken from elastic storage. There will be no drainage from the intergranular or fracture space unless the potentiometric surface falls below the top of the unit. For a unit decrease of head, the water that is released from elastic storage in a unit volume of formation (as a result of water expansion and formation compaction) is termed the specific storage, S_s, which is given by:

$$S_s = \rho g(\alpha + n\beta) \qquad \text{(eqn 1.1)}$$

where:

n = porosity
ρ = the density of water
α = the compressibility of the rock mass
β = the compressibility of water (typically around 4.6×10^{-10} Pa^{-1}).

Compressibility (strain/stress) is the reciprocal of the bulk modulus of elasticity (stress/strain), symbol κ (Greek kappa, which is not to be confused with K or k). Bulk modulus is related to Young's modulus (E) and Poisson's ratio (v) by the equation:

$$E = 3\kappa(1 - 2v) \qquad \text{(eqn 1.2)}$$

The water released from a column of unit cross-sectional area extending the full thickness, b, of the formation for a unit fall of head is termed the storage coefficient, S, so that $S = S_s b$ and

$$S = \rho g b(\alpha + n\beta) \qquad \text{(eqn 1.3)}$$

The definition of storage coefficient means that it can also refer to unconfined conditions. When water drains from intergranular or fracture space in response to a fall in the water table, there will also be a small amount of water released from elastic storage below the water table. In the case of an unconfined formation, the storage is usually contributed almost entirely by gravity drainage and specific yield is usually several orders of magnitude larger than specific storage.

Elastic storage coefficient can be determined from a non-steady-state pumping test with observation piezometers. Specific storage can also be determined from knowledge of the porosity and compressibility of the rock material; these properties can be measured on samples in a laboratory or determined from appropriate geophysical logs.

In the case of dual-porosity formations, both the matrix and fracture space can contribute to elastic storage. In this case the appropriate equation is:

$$S_s = \rho g\{\alpha_f + \alpha_m + (n_f + n_m)\beta\} \qquad \text{(eqn 1.4)}$$

where the subscript f refers to fracture properties and m to matric properties. In these cases, S is best determined from a pumping test but there may be a complex response (similar to that of an unconfined aquifer) as the two components of porosity release water at different rates. The method also involves assumptions about the fracture pattern and the distribution of head within the inter-fracture blocks (Kruseman & de Ridder, 1990).

It is also theoretically possible to determine S from the barometric efficiency, B, of the aquifer (Jacob, 1940):

$$B = \frac{\rho g b n \beta}{\rho g b (\alpha + n \beta)} = \frac{\rho g n \beta}{S_s}$$

$$\text{so that } S_s = \frac{\rho g n \beta}{B} \qquad \text{(eqn 1.5)}$$

In addition to its importance in estimating the quantity of water that will need to be drained or pumped from a formation to achieve a given reduction in water table elevation or pore pressure, the storage property is an important factor in determining the speed of propagation of a head change through the formation. This aspect is dealt with in Section 1.1.2.8 and is discussed further in Section 3.

1.1.2 Permeability and transport properties

1.1.2.1 Permeability, hydraulic conductivity and transmissivity

Permeability is the property of a porous medium such as rock or soil that controls its ability to transmit a fluid under the influence of an energy gradient. It is dependent on porosity to the extent that permeability cannot exist in the complete absence of any pore space, but is not directly proportional to it, so that highly porous materials may have low permeabilities (for example, clays).

Flow of fluid through a porous medium is described by Darcy's law. More details are given in Appendix 1, but Darcy's law is essentially a statement that as fluid moves through a porous medium it loses energy; it moves from a region where it possesses more energy to one where it possesses less; and the rate at which the fluid moves is proportional to the gradient of head in the medium, where head is a measure of the potential energy of the fluid.

For fluid flow through a homogenous, isotropic medium Darcy's law can be written as follows:

$$q = -K \frac{\Delta h}{\Delta l} \qquad \text{(eqn 1.6)}$$

where q is the rate of flow through a unit area of the medium at right angles to the direction of flow and Δh is the change in head over the distance Δl measured in the flow direction. The quantity q is called the specific discharge and is equal to Q/A, where Q is the total flow through an area A at right angles to the flow direction. The constant of proportionality, K, is called the hydraulic conductivity, and is a property of both the medium and of the fluid that is flowing through it and filling its pores. If that fluid is changed, say from water to oil, but the hydraulic gradient remains the same, the rate of flow will change and so K must change. This change depends on the viscosity and density of the fluids involved.

The quantity

$$\left(-\frac{\Delta h}{\Delta l} \right)$$

is called the hydraulic gradient and is often given the symbol i. The minus sign indicates that flow takes place in the direction in which head is decreasing.

To describe the properties of the medium (such as a rock) independent of the nature of the fluid, the term intrinsic permeability (k) is used. Hydraulic conductivity and intrinsic permeability are related by:

$$K = \frac{k \rho g}{\mu} = \frac{kg}{\upsilon} \qquad \text{(eqn 1.7)}$$

where:

g = acceleration due to gravity

ρ = density

μ = dynamic (absolute) viscosity

υ = kinematic viscosity

K has the dimensions of length over time (L/T) and in SI units is expressed in m/s. k (intrinsic permeability) has dimensions of L^2 and the SI unit should be (metre²) but this unit is many orders too large for practical purposes. The logical unit should then be the square micrometre ($1 \text{ m}^2 = 10^{12} \text{ μm}^2$) but the oil industry adopted a practical unit called the darcy (D) which is defined as the permeability that permits a flow of 1 ml of fluid of 1 centipoise viscosity completely filling the pores of the medium to flow in 1 second through a cross-sectional area of 1 cm² under a gradient of 1 atmosphere/cm of flow path. The darcy has become the standard unit of permeability in most commercial core laboratories, so laboratory measurements are usually reported in darcys or millidarcys. For conversion to SI units, $1 \text{ D} = 0.987 \text{ μm}^2$. The conversion from intrinsic permeability to hydraulic conductivity depends on the fluid properties; in the case of relatively pure water this comes down to temperature. For water at 20°C, 1 D is equivalent to 9.6×10^{-6} m/s, or in order of magnitude terms, 1 D or 1 μm² can be considered equivalent to 10^{-5} m/s.

The term 'permeability' strictly refers to intrinsic permeability, and enables two rock or soil types to be compared without consideration of the fluid they contain. The distinction between the two concepts is most important where the contained fluid can vary greatly in viscosity and/or density, as in the oil industry. In engineering work, K is often referred to informally as permeability. In most mining situations, the contained fluid at any one site is water of relatively constant density and viscosity. Thus, although there will be differences between sites such as Diavik (cold, therefore dense and relatively viscous water) and Whaleback (warmer, therefore less viscous water), such that viscosity at Diavik may be double that at Whaleback, the water properties at any one site will be relatively constant. For this reason, the terms 'permeability' and 'hydraulic conductivity' are used interchangeably in this book, with values usually expressed in units of hydraulic conductivity (m/s).

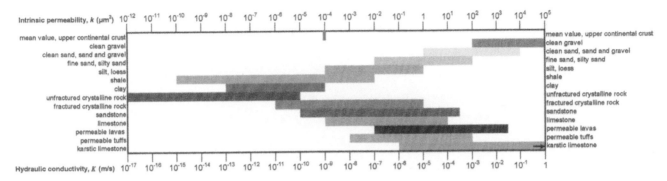

Figure 1.4: Typical permeability ranges for materials commonly encountered around mining operations. The conversion from intrinsic permeability to hydraulic conductivity assumes that 1 μm² is equivalent to 1 × 10⁻⁵ m/s. (Compiled from various sources, including: Allen *et al.*, 1997; Bear, 1972; Brahana *et al.*, 1988; Deming, 2002; Freeze & Cherry, 1979; Trainer, 1988; Wood & Fernandez, 1988)

When considering regional flow through a lithological unit, or flow to a well, the property that controls the water-bearing capacity of the unit is the product of the hydraulic conductivity and the saturated thickness. This product is termed the transmissivity (T) and is described in more detail in Appendix 1.

Hydraulic conductivity, like other conductivity properties (thermal, electrical) can vary enormously. It is probable that the range between the most permeable natural materials (such as clean gravels) and least permeable (unfractured granite, such as that from Ailsa Craig (Odling *et al.*, 2007)) covers at least 15 orders of magnitude. Figure 1.4 gives some typical values.

1.1.2.2 Speed of groundwater movement

Darcy's law is an empirical law, based on observation. It does not attempt to describe the individual flows of water molecules through pore spaces or narrow fractures, but treats the mass of rock as a continuum to describe the bulk movement of water. This leads to an important point when considering the speed of groundwater movement. The specific discharge, q, in Equation 1.6 has the dimensions of L/T, the same as speed or velocity, but q is not the true speed of the groundwater. This is because A, the cross-

sectional area that is used to calculate q, is the cross-sectional area of the whole of the rock unit whereas the water is moving through only part of that rock unit – the pore space. By the law of continuity, for the same flow rate to occur through a smaller cross-sectional area, the flow speed must increase. As shown in Figure 1.5, if the flow speed through area A_1 is q when the total area A_1 is available for flow, then the flow speed through the reduced area A_2 will have to be v, where

$$Q = qA_1 = vA_2 \qquad \text{(eqn 1.8)}$$

As a first approximation, the ratio of the area available for flow to the area of the whole mass of the rock is the same as the ratio of the volume of the pores to the total volume – in other words it is the same as the porosity. Hence;

$$v = \frac{q}{n} \qquad \text{(eqn 1.9)}$$

However, not all of the pores will be used as flow paths – some of them will be 'dead ends'. Also, the water will not make use of the full width of the pore channels; there will be an immobile layer – the boundary layer – around the grains or along the fracture faces (Figure 1.6). The effect of the variable size of the openings means that the

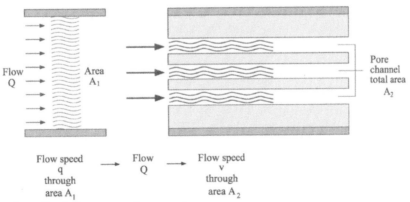

Figure 1.5: Illustration of specific discharge, q, vs flow speed

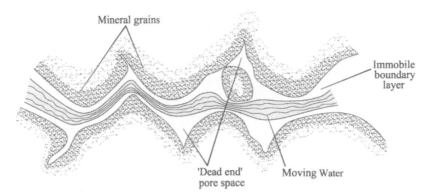

Figure 1.6: Kinematic porosity (After Price, 1996)

speed to be calculated is a mean flow speed – the flow speed that would occur if the pores, while giving rise to the same porosity and permeability, were all of the same size. The effect of the non-contributory pores and the boundary layer is that the figure for porosity that must be used in calculations of flow speed is smaller than the total or the interconnected porosity. The equation then becomes:

$$v_i = \frac{q}{n_k} = -\frac{K\Delta h}{n_k \Delta l} \qquad \text{(eqn 1.10)}$$

where v_i is the average speed of groundwater movement and n_k is called the kinematic porosity, the fraction of the rock volume that is involved in water movement. Although n_k is not easy to measure, it is often reasonably close to the specific yield, which can be used as a good approximation for n_k in Equation 1.10.

The speed v_i is not the true speed of the molecules, which must take a longer, more tortuous path around grains or along fractures; it is a means of estimating the time that a water molecule, or a molecule of some other substance such as a contaminant or tracer moving with the water, will take to travel a known distance. As Figure 1.7 shows, the speed of a molecule along an individual tortuous flow path will be v_{it} and this will be even greater than the effective mean speed, v_i.

The other complication to any form of solute transport from a source (such as a leaking tailing facility) is dispersion (Figure 1.8). The multitude of pathways available through either a granular material (Figure 1.8a) or a fracture network (Section 1.1.2.5, Figure 1.12) means that any solute introduced into the system will not move as a perfectly uniform 'front' but will be dispersed (spread out), both across the direction of flow (lateral dispersion) because of branching pathways and along the direction of flow (longitudinal dispersion) because of the varying permeabilities of individual flow paths or fractures. Dispersion is not significant for water transport but can be very significant when considering the movement of contaminants or when tracer testing is involved.

An important point to note is that the speed of transport of a molecule of water or dissolved solute is very much lower than the speed of propagation of a change of head or pore pressure through the same formation (Section 1.1.2.8).

1.1.2.3 Anisotropy and heterogeneity

Very few, if any, rocks or soils are homogeneous and isotropic. Natural variability as a result of sedimentary processes or of mineral segregation in magmas can lead to variations on several scales. Permeability can vary in direction, causing anisotropy, or vary in position, causing heterogeneity. A rock unit is homogeneous with respect to any property such as porosity or permeability if the value

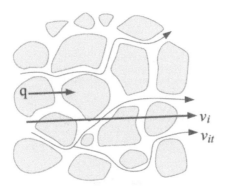

Figure 1.7: Tortuosity and mean groundwater speed

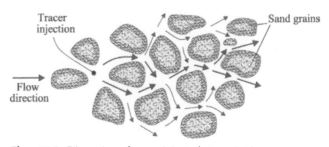

Figure 1.8: Dispersion of tracer injected at a point in an intergranular flow pattern

of that property does not vary from place to place within the unit. Because even a uniform rock consists of grains and pore spaces, or crystals of different minerals, it is necessary to compare volumes of a reasonable size when making this comparison. The smallest volume of rock that can be compared in this way is called the representative elementary volume (REV; Bear, 1972), which is described further for a fractured rock in Section 3.6.5.1.

Hydraulic gradient is a vector quantity, it has both magnitude and direction. In a perfectly homogeneous, isotropic porous medium, such as a collection of spherical grains with tetrahedral packing, permeability would be non-directional and could be treated as a scalar quantity. From Equation 1.6, the direction of the specific discharge would therefore be controlled by the direction of the hydraulic gradient, and its magnitude controlled by the magnitudes of both the gradient and the hydraulic conductivity.

Frequently, permeability is greater in one direction, making the material anisotropic. In these cases, the flow direction will try to follow the most permeable path. This will generally result in a flow direction between that of the hydraulic gradient and the direction of maximum permeability (more details are given in Appendix 1). The degree to which this is important varies with the geological setting. In some compartmentalised, fractured rock settings, the main flow direction can be almost perpendicular to the general hydraulic gradient.

The two causes of anisotropy in permeability can be summed up as (a) preferred lithological orientation and (b) significant anisotropy of *in-situ* stresses. The first is most prominent in unfractured materials such as bedded sediments and some foliated metamorphic rocks, and the second is prominent in fractured materials.

In layered sedimentary and metamorphic units, maximum permeability generally occurs along stratification or structural features, while the minimum occurs at right angles to these features; these represent two of the principal directions of permeability. The third principal direction is at right angles to both of them. In a layered system it is common to find that the minimum value is at right angles to the bedding or foliation; and that the permeability values in the two directions at right angles to the minimum value (roughly parallel to the bedding or layering) are approximately equal; such a system is said to be transversely isotropic

For crystalline rocks with an existing fracture network, anisotropic permeability can stem from significant anisotropy in the *in-situ* stress field, because this controls the direction of fracturing. In such cases, maximum permeability can often occur in the direction of maximum principal stress or stresses, and the minimum value of permeability can occur in the direction of minimum principal stress.

Permeability values vary throughout all geological settings. These variations exist even when the measurements are taken in the same lithological material and over a limited geographical area. Several forms of heterogeneous distribution can be recognised (Freeze & Cherry, 1979). Discontinuous heterogeneity is caused by the presence of faults or other large-scale geological features such as unconformities that bring two different formations into contact; good examples in mine settings are at major faults or at the junction between overburden and bedrock. Two forms of what might be called organised heterogeneity are layered heterogeneity (discussed below) and trending heterogeneity. One of the most important forms of trending heterogeneity is the decrease of permeability with depth, which occurs in many formations as a result of the greater total stress acting on the system, but is especially relevant in crystalline rocks (Section 1.1.2.5).

Random heterogeneity can occur within a single unit. It is now generally recognised that this variation in permeability often follows a log-normal distribution, for which the most suitable method of averaging is the geometric mean. This is obtained by multiplying n values of permeability together and taking the nth root of the product, so that

$$K_{gmean} = \sqrt[n]{\prod_{i=1}^{n} K_i} \qquad \text{(eqn 1.11)}$$

where K_{gmean} is the geometric mean permeability and

$$\prod_{i=1}^{n} x_i = x_1 \times x_2 \times \dots \times x_n.$$

Because this random variation occurs within virtually all geological units, trending heterogeneity is best thought of as a trend in the geometric mean value of permeability. There will be a range of permeability values in all parts of the unit, but the trending means that there is an increase in the total range of values (Freeze & Cherry, 1979). Using this approach, a homogeneous formation is one that may show variations in space but in which the mean value of K remains constant throughout.

If there is evidence of spatial distribution, then techniques can be applied to estimate the values of permeability at unmeasured locations. If it is assumed that the variation of permeability between measuring points is linear, then the geometric mean of discrete data points will coincide with the geometric mean of the continuous field.

In a clastic sediment, such as a sandstone, the grains will usually be slightly flattened in one direction and will be more likely to be deposited with the shorter axis vertical; this is especially true for clay particles. This means that flow will occur more easily in a direction parallel to the bedding than across it, making the

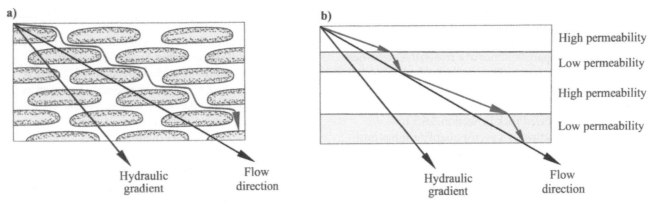

Figure 1.9: Illustration of anisotropy: (a) intergranular flow; (b) layered heterogeneity

horizontal permeability greater than the vertical (Figure 1.9a). This anisotropy will usually be relatively minor, with the ratio of K_h:K_v rarely more than 3 or 4 to 1. More significant is that most sediments exhibit various scales of layering, with coarser and finer material interbedded, so that on a larger scale what is described as bulk anisotropy is really the result of layered heterogeneity (Figure 1.9b).

A sequence of isotropic or nearly isotropic layers of different permeabilities can be shown to behave in bulk like a single homogeneous isotropic layer, with the flow direction diverted away from that of the hydraulic gradient. Figure 1.10 shows a theoretical layered system in which each individual layer is homogenous, isotropic and of uniform thickness. Thicknesses and hydraulic conductivities are as shown. For such a system, the water-bearing capacity or transmissivity is the sum of the transmissivities of the beds, so that the total transmissivity T_t is given by:

$$T_t = T_1 + T_2 + \dots + T_n$$
$$= K_1 b_1 + K_2 b_2 + \dots + K_n b_n \quad \text{(eqn 1.12)}$$
$$= \sum_{i=1}^{n} K_i b_i$$

An equivalent hydraulic conductivity for this system could be obtained by assuming that the total transmissivity T_t is the product of the total thickness b_t and the equivalent horizontal hydraulic conductivity K_x, so that:

$$T_t = K_x b_t = K_1 b_1 + K_2 b_2 + \dots + K_n b_n \quad \text{(eqn 1.13)}$$

and

$$K_x = \frac{K_1 B_1 + K_2 B_2 + \dots}{b_1 + b_2 + \dots} = \frac{\sum_{i-1}^{n} K_i B_i}{\sum_{i=1}^{n} b_i} \quad \text{(eqn 1.14)}$$

which is the sum of the transmissivities divided by the sum of the thicknesses. When all the layers have the same thickness so that $b_1 = b_2 = b_n = b$, Equation 1.14 can be simplified to:

$$K_x = \frac{K_1 b + K_2 b + \dots + K_n b}{nb} \quad \text{(eqn 1.15)}$$

which reduces to

$$K_x = \frac{\sum_{i=1}^{n} K_i b}{nb} = \frac{\sum_{i=1}^{n} K_i}{n} \quad \text{(eqn 1.16)}$$

so that, for a system of layers of equal thickness, the equivalent hydraulic conductivity parallel to the layering is the sum of the individual hydraulic conductivities divided by the number of beds. This is the arithmetic mean of the conductivities or permeabilities.

Flow parallel to the layering or bedding, like that discussed above, will be dominated by the most transmissive layer. Flow across the layering will be dominated by the layer with the greatest hydraulic resistance, where hydraulic resistance is defined as b/K, the thickness divided by the hydraulic conductivity. (If the layer itself is anisotropic, the K value to be used is in the direction at right angles to the layering; for bedded sediments this will be K_v.)

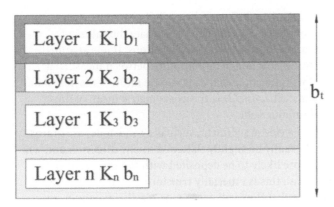

Figure 1.10: Layered system of beds to illustrate statistical means

A mean value for the vertical conductivity, K_z, can also be derived for the layered system shown in Figure 1.10. Following the argument of Freeze and Cherry (1979), it is assumed that vertical flow is taking place through the beds in Figure 1.10 with a specific discharge of q_z. This flow requires a head loss of Δh_1 across Bed 1, Δh_2 across Bed 2, and so on, with a total vertical head loss across the layered system of Δh_t. From Darcy's law:

$$q_z = \frac{K_1 \Delta h_i}{b_1} = \frac{K_2 \Delta h_2}{b_2} = \frac{K_n \Delta h_n}{b_n} \quad \text{(eqn 1.17)}$$

But over the whole thickness, also

$$q_z = \frac{K_z \Delta h_t}{b_t} \quad \text{(eqn 1.18)}$$

so that

$$K_z = \frac{b_t q_z}{\Delta h_t} = \frac{b_t q_z}{\Delta h_1 + \Delta h_2 + \ldots + \Delta h_n} \quad \text{(eqn 1.19)}$$

but

$$\Delta h_1 = \frac{b_1 q_z}{K_1}, \; \Delta h_2 = \frac{b_2 q_z}{K_2} \ldots \quad \text{(eqn 1.20)}$$

and so on, so that

$$K = \frac{b_t q_z}{\dfrac{b_1 q_z}{K_1} + \dfrac{b_2 q_z}{K_2} + \ldots + \dfrac{b_n q_z}{K_n}}$$

$$= \frac{b_t}{\displaystyle\sum_{i=1}^{n} \dfrac{b_i}{K_i}} \quad \text{(eqn 1.21)}$$

Therefore, the equivalent vertical hydraulic conductivity of the layered system is the total thickness divided by the sum of the hydraulic resistances. As for flow along the layers, when all the n layers are of equal thickness b the expression for flow across the layers (Equation 1.17) can be simplified to:

$$K_z = \frac{nb}{\displaystyle\sum_{i=1}^{n} \dfrac{b}{K_i}} = \frac{n}{\displaystyle\sum_{i=1}^{n} \dfrac{1}{K_i}} \quad \text{(eqn 1.22)}$$

The expression

$$\frac{n}{\displaystyle\sum_{i=1}^{n} \dfrac{1}{K_i}}$$

is the harmonic mean of the hydraulic conductivities. Therefore, for a system of layered rocks like that in Figure 1.10, with all the layers of equal thickness, the equivalent permeability parallel to the layers is the arithmetic mean of the permeabilities and the equivalent permeability across the layering is the harmonic mean. As the harmonic mean is always less than the arithmetic mean, the permeability is always greater parallel to the layers than across them. Therefore, a sequence of units that have different permeabilities but are individually isotropic will behave *en masse* like a single uniform anisotropic unit. If the layers are themselves anisotropic with $K_h > K_v$ as is usual in sediments, the bulk anisotropy will be enhanced.

Although such a sequence can often be represented in a regional or site-scale setting as a single homogeneous anisotropic unit, this approach may not be valid at the sector or slope scale, especially when layer thicknesses are comparable to inter ramp slope heights. In these cases, if geological knowledge points to the existence of a layered structure, then modelling the situation with an anisotropic homogeneous field by adopting values of equivalent permeabilities is not recommended. Instead, where the exact location of pathways of highly conductive material may become important, such features should be modelled explicitly.

The geometric mean always lies between the arithmetic mean and the harmonic mean. In addition to the statistical evidence that permeability often shows a log-normal distribution, this is perhaps an additional reason for using the geometric mean to provide a working average of permeability in settings where there is no clear layered or trending structure.

1.1.2.4 Intergranular ('porous medium') formations

The permeability of sedimentary materials is controlled mainly by the ease of flow through the pore throats, and the size of these is influenced by several factors, principally grain size, sorting, rounding and cementation. Other things being equal, coarser materials will possess larger pore interconnections and are therefore more permeable. However, pore sizes and throat apertures will be reduced if the grains are angular (so that they pack more tightly together) and by the presence of smaller grains (which fill the spaces between the larger ones, reducing both porosity and permeability). It has been found empirically that the D_{90} retained size (finest 10 percentile) from sieve analyses tends to have a major control on permeability. The most permeable sedimentary materials are therefore those that are well sorted with large, rounded particles. Cementation will usually close pores, but cemented material may fracture, opening up other pathways.

Figure 1.11a shows this sort of variability on a small (hand-specimen) scale; Figure 1.11b shows it at a larger (outcrop) scale. The anisotropy ratio (K_h:K_v) in a mixed bed of sandstone may be around 10:1 or 20:1. In a thick sequence of sediments, or on a regional scale, that ratio may be in excess of 10 000:1.

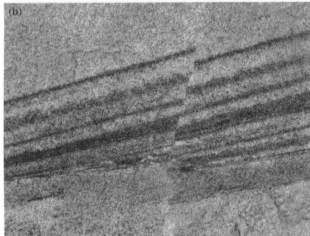

Figure 1.11: (a) Hand specimen of sandstone showing cross bedding and minor textural variations; (b) Porous sandstone aquifer illustrating inhomogeneity (from Timna mines, Israel).

1.1.2.5   Fractured rocks

Although mine locations may be bordered or overlain by unconsolidated sediments, most hard rock mines are hosted by igneous or metamorphic rocks or highly-indurated sediments. All of these rock types are prone to fracturing and the fractures will frequently contribute most, often virtually all, of the permeability.

In the case of a fractured medium, the groundwater flow occurs through interconnected fractures, joints and other discontinuities, as illustrated in two dimensions in Figure 1.12.

The local-scale movement of groundwater is influenced by the conditions within individual fractures and by the interconnectivity of the fractures. In addition to the main faults and high-order fracture zones, most hard rock types usually contain abundant low-order, small-aperture fracture and joint sets throughout the rock mass. In this

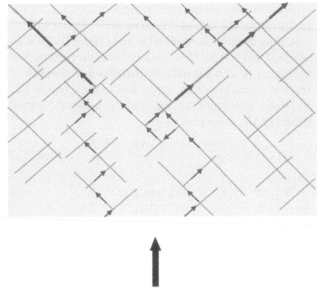

Hydraulic gradient

Figure: 1.12: Preferential and dispersive flow in a fractured rock unit

context, 'high-order' refers to the first order or largest fractures and 'low-order' to the narrower fractures.

Preferential flow is an inherent characteristic of virtually all fractured rock settings because of the preferential development of fracture patterns, alignments and aperture widths. However, because most rock is pervasively fractured at a scale of several metres, it is usual that some flow can occur at all scales along particular interconnected fracture channels. In the same way as in a porous medium, there will usually be preferential flow on a larger scale (say >10 m) due to the presence of larger geological structures or fractures with a higher and more consistent aperture, and due to variations in the lithology, alteration and mineralisation of the rock mass.

A fracture can be thought of as an extreme example of a layer of high permeability (the term **fracture** is used here to include all rock discontinuities through which flow can occur). Theoretical studies of flow in fractures are usually based on the starting, if unlikely, assumption that the fracture can be treated as an opening bounded by smooth, plane, parallel plates. It is therefore of uniform aperture. For this ideal case, the flow Q that can occur through the fracture is proportional to b^3, where b is the plate separation or idealised fracture aperture (Figure 1.13).

Figure 1.13: Idealised 'parallel-plate' fracture

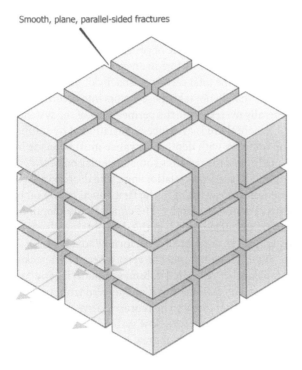

Smooth, plane, parallel-sided fractures

Figure 1.14: Flow through a rock mass containing three idealised, mutually orthogonal fracture sets (After Snow, 1968)

The hydraulic conductivity, K_f, of the fracture is given by:

$$K_f = \frac{gb^2}{12\upsilon} \qquad \text{(eqn 1.23)}$$

and its transmissivity, T_f, by

$$T_f = \frac{gb^3}{12\upsilon} \qquad \text{(eqn 1.24)}$$

where:

g = acceleration due to gravity

b = plate separation (idealised fracture aperture)

υ = kinematic viscosity of water

For pure water at 10°C (υ = 1.3 mm²/s), with b in mm, $T_f = 6.3 \times 10^{-4} \, b^3$ m²/s. Snow (1968) showed that for a system with three sets of such identical fractures, mutually at right angles and with regular spacing (Figure 1.14), the porosity and hydraulic conductivity will be as in Figure 1.15. For 'real' fractures, roughness decreases conductivity (but does not really affect porosity). This means that porosities tend to be higher for a given permeability than predicted by Figure 1.15.

In hard-rock areas, a common form of trending heterogeneity is depth-dependency, with permeability decreasing with depth. Evidence for this is seen in the decline of the specific capacities of wells (Davis & DeWiest, 1965, Figure 9.4). There can be several explanations for this, such as the increased action of weathering leading to higher permeabilities near the surface, but one of the most obvious is the increase of total vertical stress with depth of burial and the resulting tendency for fractures to be less open with depth (see also Section 1.1.1.3).

There is increasing evidence that, in general, fractures are not simple, parallel-sided planar systems. They have

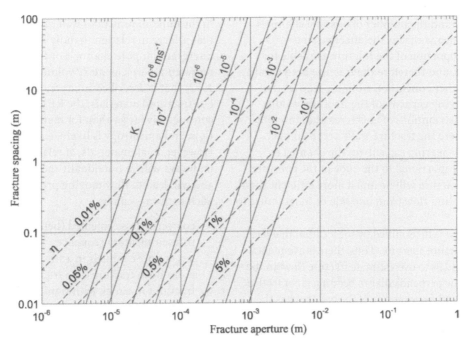

Figure 1.15: Values of porosity (per cent) and hydraulic conductivity (m/s) for fracture flow under the ideal conditions of Figure 1.14. Hydraulic conductivity values assume the fluid is pure water at 10°C (After Snow, 1968)

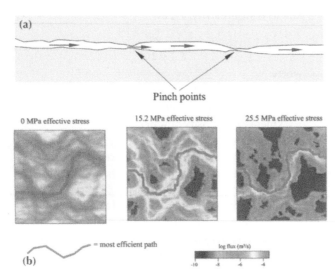

Figure 1.16: (a) 'Pinch points' in a fracture have a major control on the fracture permeability. (b) Fracture channelling and response to changes in effective stress (Source: Jeong *et al.*, 2006). The widest and most permeable pathways through the fracture will usually not be linear but will follow a tortuous or anastomosing route.

rough walls that are partially in close contact, giving rise to a constantly varying aperture and a degree of channelling (Black *et al.*, 2007). This channelling gives rise to a preferred directional permeability in the plane of the fracture. Figure 1.16 shows an illustration of such a channel with behaviour modelled at various values of effective stress, which in effect corresponds to varying depths of burial. It will be seen that, with the increase in stress, the area of fracture that is effectively 'closed' increases rapidly and the number of possible channels decreases. This means that as the effective stress increases, the proportion of the fracture available for water movement, and therefore the fracture permeability, decreases. The other conclusion from the work is that variation in carrying capacity of the fractures results from different distributions of apertures and different relative areas where the fracture walls are in contact. Because the flow-bearing capacity of the channel is approximately proportional to the cube of the aperture, one part of the fracture will be much more efficient than all others at enabling flow from one side of the fracture to the other.

In general, flow is determined by the location that has the minimum fracture aperture. Thus, there is a tendency for theoretical models to overestimate fracture flow. In the same way that flow perpendicular to bedding is controlled by the thickness and permeability of the bed that has the greatest hydraulic resistance (Section 1.1.2.3), so flow along non-uniform fractures will be dominated by the sections that have the lowest permeability. These 'choke points' may occur because of roughness, reduction in aperture, or where the most permeable channels (like that in Figure 1.16) are

reduced in carrying capacity possibly by increased effective stress, infilling by mineral deposition or weathering deposits, or the presence of gouge. These regions will act like higher-value resistors in an electrical network, controlling the total carrying capacity of the system. Simple models that do not take these factors into account will generally overestimate the permeability of the system.

Where there is evidence of variation of hydraulic conductivity with depth, allowance must be made in modelling. This requires some formulation of the conductivity–depth relationship. The decreasing trend is often more significant in the first 100 m to 300 m, but can extend to greater depths. The conductivity–depth curve can be estimated based on the data obtained from the usual packer tests that are performed to a depth of 200 m to 300 m, though it must be remembered that packer tests sample rather restricted volumes of material and 'point' conductivity estimates that are derived from packer testing are not likely to be representative of the overall interconnected fracture flow system (Section 1.1.2.7). If data from packer tests at greater depths are not available, information on the permeability–depth or permeability–stress relations of the specific deposit or the general rock type should be sought from literature.

1.1.2.6 Preferential flow

Darcy's law (Appendix 1) is an empirical law, based on observation. It does not attempt to describe the pattern of flow of water through the pore channels between sand grains (Figure 1.8) or through individual fractures (Figure 1.16), but treats the porous or fractured medium as a continuum. This works as long as the interest is only in the bulk movement of water through a volume of rock that is large in relation to individual flow paths, whether intergranular pore channels or fractures. This is the representative elementary volume (REV) concept (Bear, 1972) that is described in more detail in Section 3.6.5.1. For fractured materials, the REV will need to be considerably larger than for materials with intergranular flow, and anisotropy is likely to be more important. However, in all materials, at relatively small scales, some pathways will be fractionally more permeable than others and will lead to flow moving preferentially along them. Such pathways can be:

- coarser or less cemented (hence more permeable) layers in a sedimentary sequence;
- larger pore channels in a single sandstone bed;
- more-open fractures in a fractured rock;
- fractures in a rock with significant intergranular permeability.

The presence of these preferential pathways leads to longitudinal dispersion in flow.

Preferential flow is clearly a major consideration in materials that possess dual porosity or dual permeability, whether because of the presence of fractures or because of seams of coarser, less cemented or broken material.

1.1.2.7 Determination of permeability

Methods of assessing permeability essentially fall into five groups.

1. Estimation from literature, comparisons with similar known materials, or other sources.
2. Estimation from other properties (e.g. from grain-size distributions for granular materials; from fracture aperture and spacing for fractured materials).
3. Measurement on samples in a laboratory.
4. Measurement *in situ* using single well tests ('point' measurements).
5. Measurements *in situ* using cross-hole or multiple-hole tests ('bulk-scale' measurements).

Methods 1 and 2 may be adequate for preliminary assessments and sometimes for Conceptual level studies (Level 1), but not for planning of a general dewatering or slope depressurisation program. Laboratory methods (3) offer potentially the most accurate measurements (e.g. controlled conditions, known geometry of sample), but suffer from the disadvantage that the samples may not be representative and may be disturbed by collection, transport or handling. Laboratory measurements on core samples usually indicate lower values, often by an order of magnitude or more, than field tests. There are two principal reasons for this.

1. When coring, core loss or deterioration usually occurs in friable or closely fractured material, so that the most permeable material is not available for laboratory testing.
2. It is virtually impossible to bring back representative fractured material for testing with the apertures preserved, even from outcrop material, so laboratory tests on fractured materials will underestimate *in-situ* values.

In-situ tests deal with larger volumes but lack the precision offered by a laboratory. Standard methods for large open pit field measurements include slug tests, packer tests and pumping tests.

Slug and packer tests (Section 2.2.5) provide what essentially are 'point' measurements, yielding an estimate of the hydraulic conductivity in the immediate vicinity of the test well. In this context it must also be recognised that the measured 'transmissivity' or 'permeability' of a fracture is likely to be controlled by the interconnection with other fractures or permeable material rather than by the fracture aperture. Thus, cross-hole or multiple-hole tests (Section 2.2.8) that provide a 'bulk scale' average value for the area covered by the observation wells are more representative of the flow system. In many cases the 'bulk scale' area tested will still represent only a small part of the region to be modelled and will still face sampling difficulties in that head changes can be known only where piezometers are available. Even so, the use of 'point' permeability measurements rather than a program of 'bulk scale' multiple well testing, interpretation and judgement of the overall groundwater flow system represents a common source of error for many mine dewatering assessments.

The best approach is often a combination of methods. For example, in dual-porosity materials, it may be appropriate to test the bulk properties on site to characterise the fracture response and to use laboratory methods to help understand the behaviour of the inter-fracture blocks. This synergetic approach is especially useful for characterising the diffusive behaviour of the different (matric and fracture) components. It must be remembered that, in most cases, the ultimate control on water entry into a pit is not the permeability of the material forming the pit walls, but the transmissivity of the materials some distance away that form a barrier to water replenishment and therefore control the availability of the water to reach the pit. Thus, in many general dewatering and slope depressurisation assessments, a knowledge of the 'water availability' (from an assessment of flow barriers and recharge) is often more pertinent than a knowledge of the local-scale permeability.

1.1.2.8 Response time

The ratio of hydraulic conductivity to specific storage, or of transmissivity to storage coefficient or specific yield, is termed the hydraulic diffusivity, D, of the formation.

$$D = \frac{T}{S} = \frac{K}{S_s} \qquad \text{(eqn 1.25)}$$

This is an important parameter for transient analysis. In particular it controls the speed of propagation of a head perturbation through a rock unit.

The order of magnitude of time taken for a 'front' of head perturbation to travel a distance l (diffusion process) may be approximately estimated using the following formula:

$$t = \frac{l^2}{D} = \frac{l^2 S_s}{K} \qquad \text{(eqn 1.26)}$$

where K is hydraulic conductivity. Some workers use the form:

$$t = \frac{l^2}{2ND} \qquad \text{(eqn 1.27)}$$

where N is the number of dimensions involved. So for the movement of a one-dimensional front (planar equipotentials) $N = 1$; for two dimensions (cylindrical equipotentials) $N = 2$; for three dimensions (spherical equipotentials) $N = 3$.

Equation 1.27 can be used, for example, to estimate how long it will take for the effect of pumping from a production well to be detected at a piezometer or to influence the flow of a spring. It should be noted that this time is generally very much shorter than the time taken for a molecule of water or dissolved solute to travel the same distance (Section 1.1.2.2). In dual-porosity units, it can also be used to estimate the time for the pore water in a matric block to reach equilibrium with water in a fracture when the fracture system has been depressurised.

Clearly, response times will be shorter in highly permeable formations than in poorly permeable units and will be longer in units with large storage coefficients. In groundwater units where the pore space is emptying or filling (where the storage coefficient is effectively the specific yield), the response time will be especially long. Typically volcanic or intrusive rocks with pervasive and clean fracturing (and low porosities) tend to have very high diffusivities and very short response times.

1.1.3 Pore pressure

Pore pressure, p, is defined as the pressure of the groundwater occurring within the pore spaces of the rock or soil. At any given point below the water table, the pore pressure is the result of the weight of the column of water acting on that point through the interconnected grains or fractures. Pore pressure is positive below the water table, zero at the water table, and negative above the water table, as shown in Figure 1.17.

Above the water table, the fractures and pore spaces are partially drained, and the rock mass is not fully saturated. The water that remains is held in place within fractures and pore spaces by surface tension, so there is a negative pore pressure (suction) associated with these zones. In most instances, the magnitude of the negative pore pressure is too small to be of significance in holding the rock mass together. Significant negative pore pressures

usually occur only in fine-grained materials (such as clays) or hard rock settings where the fracture systems are infrequent, of small aperture and poorly interconnected. In these poorly permeable materials, where rebound resulting from the excavation increases the porosity of the rock but the permeability (hydraulic conductivity) is too low for the water pressures to respond (stabilise) immediately, it is possible that negative pore pressures, usually of a transient nature, can occur.

In practice, the field measurement of negative pore pressures in a fractured rock medium around an open pit is rare. Negative pore pressures can result from (i) drainage (creating pore water suction in the material that remains above the water table), (ii) removal of water by plant roots or by direct evaporation, and (iii) reduction of pressure because of expansion of pore volume. Most experience suggests that, in terms of 'drainage', clay soils may drain to 'field capacity' at suction pressures of about –5 m (about –50 kPa), while sandy soils may drain to field capacity at about –1 m (about –10 kPa). Negative pressures of –2 to –4 m (about –20 to –40 Kpa) are measured in leach pads following active drainage cycles. There are a number of literature case studies that indicate measured values well below 50 m suction (about –500 kPa pressure) in the soil region due to evapotranspiration, and values below 10 m suction (about –100 kPa pressure) at some depth in the profile. The most-negative pore pressures that result from expansion of pore volume are usually measured in plastic clays, below or surrounding an excavation, in formations that are expanding as a result of a reduction in the weight of the overlying material. In these cases, negative pressures below –20 m (–200 kPa) have been reported. The pore pressure deficit usually dissipates with time but, in low-permeability clays

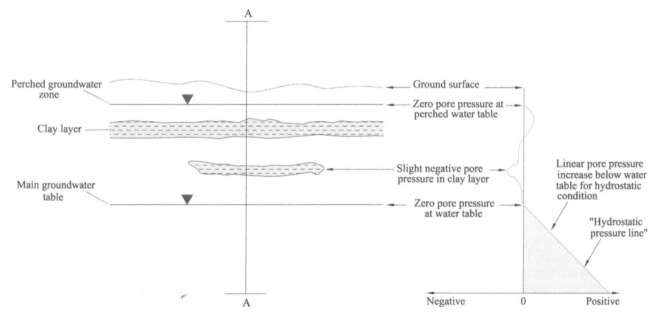

Figure 1.17: Pore pressure relative to the water table

(<10^{-9} m/s), it is possible that negative pressures may exist for several years, or even decades. A similar situation may potentially occur in hard rock at significant depth, with low permeability and poorly interconnected fracture apertures, for example in an area exhibiting isostatic rebound (see also Section 3.6.6.2).

Under porous-medium conditions, the pore pressure occurs in the interstitial spaces between grains and acts on the grains themselves. Under fracture flow conditions, the pore pressure occurs within the fractures and acts on the surface of the rock mass surrounding the individual fractures. It should be noted however, that the majority of hard rock settings associated with open pits tend to have pervasive fracturing and the pore pressure acts in the abundant low-order micro-fractures and joint sets in addition to the larger, high-order fractures. For the transitional rock types, the pore pressure occurs both in the interstitial spaces between grains and in the fractures.

Below the water table, pore pressure is determined by measuring the height of a column of water at a given point (depth and location) within the rock mass. In general, the deeper the point below the water table, the higher the pore pressure.

1.1.4 Head and pressure conditions

1.1.4.1 Steady-state conditions
Prior to the start of any mine excavation, the pressure field acting within the rock mass is usually in a state of equilibrium, provided there are no other major groundwater extractions in the area. The pore pressure field has typically evolved to near steady-state over a long period of time, and is the result of the natural hydraulic stresses that act on the groundwater system (recharge and discharge) and the hydraulic properties of the local materials. In a natural groundwater system, recharge enters through infiltration of precipitation, runoff or surface water at one elevation, and discharge occurs as a

result of evaporation or flow into a surface water body (or another groundwater unit) at a lower elevation.

1.1.4.2 Lateral head and pore pressure gradients
The magnitude of the hydraulic stresses (recharge and discharge) and the resistance to flow along the flow path (determined by the permeability of the materials and, in particular, the hydraulic barriers) combine to create a lateral hydraulic gradient (slope of the potentiometric surface) between the recharge area and the discharge area. This is diagrammatically illustrated in Figure 1.18 as the pre-mining flow field through an ore body.

For rock masses characterised by interconnected granular or fracture pore spaces with no impediments to groundwater flow, variation in pore pressure at any depth can be represented by a single potentiometric surface. The slope of the potentiometric surface in Figure 1.18 implies that there is a horizontal or lateral hydraulic gradient. Along any horizontal line, the change in head indicated by such a gradient is caused solely by a change in pore pressure, so that there is a horizontal pressure gradient between P and P′. At point P on the east side of the ore body, the potentiometric surface elevation is 1000 m and the elevation of point P is 900 m, so the pore pressure at P is 100 m water head, or ~1000 kPa. At point P′ on the west side of the ore body, the potentiometric surface elevation is at 950 m. The pore pressure at point at P′ is 50 m water head, or ~500 kPa. The pore pressure gradient means that the pressure on the upstream side of every mineral grain or block is greater than the pressure on the downstream side. Thus, there is a force (the seepage force) attempting to move each grain or block in the direction of decreasing pressure.

In natural situations, the properties of the materials along the flow path are infinitely variable, and the variations act to create lateral differences in the distribution of pore pressure along the flow path. Large-scale variations in head (>10 m) invariably occur due to

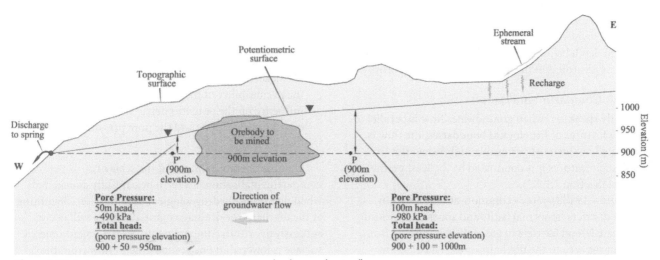

Figure 1.18: Lateral variation in pore pressure as a result of groundwater flow

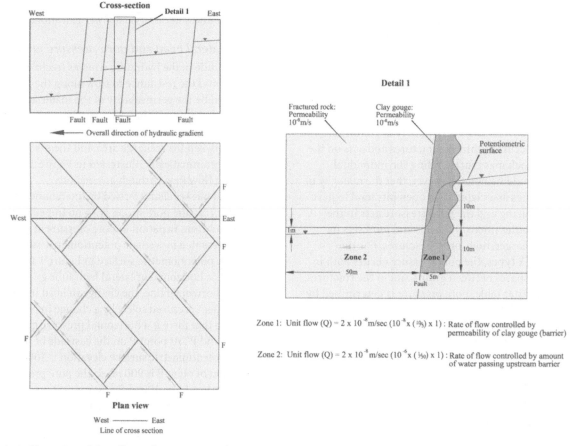

Figure 1.19: Illustration of the effects of compartmentalisation

structural zones perpendicular to the flow path (usually flow barriers), structural zones parallel to the flow path (often flow conduits) and lithological changes that occur laterally or vertically along the flow path (often associated with weathering, alteration, or mineralisation in mining districts). Small-scale variations (<10 m) also occur because of inhomogeneity and anisotropy, as described in Section 1.1.2.3 and Appendix 1.

Under natural conditions, as illustrated in Figure 1.18, the pressure distribution is not directly related to the porosity. Provided the fracture sets are interconnected, at any given depth below the water table the same pore pressure will occur regardless of the fracture aperture and porosity.

1.1.4.3   Compartmentalisation

Generally speaking, when groundwater flow is parallel to bedding features or lithological boundaries, the flow is dominated by the most permeable unit. When flow is across such features it is dominated by the least permeable material (Section 1.1.2.3).

Figure 1.19 illustrates a common situation within mining districts of normal faults and zones of alteration that create lateral barriers to groundwater moving from the recharge area to the discharge area. The diagram shows that the magnitude of flow and the shape of the

potentiometric surface are controlled by the presence of the discrete barriers, not by the properties of the intervening rock. This condition is commonly termed *lateral compartmentalisation* and is notably found in fractured rock systems. Frequently, the key features of a compartmentalised groundwater system are:

- the potentiometric surface is relatively flat between the barriers (within the compartment) and is 'stair-stepped' across the barriers;
- the ability of the system to transmit groundwater is controlled by the permeability of the barriers (the least permeable part of the flow field) rather than by the permeability of the rock mass itself;
- the permeability of the rock itself has only a minor influence on the potentiometric surface, the magnitude of groundwater flow and the pore pressure distribution.

Compartmentalisation, and the scale of compartmentalisation, needs to be carefully considered when planning a hydrogeological investigation. Beginning at the discharge or drainage point, drainage will occur successively within compartments. As the potentiometric surface is lowered in one compartment, flow from the upgradient compartment will result in the potentiometric surface being lowered in that compartment, and so on. Within an individual compartment, the fall in head will occur smoothly only if the material within it is homogeneous. If it is fractured, for example, the head in the main (high-order) fracture zones will decline first and there will be a gradual drainage of the lower order fracture zones throughout the rock mass towards the main flow paths in the high-order fracture zones. Piezometers sited only in the main fracture zones or only in the intervening, less permeable rock mass, may therefore give a misleading impression of what is happening in the unit as a whole. The presence of clay alteration along the flow path creates a further inhomogeneity to the groundwater flow pattern as a result of the rock becoming softer and the fractures becoming healed within the clay-altered zone.

The measurement and testing of the rock mass between the structures can provide a misleading representation of the magnitude of groundwater flow and the distribution of the potentiometric surface. In order to provide a representative characterisation of the groundwater system, it is necessary to apply additional temporary stress at one location and to measure the resulting hydraulic response surrounding that stress point.

In some settings, horizontal or sub-horizontal layers can create a degree of **vertical compartmentalisation**. An example of this is shown in Figure 1.20. Here, a sequence of relatively permeable sand units is interbedded with less permeable silty clay units. Under pre-mining conditions, there would have been hydrostatic conditions or possibly a small downward component to the hydraulic gradient, with steeper vertical gradients across the less permeable beds and shallower vertical gradients across the sand units. The creation of the pit void has provided a path for groundwater to drain from the system to the pit face. This has led to the drainage of the lowermost, more permeable materials and the creation of a perched water table in the upper part of the sequence, where groundwater in a sandy layer is supported by the less permeable material beneath and is unable to drain vertically downward. Both the upper sandy bed and the permeable unit beneath are now draining to seepage faces in the pit wall. The presence of high-level seepage faces, in addition to the nuisance factor, can lead to partial saturation of the material below, potentially reducing the stability of the pit slope.

1.1.4.4 Vertical head and pore pressure gradients

At the pit scale, prior to mining, most natural groundwater systems have no significant component of vertical hydraulic gradient. They are in the hydrostatic condition,

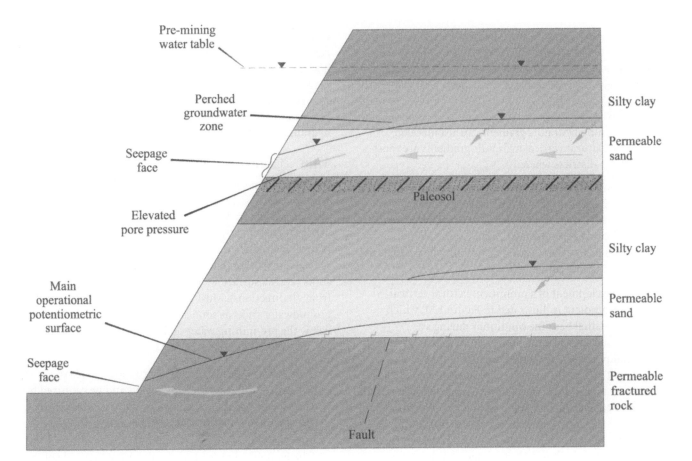

Figure 1.20: Vertical and lateral compartmentalisation with seepage faces

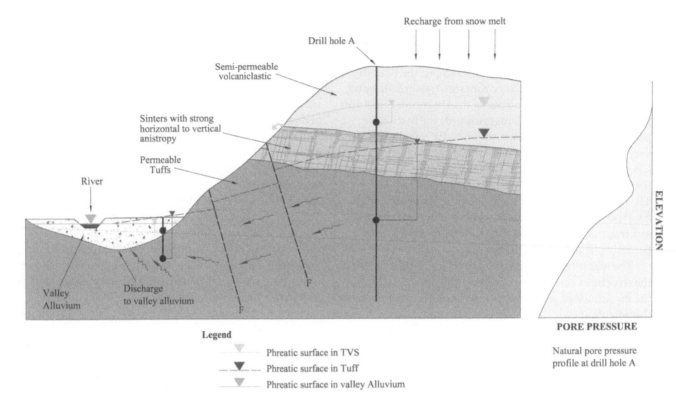

Figure 1.21: Example of natural vertical compartmentalisation with a vertical hydraulic gradient: from Lewis and Clark County, Montana, USA

where the water pressure 10 m below the water table is 10 m head (98 kPa), the water pressure 20 m below the water table is 20 m head (196 kPa), the water pressure 100 m below the water table is 100 m head (980 kPa), and so on.

However, in certain situations, a vertical component of the pre-mining hydraulic gradient can also develop naturally. Natural vertical hydraulic gradients are most common in districts where:

- there are large variations in topographic relief (between the recharge source and the discharge zone);
- there are high rates of recharge and discharge (i.e. a large natural groundwater flux);
- there is a large contrast in permeability in the material along the flow path.

The development of a prominent natural vertical hydraulic gradient is illustrated in Figure 1.21, which shows the actual baseline conditions during an investigation for a mining project in Montana, USA. In this example, direct recharge from precipitation (snowmelt) occurs to a relatively permeable volcaniclastic unit in an area of high topography. The groundwater in the upper volcaniclastic unit is trying to move towards a lower volcanic (tuff) unit, from where it discharges into a permeable alluvial groundwater system. However, there is an intervening shallow dipping sinter unit separating the two main volcanic units. Although the sinter unit contains

extremely hard and fractured bands, and itself exhibits a very high lateral permeability, it also has a very strong horizontal (k_h) to vertical (k_v) ratio in permeability, and the permeability of the unit in the direction of flow is very low. As a result, it forms a very strong barrier across the main groundwater flow path, and a pronounced vertical hydraulic gradient has developed across the unit.

1.1.5 Controls on pore pressure

As soon as mining-induced stresses are placed on the groundwater system, it is usual that steeper hydraulic gradients begin to develop, often with associated strong vertical components. The creation of a pit void disrupts the natural hydraulic gradient and groundwater flow paths. All open pits other than side-hill cuts are surface water sinks (i.e. the surface water flows towards the pit floor from all directions). Most active pits also become groundwater sinks as soon as the pit floor is excavated below the pre-mining water table and the water level in the pit is lowered. The consequences of this are as follows.

- All groundwater within the surrounding 'area of influence' will flow towards the pit and be captured by the pit (the area of influence normally gets larger as the pit is excavated progressively deeper below the water table).
- The hydraulic gradient on the downgradient side of the pit becomes reversed for some distance outside the pit.

- Groundwater flow is towards the pit from all directions, irrespective of the pre-mining groundwater gradient.

The hydrological sink condition often extends well into the mine closure period, as the pit void fills with water. In arid climates, where the mean annual evaporation rate is higher than the mean annual precipitation rate, the hydrological sink condition is often permanent. If this is the case, all the water within the pit lake remains permanently in the pit and is not transported into the surrounding groundwater system. This aspect often has significant consequences for planning of the mine closure program.

In the example given in Figure 1.22, a mine excavation has progressed below the elevation of the 'natural' water table, creating a zone of discharge and causing groundwater to flow towards the mining void from outside the crest of the pit slope. The figure shows a typical flow net that may occur in a homogenous pit slope where groundwater is trying to move from a recharge area outside the crest of the pit to an induced discharge zone at the toe of the slope. The figure shows that:

- a downward component of the hydraulic gradient has developed above the pit crest and in the upper part of the slope because water is trying to move away from the recharge source;
- an upward component of hydraulic gradient has developed at the toe of the slope because groundwater is trying to move towards the discharge zone;

- the downward component of the hydraulic gradient and the upward component of the hydraulic gradient become reversed somewhere below the middle part of the slope.

Figure 1.23 illustrates the relationship of pore pressure and depth below the water table for the flow net shown in Figure 1.22. Three distinct zones can be recognised, as follows.

1. A downward gradient below the crest of the slope (Zone A on Figures 1.22 and 1.23). In this zone, the rate of increase of pore pressure below the water table is less than hydrostatic. In the example shown, the water pressure 10 m below the water table is about 9 m head (88 kPa) and the water pressure 20 m below the water table is about 18 m head.
2. An upward gradient below the lower part of the slope (Zone C on Figures 1.22 and 1.23). In this zone, the rate of rise of pore pressure below the water table is greater than hydrostatic. In the example shown, the water pressure 10 m below the water table is about 11.5 m head (113 kPa), and the water pressure 20 m below the water table is about 23 m head. Potentially, this situation could destabilise the slope.
3. No vertical component of gradient in the central part of the slope (Zone B on Figures 1.22 and 1.23). In this narrow zone, groundwater flow is essentially horizontal and the system shows a condition with no significant vertical component of hydraulic gradient.

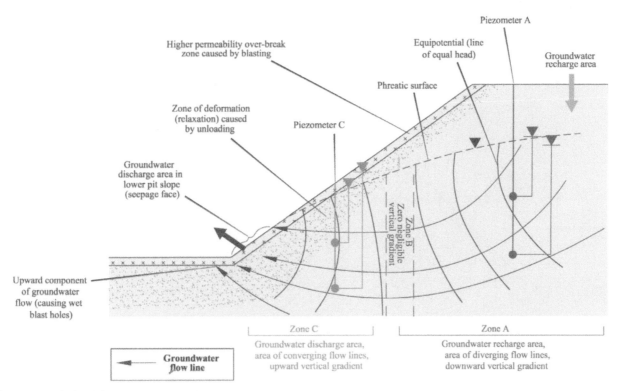

Figure 1.22: Idealised flow system in a pit slope (Source: Read & Stacey, 2009)

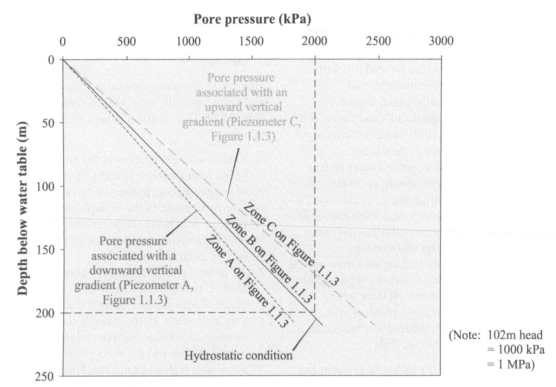

Figure 1.23: Illustration of pore pressure vs depth (Source: Read & Stacey, 2009)

Figure 1.24 is a flow net with some of the common features that are encountered in many pit slopes. In this figure, the simple groundwater flow net has been disrupted by the presence of sub-vertical geological structures that create compartmentalisation and prevent free groundwater movement from the crest of the pit towards the toe of the slope. In addition, the zone of clay alteration provides further resistance along the flow path, while the zone of enhanced interconnected permeability in the lower slope (the over-break zone) provides a

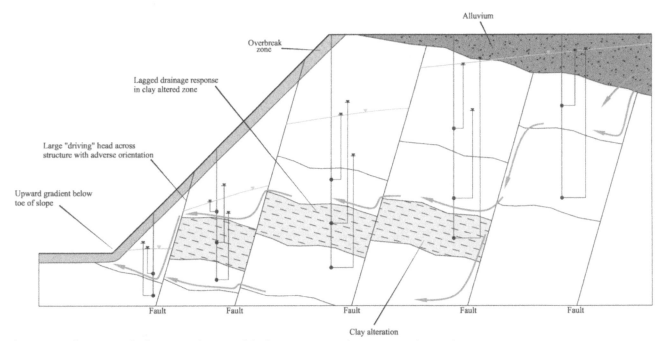

Figure 1.24: Illustration of a flow net with some of the features commonly encountered in pit slopes

discharge point and allows for improved drainage of the toe area.

The discussion above is for steady-state conditions, so the only factors controlling the head distributions and localised hydraulic gradients are:

a. the elevation of the recharge source;
b. the elevation of the discharge zone;
c. the permeability of the materials along the flow path (isotropic conditions are assumed in Figure 1.22).

However, for a dynamic pit slope, it is typically necessary to employ a transient analysis (Section 4.2.5). Therefore, the following additional factors must be considered.

- Changes in the recharge rate.
- Variations in permeability along the flow path.
- Drainable porosity and variations in drainable porosity along the flow path.
- Rate of mine excavation.
- Changes in the location of the discharge area as a result of the dynamic mine excavation.
- Implementation of drainage measures.
- Time.

Figure 1.25 illustrates the dynamic conditions behind a pushback in a homogenous pit slope. The pushback causes the groundwater head (and pore pressure) at Point P behind the pit slope to reduce because groundwater is draining through the pit walls into the excavation. However, although the absolute pore pressure has reduced, an increase in the pressure gradient has occurred from the materials behind the pit slope (Point P) to the newly exposed toe of the slope because the head loss (between recharge area and discharge point) is now occurring over a shorter distance. The significance of the pressure change for seepage force was discussed in Section 1.1.4.2; the significance of the increase in pore pressure is discussed in Section 1.1.6.2.

For a mature mining operation, it is often of benefit to evaluate the current groundwater levels in the context of the pre-mining water levels to evaluate total progress of the dewatering or slope depressurisation process. Many operations only assess achievements over recent years rather than looking at the total progress, which can often lead to a misinterpretation of results relative to pre-mining conditions.

1.1.6 The role of water pressure in slope stability

1.1.6.1 Geotechnical controls

Pore pressure is a major control on the stability of slopes in rock and soil. Moreover, in the case of large open pits, it is usually the only property that can be easily varied. Reduction in pore pressure causes an increase in effective stress and therefore an increase in the shear strength of the slope materials. Slope depressurisation is therefore a method that is readily available to the mine operator to improve slope stability and achieve a more economical slope design.

The evaluation of pore pressure is an integral part of any slope engineering assessment. A change in pore pressure exerts the following geotechnical controls:

- it changes the effective stress and hence the shear strength of the materials within the slope;
- it can cause a volumetric change in the slope materials;
- it can cause a change in hydrostatic loading.

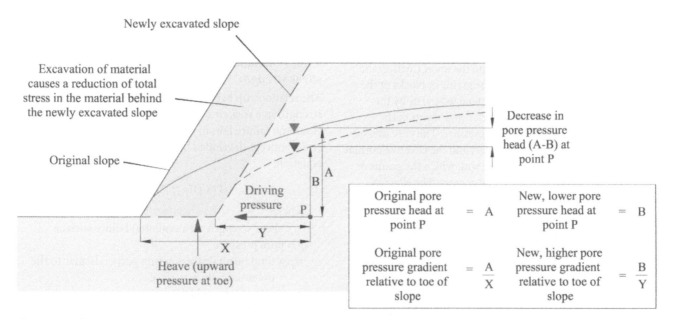

Figure 1.25: Illustration of dynamic groundwater flow system behind a pushback (Source: Read & Stacey, 2009)

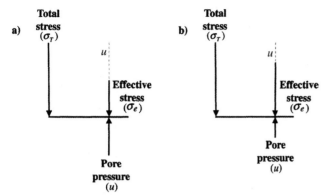

Figure 1.26: The relationship between pore pressure and effective stress (a) prior to depressurisation and (b) after depressurisation. Stresses and pressures act equally in all directions. Pressure and stress are force per unit area; force is pressure multiplied by area. If the horizontal line in the figure above represents a section through a planar surface, then consider the forces acting on a unit area of that horizontal surface. On the top surface, there will be a force, acting vertically downwards, equal to total stress (σ_T) × 1. On the underside there will be an upward force equal to (u × 1) plus the resistance of the formation to the downward weight (σ_T). The effective stress (σ_e) is ($\sigma_T - u$). The stresses act equally all around, but the forces on the horizontal plane resulting from those stresses act upwards and downwards.

1.1.6.2   Relationship between pore pressure and effective stress

For rock strength and pit slope performance, the role of pore pressure in effective stress is by far the most important consideration. Effective stress is defined as the difference between the total stress (the total pressure experienced as a result of the weight of the overlying material, including water) and the pore pressure.

The relationship between pore pressure and effective stress in a rock or soil is illustrated in Figure 1.26, in which the total normal stress (σ_T) is the pressure exerted on the rock mass (the grains or blocks of the formation or the potential failure surface) caused by the weight of both the overlying rock (lithostatic load) and the water (hydrostatic load), and is trying to compress the grains or blocks of the formation. The total stress is counteracted partly by the interaction of the granular or block components of the formation and partly by the pore pressure. The pore pressure (p) is trying to push the grains apart and the effective normal stress (σ_e) is the resulting pressure with which the grains or blocks of the formation are held in contact. The effective stress is therefore the portion of the total stress that is actually applied to the grains or blocks of the formation.

If the pore pressure is reduced by lowering the potentiometric surface, the effective stress will increase. If the total stress is reduced by removal of the weight of the overlying material, the effective stress will decrease.

If the total stress on the material in a pit wall is reduced by mining the slopes above, and if the pore pressure is not changed by an equivalent amount and at an equivalent rate,

the effective stress of the material in the slope will reduce, leading to a loss of strength. An extreme case of this would be when rapid mining (and lithostatic unloading) occurs without an accompanying reduction in pore pressure, such that the downward force caused by the weight of overlying material is reduced to near the upward force created by the pore pressure. Such a situation would potentially lead to an occurrence of floor heave (or wall heave).

At Point P on Figure 1.25 for example, the total stress due to the weight of overlying material has been reduced by the pushback and the pore pressure has been reduced by groundwater drainage into the excavation. However, the rate at which the total stress has been reduced is greater than the rate at which the pore pressure has been reduced, so the effective stress and strength of the materials at Point P have reduced.

Figure 1.27 shows an example where the mining operator has calculated both the lithostatic pressure line (shown in red) and the hydrostatic pressure line (shown in blue) to evaluate the potential for floor heave. Calculation of the lithostatic pressure line involved assumptions of the density of the material below the pit floor and the conversion of the increasing rock thickness (in metres depth) to increasing pressure (psi). Calculation of the hydrostatic pressure line used a simple 1:1 ratio of depth below the level of the pit floor (metres) to increase in pressure head (metres; converted to psi). Each of the points on the figure represents an installed vibrating wire piezometer sensor. The figure shows that many of the piezometer sensors in the lower slope and below the pit floor were above the hydrostatic condition, but none of the sensors approached the lithostatic pressure line, even though some sensors recorded heads up to 60 m (85 psi; ~588 kPa) above the pit floor. Therefore, the analysis indicated that heave of the pit floor and lower slope was unlikely.

1.1.6.3   Relationship between pore pressure and shear strength

The relationship between the pore pressure and the shear strength in a rock or soil mass is expressed in the Mohr-Coulomb failure law, in combination with the effective stress concept developed by Terzaghi, in the form of the equation:

$$\tau = (\sigma_T - p)\tan\phi + c \qquad \text{(eqn 1.28)}$$

where:

τ = shear strength on a potential failure surface
p = pore pressure
σ_T = total normal stress acting perpendicular to the potential failure surface
ϕ = angle of internal friction
c = cohesion available along the potential failure surface

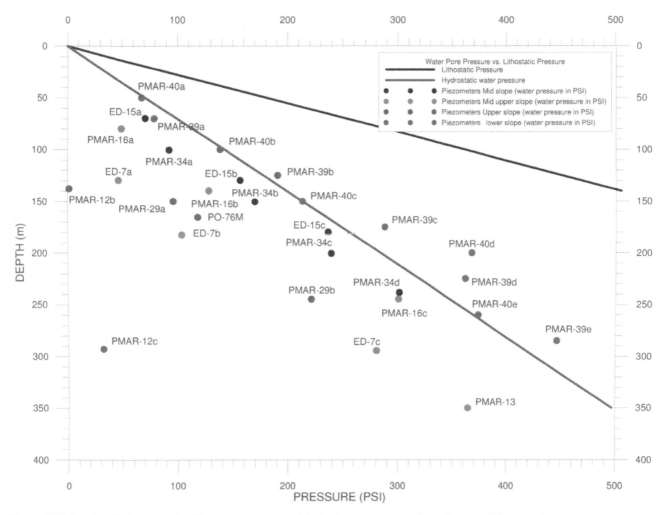

Figure 1.27: Relationship between the lithostatic pressure and the hydrostatic pressure lines (Source: Rick Tunney)

The angle of internal friction and cohesion are strength properties of the material at any point on the potential failure surface. Because the effective normal stress ($\sigma_e = \sigma_T - p$) is the portion of the total stress that is actually applied to the grains or blocks of the formation, it follows that the pore pressure can have a significant influence on the shear strength of the material.

In accordance with the modified Mohr-Coulomb expression, effective normal stress mobilises a frictional component of the shear strength of the soil or rock mass. If the pore pressure is decreased, with no change in total stress, it will lead to an increase in effective stress and hence an increase in shear strength on potential failure surfaces, with an improvement in slope stability. This is illustrated diagrammatically in Figure 1.28.

1.1.6.4 Volumetric changes

Because of the low porosity and low compressibility of most fractured rock materials, the volumetric change that results from removing water from the materials is typically a minor consideration for pit slope design and performance. Some poorly cemented silts or clays can experience minor volumetric reduction (consolidation) as a result of depressurisation but the amount of volume change for these fine-grained materials is typically small unless they occur in a thick sequence. Residual soils such as saprolite may have large porosities and are potentially subject to relatively large volumetric changes in response to changes in groundwater conditions and/or mining-induced consolidation or deformation.

Certain clays that occur above the water table and have low initial water content can experience swelling if exposed in the slope and subjected to recharge. Again, this is not common in most mining situations and typically produces only localised effects, but it needs to be recognised on a case-by-case basis. Volumetric changes in soils are not covered in this book

Volumetric change resulting from water removal should not be confused with volumetric change in the slope materials that results from mining-induced stress changes. Mining-induced stress changes and associated deformation of the slope materials can, in some

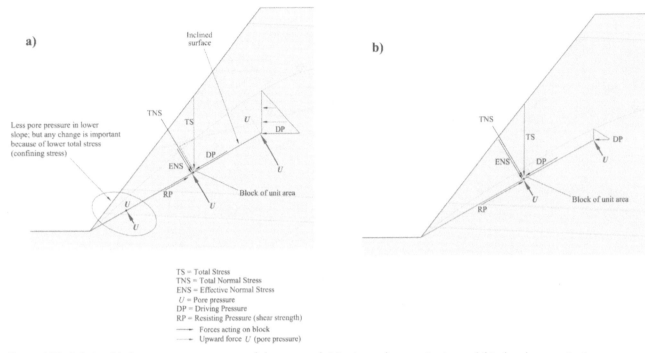

TS = Total Stress
TNS = Total Normal Stress
ENS = Effective Normal Stress
U = Pore pressure
DP = Driving Pressure
RP = Resisting Pressure (shear strength)
⟶ Forces acting on block
— Upward force U (pore pressure)

Figure 1.28: Relationship between pore pressure and shear strength (a) prior to depressurisation and (b) after depressurisation

circumstances, have a major impact on the pore pressure field. This is discussed in Section 3.6.6.

1.1.6.5 Hydrostatic unloading

Hydrostatic unloading is the reduction in total stress (σ_T) acting on the rock mass caused by the removal of the weight of water in the overlying formations (dewatering). Where significant thicknesses of sands or silts occur in the upper levels of the slope, the dewatering of these can cause a significant stress reduction. However, the stress reduction caused by the removal of water (hydrostatic unloading) is typically small when compared with the stress reduction caused by the

removal of material due to the actual mining of the slope (lithostatic unloading).

The relationship between hydrostatic and lithostatic unloading effects is illustrated in Figure 1.29. The figure illustrates that, although hydrostatic unloading can cause deformation and changes in pore pressures in the underlying materials, its role is normally small compared with other processes that are affecting the pore pressure field. Again, the effects of hydrostatic unloading on slope stability should be assessed on a case-by-case basis, especially where the mining process involves dewatering of a thick sand sequence or other materials that have a high drainable porosity (specific yield).

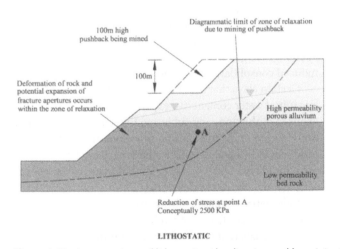

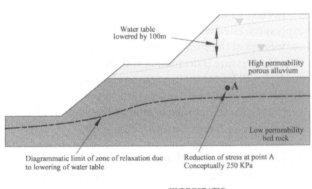

Figure 1.29: A comparison of lithostatic unloading (caused by mining) and hydrostatic unloading (caused by drainage)

1.2   The hydrogeological model

1.2.1   Basic regimes

In any hydrogeological project, three sets of physical factors or *regimes* have to be considered (Price, 1985).

1. Geology – the framework in which the water occurs and moves.
2. Hydrology (or climate) – the control on the input of water to and the ultimate natural output of water from the geological framework.
3. Hydraulics – the interaction between the water and the rock framework.

These topics must not be considered in isolation as they interact to a considerable degree, but the three themes form a useful aid to memory when planning any project involving physical hydrogeology.

In most hydrogeological studies for water resources, the first two regimes are essentially unchanging. The geology is immutable and the large-scale hydrology will not usually change over the lifetime of the project, except for possible issues of climate change. Manipulation of the hydraulics, by use of wells to remove water and to rearrange its distribution for water supply or irrigation, is essentially the aim of most water-supply schemes.

In the case of a mine, these physical regimes still form the basis for study, but the same assumptions may not be strictly valid. Although the excavation of the mine will not alter the lithology of the rocks, it will effectively remove part of the geological framework and may make significant differences locally to the framework properties by processes such as lithostatic unloading or fracture dilation as a result of over-break (blasting effects).

Similarly, the hydrology of the greater region around the mine will not change, but there can be significant changes on a scale greater than that of the immediate mine surroundings. In an arid region, for example, it may be necessary to bring in water for ore processing from more than 100 km away. This water may contribute to the local groundwater system in the form of leakage from site facilities or controlled discharge. The opening up of the pit below the water table may change the groundwater recharge, or may lead to increased evaporation, which may change the site-scale water balance, and may potentially affect local water users and other receptors.

1.2.2   Geology

1.2.2.1   Fundamental principles and rock types

Geology provides the fundamental control on groundwater. The more the geology is understood, the better the hydrogeological model will be. The most important aspects include the following.

1. Type, thickness and orientation of lithological units.

2. Presence and characteristics of fracture network (such as dip and dip direction, and degree of connectivity).
3. Presence of macro features such as faults, their orientation and extent.
4. Magnitude, orientation and trend of variation of *in-situ* stresses.

Mines are developed in more or less all geological environments, ranging from shallow sedimentary systems to deep sedimentary, volcanic and metamorphic rocks. Many metalliferous mineral deposits are in complex settings of relatively old igneous or metamorphic rocks, but these may be buried beneath an overburden of younger (and frequently permeable) sediments. Therefore, at any one site, it will usually be necessary to deal with several different lithological and structural conditions, each with very different hydrogeological properties.

This section provides a brief overview of the main rock types found in mining operations. It is followed by a brief discussion of the effects of structure. When starting work at any site, a first step should be to examine the existing geological information, which is often provided by the local geological survey for the state or country. In its absence, much general useful information, including ranges of hydraulic properties, is available in texts such as Back *et al.* (1988) which, though dealing specifically with North America, also provides useful general data applicable to other continents.

1.2.2.2   Alluvium, overburden and unconsolidated sediments

Unconsolidated sediments are rarely significant host rocks (except in placer mining), but are often found around or above ore and coal deposits. They include alluvium, colluvium, glacial till, loess, dune sands, and sands, silts and clays of marine and lacustrine origin. Loose sediments can also be found between flows in volcanic formations. The principal interest in these deposits is that they can have high permeabilities and also high specific yields (drainable porosities) so that they may feed sizeable and sustained flows of groundwater into underlying formations or into the upper pit slopes.

Permeable and less permeable deposits are frequently interbedded. When a thick interbedded sequence is being drained or depressurised, preferential flow occurs through the more permeable layers (Figure 1.20). In response to the reduction in head in these layers, water may leak into them from or through the less permeable layers, creating a leaky-confined response and helping to sustain the flow in the more permeable layers. Locally around the ore body, the groundwater flow system within the alluvium may also become influenced by the high number of geological exploration boreholes, which may potentially act to increase the vertical permeability of the total groundwater flow system. The influence of open exploration boreholes on the local hydrogeology is a factor that is often overlooked.

1.2.2.3 Consolidated sediments

Water movement in poorly cemented sedimentary rocks will usually be by intergranular flow. Cementation can produce two results. The primary porosity and permeability are greatly reduced, but fracturing can impart secondary characteristics. The rocks may show dual-porosity behaviour, with flow occurring predominantly through fractures, but with the remaining intergranular porosity providing significant storage.

As with unconsolidated sediments (Section 1.2.2.2), bedding may be an important factor, with permeable layers separated by less permeable aquitards. Flow parallel to bedding is usually dominated by the most permeable layers, while flow across bedding tends to be dominated by the least permeable layers (Section 1.1.2.3). Therefore, because bedding is often sub-horizontal, vertical hydraulic gradients and head differences are generally much greater than horizontal ones. This dictates the need for piezometers at multiple depths in layered sequences, and vertical gradients should not be taken to imply the presence of perched water tables. Where tectonic movements have resulted in bedding being steeply inclined, the presence of steeply dipping layers of low permeability can lead to horizontal compartmentalisation, such as within the steeply dipping limestone units found at Antamina (Peru) and Grasberg (Indonesia), or steeply dipping coal formations in Colombia and elsewhere.

Limestones and dolostones may also develop large openings as a result of solution. Sometimes karst development may have occurred during earlier periods with different climatic conditions and may not be continuing at the present day. In these cases there may be karst development below the water table and some or all of the voids may have been filled or partially filled with sediment. In other settings the karstic openings may have developed at a time of lower sea level or other hydrological base level and be present only at significant depth below the surface, and may therefore be unexpected. The transmissivities of these units may be extremely large and, if they are intersected without warning during excavation, can lead to very high inflows.

Some sedimentary rocks can be so well cemented that most primary porosity and permeability are destroyed by cementation, and they become subject to many of the same processes of fracturing, alteration and weathering as igneous and metamorphic rocks.

1.2.2.4 Igneous and metamorphic rocks

Metamorphic and igneous rocks (including volcanics) serve as the host rocks for many minerals. The problem these rocks often pose for mineral investigation, geotechnical studies and hydrogeological studies is their extreme variability. Unfractured igneous rocks such as granites probably represent the least permeable rocks exposed on the Earth's surface (e.g. Odling *et al.*, 2007). At the same time, some volcanic rocks, such as the basalts of the Columbia and Snake River regions in Washington, Oregon and Idaho, represent some of the most permeable (though the most permeable portions are usually the interflow horizons).

In these hard-rock settings, most of the groundwater movement occurs in joints and fractures. Volcanic and intrusive rocks typically exhibit only fracture flow, except in interflow zones and zones affected by weathering, mineralisation or clay alteration. Weathering may increase porosity and permeability in many rock types, especially if it leads to the removal of cement by dissolution or the alteration of feldspars. On the other hand, weathering of rocks such as basalts can lead to the formation of clay minerals which may reduce permeability. In zones of weathering or clay-altered bedrock, the original rock fabric often becomes destroyed, and alteration of the rock mass creates zones of interstitial porosity.

Because of their low permeability, zones of clay alteration associated with faults or within volcanic sequences can be some of the most difficult units of all for pit slope depressurisation. These zones may also be associated with a lower material strength, increasing the need for an active slope depressurisation program. Groundwater flow in these zones is often highly anisotropic. Preferential flow can occur along some relict structural features, while other relict structures may act as barriers to flow.

Within tropical areas, zones of weathered bedrock or saprolite may also be difficult to depressurise. However, the tropical weathering process is typically associated with a gradual unloading and expansion of the original rock mass over geological time. The process may mean that permeability values are somewhat higher than in zones of clay alteration that are associated with deeper mineralisation. Furthermore, sub-vertical features such as quartz-rich dykes may weather to sand and provide preferential flow conduits that enhance drainage.

1.2.2.5 Structure

Geological structure is often as, or more important than lithology in mine water control. Ideally, a mine will have a good structural model, developed by an experienced structural geologist at an early stage. This ideal can be difficult to achieve, partly because the necessary exposure may not be available until mine excavation is well under way.

The presence of high-angle geological structures is often the most important factor influencing groundwater flow within and around mine sites. Geological structures can result in enhanced permeability and groundwater flow along the direction of their strike, and in sub-parallel fractures within the adjacent rock mass. However, the

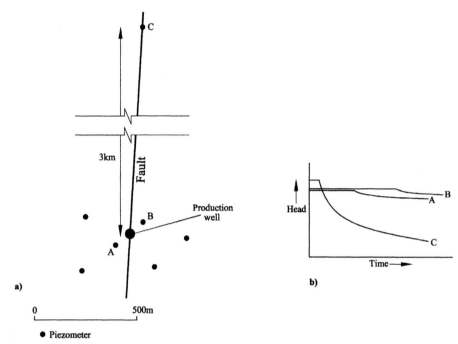

Figure 1.30: Influence of a fault on the effects of pumping (taken from the East Range, Nevada, USA); (a) Plan of wells in relation to the fault; (b) Response to pumping of observation points in (a).

same structures create barriers to flow across their strike because 1) they may offset the brittle units where fracturing and groundwater flow is occurring, and 2) they may create a zone of disturbance (often with clay gouge) of low permeability that disrupts groundwater flow. In one study, the US Geological Survey reports that clay gouge can show a permeability reduction of more than six orders of magnitude compared with the rocks on either side of the fault.[1]

In Figure 1.19 the hydraulic gradient is from east to west, but most flow is oblique to the hydraulic gradient as a result of the structural orientation imparting anisotropy to the permeability. If the north-west–south-east and north-east–south-west trending structures are contemporaneous, as is common, the structures offset each other, giving rise to the discontinuous flow system shown in the cross-section. The presence of the two structural orientations causes the groundwater system to become compartmentalised. Within each compartment, the fracturing and groundwater flow system is relatively continuous, and so the water table is relatively flat. Across each boundary, large hydraulic gradients can develop, causing the water table or potentiometric surface to be stepped and discontinuous.

Compartmentalisation of the groundwater system is a common feature of fracture flow groundwater settings. If the principal structures are widespread, the compartments can be large, sometimes greater than 1 km² in area. If the principal structures are more tightly spaced, the compartments can be less than 10 000 m² in area, and sometimes much smaller.

In certain situations, the flow system can be more continuous along a particular structural direction. As an example, a test dewatering well was installed to target a principal fault zone in the East Range, Humboldt County, Nevada. An array of six piezometers was placed radially around the pumping well, at distances ranging from 10 to 300 m. The well was pumped for 14 days at an average rate of about 35 l/s. Limited response was observed in any of the nearby piezometers, because none of them had penetrated the principal fault zone. However, an open exploration drill hole located in the principal fault zone about 3 km to the north of the pumping well responded strongly, demonstrating very discrete and rapid flow along the structural zone (Figure 1.30).

Vertical compartmentalisation of the groundwater system is common in both sedimentary and fracture flow settings. Figure 1.31 shows a cross-section through a layered volcanic sequence. Much of the groundwater flow occurs within discrete layers or interflow horizons. The intervening layers create vertical barriers to flow and allow significant vertical hydraulic gradients to develop. As in Figure 1.20, high-level seepage faces are present. The situation is made more complicated by faults causing a degree of lateral compartmentalisation in addition to the vertical compartmentalisation caused by the layering.

Sub-horizontal layering and the presence of bedding planes or paleosols can create vertical flow barriers (Figure 1.20). Although vertical jointing can increase the porosity and fracture connectivity within or between

1 (http://crustal.usgs.gov/projects/rgb/faults_gw.html accessed 9 February 2012).

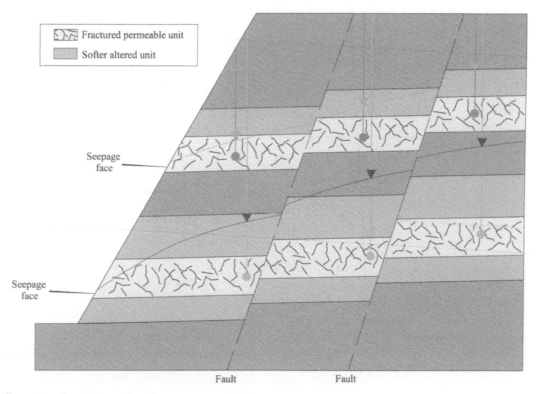

Figure 1.31: Illustration of vertical and lateral compartmentalisation

layers, continuous open vertical joints spanning wide vertical intervals are not common in mining situations. When examples do occur, they are typically found in limestone (or marble), in some sandstone (or quartzite), and very occasionally in some volcanic sequences (for example in some basalts).

1.2.3   Hydrology

1.2.3.1   Climatic data

At an early stage in mine planning, it is important to collect basic climatic information. If the area does not possess a well-developed meteorological service, one or more weather stations should be set up to monitor basic variables such as rainfall, temperature, evaporation, atmospheric pressure and wind speed. In addition to helping understand the groundwater conditions, these data provide a baseline against which to assess any potential impacts from the mining. At some locations, for example in tropical zones or areas of high topography, it may be pertinent to set up multiple rainfall stations, even if only one climatological station is planned. Also, in many areas there is (i) a marked seasonality to rainfall, temperature and evaporation, and (ii) considerable variability from year to year.

As a result, it is necessary to obtain at least several years of data to estimate the likelihood of extreme high flow or low flow events. Where existing records are available, they may need to be interpreted with caution. Early rainfall (and surface water) measurements may not have been to modern

standards. Sometimes there were interruptions when administrations changed, and sites or methods of measurement or recording may have changed without this being noted.

Statements of infrequent events, such as 1 in 10 year storms or 1 in 100 year floods, can be misinterpreted. To begin with, in remote regions it is worth checking the length of time record on which they were based and asking questions about the reliability of the measurements. Secondly, it may be necessary to ascertain whether the climate is stable or whether are there signs that recent decades have been unusually wet or dry. Thirdly, it is important to understand which methodology is best for estimating the likelihood of an event in any time period.

If an event has a return period of T years, then the probability of it occurring in any year is $1/T$, but the likelihood of it occurring at least once in any period of N years is **not** $N \times 1/T$ but

$$\left\{ 1 - \left(1 - \frac{1}{T} \right)^{N} \right\} \qquad \text{(eqn 1.29)}$$

For a 1 in 10 year event, for example, the probability P is 1/10 or 0.1. The event has a 0.1 (10 per cent) chance of occurring in any 1 year and a 65 per cent chance of occurring in 10 years. Further, there is nothing to say that two such events will not occur in successive years.

For Greenfield projects in remote areas, in the absence of any records, local inhabitants can often provide useful information. In semi-arid regions, domestic water supplies may often depend on being able to store the occasional rainfall, so local inhabitants will usually have a good idea of its frequency and reliability and their catchment structures may provide useful data.

1.2.3.2 Surface water features

Where surface water features are present, they usually need to be mapped and their discharges measured at various times of the year. Rivers flowing in from elsewhere are bringing water that might infiltrate into the ground when dewatering begins. Streams originating in the area provide evidence that there is at least part of the year when precipitation is in excess of evaporation or evapotranspiration. Springs are particularly important. Any spring or flowing stream in contact with the groundwater body may represent an 'outcrop' of the water table and so provide information on the groundwater level. However, the nature of the spring needs to be understood and it is frequently observed that:

- perennial springs are typically characteristic of the main water table;
- seasonal springs may be the result of local runoff infiltration or snowmelt that becomes perched on a local poorly permeable feature (such as the bedrock surface) and may occur above the main water table.

As natural groundwater discharge points, springs not only provide data on the magnitude of the groundwater system but are often controlled by geological boundaries or lines of structural weakness. The excess of water flowing out of the mine location over that flowing in gives some indication of the absolute minimum rate at which water will need to be removed to sustain a dewatering program.

In arid zones, dry stream beds may indicate the presence of ephemeral streams that flow in response to infrequent rainfall events. The beds of these streams should be examined to try to find evidence of the frequency of flows. In some areas, most recharge to the groundwater system may be related to extreme runoff events that may occur only every few years or even every few decades.

Other than ephemeral streams (which flow only in direct response to precipitation events), rivers are either gaining or losing streams, though some may switch between modes, perhaps seasonally, in response to changes in the level of the water table. Rivers that are gaining water from groundwater do so because their water surfaces are lower than the head in the contributing aquifer. In mine areas, this will often (though not always) be in the overburden. If the head in the aquifer is lowered as a result of mine excavation, that head condition may be reversed and the river may effectively discharge some of its flow

into the formations that contribute to the mine water. If the river is already a losing stream, the loss of water into the formations may be increased.

In either case, if the river is a major stream bringing water from outside the region, the extra inflow can be significant for dewatering. In addition to the obvious cost implications, there is also the potential impact of streamflow losses on downstream flows and riparian areas. Surface water diversions will be effective only if the river water can be isolated from the underlying permeable geological units.

1.2.4 Hydraulic controls

Most large mines extend below the local water table and therefore interrupt the natural groundwater flow regime. In simple terms, under natural conditions, groundwater flows from areas of recharge to areas of discharge, as follows.

- **Recharge areas** are topographic highs, where recharge may occur in diffuse form from precipitation and infiltration in temperate and tropical climates, but more commonly at discrete locations related to concentrated surface water (typically ephemeral stream flow or accumulating snow banks) in continental and desert climates.
- **Discharge areas** are topographic lows, where the water table either reaches the surface (as in a river valley) or comes so close that water can be discharged by evaporation or transpiration from plants. Where the discharge mechanism is a river, the process of carrying the discharge water away is much more efficient than the diffuse process of recharge from precipitation, so that discharge areas occupy a much smaller part of the groundwater flow system than recharge areas; an exception may be where the discharge area is a wetland, losing water by evapotranspiration.

This apparently simple arrangement is usually made complex by the geological setting, which may lead to the presence of multiple aquifers separated hydraulically by less permeable material, and zones or layers of high permeability that divert groundwater to alternative local discharge areas.

A discussion of the hydraulic controls on groundwater systems is provided in Section 3.6.5.

1.3 Managing water in open pit mines

1.3.1 Key factors affecting the water-management program

The control and management of water is a fundamental component of most successful large open pit mining operations. This may range from the operation of minor sumps in the lowest working area of the pit floor to the installation and operation of major dewatering wellfields that require considerable planning, investment and maintenance.

At one extreme, for some mines excavated below the water table and especially in arid regions, the evaporation of minor groundwater seepage from the pit floor or pit walls in a strong and stable rock mass can take care of all the requirements for groundwater control. At the other extreme, the largest mine dewatering operations tend to be those installed either within regional carbonate groundwater aquifers or sedimentary groundwater basins. For these operations, major pumping may be necessary, using external wells to control groundwater inflow to the pit (or to underground workings) and to lower the pore pressure in the rocks making up the pit slopes.

The factors that need to be considered for the design of an appropriate management system vary according to the regional and local hydrogeological setting, the mine plan and the geomechanical nature of the rocks that form the pit slopes. At any given site, there are five key factors that need to be considered for the water-control program.

1. The regional hydrogeological system
   - The mines that require the largest dewatering rates are generally those located in (or connected to) regionally extensive groundwater basins (see Table 1.1); either permeable carbonates or extensive sediments. Examples of these are the Goldstrike, Lone Tree and Robinson mining operations in Nevada, Vazante in Brazil, Konkola and Nchanga in Zambia, and the Ruhr, Bełchatów and Turoszów lignite basins in Germany and Poland.
   - Mines in areas of exceptionally high rainfall, or those located close to surface water bodies, may also require high-volume pumping systems. Examples of these are Grasberg and Batu Hijau in Indonesia, OK Tedi in PNG and Diavik in Canada.

2. The size of the pit, and its depth and rate of excavation below the water table
   - An obvious general rule is that mines that go deeper below the pre-mining water table require more consideration of groundwater control.
   - However, because many mine-site settings occur in low porosity crystalline rock, with low groundwater storage potential, rapid drainage is usually much easier to achieve if the host rock is not connected to a recharge source, or to a higher porosity groundwater unit in the surrounding area.
   - An example is shown in Figure 1.32, where it is much easier to generate and maintain drawdown on the west side of the pit than on the east side. The rocks in the East wall receive continuous recharge from saturated alluvial basin deposits located immediately east of the mine.

3. The prevailing climatic conditions
   - The amount of rainfall in the mine area is obviously an important factor controlling the required scale of the water control system.
   - Many tropical mining operations require considerable surface and groundwater control, even those located in rocks of low permeability. Where high-rainfall tropical mining operations are located

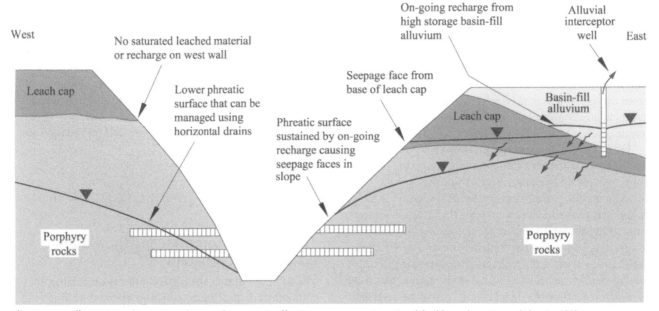

Figure 1.32: Illustration of an external groundwater unit affecting pore pressures (simplified based on Escondida pit, Chile)

Table 1.1: The role of hydrogeological settings in determining dewatering rates (including underground mine operations).

Part A: Permeable hydrogeological settings that often require large dewatering volumes	
Sedimentary basins	• Garzweiler (Germany) • La Quinua (Peru) • Naica (Mexico)
Extensive permeable limestones	• Barrick Goldstrike (Nevada, USA) • Pomorzany (Poland) • Konkola (Zambia)
High rainfall	• OK Tedi (PNG) • Grasberg (Indonesia) • Codelco Andina (Chile) • Iduapriem (Ghana)
Rapid connection to surface waters	• San Luis (Colorado, USA) • Diavik (Canada) • Lihir (PNG)
Connection to high storage Tertiary/alluvial basins	• Kori Kollo (Bolivia) • Sleeper (Nevada, USA) • Lone Tree (Nevada, USA)
Induced permeability/ connection	• Restof (New York, USA) • Boulby (UK) • Cigar Lake (Canada)
Recirculation	• Las Cruces (Spain) • McCoy/Cove (Nevada, USA) • Kori Chaca (Bolivia)

Part B: Permeable hydrogeological settings that may require lower volume water control	
Permeable volcanics, but not connected to storage	• Rosebud (Nevada, USA) • Ujina (Chile) • Quellaveco (Peru) • Asarco Mission (Arizona, USA) • Kanowna Belle (Western Australia)
Permeable limestones, but limited size blocks	• Cortez Hills (Nevada, USA) • Galmoy (Ireland) • Ruby Hill (Nevada, USA) • Murry (Nevada, USA)
Mines with strong overlying 'layer-cake' geology	• Western AG (New Mexico, USA) • Doe Run (Missouri, USA) • Red Dog (Alaska) • Olympic Dam (South Australia)

in permeable host rocks, the flow rates from the open pit can be very high, such as at the OK Tedi Mine in Papua New Guinea (Figure 1.33).

4. The hydrogeological characteristics of the rock mass in which the excavation takes place
   ▪ While the local-scale hydrogeology is important, it is often subordinate to the regional setting in terms of influence on the required pumping rate for groundwater control.
   ▪ However, the local-scale hydrogeology is clearly important for determining the type of dewatering

measures that are required and the extent to which the pit slopes will become depressurised.

5. The geomechanical characteristics of the pit slopes
   ▪ Mines with a strong and stable rock mass, such as an unaltered granite or limestone, may not require full reduction of pore pressures in the pit slopes. Mines where lower strength materials occur within the slopes are usually of more concern for slope depressurisation.
   ▪ However, in steeper, high-strength rock slopes, the potential for sudden brittle or structurally controlled failure under small mining-induced strains is reduced when the pore pressure is lowered. Hence, the best form of mitigation to reduce the risk of instability is depressurisation.
   ▪ The presence of major geological structures, and their impact on slope design and performance, is typically a major consideration that may dictate a requirement for reduction in groundwater level in order to achieve slope stability.

1.3.2 General mine dewatering

The typical focus of a general mine dewatering system is to lower the water table below the pit floor and remove residual groundwater seepage and surface water runoff inside the pit. The purpose of this section is to provide a broad background to mine dewatering within the context of pit slope stability.

There are five commonly employed methods for mine dewatering, as follows:

1. Production dewatering wells
   ▪ Most open pit mine dewatering systems use some form of pumping wells. These may be located outside the pit (in some settings by a considerable distance), at the pit crest or the upper benches, midway on the pit slopes, or in the pit floor.
2. High-volume drain holes
   ▪ Horizontal or angled drain holes can be used to dewater lateral expansions of the pit. They are mostly drilled laterally from the lower pit walls. Because they usually operate by gravity, they need to be tied into the overall pit water management system.
3. Dewatering tunnels
   ▪ Where the topography allows, dewatering tunnels can be driven underneath the mining area to drain the ore body above. It is typically necessary to drill drain holes from the tunnel to improve the dewatering efficiency and broaden the influence of the tunnel. Purpose-built dewatering tunnels are normally driven in such a way that they drain by gravity.
   ▪ Where open pits are located above underground mines, the dewatering tunnels can be driven from within the workings. Many dewatering tunnels

Figure 1.33: General view of the OK Tedi pit in PNG

installed from underground mines require the operation of a pumping system.

4. Seepage faces
   - Where open pits are installed in strong wall rocks it may be satisfactory to allow 'uncontrolled' water to seep through the wall rocks directly into the pit. However, where the seepage flows are high in volume, the water often creates a 'nuisance' factor and it is normally necessary to operate sumps below the seepage faces in conjunction with dewatering wells or drain holes.
   - Where the hydraulic connection between the pits is continuous, there are cases where seepage inflow to one pit has been used to dewater an adjacent pit.

5. Water collection systems and sumps
   - It is normal to have some form of sump system in the lower levels of most pits. Sumps often have the joint purpose of allowing surface water and residual groundwater seepage to be collected and pumped.
   - Many operations use multiple sumps, often with subordinate sumps located at higher levels so that it is not necessary to pump all the water from the pit floor against maximum head.
   - Sumps are often operated in conjunction with surface water collection trenches that collect and feed the water towards the sump. Collection trenches may also be installed to intercept groundwater seepage from alluvial deposits in the upper part of the slope, as illustrated on Figure 1.34.

Several low-permeability 'passive barrier' methods are employed as part of open pit mine dewatering systems. In particular, slurry walls are commonly applied to reduce lateral inflows in open pit coal mines. Low-permeability barriers are discussed further in Section 5.2.1.

1.3.3 Pit slope depressurisation and general mine dewatering

The focus of a pit slope depressurisation program is to locally dissipate pore pressure within the pit slope to improve slope performance (increased Factor of Safety) or reduce stripping costs by increasing the slope angle.

Pit slope depressurisation and general mine dewatering are clearly inter-related and, at many mines, both dewatering for operational efficiency and depressurisation for slope performance are required. However, there are two important differences.

Seepage collection trench
below base of alluvium

Upper slope cut into alluvium

Figure 1.34: Example of in-pit seepage collection

1. General mine dewatering programs aim to achieve a general lowering of the groundwater table in the mining area, to below the level of the working pit floor, or below some other defined target. Pit slope depressurisation programs aim to dissipate groundwater pressures more locally, generally in well-defined sectors of the mine. If any given sector of the pit slope is connected to permeable unit(s), generally it will not be possible to achieve the required depressurisation without first dewatering the permeable unit(s).

2. General mine dewatering programs often involve high-volume pumping, requiring integration of the system into the overall mine water balance and, potentially, off-site discharge. Pit slope depressurisation programs usually produce comparatively low discharges of water, meaning that downstream discharge management is often less onerous.

However, the objective of either a general mine dewatering or a targeted slope depressurisation program is the same: to achieve a lowering of the total head of water to a pre-determined target. For a general mine dewatering program, determination of the targets is often relatively straightforward because they are normally related to the elevation of the pit floor at a given time (Section 5.1.3.2). However, the required pumping rate to achieve the defined target can vary widely between different operations. For a pit slope depressurisation program, determination of the targets is usually more complex because the required pore pressure distribution is inter-related with the geomechanical behaviour of the rock mass, slope design and performance, risk tolerance, and acceptance criteria for a given mine design and operation (Section 5.1.3.3).

In some instances, a general mine dewatering system can achieve the required depressurisation of the slopes, so there is no requirement to implement focused programs aimed at depressurising the pit slopes in localised areas or in specific sectors of the pit. However, for a large number of pits, additional dedicated measures are required specifically to dissipate pore pressures for some or all slope sectors. This is particularly important when slopes are cut in a weak or poor-quality rock mass and where loss of peak shear strength due to the presence of groundwater pressure is a controlling factor in slope design and performance.

For mines located in permeable settings, it is generally not possible to achieve adequate depressurisation of the slopes without the operation of a general mine dewatering system. However, for mines in low-permeability settings or desert environments, a general mine dewatering program may not be required, and the sole purpose of the groundwater control system is to depressurise the pit slopes. In settings with a weak, deformable rock mass, groundwater inflows may be very small, but slope stability may be highly sensitive to groundwater pressure.

The relationship between general mine dewatering and slope depressurisation can be divided into five broad categories (Read & Stacey, 2009). The most important factors controlling this inter-relation are the local-scale hydrogeological setting (Item 4 in Section 1.3.1) and the strength of the material that forms the pit slopes (Item 5 in Section 1.3.1). The categories are illustrated in Figure 1.35.

Category 1: Mines excavated below the water table within permeable rocks that are hydraulically interconnected

For Category 1 (Figure 1.35a), the general mine dewatering program can adequately reduce pressure in all pit slopes, requiring no additional localised measures to dissipate pore pressure. This is because advance lowering of the water table using wells causes gravity drainage of the pore spaces within the rock mass being excavated. If the permeability of the rock mass is high and the rocks are hydraulically connected, the gradient of the water table behind the pit slope will be relatively flat as a result of nearly uniform drainage of the rock mass.

An example of this category is the Cortez Pipeline Operation in Nevada, where in excess of 1500 l/s of groundwater was pumped from peripheral and in-pit dewatering wells in a highly fractured limestone rock mass to lower the water table more than 500 m over a period of 15 years. The sole focus of the mine dewatering program was to progressively lower the water table ahead of the advancing pit floor. No localised slope drainage measures were required.

Category 2: Mines excavated below the water table that have lower-permeability rock occurring in some sectors

In Category 2 (Figure 1.35b), the rock mass may not drain fully in all sectors as the water table is lowered. The bulk of the rock mass and sectors are permeable and readily drained, but effective advance dewatering is not possible in some sectors where alteration or variations in lithology locally create lower permeability. As a result, groundwater will move very slowly and pressure within some parts of the rock mass will not dissipate quickly enough. As the mine excavation is deepened, pore pressures within all or part of the slope may need to be proactively reduced using targeted, localised depressurisation measures.

An example of this category is the North-east and South-west walls of the Sleeper Mine in Nevada, where in excess of 1300 l/s was pumped from alluvium and permeable volcanic tuffs using wells, but the drainage of argillaceous rocks in certain sectors of the pit wall was poor, requiring localised control with in-pit horizontal drain programs.

Category 3: Mines excavated below the water table containing perched groundwater zones

In Category 3 (Figure 1.35c), perched groundwater zones may develop at higher elevations as the main water table is lowered. These perched zones may be associated with less permeable structures, bedding or alteration that impede vertical groundwater movement and result in pore pressures that are 'decoupled' from the dewatered groundwater system. This situation occurs to some extent at many dewatered open pit mines.

An example of this category is the North wall of the Whaleback iron ore pit in the Pilbara, Western Australia. Dewatering of the permeable hematite formations is relatively straightforward using dewatering wells. However, shales and clay-altered materials in the upper slopes do not respond to the hematite pumping and retain elevated pore pressures.

Category 4: Mines excavated in a fractured rock mass below the water table where geological structures form barriers to groundwater flow

In Category 4 (Figure 1.35d), the rock mass may not drain because geological structures act as impediments to groundwater flow, creating compartments of trapped water with high pore pressure. As a result, the general mine dewatering program does not dissipate pressure in all pit slope sectors.

As the excavation is extended and approaches the structural compartments, localised measures become necessary to penetrate the structures and drain the water behind them. Most large open pit mines in hard rock settings encounter structural compartments that influence groundwater heads and movement to some extent. Examples of mines with structural compartments that require proactive drainage measures to support pit slope performance include the South wall of the Chino Mine in New Mexico and RioTinto Minerals' Boron operations in California.

Category 5: Mines excavated above the water table where seasonal precipitation leads to elevated groundwater heads in upper stratigraphic intervals

For Category 5 (Figure 1.35e), control of groundwater pressure is required to support the slope performance, even though the open pit is entirely above the water table. Localised infiltration of precipitation can build up on low-permeability layers and form high-level zones of perched groundwater, leading to locally high pore pressures in the pit slopes.

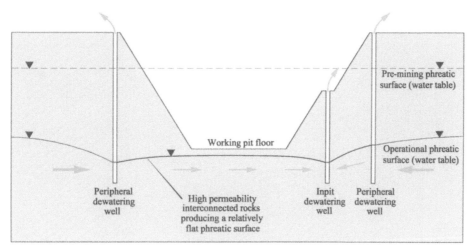

Figure 1.35a: Category 1: Mines excavated below the water table in permeable and interconnected groundwater settings

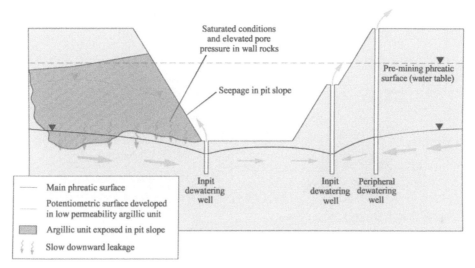

Figure 1.35b: Category 2: Mines with low-permeability rock in part of the pit walls

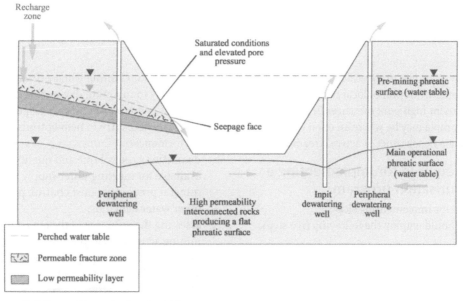

Figure 1.35c: Category 3: Mines with perched water tables

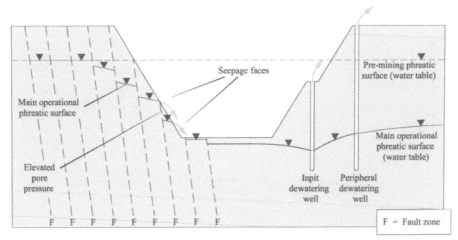

Figure 1.35d: Category 4: Mines where structural compartments impede depressurisation of the pit slope

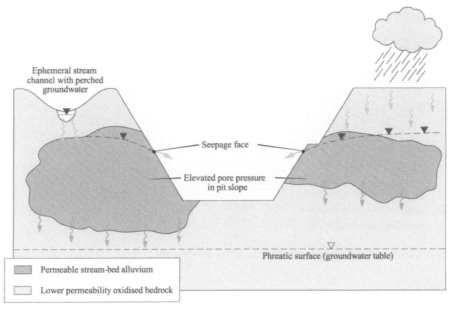

Figure 1.35e: Category 5: Mines occurring above the groundwater table

There are many examples of this category in tropical mine settings, where infiltration of local rainfall can lead to transient or permanent high pore pressures in the wall rocks. Another example may be where an open pit is excavated close to a paleochannel or active stream deposits.

1.3.4 Steps required for implementing a slope depressurisation program

A typical program for implementing slope depressurisation should employ the following five steps.

Step 1: Data collection and analysis.
All mines require a general knowledge of hydrogeological conditions to allow:

- forecast and planning when a dewatering system is necessary to maintain dry working conditions;
- forecast and planning when a slope depressurisation system is necessary to help optimise pit wall design and performance;
- quantifying the volume and quality of water that is available for meeting water supply requirements for mineral processing, dust control, potable water and other water demands;
- assessing the influence of the regional water cycle on the mining activity, an important aspect of which is the identification of the source of recharge for groundwater;
- assessing the feasibility and potential flaws of the planned mining activity;

- assessing the potential impact of the mining activity on the regional water cycle and surrounding environment, and the design of any required mitigation measures;
- satisfying the requirements of regulators and other stakeholders, and underpinning the business's licence to operate.

Collection of the required hydrogeological data and their integration with geological, geotechnical and mine planning data is an essential part of any mine development or expansion.

Once the general hydrogeological system around the mine site has been characterised, the need for a dewatering and/or pit slope depressurisation program can be assessed. A focused data collection program can then be implemented for the pit slopes themselves, as required.

Collection of the required data and characterising the hydrogeological system is described in Section 2 of these guidelines.

Step 2: Preparation of a conceptual hydrogeological model.

Preparation of a good conceptual model is the key part of any investigation of dewatering or slope depressurisation. A properly focused conceptual model is essential for underpinning the subsequent numerical modelling work and design of the depressurisation system, and is often required by regulatory agencies.

The development of the conceptual model is described in Section 3 of these guidelines. A survey of 50 operating mine sites was used as the basis for discussing the conceptual model development and for determining which components of the model are typically most important.

Step 3: Development of a numerical model of the pit slope, as required.

It is often beneficial to support the hydrogeological characterisation of the pit slopes with a numerical model. However, it is necessary to decide the type of model required and the information needed to implement its execution. It is important to understand that a numerical model is merely a representation of the conceptual model. If the conceptual model is not correct, the numerical model will not be correct. An understanding of the underlying assumptions of the groundwater flow system is therefore essential for interpreting the numerical model results.

Development of a numerical model is described in Section 4 of these guidelines. The need for, and the goals of, a numerical model are discussed in Section 4.1. The implementation of the model is discussed in Section 4.2. The input of the pore pressures to the geotechnical model is discussed in Section 4.3.

Step 4: Design and implementation of a slope depressurisation program.

A properly focused analysis and hydrogeological model will provide the design basis and required timing of the slope depressurisation system. It is usual for the depressurisation system to be installed progressively as the mining operation expands. Important mine planning considerations for the hydrogeological study are the pit floor (or pushback) excavation rate (i.e. the change in pit floor or mining bench level with time), the position of the excavation faces over the mine life, and the hydrostratigraphical units that will be exposed by the excavation. The rate of excavation often has a direct influence on the required general mine dewatering and slope depressurisation rates, and also on what measures may be practical.

Implementation of a slope depressurisation system is described in Section 5. Guidelines for defining the targets for the slope depressurisation program are described in Section 5.1. The measures typically used for groundwater control are described in Section 5.2. The measures used for surface water control are described in Section 5.3.

Step 5: Monitoring and performance assessment of the slope depressurisation program.

Because the slope depressurisation program will be implemented gradually in phases, the key to success is monitoring the performance, interactively reassessing the risks and adjusting the program as necessary, based on the continuing monitoring results.

Criteria for monitoring and performance assessment of the slope depressurisation program are described in Section 6 of these guidelines. The monitoring program is discussed in Section 6.1. Performance assessment is discussed in Section 6.2. A risk assessment related to pore pressures in pit slopes is presented in Section 6.3.

1.3.5   Mine water balance

Estimates or measurements of water inflows to the open pit are typically fed into the site-wide water balance. A typical simplified water balance diagram for an operating mine site is shown in Figure 1.36. For most mine sites, the main areas of water consumption are the process plant (where consumed water becomes 'entrapped' in the tailing stream) and the heap leach pads. In warm, arid climates a considerable amount of water can be lost from the heap leach circuit, process ponds and supernatant pond on the tailing facility as a result of evaporation. In wet climates, where incident precipitation is greater than evaporation, the leach pads, process ponds and tailing impoundments will generate water. Most climatic settings also have a strong seasonal component, usually with a wet season and dry season, and/or a high evaporation and low evaporation

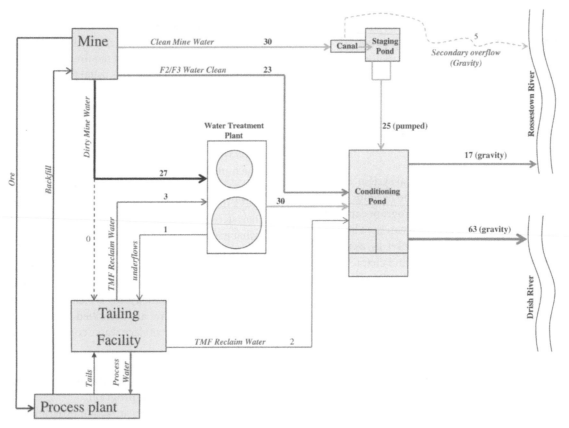

Figure 1.36: Typical mine water balance circuit (Courtesy Lisheen Mine, Ireland)

season. Mine water management is often more challenging during certain periods of the year.

In cases where the water produced from the dewatering system exceeds the total consumptive use and net evaporative losses from the water balance circuit, it becomes necessary to discharge water to the off-site environment. The requirement for water treatment and/or off-site discharge may occur throughout the entire period of the mine operation, or for a limited number of years (for example when dewatering is required for pre-stripping but the site facilities are not yet constructed). For some sites, the requirement for off-site discharge may be seasonal (for example during the wet season or winter months), or may be only occasional following extreme precipitation events. In an increasing number of situations, the 'downstream' engineering and cost implications required for water management and discharge of the open pit inflow water are greater than the cost of the pit dewatering system itself.

The influence of the pit slope depressurisation system on the site-wide water balance and mine discharge requirements is generally low, because most dedicated slope depressurisation systems produce relatively small volumes of water. As a result, it is not usually necessary to consider downstream costs when planning a slope depressurisation program.

In certain situations, where the mine operation requires additional water supplies, it is possible to create an expanded pit dewatering program, or to pre-dewater a long way ahead of the advancing pit floor, in order to create a greater flow volume. These situations are invariably beneficial for slope depressurisation, particularly when it comes to securing the required budget for the depressurisation program.

1.3.6 Mine closure considerations

The assessment of mine closure has become integral to the planning process for new mining operations and for expansions of existing mines. The generic sequence of groundwater behaviour around a pit, from pre-mining conditions, through pit operation with dewatering, to post-closure recovery of water level, is illustrated in Figure 1.37.

In addition to predicting the recovery of water levels within the open pit (or within backfill placed in the pit), it is also necessary to consider the rebound of water pressures in the wall rocks behind the pit slopes, following shut down of the active operational dewatering system. If the rebound of pore pressures in the wall rocks occurs at a faster rate than the water level rise within the pit itself, it can have important consequences for the post-closure

A. Pre - Mining

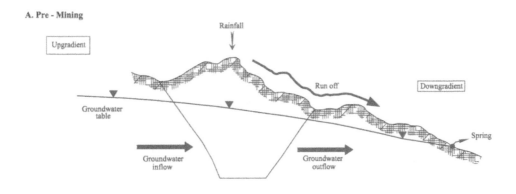

B. Operational

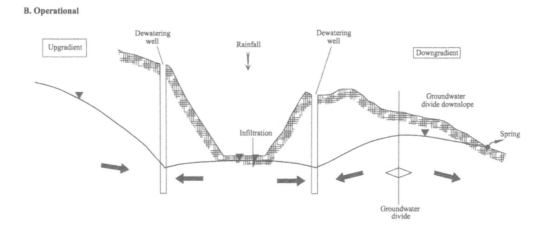

C. Post - Mining

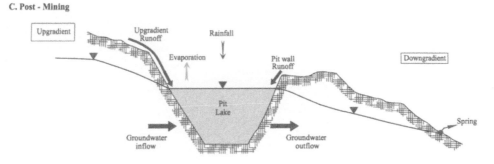

Figure 1.37: Sequence of groundwater behaviour around a pit, from pre-mining to post-closure

strength properties of the wall rocks. This is often a consideration only for short-term closure. It may become less of an issue in the long term if the pore pressures in the wall rocks become equalised by the water within the pit itself, once the hydrological system has reached post-closure equilibrium.

Table 5 of the Introduction section to this book describes considerations for both short-term (active) mine closure and long term (post-closure). The following possible consequences would form part of any evaluation of water in post-mining slope stability.

- Whether the pore pressure rebound may lead to a reduction in slope performance such that it poses a risk to infrastructure or natural features located above the crest of the pit.
- Whether the reduced performance may lead to loss of in-pit access and whether this may impact on the ability to achieve implementation of the closure plan.
- Whether pore pressure may cause instability of the slope so as to create the potential for additional mineral acidity and other oxidation products to enter the pit water.

■ Whether slope instability above the lake level may create rock falls that rapidly displace water in the pit lake and create a displacement wave (often termed a 'seiche').

In addition to evaluating the rebound of pore pressures, it is usually necessary to consider the impact of in-pit surface water runoff and the potential for erosion and/or back-cutting of the pit slopes to impact post-mining slope stability.

Example – closure of the Kori Kollo pit (Bolivia)

The Kori Kollo pit is located on the Bolivian Altiplano. The pit was mined between 1990 and 2002. The pit was mined in dacite rocks and required the operation of a large dewatering system, with more than 50 wells producing over 1000 l/s. Up to 40 m of alluvial sands and silts occurred in the upper pit walls. The dewatering wellfield consisted of a ring of alluvial wells around the pit, bedrock wells in the upper benches of the pit, and in-pit bedrock wells at deeper levels in the pit. The final depth of the pit was about 280 m.

The North-east wall of the Kori Kollo pit is located about 75 m from the Rio Desaguadero (Figure 1.38). The Desaguadero is the largest river on the Bolivian Altiplano and is the natural outlet of Lake Titicaca. A 5 km long flood protection dike was required to minimise the risk of inundation during periods of wet-season high river flow. The natural groundwater level was within 1 m of the ground surface around the entire pit crest. The presence of high pore pressure in the wall rocks was a key issue throughout the entire period of mining.

Upon shut down of the operational dewatering system, it was known that pore pressures in most of the pit walls would rebound rapidly. In particular, the North and North-east wall, along the alignment of the river, was of particular concern. Because of this, mine closure included a rapid filling program for the pit. The rapid filling program involved the construction of a controlled river intake and a 35 m high drop structure constructed to convey the water over erodible alluvial materials in the upper slope and on to the stable rocks of the haul ramp below (Figures 1.39 and 1.40).

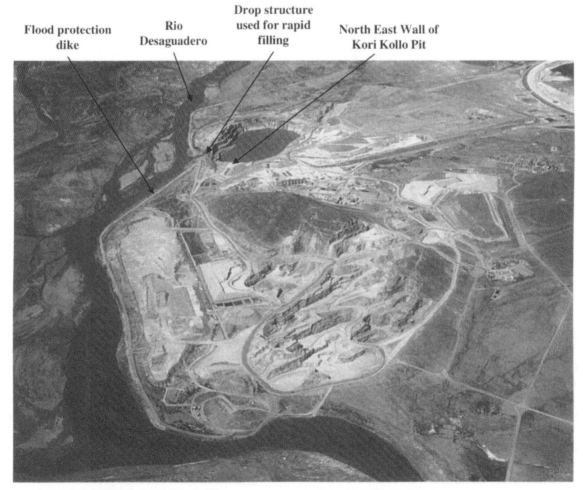

Figure 1.38: Photo of Kori Kollo during active pit filling

Figure 1.39: Drop structure used for rapid filling of Kori Kollo pit (Source: Joost Reidel and Marwin Columba)

Figure 1.40: Side view of drop structure at the Kori Kollo pit (Source: Joost Reidel and Marwin Columba)

Figure 1.41: Cracking of the North and North-east wall of the Kori Kollo pit due to pore pressure at closure (Source: Joost Reidel and Marwin Columba)

Mine closure and rapid filling was carried out during the period of high river flows in order to avoid any downstream impacts. Water was allowed to enter the intake and fill the pit at a rate of about 25 000 l/s. The goal was to fill the pit completely within a three-month period. The rapid rise of water level within the pit was sufficient to equalise pore pressures in the walls. A temporary interruption of the rapid filling program because of local social issues started to cause cracking in the slope (Figure 1.41), but the rapid filling was re-started before any significant movement of the slope occurred.

Example – closure planning for iron ore pit, Western Australia

Closure planning is being carried out for a completed iron ore pit in Western Australia. The pit extracted iron ore material from the Brockman Formation. The north side of the pit extends into saturated detrital materials which consist mostly of layered calcrete, sands, silts and clays, with a combined overburden thickness of about 40 m

along the North wall. A railroad is located above the crest of the wall. The total dewatering rate during operations was about 250 l/s from up to about 10 dewatering wells plus several phases of horizontal drains. About 30 per cent of the dewatering flow rate was derived from calcrete and sand layers in the detrital materials, and several of the dewatering wells along the crest of the North wall were installed only in the detritals. In the North sector of the pit, the detritals are underlain by mostly shale bedrock. The final depth of the pit was about 240 m (Figure 1.42).

Closure planning considers a staged shut down of the dewatering wells. The plan considers the option of initially shutting down the bedrock dewatering wells, allowing a rise in groundwater levels in the pit and in the surrounding bedrock. Continued pumping from the wells within the detrital materials would prevent a rebound of pore pressure in the softer materials of the upper wall. The wells in the detrital materials may eventually be shut down in a progressive manner, so that the rise in pore pressure within the overburden materials behind the upper wall

Figure 1.42: View of Western Australia iron ore pit

could be balanced by a commensurate rise in the equilibrating pressure in the pit lake.

Example – closure planning for open pits in the Basin and Range province, USA

Several mines in the Basin and Range province require depressurisation of prominent structural zones in their high walls. Upon closure, shut down of the dewatering system would cause the pore pressure within the footwall of the structures to rapidly rebound, which would reduce the calculated stability of the slope. For certain pits, the final pit slope extends into natural topography along the range front, well above the level of any future pit lake or equalising pore pressure in the hanging wall of the fault. When permitting pit expansions, mining operators are increasingly faced with the challenge of demonstrating permanently stable slopes, particularly in cases where long-term instability may affect public infrastructure.

Range front fault zones in the Basin and Range province typically act as natural barriers to groundwater flow across their strike. The rocks that form the hydraulically upgradient (footwall) side of the faults can receive recharge as a result of surface water runoff from the steep slopes of the range above. As a result, the natural water table on the footwall side of the fault may be considerably higher than the water table on the hanging wall (downslope) side of the fault (Figure 1.43). If the pit slope is mined down the hanging wall, depressurisation of the structure is required during operations, but it may also be necessary to consider the requirement to maintain similar levels of depressurisation during closure.

During operations, dewatering of the fault zones is usually achieved by means of dewatering wells drilled from the upper part of the slope, as shown in Figure 1.43. If the fault plane is relatively shallow dipping, it may be necessary to augment the dewatering wells with horizontal drains. The goal of the dewatering program is to reduce pore pressures in both the footwall and hanging wall side of the fault zone, and to reduce the inward hydraulic gradient across the fault zone.

Upon closure of the mining operation and shut down of the dewatering wells, there would be a post-mining

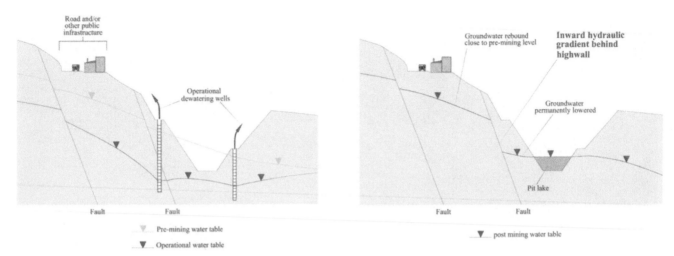

Figure 1.43: Example of inward post-mining hydraulic gradient behind high wall

rebound of groundwater in the footwall rocks which may eventually recover close to the pre-mining level. However, the post-closure rebound of water levels is much less on the hanging wall side of the fault, and the effect of the large inward hydraulic gradient may be compounded by the loss of confinement created by mining of the high wall. Therefore, to maintain stable post-closure conditions, it may be necessary to implement a system that provides permanent depressurisation of both the fault zone itself and the footwall rocks.

2

SITE CHARACTERISATION

Greg Doubek, Ashley Creighton, Jeremy Dowling, Michael Price and Mark Hawley

This section of the guidelines outlines how the hydrogeological features of a mine site are characterised and reported. Section 2.1 describes how field investigation programs are planned, Section 2.2 describes how they are implemented, and Section 2.3 describes how the collected data are analysed and reported. Figure 2.1 illustrates how each of these tasks falls within the hydrogeological pathway in the complete slope design process.

2.1 Planning field programs

2.1.1 Introduction

2.1.1.1 Objectives

Prior to commencing a hydrogeological study and field program, it is necessary to define the following.

1. The overall strategic objectives of the study and the levels of effort required to achieve the mine design (Section 2.1.5), whether it be in support of:
   i. a Greenfield site (Section 2.1.6);
   ii. a Brownfield or existing open pit expansion (Section 2.1.7);
   iii. an optimisation program for an existing operation;
   iv. a mitigation program for an identified risk within an existing operation.
2. The key demands to be placed on the hydrogeological model, and the level of supporting data and confidence necessary to underpin the key decisions that will be made using the model.
3. The level of existing hydrogeological data and knowledge at the site.
4. The timing for delivery of the results of the study.
5. The role, responsibility and accountability of the hydrogeologist as a member of the mine planning team.

2.1.1.2 Project team structure for development studies

The hydrogeology study will usually form part of a multidisciplinary and integrated program. A typical organisation structure supporting mine planning is presented in Figure 2.2. Each technical leader will provide input into defining the key objectives, co-dependencies, risks and opportunities for their discipline. In terms of the hydrogeology discipline, this will set the framework for developing a detailed work scope.

This planning team may be led by the Senior Mine Planner, Chief Engineer, Technical Services Manager, Mine Superintendent, or Mine Manager. In addition to the key defined functions shown in Figure 2.2, the team leader is also supported by the Mine Operations Group and the Process Plant Engineers. Of particular relevance to the hydrogeology program is a good interaction with the environment and community departments.

The typical role accountabilities for each key technical function are as follows:

- Team Leader (Project Manager)
  → Provides strategic direction and overall technical leadership of the project, with accountability for ensuring that the project meets:
    ▷ the corporate business strategy;
    ▷ relevant corporate policies, standards and guidelines;
    ▷ target levels of confidence, with respect to estimates of capital and operational expenditure;
    ▷ corporate methods and schedules of mine development (the mine plan) and mineral processing;
    ▷ acceptable criteria relating to project risks and opportunities, particularly regarding water;
    ▷ the required time lines for the project.
- Mine Planning Engineer
  → Coordinates the economic assessment and selection of the mining method, bringing together inputs from the hydrogeologist, geotechnical engineer, resource geologist, processing, finance, environment and community, using internal and/or external specialist consultants as needed.

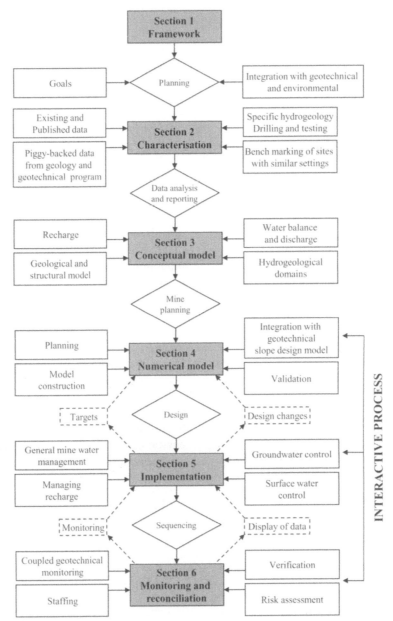

Figure 2.1: The hydrogeological pathway in the slope design

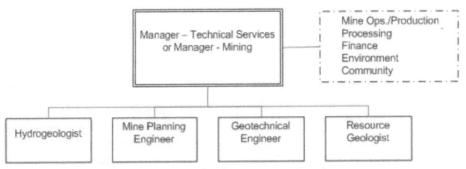

Figure 2.2: Typical organisation structure supporting mine planning at an operating mine

- → Designs the open pit void, mining method, production schedule, infrastructure layout, utilities and services.
- Hydrogeologist
  - → Leads the studies to support the water management plan for the project, which may involve a pit dewatering and/or slope depressurisation program, a water resource development and water supply plan, and an overall strategy for mine water management.
  - → Scopes the field investigations to support the development of the hydrogeological model. This could be at different scales, as described in Section 2.1.2 below.
  - → Provides estimates of the methods, designs and pumping rates for the general dewatering system, taking account of the mine void geometry and mining development rate.
  - → Provides estimates of the pumped water quality from the general dewatering system and the required management plan for the pumped water.
  - → Provides estimates of the pore pressures within the rock mass for the evolving pit slopes in each sector of the pit.
  - → Designs the methods for dissipating pore pressures.
- Geotechnical Engineer
  - → Leads the studies to support the design of the pit slopes.
  - → Scopes the field investigations to support the development of the geotechnical model. This requires close liaison with the mine planner, structural geologist, mine geologist and hydrogeologist. Co-dependencies includes delivery of an economic open pit shell and mining schedule, ore and waste rock block models, and estimates of pore pressures.
  - → Assessment of the stability of the pit slopes and provision of the slope designs, given the geology, structure and hydrogeology.
- Resource Geologist
  - → Owner of the block model, which includes the lithology, alteration, mineralisation and structural models, all of which are fundamental components of the hydrogeological and geotechnical models.
  - → Coordinates the drilling logistics on which both the hydrogeological and geotechnical field programs are based, maximising the amount of geological, hydrogeological and geotechnical information being collected from each hole drilled as part of the total program.

Specialist technical support (including management of the project) may be outsourced to consultants or to corporate staff of the mining company if this capacity does not exist within the client's organisational structure. This approach can be adopted at any stage of a mining project,

although outsourcing is most common when it is not possible to draw resources away from a mine's day-to-day operations. The hydrogeology program is closely related to the environmental program, particularly for obtaining baseline and regional information. It is therefore important that the hydrogeologist involved in the mining study also shares drilling and other resources with the environmental group.

2.1.1.3 Hydrogeology group for mine operations

Most large mining operations that require dewatering or slope depressurisation employ a dedicated hydrogeology group whose focus is to characterise the groundwater system and prepare the dewatering and slope depressurisation plans. The hydrogeology group is generally formed during the early construction phase of the mine but, for larger projects, may be formed sooner. At earlier development stages (through to Feasibility), it is normal for the mining company to utilise corporate personnel (or staff from other operations) or external consultants to carry out the hydrogeology studies.

The dedicated mine hydrogeology group may be required to control large and continuing field programs with multiple drilling rigs. Some mines have both surface and underground mining operations and may operate separate hydrogeology groups for each. However, as specific goals are often inter-linked, it is usually better to integrate the programs. The duties of the mine hydrogeology group usually include:

- installation of a monitoring or piezometer network in order to support continued hydrogeological characterisation, and to monitor the performance of dewatering and pit slope depressurisation measures;
- locating suitable dewatering well sites by drilling and testing pilot holes;
- designing and supervising the installation of production dewatering wells;
- installation of horizontal and vertical drains;
- performing aquifer tests;
- sizing pumping equipment;
- designing and supervising the installation of distribution pipelines;
- conceptualising, designing, and supervising the installation of water treatment and disposal facilities;
- securing and protecting the associated infrastructure;
- collecting data and maintaining a robust database of water levels, pumped volumes, flow rates, and water chemistry;
- obtaining permits for various operations from regulating agencies;
- securing and maintaining water rights;
- prediction and monitoring of site-scale and regional-scale impacts;
- reporting to regulating agencies;

▦ reporting dewatering progress to mine management and interfacing with other disciplines at the operation.

The variability of hydrogeological conditions that is commonly encountered around mine sites means that the methods applied to characterise the hydrogeology usually need to be tailored to the site-specific conditions. As a general rule of thumb, between 60 per cent and 80 per cent of the total hydrogeology budget is typically spent on drilling, well construction materials and pumping equipment.

2.1.2 Scale of the investigation

2.1.2.1 Introduction

An understanding of the wider-scale hydrogeology is essential for most open pits. For the majority of pit slopes, the main control on pore pressure is the wider-scale groundwater flow system and how this interacts with the pit slope geology. The hydrogeological field program must therefore allow for adequate characterisation of the groundwater system and hydraulic properties within the mine area and the surrounding district.

There are usually three scales of data collection, analysis and modelling.

1. Regional (or district) scale.
2. Mine (or site) scale.
3. Slope (or sector) scale.

2.1.2.2 Regional scale (district scale)

At a regional scale, the field program must provide the following characterisation.

▦ Units or domains with significant storage that may provide a permanent source of groundwater to the open pit, once the mine is opened up.
  → An example would be a large sedimentary basin in the immediate vicinity of a mine. Many of the gold mines in the basin and range province of Nevada are examples of where the open pit is centred on bedrock but has slopes excavated through (or hydraulically connected to) regional sedimentary basin fill materials that contain very large quantities of groundwater.
▦ Units or domains with high hydraulic conductivity that may conduct significant groundwater at a regional scale and therefore provide hydraulic communication from the pit area to a significant source of storage, resulting in the potential for sustained high rates of groundwater inflow.
  → A good example is the Diavik Mine (Section 3.3) where a fault controlled fracture zone provides a hydraulic connection between the pit and the adjacent lake, and results in a significant inflow that is mostly transmitted along a single structural zone.

It is necessary that the regional characterisation not only provides an assessment of the potential for flows to affect the open pit, but also on the potential for the open pit and other mine facilities to affect the regional groundwater system, and therefore any mitigation measures that may be required. Environmental and community aspects, and also water supply, are typically addressed at the regional scale.

2.1.2.3 Mine scale (site scale)

At the mine scale, the field program must lead to definition of the following key units.

▦ Units within or near the pit shell (the ore and waste rock) that have significant water storage and permeability, and may produce significant inflow and dewatering pumping requirements, particularly during phases of pit deepening.
  → An example would be where alteration events associated with ore deposit emplacement and mineralisation have created pervasive fracturing and high permeability that is geometrically limited to the zones of alteration. The Yanacocha Mine in Peru is an example where high permeability and storage is associated with the silica-altered rock mass centred on the pit shells, and creates the requirement for high dewatering flow rates as the pits are deepened.
▦ Units with low hydraulic conductivity that lack the ability to allow groundwater movement, but may consequently cause significant groundwater pressures in the pit slopes as they are mined.
▦ Geological structures or structural zones that may create groundwater barriers and inhibit flow and create pressure build up or, alternatively, may provide conduits for flow and recharge to the pit slopes.

The focus at the mine scale is typically mine dewatering or an assessment of the performance of other key facilities (e.g. tailing facilities and waste rock areas).

2.1.2.4 Slope scale (sector scale)

At the slope scale, the field program must provide details of the existing pore pressure profiles and the key factors that will control the future pore pressures in each slope sector as the pit is excavated. Important considerations are usually as follows.

▦ Characterisation of vertical head gradients within units and between units, and characterisation of the pore pressure responses to influences outside the slope domain.
  → For many pit slopes, the vertical head gradient that develops during excavation is often equivalent to or greater than the lateral head gradient. As a result, it

is necessary to use the hydrogeotechnical cross-sections to help to define the location and number of instruments required for the field program.

■ Characterisation of the amount of recharge that may enter the slope domain and the influence of the pore pressures. In Section 3.6.3, recharge is identified as the single-most important factor controlling pore pressures in the pit slope. It is therefore essential that the field investigation is set up to characterise the influence of all potential recharge sources to the slope domain.

→ The presence of sustained recharge usually has a particular influence on weak, deformable materials because these types of materials are often associated with clay alteration and therefore high porosity and low permeability. If recharge enters the slope domain over a wide surface area (for example, infiltration into the slope, or contact with an adjacent permeable unit outside the slope domain), it is often not possible to install a sufficient number of drains to remove the water from the slope at a greater rate than the recharge can enter it, meaning that the slope will not depressurise.

■ Identification of units or structural zones (for example gouge or clay zones) that are difficult and may take time to depressurise. In this case, suitable control measures would need to be implemented well in advance of excavating the slope.

→ The lowest permeability units in the slope are often the most important units for slope stability because of their low material strengths and deformable nature.

■ Definition of units within the slope that may be subject to changes in porosity as a result of deformation in the zone of relaxation behind the slope or beneath the pit floor, and how deformation of the materials may affect the pore pressure profile.

→ Pore pressure changes resulting from deformation typically affect only the lower-permeability units within the slope and are most apparent when recharge to the slope (or recharge to the particular unit) has been substantially reduced or eliminated.

2.1.3 Early-stage investigation

An initial walk-over and reconnaissance survey should be carried out as early as possible in the project investigation, particularly for a Greenfield site. All possible information on existing drill holes and/or wells should be recorded. The reconnaissance survey should also include basic measurements (or estimates) of stream flows, water levels and water quality samples, together with preliminary mapping of springs, seeps and wetlands.

The early-stage investigation should provide an understanding of the basic 'plumbing' of the hydrogeological system and hence the targets for the subsequent investigations and field work. The hydrogeological elements normally considered for early-stage planning are outlined in Table 2.1.

The levels of confidence required by the overall project and the hydrogeological model will directly influence the spatial resolution and scope of the field program. Close liaison between the members of the 'mine planning team' is required to ensure that there is consistency in approach in meeting the mine planning objectives.

2.1.4 Integrating the design process

2.1.4.1 General

For supporting the pit slope design, the hydrogeology program must be closely coordinated with the geotechnical program. The first objective of the hydrogeological program is to predict the dewatering requirements. The second objective is to predict the pore pressures behind the pit walls over the life of mine, or at any critical phase of pit development. The third is to determine whether a specific pit slope depressurisation program is required (Section 1.3.3). The integration of the basic hydrogeological and geotechnical elements for a typical pit slope design investigation is outlined in Figure 2.3.

2.1.4.2 Integrating the field programs

Ideally, drilling programs should always be optimised so that the data collected can benefit all disciplines involved (hydrogeology, geology, geotechnical and environmental). The obvious benefits of this include:

1. saving on drill rig mobilisation, and drilling costs and time;
2. maximising the information gained from the hole;
3. avoiding the need to re-enter the hole (or drill a new hole) to carry out hydrogeological testing;
4. allowing the hydrogeological data to be directly integrated with the core logging, geotechnical testing, and geotechnical installations from the same hole.

During the early stages of the investigation, it is normal practice for the geotechnical engineer to take the lead on locating the drill holes for an integrated program to assess pore pressures in the pit slopes, with hydrogeological data collection 'piggy-backed' on the geotechnical holes. The holes will be designed to assess the rock mass and key structures based on:

■ the mine plan and options being considered;
■ lithological or structural information that is available from field mapping and/or reconnaissance;
■ any geotechnical hazards that the geology may impose on the open pit development;

Table 2.1: Hydrological elements for early-stage planning of a project

Element	Information source	Comments
Climate	• Rainfall, evaporation and seasonal temperature records	• Required for establishing recharge and the initial project water balance
Springs, seeps and other key riparian features	• Published maps • Aerial photography • Walkover surveys	• Identification of recharge and discharge zones • Evaluation of key receptors in the area of potential mine impact
Surface water	• Site-scale and regional streams, rivers and lakes • Streamflow records • Surface water chemistry data	• Important contribution to the early site water balance and establishing groundwater recharge and discharge areas • For projects located in coastal areas, it will also be necessary to obtain information on the tidal interaction and seawater influence on the surface water and groundwater.
Topography	• Published maps • Aerial ortho-photogrammetry • Airborne LiDAR surveys • Detailed topographic surveys • Bathymetry (if project is near the sea or a lake)	• Identification of catchment divides • Identification of perennial or ephemeral surface water • Required to help establish hydraulic gradients
Regional-scale stratigraphy (lithology)	• Published regional mapping (Geological Survey reports and maps) • Published drilling reports • Previous project data	• Required for establishing the site groundwater regime in a regional context • Important for delineating boundary conditions for simulation models
Mine-scale stratigraphy (lithology)	• Published drilling or other reports • Digital geological model (as available from previous studies)	• Basis for delineating the key hydrostratigraphical units within the project area
Regional- and mine-scale structure and lineaments	• Published regional mapping (Geological Survey) • Project-specific structure mapping • Aerial photography or satellite image interpretation	• Basis for estimating the role of structure for groundwater flow • Important for helping to link the local groundwater system into the regional setting
Rock mass fabric	• Project-specific fabric/joint mapping • Fracture frequency from diamond drilling logs or downhole geophysics	• Provides a first order estimate of permeability/hydraulic conductivity/anisotropy for each lithological unit
Groundwater levels	• Hydrogeological reports • Published monitoring records • Initial field surveys	• Basic information for determining the continuity of the groundwater flow system; regionally and locally • Most important input to the preliminary hydrogeological model
Results of hydraulic testing	• Pumping tests on existing regional wells • Downhole testing from previous studies	• Provides a first order estimate of the required magnitude of dewatering and slope depressurisation
Groundwater chemistry	• Hydrogeology and geochemistry reports • Published monitoring records • Initial field sampling	• Supports the overall hydrogeological analysis and provides a first order estimate of potential environmental or regulatory concerns

- whether any downhole geophysics is being considered for the geotechnical program (e.g. acoustic or optical televiewer logging);
- geotechnical sampling requirements;
- hydrogeological testing requirements;
- the overall project schedule to ensure the field program will deliver outcomes within the expected time frame and budget.

The information required for the wider-scale hydrogeology will often be derived from the resource definition and condemnation (sterilisation) drilling programs and other regional data. Inspection and logging of the core is valuable to help define the key hydrogeological units, including zones which may form enhanced flow conduits and zones which may form barriers. The logs can be used to establish the location of downhole tests such as packer tests, slug tests, falling or rising-head tests (Section 2.2), and the position of instruments in each hole. Information on the wider-scale hydrogeology will also be derived from the environmental drilling program being carried out for baseline studies or general site environmental characterisation. It is normal to closely integrate the hydrogeology and environmental programs, particularly during the early stages of a project. There is a joint benefit to determine:

- how the regional groundwater flow system may interact with the mine-site groundwater system (e.g. impacts to planned mining activities) and whether the setting is one of general dewatering or slope depressurisation (or both);

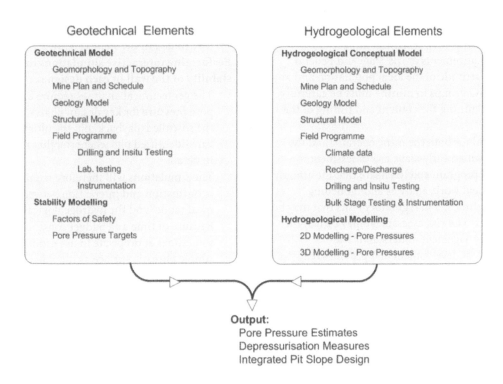

Figure 2.3: Integrated geotechnical and hydrogeological elements in the pit slope design process

▪ how the mine-site groundwater system may interact with the regional groundwater flow system and definition of where potential impacts may require mitigation.

'Piggy-backing' of data collection can provide considerable cost savings if the geology, geotechnical, hydrogeology and environmental disciplines work together. For example, a drilling program for ore reserve delineation purposes will benefit the other three disciplines as a result of the better understanding of the geology. Conversely, a drilling program for hydrogeology will provide further information about formation contacts and structure, and any collected geological samples can be assayed if required. Holes drilled for hydrogeological investigation are often of interest to the geological group, as the hydrogeological investigation normally extends outside the immediate footprint area of planned open pit development to where the geological data and knowledge may be less reliable. A number of new ore bodies have been discovered and the delineation of existing ore bodies extended following the implementation of specific hydrogeological drilling programs.

2.1.4.3 Field logistics

Projects in remote areas will usually require a greater lead time to procure and mobilise supplies and equipment and to carry out initial works prior to excavation of the pit. Import procedures for certain countries may also be a significant factor. Training of personnel in developing countries must also be factored into the planning process.

Given the required lead time, the planning process for projects in remote areas may need to occur a year or more ahead of projects in established mining districts.

However, during times of mining 'boom', even projects in well-established mining districts may also require advanced planning because of the industry-wide demands on equipment and personnel. The acquisition of drilling equipment, well construction materials and pumping systems may involve many months of advanced lead time and planning. For some operations, the ability to free-up equipment and/or gain site access for 'investigations' while avoiding conflict with 'mine operations' is a major consideration for planning.

For many Greenfield projects and major Brownfield expansions, the most important factor controlling the time for project development is often the permitting and/or the ability to gain public confidence. In many established mining districts, the time taken to achieve regulatory approval has increased significantly over the past 10 years or so, and a reasonable time frame to obtain all necessary permitting is now three to five years. For large Greenfield projects away from established mining districts, the time to receive regulatory approval for mining may be longer than five years in some countries. However, permitting times in some developing countries may be much shorter where external investment and infrastructure-building is actively encouraged.

As a result of the increasing requirements for permitting, it is often necessary to 'front-load' the early-phase studies (Conceptual and Pre-Feasibility level)

to provide sufficient technical support for the required permit applications and environmental impact assessment. The requirements of the State and Federal regulators and the attitude of the local population and any special interest NGO groups to mining must be carefully considered when planning the content and timing of the program.

It is often helpful for both the mine operator and the regulator to carry out an early-stage risk assessment to help focus the field program and to ensure that the amount and timing of the field work, analysis and modelling reflects both the technical and regulatory risks that may be posed by the project. The risk assessment may be helpful for ensuring that management understands the implications of project development and provides realistic budgeting and financial resources to perform the required investigations.

2.1.4.4 Integrating the analysis and modelling

Integration of the hydrogeological analysis and modelling into the pit slope design process typically involves the following steps.

1. **Development of a regional-scale hydrogeological model.**
   - It is important that the wider-scale hydrogeology is understood as early as possible as this provides the basis for determining the external factors (boundary conditions) influencing the pit hydrogeology and for planning the dewatering and slope depressurisation programs.

2. **Development of a mine-scale geotechnical model** that incorporates hydrogeology throughout the model domain.
   - The mine-scale hydrogeological model must be at a sufficient level of detail to provide the required input to the geotechnical model.
   - It is important that both the geotechnical and hydrogeological models are integrated with the geological and structural models (and that they both use the same geological assumptions).

3. **Identification of the geotechnical design sectors,** based on factors including the orientation of the wall relative to the geotechnical model and the potential slope failure modes.
   - Most geotechnical modelling to support the slope designs is carried out at a sector-scale (usually two-dimensional (2D) modelling but sometimes supplemented by sector-scale three-dimensional (3D) modelling). To support the sector-scale numerical modelling, it is important there is consistency between the geotechnical and hydrogeological model domains for each sector.
   - Hydrogeotechnical cross-sections should be prepared as early in the program as possible to show the geology, the important structures, the potential weak rock zones and how the groundwater is distributed in the slope (Figure 2.4).

4. **Performing interactive simulations to assess the stability of the wall in each design sector.**
   - The geotechnical analyses require estimates of the pore pressure for key design phases, which may be the so-called 'payback' pit, the ultimate pit, or other time-identified pits where stability must be assessed in detail.
   - The simulations will therefore require close coordination and interaction between the geotechnical model and the hydrogeological model.
   - Because of time and budget constraints, it is sometimes appropriate to use estimates of the simple phreatic surface as input to early-stage geotechnical models, rather than pore pressure fields. However, this needs to be assessed on a case-by-case basis and is discussed in Section 4.

5. **Finalising the slope design.**
   - The iterative design process typically involves a number of geotechnical model simulations for various levels of active depressurisation effort. This allows the planned active depressurisation measures to be focused on the identified pit slope failure mechanisms.
   - For the majority of large pits, the final slope designs will require consideration of a pore pressure field rather than the use of a simple phreatic surface but, again, this needs to be assessed on a case-by-case basis and is discussed in Section 4.

Most initial pit slope designs are based on analyses undertaken using 2D sections located to intersect the wall in critical parts of each design sector (e.g. Figure 2.4). However, for more detailed design studies, and where there is a need to analyse detailed structural orientations and/or the influence of a curved pit wall, the 2D analyses are often supplemented with a 3D analysis, which is carried out to address specific geotechnical issues. The pit slope engineer must define the hydrogeological inputs required to satisfy the stability analysis and design process for both the 2D and 3D models.

2.1.4.5 Parameter variability

Most (but not all) mineral deposits, particularly those hosted in fracture volcanic and/or metamorphic rocks, contain significant variability in geological, geotechnical and hydrogeological conditions. It is commonly observed that drill holes a few metres apart can exhibit vastly different fracturing and groundwater conditions. It is common to measure hydraulic conductivity values that are many orders of magnitude different in neighbouring boreholes of similar geology because multiple geological events have combined to create a complex distribution of fracturing and hydraulic property values.

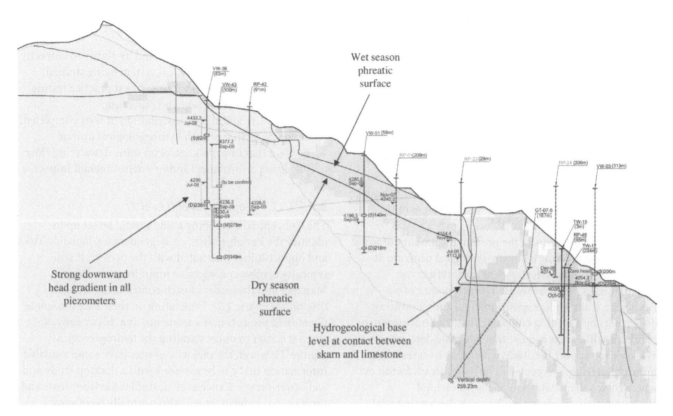

Figure 2.4: Integrated hydrogeotechnical section (Source: Antamina)

The variability often appears irregular and unpredictable. A wide spread of measured parameter values applied to a groundwater inflow analysis or a prediction of pit slope pore pressure will produce radically different results. Therefore, a key challenge is to develop a set of field test data that produces representative results for the study, in terms of pit inflows, pit slope pore pressures or potential impacts.

There are several ways that variability can be dealt with in a study.

- Understanding the scale at which variability is of consequence. For example, the hydraulic properties of a fault system present at a sector scale may have minimal impact on estimates of general groundwater inflow rates to an open pit, but may be of high consequence to pore pressure distribution in the inter-ramp or general slope stability.
- Understanding the most important hydrogeological units and features in the system, and targeting these for the majority of testing and monitoring.
- Defining an adequate test program for each important hydrogeological unit, such that statistically valid parameter distributions can be developed.
- Completion of prolonged cross-hole or multi-hole tests (with response monitoring at multiple locations) rather than reliance on 'single-point' formation testing (packer tests, drill-stem tests or falling-head tests).

- Integrating and correlating the hydraulic test data with the geological model, the structural model and the geotechnical model, so that the geological controls on parameter distribution and variability are understood to the maximum extent possible.
- For Brownfield sites, the use of practical analogues to ground-truth measured hydraulic parameter ranges. For example, at Bingham Canyon, the hydraulic conductivity ranges for the main rock types established from field testing correlate with observed horizontal drain flow and underground drain flow data from past slope depressurisation programs.
- For Greenfield sites, the use of benchmarking and the review of the hydrogeological model at similar analogue sites.

2.1.5 Required effort based on project level

2.1.5.1 Introduction

The levels of effort required to achieve the mine design at each of the project levels defined in Table 1 of the Introduction (Conceptual, Level 1; Pre-Feasibility, Level 2; Feasibility, Level 3; Design and Construction, Level 4; and Operations, Level 5) are outlined below. Specific considerations for a Greenfield site are discussed in Sections 2.1.6. Specific considerations for a Brownfield site are discussed in Section 2.1.7).

Any given field program is normally conducted in phases. Hydraulic testing normally commences as part of the Conceptual (scoping) study and is expanded as the progress moves through the various stages of the overall project development. The results of the early-phase work are used to develop and refine the site conceptual model and predict the mine dewatering and/or slope depressurisation requirements. A typical approach involves a relatively inexpensive program during the Conceptual study when most of the data can be obtained by 'piggy-backing', proceeding into more extensive and specific testing during the Pre-Feasibility and Feasibility programs, guided and defined by earlier stages of assessment.

As with all elements of the project, the amount of work and effort required for the hydrogeological program at each stage of project development must reflect the potential risks posed by water. The acceptance of a Feasibility study and/or environmental and permitting documents by a mining company, financial institution, or regulator will always require that adequate data are available to support the study. Therefore, in cases where mine dewatering is expected to be a significant factor, or where robust demonstration of environmental performance is necessary, a pilot-scale dewatering trial may be important for helping to demonstrate to the regulatory authorities the feasibility of dewatering with minimal impact to the surrounding hydrological system or to other local water users or receptors. A dewatering trial would normally be carried out as part of the Feasibility Study (Level 3), but may become part of the Pre-Feasibility Study (Level 2) in cases where it is necessary to increase the level of confidence to support the permitting documents.

2.1.5.2   Confidence levels

An important aspect for planning a field program is the required levels of confidence to support the ore body resource and/or reserve estimates, the slope designs, the mine plan and the permitting. Whether the project involves a Greenfield venture or an expansion of existing Brownfield site, the level of confidence in the hydrogeological model must be consistent with the level of confidence reported for the resource (i.e. inferred, indicated or measured) or reserve (probable or proved), and with the levels of effort and data confidence defined by Read and Stacey, 2009 (Introduction, Section 1, Table 1 and Figure 2.5).

Many mining houses have their own internal standards and requirements in terms of level of detail, confidence and certainty in results at each stage of project development, often determined as a function of the risk posed by various aspects of the project. In all cases, the

quality of field data will be carefully scrutinised with respect to:

- the type of testing completed and its ability to correctly quantify the hydrogeological system being studied;
- the quality of testing, in terms of best practice testing methods, data collection and reporting;
- the quantity and spatial distribution of tests completed;
- adequate testing of each hydrogeological unit or domain that imparts control on mine dewatering flows, pit slope performance, and/or environmental impact.

2.1.5.3   Conceptual level (Level 1)

The Conceptual or scoping study should be set up to identify the key objectives, co-dependencies, logistics, risks and opportunities associated with the project. It will typically involve considerable input from the Project Manager and the leader of each contributing technical discipline (Figure 2.2). Depending on their location, some Greenfield projects may commence as a 'blank canvas' when it comes to understanding the hydrogeological regime. However, the majority of sites have some available information that can be assessed with a desktop study and walk-over survey. Experience in similar environments and professional judgement will also normally be of great benefit when scoping potential groundwater yields and the response of the groundwater system to mining.

Additionally, it will often be possible to obtain the hydrogeological information by 'piggy-backing' onto the geological resource drilling program to establish sufficient basic data to support a first generation conceptual model. Depending on the scale of the mineral exploration program and on the inferred water-related risks, a dedicated hydrological field program may not be required; rather hydraulic testing and piezometer construction can be carried out in a selected sub-set of mineral exploration boreholes. A typical Level 1 program may include the following.

- Obtaining all published information for the mine site and regional area (climate, geology, surface water and

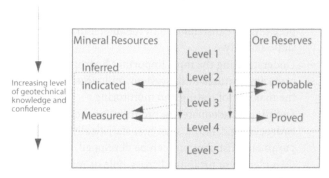

Figure 2.5: Geotechnical levels of confidence relative to the JORC code (Read & Stacey, 2009)

groundwater information, and details of other local groundwater and surface water abstractions, as outlined in Table 2.1).

- Field reconnaissance surveys, including measurements and monitoring of surface water levels and flows, and spring flows.
- Geological data gathering from the resource exploration drilling program (RC and/or diamond drill holes).
- Measurement of airlift production rates at the completion of each drill pipe and at the end of the hole (RC and/or air-drilled holes), with groundwater production rates correlated to rock type, alteration or structural zones.
- Documentation of zones of fluid circulation loss (diamond drill holes); with the zones of fluid loss correlated to rock type, alteration or structural zones.
- Measurement of groundwater levels in RC and diamond drill holes at the end of the hole, and preferably also at the start of each shift.
- Installation of piezometers in a small sub-set of the mineral exploration boreholes. A typical number would be six to 12, spread across the extent of the potential mining area. However, when feasible, it is often of benefit to install perforated pipe in a greater number of resource exploration or condemnation (sterilisation) drill holes.
- Measurement of water levels in adjacent holes during drilling and airlifting of RC drill holes.
- Initial water chemistry measurements taken from airlifting of air-drilled holes, any existing wells in the area, surface waters and spring flows.

As part of the Level 1 studies, it is important to gain as good an understanding as possible of the regional groundwater flow system as this will be fundamental for determining the dewatering system, the slope depressurisation program and the potential magnitude and extent of hydrogeological impacts. This should include:

- the key hydrostratigraphical units at a regional and mine scale;
- the elevation of the water table and the general hydraulic gradients across the site prior to mining, and the potential broad-scale compartmentalisation within each identified unit;
- the nature of the recharge boundaries (such as hydraulically connected rivers) or groundwater flow barriers (such as less permeable units), some of which may come into effect only when mine dewatering begins;
- the likely permeability and groundwater transmitting characteristics within each identified unit, and between units;
- the drainable porosity of each identified unit;

- the hydrological characteristics of major structures (e.g. faults, dykes, sills);
- the nature of poorly permeable materials where pore pressures may be of concern for the slope design;
- pre-mining water chemistry within the most important identified units.

As the data are collected, the scoping study should identify significant hydrological conditions and risks that may impact the mine design, operation, permitting or economic viability. The scoping study should also include a gap analysis and should provide a detailed work plan for the Level 2 studies.

2.1.5.4 Pre-Feasibility (Level 2)

The hydrogeological testing program for the Pre-Feasibility study needs to be adequate to support a first generation conceptual and numerical groundwater model, prediction of groundwater inflow rates and pit slope pore pressures. It also needs to support a preliminary impact assessment that defines the location and order of magnitude of potential impact.

Given these requirements, it is necessary to generate adequate test data that quantify hydraulic conductivity ranges and bounds for (i) the main hydrogeological units within the pit shell, (ii) the district geology units that contact and extend beyond the ore deposit margins, and (iii) major structures or structural zones continuous at or greater than the scale of the proposed pit in order to estimate the dewatering and slope depressurisation requirements, and also the overall project water balance. The Pre-Feasibility study would also need to provide a sufficient understanding of the hydrochemistry to provide an engineering plan and costs for any downstream treatment and discharge of excess water from the dewatering system.

A typical Level 2 program would include the following.

- Establishing a full climatological station on site (potentially several full stations and/or multiple rain gauges in areas of greater topographic relief and/or expected high rainfall variations across the site).
- Establishing surface water gauging stations specifically located in the hydrogeological catchment of the mine to support the site engineering designs, and also for baseline groundwater level and chemistry information.
- Establishing a water chemistry sampling program from surface water bodies, drill holes and monitoring wells to ascertain pre-mining water chemistry and intrinsic characteristics of each hydrogeological unit to support engineering and environmental studies.
- Conducting baseline studies for surface and groundwater (typically a 12-month monitoring period, flows,

levels, chemistry) in the hydrological catchment of the mine (Section 2.1.8).

- Drilling of eight to 15 dedicated hydrogeological test boreholes to the depth of planned mining, targeting all of the main hydrogeological units identified in Level 1. In hard rock settings, it is normally preferable to drill the test holes using the dual-tube reverse-circulation (RC) method, as described in Section 2.2.3. It will be necessary to carefully monitor the results of the drilling program as it progresses and, if necessary, to interactively adjust the program based on increasing confidence in conceptual model or changes in conceptual model as data are collected and analysed.
- Dedicated hydraulic testing of each test borehole during drilling, including injection tests and airlift yield tests at multiple depth intervals.
- Packer testing in a sub-set of geotechnical core holes, with six to 10 holes and five to eight discrete vertical tests per hole being a general guideline for a site that includes low-permeability wall rocks.
- Installation of multiple-level vibrating wire piezometers in at least 80 per cent of the RC hydrogeological boreholes and at least some of the geotechnical holes, with transducers set in specific pre-determined hydrogeological units and on opposing sides of major structures. For projects in sedimentary host rocks, or where there are overlying saturated alluvial deposits, some of the holes can be completed as standpipe piezometers.
- Construction of at least one test pumping well, targeting the most productive hydrogeological units in the system, using the hydrogeological test boreholes to identify the productive zones and optimise the location of the pumping well. As described above, it is necessary to consider a test pumping well even in very low-permeability settings, in order to confirm conditions and to provide data that defend a conceptual model involving a poorly permeable system. Where the test drilling indicates that significant dewatering will be required, more than one test production well should be considered for the Pre-Feasibility study.
- Completion of a constant rate pumping test(s) in the test well(s) and monitoring of hydrogeological system response in the installed piezometers. The duration of the test would normally be not less than 24 hours and preferably extended to at least one week. A longer period of testing may also be advisable in highly permeable settings where high-volume mine dewatering is anticipated.
- Definition of each major hydrogeological unit and/or identified hydrogeological domain. This is typically based on the geological and regional structure model, and the relative hydrogeological characteristics of each unit.
- Establishment of recharge zones (groundwater catchments) and discharge zones via field reconnaissance of streams, stream flow measurements, spring and seep locations and flow rates, water chemistry, and measurement of groundwater level and hydraulic gradients.
- Preparation of a conceptual hydrogeological model including surface water, groundwater and pore pressure distribution, and surface-groundwater interaction.
- Construction of a first-generation 3D groundwater model to test the main elements of the hydrogeological conceptual model, develop order of magnitude pit inflow estimates, and define key data gaps and uncertainties.
- Preparation of a site water balance, including water supplies, water management for individual mine facilities, and the requirement for off-site discharge of excess water.
- Preparation of surface water and stormwater management plans.
- Preparation of an updated and more focused 'gap-analysis' to help focus data requirements for the subsequent Feasibility study (Level 3) and re-assessment of water-related risks based on the improved understanding of site conditions.

2.1.5.5 Feasibility (Level 3)

The Feasibility-level program needs to provide sufficient data to support (a) a viable pit slope design, (b) a pit dewatering and slope depressurisation plan, and (c) a site-wide water management plan, including potential discharge of excess water. At this stage, confidence in the hydrogeological model should have improved to 40 to 65 per cent (Introduction, Table 1).

The testing program would be designed to quantify each of the hydrogeological units and major structures important at the scale of a new mine or a major pushback of an existing mine. The data will also need to support an environmental impact assessment and permitting documents, and will need to be sufficient to support internal risk assessments and an analysis of potential fatal flaws.

The potential content of a Level 3 study and the scale of hydrogeological testing required will be specific to an individual project, but general guidelines are as follows.

- Additional climatological stations and detailed precipitation and evaporation stations, sufficient to support a detailed assessment of groundwater recharge, the surface water management plan, the stormwater management plan, and a downstream impact assessment.
- In-fill studies for surface water and groundwater baseline.

- An additional six to 12 dedicated hydrogeological boreholes and multiple-level vibrating wire piezometers designed to (i) provide quantification of the main hydrogeological units present at the pushback scale, (ii) provide characterisation of structures present at the pushback scale, and (iii) infill identified data gaps, with a prolonged period of monitoring each transducer in each piezometer string to build a continuous record of natural seasonal variation in groundwater levels and response to pumping and drainage trials.

- Installation of additional test production pumping wells, or drainage trials using horizontal or vertical pilot holes, depending on site conditions. For most projects, some form of multi-hole testing or a pilot-scale dewatering or drainage program is essential at the Feasibility level.

- For permeable systems, pumping trials with at least 10 per cent of the expected production dewatering rate is a good rule of thumb for a feasibility study, and this may potentially require a pilot water treatment plant and/or a strategy for temporary discharge to the environment. Normally, the pumping trials would be planned so that wells and other infrastructure can be retained as long-term assets for the project operations. The pumping trials would involve (i) monitoring of groundwater levels and responses in all constructed piezometers, and (ii) monitoring of surface water flows in rivers, springs and other features identified within the potential area of impact of mine dewatering.

- A further program of local-scale packer testing in geotechnical drill holes, focused on specific sectors where poorly permeable units have been identified. At the Feasibility level, it is essential that the main geotechnical domains of strength and rock quality will have been identified. Slope sectors of potential concern for design and performance, with respect to rock mass or major structural control, will be defined. The packer testing program will focus on defining hydraulic properties for rock units, rock mass and major structure, where geotechnical design and performance is a concern, and consequently where slope depressurisation may be important.

- The field data collection for the Level 3 study will need to support the detailed hydrogeological conceptual model, predictive numerical modelling, dewatering system and slope depressurisation design with:
  → of the order of 100 measurements of permeability via single borehole measurements (RC and packer testing), focused on the lower-permeability sectors of the pit that may be slow to drain;
  → a prolonged dewatering production pumping trial (where applicable), potentially with multiple pumping wells, with measurements of the response

to pumping from all installed piezometers within the pit shell and in the surrounding area (project specific). The pumping trial should aim to monitor the (potentially rapid) response in the more permeable sectors, the response across structures that may potentially act as hydrogeological boundaries, and any (potentially delayed) response in the poorly permeable sectors.

- Refinement of the conceptual hydrogeological model including surface water, groundwater, and surface-groundwater interaction (Section 3).

- Refinement of the 3D groundwater model using the additional field data and responses to pumping and drainage trials for transient model verification (Section 4.2).

- Prediction of groundwater inflow rates to the mine on annual increments using analytical methods and the groundwater model (as appropriate). Determination of the required mine dewatering strategy.

- Development of the dewatering plan and input of the dewatering volumes to the site water balance model, including water supplies, and water management for individual mine facilities. Planning and design of off-site water treatment and discharge systems, as required.

- Development of pit slope pore pressure profiles for key years of future project development using the 3D model and 2D sector specific models, with one-way coupling of results into the geotechnical design study models (Section 4.3).

- Design of the required systems for slope depressurisation.

- Use of the additional data to finalise the surface water and stormwater management plans.

- The Feasibility study will need to include an assessment of potential hydrological impacts at both the site scale and the regional scale, with a detailed mitigation and monitoring plan. In some cases, where this involves discharge to a surface water system, a solute loading analysis and/or a mass loading model may be required.

- The Feasibility study will also need to include an engineering design for the dewatering and slope depressurisation systems and associated infrastructure, and integrate the required infrastructure into the overall mine plan. This can potentially be a significant challenge when a large number of horizontal drain holes are required and the planned sinking rate for all or part of the pit is rapid.

- The Feasibility study will also need to include a further gap analysis and risk assessment to identify any additional needs for the detailed design and execution phases.

- The Feasibility study will need to be sufficient to support CAPEX and OPEX for the dewatering and slope depressurisation plan for the break-even pit and for the dewatering discharge management system.

2.1.5.6 Design and Construction (Level 4)

The detailed design phase needs to provide a fully functional water control plan that may be handed over to the project construction group. The program should focus on the gradual ramping up of the system into the production phase and should include the following.

- Pilot hole drilling for optimisation of the design and placement of the production dewatering wells. Ideally, this should utilise RC equipment when drilling in fractured rock areas.
- Long-term production pumping trials, which should run for at least two weeks and preferably longer.
- Development of a detailed dewatering plan for the coming five years that is consistent with the overall dewatering strategy prepared during the Feasibility study.
- Construction of 30 per cent of the start-up production well requirements, with well placements tied to the start-up mine plan, which may include the pre-stripping and/or a large part of the Phase 1 mining.
- Testing for horizontal drains or other selected slope depressurisation measures.
- Long-term pilot drainage trials at several locations, with a view to expanding these into a production drainage system.
- Installation of specific multi-level VWPs for monitoring of test pumping and slope drainage trials, with emphasis on the starter pit.
- Design of pit perimeter surface water diversions and in-pit surface water and drainage pumping systems.
- Updating of the conceptual groundwater model based on the production dewatering trials.
- Development of detailed pore pressure models that are integrated with slope design models, with emphasis on the 'break-even' pit slopes.
- Re-evaluation of site-wide water balance, the need for off-site discharge of excess water, and the impact analysis.
- Updating of the CAPEX and OPEX for the dewatering and slope depressurisation plan for the 'break-even' pit as determined in the Feasibility study.
- Updating of the CAPEX and OPEX of the dewatering discharge management system as determined in the Feasibility study.

2.1.5.7 Operations (Level 5)

The following will be required once the detailed design phase has been completed and the project has moved into the operating stage (see also Section 6.2).

- Definition of dewatering targets and slope depressurisation targets for key slope sectors, tied to the mine plan.
- A phased implementation plan for the defined programs.
- Progressive construction of dewatering and slope depressurisation measures and surface water controls.
- Installation of the required monitoring points to allow interactive monitoring of drawdown and slope depressurisation towards the defined targets.
- Monitoring of the individual well and pump efficiency and performance leading to design changes as appropriate.
- Interactive monitoring of the dewatering system performance as a whole.
- Implementation of milestones for update of the groundwater model and planning of the dewatering system and impact assessment based on system performance.
- Continuous monitoring of the dewatering discharge management system.
- Mine water balance evaluation, model development and predictive simulations for future water management requirements.

2.1.6 Planning for a Greenfield mine development

The discovery of an ore body, the definition of a resource, and the economic and technical viability assessment for subsequent mine development follows a logical, 'stage-gate' study path. At each stage, it is necessary to (i) identify opportunities, (ii) reduce uncertainty and improve confidence, (iii) reduce risk, and (iv) identify a preferred development option. At all stages the hydrogeological program must mirror the overarching project objectives outlined in Section 2.1.5 and summarised in Figure 2.6.

It is important to understand that the 'stage-gate' process may vary between mine sites. It is to be appreciated that the actual level of effort depends on the hydrogeological conditions encountered, the water-related risks, and the applicable regulatory controls. In addition, the scope may depend on the individual guidelines and standards of the particular mining company in terms of detail and confidence required at each stage of project development.

The following are examples of situations where an extensive program of hydrogeological testing would need to be initiated at an early stage and completed by the time the project reached Feasibility status.

- Where high-volume dewatering is indentified to be critical to successful mine operation and project economics.

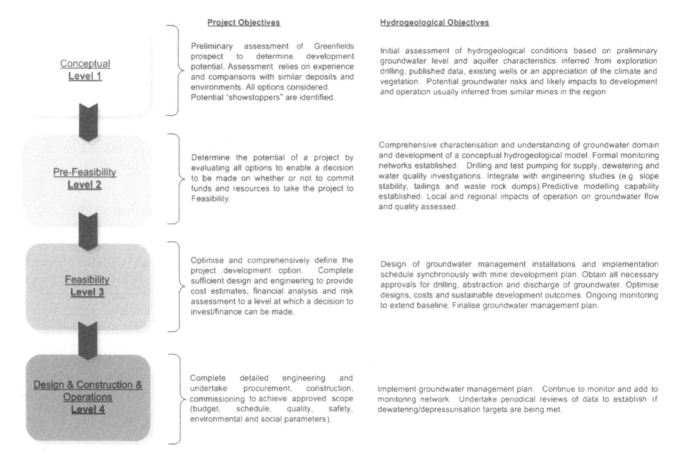

Figure 2.6: Study pathway for a Greenfield project

- Where the site-wide water balance raises sensitive issues and is critical to project feasibility, for example because of low water availability in an arid setting, or where excess water flow may lead to treatment and discharge.
- Where the economics are sensitive to overall slope angles, and where groundwater pressure is a significant factor in the slope design and for the accurate declaration of resources.
- Where dewatering may potentially create environmental impacts in the form of drawdown or changes in water chemistry. In this case, regulatory process and public scrutiny is likely to be a critical path item and the hydrogeology program may become the controlling factor for mine development.

If the project is hosted mostly in low-permeability rocks, then it may not be necessary to consider the installation of a test dewatering well. However, even if the rocks are of low permeability, it is important that some form of groundwater stress testing is included in the program as early as possible. An integrated approach to the assessment of groundwater pressure and its influence on mine geotechnical design and performance is common practice from the Pre-Feasibility stage onwards.

2.1.7 Planning for a Brownfield site development

A Brownfield development may encompass an existing dormant mine site or the expansion of an existing active mining operation. In some cases, the dormant mine site may have been partially or fully reclaimed, but its re-development may have been made economic by increasing commodity prices. The expansion of an existing active operation may involve a full pit expansion and deepening, a pushback (cutback) of one or more of the pit walls, or the development of a satellite ore body on the mining lease.

The main differences between a Greenfield and Brownfield site are:

- a Brownfield expansion often involves mining within a rock mass that is familiar, and in which groundwater response may already be at least partly understood;
- background data may already exist for the site;
- a hydrogeological baseline may already exist and only supplementary additions may be required;
- there may be existing infrastructure and support.

Notwithstanding these differences, an assessment of the potential for reactivation or expansion of any mining operation should follow the same general stages as outlined in Sections 2.1.5. Sufficient studies should be performed to confirm the dewatering and pit slope depressurisation steps needed to support each stage of the expansion project.

Significant mine expansion and/or deepening may change the future dewatering requirements, compared with historic operations at the site. Groundwater pumping rates, pumping locations and pumping heads may change. Revision of the slope angles, increases in the total slope heights, and changes of ramp configurations as part of an expansion can create a significant increase in the depressurisation efforts that are required to support the geomechanical performance and slope design.

In a mature mine, expansion studies can also be supported by long-term production pumping and slope depressurisation experience within the existing mine. Further, the exposed pit slopes provide access to conduct drainage trials. In many mine expansion situations, work to define dewatering and slope depressurisation can commence at Level 4. However, the emphasis remains on developing certainty and confidence through the break-even period of the mine expansion project.

Figure 2.7 shows a typical study path of a Brownfield expansion. The objectives of the regional-scale and site-scale field investigation normally change focus from 'discovery' of the hydrogeological conditions in a Greenfield study to 'confirmation' (or otherwise) of conditions encountered with the existing mine operations. Confidence levels may be higher for a Brownfield operation but, at the slope scale, the characterisation required for a pit deepening or slope pushback may be almost equal to that required for a Greenfield development.

2.1.8   Environmental baseline studies

It is not the purpose of these guidelines to detail the procedures and data requirements for water chemistry sampling and testing or environmental studies, for which numerous references are available in the public domain. For Greenfield sites, a formal environmental baseline monitoring program is normally initiated at the time of the Pre-Feasibility study (often before, particularly in environmentally sensitive areas).

The baseline hydrogeology monitoring program will need to characterise the following aspects.

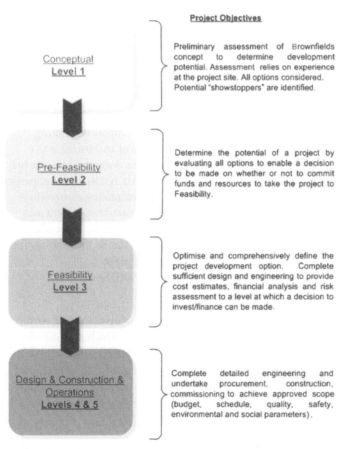

Figure 2.7: Study pathway for a Brownfield site

- Climatic variables.
- Surface water levels, flow and chemistry (including rivers, streams, lakes, springs, seeps and wetland areas).
- Groundwater levels and chemistry for each key hydrogeological unit.
- Mine-scale and regional-scale interaction between the groundwater and surface water systems.

The main goal of the baseline studies is to define pre-mining background conditions in the proposed project area. The ambient groundwater chemistry in areas upgradient and downgradient of the mine itself (and the individual mine facilities) is typically used as the benchmark for defining water quality impacts. It is therefore important that the baseline studies are focused to quantify the natural variability of the hydrological system, paying particular attention to naturally elevated chemical constituents and/or pre-existing conditions, and also to the natural extremes of surface water flow, spring flow and groundwater levels.

The natural water chemistry associated with many ore bodies often contains elevated level of trace metals and other important parameters, and these may also occur in the downgradient groundwater and surface water systems. It is therefore important that the baseline study properly characterises these ranges and uses the information to help determine groundwater and surface water compliance levels during mine operations and for mine closure. Mine operators are being increasingly required to demonstrate that changes or trends in water chemistry and/or water levels may be the result of the natural variability of the system or other external sources, rather than due to impacts from the mining operation itself.

The number of baseline monitoring stations required depends on (i) the area, topography and location of the site, (ii) the type of climate and seasonality, (iii) the natural variability of the hydrological system and the presence of elevated chemical parameters, (iv) the mine plan relative to surface water bodies and groundwater levels, (v) the number and type of local and downstream receptors, (vi) the interest of NGOs and other public groups, (vii) local regulatory requirements and the needs of regulatory bodies, and (viii) internal mining company standards. The standard baseline monitoring program is normally a minimum of 12 months, but may be longer, depending on the site characteristics and seasonal variations. The potential for seasonal variations needs to be considered when designing the baseline sampling program, and it is often appropriate to use a reduced sampling interval during any wet season.

In addition to providing a regulatory baseline, the characterisation of pre-mining surface water and groundwater chemistry has the following objectives.

- To supplement the interpretation of the hydrogeological system and support the conceptual model. For example, groundwater chemistry can help define hydrogeological domains and compartments, based on spatial variation in parameter values. Zones that are potentially connected or compartmentalised can be defined and locations of groundwater recharge can be identified.
- To help determine appropriate construction materials for dewatering system infrastructure. For example, a corrosive groundwater chemistry may require stainless steel well construction and pumping system components, which may affect project costs.
- To characterise the chemistry of the dewatering flows, and changes in the chemistry with time, particularly where discharge to the environment is expected and a permit to discharge is required. For larger dewatering projects, a solute balance and/or mass loading model may be required.
- To predict the consequence of mixing and blending different water types that may be produced in the process of dewatering.
- To determine an appropriate strategy for accommodating mine dewatering flows within the mine-wide water management system (for some operations, for example, the mine dewatering chemistry makes it unsuitable for use as make-up water for the mineral process plant).
- To determine the potential need to treat the mine dewatering flows, in order to either utilise the water within the mine water balance or discharge to the environment.
- To predict potential groundwater chemistry impacts resulting from mine dewatering and, eventually, open pit closure and shut down of the dewatering system.
- To ensure compliance with environmental monitoring and reporting regulations. Ambient groundwater chemistry and natural ranges of variability can be used as the basis for establishing compliance for discharge to the environment.
- Prediction of viable closure and reclamation options for the open pit. The chemistry of groundwater inflow to the pit after closure will affect the pit water quality and potential options for permanent closure of the mine.

2.1.9 Water management practices during the field investigation program

2.1.9.1 General
The protection of local surface and groundwater resources is also an important planning component for any hydrogeology field program. Consideration needs to be given to the following aspects.

- Water produced from exploration holes and/or fluid circulation losses to the local groundwater system during the drilling of exploration holes.
- Management and discharge of larger water volumes that may be produced during test pumping of production wells.

2.1.9.2 *Exploration drilling and single hole testing*

Many countries now require an assessment of how exploration drilling activities may affect local surface and groundwater resources. This may be different for 'wildcat exploration' involving relatively few holes and 'ore body definition' involving tens or hundreds of holes. The first step is often a screening-level assessment to determine the potential risks. This may comprise an assessment of the distance to the nearest known water supply well or other sensitive groundwater feature (e.g. a spring). Those exploration programs that do not pass the screening-level assessment should be subject to a more detailed risk assessment, which may include an assessment of whether groundwater receptors are upgradient or downgradient of the exploration site, whether there is potential for losing a reasonably large amount of fluid into the groundwater system, and the need to sample the make-up water for the drilling process. Most drilling fluid additives are inert but many local regulations provide guidance on what additives are acceptable.

Local regulations also provide guidance on:

- containment of water produced from the drill hole and when this water may be discharged to nearby surface water courses;
- the need to case off shallow groundwater sources that may overlie the deeper mineralised bedrock;
- requirements for plugging and abandonment of the drill hole once exploration, logging and testing activities are complete.

2.1.9.3 *Pumping tests*

Many hydrogeological field testing programs eventually involve pumping tests that produce groundwater at flow rates that may require a management and discharge plan. A key challenge at many Greenfield sites is that infrastructure may not exist for handling, treatment or discharge to the environment of the water produced during pumping tests. Another reason why adequate early-stage (Level 1) characterisation is beneficial is so that subsequent hydrogeological testing requirements can be anticipated and planned well in advance.

In most countries, the drilling and construction of pumping wells and piezometers, even for temporary use, requires some form of permitting. Establishing a temporary water management plan, including a permitting program, can be a drawn-out process and may affect the critical path for the overall mine study. A pumping test program may require a temporary discharge permit if release of water to the environment is necessary. This will usually involve a constraint on water quality and rate of flow.

A program of water quality monitoring during testing will normally be required since the chemistry of the water may not be proven at the time that test pumping is initiated. Indicator parameters including Total Dissolved Solids (TDS), pH, Electrical Conductivity (EC) and Dissolved Oxygen (DO) are used to quickly identify possible changes in water chemistry as the test progresses. In many regions, regulators will allow exemption of groundwater abstraction itself if the testing is temporary and designed for the purpose of groundwater quantification and impact assessment. However, a process of engaging the regulator is required and, in regions where there is social sensitivity and an indigenous presence, a prolonged process of social engagement and education may be necessary as a precursor to test pumping.

It is usually not advisable to allow discharge of pumped flow to occur within the expected area of influence of pumping and, in particular, in the vicinity of test pumping wells or piezometers. Rapid recharge and re-circulation of discharge water into the groundwater system can bias results. This is particularly important where there may be permeable materials and therefore a high infiltration rate at the surface.

For Brownfield sites, the burden of planning and permitting is normally less onerous. Infrastructure at the mine will probably exist for the management (storage, treatment or use) of flow produced from testing. Permitting of dewatering wells and piezometers located within existing operating areas may involve only a minimal process.

2.2 Implementing field programs

2.2.1 Background

This section discusses procedures for data collection in the field. Many of the field procedures are common to both the investigation and planning phase (Section 2.1) and the operation or implementation phase (Section 5) of a mine project.

Mine hydrogeology regimes are as variable as the ore deposits themselves. They may include saturated alluvial units, poorly permeable units (perhaps with highly permeable zones or discrete flow conduits), highly permeable units (often with discrete flow barriers), units with only interstitial flow, units with only fracture flow, units with dual-porosity or dual-permeability characteristics, units that have karst-type solution cavities, or a combination of some or all of these features. As described in Section 2.1, a thorough understanding of the geology (lithology, alteration, mineralisation and structure) should form the basis for developing an early-stage understanding of the regional hydrogeology and for preparing a conceptual model to provide focus for the subsequent field program.

For small scale excavations, such as for construction projects, dewatering is generally completed *prior* to commencing excavation. However, in the case of open pit mining, where the excavations are much larger, dewatering and slope depressurisation is most often done *concurrently* with the mining process itself, with the goal being to keep the dewatering and pore pressure targets ahead of the mining schedule. Except for the smallest of operations, it is usually impractical and seldom cost-effective to dewater prior to mine development. Therefore, as described in Section 2.1.5, the planning and implementation of mine dewatering and slope depressurisation for a large project is best accomplished through a phased approach, progressing from the Conceptual study (Level 1) through to active operations (Level 5).

2.2.2 Drilling methods

Many drilling methods are used for hydrogeological investigations. It is beyond the scope of these guidelines to detail all the methods and procedures. Pertinent reference texts with detailed descriptions of drilling and well construction methods include Roscoe Moss Company (1990), Groundwater and Wells (3rd edition) (Johnson Screens, 2008), and the Australian Drilling Industry Training Committee Ltd (1997). Drilling methods typically used in hydrogeological field investigations are summarised in Appendix 2.

2.2.3 'Piggy-backing' of data collection

2.2.3.1 Data from drilling programs

During the development of any new project, hundreds of boreholes may be drilled across the mine exploration concession area. Many of the holes are often concentrated within the final footprint area of the open pit. Although the primary objective of these boreholes is to assess and delineate the mineral resource and host rock material, these boreholes provide an important opportunity to 'piggy-back' the collection of both hydrogeological and geotechnical data. Furthermore, condemnation (sterilisation) holes drilled for the waste rock facilities and other site infrastructure are usually spread out from the main area of the ore body and therefore provide the opportunity to help characterise the wider-scale groundwater system. It is often possible to collect a considerable amount of the required

hydrogeological information from the mineral exploration and condemnation (sterilisation) drilling program, particularly during the early stages of project development; equally, it is easy for the opportunity to be missed unless preparation is made for it.

Most exploration drilling programs use a combination of rotary mud techniques (such as core drilling and conventional mud-drilled boreholes) and reverse circulation (RC) utilising air as the main circulation medium. 'Piggy-backed' data can also be collected from drill holes for the geotechnical and environmental programs.

2.2.3.2 Supplementary hydraulic testing

Hydraulic tests can also be performed on exploration boreholes upon completion. These tests may add to the total cost of the drilling program, as there is often a significant amount of time involved in conducting the tests. However, the cost is relatively small compared with the cost of having to subsequently drill a 'twin' investigation hole for collection of required hydrogeological data.

Hydraulic tests that can be performed within mineral exploration and geotechnical investigation holes include injection tests, airlift recovery tests, variable head (slug) tests, or extended airlift tests. These testing methods are described in Section 2.2.5.

2.2.4 Dedicated hydrogeological drilling programs

2.2.4.1 Introduction

Although much of the background data for the hydrogeology program can be obtained by 'piggy-backing' (particularly for Level 1 and Level 2), most operations that require general mine dewatering or pit slope depressurisation will require dedicated hydrogeology drilling programs from Level 2 or Level 3 onwards. For larger mining operations, the dedicated hydrogeology programs can be extensive. They will frequently include the following.

- Groundwater exploration drilling, which is usually best carried out using reverse-circulation (RC) or alternative air-flush methods (Section 2.2.4.2).
- Surface geophysical programs to support the general geological and hydrogeological characterisation, typically at a regional scale, and in particular to provide definition of alluvial or sedimentary basins or to define regional structural zones. The application of surface geophysics to mine hydrogeology is not covered by these guidelines.
- Pilot hole drilling to optimise sites for high-cost production dewatering (or water supply) wells. Again, in hard-rock settings, the pilot holes are usually best

carried out using RC or alternative air-flush drilling methods (Section 2.2.4).

- Single-hole testing (Section 2.2.5).
- Installation of observation wells and piezometers, which would typically use the groundwater exploration holes plus other dedicated holes (Section 2.2.6).
- The installation of test dewatering (or water supply) wells, which is discussed in Section 5.2.6.
- Downhole geophysical logging (Section 2.2.7).
- Cross-hole and multi-hole testing (Section 2.2.8).
- Water chemistry analysis (Section 2.2.9).
- The setting up of pilot drainage trials (Section 2.2.10).

The management of surface water in the slope design process is discussed in Section 5.3. Most mines located in wet or seasonal environments carry out a significant amount of climate and surface water monitoring. Suitable references covering surface water flow monitoring include:

- *Water Measurement with Flumes and Weirs*, A. Clemens, T. Whal, M. Bos, J. Replogle, ILRI, 2001;
- *Water Measurement Manual*, USBR, 2001, <http://www.usbr.gov/pmts/hydraulics_lab/pubs/wmm/>.

References covering the analysis and application of surface water hydrology information include:

- *Physical Hydrology*, Lawrence Dingman, Waveland Press, 2002;
- *Handbook of Hydrology*, David Maidment (Editor) McGraw Hill, 1992;
- *Design Hydrology and Sedimentology for Small Catchments*, C. Haan, B. Barfield and J. Hayes, Academic Press, 1994;
- *Applied Hydrology*, Ven Te Chow, D. Maidment, L. Mays, McGraw Hill, 2013 (last edition).

2.2.4.2 Groundwater exploration drilling using RC drill holes

In most hard rock settings, a dedicated groundwater exploration drilling program is best carried out using air-flush drilling methods. RC drilling (Figure 2.8) is ideal for groundwater exploration and characterisation in hard-rock geological units and much of the discussion on data collection contained within these guidelines is focused around the RC drilling method. However, most of the techniques can be adapted to other air-flush drilling methods. RC is the most common drilling method used for mineral exploration worldwide so, in many cases, the dedicated hydrogeology program can be closely integrated with, and added on to, the mineral exploration program. The other common drilling method for mineral exploration is diamond drilling and notes for hydrogeological data collection in cored holes are included in Section 2.2.4.3.

The RC drilling method (and other air-flush methods) are much less suitable for groundwater exploration in alluvial and other soft rock settings where there is a tendency for the formation to collapse around the drill pipe and reduce the downward flow of water in the hole to the drill bit. In soft rock settings, it is often necessary to employ casing advance methods or mud-based drilling where the opportunities to collect hydrogeological data as the hole is advanced are typically more limited.

The following hydrogeological data can usually be collected at minimal incremental cost during the drilling of RC and other air-flush holes.

- Airlift flow rate, after drilling down each 6 m (20 ft) drill pipe. A good record of airlift flows enables the volume of water entering the hole, and the depth at which it enters, to be characterised.
- Air pressure during drilling, taken after drilling down each section of drill pipe. This provides supplementary data to quantify the depth of the inflow zones.
- Air pressure required to unload or 'kick-off' the hole when re-starting drilling each time a new length of drill pipe has been added. This allows assessment of the recovery of the water level during the pause in drilling.
- Quantitative information on fracture location and intensity.
- Quantitative information on the degree of clay alteration.
- The temperature of the airlifted water.
- Penetration rate (for the bit type in use).
- General notes about the drilling process itself (for problems and events).
- Static water level, measured down the inside of the drill pipe at the start of each drill shift (or after shut-downs of long durations such as rig repairs).

With the correct set-up, the hydrogeological data collection should not slow the drilling program down by more than a few minutes each day. Ideally, a standardised field data form should be utilised to allow the data to be systematically recorded. An example of a typical hydrogeological logging form and notes for collecting data during drilling are included in Appendix 2.

The data that are important to record can vary from one project to the next. The users of these guidelines need to assess what will be important for their individual projects. For example, the occurrence of groundwater may be associated with a specific type of lithology, alteration, and/or mineralisation. Therefore, a column may be added to the form that would be used to record this specific information. Alternatively, measurements of water chemistry (pH, TDS, and EC) might be important, and columns could be added for these parameters.

If there is continuous supervision of the drilling program, the geologist can be made responsible for

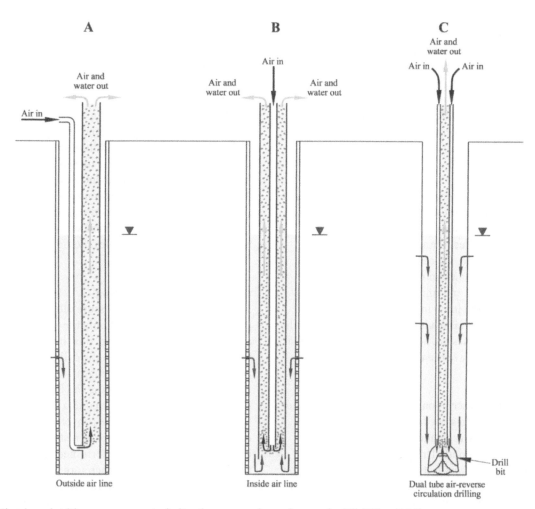

Figure 2.8: Sketches of airlift arrangements, including the commonly used set-up for RC drilling (2.8C)

recording the hydrogeological data. More commonly, the drilling rigs are visited by the supervising geologist once or twice daily, so the responsibility of recording the hydrogeological data falls to the driller. Most drillers will provide good cooperation with data collection provided an appropriate level of guidance and the right incentives are offered beforehand. The driller will benefit because the hydrogeology data collection process helps provide a better understanding of drilling conditions in the hole.

The common RC method uses an airlift arrangement to drive the reverse-circulation process. Compressed air is introduced into the column of water and creates a column of aerated fluid ('froth'). The density of the aerated fluid is less than that of the water in the adjacent hole, so that a longer column of aerated fluid is needed to balance the pressure of the water. If the delivery pipe containing the aerated fluid is too short for that balance to be achieved, aerated fluid flows from the top of the delivery pipe and is replaced by more water being drawn in from the bottom. The airlift arrangement commonly used in RC drilling is shown in Figure 2.8C.

Penetration rate
The penetration rate can be recorded in several ways but it is probably easiest (and best) to record start and stop times for each 6 m (20 ft) drill rod. It can also be recorded as minutes per rod or minutes per metre (or per foot). This information can be helpful for identifying hard or soft rock, or easy or difficult drilling.

Unloading ('kick off') pressure
The unloading pressure is the pressure required to 'unload the hole' and get the water circulating after adding another length of drilling pipe. After a length of pipe has been drilled all the way down, the air is shut off in order to add another length of drill pipe. When this happens, formation water will continue to enter the borehole and will begin to fill up the hole and the drill pipe. How much the hole (and pipe) fill up is a product of the amount of time the air is off and the inflow rate to the hole. After the pipe is added, the driller turns the air back on and the pressure gauge on the control panel will rise to a high point and then drop back somewhat before it stabilises. The highest pressure reached

is recorded on the form. This pressure depends on the height that the water has recovered above the drill bit, and can therefore be used to estimate the inflow rate to the hole. It can be used to help determine when additional productive fractures have been encountered by the borehole. It can also be useful for determining whether a 'boot' has formed around the drill pipe, thus impeding some of the water that is entering the borehole from reaching the bit.

Bit type

This is simply meant to record whether a tri-cone bit is in use or whether a down-the-hole hammer is in use. The type of bit is important as it can affect the pressure readings and it can affect the amount of water being airlifted out of the hole. Tri-cone bits often allow more water to be airlifted than hammer bits. Many holes are started with hammers and finished with tri-cones. It is likely that an apparent jump in water production will occur where the switch is made. The increased flow to the surface is real but it does not necessarily mean additional productive fractures have been encountered.

Degree of fracturing

This is a subjective description normally recorded by the driller on how he 'judges' the fracturing of the rock. Sometimes fracturing is obviously non-existent and drilling is slow but smooth. Drilling speed in highly fractured rock can be very fast or very slow depending on how the bit copes with the fractured blocks, the cuttings may normally increase significantly in size, the water colour may change, the drill rods may tend to lock up or the drilling may just be extremely rough and 'jumpy'. The driller may write down a number from 0 to 3 or 0 to 5 (depending on the convention chosen) where 0 means no fracturing and 3 (or 5) means extremely fractured.

Clay content

The description of the clay content is also somewhat subjective and is normally recorded by the driller. Again the scale can be 0 to 3 (or 5). Zero would indicate no clay present and a 3 (or 5) might indicate that the entire sample was in the form of clay balls. When drilling RC, it is not uncommon for minor amounts of clay to be completely washed away and go unnoticed.

Airlift flow rate

The airlift flow rate is the most important data to be recorded while drilling. This is simply a record of how much water is being airlifted to the surface by the drilling system. It is typically measured at the end of drilling each drill pipe and is obtained from timing how long it takes to fill a container of known volume. A five-gallon (US) bucket (19 litres) can be used up to flows of roughly 9.5 l/s

(150 US gpm), but a larger vessel should be used for flows greater than this to retain a reasonable amount of accuracy. The volume of the container should be accurately calibrated and marked.

The airlift flow measurement should be obtained after the drilling has ceased, after the discharge has 'cleaned up' (relatively few cuttings coming up), with the pipe near the bottom of the hole, and with the drill pipe slowly turning. If the driller was injecting additional water while drilling, the injection pump should be turned off a minute or two before taking this measurement. Measuring a flow rate of 0.3 l/s (5 US gpm) would obviously add one minute of time to drilling each pipe. It takes two seconds to measure a flow rate of 9.5 l/s (150 US gpm) so the cost in added time is small. Fairly high flow rates should be measured several times and then averaged.

Air pressure 'during drilling'

More accurately, the air pressure during drilling is the air pressure required to airlift only the water (not water and cuttings) out of the borehole. It is the air pressure that is observed from the gauge on the driller's control console at the end of drilling each length of drill pipe, usually just before or just after performing the flow measurement. This pressure can be converted to a head of water above the bit, which in turn can be converted to a value of drawdown if the initial static or rest water level (RWL) is known (the RWL can be determined approximately from the 'kick-off' pressure). It therefore provides supplementary data on new inflow zones.

Most pressure gauges on rigs are calibrated in pounds-force per square inch (lbf/in² or psi). For fresh water, 1 psi is equivalent to 0.703 m head of water: hence the depth to water level is given approximately by:

$$d_w = d_b - 0.703 p_g \qquad \text{(eqn 2.1)}$$

where:
d_w = depth to water from driller's datum level
d_b = depth to drill bit from driller's datum level
p_g = air pressure indicated by gauge.

It should be noted that using the pressure indicated by a gauge at surface ignores the weight of the air in the pipe between the gauge and the drill bit. As the density of air is very low and the estimation of water level by this method is only intended to be approximate, the resulting error is insignificant at relatively shallow depths but it may become important in very deep holes. Errors may also be introduced as a result of friction losses as the air travels to the bit. More details of airlift principles and arrangements are given in Appendix 2.

Additional observations

Several other columns may be added to the data collection form on an as-needed or site-specific basis. Other data that

Figure 2.9: Diamond (core) drilling (HQ size)

might be important for characterising the local hydrogeological regime include:

- water temperature;
- simple field water chemistry such as pH and TDS;
- water colour;
- simple geological notes like formation changes;
- the location of 'bridges' or 'boots'.

A bridge is best identified by the driller while tripping out the drill pipe after a bit change and is where the formation has collapsed into the borehole and blocks the re-entry process. A boot is when the formation has collapsed or squeezed (in the case of some clays) around the drill pipe which may impede the downward flow of water in the annulus of the hole from reaching the drilling bit. A boot may or may not require a significant amount of additional torque to be applied to maintain the speed of rotation of the drilling tools and may become a bridge upon the removal of the drilling pipe. Boots can be short or long and they can partially or totally block water movement.

Ample space should be left on the data collection form for the driller to make any additional comments about the drilling process. Records of the static water levels obtained from the hole can be recorded in this column as well as notes pertaining to problems with hole stability, mud use, abnormal torque, notes about rig maintenance and downtime, changes to the air package, or other information that may be deemed important.

2.2.4.3 Groundwater data collection from diamond drill holes

Diamond (core) drilling (Figure 2.9) is not as adaptable to groundwater exploration as RC drilling, but allows the collection of the following information.

- Static fluid level, measured down the inside of the drill pipe at the start of each drill shift.
- Depth of circulation or fluid loss in the hole, which can be correlated with permeable zones.

It is also common practice to utilise diamond drill holes for conversion to piezometers or groundwater monitoring wells, and they are particularly valuable for the installation of vibrating wire piezometers because the depth of each piezometer sensor can be selected following visual examination of the drill core.

2.2.5 Single-hole testing methods

2.2.5.1 Introduction: types of single-hole tests

While a considerable amount of information can be gained from the interpretation of airlift flows, it is usually beneficial to perform additional tests either during the drilling process, upon completion of the open borehole, or after a cased monitoring well or piezometer has been completed. There are several short-term test procedures that can be performed on a single borehole to provide an estimate of 'point' permeability and to help estimate the potential pumping yield from the hole. These test methods include:

- drill-stem injection tests (not to be confused with oil-industry type DSTs);
- variable (rising or falling head ('slug')) and constant-head tests;
- packer (Lugeon) tests;
- single hole airlift and recovery tests.

In addition, somewhat longer-term, multi-hole tests can be performed easily and economically by airlifting a newly drilled borehole.

2.2.5.2 Drill-stem injection tests

Injection tests can be carried out in RC holes to help characterise the fracture zones penetrated during drilling. A sketch of the set-up is shown in Figure 2.10 and procedures for injection testing are included in Appendix 2. Water is pumped down the annular space between the inner and outer tubes of the RC drill pipe using a small trash pump. The resultant rise in water level in the inner tube is measured using a sounding probe. Manual sounders are usually adequate for measuring the response when the depth to water is shallow but when the depth to water is greater than about 100 m, pressure transducers and data loggers may be preferable. The data allow characterisation of each main fracture zone. The data also allow for the calculation of the specific capacity of each hole.

Fabrication of a suitable sub-adapter head for injection can easily be carried out by the drilling contractor. A 50 mm diameter flow meter and 3–5 kW trash pump are also required. Each test may last approximately 20 to 30 minutes. Typically, up to three tests may be carried out (at various depths) on holes that show good water-bearing fractures. Experience has shown that better results are usually obtained by pulling the bit up so that it is just above productive fracture zones that were identified through the drilling process. These tests can be performed as the drill hole is advanced or may be carried out after the

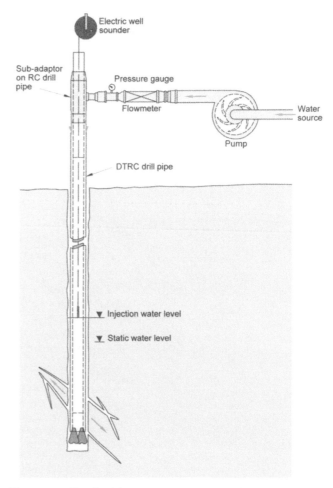

Figure 2.10: Sketch of the arrangement commonly used for RC drill-stem injection testing

hole has been drilled to its total depth and the bit is being pulled out. The former method is best but will add additional rig standing time.

Drill-stem injection tests provide a good method of characterising different intervals and different fracture zones within a drill hole. They are most often used for comparing the performance and specific capacity of individual fracture zones, but the data can be used to estimate permeability according to the procedures described in Appendix 2. Apart from minor equipment costs, the cost of the testing is limited to the cost of standing time for the drill rig. For a typical hole, two hours standing time may be required for three test intervals. The method is therefore attractive for rapidly obtaining quantitative data from many drill holes.

It may not be possible to perform this test through some makes (manufacturer brands) of down-the-hole-hammers as the springs behind the dart valves may be too tight to allow the valves to open and let the injected water through the hammer and into the borehole. This is dependent on the design of the hammer, the depth to water, and the capabilities of the trash pump (i.e. the head that can be developed to open the valves).

2.2.5.3 Variable-head ('slug') tests

Variable-head tests are also known as rising or falling-head tests, slug tests or Le Franc tests. They offer the simplest and lowest cost method of *in-situ* measurement of hydraulic conductivity, either in an existing hole, or in a hole that is actively being drilled. Tests can often be rapidly carried out in many holes to provide a basic characterisation of the hydraulic conductivity of the materials present at the mine site. They can be carried out in an open hole during or following drilling when formation stability is suitable, but are more often performed immediately following construction and within a standpipe piezometer. Such tests may also be performed in newly constructed dewatering wells in advance of more rigorous pumping tests. In some circumstances, a value of storage coefficient may also be estimated from test data.

Test procedures and analytical methods for variable-head tests are described in numerous technical papers and publications, including Freeze and Cherry (1979),

Kruseman and de Ridder (1990), Brassington (1998), and Domenico and Schwartz (1998). Many of these refer to the detailed descriptions provided in the standard methods publications by the British Standards Institution BS 5930 (1999) and the American Standard and Testing Methods ASTM D 4044-91 (1991). A summary of the field procedures is provided in Appendix 2.

The aim of the falling-head test (Figure 2.11) is to cause an instantaneous increase in head of water in the test hole and then observe the rate at which the head declines ('falls') back towards its original level, using either a pressure transducer and data logger or an electric sounding line ('dipper') and stop watch. The rate at which the water level returns to its original depth is related to the hydraulic conductivity of the formation tested, a faster recovery indicating a higher value of *in-situ* hydraulic conductivity.

Although the test is often attempted by simply pouring a quantity of water into the hole, this has the disadvantage that the water will not all arrive at the same time, especially when the depth to water is more than a few metres. For permeable formations, the result is that, while the decline

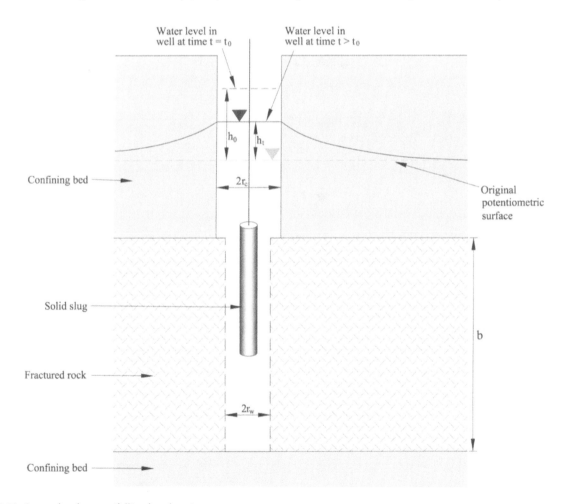

Figure 2.11: Set-up for slug test (falling-head test)

in head is under way after the arrival of the main slug, some of the water will still be trickling down the sides of the hole, making reliable interpretation difficult. However, for lower-permeability settings, the decline in head may take minutes or hours, so the error caused by the 'non-instantaneous' arrival of the water slug is negligible, particularly when considering that 'point' permeability data are more valuable for characterising low-permeability settings than high permeability flow systems. Adding water into the hole from the surface has the advantage of making it easier to apply a greater initial head.

An alternative approach is to raise the water level instantaneously by inserting a solid 'slug' on a nylon or steel line below the water level to displace water and cause the rise in head. The slug can be a specially made solid cylinder of dense plastic (such slugs are available commercially) or similar article fabricated locally; obviously, the average density must be greater than that of water. In the absence of such a device a bailer with a closed valve and filled with water can be used. Plastic materials have the advantage that in the unfortunate event of a slug becoming stuck in the hole, they are more readily drilled out.

A suitable procedure is:

1. place a pressure transducer with data logger into the hole below the water level; the data logger should be set to record at short intervals;
2. lower the slug to just above the water level;
3. when the water level is known to be stable, rapidly (within a few seconds at most) lower the slug to submerge it completely below the original water level;
4. record the changes in water level against time until the water level is again stable.

When the water level is again stable, a rising-head test can be performed by rapidly removing the slug to cause a fall in water level and monitoring the subsequent recovery (Figure 2.12).

Rising-head tests can also be carried out independently of falling-head tests. Depending on the hole diameter and the depth to water in the hole, the water level may be reduced by introducing a bailer to remove smaller volumes from narrower diameter monitoring wells or by pumping to remove larger volumes in larger diameter monitoring wells, with the use of a manual inertial pump like the

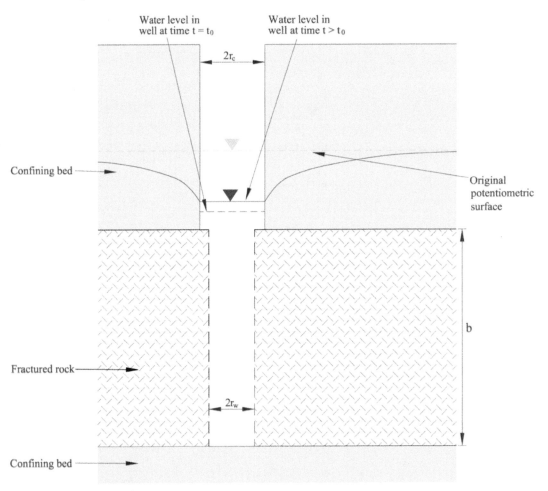

Figure 2.12: Set-up for slug test (rising-head test)

Waterra (In-Situ Inc.) or by installing a small electrical submersible pump such as the Grundfos MP1. Similarly, these tests may be conducted by removing a measured volume of water from the hole with an airlift system. Again, these will not remove an 'instantaneous' slug of water, so the results obtained may need to be interpreted with caution in higher permeability settings.

Manual measurements of initial water levels and levels during the test, commonly at 0.5 to 1 minute intervals, can be made with an electrical sounding line. However, pressure transducer/data-logger sensors should preferably be installed. Sensors that are widely used for this type of application include the Diver (Schlumberger Water Services), MiniTroll (In-Situ Inc.), or Level Logger (Solinst). The advantage of using a pressure transducer for variable-head tests is that it can be pre-programmed to record measurements at fixed or variable frequency. Important early-time measurements can be collected by programming the system to record data at 10 second intervals, which is important for more permeable formations, where the rate of water level recovery may be rapid and difficult to measure manually. Conversely, where the formation permeability is expected to be relatively low, the test may take several hours to complete and longer data intervals may be more appropriate, and it may only be practicable to measure the early part of the water-level decay curve. In this case, it is usually beneficial to conduct tests in different holes simultaneously.

2.2.5.4 Constant-head tests

Constant-head tests typically involve a higher rate of flow than variable-head tests and may therefore be useful for characterising a greater volume of the formation surrounding the drill hole. The method involves introducing a flow of water to the hole (or removing water from the hole) and adjusting the flow rate until a constant head or column of water is maintained above (or below) the pre-test water level. A description of the method can be found in BSI (1999). A typical configuration for a constant-head test in a monitoring well is shown in Figure 2.13.

For a constant-rate test involving injection, an ample supply of clean water is needed, and this may be pumped or gravity fed via a header tank with a valve for controlling the rate of flow. The method also requires that a dip tube (sounding tube) is installed within the test hole to a depth of several metres below the pre-test water level to allow an undisturbed measurement of water levels. As for the variable-head tests, the water levels may be measured within the dip tube using either a dip meter (electrical sounding device) for manual measurement or a pressure transducer/data logger for automatic measurement.

The rates of flow required to establish and then maintain the constant head over fixed intervals of time are recorded and analysed to determine the formation permeability. The most common empirical method and associated assumptions for deriving estimates of permeability from constant-head test results are described by Gibson (1963). It is important to note that, in many stratified sediments or fractured rock settings, permeability estimates derived from a constant-head test may be representative only of a discrete single sedimentary layer or single fracture zone, rather than the entire sequence. In addition, for tests involving injection of water, it is necessary that the zone of water level rise (above the

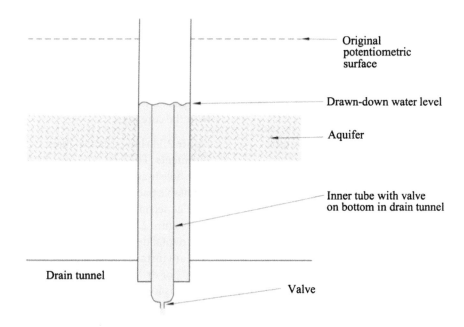

Figure 2.13: Set-up for a constant-head test

pre-test water level) is of several orders of magnitude lower permeability than the interval being tested.

Although fairly simple, both variable and constant-head tests need to be carried out with care when using the data to calculate permeability, as small errors in the measurement process can lead to significant differences in the analytical results. Repeating the tests can improve the level of accuracy and confidence and, as for all field test methods, careful review of the testing assumptions and the data obtained is required. However, the tests are often simple to perform, and are therefore suitable for providing information at multiple locations around a site.

2.2.5.5 Injection tests

In addition to the test procedure for RC drill holes described in Section 2.2.5.2, the use of injection tests can provide a rapid method to determine the specific capacity of a drill hole and therefore to determine the suitability of the hole for reaming out and conversion into a production well. The method also involves the installation of a dip tube to just below the pre-test level. A steady stream of water is allowed to flow (or is pumped) into the hole under controlled conditions. The rise in water level in the hole is measured within the dip tube.

While injection tests are often best suited for providing information to support the cost of reaming out the hole into a production well, the results can be used to calculate the permeability of the formation, provided the test is carried out under controlled conditions (ideally at a constant rate). Again, the results need to be interpreted carefully, with particular consideration given to the permeability (and any fracturing) within the zone of water level rise (above the pre-test water level) as this will have a strong influence on the results.

In remote areas, where the availability of test pumping equipment is poor or non-existent, it may be possible to carry out an entire characterisation program using injection tests rather than pumping tests. It is possible to undertake the tests under controlled conditions and to carry out measurements in nearby observation wells and piezometers (Section 2.2.8). The use of injection tests may also be appropriate for carrying out cross-hole or multi-hole testing in low-permeability formations, where the response in a nearby piezometer to injection may provide much more valuable information than permeability estimates derived from single-hole 'point' testing. However, if high-volume injection of water into a formation is to be considered, it is necessary to consider any local regulations that may apply.

2.2.5.6 Packer tests

Packer tests (also known as Lugeon tests) are a widely used field method for measuring *in-situ* hydraulic conductivity of consolidated sediments and fractured rock at discrete depth intervals in drill holes. Packers are temporary plugs, usually consisting essentially of inflatable sleeves expanded by compressed gas or a fluid. Packer tests involve isolating a length of a borehole between two packers or between one packer and the base of the borehole and then injecting water into or pumping water out of the interval. Methods are described in various texts and papers, including Muir-Wood and Caste, 1970; Pearson and Money, 1977; Price *et al.*, 1977 and 1982; Daw and Scott, 1983; Brassington and Walthall, 1985. Although there are advantages to pumping water from, rather than into, the test interval, for practical reasons tests at mine sites usually involve injection.

Packer tests may be carried in both vertical and inclined holes, but it is generally not feasible to run packer tests in inclined holes that are greater than 35–40° from the vertical. If the tests are run properly, the angle of the hole is not usually important for achieving good results. Single-packer testing systems may be easier to install and remove from inclined holes than double-packer systems. In theory, packer tests may be conducted in 50 mm (2 inch) to 533 mm (21 inch) diameter holes given the current range of available inflatable packer equipment. They also may be performed in a variety of drill hole types (mud rotary, RC, diamond core). In practice, and particularly in mining, it is usually preferable to carry out packer tests in diamond core holes of PQ (122 mm), HQ (96 mm) or NQ (75 mm) diameter, for the following reasons.

- It is generally easier to obtain a reliable seal with the inflatable packers in core holes given the smoother wall conditions.
- The inflatable packer assembly can be installed through the core barrel on the wireline system, with less risk of damage to the packers during installation.
- The standard drill diameter sizes mean that packers and testing equipment are widely available throughout the world.
- The packer test information can be readily correlated with the core log.

Figure 2.14 shows the two main types of packer system arrangements. The first type uses a single packer, run as the core hole is advanced, to isolate a test interval between the inflatable packer and the base of the hole (Figure 2.14a). The second type is a double (straddle) packer system (Figure 2.14b), typically run after the borehole is drilled to total depth and where upper and lower inflatable packers are installed to isolate the test section. In the oil industry (for drill-stem tests) or hydrogeological applications for water-resource evaluation or sampling, packers are normally run in open hole on drill pipe or other steel tubing. For mining applications, they are frequently deployed on wireline.

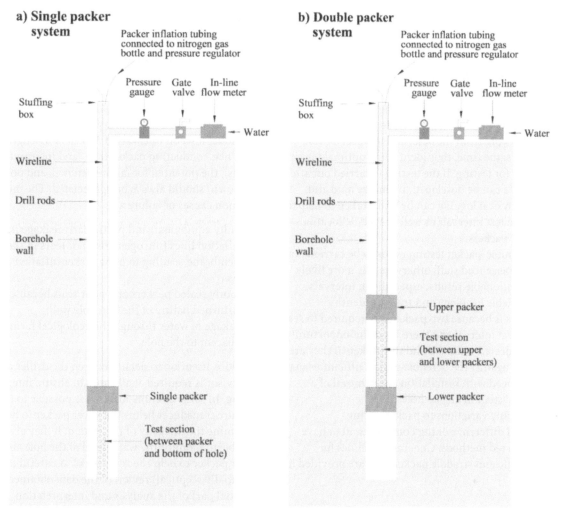

Figure 2.14: Types of packer system arrangement

Early practice in so-called Lugeon tests appears to have been to test at only one flow rate and test pressure; this involves the tacit assumption that flow is conforming to Darcy's law and that the test procedure does not alter the natural conditions. A better procedure is to:

1. inflate the packers, ensuring that the inflation pressure is not high enough to damage or fracture the rock;
2. allow the head in the isolated test interval to come into equilibrium with the head in the formation;
3. inject water at five (or more) flow rates/injection pressures in a sequence increasing to a maximum and then decreasing to repeat (approximately) the lower flow rates; the sequence is thus *a-b-c-b-a*. The maximum pressure in the interval (corresponding to *c*) must not be so great as to risk fracturing the rock.

At each flow rate/pressure step, the flow and pressure should be monitored over five-minute intervals. When two successive five-minute readings are essentially the same, the flow rate is increased (or decreased) to the next step.

Single-packer tests

The single-packer system has the advantage that it is only necessary to seat one packer, instead of the two required for the straddle system. For the single test procedure, the bottom seal is provided by the bottom of the hole at the time of the test. The single-packer test is also easier to run because there is less risk of installation/removal problems for the packer equipment if the hole is unstable.

If drilling conditions dictate the use of mud to enhance hole stability, the use of a single-packer system is preferable, as mud in the test interval may first be flushed out or broken down with clean water and/or an agent such as hypochlorite. The packer system is lowered to the selected interval through the wireline core barrel, with limited exposure to the open hole.

A disadvantage of using a single-packer system is the additional standing time required as drilling is stopped to allow installation of equipment and running of each individual test. Additionally, a technician or geologist may be required at the drill site for a longer period of time to log core, select suitable test intervals, and direct the tests concurrently with drilling.

Double-packer tests

Advantages of the double (or straddle) packer method include the ability to do the tests sequentially while tripping out of the borehole after drilling, minimising rig standing. It allows the field supervisor to observe the entire profile of core from the hole, and to select all test intervals at the same time, thus identifying optimal target depth intervals for testing. If the testing is carried out after drilling, the hole can be developed to remove mud and infill and geophysical logging can be undertaken to help to identify suitable test intervals as well as specific locations at which to seat packers.

However, double-packer testing needs to be carried out with care by experienced staff, otherwise it is more likely to generate questionable results, especially in intervals with highly variable lithology and fracture-density conditions. This is because two packers are required to seal off the packer test interval and there is greater opportunity for by pass of injected water around the packers if they are not fully sealed against the borehole walls. Difficulties may also be experienced with installation and removal of a double-packer system in angled holes.

There are many variations to packer testing procedures, and different mining companies often have their own preferred methods. General guidelines for conducting single and straddle packer tests are provided in Appendix 2

Considerations for packer testing

Both the depth of the target zones and length of the test interval for the packer tests should be determined based on analysis and interpretation of the geological (or core) log of the hole, any downhole geophysics, and the requirements of the hydrogeological or geotechnical investigation. Information from geophysical logs is often also available, including acoustic and optical televiewer logs, which provide valuable complementary data to the geological and geotechnical core logs, particularly for confirming depth and *in-situ* condition of water-bearing discontinuities (joints, fissures, fractures and faults) and identifying suitable depths for placement and seating of the packers.

The majority of packer testing in the mining industry is carried out in core holes, so the target zones (and packer locations) can be determined based on fractures, lithology and alteration types, and the known condition of the borehole wall. Packer testing is also regularly carried out in RC holes but, in this situation, the detailed control on the geology is often less, and there is more risk of poor seating of the packers. It is critical that locations for packer placement in the core hole are picked where there is a smooth borehole wall to provide the greatest seating

(formation sealing) characteristics. This is easier for the single-packer system, since there is only one packer to seat.

When conducted correctly, it is often possible to correlate the results with the geological and geotechnical properties of the rock, including lithological and alteration types, discontinuities, fracture frequency and Rock Quality Designation (RQD). For both single and double-packer methods, the best results are usually obtained by running individual tests on intervals of 3 to 10 m in length.

When conducting packer tests (and interpreting the results), the potential for failure (known, and potentially unknown) should always be appreciated. The most common causes of failure are:

- faulty equipment, and particularly leakage in the inflation lines (nitrogen hoses) or tears in the packer membrane, leading to poor packer inflation and a poor seal;
- poorly seated packers and poor seals because of the disturbed nature of the borehole wall;
- leakage of water through the geological formation adjacent to the hole.

Close attention to detail between the driller and field supervisor is required at all times to ensure integrity of the testing. In some circumstances, it is possible to install a pressure transducer below the lower packer to help determine the integrity of the system. It may also be possible to monitor the water level of the hole above the upper packer to help check for leaks. A careful (and potentially sceptical) review of the data obtained is an essential part of the analysis and interpretation. Methods for analysis are described in Section 2.3.

Gas and water inflatable packer testing equipment

There are two main types of packer testing equipment, those with gas inflatable (pneumatic) packers and those with water inflatable (hydraulic) packers.

Gas inflatable packers have been used more regularly in the past. Generally, an inert gas such as nitrogen is used for safety reasons. A disadvantage of compressed gas or the pneumatic packer system in deep testing applications is the inherent safety concern of bursting an inflation line or fittings near to the testing crew at ground surface, due to the high working pressures required.

Alternatively, for deeper testing, water or hydraulic fluid/antifreeze can be used to inflate the packers. The advantage is that water is essentially inelastic so the inflation pressure does not have to be as high and the packer inflation lines do not have the same degree of stored potential energy as the elastic gas filled lines. Rupture of water filled inflation lines therefore poses a lower hazard than for pneumatic lines. A further advantage of the water inflated system in deeper tests is that the natural hydrostatic pressure of the water in the inflation line will equal, or exceed (if static water level is below ground surface), that of the adjacent water in the formation, and therefore the only pressure required at the surface is the incremental pressure to inflate the packer itself.

It should be noted that both pneumatic and water inflated systems are prone to freezing in cold conditions. Antifreeze or hydraulic fluid can be used to overcome this problem, but may present an unacceptable environmental risk. Brine solutions may also be used to reduce freezing but this may cause corrosion that in the longer term may be problematic.

Inflation fluids must also be assessed for compatibility with the packer gland material.

Other types of testing using packers

Packers are merely a method for isolating a selected vertical interval of the borehole. Therefore, where there is sufficient diameter, any type of hydraulic test may be carried out inside a packer system. In some cases, for larger diameter holes with more permeable zones, extended airlift or pumping tests can be run within packer systems (Figure 2.15a), with the pumping response monitored in

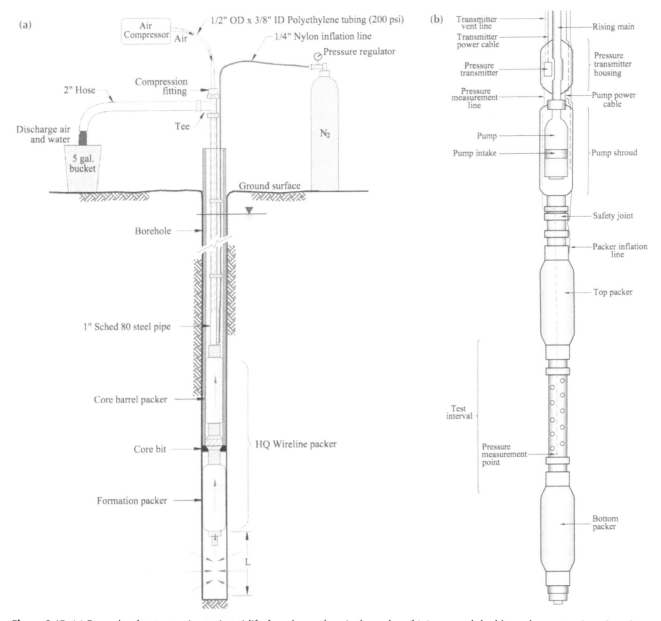

Figure 2.15: (a) Example of test pumping, using airlift, from beneath a single packer; (b) A pumped double-packer system incorporating a submersible pump and pressure monitoring system. The safety joint provides a 'weak link' designed to fail at a particular point if the packers become stuck, so allowing easy retrieval of the pump, transducer and production tubing, and providing a large shoulder for 'fishing' purposes (After Price & Williams, 1993).

multi-level piezometers in adjacent holes. The tests can be repeated with different sections of the airlift or pumping hole isolated by the packers. In this way, a vertical profile of the groundwater system can be developed, using multi-hole measurements, rather than the 'point' measurements that are obtained using the conventional packer test methods. High-volume injection tests can also be considered using packer systems (depending on any local regulations concerning injection of large volumes of water into a formation). The upper packer can be positioned so as to avoid any water level rise within the unsaturated zone.

A testing system that incorporates a submersible pump to lift water from the packer interval is described by Price and Williams (1993). This type of system also obviates potential problems related to the clogging of the test interval with suspended solids in the injection water, or changing the natural chemistry of the formation (if this is of concern). A typical set-up for this type of packer test system (which is run on drill pipe or tubing) is shown in Figure 2.15b.

2.2.5.7 Airlift and recovery tests

Upon completion of drilling (or after penetration of key productive fracture zones during drilling), controlled airlift testing can be carried out in air-drilled holes. Provided the hole is stable, it can be airlifted for periods of 30 minutes to four hours to allow the potential steady-state yield of the hole to be determined. The drawdown in the hole can be estimated from comparison of the unloading or kick-off pressure (required to begin the airlift) and the running pressure.

When the airlift system is shut down, a water level sounding device can be run into the hole, and the water level recovery can be measured inside the drill pipe. The recovery data can be analysed to estimate the permeability of the interval.

This testing method is best suited to boreholes in formations of relatively low permeability. This is because it usually takes a minute or two for the driller to shut off the air and 'break' the pipe at the surface to allow a well sounder to be introduced into the drilling pipe and run down to the water table. In formations of higher permeability when the borehole makes a lot of water, most of the recovery will have occurred before the first data point has been obtained. However, while the data may not allow an accurate calculation of the permeability, they do provide good information on the producing capacity of the hole.

2.2.5.8 Surface pumping

In situations where the water depth in the hole is less than about 5 m or so, consideration can be given to installing a surface suction pump to carry out hydraulic testing. Again, the pumping rate from the hole and the water level

would be carefully measured. The installation of a monitoring pipe is normally beneficial to facilitate the measurement of water levels.

2.2.6 Monitoring installations

2.2.6.1 Overview

A groundwater monitoring network is required for all mining operations, whether or not the mine workings extend below the water table. As outlined in Section 6.1, a typical monitoring network may consist of several types of installations.

- Boreholes that are temporarily left open (with or without liner pipe).
- Standpipe piezometers or observation wells (single or multi-level).
- Monitoring wells (for groundwater chemistry).
- Grouted-in vibrating wire piezometers (VWPs).
- Installations with a combination of these methods.

In addition, monitoring should also be carried out in inactive (off line) pumping wells and, depending on local circumstances and access considerations, any surrounding third party observation wells and pumping boreholes.

Initially, monitoring networks should be planned to characterise both the site-scale and regional-scale hydrogeology. Monitoring networks necessarily evolve after mining begins and a greater understanding of the hydrogeological regime is attained. For mines with combined surface and underground operations, it is not uncommon to add water level or chemistry monitoring systems from underground once the workings become established. In complex hydrogeological regimes, every borehole drilled at the site should be considered as a possible candidate to be completed as an observation well, monitoring well or piezometer. Open holes, standpipe piezometers and 'grouted-in' VWPs all have their part to play in helping to understand water level behaviour and head conditions around a mine site.

The static head at any point in a formation, such as the measuring element of a piezometer, consists of two components – the elevation head, which is the height of the point above the datum level, and the pressure head, which is the pressure at the point divided by the specific weight of the water column above that point. A transducer will make a direct measurement of pressure (though it may need adjustment for changes in atmospheric pressure) and, in a static system, that pressure will not change. If the measurement is made by measuring the depth to water level below the top of the standpipe or well, and the density changes, the water level will change even though the pressure does not. Care therefore needs to be exercised when comparing head values derived from a combination of measurement methods.

2.2.6.2 Open observation holes

The term 'open observation hole' is used to describe an open drill hole, which is unlined except perhaps for the upper few metres, and is used for measuring groundwater levels (Figure 2.16a). Open observation holes may have small diameter monitoring pipe installed to provide stability against collapse of the hole, but have no impermeable seal.

Measurement of groundwater levels in open observation holes (with or without slotted pipe installed) can allow rapid initial characterisation of the piezometric surface at low cost. The water level recorded in an open hole represents an average of the head over the entire saturated interval. This is useful in the early stages of project development, where there has been little stress on the groundwater system, and where significant vertical head gradients may have not yet developed. Where vertical gradients are present, open holes, or lined holes with long open or screened intervals, can provide a cross-connection between formations or parts of formations where the heads vary with depth. Around large open pits, the vertical hydraulic gradients are often much steeper than the horizontal hydraulic gradients and, by providing an easy connection for water to flow vertically through a borehole or standpipe, such holes may be altering the very conditions that they were installed to measure. In some countries, in order to protect against cross-connection of aquifer units, there is a legal requirement to seal open drill holes within a specified time period.

2.2.6.3 Standpipe piezometers

The traditional piezometer is a standpipe allowing the water pressure above the sealed interval to be measured as a head of water in a pipe. Historically, 'Casagrande' piezometer tips were installed, but the piezometer may consist of any filter tip or perforated interval of pipe of a desired length attached to the bottom of impermeable casing that rises to the surface.

All standpipe piezometers require a sand pack (or gravel pack) to be emplaced in the annulus adjacent to the perforated interval (or filter tip), usually with a bentonite seal in the annulus 3 m to 5 m above the sand pack, and with the annulus grouted above the bentonite seal to the surface. It is usually necessary to install the sand pack, bentonite seal and grout using a tremie pipe that is progressively raised in the hole as the materials are emplaced in the annulus, and is eventually removed from the hole once the entire piezometer completion has been constructed. Figure 2.16b shows a typical single-completion standpipe piezometer. The level to which the water rises in the pipe is indicative of the average head over the perforated interval.

General guidelines on the installation of piezometers are given in Appendix 2. There are numerous standard

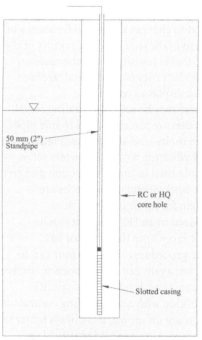

a) **Simple standpipe**
Slotted pipe installed in geology
exploration drill hole.

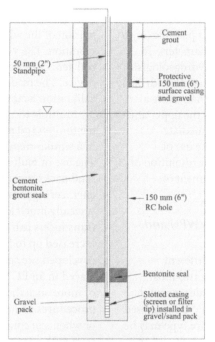

b) **Sealed standpipe piezometer**
Smaller perforated interval.
Sealed below the water table.
Installed in RC drill hole.

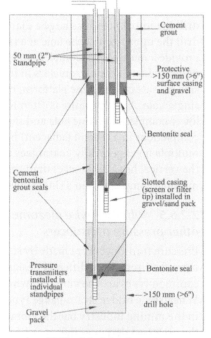

c) **Multilevel standpipe piezometer**
Using required larger drill hole.
Necessary to use tremie pipe
to place ground packs and seals.

Figure 2.16: Types of standpipe piezometer completion (see text for details)

hydrogeological texts that describe piezometer construction in detail, including Freeze and Cherry (1979), Kruseman and de Ridder (1990), Domemico and Schwartz (1998). In the early stages of groundwater investigation, if piezometers are used in preference to open holes, the perforated interval of the piezometer may be long (say 20 m or more), so the average head of a formation is measured. As the investigation becomes more focused, a shorter (6 m or less) perforated interval may be installed in subsequent holes to measure the water pressure at a discrete depth, to allow a better assessment of vertical head gradients.

2.2.6.4 Multi-level standpipe piezometers

It is common practice to install multiple standpipe piezometers in a single borehole to allow the water pressure to be measured over several depth intervals in the hole. Where multiple standpipes are installed, the grout placed above the lower piezometer is brought up to just below the base of the selected interval for the second piezometer, and a bentonite layer is placed above the grout prior to the installation of the second pipe. Figure 2.16c shows a typical multiple-completion standpipe piezometer (three standpipes).

Multiple-level standpipe piezometers may include two, three or four installations at various depths in the hole. However, in cases where the hole condition is unstable, or the hole diameter is small, it may be difficult to place the seals correctly in the annulus, meaning that the subsequent monitoring data are questionable. When considering a multiple completion, it is usually better to drill the initial hole at as large a diameter as possible (or to drill the upper part of the hole at a larger diameter). To provide a more robust completion, many hydrogeologists prefer to install the piezometers in individual holes drilled side-by-side off the same platform, rather than using a single hole. A compromise is often to place the upper and lower completion in one hole and the intermediate completions in a second (adjacent) hole. The use of multiple holes generally guarantees a better resolution of the vertical head differences than the placement of multiple completions in a single hole.

2.2.6.5 Vibrating wire piezometers (VWPs) and other pressure transducers

Pressure transducers are made by several different manufacturers using different designs. The two most common types are the vibrating wire transducer and the Wheatstone bridge strain-gauge type. Both types are used in the mining industry but the vibrating wire type may be the preferred design as they may be more robust and are usually more economic.

Either type may be vented or unvented to the atmosphere. Vented transducers have a small tube that is bundled into the cable sheath. The tube runs from the body of the transducer to the surface so that atmospheric pressure changes do not affect the pressure measurements, but if the water levels in the hole are subject to changes due to atmospheric pressure (Jacob, 1940) the venting will not remove these effects (Price, 2009). At the surface, the tube usually terminates in a pod that contains desiccant capsules to keep moisture out of the tube. The desiccant capsules require regular maintenance. Unvented pressure transducers may be affected by barometric pressure changes. The use of unvented transducers during investigations that require a high level of sensitivity may mean that an additional barometric pressure transducer needs to be set up at a base station so that the piezometer readings can be corrected for changes in barometric pressure. However, if the aquifer has a barometric efficiency this is true whichever type of transducer is used and also true if the water levels are 'dipped' by hand.

Vibrating wire pressure transducers provide a good and well-established method of collecting pressure data and, in particular, for characterising vertical hydraulic gradients around active pit slopes. The system comprises a vibrating-wire sensing element enclosed in a protective steel housing. The sensing element consists of a tensioned steel wire that is fixed to a hollow cylindrical body at one end and a sliding diaphragm at the other. The vibrating wire transducer or sensor converts water pressure to a frequency signal via the diaphragm, the tensioned steel wire, and an electromagnetic coil. When excited by the electromagnetic coil, the wire vibrates at its natural frequency (like a guitar or piano string). The transducer is designed so that a change in pressure on the diaphragm causes a change in tension of the wire and so changes its natural frequency of vibration. The vibration of the wire in the proximity of the coil generates a signal that is transmitted to the readout device. The readout device processes the signal, applies calibration factors, and displays a reading.

Multiple transducers may be set in alternating sand/bentonite/sand packs that are placed using a tremie pipe, in a similar manner to multi-level standpipes (Figure 2.17). The use of multi-level vibrating wire piezometers set in sand packs is commonly used in the construction and civil engineering industry, where installation depths are typically much less than for mining. Two to three transducers can be placed in an HQ hole, but can be increased up to five or more once the operator has developed site specific procedures. Three to four can be placed in an RC hole, but again can be increased up to five or more once the operator has developed site specific procedures. When working with a new drilling contractor or when site conditions are unknown, it is always better to start with a fewer number of sensors and to provide a robust grout seal. The number of sensors can be increased as familiarity of site staff and site conditions increases.

The advantage of using sand packs and seals is that discrete, pre-determined, known intervals can be monitored. When installed in this manner, the piezometer sensors and cable are normally strapped to the outside of a non-perforated sacrificial pipe which is left in the hole, and a separate tremie pipe is used to emplace the sand packs and seals in the borehole. Difficulties in placing the seals may be less compared with the standpipe installations because there are fewer pipes (and therefore more room) in the hole. However, poor and unreliable seals are frequently encountered, particularly in deeper holes, or in holes where the walls are unstable.

2.2.6.6 Grouted-in, multi-level vibrating wire piezometers

Since the late 1990s, the mining industry has pioneered the application of grouted-in multi-level vibrating wire

piezometers. A string of vibrating wire piezometer instruments is grouted directly into the hole without the use of a filter pack. This provides a rapid way to install multiple piezometers in one borehole, to install piezometers in holes with inclinometers or other geotechnical instruments, and to monitor pressures at discrete 'points'. Key technical references that describe the theory and practice of installing grouted-in multi-level vibrating wire piezometers (VWPs) include McKenna (1995) and Mikkelsen *et al.* (2000).

Figure 2.18 shows an example of multiple grouted-in vibrating wire installations. Typically two to five sensors are placed in the hole but, in some cases, it is possible to install six or more instruments in HQ or RC holes. Grouted-in vibrating wire installations are frequently made by piggy-backing on mineral exploration or geotechnical drill holes. Consequently, the cost is low, and it is often not necessary to drill a dedicated hole. Many installations can therefore be made rapidly at any given mine site.

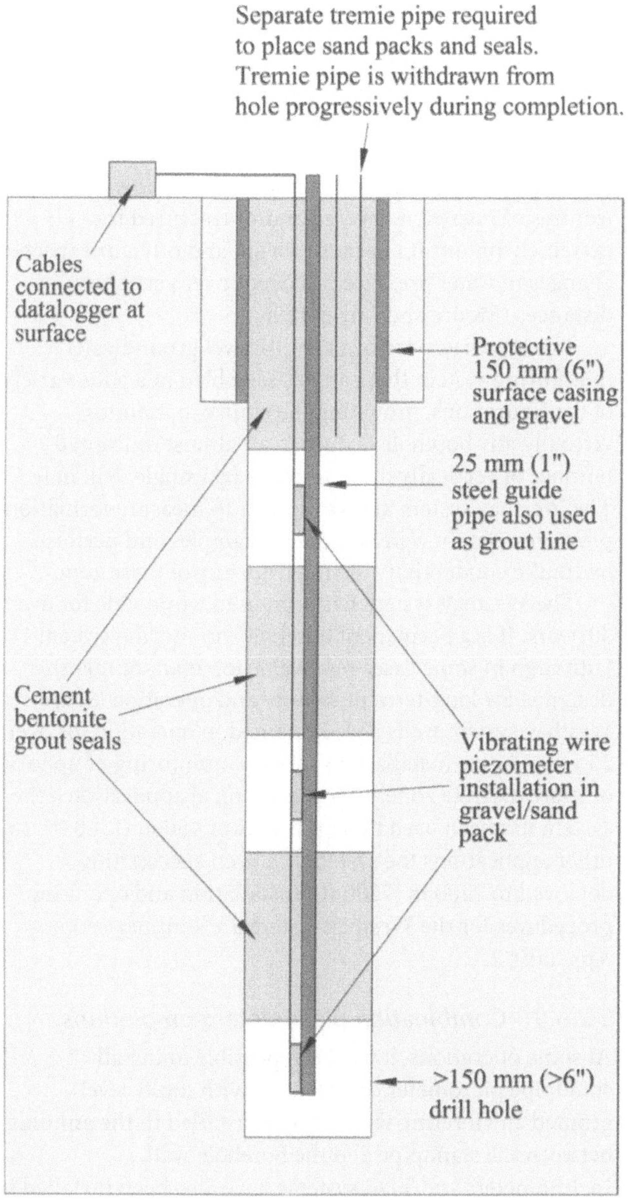

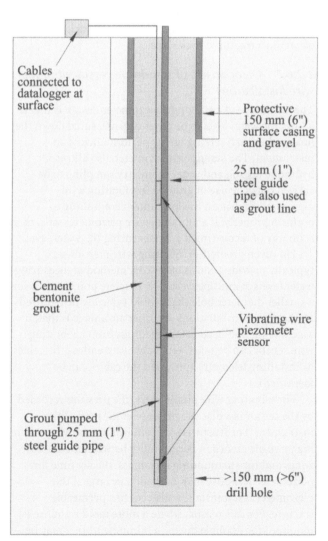

Figure 2.17: Setting of vibrating wire transducers in sand packs

Figure 2.18: Example of multiple grouted-in vibrating wire piezometers

There are a number of different installation methods currently practised; the selection of the most appropriate method depends on anticipated ground conditions; hole diameters, depth and angle, and partly on the type of vibrating wire instruments deployed. The ease of installation has increased as the diameter of the instruments has got smaller, and as the use of a single cable for multiple sensors has become more commonplace. However, where field experience at a given site is limited, it is usually sensible to start off with fewer instruments in the hole (say two to three) and to slowly increase this once site-specific conditions are better understood, and when site-specific installation procedures are developed by the mine operator. Installation methods for grouted-in VWP strings are included in Appendix 2.

The increasing use of the multi-level grouted-in piezometer over the past 10 to 15 years has led to major improvements in the understanding of pore pressures and, in particular, how vertical pore pressure gradients develop around pit slopes. It is considered to be the single-most important factor contributing to the rapidly improving and practical understanding of how depressurisation occurs in a fractured rock slope.

2.2.6.7 Comparison of standpipe versus vibrating-wire installations

The advantage of a standpipe piezometer design is that it allows the water level to be physically measured down the hole, rather than relying on the performance of an instrument. The standpipe piezometer also allows for hydraulic testing and water chemistry sampling to be carried out. The use of grouted-in vibrating wire piezometers relies on the instrument continuing to perform properly. If a vibrating wire piezometer fails, there is no way of recovering it and measuring the water level.

However, grouted-in vibrating wire piezometers typically provide a much lower cost method of measuring water levels at multiple vertically discrete points, and using a smaller diameter hole than would typically be required for standpipe installations. Another advantage is that, if access to the wellhead is lost, readings from a vibrating wire sensor can be taken remotely by extending the cables or installing telemetry (provided the cables can be identified).

For vibrating wire piezometers, the pressure recorded by the sensor is a discrete pressure at the level of the instrument. For fractured rock environments, where there is a large difference in permeability between the fracture zones and the surrounding rock mass, the lag time for piezometer response may, in theory, be large if the piezometer is not installed adjacent to a permeable fracture. For this reason, where a more rapid response is required (for example during a pumping test), it may be better to use sand pack intervals rather than the grouted-in

system. However, the general experience to date when using grouted-in vibrating wire piezometers is that rapid responses to pumping are commonly observed in many fractured rock settings even when the instrument is not positioned near to a large facture zone.

Grouted-in vibrating wire piezometer installations are also beneficial in settings with low permeability. Because there is no standpipe, and therefore no water in the completed hole, there are no wellbore storage effects. Therefore, the response time of a vibrating wire piezometer in a low-permeability setting is much more rapid than a conventional standpipe.

2.2.6.8 Westbay piezometers

Another option for multi-level groundwater monitoring is the Westbay system (Figure 2.19). The system was originally developed in the 1980s for geotechnical applications, based on the premise that densely spaced pore pressure monitoring is required to correctly understand the complex heterogeneities and compartmentalisation common to large rock slopes. In addition, establishing short completion zones within the features of interest in layered and/or fractured rocks is extremely important because abrupt and often unexpected changes in water pressure can occur over very short distances (Moore and Imrie, 1995).

The Westbay system is a multi-level groundwater monitoring system that can be assembled in a wide variety of configurations, providing the ability to monitor virtually any borehole feature in an almost unlimited number of vertically discrete zones in a single drill hole. The Westbay system allows the user to measure formation pressures, obtain water chemistry samples and perform hydraulic conductivity testing from any of these zones.

The Westbay system has been used worldwide for over 30 years. It is a permanent borehole monitoring system (although in some cases removable for abandonment) designed for long-term durability and operation. Some Westbay system wells have remained in operation for over 25 years. Some installations include monitoring of up to 50 or more discrete zones. For geotechnical applications, the system has been used to depths of over 450 m (1500 ft). In other applications the system has been successfully deployed to 2200 m (7200 ft). Installation and operating procedures for the Westbay system are summarised in Appendix 2.

2.2.6.9 Combination piezometer completions

At some operations, it has been possible to install standpipe piezometer completions with multi-level grouted-in vibrating wire sensors installed in the annulus between the standpipe and the borehole wall. Inclinometers and TDR systems have also been installed in the same manner (Section 2.2.6.12).

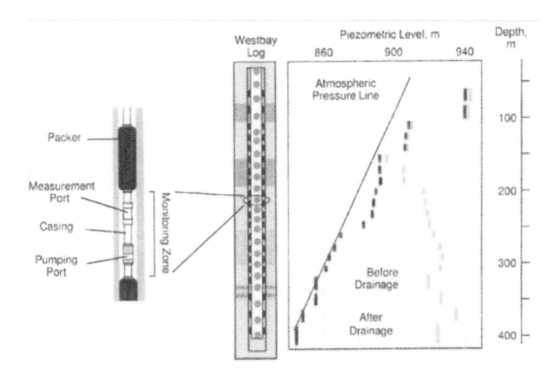

Figure 2.19: Example of slope-depressurisation data collected using the Westbay system

These designs involve building a standpipe well with slotted pipe on the bottom and attaching pressure transducers to the blank casing, at various levels, as the casing is run into the hole. With this method, the casing provides a positive way to ensure that the transducers are placed exactly where intended. The well is then completed by placing sand pack around the well screen and grouting the annulus of the hole outside the casing above the screened interval, so that the vibrating wire sensors become grouted in. It is often advisable to employ centralisers on the casing when using this method, provided there are no broken zones within the borehole wall where the centralisers may become hung up. Grouting of the annulus is best carried out using a separate tremie pipe but this is not always possible and, in some situations, it may be necessary to grout from the surface.

2.2.6.10 Horizontal piezometers
During active mining operations, piezometers can be installed horizontally from the pit slope face, often as part of a horizontal drain drilling program. However, as for vertical piezometers, the target zone for pressure monitoring needs to be isolated and sealed. There are two practical ways to achieve this. The first is to use a packer system, and to grout the annulus inward from the packer, as shown in Figure 2.20. The second is to grout vibrating wire sensors into place, as shown in Figure 2.21. The second option allows multiple sensors to be installed within the hole.

Horizontal piezometers can be drilled at a slight downward or upward angle. For down-drilled holes, the grout is pumped into the hole using a sacrificial tremie pipe, and the hole is grouted from the bottom up. For holes drilled at an upward angle, the piezometers are attached to the sacrificial tremie pipe in a similar way, but the pipe needs to be pushed into the hole, and the method of grouting needs to be modified. A second short grout pipe is placed from the collar and the collar is sealed. The grout is then pumped through the short pipe and the hole is filled from the collar upward. The displaced air and water flows out of the hole through the deeper tremie pipe, and the grouting is complete when grout returns are observed from the deeper pipe.

Experience has shown that a hole drilled at a slight upward angle will often deviate downward such that it approaches the horizontal or a slight downward angle when total depth has been attained (Section 5.2.3.3). Therefore, because knowledge of the exact sensor location is essential for determining the static head, all horizontal and angled holes should be surveyed for deviation; this is discussed further in Section 5.2.3.6.

2.2.6.11 Piezometers installed in drainage tunnels
The designs shown in Figures 2.20 and 2.21 can also be applied to piezometers installed in underground drainage tunnels. Depending on the capability of the underground drilling set-up, such piezometers can be installed at any angle to best target specific zones. Most underground drill

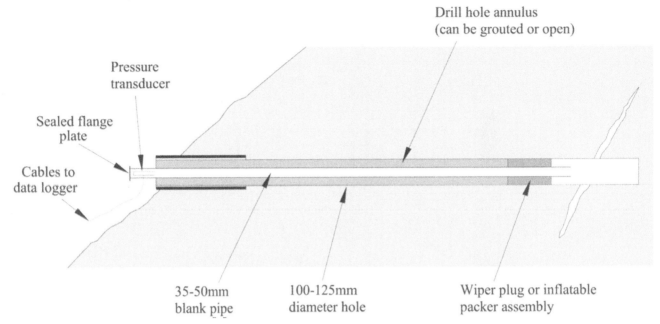

Figure 2.20: Horizontal piezometer completed using a packer

holes are drilled at a relatively narrow diameter compared with surface drilled holes. Although the piezometer string is best placed in holes of HQ size or larger, it is perfectly feasible to install multiple piezometer sensors in NQ size holes drilled from underground. Figure 2.22 shows an example of an underground drilling set-up). Figures 2.23 and 2.24 show drain holes and piezometers installed in underground drainage galleries (see also Introduction, Section 3.4). The procedures for underground installation are also included in Appendix 2.

2.2.6.12 Integrated hydrogeological and geotechnical installations

The relatively small diameter of the VWP sensors and cables leaves space within the borehole for additional instruments, such as inclinometer casing, Time Domain Reflectometry (TDR) cables, accelerometer arrays, and borehole extensometers. Examples of these types of combined instrumentation systems and some practical suggestions for installation are outlined below.

Combined VWP and inclinometer installations

Conventional inclinometer systems typically comprise a series of ABS (and in some cases aluminium) casing segments that are coupled together with a sealed cap on the end and inserted into a borehole. The annulus between the borehole wall and the casing is typically grouted with a cement-bentonite mixture. A portable survey probe (torpedo) with a sensitive, mobile accelerometer array attached to a cable is fitted into aligned, machined grooves on the inside of the casing and lowered down the hole. The orientation (tilt) of the probe is recorded at prescribed

depth intervals. Repeated hole surveys can be compared against a baseline survey to detect deflections or distortions of the casing over time. Using this system, both creep and discrete movements and movement rates down the hole can be determined over time.

Depending on the relative size of the borehole and the inclinometer casing, it may be possible to install one or more VWPs in the annular space between the borehole wall and the outside of the inclinometer casing. The VWP tips and readout cables, and a disposable PVC or steel grout tremie pipe are secured (taped) to the outside of the inclinometer casing as it is lowered into the hole. Once the assembly is fully inserted, the annulus is grouted from the bottom of the hole using the tremie pipe. The grout, which is typically composed of a lean mixture of Portland cement and bentonite, encapsulates the VWP tips and stabilises the inclinometer casing. Depending on the depth of water in the hole, it may be necessary to fill the inside of the inclinometer casing with water to weigh it down. For deep installations, it may also be necessary to grout in stages, in which case more than one tremie pipe may be required. For installations where the water level is very deep, it may be necessary to support the weight of the assembly using a disposable wireline so that the stress on the inclinometer casing couplings stays within tolerance. In holes subject to caving, it may be necessary to install the assembly through drill rods or steel drill casing, in which case it may be possible to use the rods or casing to inject the grout and eliminate one or more of the tremie pipes.

The key to a successful installation is in designing an instrument bundle that is appropriate for the effective hole diameter. There must be enough annular clearance to

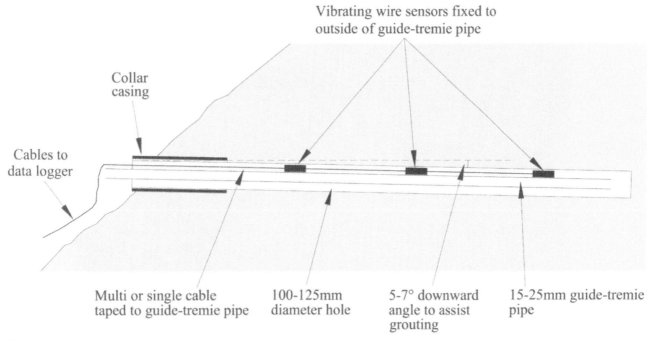

Figure 2.21: Horizontal piezometer string completed with vibrating wire piezometers

Figure 2.22: Drill set-up used for underground drain hole and piezometer installations

Figure 2.23: Array of drain holes and piezometers installed in an underground drainage gallery

Figure 2.24: Array of underground drain holes and upward piezometers installed in the Highland Boy drainage gallery at Bingham Canyon Mine

accommodate the VWP tips and cables and other instruments, as well as any tremie pipes, plus sufficient space to ensure that the bundled instruments are fully encapsulated in grout. VWP tips should be spaced well apart down the hole to minimise the potential for shear or extensional movements along the inclinometer casing to cause hydraulic short-circuiting. Small diameter VWP tips (typically 21 mm diameter) are generally preferred as they provide the most flexibility, and the use of a single cable for multiple VWP sensors is becoming increasingly common. Small diameter diamond drill holes (i.e. HQ – 96 mm diameter or smaller) are generally not large enough to accommodate combined inclinometer and VWP installations; consequently, larger diameter (e.g. PQ – 122 mm diameter) diamond drill holes or conventional rotary or RC holes are usually required. Note that because the inside diameter of drill rods or casing is smaller than that of the hole, installing the instrument package through the drill rods or casing may limit the number of instruments that can be bundled together.

Most conventional inclinometer systems are installed in vertical or near vertical boreholes. However, provided the hole is stable or can be stabilised with casing, installation in inclined holes is also possible.

Schematic illustrations of a bundled inclinometer and VWP installation are shown in Figures 2.25 and 2.26.

Combined VW piezometer and TDR installations
The use of Time Domain Reflectometry (TDR) in addition to, or in place of, conventional inclinometer systems is gaining popularity in open pit mining applications. TDR borehole systems consist of a coaxial cable that is grouted into a borehole. A cable tester (signal generator) is attached to the cable at the collar of the borehole and sends pulsed

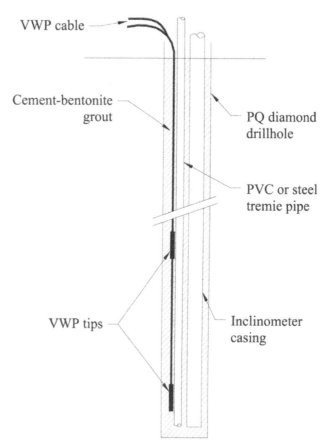

Figure 2.25: Profile of inclinometer and VWP installation

signals along the cable. These signals are reflected at the end of the cable and travel back along the cable to the surface where their arrival time is recorded. Reflections are also triggered along the cable at places where the impedance (resistance) of the cable changes. The travel

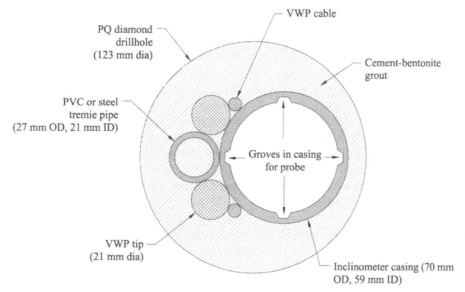

Figure 2.26: Bundled inclinometer and VWP installation

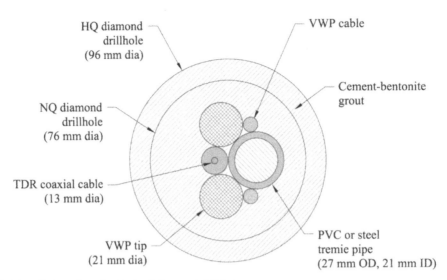

Figure 2.27: VWP/TDR bundle in HQ and NQ-sized diamond drill holes

time for the reflected signals to return to the collar of the borehole is proportional to the depth at which the reflection is generated. As for conventional inclinometers, repeated readings can be compared against a baseline reading to detect the depths of impedance changes in the cable over time that may indicate occurrences of shearing, or deflection of the hole. The capital cost for a TDR system is much less than for a conventional inclinometer system, and they are much easier to install and read. Another feature is that they can be read remotely, and hence can provide essentially real time monitoring, which may be useful in cases where active deformation is occurring. Unlike inclinometers, TDR cables cannot be used to measure directional movement, or be used for movement trend analysis.

Because the only downhole component is a single coaxial cable, TDRs can be easily installed in tandem with a conventional inclinometer, or even bundled with a combined inclinometer/VWP installation, in which case the TDR would provide an inexpensive redundancy for the inclinometer. TDR/VWP bundles (without the conventional inclinometer casing) can also be installed in smaller diameter drill holes. In this application, both the VWP tips and cables, and the TDR cable, would be taped to a tremie pipe. The number of VWPs that could be accommodated would depend on the size of the hole and the size of the tremie pipe. Figure 2.27 schematically illustrates one possible configuration of a VWP/TDR bundle in HQ and NQ-sized diamond drill holes.

Combined VWP and borehole accelerometer array installations

A relatively new concept in geotechnical instrumentation in open pit mines is the borehole accelerometer array. This instrument consists of a series of mutually perpendicular accelerometers that are spaced evenly along a flexible or articulated cable. This cable can be lowered down a casing or conduit in a borehole (portable system), or alternatively permanently grouted into a borehole. Signal cables communicate the tilt of each accelerometer group to a readout box at the collar. Once processed, the data can be used to define the inclination (tilt) of the borehole in a similar manner as a conventional inclinometer, and repeated readings can be compared against a baseline to detect deflections or distortions of the borehole over time. Sensitive borehole accelerometer arrays may also be useful for monitoring microseismic activity or blast damage. Like TDRs, they can be read remotely and provide essentially real-time monitoring.

Borehole accelerometer arrays could be bundled with VWPs in boreholes in the same way as TDRs as illustrated schematically in Figure 2.28. Unfortunately, these systems are still relatively expensive and have yet to gain wide acceptance in practical open pit mining applications.

Combined VWP and borehole extensometer installations

While inclinometer systems are traditionally employed to monitor and measure shear displacements at depth, borehole extensometers are designed to monitor extensional deformations or settlement. Borehole extensometers can be single or multipoint, and may be constructed from solid rods or wire cables. In a single point extensometer, the end of the rod or wire is anchored in the bottom of the hole, and the rod or wire extends to the collar of the hole. Multipoint extensometers have multiple rods or wires anchored at various depths in the borehole. Deformations of rod extensometers are measured against a reference plate at the collar of the hole. In the case of wire extensometers, a variety of deformation measurement techniques are used depending on the system design.

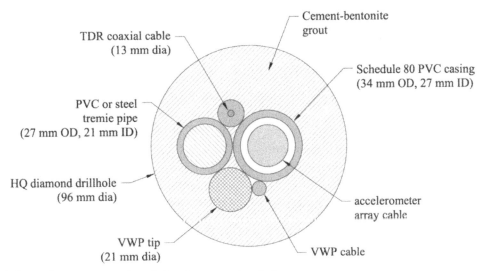

Figure 2.28: Borehole accelerometer arrays bundled with VWPs in boreholes

The key requirement of a borehole extensometer is that the portion of the rod or wire that extends between the anchor and the collar of the hole must be free to move; consequently, the hole cannot be fully grouted unless the rods or wires are encased within a conduit.

Single borehole extensometers can be bundled with conventional inclinometers provided there is sufficient space to accommodate both the rod/wire and the conduit. In this case, both extensional and shear displacement could be monitored. Likewise, borehole extensometers could be bundled with VWPs and/or TDRs, subject only to the available annular space in a given borehole. A schematic illustration of a simple rod borehole extensometer, TDR and VWP bundled in an HQ borehole is shown in Figure 2.29. It should be noted that rod extensometers are practically limited to about 100 m in length.

2.2.7 Downhole geophysical logging

2.2.7.1 General

Downhole geophysical logging is carried out worldwide for all types of projects. There is a general paucity of reference books applied to water and mining but descriptions of types of logs and their applications can be found in the following texts.

- Bloetscher F, Muniz A and Largey J (2007) *Siting, Drilling and Construction of Water Supply Wells (1st Edition)*, American Water Works Association (AWWA), Chapter 4: Geophysical Logging and Field Testing.
- Kobr M, Mareš S and Paillet FL (2005) Hydrogeophysics, Water Science and Technology Library, Volume 50, Chapter 10: Geophysical Well Logging, pp. 291–331.
- Hearst JR, Nelson PH and Paillet FL (2000) *Well Logging for Physical Properties: A Handbook for*

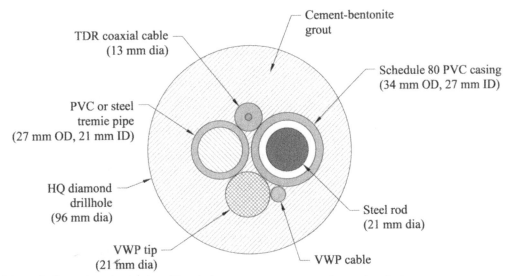

Figure 2.29: Schematic illustration of a simple rod borehole extensometer, TDR and VWP bundled in an HQ borehole

Geophysicists, Geologists, and Engineers (2nd Edition). John Wiley & Sons Ltd.

Standard geophysical logs used in groundwater investigations consist of spontaneous potential (SP), resistivity (e.g. long and short normal), natural gamma ray, sonic, neutron porosity, and optical/acoustic televiewer. The majority of these instruments are typically designed for use only in uncased boreholes. For specific applications to dewatering and pit slope depressurisation studies, the following types of down-hole logging are often the most useful.

- Down-hole surveys to measure the deviation and alignment of drill holes and wells.
- Formation logs to characterise the geotechnical properties of the formation, and specifically to characterise fractures.
- Fluid logs to estimate and quantify inflows to a test well or production wells.
- Down-hole video methods to allow inspection of casing and equipment.

Geophysical logging is typically undertaken after the drill string has been partially or totally withdrawn from the hole. Logs that may be of mutual interest to both the hydrogeologist and geotechnical engineer are summarised in Table 2.2. Figure 2.30 illustrates an in-pit geophysical logging set-up.

Advanced geophysical logging systems that are regularly used by the oil and gas industry are also now being applied in mining situations to provide quantitative information on:

- fine-scale variation and heterogeneity in rock/sediment matrix porosity and permeability;
- pore size distribution and water/moisture content;

Table 2.2: Geophysical logs frequently used in diamond drill holes

Geophysical test	Application
Acoustic Televiewer	Fracture detection, orientation
Optical Televiewer	Fracture detection and orientation, casing or screen inspection
Full Waveform Sonic	Lithology, rock UCS, porosity, Poisson's ratio, fracture aperture and permeability index in hard rock
Multi Parameter Resistivity	Permeable zones, porosity, alteration and fracture zones, lithology
Compensated Neutron	Aquifer, aquitard, porosity, lithology, moisture content
Neutron Tool – Single Point	Aquifer, aquitard, porosity, lithology, moisture content
Dual Density	Bulk density, lithology, rock strength and elasticity (with sonic log)
Conductivity	Water table, detection of flow horizons

- fracture systems (fracture orientation, aperture, fill, intensity/density, porosity);
- secondary porosity (vugs, cavities);
- sedimentary features, geological structure (type and orientation);
- rock geochemistry and mineralogy;
- geomechanics (elastic properties and rock strength, state of stress).

The precision of the measurements, in concert with the potential to measure the same volume repeatedly (i.e. re-enter the same well), enhances the use of logging systems for monitoring (e.g. variations in moisture content with time, changes in fracture aperture with time, and bulk chemistry changes such as copper concentration).

2.2.7.2 Advanced hydrogeological and geotechnical characterisation

Advanced wireline geophysical logging is being increasingly used to assist with detailed hydrogeology and geomechanical studies. Techniques including micro-resistivity image logs, magnetic resonance and full waveform dipole sonic logging can provide high-resolution, continuous *in-situ* information about key rock properties across the entire geological section. The logs can provide:

- comprehensive fracture and geological structural evaluation with micro-resistivity imaging and dipole sonic: detailed delineation, tracing, orientation (strike and dip),and classification of fractures (including fracture fill material) and other geological and bedding features; calculation of discrete fracture apertures and bulk properties (fracture density and porosity); estimation of fracture transmissivity;
- matrix evaluation with the magnetic resonance, dipole sonic, and induction resistivity: total porosity, effective and isolated/bound porosity; estimated rock matrix hydraulic conductivity, specific storage;
- *in-situ* rock geomechanics evaluation with the dipole sonic and integrated log analysis: compressional and shear velocity, rock elastic properties, estimated *in-situ* rock strength and state of stress;
- lithological evaluation through integrated log analysis – dry weight and volumetric percentages of rock-forming minerals (including carbonates, clastics, and volcanics).

Figure 2.31 shows a 10 m section of a processed and analysed micro-resistivity image log from a 405 m deep hydrogeological and geotechnical test borehole in fractured volcanic tuff. Displayed in each track are (left to right): auto-picked fracture traces overlaying oriented resistivity image (true north on left and right edges of 'unrolled' borehole wall image); fracture orientations displayed as tadpoles with the heads indicating dip angle from horizontal

Figure 2.30: In-pit geophysical logging set-up (Courtesy Schlumberger Water Services)

(on scale of 0 to 90 degrees, left to right) and tail pointing in the direction of dip azimuth (direction the fracture is dipping towards); fracture density, trace length and porosity; fracture apertures; and 3D 'electrical core' image.

Micro-resistivity image logging

Micro-resistivity imaging produces a fully oriented, high resolution electrical resistivity map of the rock at the borehole wall. The technique is operable in most fluids and is relatively insensitive to borehole size and condition. It can delineate lithological changes and bedding, as well as geological texture variations (grain size, sorting, secondary porosity). Also, because the micro-resistivity measurement is very sensitive to electrically conductive, fluid-filled fractures and the measurement response is well characterised, the aperture of individual fracture traces can be reliably computed.

The Fullbore Formation Micro-Imager (FMI; developed by Schlumberger) creates an oriented 'electrical core' picture by mapping the electrical resistivity measured by an array of 192 small, pad-mounted button electrodes to provide an electrical image of the borehole wall with a physical resolution of 5 mm (Ekstrom *et al.*, 1986). A triaxial accelerometer permits determination of tool

position, and three magnetometers allow determination of tool orientation, in turn used to orient the produced image and determine borehole profile (Figure 2.32). The FMI requires an uncased, fluid-filled borehole with a minimum diameter of 150 mm and maximum diameter of 530 mm.

The orientation (strike and dip relative to true north) of any planar features that intersect the borehole wall (e.g. bedding and fractures/faults) is calculated as it is manually or automatically picked on the image. Fracture aperture is estimated when electrically conductive fracture traces are picked.

Figure 2.33 shows the processed FMI image log, with automated analysis results, for the bottom 8 m of a blast hole, along with the processed bulk porosity logs from the other tools and breakout analysis from the FMI oriented dual caliper. Displayed in each track are (left to right): borehole diameter measured along perpendicular axes, with tadpoles indicating where there are borehole breakouts detected from oriented caliper breakout analysis (tadpole tail points in the direction of the breakout strike, often indicative of the minimum horizontal *in situ* stress); processed oriented resistivity image; porosity derived from sonic (dark shading) and bulk density, fracture density (hashed) and porosity (light shading); processed image

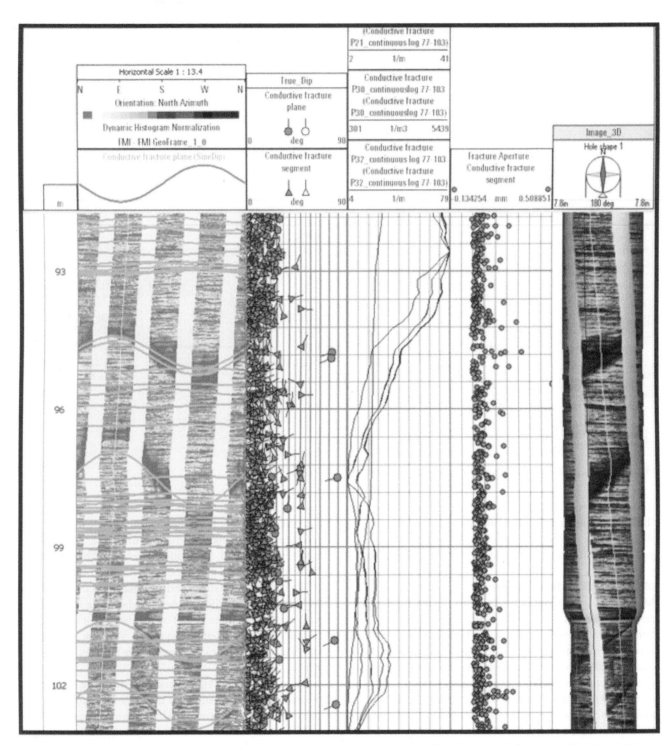

Figure 2.31: Processed borehole electrical imaging log in volcanic tuff with fracture analysis performed (Courtesy Schlumberger Water Services)

with auto-picked fracture traces overlain; major fracture strike and dip.

Magnetic resonance logging

Nuclear magnetic resonance (NMR) measures total fluid-filled porosity and pore size distribution in the rock matrix, from which retained and drainable porosity (specific yield) can be estimated, together with fluid volumes and a continuous permeability log. The Combinable Magnetic Resonance (CMR) and MR Scanner tools use the NMR technique to provide a measure of pore-size distribution independent of

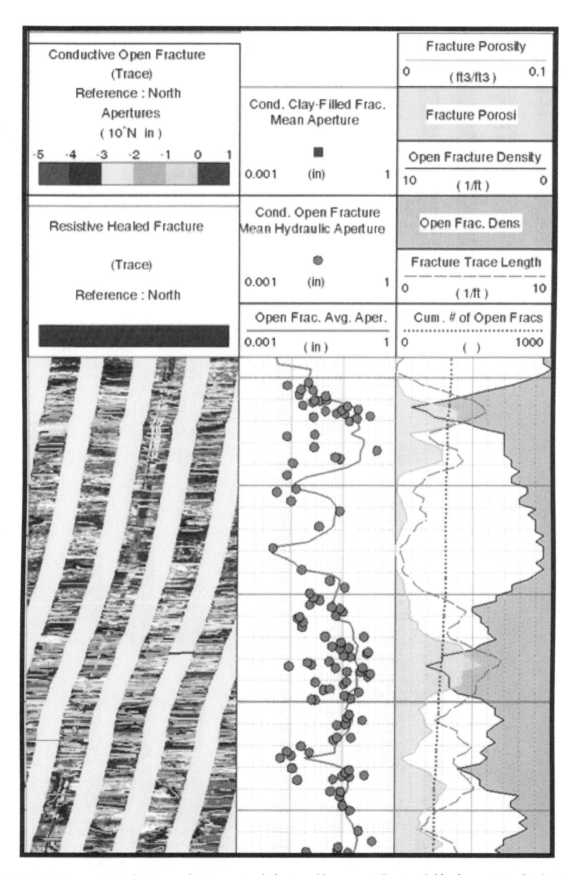

Figure 2.32: FMI micro-resistivity fracture analysis in a rotary hole at a gabbro quarry (Courtesy Schlumberger Water Services)

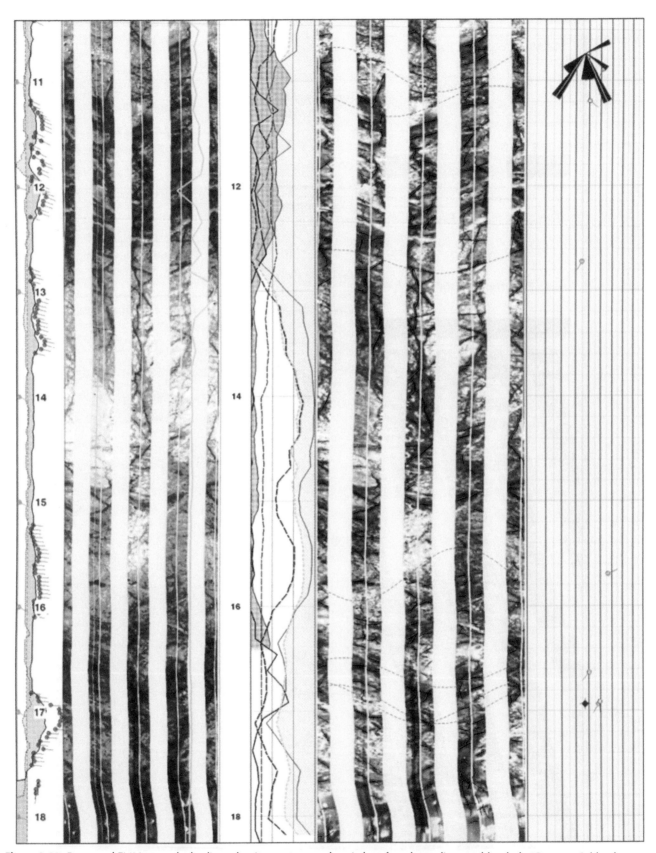

Figure 2.33: Processed FMI image, dual caliper, density, neutron, and sonic logs from large-diameter blast hole (Courtesy Schlumberger Water Services)

lithology, without the requirement of a radioactive source.

The technique has a number of applications, including the estimation of:

- total bulk water volume present in rock (drainable and retained);
- pore and grain size distribution in saturated clastic material.

A limitation of the NMR method is that valid measurements require a porosity of greater than about 2 per cent of total rock volume and thus measurements are typically insensitive to fractures. NMR hydraulic conductivity values thus largely reflect the hydraulic conductivity of the rock matrix. Both the CMR and MR Scanner require a hole size greater than 152.5 mm (there is no maximum hole size as long as the tool can be properly pushed against the side of the hole), and the system can be run in fluid- or air-filled holes. Figure 2.34 shows the CMR log for a deep (400 m) water supply well in central Texas, USA, with permeability and porosity measured from side wall core overlain for comparison.

2.2.7.3 Deviation logs

The use of deviation logs to survey the alignment of boreholes has become common practice for mineral exploration drill holes. Directional survey equipment is now available in all countries that have an active mining industry, and in many other locations. For hydrogeology and dewatering studies, deviation logs are necessary for the following applications.

- Surveying of pilot holes in order to determine the exact location of productive fracture zones so that these can be targeted by the subsequent drilling of production wells.
- Surveying of 'vertical' or angled boreholes where vibrating wire piezometers will be installed. Vibrating wire piezometers (and other types of pressure transducer) measure the water pressure above the instrument. To allow an accurate calculation of static head, it is necessary to know the exact position of the piezometer sensor. Therefore, a directional survey of the borehole is usually necessary.
- Surveying of horizontal drains and piezometers to allow the exact alignment of the hole to be determined relative to fault zones or other important features (Section 5.2.3.6). Again, for 'horizontal' piezometers, it is necessary to know the exact position of the piezometer sensor to allow a calculation of static head.
- Surveying of production wells and standpipe piezometers, particularly those drilled to greater depths and those in formations where there may be a tendency for the holes to deviate. Production wells and piezometers that penetrate formations with steeply dipping bedding

planes or steeply dipping fault planes may have the greatest tendency to deviate. For instance, at a mine located in a range front setting in Nevada, RC pilot holes initiated vertically for water exploration and piezometer installation routinely ended up at between 36 and 46 degrees from the horizontal at 350 m depth as a result of drilling through a layered volcanic rock sequence with well-defined bedding planes.

2.2.7.4 Fluid logs

For pilot holes and production wells, knowledge of the depth of the groundwater ingress zones in a drill hole is as important as the knowledge of the hydraulic behaviour of the ingress zone (Section 5.2.5.2). In dewatering applications, and for water supply wells where continuous drawdown of the water table is expected, production wells that rely on water ingress from shallow levels in the hole will often show a rapidly declining yield and may ultimately 'dewater themselves'.

Logging instruments to detect inflow horizons (especially fractures or fracture zones) can be divided broadly into two groups.

- Temperature and conductivity (fluid) logs: which measure changes in fluid properties to detect inflow horizons.
- Fluid velocity logs: which detect inflow horizons by observing changes in the velocity of the water column.

Instruments from both groups can operate under 'static' head conditions. However, better results can often be obtained when the hole is being pumped; abstraction of water by pumping increases inflows and emphasises the differences in properties, making inflow horizons easier to detect. In observation wells, the abstraction may be achieved by placing a small pump with its intake just below the water level. Abstraction from the well may reverse normal gradients and cause inflows along fractures that are naturally outflow horizons, so making them detectable.

Unlike most geophysical logs, which are generally run from the bottom of the hole upwards, fluid logs run under natural conditions are usually run from the top downwards, to avoid stirring up the water column and making minor inflow features harder to detect.

Temperature and conductivity logs

Temperature and conductivity logs (fluid logs) work on the principle that groundwater at different depths in a formation will usually have slightly different electrical conductivity and temperature (the latter, especially, tending to increase with depth under the influence of the geothermal gradient). Therefore, a significant inflow to a borehole will usually consist of water that has slightly

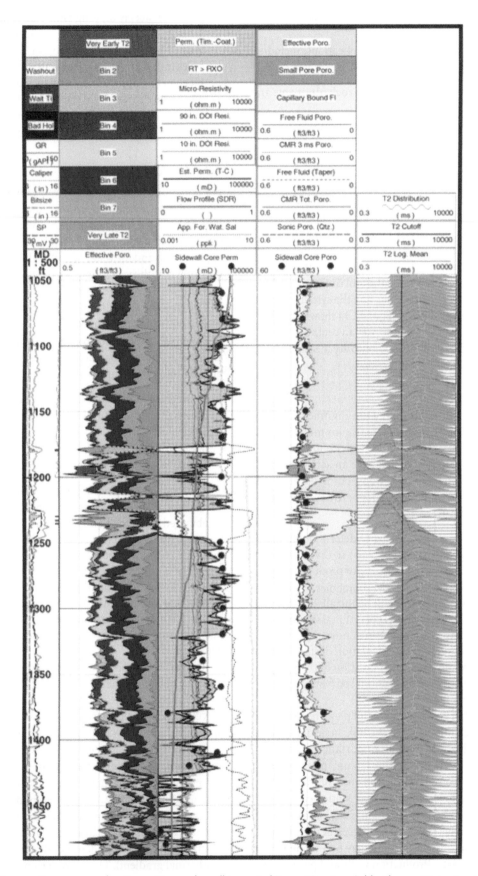

Figure 2.34: CMR magnetic resonance log in a water supply well in central Texas (Courtesy Schlumberger Water Services)

different temperature and/or conductivity from the column of water that is already in the hole. Temperature sensors are usually based on thermistors and conductivity probes use a simple conductivity cell. Both types can record either the actual value of the property, or have two similar sensors a short distance apart on the logging sonde to record the local differences.

Results of an electrical conductivity log from a borehole in fractured sandstone are included in Section 2.3.3.3. Multiple minor fluctuations in conductivity at around 20 m depth indicate the presence of several minor fractures, each contributing small inflows of slightly 'different' water to the borehole water column. Below 25 m, the log is essentially smooth for a few metres but then there is an abrupt decrease in conductivity at around 29.5 m, indicating a major inflow of 'fresher' water. The straight nature of the log from that point to the bottom of the hole suggests that water of that conductivity is moving down the hole and 'swamping' any inflows of different conductivity; it also suggests that there must be an exit for the water near the bottom of the hole. Outflows from the hole generally do not cause any variation of the properties of the fluid column, so that they cannot usually be detected except by inference. These conclusions were borne out by the results of double-packer testing and of borehole television inspection. One of the first papers to discuss the use of fluid logging in studies of fissure flow was by Tate *et al.* (1970); these authors emphasise that no single technique is likely to prove satisfactory in isolation and that a combination of techniques is generally much more useful.

Fluid velocity logs
Fluid velocity logs record the speed and direction of flow (upward or downward) of the water in a borehole. There have been experiments with tools that will measure velocities across a hole but the common types operate along the borehole axis; the changes in velocity are interpreted to indicate and quantify the inflow horizons in boreholes. The earliest and most common types use an impeller or 'spinner' tool on a wireline; the tool has a high-precision impeller that spins due to tool and water movement, and therefore in effect measures the relative speed of the tool relative to the water. Compensation must be made for the line speed when deriving the true water velocity but this is normally done by the logging software. Impeller flow meters are mostly used in cased water wells but can be run in open holes where hole stability is not an issue. The well or hole must be free of drilling mud and contain fairly clean water.

The tool can be used for checking 'static' conditions, where it is lowered into the hole and stopped at several locations in the borehole to check for 'natural' upward or downward flows between groundwater units that have been intersected by the borehole. However, more normally,

the tool is run under dynamic pumping conditions to produce a profile of the entire wellbore (below water level), identifying the zones of water inflow. The well is stressed by either pumping water out of the well or injecting water into the well while the tool is either raised or lowered at a constant rate. For cases where water is injected, the injected water is transferred to below the fluid level by way of a small diameter tremie pipe.

Figure 2.35 shows an example of a 'spinner' log that identifies the depth of the producing fracture zones and the quantity of inflow from each zone. The log is from a fractured intrusive rock and the well was being injected during the logging period. Other logs of the same well were made during pumping conditions using a small pump placed immediately below the static water table, and the two sets of logs were compared. As with temperature or conductivity logging, the injection or pumping rate during the test needs to be sufficiently high that a measurable downflow or upflow in the hole is produced. The amount of downflow/upflow required, and therefore the injection/pumping rate, needs to be calculated for each individual well based on the sensitivity (the minimum response velocity) of the tool being used.

Normally, when pumping a well for spinner logging, a 50 mm diameter guide pipe is run to a depth immediately below the pump, and the spinner tool is run through the guide pipe and into the well below. As the tool is lowered down the hole, the upflow velocity becomes progressively lower when the tool passes below each of the contributing inflow horizons. The inflow rate from each inflow zone can therefore be calculated. Most geophysical logging companies that offer this service have software that will calculate the contribution of each production zone as a percentage of the total.

In non-pumping wells such as observation wells, the natural flow is generally too slow to be detected by an impeller flow meter. In these cases, a pump can be used to generate artificial flows but if there is a need to know the flow under natural conditions a heat-pulse flow meter can be used. Heat-pulse flow meters consist of a small heater grid with thermistors mounted a fixed distance (usually a few centimetres) above and below it. The flow meter is suspended in the hole and, when it is judged that any disturbance caused by its entry has subsided, a short pulse of current is applied to the grid. The warmed water is carried by the natural flow and detected at one or other of the thermistors. Although there is a tendency for the warmer water to rise, this problem is overcome by calibrating the meter in tubes of varying diameter (to match the sizes of borehole being investigated) through which water is passed at varying speeds. An example of the output from a heat-pulse flow meter in also included in Section 2.3.3.3. Heat pulse flow meters are more sensitive than impeller tools, but they have the disadvantage that

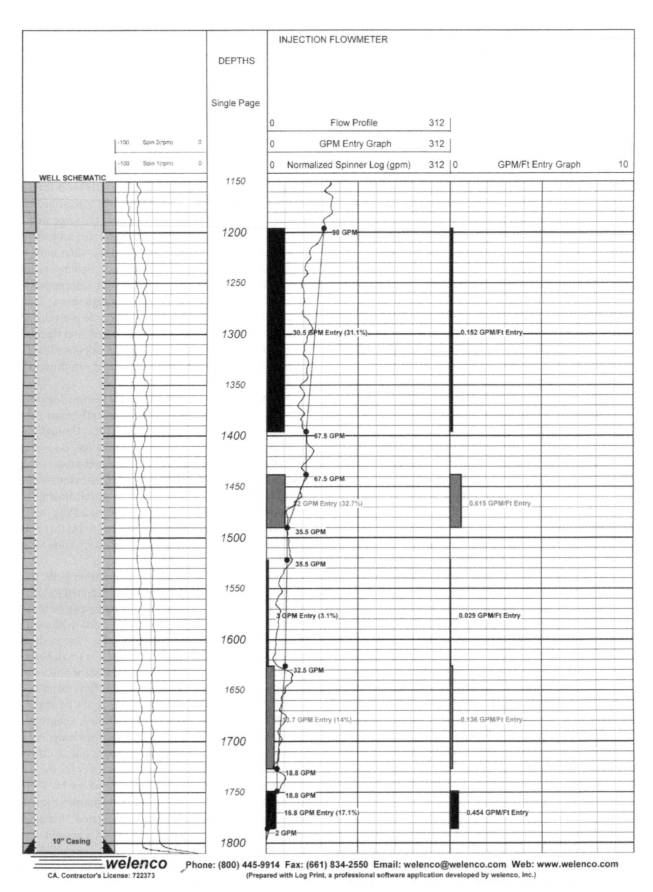

Figure 2.35: A fluid velocity log made with an impeller-type ('spinner') flowmeter

they must be suspended at one level while conditions stabilise and the measurement is taken, so only point readings (rather than a continuous log) can be taken.

Another technique for measuring natural or low flows is oxygen activation tracer logging, where a pulsed neutron generator activates the oxygen in the water and gamma ray detectors above and below the generator measure the emitted radiation as the activated water molecules move by. By analysing the gamma ray profiles through time the velocity of the water can be accurately determined. As with heat pulse flow meters, a disadvantage of oxygen activation tracer logging is that it requires a stationary measurement.

2.2.7.5   A case study on HydroPhysical logging

HydroPhysical Logging (HPL) is an increasingly common method for evaluating the vertical distribution of permeable fracture locations, inflow rates and water quality in wells and boreholes. The method has several variations but normally deionised water is injected into the borehole while a conductivity/temperature tool is moved through the water column in the same borehole or one nearby. In this way, flow zones are identified and permeability can be estimated. The method characterises flow dynamics under ambient conditions in the tested borehole and/or when the tested borehole or a nearby well is being pumped.

At a project area in Arizona, USA, a detailed HPL program was performed in wells completed in a volcanic tuff sequence under static and transient pumping conditions to characterise the formation waters and cross-borehole hydraulic connection of identified fractures and features intersecting four specific boreholes (Figure 2.36). The objectives of the investigation were to:

1. evaluate temperature and fluid electrical conductivity under pre-testing conditions;
2. characterise and quantify flow in the borehole under both non-stressed (ambient) and stressed (pumping) conditions;
3. characterise and quantify flow in the borehole under cross-borehole pumping conditions with a single pumping well;
4. evaluate the vertical distribution of flow and interval-specific permeability for all identified water producing fractures or intervals.

The four boreholes hydrophysically logged were PHRES-1, PHRES-2, PHRES-3 and PHRES-4. The pumping well during cross-borehole testing was HRES-9. The hydrophysical investigations were typically performed in approximately 157 mm (6.2 inch) diameter open boreholes up to 360 m (1180 ft) in total depth. The four wellbores were tested under both non-stressed (ambient) conditions and stressed (pumping) conditions, as well as cross-borehole pumping conditions. Prior to testing, the original fluid in each well was displaced with water of low electrical conductivity. Each well was then tested during cross-hole pumping twice: once early during the pumping of HRES-9 and once later when water levels stabilised, with the exception of PHRES-1 which was cross-borehole tested only once late in the pumping of HRES-9.

The testing identified upward vertical gradients, downward vertical gradients and horizontal flow during ambient and cross-borehole testing. The HPL results

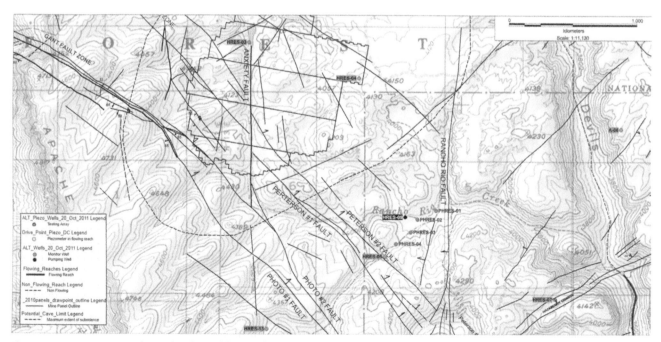

Figure 2.36: Location map for HydroPhysical logging holes

identified significant changes in flow characteristics between ambient wellfield conditions and stressed wellfield (cross-hole testing) conditions, suggesting hydraulic communication at these respective intervals with the pumping well HRES-9. These hydraulic characterisations in each wellbore were identified with HydroPhysical ambient, stress and cross-hole testing. In general, at first inspection of the data, three of the four subject wellbores (PHRES-2, PHRES-3 and PHRES-4) appear to have similar flow regimes. Upon closer inspection, the details of the flow results for three of the four wellbores show heterogeneity (Figure 2.37).

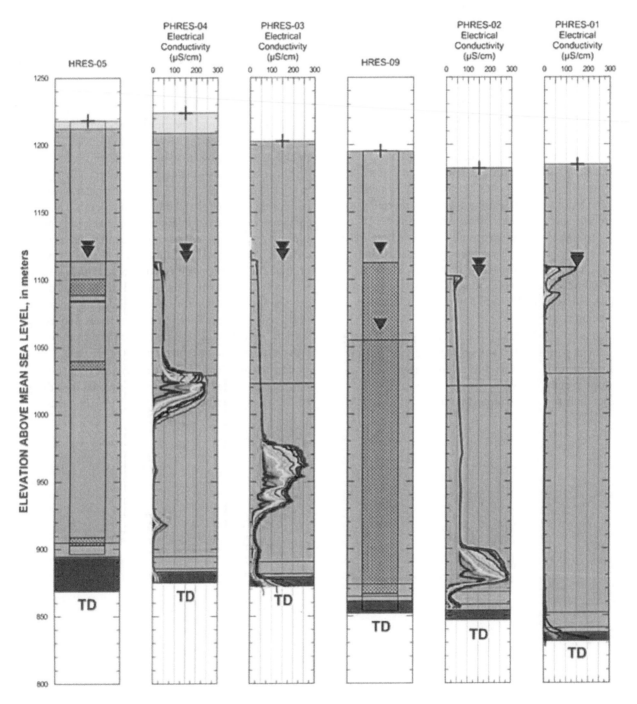

S:\projects\605\605.4\605.411\ApacheLeapSpeedholes\PHRES Piezo Schematics\Hpl_Logs.grf

Figure 2.37: Example of HydroPhysical Logs. The coloured lines represent fluid electrical conductivity as a function of time. Red is early time; indigo is late time. The total time lapse for the log is in the order of an hour from the far left (red) to far right (indigo) curves. Therefore, what one sees is the influx of groundwater into the well which is full of de-ionised water at the start. The figure shows a productive fracture high in PHRES-04 dipping to the east and occurring deep in PHRES-01.

The results were summarised as follows.

- Water was observed to enter the borehole from above the water level in PHRES-1 and PHRES-2 during ambient, stress and cross-hole testing. Water was not observed to enter the borehole from above the water level in either PHRES-3 or PHRES-4.
- The specific capacity is comparatively low in PHRES-1 (0.56 gpm/ft) compared to PHRES-2, 3 and 4 (11.0–36.1 gpm/ft).
- Most of the fracture permeability in PHRES-2, 3 and 4 occurs in the lower half of the saturated thickness.
- Data from PHRES-2 and PHRES-3 suggest significant vertical hydraulic communication between dominant fractures in the immediate vicinity of the borehole.
- Data from PHRES-1 suggest little to no vertical hydraulic communication between fractures (data from PHRES-4 are inconclusive in this regard).
- Regarding elevations of the four wellheads, PHRES-1 is the lowest in elevation, PHRES-2 the next lowest, PHRES-3 the second highest and PHRES-4 has the highest wellhead elevation. This would suggest any marker flow intervals, such as a dominant outflow zone connected to the pumping well HRES-9 would be the shallowest in PHRES-1 and the deepest in PHRES-4, all other considerations being equal. However, the opposite is observed with the exception of PHRES-1 which, by nature of having the least hydraulic connection to the pumping well HRES-9, did not exhibit similar flow characteristics to the remaining three tested wellbores. The dominant outflow zone in PHRES-2 during cross-hole testing was observed at approximately 229 m (750 ft), PHRES-3 at approximately 283 m (930 ft) and PHRES-4 at approximately 200 m (655 ft).

Data acquired from this analysis were used to refine the understanding of flow dynamics in the tuff sequence and to determine whether flow anisotropy exists and is compartmentalised. The nature of the flow paths were shown to substantially affect the magnitude and timing of drawdown estimates during predictive pumping simulations.

2.2.7.6 Downhole video inspection

Downhole 'video' (also called closed circuit television or borehole CCTV) is regularly used for inspections of test holes and production wells around mine sites. Most drilling contractors carry downhole inspection equipment, as do many consulting and engineering companies. Many mines that operate multiple wells carry the equipment in house.

Video equipment can be used to inspect open pilot holes for water movement and fracture locations. The water in the hole must be reasonably clear. The inspection can be done under ambient (static) conditions or under dynamic conditions where water is being injected into the borehole. The video log can be run in stages by running RC drill pipe into the hole and lowering the camera down the centre tube while injecting water down through the annular space normally used to convey air. The sub-adaptor used for injection testing is ideal for this. This technique is often used in karst limestone formations where cavity zones with natural flow have been encountered. It can also be used to identify perched ingress zones above the main water table.

It is commonplace to use downhole video to assess the condition of casing and screen in a completed well. Usually, the inspections are related to corrosion or damage to the casing and well screen. Inspection of the pump and rising main can also be carried out, although care must be taken not to run the camera system in a hole containing loose power or guide cables for the pump. The camera is lowered down the hole and a continuous video log of the hole is produced. The camera can be stopped for detailed inspection of specific zones, and most cameras have the option of an axial view (looking straight down the hole below the camera) or a radial view (where the camera is directed at the borehole wall or lining and can usually be rotated to view the entire circumference).

Figures 2.38 and 2.39 show examples of a video inspection of slope depressurisation wells, where the wells had been drilled through the shear plane of a slope failure. Figure 2.38 is an example showing damage to blank casing. Figure 2.39 is an example showing how the well screen may become damaged. For both of these wells, it was possible to remove the pump and riser pipe but it was not possible to rehabilitate either well. For the well shown in Figure 2.38, it was necessary to drill a replacement well at a nearby location. For the well shown in Figure 2.39, a judgement was made that existing nearby wells could produce sufficient drawdown to maintain the required pore pressure along the failure surface, and the well was not replaced. In the case of depressurisation wells that are drilled through a failure surface, continuous operation of the well will maintain a minimum effective stress along the failure plane and will typically extend the operating life of the well. However, when pumping from the well is stopped, the water level in the wellbore will rise and the effective stress on the failure surface will therefore decrease, which may increase the potential for shearing of the well casing.

2.2.8 Cross-hole and multi-hole testing

2.2.8.1 Introduction

Short-duration, single-hole tests typically provide information only for the local-scale in-situ permeability of the material within the borehole itself, or the formation or fracture systems immediately surrounding the wellbore. However, the local-scale 'point' permeability is usually not the primary control on the 'bulk-scale' groundwater flow, particularly in a fractured rock environment.

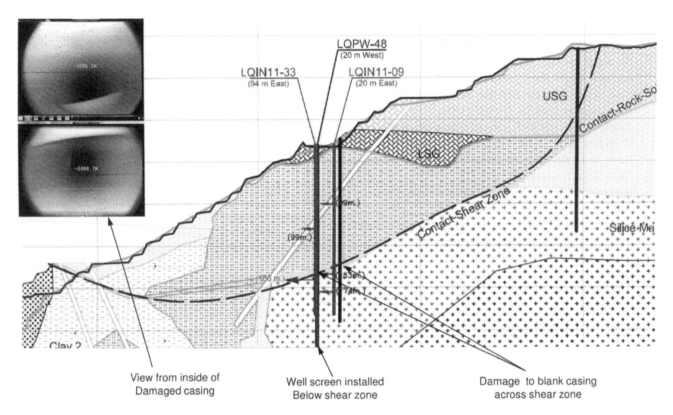

View from inside of
Damaged casing

Well screen installed
Below shear zone

Damage to blank casing
across shear zone

Figure 2.38: Example of casing damage caused by shearing (Source: David Rios, MYSRL)

The groundwater flow system within many bedrock units is controlled by the broader-scale bounding features, and particularly geological structures, which are not possible to characterise in the field by single-hole testing. The ability to dewater a pit and/or depressurise the pit slopes is mostly determined by the general inter-connectivity of the fracture systems rather than by the 'local-scale' properties of fractures. Therefore, while they are useful during the early stages of a study and for

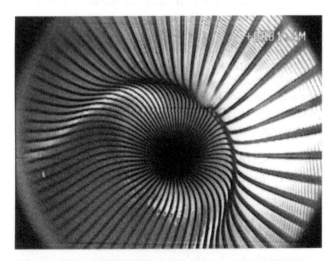

Figure 2.39: Example of a borehole video inspection showing damage to wire-wrap well screen caused by shearing along a failure surface in an active pit slope

providing information at multiple locations around the site (particularly for the lower-permeability materials), it is usually not representative to apply 'single-point' information to characterise the 'bulk-scale' behaviour of the groundwater system.

Knowledge of the complete flow characteristics of the groundwater system can only be determined from a testing program that places a prolonged hydraulic stress on the groundwater system at one (or more) location(s) and allows the response to that stress to be measured at other surrounding observation points. Multi-hole tests typically involve pumping, airlift or injection in one hole, and measuring the response in at least one adjacent observation hole, and preferably at multiple horizons within the observation holes. For general dewatering programs, the information is normally gained by using vertical hole programs. However, for pit slope depressurisation programs, where pit walls already exist, the information can be gained by horizontal programs. It is only through the use of multi-hole programs that the true groundwater flow system can be characterised, including the influence of anisotropy, enhanced flow features, and compartmentalising features (such as structures and dykes).

Some form of cross-hole testing is essential for any feasibility study. It is also advisable for the majority of pre-feasibility studies where dewatering or slope depressurisation is known to be a significant issue. Cross-hole testing may be carried out between two holes but it is normally preferable to use multiple holes. It may include:

- prolonged airlift testing while measuring the response in nearby holes;
- extended pumping or injection, measuring the response in adjacent or nearby holes;
- prolonged testing of a selected interval between packers, with the response measured at multiple intervals in an adjacent hole;
- cross-hole testing using Westbay systems;
- longer duration pumping tests with multiple observation wells.

2.2.8.2 Longer duration airlift tests

Prolonged airlift testing can be carried out in pilot holes to allow the response in surrounding piezometers to be observed. If there are other open drill holes or piezometers installed near the hole being airlifted, it is possible to monitor the groundwater level response in these 'observation wells'.

By using a carefully controlled airlift test with a yield of about 1 l/s or more carried out for a prolonged period (say at least four hours), it is often possible to analyse the response in the 'observation wells' in the same quantitative manner as for a conventional pumping test. This is an excellent and cost effective method for characterising an aquifer over a greater areal expanse than say single-well injection tests. In less permeable settings (between 10^{-6} and 10^{-7} m/s), the tests may be carried out in a similar manner but, because of the lower airlift rate, it may be necessary to use closer observation holes (say 10–30 m distant) and to extend the airlift period for eight hours or more.

In this way, airlift testing in small diameter RC holes can be important for helping to characterise the groundwater system in areas where the expenditure associated with drilling a large-diameter pumping well is not justified, or for projects where the remote or inaccessible nature of the site makes the mobilisation of a larger water-well drilling rig impractical.

For example, as part of the initial site characterisation work for the remote Agua Rica copper project in Argentina, it was not feasible to mobilise a drilling rig large enough to drill water wells. Detailed testing information was obtained by controlled airlifting of RC holes for a 48-hour period, monitoring the response to airlift pumping in several purpose-installed piezometers drilled nearby. An example of the airlift data from the Agua Rica testing program is provided in Figure 2.40.

2.2.8.3 Extended pumping or injection tests

Prolonged pumping or injection testing of pilot holes can be carried out in a similar manner while monitoring the response in surrounding piezometers and other observation points. Again, in poorly permeable formations the tests may require a lower flow rate, using closer observation holes and extending the test period for eight hours or more.

Injection tests provide quantitative data on the response of the groundwater system (Section 2.2.5.5). In practical terms, the hydraulic behaviour and response is similar to a pumping test, so the test results can be used to calculate aquifer parameters as well as to provide empirical estimates. However, when carrying out an injection test, it is necessary to interpret the data considering the possibility that some or all of the injected flow may have

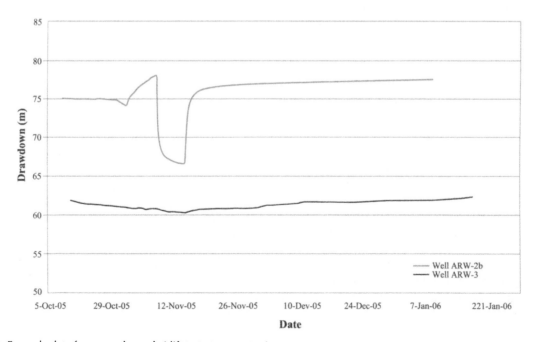

Figure 2.40: Example data from a prolonged airlift test at a remote site

occurred within fracture systems above the water table, which will clearly not be representative of any saturated aquifer response, and would not contribute to any dewatering or slope depressurisation program.

Therefore, if possible, pumping tests are preferable to injection tests. It is also necessary to check any local regulations that may be applicable for longer-duration injection testing. However, injection tests are often logistically easier to carry out in small diameter holes and do not require a submersible pump and power source.

2.2.8.4 Prolonged testing of a selected interval between packers

Prolonged pumping or injection (or airlift) testing may be carried out using packer systems to isolate the contributing inflow zone to a discrete test interval. The procedure may be used to help characterise vertical 'slices' of the groundwater system. Observation holes or piezometers can be used to measure the response at discrete monitoring horizons and therefore provide data to help assess the vertical component of groundwater flow and/or the amount of leakage occurring between vertically discrete stratigraphical layers or fracture zones. If a straddle packer system is used, the test may be repeated across other vertical intervals.

2.2.8.5 Cross-hole testing using Westbay systems

Prolonged cross-hole testing may also be carried out using Westbay systems (Section 2.2.6.8). The tests may be carried out using a single Westbay system, either for injection or measurement of the multi-interval response, or using two Westbay systems.

Figure 2.41 shows the results of multi-level cross-hole testing using a single Westbay system for a deep site characterisation in Alaska, USA. The data were obtained by measuring the multi-level response in the Westbay system to airlifting from a straddle packer system in a nearby hole. The airlifting was carried out at various levels in the hole using the straddle packers, and the response in all of the Westbay ports was monitored. Below about 1500 ft (about 450 m) depth, the pressure drops over the record, even though there is some hydraulic conductivity. The only reasonable interpretation is that the drilling has loaded the compartment with water and it is slowly recovering. The testing indicates moderate near-hole permeability values, and this response indicates relatively low permeability boundaries to the compartment. The interpretation of individual tests can be interpreted in the same way, with lower heads after pumping.

Figure 2.42 shows the results of multi-level cross-hole testing using two Westbay systems to characterise conditions across a vertical normal fault. The figure illustrates the level of detail that can be obtained using a detailed multi-level cross-hole program. Faults zones are

normally one of the most important factors controlling groundwater flow in fractured rock settings, as discussed in Section 1.2.2.5, and their hydraulic behaviour can normally be characterised only by using cross-hole testing. While it is normally not practical for a mining operation to characterise a fault zone at the level of detail shown, Figure 2.42 does illustrate the quality of data that can be achieved using a detailed cross-hole testing program.

2.2.8.6 Pumping tests

General

Pumping tests are one of the standard hydrogeological testing procedures for characterising the 'bulk-scale' behaviour of groundwater systems. They are commonly carried out all over the world in many types of hydrogeological setting. There are a large numbers of texts and references describing pumping test procedures, including Freeze and Cherry (1979), Kruseman and de Ridder (1990) and Domemico and Schwartz (1998). These guidelines provide a background discussion on the application of pumping tests to mine hydrogeology, but it is necessary to refer to one of the standard texts for the detailed procedures.

The test procedure involves pumping water out of a single well or borehole in a controlled manner, while monitoring the flow in the pumping well, the water level in the pumping well, and the water level in surrounding observation wells and piezometers. Pumping from a production well creates a 'cone of depression' in the potentiometric surface (Figure 2.43). The shape and rate of expansion of this 'drawdown cone' depend on the hydraulic properties of the formation. By monitoring the way an area of drawdown expands, it is possible to derive information about those properties. In a typical pumping test, a larger diameter well is pumped, and piezometers installed within the area of pumping influence are monitored for drawdown response.

Pumping tests are most typically undertaken in permeable settings, where it is possible to pump at significant rates from wells. Thus, for mining applications, pumping tests are more applicable for general mine dewatering evaluations. However, in many situations involving sector-scale slope depressurisation, where the permeability of the materials may be low, tests can be successfully carried out by pumping at low rates, or by airlifting, as long as closely spaced piezometers are available to allow discrete measurements at different lateral distances and vertical horizons.

The testing provides information to characterise the pumping well itself and the surrounding groundwater system and may be used to:

- assess the hydraulic performance of the pumping well itself;

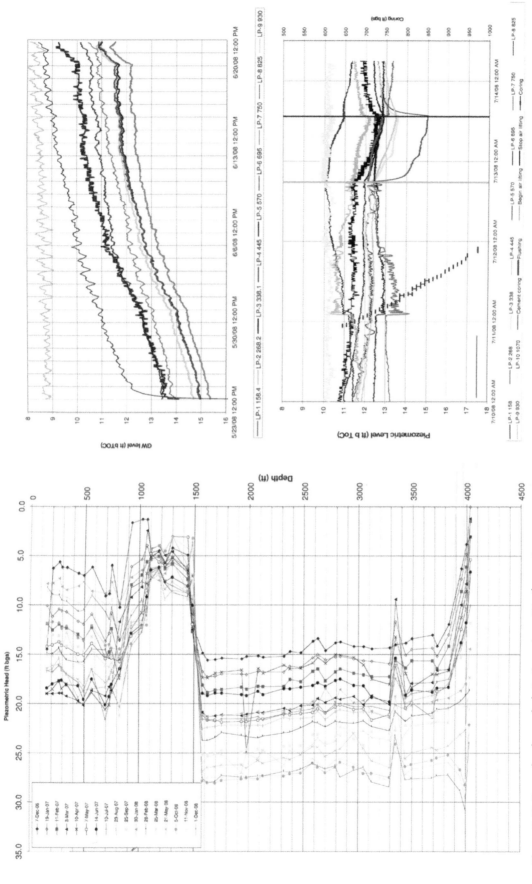

Figure 2.41: Example of cross-hole testing using a Westbay system

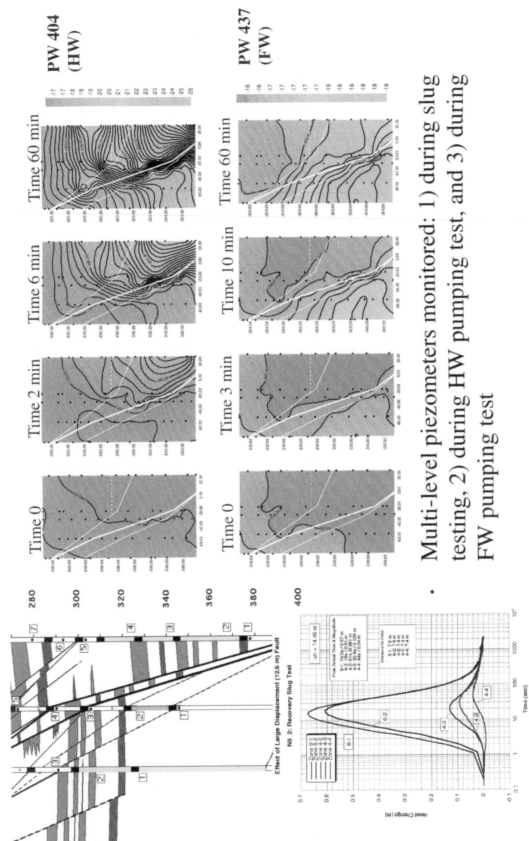

- Multi-level piezometers monitored: 1) during slug testing, 2) during HW pumping test, and 3) during FW pumping test

Figure 2.42: Characterisation of a large sub-vertical structure using cross-hole testing (Courtesy of Texas A&M University)

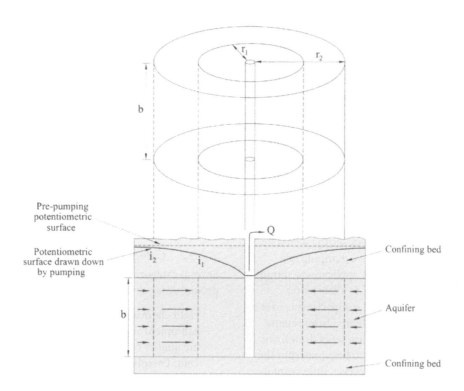

Figure 2.43: 'Cone of depression' around a pumping well (After Price, 1985 and 1996)

- test the operation of the pumping and monitoring equipment;
- derive the hydraulic properties of the surrounding groundwater system;
- determine the potential connection to remote permeable formations that may provide ongoing recharge to the dewatering or slope depressurisation system;
- determine the cumulative interference effects on nearby wells operated by the mine (other dewatering wells or water supply wells);
- determine the potential impacts on the surrounding groundwater system, including nearby wells operated by other local groundwater users;
- determine the environmental impacts of the abstraction, including impacts to springs and seeps, wetlands and surface waters;
- provide information on groundwater quality;
- optimise the design of the operational dewatering system.

There are many types of test, but the most commonly applied are the step drawdown test and the constant rate discharge test. A period of trial pumping and monitoring is normally carried out prior to the start of testing to provide an estimate on the maximum short-term yield of the well, which is then used to determine the pumping rates for the step test. The discharge arrangements for the pumped water are also important and the discharge system should be designed such that re-infiltration of the pumped water back to the groundwater system does not occur within the planned period of testing. This is particularly important when working in an unconfined groundwater setting.

Step test
The step drawdown test is used to establish a short-term yield-drawdown relationship to determine the performance of the pumping well itself. Analysis of the data provides information on the expected short-term yield of the pumping well and also head losses (local-scale formation losses, well losses, laminar flow vs turbulent flow). The test involves pumping in a series of steps of roughly equal duration. Normally, there are at least four steps, usually consecutive, with the lowest discharge rate first, increasing to the highest discharge rate, which should be close to the anticipated maximum short-term yield of the well. Typically, for high-yielding wells, steps of two hours are typical.

Constant-rate test
The purpose of the constant-rate test is to obtain information on the hydraulic characteristics of the groundwater system within the radius of influence of the pumping well. The distance to which the 'cone of depression' will expand (essentially the 'radius of influence' of the well) depends on the properties of the

formation and on the time for which the well is pumped, but not on the pumping rate. (A higher pumping rate causes the cone of depression to deepen but not to extend further laterally.) The testing involves pumping at constant rate for a period of time, with the pumping rate usually determined based on the results of the step test. To provide the best data for calculation of aquifer characteristics, the pumping rate should be maintained as constant as possible throughout the duration of the testing period. In addition to estimating transmissivity and hence hydraulic conductivity, the data from the observation wells can be analysed for storage coefficient (Appendix 2).

The constant-rate test should be run for as long a time period as practicable. A typical time for achieving a reasonable response in a fractured rock setting would be three days to two weeks for a moderate pumping rate (10–100 l/s). However, for dewatering applications, pumping for a longer period is often beneficial to allow better characterisation of flow barriers and compartments over a wider area, to assess the recharge potential from remote permeable formations, and to provide a set of drawdown data over as wide an area as possible to help underpin the feasibility of dewatering. There is, however, little point in extending the test beyond the time when drawdown becomes stable at a constant pumping rate. This may occur, for example, when a low storage fractured rock groundwater system demonstrates a good hydraulic connection with a high storage alluvial system (recharge source).

Monitoring of the recovery period
On completion of a constant rate pumping test, it is normally advisable to continue detailed monitoring of the recovery period following pumping. The recovery monitoring may be used to check values of aquifer parameters derived from the discharge portion of the test. In fractured rock or compartmentalised groundwater settings, the rate of recovery, and the degree to which recovery occurs relative to the initial pre-test water level, can be used to help identify when permanent (or semi-permanent) storage has been removed from a groundwater compartment or the particular set of interconnected fractures. If there is an incomplete recovery of water levels after an equivalent time to the constant-rate test, it may indicate that ongoing recharge to the groundwater system is low, which is usually favourable for dewatering. Proper interpretation of the rate of recovery can provide a good indication of how sustained the future operational dewatering rate may be.

2.2.9 Water quality testing

2.2.9.1 Introduction
The typical goals of a water quality testing program were listed in Section 2.1.8. It is not the intent of these guidelines to specify water quality sampling for environmental programs, but rather where chemistry studies are required for dewatering and slope depressurisation programs. The water quality testing program will provide an indication of the corrosion potential of the water (for well and pump design), how the water may need to be managed within the site-wide water balance, and potential downstream treatment and management costs if there is requirement for off-site discharge. An understanding of the spatial variations in groundwater chemistry, and how the chemistry changes with time during pumping, is also important for supporting the conceptual hydrogeological model and, in particular, for helping to determine the wider sources of recharge for the water pumped from the pit area.

2.2.9.2 Categories of analysis
Several categories of analysis can be performed for groundwater chemistry characterisation.

- **Field chemistry parameters**: measured with portable equipment in the field, providing a quick indication of the quality of water. Parameters normally include pH, Total Dissolved Solids (TDS), Electrical Conductivity (EC), Dissolved Oxygen (DO) and in some cases Total Suspended Solids (TSS). Field chemistry testing provides a rapid and low cost measurement of multiple samples to allow a basic characterisation of groundwater system chemistry, and identification of temporal changes or areal variations.

- **Dissolved major ions**: dissolved major cations (calcium, magnesium, sodium, potassium) and anions (chloride, bicarbonate, sulphate and nitrate/nitrite) are measured via laboratory analysis. The composition and relative concentration of each dissolved ion is used as a basis to group and categorise water 'types' and to understand the chemical interactions that occur as groundwater moves through a soil or rock.

- **Dissolved metals**: also measured via laboratory analysis. Dissolved metals are often naturally elevated in groundwater in the vicinity of ore deposits due to the interaction of groundwater with the ore deposit. For any mine dewatering projects, it is essential to define the natural background concentrations of dissolved metals and to use these as the basis for determining background and potential compliance standards, where existing standards do not already exist. Further, understanding the ranges of dissolved metals in the pumped flow and the potential need for treatment is also critical if off-site discharge and associated permitting are anticipated.

For hydrogeology and engineering studies, it is normally of benefit to define three sampling suites which would normally include (i) field parameters which can be sampled from many sources on a routine basis, (ii) an indicator parameter suite consisting of those parameters which may be considered important for the particular site being investigated, and (iii) a full parameter suite which allows detailed analysis and 'typing' and characterisation of the groundwater system. For special 'one-time' studies that may involved the age-dating of water for characterising recharge sources, it may be advantageous to carry out isotope sampling and analysis.

2.2.9.3   Typical program for water quality testing

At the scoping level (Level 1), a simple program of field parameter measurement supplemented with a limited number of full parameter samples should be carried out opportunistically using groundwater from exploration drill holes and surface water flows or springs in the hydrogeological catchment of the mine. It is noted, however, that field measurement of water produced during exploration drilling requires careful judgement since drilling fluid or rock cuttings may influence the samples. The data will contribute to the first generation conceptual model, and will be used to define possible water chemistry zones and domains that require more in-depth sampling and analysis in later project phases.

At the Pre-Feasibility stage (Level 2) it becomes important to establish the baseline mine-wide water chemistry monitoring, at locations that can support the permitting process. Quarterly sampling (or other frequency specified by site-specific conditions, climate or local regulations) will be initiated and continued for at least a year and sometimes longer, potentially until the time that the mine project is taken into production. It is normal to supplement the baseline dataset with additional samples that may include vertical or lateral groundwater profiles or synoptic sampling along water courses. For the open pit dewatering and slope depressurisation program, it becomes important to establish ranges of groundwater chemistry that will allow for:

1. zoning of groundwater chemistry domains in the pit shell and surrounding district system;
2. identification of dewatering system construction materials for in-pit and perimeter dewatering systems;
3. prediction of groundwater chemistry impacts that may occur as a consequence of mine dewatering and later open pit closure;
4. the potential requirement for of water treatment for either discharge to the environment or input to the mine water circuit.

In addition to permit-related baseline monitoring, the following would be typical at the Pre-Feasibility level.

- Single samples would normally be collected from the dedicated RC hydrogeological holes on completion of

drilling (drill fluid permitting) for field parameter testing and laboratory analysis.
- At least one sample would be collected from each major rock type within the pit shell and surrounding country rock that may become a target for dewatering pumping.
- Repeat samples would be collected from any 'first phase' pumping test well(s) for field parameter testing and laboratory analysis.
    - → Typically one sample is collected at initiation of pumping, and then repeat samples taken at 30 per cent, 60 per cent and 100 per cent of the test duration (four samples in total).
- In some circumstances, borehole wireline logging of water chemistry, commonly including TDS, EC, and temperature, may be completed as part of the process of defining groundwater flow horizons within the geological sequence.

For a Feasibility study (Level 3), adequate groundwater chemistry data are needed to define potential groundwater quality impacts during mine operations and closure, at a level that supports the environmental impact analysis and permitting. A comprehensive understanding of water chemistry domains and predictive modelling of mixing, transport and impacts is required. In addition, the potential requirement for water treatment needs to be defined to the point that a viable treatment system is defined, and the cost implications of water chemistry are defined with +/– 20 per cent confidence through the payback period of the mine.

In addition to permit-related baseline monitoring, the following would be typical for a Feasibility study.

- Additional samples would be collected from RC hydrogeological holes on completion of drilling.
- At least three samples would be collected from each major rock type within the pit shell and surrounding country rock that may become a target for dewatering pumping.
- Repeat samples would be collected from any pumping test well(s).
    - → Field parameter testing is carried out at regular intervals and a full laboratory sample is collected at initiation of pumping from each well and then repeat samples taken at 30 per cent, 60 per cent and 100 per cent of the test duration (four samples in total). However, if the testing is extended for a period of many weeks, then a typical protocol is to collect weekly samples for the duration of testing.
- For situations where treatment of the dewatering flow is important, bench-scale and potentially pilot-scale testing of the treatment alternatives is needed that can be used to support the design of an operational water treatment system.

- Additional borehole wireline logging of water chemistry may be completed as part of the process of defining groundwater flow horizons within the geological sequence.

2.2.10 Pilot drainage trials

The final step in the investigation program is often the implementation of an instrumented pilot trial area, which can be progressively expanded into a full dewatering or slope depressurisation system. The following types of pilot trial may be considered.

- **An extended pumping trial** for the dewatering system, running for a period of several months, and potentially using multiple wells, and with variable pumping rates. The use of multiple wells may complicate the data for analysis and quantification of aquifer properties. However, it often produces a better 'site-wide' response and may provide an early indication of sectors or the pit where an increased level of general dewatering or a specific program of pit slope depressurisation may be required. The trial is usually carried out at Level 3 for a project where dewatering is a major factor for project planning, or as part of the detailed design studies in Level 4, from where the trial is expanded into the initial production dewatering system.

- **Horizontal drain trials** where multi-level piezometers are first put in place, and a trial network of horizontal drains is drilled close to the piezometers. The initial trial would typically use two multi-level piezometer strings and four to six horizontal drain holes installed to target the piezometers. It is often preferable to position the piezometer sensors so the drains can target the middle or deepest sensor and also to monitor the response of the overlying and underlying sensors, but this will depend on site specific conditions. The flow rate of the drains and the response of the piezometers are monitored. It is clearly necessary for a pit slope to be in place prior to carrying out the trial. However, for a Greenfield site, excellent response data can often be obtained by installing the piezometers and carrying out extended airlift trials in nearby vertical (or angled) pilot holes. The conceptual layout for a horizontal drain trial for an existing pit expansion is shown in Figure 2.44.

- **Vertical drain trials** where there are known to be vertical gradients already, and a trial can be designed to provide a vertical cross-connection of different units. Again, multi-level piezometers are first put in place, a trial network of vertical drains is drilled close to the piezometers, and the response of the piezometers is monitored. The initial trial would typically use one to two piezometer strings and four to six vertical drain holes installed adjacent to the piezometers. It is

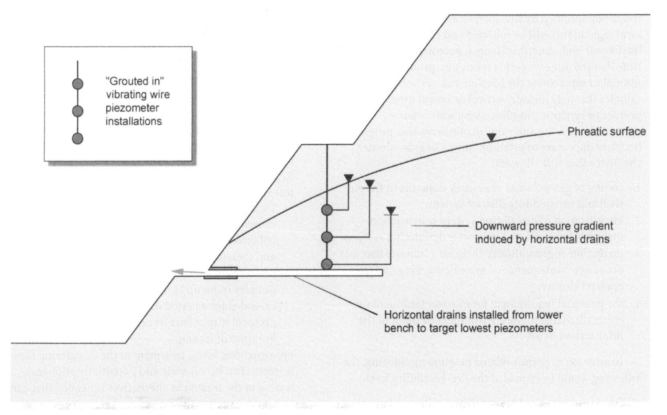

Figure 2.44: Conceptual set-up for horizontal drain trial

necessary to have a lower 'already-depressurised' unit for the trial which can be carried out for any site where significant vertical head gradients have been identified. The conceptual layout for a vertical drain trial is shown in Figure 2.45.

In the example shown in Figure 2.44, three strings of vibrating wire piezometers (with three sensors per string) were installed in a pre-selected area of the pit slope. Four horizontal drains were drilled, with at least one drain oriented to target the level of the deepest piezometer in each string. The flow rate of the drains and the response of the piezometers were monitored. The piezometers at the level of the drains showed the highest response, with the shallow piezometers in each string showing a lesser response. The pilot test allowed detailed design of the depressurisation system for the slope.

Figure 2.45 shows a vertical drain trial at the Round Mountain Mine in Nevada, USA. The drains were drilled to depressurise a sequence of lake bed clays and clay-altered volcanic layers (the Stebbins Hill Formation). The small amount of water in the Stebbins Hill flowed down the drains and entered the underlying fractured volcanic rocks, which had already been depressurised using dewatering wells. The pilot test consisted of six test drains and two piezometers. Based on the results from

the pilot area, the drain array was expanded into a field of 45 drains, each drilled to a depth of about 75 m. Vertical drains are often most effective when the horizontal permeability is greater than the vertical permeability so that the water can flow along the bedding towards the drains and then flow vertically within the drains.

2.3 Presentation, analysis and storage of data

2.3.1 Types of data

2.3.1.1 General
The data obtained from the hydrogeological field program can broadly be divided into two groups.

1. 'Time-series' data (collected on a continuing basis).
2. 'One-off' data (from specific characterisation or testing programs).

2.3.1.2 Time-series data
Time-series or monitoring data include meteorological data, surface water flow and chemistry data, flow rates (from pumping wells, sumps, seepage faces and horizontal

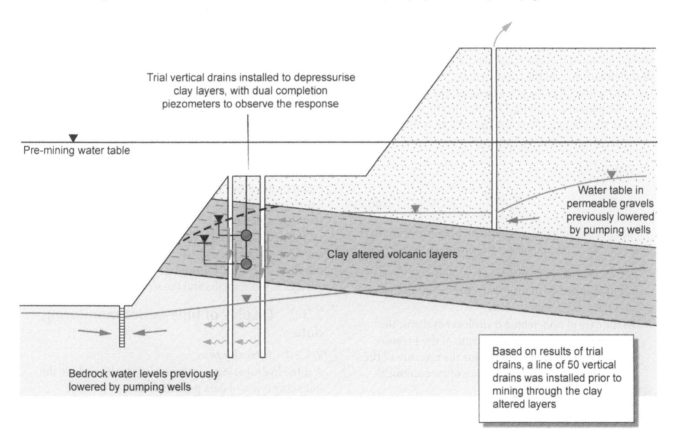

Figure 2.45: Conceptual set-up for vertical drain trial (based on Round Mountain Mine, Nevada, USA)

drains) and, perhaps most importantly, piezometric data on groundwater heads. The time-series data are used in all project planning stages and usually form the basis of judging the actual performance of the mine dewatering and slope depressurisation systems.

The aim of the time-series monitoring is usually to provide regular data from as many consistent monitoring points as possible. However, achieving continuity of monitoring stations around an active mine site is often not practicable, and it usually becomes necessary to replace wells and piezometers that are located in and around the immediate pit area on a continuing basis. Because of this, the presentation of data should ideally provide an overlap of older monitoring points and newer (replacement) points.

Water-level data from piezometers (whether standpipes or VWPs) and other observation wells are time-series data. They need to be corrected, where necessary, for atmospheric pressure changes (and potentially other external variables) and converted to static head (water level elevation) measurements above the usual mine datum, usually mean sea level or the local equivalent such as Australian Height Datum (AHD). It is always advisable to record the height of any temporary monitoring reference point used for the measurement (for example the top of the casing, or ground level), and it must always be borne in mind that piezometer locations and monitoring reference points may become disturbed or destroyed by subsequent mining operations.

Where VWPs or other pressure transducers are used, it should be noted that the static head at any point is the sum of the elevation, z, of the measurement instrument above the datum level plus the pressure head; the pressure head, h_p, is the pressure, p, divided by the specific weight of the water in the column above the transducer, so that:

$$h = z + h_p \qquad \text{(eqn 2.2)}$$

and

$$h_p = \frac{p}{\rho g} \qquad \text{(eqn 2.3)}$$

When comparing heads between piezometers or over time, it is therefore important to remember that:

- the comparison assumes that the density of the water in the standpipes or rock units is both constant in time and the same across the site;
- in the case of non-vented transducer systems, the pressure being recorded is the sum of the pressure exerted by the water column plus the pressure of the atmosphere at the time and place of measurement.

The atmospheric pressure change is usually small but may be significant when dealing with shallow hydraulic gradients or small changes in head during pumping tests. When dealing with confined systems, or those that are nominally unconfined but where material of low permeability is present above the water table, the water level in a standpipe may be affected by changes in atmospheric pressure whether or not the transducer is vented (Price, 2009). Such changes are usually small but may be significant when monitoring the short-term response to new pumping stress. In some confined groundwater systems, it may also be necessary to consider the influence of earth tides, for example in low porosity sub-permafrost groundwater systems.

2.3.1.3 One-off data

One-off data include specific tests and measurements to determine permeability and porosity (both *in-situ* and in laboratories), geological and geophysical logs, specific groundwater measurements, and synoptic surface water surveys. The one-off data are used for system characterisation and provide the basis for initially determining the hydraulic properties of the groundwater units, and for characterising the hydrological system. In many cases, the results and conclusions derived from the one-off data are confirmed or modified using the time-series monitoring data.

Most tests carried out to determine permeability or other formation properties will involve data collection (usually of flow rates and heads or pressures) over time periods varying from seconds or minutes to many days or weeks. Although some tests or logs may be repeated or augmented, they will not generally provide long-term continuous monitoring information.

2.3.1.4 Representation of the data

Like other forms of geological data, the hydrogeological data from mine sites represent a three-dimensional reality that conventionally has to be conveyed in two-dimensional form as maps, sections, graphs or tables. While the increasing use of GIS systems has become commonplace so that data are often now viewed in three dimensions, the basic components for hydrogeological analysis are still two-dimensional.

All data should be entered into a database that ideally is coupled to some form of GIS system for analysis and display purposes (Section 2.3.5), and for integration with the geological block model and the structural geology model.

2.3.2 Display of time-series monitoring data

2.3.2.1 Introduction

A mine hydrogeology analysis will usually include the following types of data presentation.

1. A base map.
2. Water level maps in various forms.
3. Maps of airlift yields from pilot holes.

4. Maps of horizontal drain flows.
5. Water chemistry maps.
6. Hydrographs of piezometers and observation wells.
7. Graphs showing well yields or drain yields with time.
8. Graphs showing changes in water chemistry parameters with time.
9. Hydrogeological (hydrogeotechnical) cross-sections.

Each of these is discussed in the following sections.

2.3.2.2 Base map

A base map should normally convey as much of the following 'fixed' information as possible.

- Topography in the form of contours.
- Surface geology.
- Principal geological structures.
- Surface water features including rivers, lakes and prominent springs.
- Components of the mine site such as buildings, roads, waste dumps, and tailing impoundments.

All data points pertaining to the type of analysis should also be shown. This may include hydrogeological test holes, pumping wells, drain holes, piezometers, climatological stations and surface water monitoring locations. The actual monitoring data (e.g. water levels, airlift yields, horizontal drain yields, water chemistry data) can be added to this map or presented as overlays as described below.

Base maps normally show the 'fixed' information for the ground surface. However, in some instances, it may be pertinent to show 'slice' maps at a given elevation below the surface. In particular, for underground mines, it is common to show the mine infrastructure, lithology and geological structure at a particular mine level. Sector-scale maps can be prepared for pit slopes, for instance when defining targets for horizontal drain hole programs.

2.3.2.3 Water level maps

Groundwater level maps may show either past recorded levels, currently recorded levels, predicted levels indicated by modelling for some future date; or past or predicted future changes in levels over a defined time period. To aid data interpretation, it is normal that all elevations (such as groundwater heads) should be plotted relative to mine datum (usually mean sea level or local equivalent). The groundwater maps are often contoured, but this may not always be appropriate, for example where groundwater levels become strongly 'stair-stepped' across geological structures or between geological units.

Maps that are contoured should also show the actual data points as there is often little value in showing groundwater contours without illustrating the controlling features. There will often be a sequence of maps, which may show either:

1. static head (groundwater elevation) at a given point in time;
2. change in static head or pore pressure over a given time period.

Vertical differences in head can be represented by colour coding data from vertically separated piezometers. However, strong vertical head gradients are associated with many pit slopes and may, in some cases, become larger than the horizontal gradients. Therefore, the use of 2D groundwater maps may be more applicable for presenting the site-scale data rather than the sector-scale data (although this is not always the case).

At the Grasberg pit in Indonesia, there is high rainfall and rapid infiltration at the surface and the groundwater moves rapidly downwards through the hydrogeological units towards the underground dewatering system below the level of the pit floor. The groundwater heads measured in the Grasberg area are more a reflection of the depth of the piezometer than its lateral position, so in this case the preparation of groundwater maps has limited value and can be misleading, and it is important to use cross-sections for data presentation and interpretation.

For some pit slopes, rather than plotting data points that are referenced to a common datum, it is useful to show the depth of the water table below the slope (or below the pit floor). This may help illustrate zones where a particular mine expansion or slope pushback may become wet. However, rather than showing contours of both the water table and the ground surface, it may be more effective to present contours or colour shading of the difference.

It is usually beneficial for points of hydraulic stress to be included on the groundwater level maps. These are most commonly pumping wells, where the flow rate from the well (usually at the time the groundwater levels were collected) may be added to the map. Flows from horizontal drains and prominent seeps can also be added to the maps where appropriate.

Natural hydraulic stress may also be important in wetter climatic zones. Figure 2.46 shows a pre-mining groundwater map developed for a sedimentary ore body where recharge to the formation flows laterally through the ore body and discharges as natural springs. The map includes the geological structures and lithology, contoured groundwater levels, and spring locations and flow rates as an aid to identifying the groundwater flow paths and groundwater discharge points.

2.3.2.4 Maps of airlift yields from pilot holes

Maps showing the distribution of airlift yields from pilot holes may be useful for determining general areas of potentially higher or lower yields and for correlating yields with geological structures or other features. Where there is

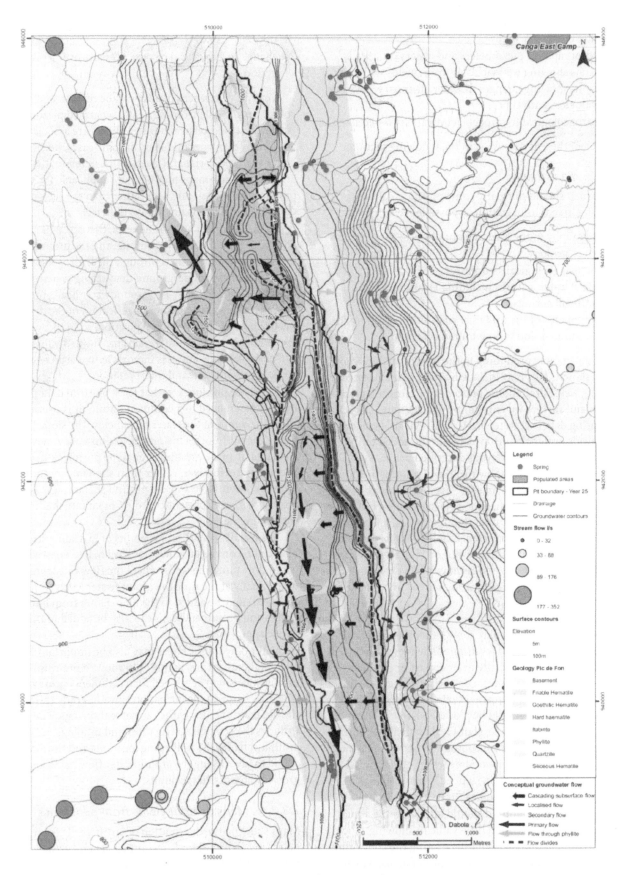

Figure 2.46: Example of a groundwater map including surface water discharge points

a relatively small number of pilot holes, the data may be added to the groundwater map. Where there is a lot of information, it is usually beneficial to create a separate map or overlay that may be colour-coded with 'bands' of yields (for example <1 l/s, 1–3 l/s, 3–5 l/s, >5 l/s, and so on). It is sometimes useful to create a map where the depth of the main groundwater ingress zone in the pilot hole may be colour-coded.

2.3.2.5 Maps of horizontal drain flows

Maps showing the distribution of horizontal drain flows provide an excellent way of rapidly reviewing drain hole results. The trace of the drain hole is shown on the map and is colour-coded to show the initial flow rate of the drain (Figure 2.47). The position of the first water strike in each drain hole can be added to the map to provide an indication of the depth of water behind the slope. In some cases, to illustrate the location of the most prominent inflow zones, the colour along the drain trace can be varied according to changes in the inflow rate as the hole is drilled.

The maps allow the areas of the highest drain yields to be easily determined and can also assist with the interpretation of the most favourable orientation for drilling of subsequent drains. If the geological structures are included on the base map, a correlation can often be made between water strikes or 'jumps' in drain flow and the location of structures. This correlation can often be better achieved by entering the colour-coded drain hole results into the geological block model and viewing the drain traces and colour-coded yields in three dimensions.

It may also be appropriate to create sector-scale maps for particular pit slopes that show the horizontal drain results together with the piezometer information. However, if such maps are used, it is usually necessary to consider the vertical gradients and to discretise the piezometer information using colour codes.

2.3.2.6 Water chemistry maps and diagrams

Water chemistry maps

Maps of selected water chemistry parameters can be used to support the conceptual hydrogeological model by showing the variation in water quality across the site (Figure 2.48). It may be possible to correlate the water chemistry values with different geological units or across different structural zones. By viewing maps for a number of different time intervals as pumping progresses, it may be possible to make a qualitative and sometimes a quantitative analysis of the direction of groundwater flow, or the way that groundwaters from different zones are mixing. The maps may also be useful for helping to identify recharge and discharge zones, particularly if surface water chemistry parameters are included in addition to the groundwater chemistry values.

When preparing water chemistry maps to support the conceptual hydrogeological model, it is generally appropriate to select those parameters that are chemically conservative (inert) and are less likely to react with the geological materials or other waters as they move down the flow path. As such, groundwater maps often include total dissolved solids (or electrical conductivity), the major cations (sodium, calcium, potassium, magnesium) and the major anions (sulphate, chloride, nitrate and those causing alkalinity, which are generally assumed to be bicarbonate (hydrogen carbonate) and carbonate). It may also be beneficial to plot the ratios between certain major ions. In mines close to agricultural areas, maps showing nitrate or ammonia results may help illustrate the difference between external sources and internal sources (both agriculture and blasting agents are commonly a source of nitrate and ammonia in some waters).

Water chemistry diagrams

Other methods of graphical water quality representation are described in standard texts such as Freeze and Cherry (1979), Hem (1992), Appelo and Postma (2005) and Drever (1997). The presentations fall into two main groups.

1. Those intended to give a pictorial representation of total solute concentration and the proportions assigned to each ionic species.
2. Those designed to help detect and identify mixing of waters of different compositions and to identify some of the chemical processes that may operate as the waters are drawn towards the dewatering system.

The unit used for the majority of water chemistry maps is milligrams per litre, which is the standard reporting unit for most chemical analyses. However, for many of the methods for graphical representation, the unit in which concentrations are usually expressed is milliequivalents per litre.

A simple bar diagram provides a useful means of representing analyses on hydrogeological maps and sections. Each analysis is represented by a vertical bar graph with a total height proportional to the concentration of the determined anions or cations, in milliequivalents per litre (Hem, 1992). The bar is divided vertically with the left half representing cations and the right half representing anions (Figure 2.49). Assuming that the analysis has achieved an ionic balance, the two bars should be of the same height. Figure 2.49 shows analyses of two water samples from a carbonate aquifer. The first, from the unconfined (outcrop) area, shows a typical calcium bicarbonate-rich water. The second sample is from downdip, below a clay-rich confining layer; here it is believed that natural ion exchange has reduced the calcium

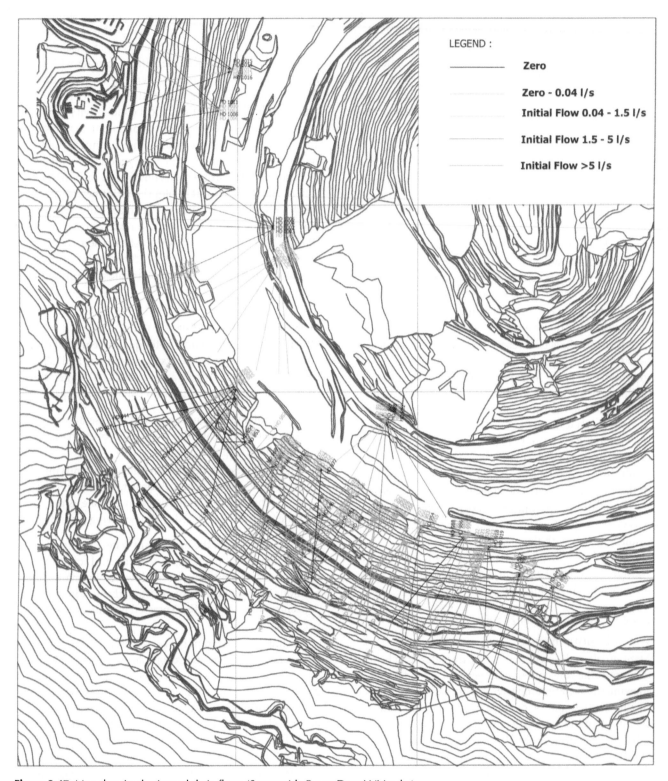

Figure 2.47: Map showing horizontal drain flows (Source: Ida Bagus Donni Viriyatha)

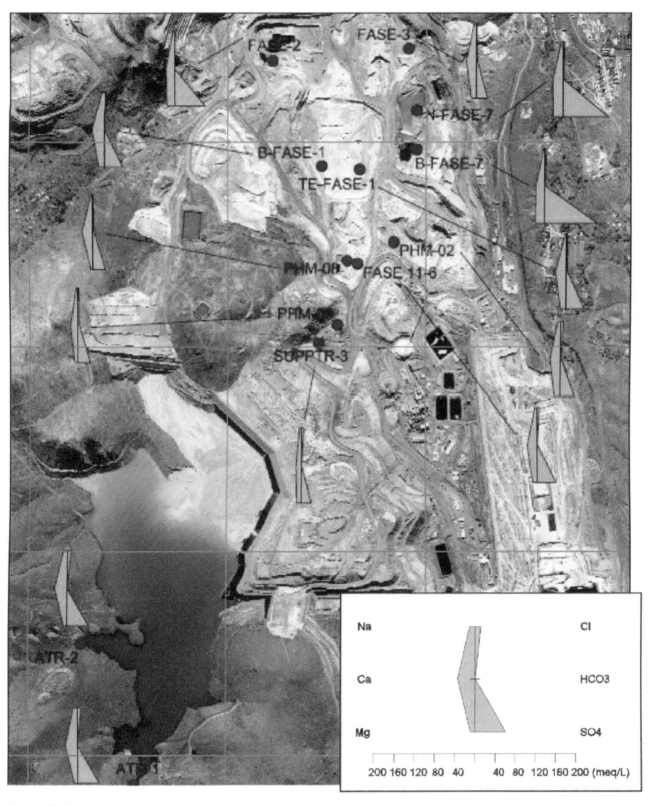

Figure 2.48: Example water chemistry map (Source: Mauricio Tapia and Marcelo Andres Kong Castro)

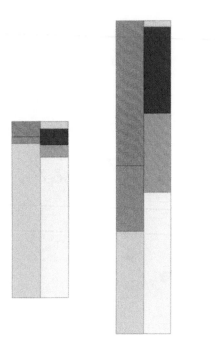

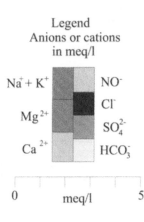

Legend
Anions or cations
in meq/l

Na$^+$ + K$^+$ NO$^-$

Mg^{2+} Cl$^-$

Ca^{2+} SO$_4^{2-}$

HCO$_3^-$

0   meq/l   5

Figure 2.49: Example of a Collins plot

content and replaced it with sodium and magnesium; the chloride level has also increased.

An alternative form of representation, which has been widely used for maps and sections, is the Stiff pattern diagram. This uses four parallel horizontal axes extending each side of a vertical zero-concentration axis. Concentrations of four cations and four anions can be plotted, the cations on the left and the anions on the right, in units of milliequivalents per litre, and the resulting points are connected to give an irregular polygonal shape (Hem, 1992). Figure 2.50 shows the water analyses of Figure 2.49 represented in this form; the change in dominant chemistry between the two samples is immediately obvious.

Stiff diagrams are useful for representing concentrations of major ionic species to allow comparisons of waters over the area covered by a map or section. However, each diagram represents only one 'snapshot' analysis and does not provide any means to explain how the various concentrations may have occurred. By combining the major cations and anions into three groups (calcium, magnesium, and sodium with potassium for the cations, and bicarbonate, sulphate, and chloride with nitrate and fluoride), and expressing the concentrations as percentages of the total milliequivalents per litre of cations and of anions, the composition of most waters can be represented by a trilinear plotting technique. Trilinear plots have taken various forms but the Piper or Piper-Durov diagrams are most common. The diagrams consist of two triangles, one for cations and one for anions, side by side. The cations and anions for each analysis are plotted

on the triangles and the relevant points are then projected into a diamond shaped field (Figure 2.51). Trilinear diagrams can be used to identify trends such as mixing of waters; the analysis of any water which is a mixture of two others will lie on a straight line joining the other two in the diamond plotting field. Their disadvantage is that the points represent only relative concentrations, not strengths of solutions; symbols of different diameters can be used to indicate absolute concentrations but this method becomes impractical for more than a few samples.

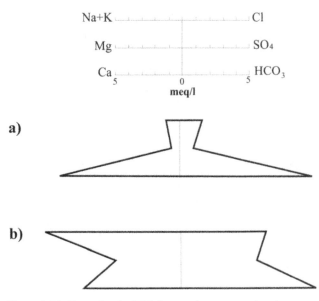

Na+K   Cl

Mg   SO$_4$

Ca   HCO$_3$

5   0   5

meq/l

a)

b)

Figure 2.50: Example of a Stiff diagram for representing the results of chemical analyses

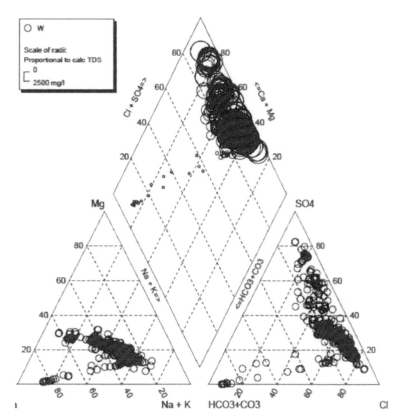

Figure 2.51: Example of a trilinear diagram showing mixing of waters of two end members

2.3.2.7 Hydrographs of piezometers and observation wells

Hydrographs of piezometers and observation wells are an essential component of any hydrogeological analysis. Multiple records should be included on the same plots to allow correlations to be made. The hydrographs should be annotated with as much pertinent information as possible to allow the influence of external variables to be determined and to help build up an understanding of the principal factors that influence piezometric levels in each sector of the pit.

In addition to annotations showing pumping rates and horizontal drain flows, the annotations may include such factors as the start of deposition of tailing in a nearby facility or the operation of process tanks and pipelines that may have a component of leakage affecting the groundwater system. In lower-permeability settings, a correlation of the mining rate or bench level with piezometers in a particular slope sector can help determine where pore pressure dissipation may be occurring due to deformation of the rock mass.

Piezometer plots are usually grouped by their physical location within the pit, but may also be grouped by hydrogeological unit. Records from multiple piezometers in each sector, particularly vertically discretised piezometers, should ideally be included on a single plot

(Figure 2.52) provided the amount of information doesn't become overwhelming, in which case several plots may be required. Plots of vertically discretised piezometric data against time, with the piezometers 'hung' on the plot at the appropriate depth, illustrate the way in which vertical heads are altering with time; they can also indicate the effects of vertical compartmentalisation.

It is usually important to include the pumping rate on the piezometer hydrographs; for each key well, for all wells in a particular sector, or for the total pumping rate for the site. Where appropriate, the piezometer hydrographs can also be annotated with horizontal drain flows, normally by sector, but sometimes the total flow from all drains. It is useful to annotate the hydrographs to show the start up (or shut down) of key wells or the timing of various horizontal drain programs. A histogram of daily, monthly or accumulated precipitation (depending on the time scale of the hydrograph) can be useful for identifying key recharge events, or for identifying longer periods of enhanced recharge. In some circumstances, correction of the data for atmospheric pressure or other variables may be required to illustrate observed changes with time in the piezometric levels.

It is normally good practice to use consistent scales between plots. However, this may not be practical where there is a large elevation range in total head values

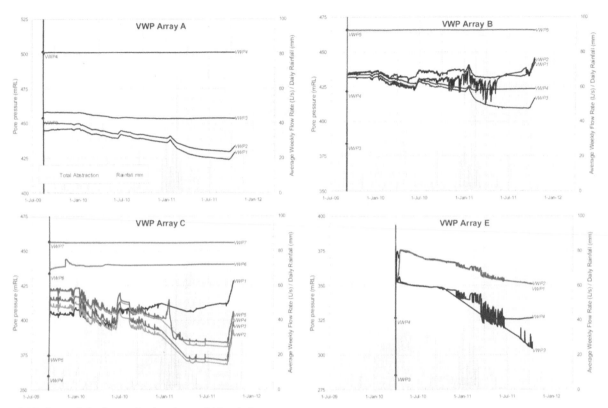

Figure 2.52: Example hydrographs showing multi-level piezometers

between piezometers, or where there is a large difference in the magnitude between the observed responses. Where it is necessary to illustrate a correlation between two or more hydrographs, it is acceptable to use multiple scales on the same plot, provided they are clearly labelled. For example, if there is a correlation between two piezometers that have 100 m difference in total head, then the piezometers can be plotted together but with different elevation ranges. Similarly, if one piezometer shows a response of 0.1 m to a pumping stress and a nearby piezometer shows a response of say 50 m, then the piezometers can be plotted together but using difference scales.

The development of dewatering and pore pressure targets is discussed in Section 5.1. The targets may be included on the hydrograph plots, as illustrated in Figure 2.53. The purpose of such plots is usually to illustrate the progress of the dewatering program to mine operations or management. As such, it often pays to keep plots showing targets as focused and easy to understand as possible.

2.3.2.8 Graphs showing flow rates from pumping wells or drains with time

Flow measurements from seepage faces, in-pit sumps, horizontal drains, or pumping wells are an integral part of any monitoring program. Graphs showing flow rates from pumping wells or drains with time are fundamental to any

hydrogeological analysis. It is normal to integrate the plots of the flow rate with hydrographs showing piezometric responses.

For vertical pumping wells, installations from each sector of the pit may be included on a single plot, with the total flow rate from all the wells in the sector also shown. The plots of flow rates from wells may also be grouped by their hydrogeological unit.

Plots showing the variation of specific capacity (discharge rate divided by drawdown) with time may be useful for certain hydrogeological units, or for the system as a whole. This may help illustrate the reduction in well performance as dewatering proceeds and may help with the scheduling of replacement wells. In strongly bounded settings, a plot of the specific drawdown rate (for example in metres per month per litre per second pumping rate) with time for selected piezometers may help in the understanding of how conditions are changing as dewatering proceeds and as the inward head difference across boundaries increases. This may help determine how the magnitude of inward leakage across boundaries is increasing as more drawdown is placed on the groundwater system within the pit.

Plots of horizontal drain flows against time can be useful for determining whether a slope sector is draining and depressurising or whether there is continuing recharge. Types of horizontal drain responses are

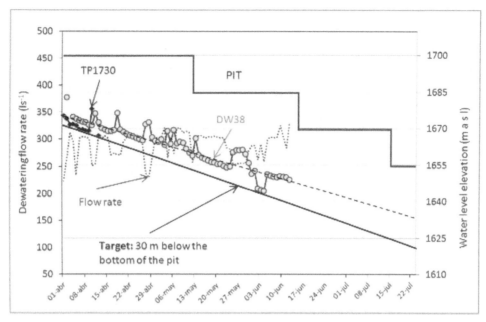

Figure 2.53: Example of plot showing target elevations on water level hydrographs (Courtesy Minera Peñasquito)

discussed in Section 5.2.3.7. Multiple drains from each sector of the slope may be included on a single plot, with the total flow rate from all the drains in the sector shown on the plot (Figure 2.54). Again, drain flow plots are usually grouped by their physical location within each pit sector, but may also be grouped by hydrogeological unit.

It is highly desirable to create composite plots that show changes in head or pore pressure combined with measurements of discharge or pumping rates. It is usually

beneficial to annotate such plots with dates where new wells or drains came on line, dates where a pushback was mined in a particular slope sector, dates of high precipitation events, and other factors that may have influenced the observed hydrograph responses.

Hydrographs showing the total dewatering flow from the mine site may also be useful for determining long-term trends. Figure 2.55 shows the pumping rate for three carbonate-hosted mines in Ireland plotted together to

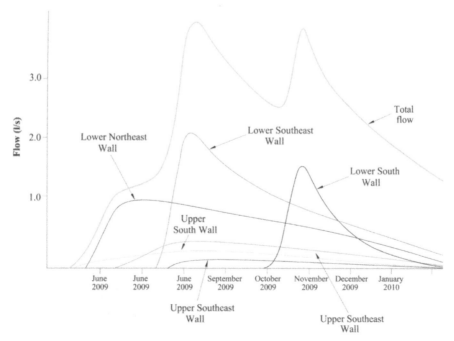

Figure 2.54: Example of horizontal drain response showing flow from drains by sector

illustrate the effect of precipitation and recharge cycles on the pumping rate.

2.3.2.9 *Graphs showing changes in water chemistry parameters with time*

Graphs showing changes in water chemistry parameters with time are commonly used to illustrate how the groundwater system is changing as pit dewatering progresses. They may be used to help illustrate how groundwater is moving between hydrogeological units, or used to help quantify the influence of recharge either from precipitation or snowmelt, discrete surface water sources, or from leaky site facilities.

Water chemistry graphs may also be used as part of the analysis to predict the long-term pumped water quality from individual wells or for the site as a whole, which may assist in the planning of downstream engineering, water treatment and discharge facilities. It is normally beneficial to include several different parameters on the same plot and, as for all other types of plots, it may be beneficial to plot chemistry alongside graphs of water level or pumping rate.

2.3.2.10 *Cross-sections*

Sector-scale cross-sections are the basic component of any pore pressure analysis (Figure 2.56). It is normal for the cross-sections to include lithological, alteration and structural information. Also, it is usually beneficial to make the hydrogeological cross-sections coincident with the geotechnical design sections. The sections may be annotated to include geotechnical information as well as hydrogeological information, provided that this does not overload the diagrams and prejudice clarity.

Historical and current piezometer information can be plotted onto the sections to show the phreatic surface or to show 2D groundwater head (or pore pressure) contours for a given time period based on vertically discretised piezometers, and/or changes of the phreatic surface or head distribution with time. The sections can also be annotated with known fracture or groundwater-flow zones and other pertinent features.

For the majority of mine sites, it is not usual that the boreholes or data points will lie exactly on the line of the section, and it is normal to project data from nearby measurement points onto the section. When this is done, the points should always be clearly identified and, if there are anomalies, the data that are furthest from the section line should be the first to be questioned. The lateral 'search distance' for adding piezometers to the section is typically 100–200 m, but may be greater than this where the density of data is low or where the geology of the slope is relatively consistent away from the section lines. It is usually of benefit to annotate the distance of the piezometer from the section line.

In addition to the sector-scale cross-sections, regional and site-scale cross-sections commonly form part of the overall hydrogeological analysis. Fence diagrams that link boreholes or piezometers parallel, as well as perpendicular, to the slope may be useful in addition to cross-sections for helping to interpret the hydrogeological data in the context of the geology.

2.3.2.11 *Composite data presentation*

The more that the different types of data can be integrated onto composite maps, data plots and cross-sections, the more useful the analysis will be for understanding the behaviour of the hydrogeological system. The integration of the geological and hydrogeological data onto maps and cross-sections is essential for providing a meaningful interpretation and for subsequently building the conceptual model.

It is common to integrate plots of water level, pumping rate and chemistry onto a single display to show how the various measured parameters are linked. As discussed above, it is also beneficial to annotate the composite plots with key events, such as dates when new wells came on line, dates when a pushback was mined in a particular slope sector, dates of high precipitation events, and other factors that may have influenced the observed hydrograph response.

There are many types of GIS system where piezometer hydrographs, pumping rate plots or water chemistry plots of observation wells or piezometers can be plotted as 'windows' onto a water level map. This may allow the various groups of piezometer responses to be correlated with water level and water level change. Most GIS systems now allow this to be done digitally by clicking on an individual well on the base map and pulling up all pertinent information and data plots for that well.

Section 2.2.6.12 illustrated some techniques for installing vibrating wire piezometers and geotechnical instrumentation in the same borehole. Clearly, to obtain an understanding of how pore pressures may be influencing the movement of the slope, every effort should be made to compile combined plots of piezometer hydrographs with data from slope monitoring prisms, inclinometers or TDR systems. A number of such examples are included in later sections of these guidelines although, to date, there are relatively few examples from the mining industry. In the future, it is expected that an increasing number of joint monitoring programs will provide an expanding worldwide database to illustrate the relationship of pore pressure and slope movement for various settings.

2.3.3 Analysis of one-off data

2.3.3.1 *Introduction*

One-off data are typically used for general system characterisation rather than continuous monitoring. As

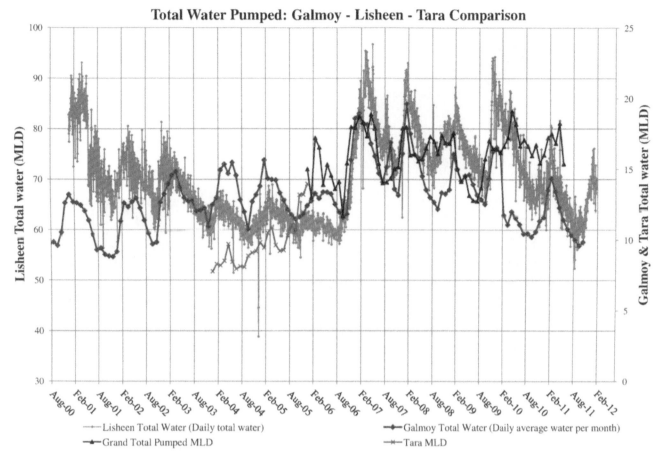

Figure 2.55: Combined plot of the total dewatering rate for three carbonate-hosted mines in Ireland

such, they usually form part of reporting for a specific investigation level rather than data presentations that are continually updated. Examples are as follows.

- Drill logs and downhole geophysical logs to provide geological information.
- Surface geophysical surveys to help delineate geological or hydrogeological units.
- Single-hole, cross-hole or multiple-hole tests to determine aquifer parameters and to characterise the groundwater system.
- 'Snapshot' (synoptic) groundwater surveys to delineate specific groundwater units, seeps or springs (although this may require several measurements to quantify seasonal variations).
- 'Snapshot' (synoptic) streamflow surveys to determine lateral changes along surface water courses (which may also be repeated to investigate seasonal changes).

The results of these test programs or surveys are normally contained in specific reports that are prepared at the time of testing (or shortly thereafter). The information is also usually entered into the database. The 'one-off' testing results contribute to a developing understanding of the groundwater system as an increasing number of test programs and surveys are carried out.

2.3.3.2 Development of hydraulic parameters

Field methods for hydraulic testing (single-hole, cross-hole or multiple-hole tests) were discussed in Sections 2.2.5, and 2.2.8. For a mining project, the data from these tests are normally used as follows.

- To determine lateral and vertical distributions of hydraulic parameters throughout the groundwater system, using data from different holes, and from different down-hole test intervals.
- To determine the hydraulic parameters (permeability, transmissivity, specific capacity and storage) and their lateral and vertical distribution within the defined hydrogeological units. Correlation of the testing results with the defined hydrogeological units forms the basis for developing the conceptual model (Section 3).
- To provide the input to analytical calculations which may be used to predict future groundwater flow rates (or future drawdown rates) and/or as input to the numerical model (Section 4).

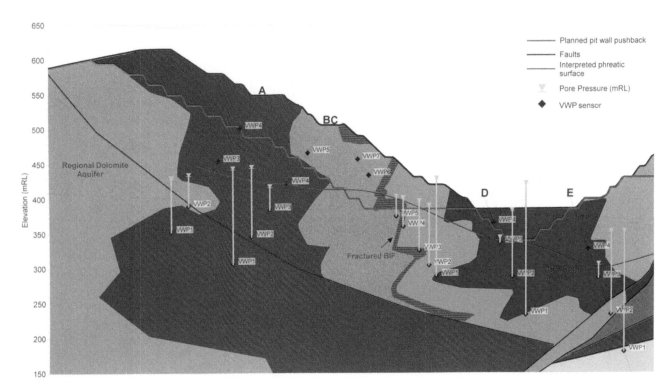

Figure 2.56: Illustration of a pit slope cross-section

- To provide 'empirical' response data to help determine the interconnection of fracture systems within given hydrogeological units or the behaviour of the groundwater system as a whole (for cross-hole and multiple-hole tests).

The methods used to analyse the various test programs are discussed in Appendix 2. It is essential to analyse the data within the context of the known geology and to consider that the physical nature of most formations (and therefore the hydraulic parameters) are likely to vary laterally and vertically over short distances, so some form of data 'averaging' is required, particularly for single-hole test results. As discussed in Section 1.1.2, in layered sequences, the use of an arithmetic mean, weighted to allow for variations in thickness, is usually the best 'average' to use for water movement parallel to the layering, and a weighted harmonic mean is most appropriate for flow across the layering. This applies regardless of whether the layering is horizontal or arises from some form of compartmentalisation with individual units that are aligned sub-vertically or are dipping at an angle. Where there is no layered structure, the most suitable average is the geometric mean.

However, as discussed in Section 1.1.2, although these averages may be useful for estimating bulk rates of groundwater flow, they will not generally be adequate for understanding pore pressure gradients (which will usually

be steeper within less permeable layers) or the speed of groundwater flow or transport (which will be greater through more permeable and/or less porous units, especially if they are fractured). Hydraulic gradients can be especially steep across vertical barriers such as faults containing clay gouge or across sub-vertical dykes, and significant pore pressures can be encountered on their upgradient sides. Therefore, while assessments based on bulk values of permeability averaged over large blocks may be adequate for early-stage assessments, they are unlikely to be suitable for predicting the widespread response of the groundwater system or for detailed modelling of critical slope sectors. It is therefore desirable to carry out as much cross-hole or multi-hole testing as possible to obtain groundwater response data over as wide an area as possible.

2.3.3.3 Analysis of vertical variability

Large open pits usually impose a major stress on the groundwater system. There is an increasing tendency for vertical gradients to develop as the water moves closer to the pit. For analysis and interpretation at the sector scale, it is therefore necessary to focus on the vertical changes in property values as much as the lateral changes. The groundwater flow paths will often become deeper as the shallow formations become dewatered, which will increase the tendency for vertical gradients to develop.

Single-hole tests mostly provide an indication of the lateral hydraulic conductivity in the immediate vicinity of the borehole. However, data from discrete-interval tests (such as slug or packer tests) can be plotted against depth to provide an indication of how the hydraulic conductivity changes. It is sometimes useful to superimpose laboratory values, where available, on the same depth plot, but any laboratory values will almost invariably be lower than those measured in the field because they have a tendency to measure the disturbed matric permeability rather than that of the *in situ* fractures (Section 1.1.2.7). The results of interval-tests can be plotted as hydraulic conductivity vs depth, hydraulic conductivity vs material type, and hydraulic conductivity vs RQD of the test interval, where there are data to support this.

Figure 2.57 shows an example of a hydraulic conductivity vs depth plot from a fractured and porous limestone formation in which the fractures make an important contribution to permeability throughout the sequence, but where one major fracture, enlarged by solution, is particularly important. Figure 2.58 shows an example of plotting packer test results from a volcanic setting. In addition to indicating the distribution of hydraulic conductivity with depth, this particular plot showed a fairly broad range of hydraulic conductivity values at all depths, indicating a wide variability in the hydraulic properties of the various fracture zones that were tested.

Figure 2.59 shows an example from a borehole in fissured sandstone. A geophysical log of fluid conductivity (Section 2.2.7.4) is also included. In this case, the similarity of permeability results obtained from laboratory samples and those measured with packer tests shows that fracturing is relatively unimportant, except in two intervals at around 30 and 60 m below datum.

When dealing with large thicknesses of sediments (as in the case of overburden or coal workings) it can be useful to plot the increase of transmissivity with increasing depth below the water table; Figure 2.60 shows the data from Figure 2.59 presented in this way. However, as noted above, in situations where dewatering of the formations is required, the effects of the falling water table and thinning of the saturated aquifer unit need to be strongly considered in the data interpretation.

2.3.3.4 Analysis of storage and specific yield

Storage properties usually offer a smaller range of variability than does permeability. For most situations involving depressurisation, the system (or, at least, the permeable parts of it) will be confined and the relevant property will be the (elastic) specific storage, S_s. This will be the case, for example, when dealing with matrix blocks in a fractured formation. It will remain the case until the pore pressure is reduced to atmospheric, at which time the relevant property will be specific yield. However, by this time, the pore pressure is often of less significance in calculating effective stress and shear strength.

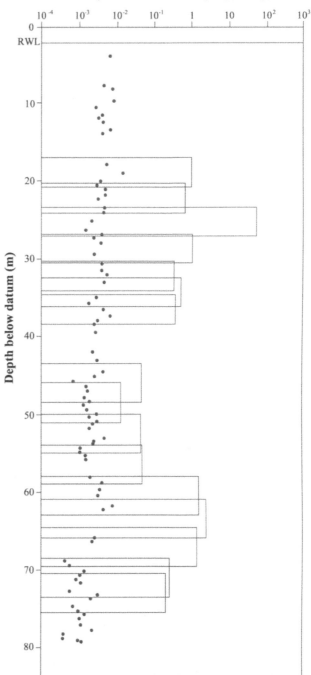

Figure 2.57: Laboratory values of permeability (circles) and permeability values obtained from packer tests (bars) combined on a permeability-depth plot (After Price *et al.,* 1982)

Specific storage can be estimated from knowledge of the porosity and compressibility of the formation (Section 1.1.1.5). If derived from laboratory measurements on small samples, the value of S_s will be relevant to the matrix; if derived from field values (probably from geophysical logs) the value should relate to the bulk behaviour of the

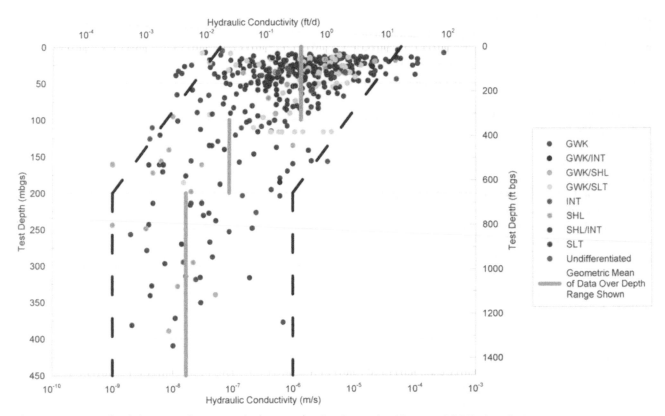

Figure 2.58: Example of plotting packer test results from a volcanic mine setting (Courtesy BGC Engineering)

formation. In dealing with dual-porosity materials, the relevant equation is Equation 1.4.

Specific yield is relevant to understanding the response of slope environments to recharge and in estimating the volume of water to be removed. When using specific yield, it is important to remember that sub-horizontal layers of fine material can have a significant impact on drainage, both by:

1. the low permeability impeding downward movement from the rock above;
2. possibly more importantly, the effect of the fine pores in the material restricting air entry and inhibiting drainage from the rock below.

For this reason, laboratory values of specific yield (obtained, for example by pressure-plate or centrifuge methods) may imply values of specific yield that are much higher than those likely to be encountered in the field in the short term. Conversely, values of specific yield obtained from short-term pumping tests may underestimate long-term drainage volumes.

2.3.3.5 Development of composite borehole logs

It is routine to present data from drill holes in the form of borehole logs showing lithological types, casing and completion details. These logs can be expanded to include drilling information such as the rate of penetration, the water level at the start and end of each shift, significant water strikes and intervals of lost circulation. The rate of water production and increases in water production with depth should also be recorded on the composite log, along with any measurements of temperature and water quality. Figure 2.61 provides an example of a composite log of a drill hole developed during the Pre-Feasibility stage of a mine hydrogeology investigation.

When lithological data are derived only from drill cuttings, the information available may be limited but when core is available much more information can be summarised on the composite log. This can include laboratory data including strength properties, porosity, permeability and other testing results.

Where geophysical logs are recorded, lithological and engineering data should be presented alongside them. Geophysical logs are of particular value where coring is not continuous or when core loss is common. In this case, the logs from areas where core has been recovered can be used to extrapolate to areas of core loss. The geophysical logs can also provide useful engineering data such as casing details and the integrity of cement seals. The most effective interpretive use of geophysical logging is the synergetic approach, where data from various types of log (for example, sonic, electrical and nuclear) are combined to provide data on porosity, permeability and conductivity of pore fluids. Geophysical logs can also be incorporated

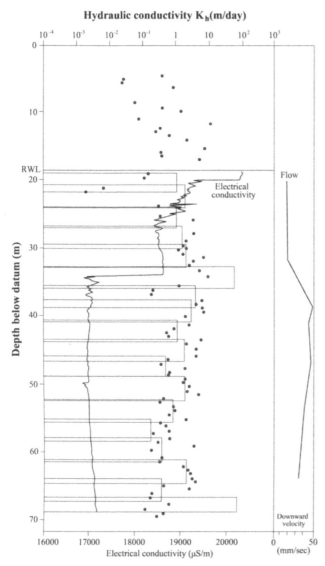

Figure 2.59: Laboratory values of permeability (circles) and permeability values obtained from packer tests (bars) combined on a permeability-depth plot. A log of fluid conductivity and a heat-pulse flow log (Section 2.2.7.4) is also shown (After Price *et al.,* 1982)

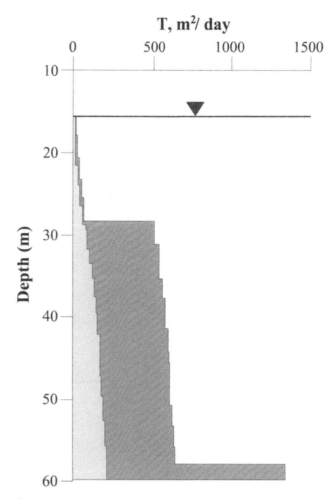

Figure 2.60: Increase in transmissivity with depth below water table in a sandstone formation. Yellow – laboratory results, showing the contribution of intergranular permeability; orange – field results from packer tests showing the contribution of both intergranular and fracture permeability. Note the large increases in total transmissivity at around 30 and 60 m caused by permeable fissures (Data from Price *et al.,* 1982)

into plots of permeability variation, as in Figure 2.59, where the abrupt change in the fluid conductivity log indicates inflow along the major fracture at around 30 m and the straight portion of the log below that level (down to 50 m) suggests downward flow in the hole.

2.3.4 Levels of data analysis for a typical development program

Hydrogeological data analysis, interpretation and reporting should proceed in a similar phased manner to that of the field program (Section 2.1.5). The results of the analysis should be interactively fed back to guide the next stage of the field program. It is also important that the

hydrogeological data are continually integrated with the geological, geotechnical, environmental and other data from the overall mine development program as it passes from one level to the next.

2.3.4.1 Level 1

At the Conceptual level (Level 1), where the density of data is often low, the analysis and interpretation can be carried out relatively quickly, often based on professional judgement and by benchmarking with analogous (proxy) sites. A conceptual hydrogeological model (Section 3) should ideally be developed as early as possible in the planning process. The conceptual model should be continually updated and refined. The field programs should be adjusted and focused in line with the evolving conceptual model.

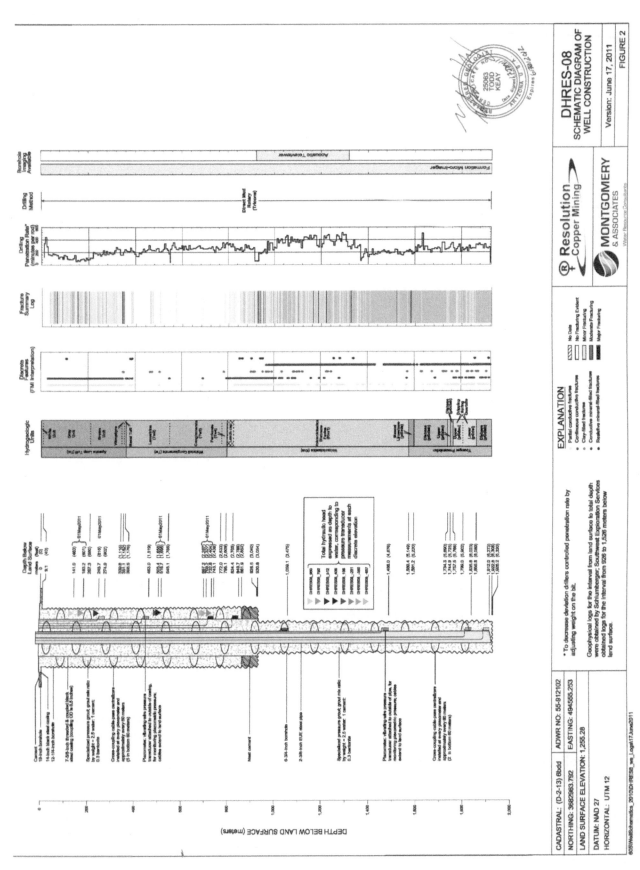

Figure 2.61: Example of composite borehole log (Courtesy Resolution Copper and Errol Montgomery and Associates)

For most groundwater programs, the most important information for early-stage planning is a general knowledge of regional and site-wide groundwater levels and correlation of these with the project geological model, the mine plan, and locations of potentially connected surface water, recharge and discharge. This information can typically be gained by implementing a program of water level measurements as early as possible using as many existing regional holes and wells, mineral exploration drill holes and condemnation (sterilisation) holes as possible.

The head contour (groundwater elevation) map is often the most important tool for initial characterisation of the groundwater system (Figure 2.62). Other important components of the early-stage analysis will typically include regional and site-scale hydrogeological cross-sections, initial estimates of hydraulic conductivity and storage for the key identified hydrogeological units, an analysis of hydrographs for any existing observation points, and an initial assessment of meteorological and surface water conditions.

The groundwater map should include the key lithological units, geological structures and the surface topography (including the planned pit and other mine facilities) (Section 2.3.2.2). If there are sufficient data, it may be appropriate to prepare a regional-scale map and a site-scale map. The maps can be used to identify zones where:

- groundwater heads are similar (flat gradients) over a wide area. This often indicates interconnection of the groundwater system, and may indicate the potential for regional-scale groundwater flow;
  - → for example, strongly interconnected carbonate groundwater systems in areas of low recharge often exhibit very similar total head values over the scale of many kilometres. This tends to indicate a requirement for a high-volume dewatering system.
  - → conversely, discontinuous groundwater levels and strong lateral gradients indicate a general poorly connected groundwater system, suggesting that the inflow rates in that sector of the project will probably be low, but increased emphasis on slope depressurisation may be important.
- 'stair-stepping' of groundwater levels can be correlated with known geological structures. This often provides good evidence of discrete flow barriers and 'compartmentalisation' of the groundwater system, which may also inhibit the overall required dewatering rate;
  - → potentially one of the most favourable conditions for groundwater control is an ore-body hosted in locally permeable rocks, but which shows strong hydrogeological compartmentalisation as a result of regional geological structures or other groundwater flow barriers.

- → an example of strong hydrogeological compartmentalisation is the Ruby Hill Mine which is hosted in permeable limestones in Nevada, USA. Good local drainage of the ore body and pit slopes can be achieved with a relatively low dewatering rate, with limited influence on adjacent groundwater systems. During the Feasibility study, it was possible to predict this situation in advance of dewatering by an assessment of the discontinuity of the total head measurements in exploration holes, and correlation of these with known geological structures.
- weak rock and low permeability occur and may indicate sectors of the mining operation that require intensive effort to depressurise and stabilise the pit slopes. These can usually be identified with monitoring of water production during the course of drilling (or by packer testing in diamond drill holes – usually commencing at Level 2).

Thus, an interpretation of the initial groundwater map can form the basis for determining the initial estimates of dewatering rate and for identification of any zones that may be difficult to drain in the future pit slopes.

2.3.4.2 Level 2

At the Pre-Feasibility stage (Level 2), the whole mine site will probably receive equal treatment with the aim of establishing supportable estimates of mine dewatering and slope depressurisation requirements. The Pre-Feasibility study will normally include a detailed groundwater map, piezometer hydrographs, a hydrochemistry assessment and a series of pit-scale hydrogeological cross-sections, in addition to a detailed assessment of meteorological and surface water conditions. Because of their importance in understanding the distribution and behaviour of pore pressures, it is important to build the knowledge of vertical hydraulic gradients from the Pre-Feasibility stage onwards.

2.3.4.3 Level 3

As work proceeds to Feasibility level (Level 3), the analysis and reporting may concentrate increasingly on the principal areas of risk for the general mine dewatering system (and the associated potential regional impacts), and for the pit slope design. The areas of higher risk for the pit slopes will usually be those areas that are being shown to be most vulnerable, for example, because of their geotechnical properties or because of greater water recharge potential from outside the pit area.

For the complete dewatering design, it is important that the Feasibility study includes pumping trials (with single or multiple test dewatering wells) that show an empirical response to pumping over a reasonably wide area, and allow important hydrogeological boundaries to be identified. For the pit slope design, the Feasibility study

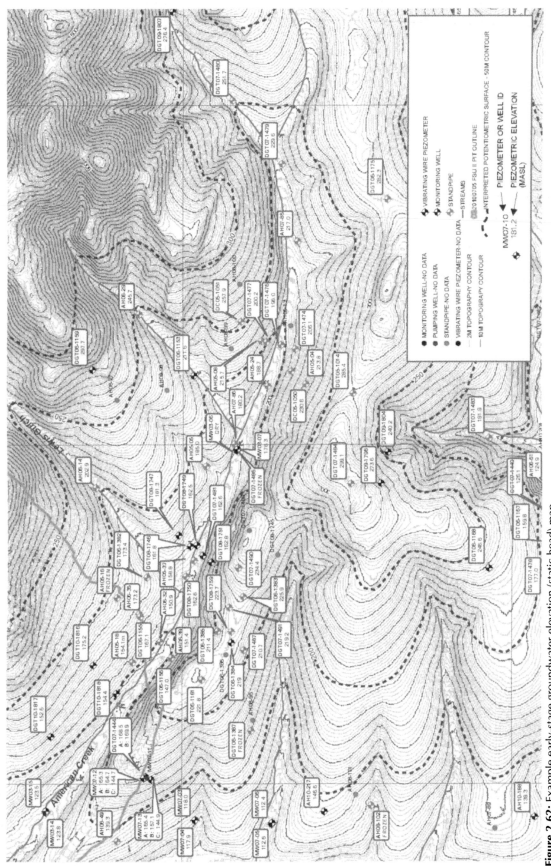

Figure 2.62: Example early-stage groundwater elevation (static head) map

must include detailed hydrogeotechnical cross-sections (Figure 2.4), together with hydrographs of piezometers in those sectors of the slope that have been identified as being most critical for stability (generally the weaker and less permeable materials). The aim is to quantify mine dewatering requirements and pore pressures in the slopes as the mine develops in order to identify and design the required measures for slope stability.

2.3.4.4   Levels 4 and 5

The level of analysis and reporting at the Detailed Design stage (Level 4) must be sufficient to provide confidence that the planned dewatering and slope depressurisation measures will allow the mine plan to be achieved with a high degree of certainty, including discharge of any excess water from the mine site, and potential environmental impacts. It normally includes all of the components of data presentation described in Section 2.3.2. During operations (Level 5), as the pit is deepened or the slopes are being pushed back, it is essential to keep track of the groundwater heads and pore pressures, and the data presentation methods described should be linked to identified target values. Continual analysis of the hydrographs may reveal trends that indicate the need for changes to the future hydrogeological program, and potentially changes to the mine plan.

2.3.5   Databases

2.3.5.1   General

The advent of data loggers means that variables such as groundwater head can be recorded as frequently as every few seconds. The resulting large quantities of data can be held on electronic databases and plotted relatively easily and in various forms using GIS systems, so that various data can be combined on maps and sections. While this offers great advantages, it can also pose a problem. It may be tempting to show, for example, heads, mine levels, drain flow rates and times on a single diagram. The aim may be to communicate the situation clearly, but the end result may be something that is so complicated that it does the opposite. It is therefore important to:

1. decide what information is most important to the identified areas of risk;
2. determine what each presentation is trying to portray;

3. consider the audience, who may not be familiar with the data collection, analytical or modelling processes.

Effective data management is fundamental to any groundwater investigation and requires investment and planning from the initiation of the project. The purpose of this section is to provide guidance for defining and implementing the data management process. The use of database software is also discussed, although it is emphasised that software solutions will only add value to a project when they are part of a well-defined, documented and rigorously implemented data management process.

2.3.5.2   Data management process

A schematic of the data management process is presented as Figure 2.63. It is necessary to provide a clear definition of what data are required at the outset of the project. The consequences of not defining and implementing the data management process are far reaching. There are numerous examples of projects where data have been lost or disregarded due to poor data management. The end user of the data will frequently be working to a tight budget or deadline and will not have the time or resources to compile, verify and validate poorly managed data. Without traceability and confidence in the data gathered, the effort and expense is wasted.

The data management system should be designed with these requirements in mind and the following questions must be considered during the definition of the data management process.

- Why are the data being collected?
- What are the parameters that need to be measured?
- At what frequency do the data need to be collected?
- What level of accuracy and precision is required?
- How are the data to be used?
- Who are the data for and how are they be accessed and delivered?
- When are the data needed?
- What level of quality assurance is required and at what stage?

The definition of the data requirements must also consider whether the data need to be integrated into an existing data management system, or link with the systems

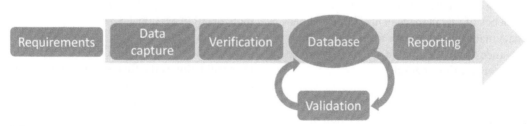

Figure 2.63: Schematic representation of the data management process

of other departments or companies. This is more likely to be a requirement at established operations where advanced data management systems may already be in place.

The basic rules of data handling, analysis and reporting should also be considered at this stage of the process. These include the following.

- Defining the coordinate system within which all data are to be recorded and presented.
- Defining the datum level.
- Agreement of a logical and consistent naming protocol for all samples and sampling locations.
- Definition of any standards that must be adhered to or methods that must be followed. These could be regulatory (as is the case with certain water quality parameters), or could relate to internal company standards or industry best practice.
- Confirmation of what Standard Operating Procedures (SOPs) will be required, and who will be responsible for preparing and implementing them.

A description of the key considerations for each step of the process is provided below.

2.3.5.3   Data capture

The first step in the process is defining how the data are going to be captured. The traditional method of data capture is by recording of measurements and notes in field note books or pro-formas. Pro-formas are often preferred as they provide a systematic method of data capture and reduce the likelihood of key data being missed or forgotten by the field operatives.

Technology is increasingly prevalent at the data capture phase. This ranges from the use of data loggers in the collection of climate, water level and water quality data through to the use of tablet PCs in the recording of data in the field. The use of technology has advantages because it can reduce the possibility of transcription errors, ensure basic data verification processes are carried out immediately, and accelerate the process of database updating and data validation.

Telemetry has also advanced considerably in recent years. Information transfer technology such as GPRS (mobile phone networks), RF (radio frequency) and Bluetooth are frequently used where available to relay remote and even real-time data. The advantages of automated logging and transfer of data are obvious with respect to efficiency of data management. The subsequent steps in the data management process must however be planned in advance due to the volume of data that can be generated in a relatively short period.

Verification

Data verification is the first stage in the validation of collected data. The purpose of the data verification step is to define the basic checks that should be conducted at the point of data capture, or immediately afterwards. This step in the process provides a means of identifying errors at the earliest opportunity.

In the case of field data, it is simply a check that all required data have been recorded in accordance with the SOP. In the case of data loggers it would be a basic check on the functionality of the equipment.

Database design

A single, centralised, validated data source is the key objective for any data management system. A basic relational database provides a means of ensuring data integrity through the definition of database rules. Basic rules such as verification of location IDs at the data entry stage, and field settings that ensure correct date and number formats, are vital to ensure the integrity of data.

In addition to the database itself, original data transfer files (e.g. field data sheets, electronic data deliverables, lab receipts, results) and auxiliary files (e.g. calibration certificates) need to be archived in a traceable manner. This can either be within an organised file structure or imported as whole documents into the database.

The process of data validation should also be integrated into the database design. A well-designed database will provide a complete audit trail from data import through data validation to the final reportable data set. This can be achieved, for example, through incorporation of date-time-user stamps on all imported data, and the incorporation of data quality flags into the database tables.

There is a large amount of material available on the internet and in reference books on the principles of data management, and how to establish a basic database. The availability of information is positive as it enables individuals to become familiar with the principles of database design. However, it can also lead to database development by relatively inexperienced individuals. This may result in an inefficient development process and a sub-standard product. A key requirement at the outset of the project is therefore to determine who is responsible for developing the project database and at what stage this database will be developed. The initial focus should be on maintaining a well-organised, quality-assured, data set. This will require documentation that details each step of the data management process.

The software options for data storage are discussed in more detail in Section 2.3.5.4.

Validation

Data validation is the final stage in the quality assurance process. It is more involved than the basic data verification described above. The data must be reviewed against historical data and anticipated trends, and experience of the potential sources of data anomalies is desirable. Data validation will often involve advanced data processing, such as the calculation of ionic balance errors on water quality data, or correction for barometric pressure fluctuations in the case of water level data recorded by pressure transducers.

A fundamental rule in any data validation process is to maintain the integrity of the original data set. Where original data are being overwritten for analysis or reporting purposes, the reasons should be documented and the corrected data stored in parallel to the original data set. The use of QA flags/codes is often the most efficient mechanism for recording the data quality. The data validation process must be clearly defined at the outset of the project so the mechanism for review and correction of poor data can be incorporated into the database design.

Reporting

The final step in the data management process is reporting of the data set. The detail and frequency of the reporting requirements must be defined, as should the level of data validation that is required prior to reporting. In some instances, data must be reported without validation if real-time reporting is required. Full validation will be required before data are reported to regulators, or submitted to engineering design teams.

Reporting formats should be standardised (where possible) to improve the efficiency of report generation. This can include standardised table formats and figures and automatic linkages to GIS systems for updating of maps. Typical reporting requirements for a mine hydrogeology program were provided in Section 2.3.2. It is important, however, to review the plots and make interactive modifications as the program progresses. For example, as the pit expands, it may be advisable to change the boundaries of the sectors or change the grouping of data.

Another component of data reporting is the method of transfer to third parties. It is important that the method of data collection and level of validation is communicated to the recipients of the data so they are fully aware of any limitations that may apply to the data set. Basic information such as locations, parameter names and units must also be communicated clearly.

2.3.5.4 Database management systems

Database management systems (DBMS) generally revolve around a central structured data store (i.e. a database) and will offer various levels of functionality, automation and accessibility depending on cost and available infrastructure. The capabilities of the database management system will lead to refinements to the data management process, but it is essential that the process is reasonably well defined before a database management

system is selected. The following discussion presents an overview of the various levels of DBMS available from relatively low-cost desktop products through to full enterprise systems.

Desktop products

The simplest and cheapest form of DBMS can be created from generic office software such as Microsoft Access. Desktop solutions are typically implemented during the early stages of a project when the expense of a fully networked or enterprise level solution may not be justifiable. A well-designed and implemented desktop product will also ensure that data can be readily transferred to a more sophisticated system at a later date.

A frequent failing of a customised DBMS is lack of documentation. This will often lead to the DBMS being abandoned when the developer moves on. It is therefore essential that time for documentation and review is built into any database development process. If the capability, time or budget required to implement a documented, quality-controlled, customised DBMS are not available an alternative solution will be required. This could involve an off the shelf package, or the simpler option of a well-organised file structure containing data in MS Excel spreadsheets. Whilst the latter option has a large number of limitations with respect to efficiency of data handling, data protection and accessibility, it is still preferable to a poorly designed DBMS. There are many examples of projects where good data have been disregarded because the end-user did not have confidence in the source of the data.

Purpose-designed desktop software offers a quicker route into database management and will give greater off-the-shelf functionality. Importing and exporting data will be automated as will various methods of reporting and visualising the data. Some products may come with GIS and procedures for linking or exporting to other software packages for data analysis. It is important to understand clearly the limitations of a purpose-built system at the outset as the complete data management process will need to accommodate these limitations. The total cost will also need to take into account maintenance and support and need to include some form of training to gain the full benefits from the system.

Locally networked systems

Locally networked systems extend the concept of purpose-designed desktop software and allow multiple users to access data simultaneously. For larger monitoring networks, this will help remove the bottleneck of data requests and distribution. The level of functionality expected would be similar to the desktop solution. Consideration may also need to be given to integration of the mine site database into a corporate database system. For plotting and analysing data, it is important that the database has sufficient GIS functionality that it can be incorporated with the geology block model and the structural geology model. This will allow, for example, the 'jumps' in flow rate during the drilling of pilot holes or horizontal drains to be viewed in 3D with lithological contacts and/or the location of major structures.

This level of system would use a networked database such as SQL Server, Oracle, MySQL or PostgreSQL. These systems may require licensing, depending on the product, and will be an addition to the cost of the DBMS software. An increased level of database administration will also be required on a locally networked system.

Enterprise
This high-end level of DBMS will typically offer web capabilities with a more automated workflow based on multiple users across multiple sites. This will more than likely combine the power of local network users with higher-level decision, review and distribution functionality via internet services. This requires greater infrastructure costs if hosting internally, although hosted services are increasingly being offered on a subscription basis.

3

PREPARING A CONCEPTUAL HYDROGEOLOGICAL MODEL

Geoff Beale, Pete Milmo, Mark Raynor, Michael Price and Frederic Donzé

3.1 Introduction

3.1.1 Background

This section of the book outlines how a conceptual hydrogeological model is prepared. The development of a conceptual model is described in general for the mine site, but much of this chapter is focused on the development of a model that is specific to pit slopes and that can form the basis for the pore pressure predictions that will ultimately be used as input to the pit slope design process. How the conceptual model fits into the hydrogeological pathway to support the slope design is illustrated in Figure 3.1.

The guidelines for development of a conceptual model for the complete mine site were prepared based on the operating experience drawn from the LOP sponsor organisations. The guidelines for developing a conceptual model specific to the pit slopes were developed following research carried out using data from the North-west wall of the A154 pit at the Diavik Diamond Mine, the case studies presented in Appendix 4, and a survey of 50 mine sites to determine the key hydrogeological factors that contributed to the observed pore pressure profiles at those sites.

The background to the process of developing a conceptual model is outlined in Section 3.1 and 3.2, including the key components of a site-wide model. The research carried out using the data from the Diavik A154 pit North-west wall is discussed in Section 3.3 with supporting information included in Appendix 3. Section 3.4 describes how Discrete Fracture Network (DFN) modelling can be used in support of the conceptual model. The results of the case studies for other mine sites are summarised in Section 3.5 and a write-up of the case studies is included in Appendix 4. Section 3.6 presents the common factors and processes that must be evaluated for developing the conceptual model specific to pit slopes. Section 3.7 presents the conclusions of the work and discusses the key issues for making decisions regarding a slope depressurisation program and addressing other water-related issues.

3.1.2 What is a conceptual model?

A conceptual groundwater model may be described as a realistic, defensible, albeit simplified representation of the hydrogeological system. At any given scale, the hydrogeological system can be broken down into eight fundamental components, as discussed in Section 3.2. For any conceptual model, it is important to gain as much information as possible firstly on the detailed geological conditions and the main hydrostratigraphical units, and secondly on the response of the hydrogeological system within the model domain to changes in hydraulic stress (recharge or discharge). This provides essential information on the link between the different hydrostratigraphical units and the magnitude of groundwater movement within the conceptual model domain.

The level of detail within the conceptual model must be commensurate with both the available data and the objectives and risks related to the study. For any given mine site, the model must incorporate the regional (or district) scale, the mine (or site) scale and the pit slope (or sector) scale, as described in Section 2.1.2.1.

3.1.3 Development of a sector-scale model

At the sector (or pit slope) scale, a fundamental objective is to answer the question 'How does a fractured rock slope depressurise?' As outlined in the introductory section to

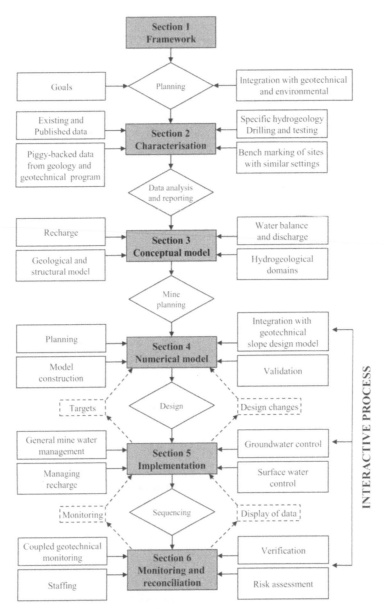

Figure 3.1: The hydrogeological pathway in the slope design process

this book, the question was addressed by carrying out a research program in four steps, as follows.

1. Literature review. A review of the available information on fractured rock hydrogeology and the behaviour of water pressure under fracture flow conditions was first carried out, mostly using information from the mining, nuclear and construction industries.

2. Analysis of the detailed structural and hydrogeological data from a well-instrumented rock slope at an open pit mine. The mine selected for the analysis was the A154 pit at the Diavik Diamond Mine, which is located in the Northwest Territories of Canada. Section 3.3 and Appendix 3 contain the outcome of the hydrogeological analysis, which was performed by staff of Schlumberger

Water Services (SWS) and Rio Tinto. The structural analysis and associated DFN modelling was performed by Golder Associates (Golder), and is also presented in Appendix 3.

3. Carrying out analyses of supplementary structural and hydrogeological data based on case histories at nine other large open pits. The mine sites used to provide the data were:

- Escondida (Chile)
- Chuquicamata (Chile)
- Radomiro Tomic (Chile)
- Antamina (Peru)
- Jwaneng (Botswana)
- Layered limestone sequence (Nevada, USA)

- Cowal (New South Wales, Australia)
- Whaleback (Western Australia)
- La Quinua (Yanacocha, Peru)

The sites represent open pits with fractured low-permeability slopes and were selected to establish the importance of varying physical, geological and climatic settings on pit slope depressurisation in the different environments. The information from these sites is presented in Appendix 4 and summarised in Section 3.5.

4. Integration of all of the data to determine the key factors for developing a sector-scale conceptual model, focussing on how depressurisation occurs in a fractured rock slope.

3.1.4 Available data

The literature review provided a large number of desktop and laboratory-scale studies regarding porosity and permeability, together with a wide range of fracture mapping and modelling studies, many of which were carried out for the nuclear waste industry. However, there were few studies containing actual pore pressure data relating to the behaviour of fracture systems due to changing hydraulic stresses. The construction industry has carried out numerous studies on groundwater flow and pore pressure associated with shallow excavations and landslides that generally support the findings of the monitoring programs carried out for large open pits. In particular, several hydroelectric power companies have monitoring data associated with the reduction of pore pressure and stabilisation of natural slopes.

Discussions were held with each of the LOP sponsor organisations to determine the amount and quality of hydrogeological data that are generally available from mine sites worldwide. While there is currently a variable standard of data collection within the industry, it is clear that the importance of water, and in particular pore pressure, as a factor for mine planning and pit design, has come increasingly to the forefront over the last five to 10 years, with a commensurate increase in both the amount and quality of data collected. The wide acceptance of the grouted-in vibrating wire piezometer and the greater reliability of these instruments has also been a key factor in the improvement of both data quantity and data quality. However, at the current time, there are still only a limited number of mine sites with well instrumented pit slopes where the response to mining has been monitored over an extended period of time. In particular, relatively few monitoring examples could be sourced to illustrate the behaviour of hydromechanical coupling, and there are also few examples where pore pressures have been monitored within an active slope failure.

It was generally agreed at the LOP Sponsor Management Committee meetings that the North-west wall of the A154 pit at Diavik represents the best example where the pore pressures in a slope have been monitored prior to and during an active mining sequence. The supplementary case studies were also selected based on discussion with the LOP Sponsors regarding the quantity and quality of data available for providing a distribution of prevailing conditions in terms of climate, geology and mineral commodity type.

The data from the vibrating wire piezometers at Diavik were considered to be of excellent quality and sufficient to describe accurately the variation in pore pressures in space and time, including:

- the longer-term patterns of depressurisation (i.e. the overall depressurisation rate as the slope was being mined);
- the medium-term pressure trends (i.e. deviations from the long-term trend that may be observed on a weekly or monthly basis, or as a result of seasonally changing climatic variables);
- detailed short-term analysis of pore pressure fluctuations (i.e. responses to hydraulic or mechanical 'events').

In the development and testing of a sector-scale conceptual model, analysis of pore pressure trends is required at these different time scales. The detail of knowledge required for the fracture hydraulics, in space and time, will also depend on the scale that the pit slope is viewed at, and may be:

- bench-scale;
- inter-ramp-scale;
- overall pit slope-scale.

The conceptual hydrogeological model of the Diavik North-west wall was developed in a general sense by the mining operation during mining and monitoring of the slope. However, the analysis and research carried out as part of this current study has enabled a more detailed evaluation of the pore pressure behaviour which is applicable to other hydrogeological settings. The conceptual hydrogeological models for the supplementary case studies have mostly been developed by the operator both prior to and concurrently with this book. It is important that the bench and inter-ramp scale analysis considers the role of transient pore pressures which may develop over short time periods or seasonally.

3.2 Components of the conceptual model

3.2.1 Components of a larger scale conceptual model

At the regional and mine scales, the conceptual hydrogeological model must include consideration of the three 'regimes' (geology, hydrology, hydraulics) described in Section 1.2.1. These regimes include the following components.

3.2.1.1 Geology

- **The larger-scale geology**; with the focus on alluvial stratigraphical units, bedrock stratigraphical units, large-scale variations within those units, the relative location and interaction of those units, and the location and nature of regional geological structures. The geology will typically form the basis for determining the main hydrogeological units.

3.2.1.2 Hydrology

- **Natural recharge to the groundwater system**, which may be derived from net infiltration of areal rainfall ('blanket recharge'), but is more commonly derived from 'concentrated water' in the form of runoff channels, temporary ponding of water, losses from ephemeral or perennial streams, rivers and lakes ('losing reaches'), and the melting of accumulated snow (for example sheltered snow banks on the lee side of topographic features).
- **Artificial recharge to the groundwater system**, which is often an important consideration around mining districts and may include leakage losses from man-made facilities (including mine site facilities), losses from nearby agricultural irrigation systems, or the melting of accumulated snow on the lee side of man-made features. It may be possible to alter the pattern of artificial recharge to a greater extent than the natural recharge sources can be changed.
- **Discharge from the groundwater system**, which will usually consist of 'passive' hydrological outflows which may occur due to evaporative losses where the perched or main water table occurs at a shallow depth; groundwater flow into streams, rivers or lakes ('gaining reaches'); or 'active' hydrological outflows that may be added in the form of pumping wells or other forms of groundwater extraction.
- **Groundwater-surface water interactions** that may occur within the overall domain of the conceptual model (for example gaining and losing stream segments).

3.2.1.3 Hydraulics

- **The hydraulic properties of the main geological units**, which will help to determine when units can be 'lumped' together, or when they must be separated or sub-divided. At a gross scale, it is important to define both the vertical and lateral (directional) permeability of the units. For example, strong hydraulic layering of

alluvial sequences is common because of the nature of their geological deposition, and many regional geological structures exhibit enhanced flow in a direction parallel to strike and reduced flow in a direction perpendicular to strike (Section 1.1.2.3).

- **The groundwater flow paths** between the recharge sources and the discharge locations, which are controlled by the permeability of each of the hydrogeological units, and the amount of flow that can occur between the units. Another important consideration is whether the hydraulic properties of individual units may have become altered by previous drilling programs that have occurred within the model domain, and this may also be a factor enhancing the cross-connection and flow between certain identified units.

3.2.1.4 Influence of the mining operation

- **The key features that disrupt the natural groundwater flow system**, of which the presence of an open pit is usually the main consideration.

The larger scale conceptual model must provide a sufficient level of detail to broadly categorise each of the above factors, to identify which of the factors are important within the conceptual model domain, and to provide a sufficient level of detail for those important factors such that the implications for the mining operation can be identified and addressed.

3.2.2 The 'A-B-C-D' concept of fracture flow

Chapter 6 in the reference book *Guidelines for Open Pit Slope Design* (Read & Stacey, 2009) provides guidance on the development of conceptual and numerical pore-pressure models for pit slope evaluation purposes. Discussion of the influence of fracturing on groundwater flow is also presented, along with a representation of how flow may occur in a fractured rock slope based on the 'A-B-C-D' concept of fracture flow (Figure 3.2), which is linked to the principle of a size-order relationship within the fracture network.

A typical fracture network might comprise a few highly transmissive and pervasive 'first-order' fractures (A), many more less transmissive second-order fractures (B) and so on for third- (C) and fourth- (D) order fractures. In some more porous rock settings or rock types containing a pervasive network of micro-fractures, 'D' may also be considered as the rock matrix. The permeability of the fracture network is dependent on the scale of this distribution.

The more permeable 'A' fractures respond first to depressurisation by drainage. An interconnected 'plumbing' system of these first-order fractures allows rapid propagation of pressure changes through the system.

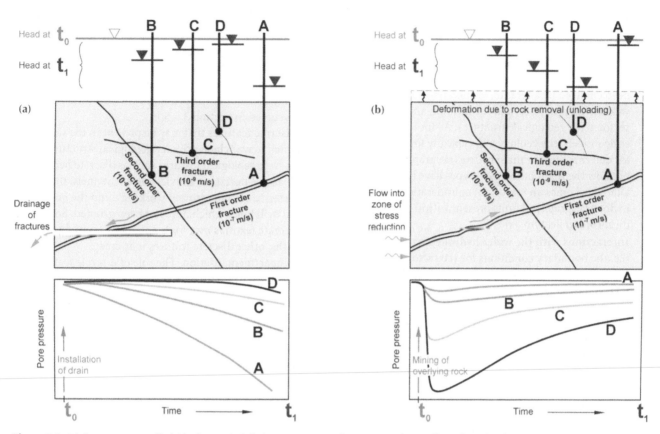

Figure 3.2: (a) Fracture-controlled (dual-porosity) drainage response due to groundwater flow; (b) unloading response in a fractured rock mass (Read & Stacey, 2009)

Typically, the greater the number of interconnected first-order fractures, the easier the rock will drain. As pressures within the first-order fractures begin to decrease, flow along the second-order, less permeable fracture sets begins to occur towards the first-order fractures. This dual-porosity response similarly occurs from third-order fracture into second-order fractures, and so on through the entire interconnected network of fractures.

Although the 'A-B-C-D' concept is simple and theoretically robust, further review of pore pressure responses in fractured-rock mining environments was required to verify its suitability at the scale of measurement required in pit slope design studies, and particularly with regard to the time dependency of the pressure responses in the lower order fracture sets and/or the matrix porosity. At all but the smallest scales, variations in the permeability of fractures over their length provide further complications to the pattern of groundwater flow. The findings of the review are presented in Section 3.3.

For many slope stability analyses, a current underlying assumption is that the effective stress principle applies across the full range of the fracture network. However, the influence of pore pressure and

effective stress may clearly be different for a brittle failure along a major structure than for a rock mass failure in pervasively weak materials. Consequently, it is important that the sector-scale conceptual hydrogeological model makes allowance for:

- the pore pressure differences that may occur between the large (A-type) fractures through to micro-scale (D-type) fractures;
- the scale of the analysis, for example whether the 'A-B-C-D' principle occurs at the scale of several hundreds of metres (potentially equivalent to the overall slope-scale), several tens of metres (potentially equivalent to the inter-ramp scale) or several metres (potentially equivalent to the bench-scale);
- the transient nature of the drainage process and the time-dependency of the pore pressure differences (which may potentially range from minutes to years).

3.2.3 Components of the sector-scale conceptual model

The 'A-B-C-D' concept is focused on the hydraulic properties only. As such, it is not a true conceptual model. For the sector-scale model, it must be supplemented with the following factors that consider also the regimes of the geology and the general hydrology.

1. **The sector-scale geology**, with the focus on lithology, how the lithological properties may themselves have been changed by alteration, mineralisation and weathering processes, and the location and nature of sector-scale geological structures. As for the larger-scale models, the geology will typically form the basis for determining the main hydrostratigraphical units, but how the main lithological types have been locally altered becomes an important consideration. The hydrostratigraphical units present within a pit slope are discussed in Section 3.6.2.

2. **Interactions with the wider hydrogeological system** (i.e. the boundary conditions for the sector-scale model), which is a fundamental consideration because it is seen that the majority of pit slopes form part of the wider hydrogeological setting.

3. **Recharge** to the slope domain itself, which includes (i) recharge to the slope from precipitation and local-scale runoff, (ii) recharge to the slope from adjacent surface water bodies, (iii) recharge to the slope from adjacent hydrogeological units, and (iv) recharge to the slope from leaky site facilities (e.g. pipes, tanks, ponds, drainage ditches). Knowledge of the recharge and how this may vary spatially, with elevation and with time, is an essential part of the understanding. Recharge is discussed in Section 3.6.3.

4. **The discharge** mechanisms for water in the slope, which are often related to the primary flow pathways, and may be 'passive' as a result of seepage to the pit face (or to nearby underground workings) or 'active' as a result of installed depressurisation measures. In rare cases there may be 'passive' outflow to other groundwater units outside the slope domain but, because most active pits are hydrological sinks, this is not common and mostly occurs when underground workings underlie the open pit. The role of discharge on slope depressurisation is discussed in Section 3.6.4.

5. **Changes to the hydrogeological system that may occur with time**, such as changes in the recharge patterns due to the evolving nature of the mine site or changes to rock properties that may result from the mining process. These may also include changes in water quality through time, for example due to exposure of pyritic rock, which may cause a deterioration and alter the physical properties of the rock mass, and will also need to be addressed in terms of site water management.

6. **The hydraulic properties** of the defined hydrogeological units and changes to the hydraulic properties that may occur in space and time (as the slope is excavated) as a result of deformation of the

materials, or by blast-induced changes. The hydraulics of the slope materials are discussed in Section 3.6.5.

7. **The primary flow pathways** within the slope, and their interconnections, and how they may be influenced by the presence of discrete features. The influence of the primary flow pathways is discussed in Section 3.6.5.1.

8. **Discrete features** that may be present in the slope, which may include primary geological structures, other discrete geological features such as discrete bedding or palaeosol horizons, alluvial palaeochannels, the presence of underground workings, and the presence of old drill holes which have not been grouted. Some discrete features may form enhanced groundwater flow paths; other discrete features may create compartmentalisation. The role of discrete features is discussed in Section 3.6.2.7.

In general, the understanding of the sector-scale conceptual model will be greatly enhanced if there is information on the response of the system to changes in hydraulic stress (recharge or discharge).

It is important that the sector-scale conceptual model considers the role of transient pore pressures (short-term pressures and pressure changes) that may occur anywhere in the slope but are often of most concern at shallow levels (and often within the over-break zone), particularly in high rainfall and/or seasonal climatic regions (Section 3.6.7). Many slope failures worldwide occur at relatively shallow depths behind the slope. As the quality of piezometer data continues to improve, it is becoming increasingly apparent that shallow failures are often associated with short-term (transient) pore pressure changes.

3.3 Research outcomes from Diavik

3.3.1 Background

One of the goals of the LOP project was to carry out a thorough analysis and interpretation of the monitoring data from an active pit slope to help answer the question 'how does a fractured rock slope depressurise'. Of the LOP Sponsor sites that were considered for this analysis, the North-west wall of the Diavik A154 pit was selected as the most suitable for the following reasons.

- It included the most closely spaced piezometer installations of any available site, with an excellent monitoring record over a period of up to five years.
- The data collected are of high quality and have been verified by previous QA programs.
- The data are influenced by a large number of hydrogeological responses, which include: (i) overall slope-scale drainage, (ii) short- and long-term testing of pumping wells, (iii) short- and long-term testing of

vertical and angled gravity drains, (iv) numerous blasting events, and (v) the seasonal freeze-thaw cycle.

- There had been an extensive program of bench face and downhole fracture mapping and analysis, with a partially developed DFN model.

The key findings of the Diavik study are summarised in this section. Further details are described in Appendix 3.

3.3.2 Diavik site setting

3.3.2.1 General site setting

Diavik is located adjacent to Lac de Gras, 300 km north-east of Yellowknife, in the Northwest Territories of Canada. The A154N and A154S diamond pipes are accessed through the A154 open pit and associated underground workings. Figure 3.3 shows that Lac de Gras encircles the north-east, north and west sides of the A154 pit. The A418 pit is located to the south-east.

The average annual temperature at Diavik is –12°C, and the A154 pit shows permafrost conditions in its South wall. The northern half of the pit is outside the influence of permafrost because of the proximity of Lac de Gras, but seasonal freezing does occur at the surface and on the pit slopes. Mean annual precipitation at Diavik is less than 300 mm and does not provide any significant recharge to the pit slopes.

The layout of the Diavik site is shown in Figure 3.4. Mining occurs in competent granitic country rocks with Rock Quality Designation (RQD) of 95 per cent (Kutchling *et al.*, 2000). Sub-horizontal metasedimentary rafts occur through the granite rock mass. These may vary from a few centimetres to several metres in thickness. Site-wide, the main hydrogeological feature at Diavik is a prominent NE–SW trending structural zone known as Dewey's Fault zone. This bisects the A154 pit. The A154N and A154S pipes also occur along this feature. Dewey's Fault zone is a zone of significant fracture development and enhanced permeability.

Due to the close proximity of the lake (elevation 416 masl), water retention dykes have been constructed along the north-east, north and west margins of the A154 and A418 pits. The crest of the North-west wall is about 130 m inside the water retention dyke. Before selecting the site for the current study, it was recognised that most piezometer responses showed a 'damped' effect because of the proximity of the recharge source. Prior to development of the underground mine, dewatering of the A154 pit was mostly achieved by pumping from sumps excavated in the lower pit benches and at the pit floor close to Dewey's Fault zone (Figure 3.4). Since late 2009, with the development of the decline beneath the A154 pit, an increasing amount of the water has been intersected underground.

Figure 3.3: Air photograph of the Diavik Mine site (looking East)

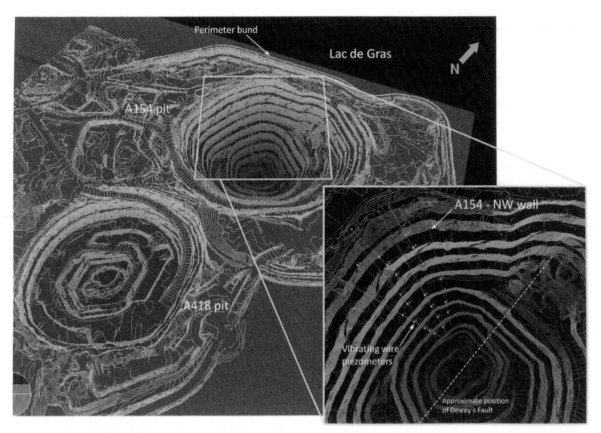

Figure 3.4: The A154 Pit NW wall at Diavik

3.3.2.2 Main fracture orientations

Fracture traces from bench mapping were used in the development of the Diavik DFN (Appendix 3). There are three principal fracture orientations.

1. Sub-vertical fractures sub-parallel to Dewey's Fault zone (NE–SW).
2. A sub-vertical fracture set oriented WNW–ESE.
3. A single sub-horizontal fracture set.

3.3.2.3 Observations during dewatering

The average dewatering rate during the period of the study was approximately 700 l/s. As the A154 pit was progressively deepened, it was noted that pore pressures in parts of the North-west wall were decreasing at a lesser rate than the mine sinking rate. Consequently, heads in the middle to lower slopes of the NW wall became artesian, with up to 80 m of head above the level of the slope.

A depressurisation program was implemented for the North-west wall because of the high artesian heads.

However, compared with the dewatering rate from the pit floor sumps, the achievable flow from the horizontal and angled drains and pumping wells installed in the slopes was very low. The total dewatering rate for the in-pit wells collared on the 280 m bench (DW01-05, DW07, DW09, DW10, DW29 & DW30) was between 5 and 12 l/s

in 2008, which was around 5–8 per cent of the total dewatering rate for the A154 pit in 2008. The total combined flow rate from the 280 m bench wells was steady at around 10 l/s during 2009. An important event in the North-west wall dewatering program was the initiation of dewatering from the underground drains illustrated in Figure 3.5.

3.3.2.4 Pore pressure monitoring

Pore pressures in the North-west wall were monitored by an array of multi-level, grouted-in vibrating wire piezometers (VWPs), with up to four grouted-in VWP sensors in each hole providing discrete pressure measurements at their position behind the pit face (Figures 3.6 and 3.7).

The VWP sensors were assigned the letter A, B, C or D, depending on their position within each piezometer string (increasing in depth from A to D). It is expected that, prior to mining, the head at the location of each of the installed VWP sensors would have been 416 masl (the elevation of the lake). Therefore, any downward departure from 416 masl head is assumed to represent a change caused by the mining operation. Example hydrographs of daily mean head are shown in Figure 3.8 for piezometers A154-MS-08, A154-MS-11 and NWPT-M-02. Also shown are certain 'timeline events' such as the installation of drains,

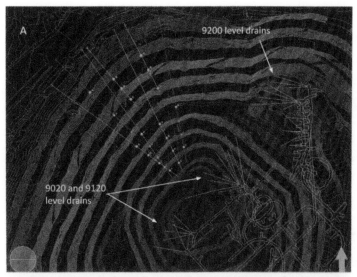

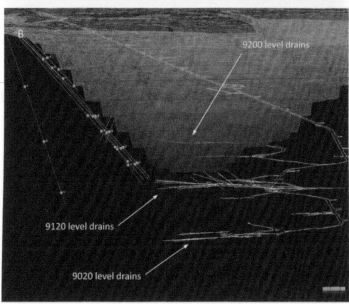

Figure 3.5: Diavik: Underground dewatering on 9200, 9120 and 9020 levels in plan view (A) and looking NNE through the NW wall in 3D (B)

activation of pumping wells, and the changing elevation at the toe of the North-west wall to demonstrate the rate of pit advance. A review of the entire long-term VWP dataset reveals the following.

- The rate of depressurisation in many of the piezometers, and variations in the rate, is a general reflection of the rate of advance of the pit.

- The increasing head difference between shallow and deep piezometers reflects the increasing downward vertical hydraulic gradient (head gradient) that progressively developed as the pit was deepened.

- Superimposed on the general trend of gradual depressurisation are a series of much sharper pressure fluctuations, which are present as either (i) short-

duration spikes (increasing or decreasing head), which quickly recover, or (ii) larger permanent displacements in pressure, which permanently affect the long-term depressurisation trend.

- The pore pressures exhibited in many of the deeper piezometers reflect heads which are above the elevation of the adjacent pit face. The heads only dropped below the elevation of the bench above following installation of the underground drains.

In general, the long-term VWP data indicate that depressurisation is mostly occurring through a process of passive drainage towards the pit floor. Localised active depressurisation measures (such as the installation of drains and pumping wells within the face) did not have a

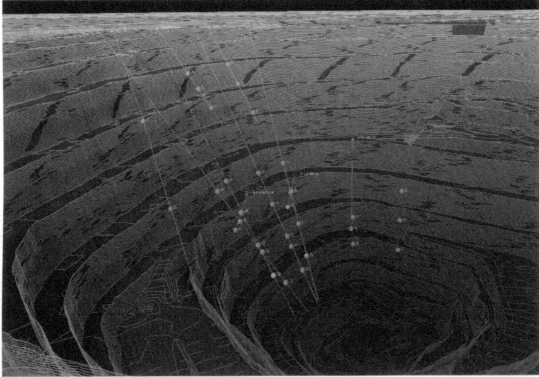

Figure 3.6: Diavik NW wall: Location of VWP arrays

significant effect on the rate of depressurisation. Shallow VWPs, which are further from the discharge area at the pit floor, but closer to the source of recharge from Lac de Gras, show much slower depressurisation (e.g. NWPT-M-02), whilst the deep VWPs show a depressurisation rate that more reflects (but was slower than) the rate of pit floor advance. The steepening of the vertical hydraulic gradients measured in each piezometer is indicative of a hydrogeological system that is controlled by strong hydraulic conditions at its boundaries (i.e. recharge from

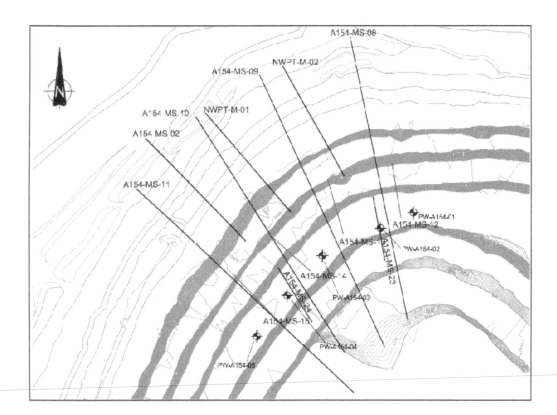

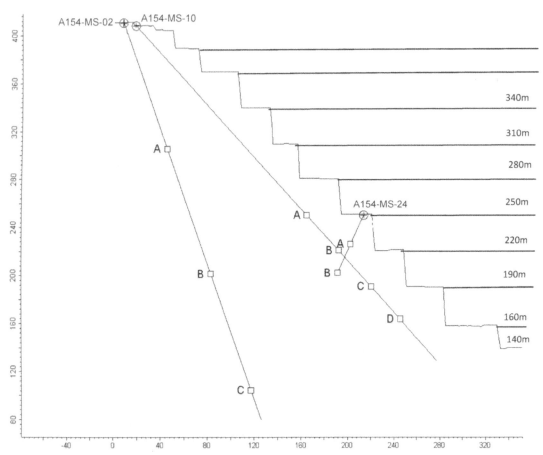

Figure 3.7: Diavik NW wall: Plan and cross-section showing VWP positions

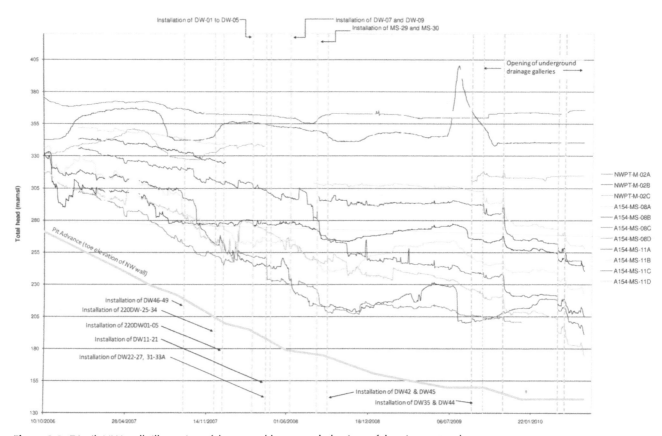

Figure 3.8: Diavik NW wall: Illustration of the general long-term behaviour of the piezometer data

Lac de Gras and discharge to the pit floor). Water from the recharge source is available to enter the fracture system behind the pit wall at a greater rate than the water can be transmitted downward along its flow pathway of interconnected fractures.

To illustrate the relationship between VWP depth and rate of depressurisation, the mean rate of depressurisation is plotted against VWP elevation in Figure 3.9. With the exception of A154-MS-02 (which is farther behind the pit face), the data show a strong correlation between sensor elevation and rate of depressurisation. There is a linear relationship between distance from the lake and the long-term mean rate of depressurisation (Figure 3.10). Pumping from the pit-floor sump reduces heads to the level of the pit floor, effectively reducing the hydrological 'base-level' for the pit slope as the pit is mined to progressively lower elevations.

The main control on the rate of depressurisation exhibited in any particular VWP is its relative position between the constant-head boundary of Lac de Gras and the changing hydrological base-level of the deepening pit floor. As illustrated in Appendix 3, the distribution of hydraulic heads was easily reproduced in a 2D numerical simulation by assigning constant-head boundary conditions to simulate Lac de Gras (recharge boundary) and the changing pit floor elevation (discharge boundary).

3.3.2.5 Monitoring of specific events

Within the 'long-term' piezometric response caused by progressive lowering of the pit and pumping of water from the pit floor, there are many 'medium-term' (on the scale of days to weeks), and 'short-term' (responding in minutes to hours) fluctuations in pore pressure.

'Short-term events' can be due to either hydraulic or mechanical stresses, as follows.

- Blasting.
- Drilling for the installation of drains or wells.
- Hydraulic testing of wells via airlifting or longer-term pumping tests.
- Power outages, resulting in the short-term interruption of pumping.
- Opening of relatively low-flow drains in the pit face.
- Opening of higher-flow drains in the underground workings.

'Medium-term events' are brought about through:

- large magnitude responses with slow (or no) recovery;
- compound effect of successive 'events';
- seasonal freezing and thawing of the slope.

In total, 37 events were examined as part of the current research. The following sections summarise the various

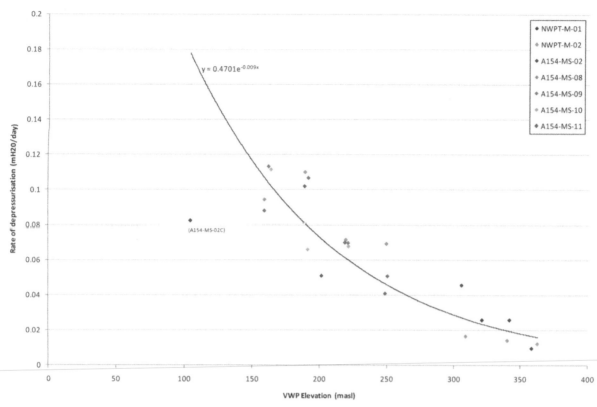

Figure 3.9: Diavik NW wall: Relationship between the rate of depressurisation and VWP elevation

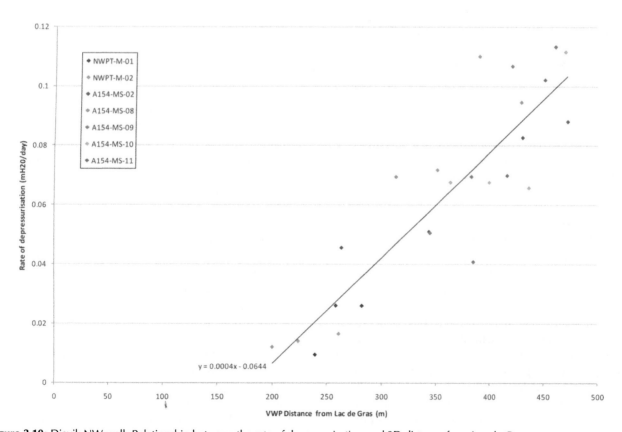

Figure 3.10: Diavik NW wall: Relationship between the rate of depressurisation and 3D distance from Lac de Gras

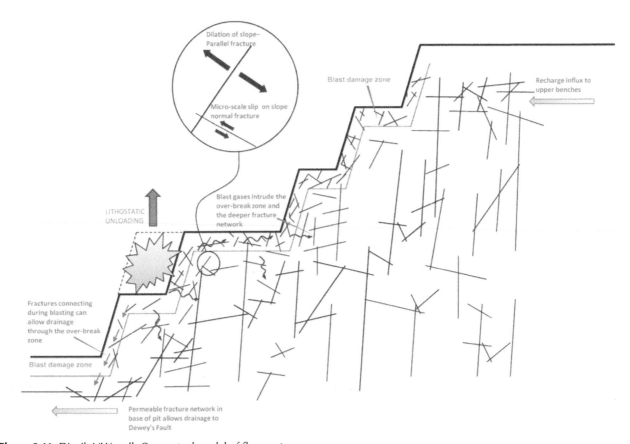

Figure 3.11: Diavik NW wall: Conceptual model of flow system

influences on pore pressure that are evident from the hydrograph analysis and present an interpretation of some of the processes involved. A schematic drawing illustrating the flow system for the North-west wall is included as Figure 3.11.

The hydrograph in Figure 3.12 is a good example of the different types of pressure responses which are observed due to blasting. The types of response are generally consistent and are summarised in the following sections. However, individual VWPs can show different responses to consecutive blast events.

3.3.3   Effects of blasting

3.3.3.1   Pressure pulses due to the intrusion of gases released from blasting

Gases released from blasting may send a pressure pulse that moves through the interconnected fracture network and causes a 'hammer' effect on pore pressures. Such spikes in pressure are frequent in the Diavik data set. Generally, the increases in pressure gradually dissipate over a number of days, depending on the scale of the displacement and many of them show a response similar to that which may be typically observed during a slug test (Section 2.2.5.3).

VWPs tend to show the largest pressure spikes from blasting as the pit advances past (below) their position in the

wall. The largest observed pressure spike in the Diavik data set is in piezometer MS10D on 15 February 2008. However, all of the near-face VWPs show responses to blasting at all levels, indicating a generally good interconnectivity in the fracture network. It is postulated that a 'fluid-to-solid' effect may occur as a result of the pressure pulse due to blasting. The blast-induced water hammer intruding the fracture network could potentially result in poro-elastic dilation (opening) of fractures. Following the blast, this dilation of fractures could subsequently cause a temporary or permanent increase in fracture interconnectivity, allowing pressure dissipation in some of the interconnected fracture sets. However, the data from Diavik suggest the rate of pressure dissipation is variable.

Although pressure increases that result from blasting do not typically affect long-term trends, 'hydraulic jacking' can occur, when successive blasts cause step-wise increases in pore pressures. At times, this has significantly retarded the progress of passive depressurisation in localised areas of the slope.

Dilation of fractures from blasting-induced pressure pulses could cause permanent changes to the fracture network, as small movements between fracture walls may create offsets in asperities, which can permanently increase the fracture aperture by preventing the walls of

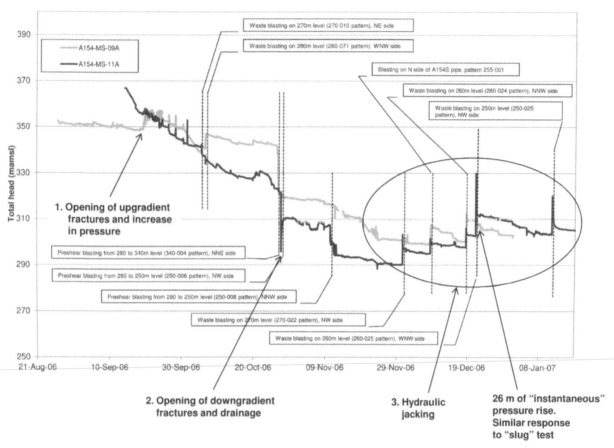

Figure 3.12: Diavik NW wall: Events for 'A' group VWPs: September 2006 to January 2007

the fracture from exactly fitting back together. The pressure response which accompanies such movement would be evident as a rise in pressure following the blast, followed by recovery to a permanently lower pressure, provided that the increase in fracture aperture is associated with improved connection to the downgradient drainage pathway.

3.3.3.2 Blast-induced fracturing

Blast-induced fracturing may be caused by the pressure pulse created by intruding gases, or by the blast shock wave itself. Whilst fracturing from the shock wave might occur only relatively close to the blast position in the face, intruding gases may potentially cause induced fracturing and deformation deeper within the interconnected fracture network.

The permeability of the fracture network will permanently increase in the affected areas, most typically within the near-face (over-break) zone. The thickness of the over-break zone behind the face is dependent on the type and scale of blasting. The controlled blasting carried out at Diavik probably results in a relatively narrow zone of blast damage (<10 m). Improvements to fracture interconnectivity will occur within this zone, but may also

extend further out into the rock mass as a result of the pressure pulse. Either case can cause improved drainage of the rock mass at greater distances behind the wall.

Numerous drainage-type responses are evident in the Diavik data following blasting. The apparent drainage 'recession curve' represents in part the re-equilibration of pressures with the more transmissive seepage pathway created from blast-induced fracturing (either by opening of new fractures and/or by dilation of existing fractures).

Blast 160-018 (Figure 3.13) caused significant depressurisation in several deep VWPs. The blast was located in the South wall of the pit, at an elevation of 155 m amsl. This position, away from the North-west wall, makes this event particularly interesting, as it appears to show the influence of a well-connected 'plumbing' system of permeable fractures at the base of the pit, through which a drainage response can propagate into the North-west wall. The most plausible explanation is that either the shock wave or the pressure pulse created by the blast caused an enhancement of one or more discrete fracture connections between the already-depressurised fracture zones surrounding Dewey's Fault zone and connected to the lower pit floor, and the location of the deep North-west wall VWPs.

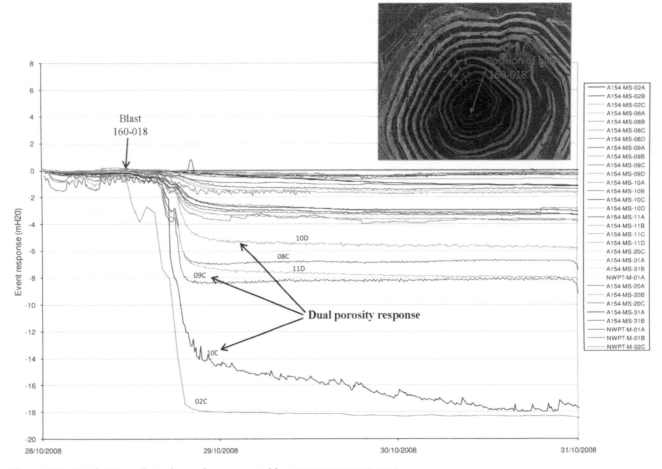

Figure 3.13: Diavik NW wall: Hydrograph response to blast 160-018 (28/10/2008)

In general, the magnitude of the drainage response due to blasting is indicative of the level of developed interconnectivity. With reference to the 'A-B-C-D' conceptual model, a small drainage response might occur following connection of two 'C' (third-order) fractures, whilst a hydraulic connection developing between two previously unconnected 'B' (second-order) fractures may result in prolonged and significant drainage and much lower equilibrium pressure conditions. The large, long-term drainage responses evident in 10C, for example, may be the result of blast-induced fracturing connecting larger aperture 'B' type fractures.

In several instances, blast-induced fracturing has caused a sustained rise in pore pressure, which is thought to result from the opening up fracture pathways that improve the hydraulic connection between the upgradient recharge source and the piezometer location. This may occur, for example, when the piezometer is located at an elevation below that of the blast. Furthermore, because recharge pathways may typically be more distant from the face, the opening up of connections to upgradient recharge may be indicative of a pressure pulse moving further into the wall.

3.3.3.3 Blast-induced deformation

In addition to the rapid opening up of new flow pathways, the sharp, instantaneous decreases in pressure that may follow a blast could be partially related to changes in porosity caused by fracture deformation. The dilation of a fracture can potentially occur from lithostatic unloading and stress relief through blasting (rapid) and the subsequent removal of the shot muck (gradual).

Blasting may initiate the release of this stress through fracture-normal opening or fracture-parallel slip. Blast gas intrusion that creates pressure pulses through the fracture network could conceivably cause small-scale slip as fracture contact area (and internal shear strength) is reduced during the short-term poro-elastic dilation of fractures. Dilation of fractures could also occur where fractures are oriented normal to tensile stresses in the pit face. Such stresses are known to develop close to the toe of high pit walls (Stacey *et al.*, 2003). However, in the case of Diavik, permanent changes in piezometer levels due to dilation of fractures may be rare because the ongoing movement of recharge from upslope quickly equalises any pressure drop. The observed decreases in pressure are

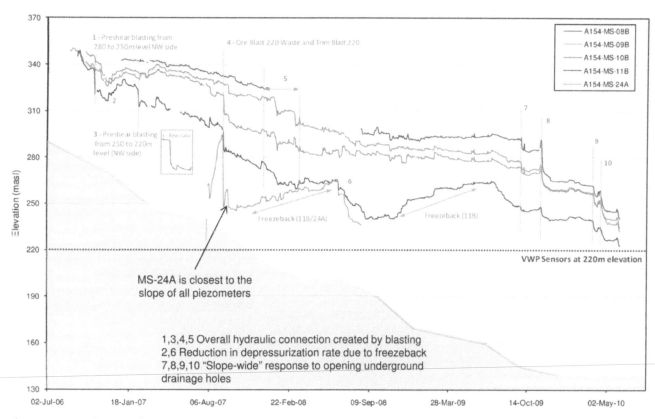

Figure 3.14: Diavik NW wall: Hydrographs for 'D' group VWPs

more likely to be caused by the opening of new flow pathways during blasting and connection to downgradient zones of lower head (Figure 3.14).

3.3.3.4 Discussion

Rapid pore pressure increases are thought to be the result of intrusion of high pressure gases into the fracture network. Four piezometric responses can be recognised from the dataset. These are shown in Figure 3.15.

- Pore pressure increases (A on Figure 3.15) or decreases (D) due to deformation of the rock mass and/or fracture network following blasting events.
- Pore pressure increases (B) as a result of blast damage and/or deformation causing an increased connection between the *recharge source* and the piezometer.
- Pore pressure decreases (C) as a result of blast damage and/or deformation causing an increased connection between the piezometer and the *discharge point*.

Rock mass deformation resulting from blasting and mining typically causes an increase in fracture aperture, and hence an opening of flow pathways. When blasting leads to removal of total stress, the dilation of fractures can cause a rapid 'direct' hydromechanical response (e.g. lithostatic unloading), with the resulting increase in fracture

porosity causing a lowering of pore pressure. The dilation of fractures may also create more permeable pathways.

In relatively permeable fracture networks, such as at Diavik, the 'direct' hydromechanical response is generally short-lived (and is mostly not noticeable in the Diavik data). However, the improved transmissivity of flow pathways can cause pore pressures to equilibrate to lower levels.

The pressure pulse from blasting that moves through the fracture network may be exaggerated at Diavik compared with other sites because of the ongoing recharge moving down from above (in effect, the water hammer effect is greater because of the two opposing hydraulic stresses). This is supported by the observation that the pressure rise tends to be greater in piezometers located above the level of the blast (i.e. between the blast and the recharge source).

3.3.4 Influence of freeze-back

Ambient air temperatures at Diavik typically fall below freezing in early October and rise above freezing at the beginning of May. 'Freeze-back' occurs when pore water freezes in the pit face and floor, causing a reduction in both seepage at the face and drainage within the over-break zone. The frozen pore water creates a hydraulic barrier at the pit face. Groundwater heads behind the face increase as

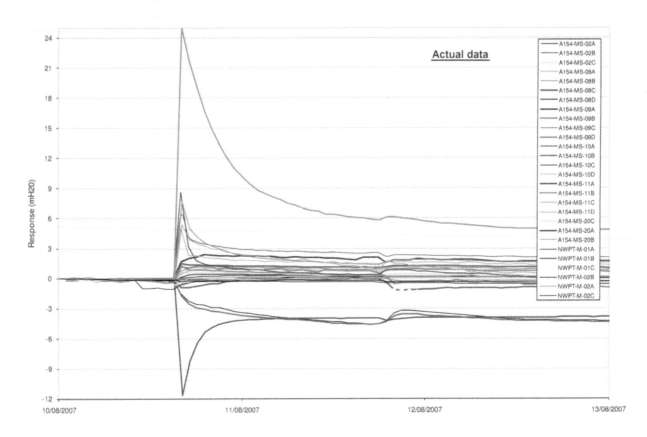

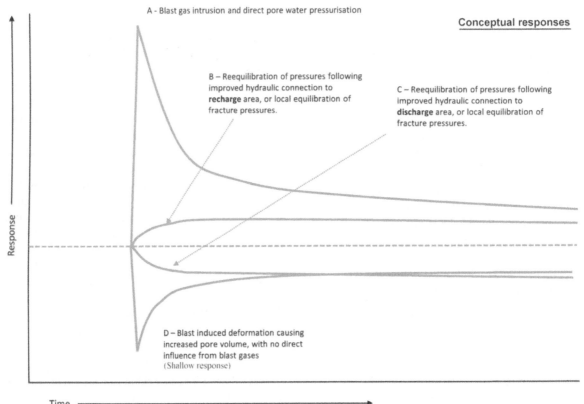

Figure 3.15: Diavik NW wall: Example blast response to waste blasting on the 230 m level (230-024 pattern)

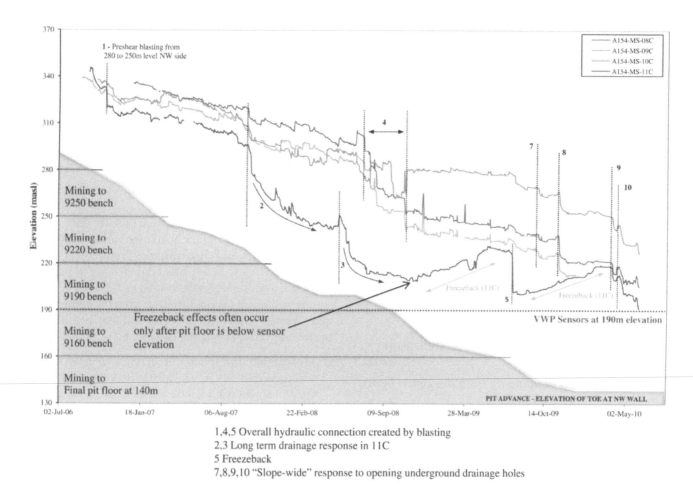

1,4,5 Overall hydraulic connection created by blasting
2,3 Long term drainage response in 11C
5 Freezeback
7,8,9,10 "Slope-wide" response to opening underground drainage holes

Figure 3.16: Diavik NW wall: Hydrographs for 'C' group VWPs

a result of the 'backing up' of continued recharge from Lac de Gras (Figure 3.16). In many of the piezometers, the freeze-back cycles are evident as a transient (seasonal) reduction in the rate of depressurisation.

In order to significantly affect the permeability of these zones, it is probable that freezing must occur in a 'skin' about five to six metres thick behind the pit face. There is thus some delay between the time that ambient air temperatures drop below zero and the increasing pore pressure measured behind the pit face. The rock mass and flow of water will also provide an element of thermal insulation and heat capacity which might delay the onset of freezing (and spring break-up) to some extent.

There is also a lag time associated with heads 'backing up' against the pit face. This is seen as a delay between the onset of freeze-back in deeper piezometers (close to the pit floor and the point of discharge) and those higher up the pit face.

Freeze-back is generally only evident in the VWPs once the pit (i.e. the point of discharge) has advanced to an elevation below the VWP, so that there is an exposed face below the VWP position, although there are exceptions to

this (for example the freeze-back response recorded in piezometers 02B and 02C during the winters of 2005/6 and 2006/7, despite their position below the pit floor at that time).

At the time of spring break-up, there is a delay in the onset of depressurisation at higher elevations in the pit face (evident in multi-level VWPs of A154-MS-11). This is due to a combination of a thermal insulation effect and the natural geothermal gradient, both of which ensure the rock mass at depth is at a higher temperature, and is less susceptible to seasonal variations in temperature.

The freeze-back process draws parallels with the *in situ* hydromechanical coupling experiments at the Coaraze test site in south-east France (Section 3.6.6.3), in which the deformation of a limestone test slope was instrumented and measured for surface tilt and changes in fracture aperture during artificial flooding through the closure of a downstream 'water gate'. The closure of the water gate induces the same 'backing up' of pressures as freeze-back. At Coaraze, the increase in the water level was about 8 m. The experiments recorded deformation which resulted in a slight 'swelling' of the slope. Freeze-back at Diavik has caused an increase of up to 30 m of pressure head in some piezometers. Based on the Coaraze analogy (Section 3.6.6.3), this is likely to influence hydraulic properties behind the face, as increasing heads cause fracture dilation and increased fracture transmissivity.

The analogy to Coaraze suggests that freezing of pore water will cause fracture apertures behind the zone of freezing to increase. Some of the fractures may not exhibit full elastic recovery to their previous positions, depending on the fracture orientation, and on the type of infilling on the fracture surfaces. Therefore, it is considered probable that the 'backing up' effect caused by repeated seasonal freeze-thaw will enhance the permeability of fractures further behind the face, depending on the elastic response of the fractures. In addition, if the VWP temperature data are reliable in indicating gradually reducing temperatures in the rock mass (Appendix 3), a pit-face permafrost layer may ultimately develop, which would permanently alter the hydrogeology of the North-west wall.

3.3.5 Responses to changes in hydraulic stress

3.3.5.1 Initiation of pumping from the 280 m bench

On 8 March 2008, wells DW01-5 on the 280 m bench (Figure 3.17) were brought on line with an initial combined flow rate of 5.9 l/s. The main depressurisation occurred in the A and B group piezometers above an elevation of 200 m amsl, which was around the elevation of the pit floor at the time (Figure 3.18). The magnitude of the response in the individual piezometers appears to be governed by general proximity to the pumping source and also by the amount of previous depressurisation that has occurred through passive drainage. VWPs that had shown more prior depressurisation generally show less incremental depressurisation because of the pumping.

The rapid re-equilibration of pressures following the start-up of pumping, and the similar lengths of time required in different VWPs to re-establish hydraulic equilibrium, is indicative of high fracture permeability, and good interconnectivity in the fracture network. This happens very quickly (within one to two days), with a relatively small drop in pressure head, and shows that the rate of pumping from the dewatering wells was not able to match the diffuse groundwater flux moving from the Lac de Gras recharge area (i.e. groundwater can move *en masse* through the fracture network at a greater rate than it can flow locally towards the pumping wells).

3.3.5.2 Response to activation of MS-29 and MS-30

MS-29 is located on the 280 m bench. Pumping from the well started on 15 August 2008 (Figure 3.19). The response to pumping was localised and most of the response occurred in VWP8C and 9C (9 m to 10 m decrease in

head) with a smaller response in VWP8D (2.5 m decrease in head). Many of the other near-face VWPs showed responses of less than 1 m.

Pumping from MS-30 (also on the 280 m bench) commenced on 20 August 2008 (Figure 3.20). As with MS-29, the depressurisation was localised in the fracture network close to the dewatering well. Depressurisation was greatest in VWP8C (10 m), 9C (9 m), 8D (6.5 m) and 8B (3.5 m). A154-MS-08 is the closest piezometer in the North-west wall to MS-30, and the greatest amount of depressurisation occurred in VWP 08C, which is closest in elevation to the base of well MS-30.

The magnitude of responses from the two wells conforms to a distance-drawdown relationship (Figure 3.21). The theoretical (Theis) distance-drawdown relationship has also been plotted for reference to show the distribution of drawdown that would be expected in a porous medium at various distances from the pumping well. The data points show a degree of scatter, but the data can broadly be separated into two groups.

1. Low magnitude responses occurring away from the well (generally >40 m) to which a theoretical Theis curve is a reasonably good fit to the upper bound of the data.
   - The distance-drawdown relationship exhibited in the first group above supports a conceptual model in which the fracture network can be treated as an equivalent porous medium (EPM) on the slope scale.
2. High magnitude responses closer to the pumping well (generally <40 m) which exceed the Theis curve fit.
   - Data falling below the Theis curve probably reflect responses from VWPs that are installed in or close to permeable fractures that are intersected by the pumping well. These fractures are in effect extending the diameter of the pumping well as it is 'seen' by the formation; the head loss along a permeable fracture will be relatively small, so that the head in the fracture will be nearer to that in the pumping well than the head in the bulk of the formation at the same distance.
   - Therefore, within the large-scale framework of an EPM response, local-scale flow hydraulics may be controlled by discrete fracture flow patterns. The presence of 'high response' parts of the fracture network may indicate the breakdown of EPM flow behaviour moving closer to the source of pumping. Alternatively, it can simply be regarded as signifying the behaviour of a well with an extended effective radius, as discussed in Appendix 2 and illustrated in the semi-logarithmic distance-drawdown plot showing the concept of effective well radius shown in that appendix.

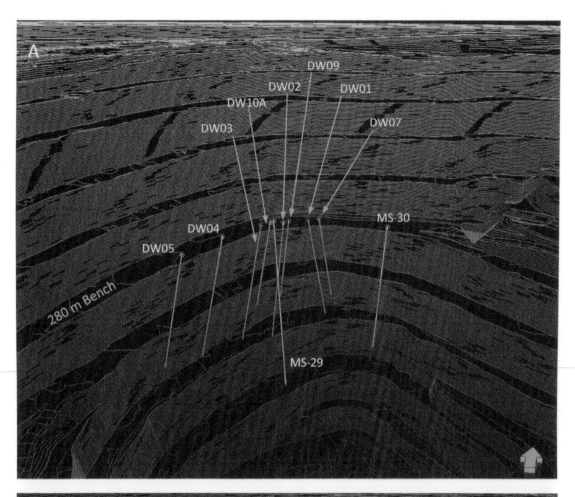

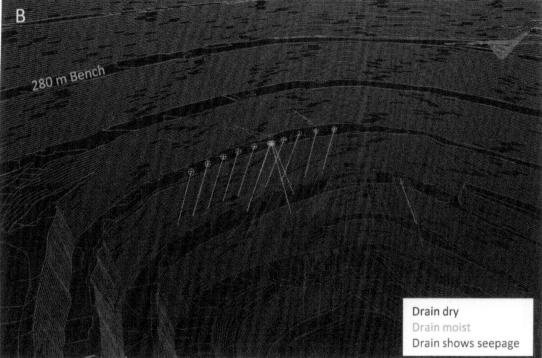

Figure 3.17: Diavik NW wall: Location of drains and wells

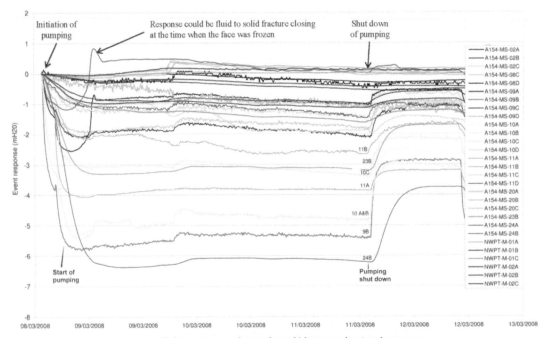

- All piezometers respond except those which are very close to recharge source
- Magnitude of response correlates to distance from pumping source
- Rapid re-equilibration of pressures (wthin 1 day) following addition of new stress
- Very little lagged ("D") response

Figure 3.18: Diavik NW wall: Hydrograph response to initiation of pumping of 280 m bench dewatering wells (08/03/2008)

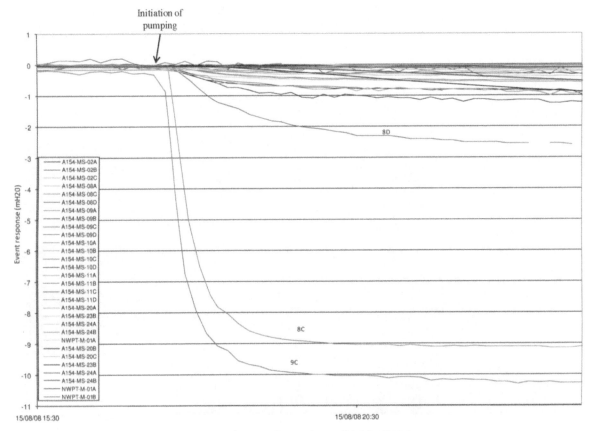

Figure 3.19: Diavik NW wall: Response to activation of MS-29 dewatering well (15/08/2008)

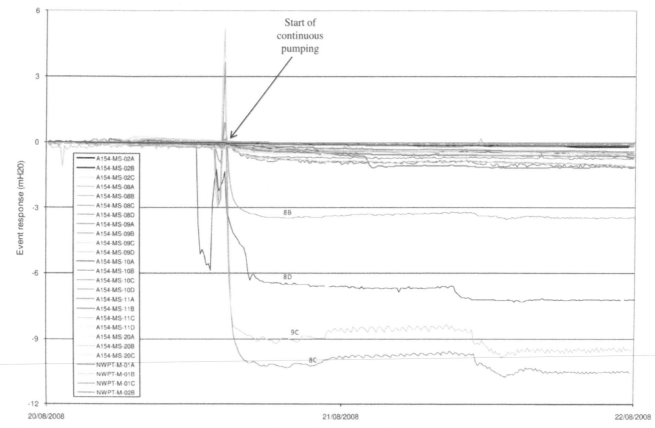

Figure 3.20: Diavik NW wall: Response to activation of MS-30 dewatering well (20/08/2008)

3.3.5.3 Response to opening of the underground drainage galleries

Figure 3.22 shows the pore pressure responses to a short-term underground flow test (28 September 2009). Figure 3.23 shows the longer-term depressurisation brought about through opening of the underground drains (September 2009 to June 2010). A rapid and more pervasive pressure response occurs following opening of the underground drainage galleries. The 'recession curve' response, which is characteristic of drainage, is actually the re-equilibration of pressures to the new, lower elevation discharge point.

The rapid equilibration of pressures is testament to the relatively high interconnectivity in the fracture network. The immediate onset of depressurisation following the opening of underground drains indicates the very high diffusivity of the fracture network. High fracture diffusivity (high transmissivity coupled with low storage) enables pressure responses to propagate very quickly through the fracture network. The main controlling factor in the magnitude and time of the response is the distance of the piezometer from the discharge source. The rapid response time is measured as a lag time of approximately one hour between the onset of the hydraulic response in VWPs of A154-MS-08 (closest to the nearest underground drain) and A154-MS-11 (furthest away) (Figures 3.22 and 3.23).

Figure 3.23 shows several phases of depressurisation associated with the opening of drains in different parts of the underground workings. In these hydrographs of event responses, sensors 02A and 02B show the greatest amount of total depressurisation. Unlike the case of the pumping wells, the combined yield of the underground drainage galleries is high relative to the rate of recharge, and is sufficient to cause widespread depressurisation. In the shallower fracture network, the resulting drainage flux exceeds the rate of recharge influx, and hence a drawdown occurs across the entire interconnected fracture system.

3.3.6 Overall interpretation of the Diavik results

The pressure responses observed in the Diavik A154 pit North-west wall support the concept of a fracture network which behaves as a hydraulically homogeneous body on the scale of the pit slope. It is postulated that the Diavik data set supports the concept that EPM flow conditions are valid at the 'slope-scale' and 'inter-ramp-scale', away from discrete points of pumping stress. At Diavik, it is considered that, if the recharge source were removed, 'free drainage' of the fracture system would occur throughout the majority of the slope.

A

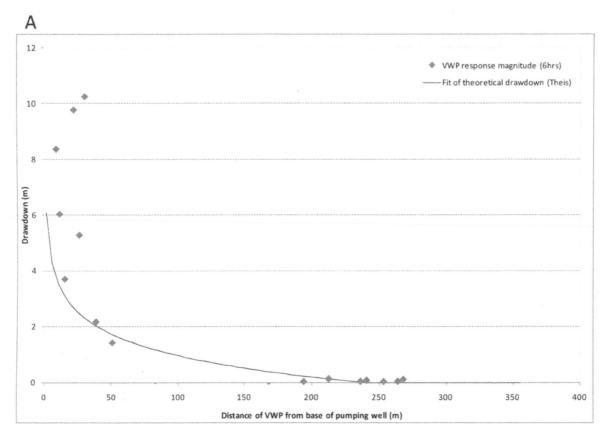

B

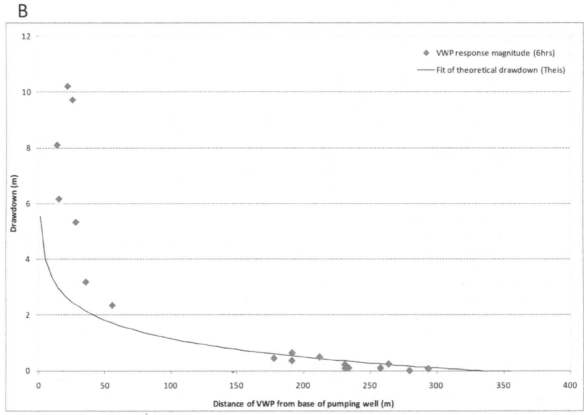

Figure 3.21: Diavik NW wall: Distance-drawdown relationships after six hours of pumping from MS-29 (A) and MS-30 (B)

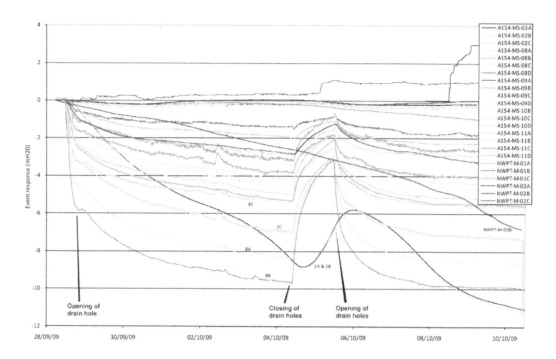

- All piezometers respond except the most shallow
- Magnitude of response correlates to distance from stress
- Rapid re-equilibration of pressures (within 6 days) following addition of new stress
- Lagged (but still fairly rapid) response in piezometers deeper behind slope

Figure 3.22: Diavik NW wall: Hydrograph response to the underground drainage gallery

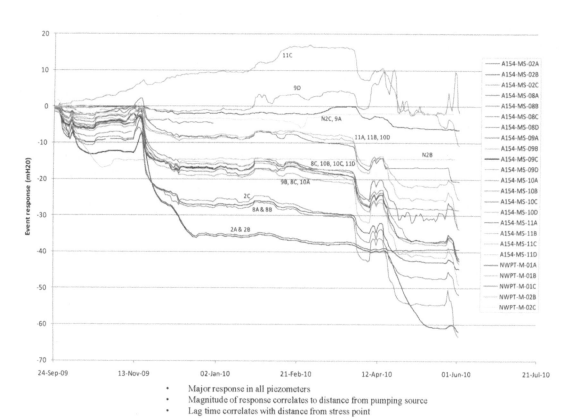

- Major response in all piezometers
- Magnitude of response correlates to distance from pumping source
- Lag time correlates with distance from stress point

Figure 3.23: Diavik NW wall: Hydrograph response to underground dewatering

Fracture flow characteristics are evident at a smaller scale, close to points of new pumping stress, and for a limited time period until pore pressures adjust to the new applied stress. This concept is illustrated in Figure 3.24. In the left hand diagram, immediately after drainage begins, the piezometers in Group 1 show relatively similar heads, with all heads significantly above the head at the new discharge point (the horizontal drain hole). In contrast, Group 2 communicates directly with the horizontal drain. Because the 'A' fracture is permeable, the head in that fracture is relatively similar to the (low) head in the drain. The head loss involved in the flow of water through the lower order fractures, and the resulting hydraulic gradient, mean that the head in the 'D' fracture will be higher than the head in the 'A' fracture.

There will be locally higher groundwater flow velocities in the discrete, more permeable 'A' fracture that is connected to, and directly feeds water into, the drain hole. The faster flow leads to an increase in kinetic energy at the expense of potential energy and therefore a further reduction in fluid pressure and static head. If, for example, flow speed is 1 m/s, head loss due to the conversion of potential energy to kinetic energy (and therefore decrease in static head) would be only 0.05 m; if 5 m/s, it would be 1.25 m. The hydraulic gradient will be further steepened if the higher flow speed also leads to a loss of total energy because of frictional losses.

The hydraulic gradient steepens as the water within the fractures moves closer to the drainage location, and as the groundwater velocities within the fractures get progressively higher due to the reducing radial area of flow. A similar transition between EPM and fracture flow conditions around point sources of pumping is likely to be a feature of many pit slopes excavated in fractured rock. The transition from fracture flow conditions to EPM conditions is likely to be gradual and, in the case of Diavik, appears to occur at less than 40 m distant from the new pumping centre (depending on the magnitude of the applied stress).

The discrete fracture flow condition close to a new point of hydraulic stress is transient following the application of the new stress point. In the right hand diagram of Figure 3.24, the general water table has fallen and the heads in the 'D' fractures have adjusted to the lower head in the 'A' fracture that is connected directly to the drain hole. Locally, an approximate equilibrium condition has developed following application of the new point of drainage stress. At Diavik, it appears that equilibrium to points of new hydraulic stress occurs within less than a week, following which time the EPM-type flow behaviour becomes re-established.

During the drilling program for the North-west wall, it was not possible to identify discrete fractures with continuous large apertures that would allow withdrawal of more than a few litres per second. The yield from individual or combined fractures encountered within wells and drain holes was not sufficient to overcome the diffuse EPM flow that was moving through the fracture network as a whole. Therefore, no significant depressurisation was achieved by the localised well and drain hole installations.

Thus, it can be postulated that the Diavik fracture network falls into a relatively tight range of middle-order fractures, for example mostly 'B'- and 'C'-type fractures. Water from the recharge zone can readily move diffusely through these interconnected fractures *en masse*, but no discrete high-aperture fractures were present that would allow water to be removed from a point source at a sufficient flow rate to create a large amount of drawdown. The combined effect of the available point sources was not sufficient to overcome the *en masse* flow. As a result, a significant amount of depressurisation only occurred when a new discharge condition created a high discharge flux in comparison with the recharge flux (which is the overriding control). The underground drainage galleries yielded the necessary flow rate to achieve this.

The observed responses to hydraulic stresses are generally very rapid, indicating that the fracture network as a whole has a high diffusivity. There is little noticeable time lag in the responses of the different VWPs. Due to the very low storage of the fractured granite, there is sufficient

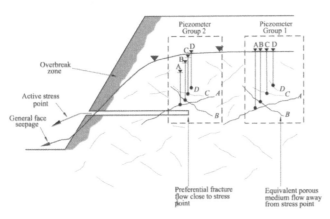

a) After one hour following drain installation

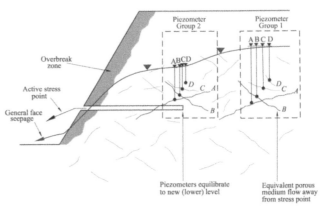

b) After one week following drain installation

Figure 3.24: Illustration of the transition between porous-medium and fracture flow conditions

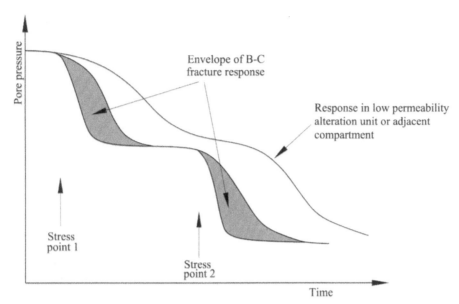

Figure 3.25: Illustration of the response to new hydraulic stresses

interconnected permeability in the fracture network to allow the small pressure changes to propagate rapidly through the system. This concept is illustrated by Figure 3.25.

There is little evidence for significant lagged response (for example 'D-type' fractures) in any of the Diavik piezometer installations. Similarly, 'A-type' fractures with larger continuous apertures, sufficient to yield significant flow to wells and drain holes, have not been encountered in the North-west wall. Such high-order ('A-type') fractures are present only in the proximity of Dewey's Fault zone and beneath the pit floor.

The over-riding control on pore pressures is the ability of groundwater recharge to enter the slope domain, and the ability of the water to flow *en masse* through the interconnected fracture network. It is considered unlikely that changes (increases) in fracture aperture that may have occurred due to deformation (unloading) as a result of mining or other processes would have had any material effect on pore pressures. The increases in fracture aperture and short-term pressure reductions would have been rapidly overcome by the *en masse* flow. However, the improved interconnectivity between fractures created by blasting did lead to improved drainage characteristics, which did influence pore pressure on a local scale (approximately bench-scale).

3.4 Discrete Fracture Network (DFN) modelling

3.4.1 DFN development

The DFN model for the North-west wall of the Diavik A154 pit presented in Appendix 3 is a stochastically generated interpretation of the fracture network. A number of realisations of the DFN can be generated based

on the same model inputs. Each of the model realisations is equally probable, but different. The following is a brief summary of the DFN and its use in the context of developing a conceptual hydrogeological model in a fractured rock slope at a sector (pit slope) scale.

For any particular pit slope, the need for a DFN model, and the objectives of a DFN, need to be carefully considered within the context of the geological and hydrogeological data that are available. At most scales used for pore pressure modelling, the majority of pit slope settings can be considered to be an EPM equivalent, so the effort and expenditure of constructing a DFN needs to be justifiable in terms of added value to the conceptual model. Even if there is sufficient geological control to construct a DFN to support the geotechnical model, uncertainties surrounding the hydraulic behaviour of individual fracture sets may reduce the value of a DFN for the purposes of pore pressure modelling.

For the DFN developed for the Diavik North-west wall, the key steps in the development process were:

- compiling and analysing fracture orientation and density data, bench-face mapping and geotechnical logs;
- determining a fracture-length distribution using a power-law function;
- determining the dependence of fracture length on orientation;
- assigning fractures a mean transmissivity based on a depth-transmissivity function developed from packer-test data;
- deriving a fracture-intensity property (known as 'P32 potential'), from cumulative fracture intensity (CFI) plots for all geotechnical logs in and around the A154 pit;

- developing a 3D block model of P32 potential values for the site through geostatistical interpolation between geotechnical boreholes;
- generating the DFN with 'FracMan' (Read & Stacey, 2009);
- conditioning fracture transmissivity based on fracture size (correcting the transmissivity-depth function to take account of fracture size).

In order to understand how the DFN can be used in the context of developing a conceptual hydrogeological model, it can be broken down into two principal components.

1. Data analysis and geostatistical component – in developing the P32P block model from fracture intensity data.
2. Stochastic component – using the P32P block model to generate discrete fractures.

The Diavik DFN appears to make the best use of the available data. However, for use as a tool in developing a conceptual hydrogeological model, the approach used needs to be evaluated in terms of uncertainty and the appropriate scale for which a stochastic solution is valid.

3.4.2   Stochastic realisations of the DFN

Early work carried out to analyse the Diavik VWP hydrographs was focused on trying to relate shared responses in different VWPs to hydraulic and mechanical 'events', and the identification of similar trends in different VWPs. It was originally expected that the Diavik DFN would represent more of a deterministic solution, and that responses in the North-west wall VWP data would be better understood through the identification of flow pathways in the DFN.

Whilst actual field data are used to constrain the orientation and fracture length distribution during the stochastic generation of discrete fractures, the actual positions of fractures of different length and orientation are different in each realisation. This has important implications for groundwater modelling as a high degree of uncertainty is associated with both the position of large structures in the slope and the hydraulic characteristics (and hydraulic continuity) of those structures.

Larger structures within the fracture network are the most important for controlling the sector-scale behaviour of the hydrogeological system. The presence or absence of large-scale flow 'conduits' determines the length of the flow path for drainage of the smaller fractures, and therefore how rapidly a slope will depressurise (and in the case of Diavik, whether adequate depressurisation of the slope can be achieved at all). In the DFN, the large structures show significant variability between visualisations, in terms of position and orientation. Therefore, in terms of modelling the actual behaviour of

the slope, and designing an appropriate slope depressurisation system, the usefulness of the DFN to address slope-scale conditions is questionable.

At Diavik, the significance of this variability is due to the fact that the largest structures in the DFN are few in number and are too large to be properly constrained in their distribution by the P32P model. Also, the orientation of these larger fractures is of greater relative importance, so a stochastically assigned orientation may not be appropriate for assisting the hydrogeological analysis. The limitations of the DFN in terms of accurately placing larger structures are discussed in detail in Appendix 3.

Smaller fractures, in comparison, can be orders of magnitude more numerous, but are less significant in controlling the slope-scale hydrogeology. For the smaller fractures, because of their greater relative density, their distribution is better constrained by the P32P block model. However, because of their relatively high density, the lower-order fractures are also most easily modelled using an EPM approach. The lower-order fractures are stochastically generated by FracMan, so zooming in on a smaller area of interest in the DFN would lead to similar uncertainties for determining actual fracture positions and actual hydraulic connections that may occur in reality in the slope.

3.4.3   The DFN as the basis for a groundwater flow model

In addition to the discussion in Sections 3.4.1 and 3.4.2, the following are also important considerations with respect to the way a DFN can be used to enhance the understanding of groundwater flow.

1. **The importance of hydraulic diffusivity:** the hydraulic property that most separates the behaviour of a fracture network from a porous medium is hydraulic diffusivity. The very low storage in a fracture network can cause pressure responses to propagate quickly through the system if the transmissivity (and hydraulic interconnection) of fractures allow this to happen. This is abundantly evident in the Diavik data. A DFN model is likely to be able to replicate this behaviour more easily than an EPM model as storage in a DFN model is limited to the fractures. Depending on how it is calculated, the storage assigned to an EPM model may be more of an average value for the rock mass as a whole, so short-term pressure responses may not be accurately reproduced. However, it should also be noted that diffusivity can be calculated on the basis of fracture porosity and applied to simulate short-term pressure responses in an EPM model.

2. **Constraining fracture transmissivity:** in the DFN, fracture transmissivity is a function of depth, and also fracture radius. The field observations at Diavik

indicate that individual fractures (both small scale and larger scale) pinch and swell (Figure 1.16). This phenomenon is commonly observed at many sites, and may encompass the full range of fracture sizes, from regional-scale faults down to the scale of fractures of a few centimetres length. This means that fractures might have a transmissivity commensurate with their total length over only a portion of the fracture. This reduces the effective hydraulic length of individual fractures, and results in a breakdown in the validity of the power-law scaling used in defining the fracture-length distribution. The effective transmissivity of the fracture is controlled by the lowest transmissivity at any point along its length (the 'pinch-points') and, because of this, needs to be determined on the basis of a harmonic mean of the transmissivity values of the different sections rather than an arithmetic or geometric mean. This is because the sections of the fracture with different 'permeabilities' are arranged in series, so the effective value must be obtained from a sum of the resistances rather than the conductances.

3. **Simulation of the variation of physical properties of fractures at a slope scale:** in any given slope, the physical properties of the fracture network, and therefore the hydraulic properties, will vary according to small scale changes in the lithology, hardness, degree of alteration, degree of oxidation, secondary chemical changes, and so on. To gain an understanding of these potential small-scale changes, it is necessary to achieve a good correlation between field hydraulic testing results and the head distribution from piezometer data, and the simulated fracture network distribution from the DFN.

4. **Dual porosity:** in most fractured rock, most drainable groundwater storage occurs within the fracture network. However, the rock matrix can have a degree of primary porosity. In granite, this is usually less than one per cent. In limestone and shale it could be significantly higher. DFN models might underestimate the storage in a fracture network if they do not have the ability to account for this.

For all practical purposes, the DFN model at Diavik was found to be of limited value for understanding how the slope would depressurise, or for helping to predict future pore pressures. If a DFN is to be used successfully for simulation of conditions within a pit slope, it must fully account for the slope-scale hydrogeology, and must represent the actual locations of the larger fractures, permeable horizons and other discrete features that transmit recharge into the slope ('sources'), and the discrete features that control drainage from the slope ('sinks').

The key benefits of a DFN model for assisting the development of a conceptual model are currently considered to be:

- gaining an understanding of the relative orientation and interaction of the various fracture sets, to improve the understanding of how the less permeable parts of the flow system may drain into the more permeable features;
- providing a basis for estimating how changes in fracture aperture may occur in response to unloading the slope (deformation) and depressurising the slope (hydrome-chanical coupling). For example, changes in aperture size are more likely to occur in fracture sets oriented parallel to the slope rather than perpendicular to it;
- providing basic data analysis and geostatistical support work to develop the model of fracture intensity potential (Component 1, as discussed above). For example, the block model of P32 potential would be a good basis for assigning zones of storage and perme-ability in an EPM model.

In considering the use of a DFN as a 'standardised' tool for developing the conceptual model, it should be realised that few sites are able to achieve the density of information that is required to support a DFN. At the current time, DFN modelling has been a part of the conceptual pore pressure model development and slope depressurisation program at few, if any, mine sites.

For field testing to provide sufficient hydrogeological data to support a DFN, a difficulty arises in the practicality of carrying out packer tests with a small enough test interval to encompass only a small number of fractures. The minimum packer test interval employed in the site investigations for most pit slopes is three metres. A 3 m test interval can encompass numerous fractures of varying orientations. Thus, it may be difficult to interpret the data in a manner which produces confidence in the input of hydrogeological parameters to the DFN. Reducing the length of the test interval is theoretically possible, but would require many more packer tests to cover the same thickness of material.

3.5 Summary of case studies

3.5.1 Introduction

The study of the North-west wall of Diavik is discussed in Section 3.3 and Appendix 3. A summary of the supplementary case study sites is included in Appendix 4. The results have been used as the basis for understanding the various controls on pore pressure distribution in a variety of settings, and how the relative importance of the various components of the hydrogeological system differs between sites.

The sections that follow summarise, in general terms, the factors that have been found to be the main influences on pore pressure at some of the sites used for case studies. The research has been further corroborated using data and observations from a total of 50 mine sites. On a qualitative level, the review of multiple sites enables the resolution of unknowns in a way that would not be possible through examining one site in isolation.

3.5.2 Diavik North-west wall: an interconnected rock mass that is strongly influenced by recharge and discharge boundaries

The key features of the Diavik North-west wall are as follows.

- The dominant control on pore pressures is recharge from Lac de Gras, which essentially forms a constant-head boundary immediately adjacent to the pit.
- Groundwater flows in a well-connected and reasonably permeable fracture network.
- Depressurisation effects from drains and wells in the face are limited because of the continuous supply of recharge.
- Pore pressure reduction occurs primarily due to the lowering of the local hydrological base level, formed initially by the pit floor and subsequently by the underground dewatering system. As the base level is progressively lowered, the groundwater system has adjusted, which has led to the creation of very steep hydraulic gradients between the lake and the pit floor.
- The opening of the underground dewatering system results in the greatest amount of slope depressurisation because it creates a large volume groundwater discharge below the lower part of the slope. In reality, this depressurisation is just the equilibration of the hydraulic gradient to the presence of a deeper discharge point, through the control of the hydraulic boundary conditions on the system.
- At the local scale (tens of metres), the influence of variability in fracture apertures causes variable yields in dewatering wells and drains.
- Discrete fractures with apertures large enough to sustain groundwater flows to wells and drains sufficient to overcome the recharge flux (high-order 'A-type' fractures) have not been encountered in the North-west wall.
- There is little evidence from the piezometer data of low-order 'D-type' fractures. At the pit slope scale, isotropic behaviour of the fractured rock causes all piezometers to show broadly consistent long-term depressurisation patterns. There are no fractures or zones of poorly permeable rock mass that show a significant lag in the drainage response time. All lags in drainage response can be attributed mostly to distance from the point of stress.

- The data indicate a relatively limited range in fracture permeability (for example 'B' and 'C' fractures, with no 'A' or 'D' fractures).
- Stress reduction and the associated increase of fracture aperture has had little effect on pore pressure because the additional fracture space is rapidly filled by water moving downward from the recharge source through the interconnected fracture system.

3.5.3 Escondida East wall: alteration in the fracture network and groundwater recharge from outside the pit crest

The important factors for controlling pore pressure in the Escondida East wall are as follows.

- A constant recharge influx occurs at relatively high levels in the slope from the basin-fill deposits of the Salar de Hamburgo.
- The upper leached zone of the porphyry creates a unit where many of the fractures have become healed, and softening of the rock has created some matric porosity and higher storage. Much of the recharge from the basin-fill deposits occurs into this unit.
- In the underlying 'fresh' porphyry, the groundwater flux moving downslope is interrupted because of barriers created by the dominant NNE–SSW trending fracture systems that occur along the length of the wall. Because the dominant fracture orientation is sub-parallel to the strike of the slope, it cannot increase the amount of downslope flow, and thus it acts to sustain elevated pore pressures.
- Successive phases of horizontal drains have released water from behind the NNE–SSW trending fracture sets that run parallel to the slope. The drains derive much of their water from a very limited number of fracture intersections where the fracture aperture is greater.
- The amount of depressurisation achieved is related to the balance between recharge from the pit crest moving down the slope and the amount of water removed ('released') by the horizontal drains.
- The recharge to the shallow leached porphyry zone creates a downward hydraulic gradient in the upper slope, while discharge to seepage faces and to drains in the unaltered porphyry creates an upward hydraulic gradient in the middle and lower slope.
- Within the fracture network of the unaltered porphyry, there are zones of sericite alteration where the rock fabric has been softened and the fractures have been healed. Removal of water from these zones is more difficult, and higher pore pressures are sustained because of the continuing recharge.
- Conversely, there are zones of potassic alteration that exhibit a higher proportion of more permeable (higher

aperture) fractures; in these areas, it is possible to remove water at a rate faster than the recharge flux, so the pore pressures tend to be lower.

3.5.4 Chuquicamata, a very low-permeability system with little recharge or discharge

The key features of the Chuquicamata pit hydrogeology are as follows.

- There is near-zero natural recharge from precipitation or from district-scale groundwater inflow. There are, however, localised recharge sources from leaky mine facilities, mostly around the Southern pit margin.
- The slope materials are the least permeable of any of the case studies. Packer test results have a geometric mean of 5×10^{-10} m/s for the East wall, with some locations as low as 2×10^{-12} m/s.
- The hydrogeology is dominated by the existing pore water in the rocks. However, because the rocks are relatively impermeable, this water does not drain readily. Static heads have been lowered mostly because of (i) deformation of the rock mass and expansion of the fracture porosity (East wall), and (ii) drains drilled from underground tunnels (lower West wall and pit floor).
- There is an extremely low combined yield in underground drain holes. Away from the influence of the South wall, the maximum combined yield of all underground drains is less than 2 l/s.
- However, the granodiorite in the West wall is an excellent example of free drainage in the absence of recharge, with a rapid depressurisation of as much as 400 m of pressure head occurring within about a week in response to underground drain installation.
- In the altered units of the East wall, typical fracture yields are an order of magnitude lower than the West wall (initial yields were typically less than 0.005–0.01 l/s). A significant component of depressurisation in the East wall has progressively occurred as a result of stress release (lithostatic unloading) during the course of mining (>25 years).
- Chuquicamata is unusual in that a downward gradient in static head is evident in all piezometers in the slope (except those within the immediate influence of the South wall recharge). It is postulated that the downward head gradient has been created by the stress release (lithostatic unloading). Even though the magnitude of rock mass deformation is less at greater depths below the active slope, the initial fracture porosity at depth is also lower, so the observed rate of pore pressure dissipation is greater. The downward head gradient is a transient condition that is sustained because of the continuing mining (and therefore the

continuing deformation) and the very low permeability, meaning that groundwater flow cannot occur at a sufficient rate to equalise the head changes caused by the continuing excavation process.

A comparison between the North-west wall of Diavik and the West wall of Chuquicamata provides a good illustration of the importance of recharge, as discussed in Section 3.6.3.5 below.

3.5.5 Antamina West wall: drainage of the slopes inhibited by structural barriers

The important controls on the pore pressure in the Antamina West wall are as follows.

- Dewatering of the more permeable intrusive and skarn rocks in the central pit floor is carried out using in-pit wells. The intrusive and skarn units are very well interconnected. As a result, the lateral and vertical head gradients in these units are extremely small, even with a large applied pumping stress.
- The pit slopes consist of moderately permeable limestone. Groundwater in all slopes drains downward toward the limestone-skarn contact, which occurs in the lower pit walls in all sectors. The limestone-skarn contact forms the hydrogeological 'base level' for pit slope depressurisation.
- The pit slopes receive seasonal recharge (annual precipitation is around 1150 mm). All piezometers in the limestones show a strong seasonal variation in head. The seasonal amplitude of recharge response generally decreases with depth below the slope, but is also controlled by the lateral position of the piezometer in the slope. A greater response is typically observed within those sectors of the slope that offer greater recharge potential (flat areas, ponding water, influences from recharge above the pit crest).
- Elevated pore pressures in the pit slopes are maintained by a combination of recharge from precipitation and the presence of low-permeability structures that form barriers to groundwater moving downslope. There is a strong downward head gradient in most sectors of the pit slopes. Therefore, although the wall rocks are saturated at a fairly shallow depth below the slope, the pore pressures below the initial saturated zone are a long way below the hydrostatic condition.
- The main potentiometric surface can be determined when two or more multi-level piezometers at any location show a similar head. This indicates they are below the level at which the hydrostatic condition occurs. In some sectors, the main potentiometric surface occurs more than 400 m below the level that groundwater is first encountered in the slope.
- Horizontal drains are used successfully to release water 'dammed up' behind structures, and therefore reduce

pore pressures in the saturated zone above the main potentiometric surface.

- All installed piezometers show a good response to recharge and to drainage. There is limited evidence of low-order fractures that show a significant lag in response time.

3.5.6 Jwaneng East wall: poorly permeable but highly interconnected shale sequence

The East wall of the Jwaneng pit shows the following hydrogeological behaviour.

- The main lithology is a poorly permeable, strongly bedded shale sequence. The shales contain a relatively well-developed fracture network that is interconnected by bedding-controlled orthogonal jointing.
- The permeability of the shale is controlled by joint sets that are perpendicular to bedding. However, because the joints are interconnected, the resultant groundwater flow is mostly parallel to bedding.
- Minor recharge at the pit crest infiltrates through the overlying unconsolidated Kalahari sands and enters the underlying shale sequence as diffuse flow. A large portion of the recharge is derived from infiltration through the base of stormwater drains and from other site facilities.
- The shale sequence possesses extremely low but well-connected storage. A single dewatering well located at the crest of the slope, pumping at a relatively low rate of 7.5 l/s, has caused widespread depressurisation to occur over time across the entire area of the wall. Many of the major structures, which are mostly sub-parallel to the wall, do not appear to act as major barriers to the spread of the drawdown response.
- Despite the structure and fracture density, the permeability remains very low due to the small-aperture fractures inherent in the shale lithology and the absence of larger permeable structures.
- None of the installed piezometers provides evidence of non-connected fracture systems and/or low matric permeability.

3.5.7 Cowal

The important features controlling the hydrogeology and pore pressures at Cowal are as follows.

- The saprolite material contains relatively high groundwater storage compared to the underlying unweathered bedrock. Regional groundwater levels in the saprolite are sustained by recharge and seasonal infiltration of surface waters.
- It is possible to under-drain the saprolite by pumping from the underlying fracture zones in the saprock and at the top of the unweathered (primary) rock. This behaviour is similar to many other saprolite groundwater settings worldwide. However, at Cowal, the obtain-

able yield from the upper fractured bedrock has generally been low.

- Similar pore pressure responses are evident for the saprock and primary rock zones, indicating a typical depressurisation rate of ~0.9 m/month as a result of drainage into the pit and pumping from wells.
- The drainage response in the saprolite (under-drainage) is typically intermediate between the fractured bedrock below and the transported sediments above. Although a net downward head gradient is observed in nearly all VWP arrays, the sapolite water levels do not become decoupled from the underlying bedrock water levels. Slow drainage of the saprolite occurs as a result of pumping and depressurisation of the underlying saprock and upper unweathered (fractured) bedrock.
- The transported sediments in the upper slope (above the saprolite) have very low vertical permeability. The depressurisation rate for the transported sediments is typically the slowest of all units at ~0.4 m/month. Water levels in the underlying saprolite have fallen at a greater rate than those in the transported sediments and have become decoupled from those in the transported sediments (with an unsaturated zone occurring below the transported sediments in many locations).

3.5.8 Layered limestone sequence in Nevada, USA

In Nevada, USA, strongly layered limestone sequences occur in a number of pits. Where water levels in the limestones are sustained by shallow recharge, the following observations are made.

- Elevated pore pressures occur in some of the limestone units because of the strong anisotropy in horizontal to vertical permeability that has developed because of the bedding.
- When the low-permeability barriers caused by bedding are penetrated by drill holes, the entire sequence shows a drainage response.
- Drainage is observed throughout the entire rock mass and there is no evidence of any small 'D-type' fractures which do not depressurise.
- In addition to the horizontal barriers, the groundwater regime is dissected by sub-vertical structures. The sequence can exhibit strong lateral compartmentalisation. However, within each compartment, the drainage response is good for the most part.
- In some instances, the degree of compartmentalisation is increased by the presence of dykes that often run sub-parallel to the main fault zones.

3.5.9 Whaleback South wall

The factors influencing the pore pressures in the South wall of the Whaleback pit are as follows.

- The strike of the geological units is sub-parallel to the wall.
- Most of the units within the wall are low-permeability shales, but there is a prominent bed of permeable iron formation inter-bedded within the shale sequence (Bruno's Band).
- Groundwater moves along the strike (sub-parallel to bedding) of Bruno's Band from a recharge source to the west.
- Pumping from Bruno's Band causes significant drawdown within the unit. The pumping also influences other shale units extending across the strike of the formations. However, depressurisation in these units is lagged compared with the pumping response in Bruno's Band.
- Pore pressure data throughout the South wall exhibit a significant recovery when pumping rates from Bruno's Band are reduced.
- The South wall piezometers do not show the same response to rainfall and runoff events as those located in the permeable iron formations that are present within the central pit area.

3.6 Factors contributing to a slope-scale conceptual model

3.6.1 Regimes

As discussed in Section 1.1.6, pore pressure is a major control on the stability of slopes in rock and soil. In the case of large open pits, it is usually the only property that can be easily varied. Reduction in pore pressure causes an increase in effective stress and an increase in the shear strength of the slope materials.

Section 1.2.1 discussed the regimes (three sets of factors) that comprise any hydrogeological framework.

1. **Geology:** the physical framework in which the water occurs and moves. The geological setting forms the basic and most important control on the hydrogeology. The geology is essentially immutable except for changes that may be caused by the activities of the mining operation itself.
2. **Hydrology:** representing the input of water to, and output of water from, the geological framework. This comprises the water entering the model domain (recharge) and water leaving the model domain (discharge).
3. **Hydraulics:** the factors that control the way that the framework constrains the water movement. This category includes consideration of:
   - the presence of *major features* which may form (i) discrete permeable pathways that provide 'conduits', (ii) discrete low-permeability barriers that act as 'dams', and (iii) a series of low-permeability barriers

that may 'compartmentalise' the slope, either in a lateral or vertical plane, or both. The major features may be geological structures, discrete stratigraphical features, weathering, mineralisation or alteration zones, the presence of underground workings, or drill holes that may remain open in the slope domain;
   - the *small-scale fracture sets and properties of the rock mass itself* and how they interact with the major features.

The three regimes need to be considered from the earliest stages of investigation at every site. These factors interact, and they are applicable to any conceptual model, regardless of scale. Their relative importance will vary from site to site.

Section 3.2.1 described the basic components of a wider-scale conceptual model (regional- or site-scale). Section 3.2.3 described the components of a sector-scale (slope-scale) conceptual model. The purpose of the research using the Diavik data, and the analysis of data from the supplementary sites, was to provide a better understanding of the components of the sector-scale model and to provide a methodology for developing a conceptual model on which to pin subsequent numerical modelling work and base a slope depressurisation program.

3.6.2 The influence of geology on the conceptual model

3.6.2.1 General

The key for developing any conceptual model is a good understanding of the geology. This needs to cover the pit slope domain, plus the entire area that may feed water into the slope domain. Therefore, the wider-scale geological evaluation should identify:

- the key lithological units; and lateral and vertical variations within those units;
- zones of alteration and/or mineralisation;
- the major structural features (which may act as flow barriers or flow conduits);
- zones of enhanced weathering and/or oxidation.

The wider-scale assessment will often need to focus more on the alluvial and basin-fill deposits because these will typically contain more groundwater in storage and have greater potential for regional-scale and site-scale groundwater flow than the bedrock (the exception often being carbonate groundwater settings).

3.6.2.2 Lithology

Lithological types commonly found in mining districts are:

- igneous rocks (e.g. Diavik and Cowal);
- metamorphic rocks (e.g. Escondida and Chuquicamata);
- cemented clastic rocks (e.g. Jwaneng);

carbonate rocks (e.g. Antamina and Nevada limestones).

In all rock types, virtually all of the drainable groundwater is contained within fractures. In fresh (unweathered) rock, the drainable porosity is often very low, particularly at increasing depth. If there is primary porosity, the intergranular openings are typically so small that the pore space is non-drainable, even after the material has been blasted.

Lithology is a principal factor controlling the hardness of the rock mass, and therefore its tendency to fracture. Lithology governs the development of the fracture network through the following:

- the amount of inherent discontinuities in a rock mass (e.g. bedding planes in sedimentary rock);
- the propensity of the rock mass to fracture in response to stress (strength, brittleness) and variations in this potential throughout the rock;
- the orientation and number of inherent weaknesses within the rock.

3.6.2.3 Alteration and mineralisation

Alteration may change the fabric of the rock mass and may therefore cause significant changes to its hydraulic properties. Silicic or potassic alteration may increase the tendency of the rock to sustain open fracturing. At Escondida, like many other porphyry settings, zones of potassic alteration generally provide greater groundwater ingress to dewatering wells or horizontal drains. Conversely, sericitic or argillic alteration tends to cause fractures to heal, reducing their potential to transmit groundwater. When argillisation leads to widespread softening of the rock and healing of fractures, it may change the nature of groundwater movement from fracture flow to intergranular (porous-medium) flow (or a combination of the two).

Mineralisation may also alter the fabric of the rock mass. The Escondida case study is again typical of many porphyry-copper settings, where leaching in the upper levels of the ore body has caused softening of the rock and healing of fractures, and groundwater movement in the leached zones can be a combination of fracture and intergranular flow. As a result of the rock softening processes, the leached zone of a porphyry copper deposit may show a greater drainable porosity than the underlying fresh rock. Locally argillic mineralised rock can also be a feature of some base metal ore bodies (e.g. Mississippi-Valley hosted lead/zinc deposits).

Mineralisation can also support open fractures at great depths, by providing structural 'bridges' to support an open aperture. Mineralisation in such a way has been cited as the cause of statistical outliers in field observations to quantify depth-permeability relationships (Rutqvist & Stephansson, 2003).

3.6.2.4 Structural influence

Geological structure is the major contributor to the distribution and alignment of fractures in any rock mass. Typically, in any given formation, two or three prominent fracture orientations can be recognised, depending on the structural setting. Certain lithological types are brittle and sustain extensive open fracturing. Other lithological types are more ductile, and open fractures are less common, or fractures have become healed and fully or partially closed. The case studies cover a broad range of geological settings. They exemplify the combined importance of lithology, structural setting and structural evolution as a control over:

- the development of the fracture network;
- the bulk hydraulic properties of the fracture network;
- the properties of individual fracture sets.

Within sedimentary (and also volcanic) rock types, variations in brittleness between layers can lead to corresponding differences in fracture density. Thus, although geological structure is the main reason for fracture development, subtle variations in lithology are often as important and sometimes more important than structure for maintaining open fractures in the rock mass, and therefore for controlling the hydrogeology. Of the case studies discussed in this report, the best demonstration of how different lithologies give rise to different fracture architecture is a comparison of Jwaneng, a stratified, cemented, clastic sedimentary rock sequence, with Diavik, a homogeneous hard rock setting.

At Jwaneng, bedding planes are present as inherent discontinuities, but the primary open orthogonal joint sets in the rock mass have developed perpendicular to the bedding planes. Nonetheless, the presence and variation in intensity of the orthogonal joint sets is controlled by the lithological differences between beds (i.e. some beds sustain a better orthogonal joint set than others). Hence, groundwater flow is primarily controlled by bedding. Bedding-controlled heterogeneity in the fracture network is common in hard rock sedimentary sequences, and often results in anisotropy. The primary flow direction is usually oriented parallel to bedding, even though the main open joint and fracture sets may not be orientated in the same direction as bedding.

At Diavik, in massive and more homogenous granite, the fracture network that has developed is more a function of the orientation and magnitude of the various regional stress fields that have existed during the rock's history. For fractures to open, one of the principal compressive stresses must be smaller than the other two, and the formation of open fractures is related to the relief of confining pressure (effective stress) (Trainer, 1988). In the North-west wall of

Diavik, the primary open fracture sets are NNE–SSW trending. Early test pumping data indicate that interconnection of fractures is most developed in this orientation.

3.6.2.5 Weathering and oxidation

Weathering also alters the physical nature of the rock. In tropical and sub-tropical areas, a feature of the upper weathered regolith or saprolite zone is that most fracturing has become healed, and intergranular flow dominates. Often, there is a transition zone near the base of the weathered zone, where some of the relict fractures remain partially open, and a combination of intergranular and fracture-controlled flow may occur. Weathering to ilmonite/goethite can also occur in iron formations along fractures/shear zones that produce zones of higher hydraulic conductivity, for example at the Carol Lake Mine in Newfoundland.

In volcanic and sandstone sequences, weathered palaeosol layers (Figure 3.26) often exhibit healing of

fractures, so these layers often form significant barriers to vertical (or across-bedding) groundwater flow. In the case studies, softer, more weathered and less permeable layers are found within some Nevada limestone formations.

3.6.2.6 Joints and faults

Fractures fall into two main groups, joints and faults. Joints show displacement perpendicular to the fracture plane, but no movement along it. Faults are characterised by movement along the plane or surface of the discontinuity. Joints below the water table are always conduits for water flow (except where healed), though they may not necessarily be aligned along the hydraulic gradient. Faults may represent either conduits or barriers, depending on their nature. As discussed in Section 1.2.2.5, a fault zone lined with gouge, for example, will act as a barrier to flow across it, but may offer a permeable pathway to flow along it. This may be due to a combination of:

- the fault offsetting permeable layers either side of it, therefore impeding flow across the fault;

Figure 3.26: Example of Palaeosol in a volcanic sequence (Gran Canaria, Spain)

- the presence of clay gouge along the footwall or hanging wall rocks, or along the fault trace itself;
- preferential fracture orientation and connectivity along the strike of the fault zone, often in the footwall rocks.

An interconnected fracture network in hard rock therefore indicates a geological history in which a number of differently oriented stress fields are likely to have been present over time. Fracture networks that have evolved over time as the result of regional stress fields should also conform to relationships between fracture size and frequency. In a body of granite, for example, few very large (first-order) discontinuities may be expected, with a significantly larger number of smaller (second-order) fractures and a much higher density of small (third-order) fractures. This appears to be the case at Diavik.

The timing of the evolution of the fracture network also needs to be considered in the context of other geological processes. The rock mass as a whole, or the fracture surfaces themselves, can undergo processes over time such as alteration, mineralisation, dissolution and weathering that may improve or degrade their transmissivity and open area. Faults and joints may have experienced hydrothermal activity or secondary mineralisation. Hot mineralised waters passing through these discontinuities can alter existing minerals to clays, so that the surfaces of the joints and *en echelon* fractures may become sealed, and fault gouge may be turned into clays with very low permeability (Banks & Robins, 2002). The younger fracture sets may also offset the older fracture sets and therefore may provide more continuity for groundwater flow. It is not uncommon for the older fractures in a rock body to have become 'healed' to a certain extent, leaving the younger fractures to sustain much of the bulk permeability of the rock mass. In a study for a sub-sea tunnel off Norway, for example, Banks *et al.* (1992) found that major fault zones were filled with secondary clay minerals and were of low transmissivity, while inflows to the tunnel tended to occur through lesser, simpler fractures not detected by the preliminary investigations. Within a given structural setting, the properties of individual fractures, such as fracture size, aperture, and wall roughness, are dependent on lithology. Lithology controls the ability of a rock to sustain open fracturing, and the potential for fractures to close or 'heal' after formation. Typically, hard rock, such as the granite at Diavik, is much more able to support larger-aperture fractures than softer rock types.

3.6.2.7 *Large-scale features*

The large-scale geological features (for example the orientation of bedding, and the position and orientation of major structures) can have a significant control over groundwater flow within the pit slope. Where they are connected to a recharge source, continuous 'slope-scale' permeable structures (e.g. highly transmissive faults or more permeable stratigraphical units) can transmit a considerable amount of water into the slope domain. At Diavik, Dewey's Fault zone is a NE–SW trending structure that bisects the A154 pit and continues through the North-east wall and beneath Lac de Gras. A large proportion of the total recharge to the pit moves from Lac de Gras along Dewey's Fault zone. At the Luce pit in Labrador City, a similar structural zone is observed to convey recharge from the adjacent Luce lake into the pit area. In the North-east wall of the Grasberg pit, discrete structures that feed water into the slope from upgradient (recharge) areas are thought to have been the cause of anomalously high piezometer readings in a sector of the slope where all other piezometer measurements showed low pore pressure.

However, where they provide a connection to the pit floor or can be 'tapped' by pumping wells or drains, continuous permeable features will, in many cases, have a positive impact on depressurisation, as they will tend to act as drainage features. Dewey's Fault zone was intersected by the underground drainage holes to create the rapid lowering of the hydrological base level below the toe of the North-west wall. At other sites, the presence of large-scale permeable features within the slope domain has provided flow pathways into which the less permeable fractured rock mass can drain. The further these primary drainage conduits extend throughout the rock mass, the shorter the flow path will be in the subordinate fractures, and the faster the rock will drain.

Where major structures are aligned sub-perpendicular to the hydraulic gradient, they will often form barriers to groundwater flow. Where they are aligned sub-parallel to the slope, they will often tend to inhibit drainage to the slope face. Such poorly aligned permeable structures often provide good targets for horizontal drains as they may transmit a considerable amount of water into the drains which are then able to convey the water out of the slope.

In the case studies, Antamina, Escondida, Jwaneng and Chuquicamata all demonstrate the importance of structures on depressurisation. The West wall of Antamina is a good example of a site where permeable geological structures (the 'valley-parallel' NE–SW trending structures) inhibit the drainage of water down-slope and into the permeable skarns in the lower slope. In addition, beds of mudstone and siltstone in the Celendin Limestone at Antamina create bedding-aligned anisotropy; hence bedding dipping away from the face also presents barriers to flow.

In the East wall of Jwaneng, the opposite is true, and the bedding orientation aligns the most permeable flow pathways towards the pit face. At Jwaneng, the monitoring results indicate that many of the major structures have no apparent major control on slope-scale groundwater flow.

This may be, in part, because of the relatively low groundwater flux in the East wall.

Large permeable structures will not be beneficial to the progress of depressurisation if those structures are well connected to a recharge source, but do not have a natural discharge point. Chuquicamata is an example where pore pressures in the RQS unit at the base of the pit were elevated due to north–south oriented structures transmitting minor artificial recharge northward from the southern end of the pit. Penetration of these structures by drains, however, caused a very large and rapid drop in pore pressure.

3.6.3 Hydrological input: recharge to the slope domain

3.6.3.1 Sources of recharge

At the mine sites analysed as part of the current study, recharge is considered to be the single-most dominant factor controlling the pore pressure distribution in the pit slopes. The case studies presented exhibit a wide range of different types and rates of recharge. Recharge can take the following forms.

1. **Infiltration of incident precipitation or local runoff into the slopes themselves** (e.g. Antamina). In particular, seasonal infiltration is usually important in tropical or high altitude settings, where intense seasonal precipitation may occur and, in some extreme cases, over 50 per cent of the precipitation may potentially infiltrate into the slope materials.

2. **Recharge from a surface water body** (e.g. Diavik). There are many open pits where adjacent surface waters (rivers or lakes) produce local infiltration and recharge at the crest of the slope or in the immediate vicinity of the pit. This water may move into the slope domain.

3. **Recharge to the slope from more permeable groundwater units outside the pit** (e.g. Escondida). Many pits that are located adjacent to alluvial basins or regional aquifers can experience a considerable amount of recharge into the slope domain from the more permeable (and often higher-storage) units that occur adjacent to the pit.

4. **Recharge from tailing facilities, leaky process facilities, tanks and pipelines adjacent to the pit** (e.g. Jwaneng). Operating mine sites often experience some water losses from their facilities or infrastructure. Losses from diversion drains and runoff collection sumps also fall into this category. This recharge source is often more obvious at sites located in low rainfall environments. It can often be very difficult to intercept this water, and the recharge can have a major impact for sustaining pore pressures within the slope domain. At any given site, recharge to the groundwater system may potentially occur from the tailing impoundment, the base of the waste rock facilities, stormwater control ditches and ponds, or the process plant facilities, such as storage tanks, pipe networks or the thickener overflow.

3.6.3.2 Mechanisms for recharge

In areas of high rainfall or low evaporation, there may be a considerable amount of areal or 'blanket' recharge caused by the incident precipitation on the site in general. To gain a general understanding of the recharge potential, it is often useful to calculate the effective rainfall, which is defined as the total rainfall minus the actual evapotranspiration. The determination of actual evapotranspiration (as opposed to pan evaporation or some other empirical measurement) can be complex but a reasonable approximation can often be made using available pan evaporation data. The assessment of the mean effective rainfall on a monthly basis can provide a useful rule-of-thumb method for determining the approximate recharge rate to the groundwater system. For many climatic settings, effective rainfall is only positive for certain months of the year, and recharge occurs only during those months. In the absence of rainfall, evapotranspiration removes water from the soil, creating a soil-moisture deficit (SMD).

Figure 3.27 shows a typical soil moisture balance and illustrates the component of rainfall that may become recharge. Depending on the intensity of the rainfall event and the antecedent conditions of the soil, a proportion of the rainfall will form runoff, and a proportion will infiltrate. The infiltrated water becomes available for evapotranspiration by plants. When the evapotranspiration demand has been met, any excess water is stored in the soil horizon. If the SMD has not been satisfied, the water remains stored in the soil horizon. However, if infiltration continues to occur, the SMD reduces until the condition known as field capacity is reached (the storage capacity is used up), and any excess water entering the soil infiltrates downward to become groundwater recharge. For steeper topographic settings, much of the effective rainfall can

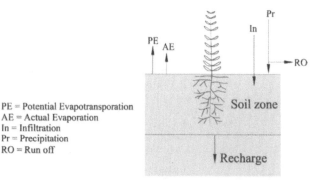

Figure 3.27: Processes occurring within the soil profile and resulting in recharge (Taken from Hulme *et al.*, 2001)

become runoff to the surface water system rather than groundwater recharge.

In higher rainfall settings, a more detailed assessment of recharge is usually required to assist either the engineering designs or the regulatory permitting. It may be necessary to install more than one on-site climatological station, possibly with multiple rain gauge sites, to help evaluate the areal distribution of the recharge. Knowledge of the near-surface soil or rock characteristics is also essential for helping to quantify the amount of precipitation that becomes recharge. Key factors in the assessment are:

- how much of the incident rainfall will runoff from the ground surface or move laterally through the soil rather than infiltrate to form potential groundwater recharge;
- how much of the infiltration will be stored in the near-surface soil zone, where the water can subsequently be withdrawn back to the atmosphere by the

root systems of plants and evapotranspiration, and therefore does not contribute to recharge.

It is only when water moves downward past the root zone and the extinction depth for evaporation that it may recharge the underlying groundwater system. For a typical hydrogeological setting, the extinction depth may be one to five metres below ground. However, in some climates where moisture is scarce, the root zone of some phreatophytic plants may extend to depths of more than 50 m.

Examples of hydrogeological settings where rainfall is high but where areal recharge is low may be:

- a tropical zone with hard lateritic near-surface soils, where much of the available water runs off the surface into the surrounding streams. In this case, the effective rainfall is high, but the vertical permeability of the soils is lower than the intensity of the rainfall, so the amount of water that can percolate downward through the soil profile is limited;

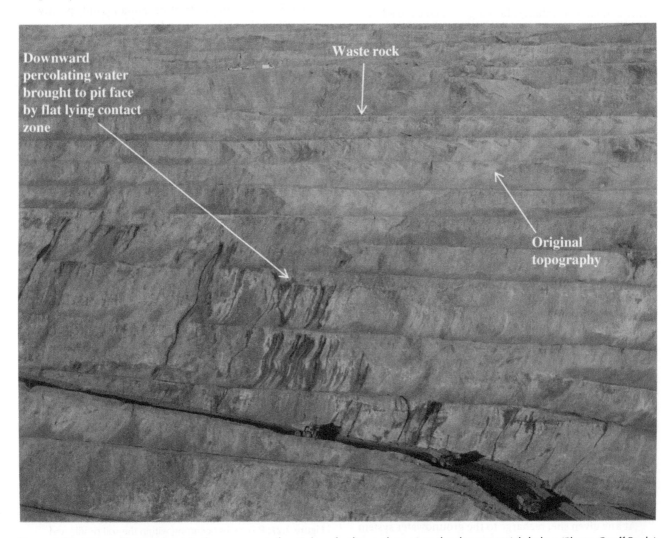

Figure 3.28: Historical waste dump in the upper part of a pit slope feeding recharge into the slope materials below (Photo: Geoff Beale)

sites with porous (high storage) soils, where rainfall and evapotranspiration are evenly distributed throughout the year, and therefore the effective rainfall is low. In this case, the soil moisture deficit is often high, so most of the water that infiltrates into the surface is stored in the active zone and subsequently removed by evapotranspiration. Little water percolates downward beyond the root zone to become recharge to the underlying groundwater system.

3.6.3.3   Recharge cycles

In temperate climates, a typical seasonal cycle of the soil moisture balance and recharge may be as follows.

- **Summer**: high rate of evapotranspiration and soil water removal; increasing SMD; rainfall events cause near-surface infiltration but the water is quickly removed from the soil profile by evapotranspiration.
- **Fall (Autumn):** high soil moisture deficit, which has been gradually built up over the summer months; infiltration from rainfall events is stored in the near-surface soils, even though evapotranspiration rates are low; little water percolates downward below the extinction depth to become recharge.
- **Winter:** the soil moisture deficit that was built up during the summer months becomes progressively replenished by continuing infiltration from precipitation events. At some point the soil moisture deficit is satisfied, breakthrough occurs, and the infiltrating water moves downward below the capture zone of the root system; the water is able to move downward below the extinction depth and become recharge to the groundwater system.
- **Spring:** the soils are fully saturated and any rainfall or snowmelt is transmitted rapidly downward below the root zone to become recharge. This may also be the period of high water availability when snowmelt occurs, so most or all of the annual recharge may occur during this period. As ambient air temperatures increase, so evapotranspiration rates also rise, and the SMD starts to build up as summer approaches.

An example of a hydrogeological setting where mean annual precipitation is low but recharge is relatively high is a temperate climate, where most of the precipitation occurs during the winter months when evapotranspiration is low and water can infiltrate downward below the root zone. However in colder climatic zones, where the near-surface soils become frozen, snowmelt often occurs prior to the thawing of the underlying soils such that the water is unable to percolate downward and most of the available water contributes to runoff.

Any assessment of recharge must also consider the variations in aspect across the site and the influence this may have on infiltration and recharge. Areas which are shaded from the sun may experience less evaporation and may therefore be slower to develop a soil moisture deficit; they may also develop a thicker snow pack which provides recharge upon melting. Areas on the leeward side of hills (and waste rock areas and heap leach pads) may also develop a thicker snow pack which, in turn, creates more recharge.

3.6.3.4   Dominant recharge controls

An over-riding control of the amount of recharge to the groundwater system is often the way the surface water becomes concentrated; either in the form of stream flow or accumulating snow. In many natural settings, much of the recharge will occur because of 'concentrated water'. Sources of 'concentrated water' may be as follows.

- **Losing (influent) rivers and streams,** which provide a head of water to cause infiltration and downward percolation of water beyond the root zone. In many arid climatic settings, virtually all of the recharge may be produced by infiltration from ephemeral streams that are fed by runoff from poorly permeable upslope bedrock areas with steeper topography. The 'concentrated' stream flow may then provide recharge to the downslope area, where the streams flow across permeable alluvial soils or rocks at a shallower topographic gradient. Larger rivers fed by glacial melt water or significant snow melt may lose water to the ground on a more-or-less continuous basis.
- **Lakes and ponds,** which also provide a constant-head water source, and may therefore cause continuous infiltration and downward percolation of water beyond the root zone. However, the downward percolation rate and the recharge from lakes are sometimes limited because of the fine grained sediments that often accumulate on the bottom.
- **Snow banks,** where considerable thickness of accumulating snow may develop in sheltered ground and melt to provide a continuous infiltration source over a period of weeks to months, sufficient to create downward percolation of water beyond the root zone.
- **Leakage from mine facilities,** which again may provide a constant water source and may therefore cause continuous infiltration and downward percolation of water beyond the root zone. Figure 3.28 shows a good example of a historical waste dump in the upper part of a pit slope feeding recharge into the slope below. The recharge from the waste dump is hitting a sub-horizontal feature immediately below an old valley that was filled with waste and the water is being brought into the slope as seepage.

At any given site, a high proportion of the recharge may result from intense precipitation events, where there is increased precipitation, extreme runoff flows and ponding of water at the surface. There are many arid hydrological environments where virtually all of the recharge occurs only from discrete precipitation events that may be related to (i) a single intense storm of short duration, (ii) a sequence of wet weather lasting for several weeks or months, or (iii) a wet year such as an El Niño event during a normal climatic cycle. Thus, all the recharge may occur in a very short time period, and there may be intervening months and sometimes years when there is no or minimal recharge.

3.6.3.5 Examples showing the influence of recharge

A good example in the case studies of the importance of recharge is the comparison between granite in the North-west wall of Diavik and the granodiorite in the West wall of Chuquicamata. Both lithologies have comparable hydraulic properties. In general, Diavik has more permeable fractures and higher individual drain yields than Chuquicamata. The sites can be compared as follows.

- At Diavik, Lac de Gras provides a constant-head source of water at the crest of the North-west wall (and indeed around much of the pit crest). Recharge from the lake can readily enter the fracture systems *en masse* above the crest of the slope. The controlling factor in the amount of recharge from the lake is the ability of the interconnected fracture systems below the pit crest to transmit the water. For depressurisation, it was not possible to intersect fractures with sufficient aperture behind the wall to allow water to be removed at a faster rate than the water could enter the slope domain as recharge. Therefore, it was not practicable to locally depressurise the slope (successful depressurisation would have required a very large number of low-yielding wells and drains).
- At Chuquicamata, where recharge to the Granodiorite Fortuna is minimal, drain holes with individual flow rates about an order of magnitude lower than those at Diavik (about 0.5 l/s) produced about 400 m of permanent depressurisation within a time period of about a week (see also Section 3.6.5.3).

Diavik is an extreme example of the influence of recharge in a relatively permeable and interconnected hard rock fracture system.

In poorly permeable fracture networks, or argillic zones within the pit slope, even relatively small rates of recharge can have a large influence on pore pressure. Where recharge to a clay unit occurs, the rate at which recharge can enter the formation over a diffuse area, and the *en masse* movement of water through the low-permeability formation is usually greater than the rate that water can be removed by drains or other point discharges. Therefore, slopes that are cut into poorly permeable formations that are sustained by recharge can be the most difficult to depressurise. They may require a very large number of drain holes with a close spacing, or it may not be practicable to drain the slope.

As will be discussed in Section 3.6.6, where the bulk permeability is less than 10^{-8} m/s deformation of the slope materials often leads to an expansion of the pore space, potentially with a significant reduction in the pore pressure. However, where the poorly permeable formations are being recharged, the reduction in pore pressure may become quickly damped by 'diffuse' recharge water entering the formation.

A good example of this can be taken from the Yanacocha Mine in northern Peru, where mean annual precipitation is between about 1100 and 1500 mm, mostly occurring in a six-month season. The North wall of the Yanacocha Sur pit and the North-east wall of the Phase 1 Maqui Maqui pit are both excavated in similar clay-rich altered materials. In the Yanacocha Sur pit, a system of lined drainage channels was installed and maintained on each catch bench, minimising recharge to the underlying slope materials. The result was that the recharge was effectively removed from the system and the slope depressurised (Figure 3.29). As a result, it was possible to maintain the slope in excellent condition (Figure 3.30). In contrast, Phase I of the Maqui Maqui pit was excavated early in the mine life without any drainage control. Recharge was able to enter the slope materials and the pore pressures were maintained, and it was difficult to maintain the integrity of the slope.

3.6.4 Hydrological output: the role of discharge in slope depressurisation

An open pit that has extended below the water table and is being drained becomes a groundwater discharge feature. Most pits are therefore hydrogeological sinks. Under equilibrium conditions, the discharge flux from the lower slope and pit floor will equal the recharge flux into the slope domain (from infiltration and site-scale groundwater flow).

It should be noted that for some open pits in low precipitation desert environments, the small seepage that does occur may not be evident. It may move downslope in the over-break zone, and may be consumed by evaporation, without creating noticeable wet points on the slope.

As discussed in Section 3.6.2.7, favourably oriented permeable structures (potentially fault zones or more permeable stratigraphical units) can act as prominent discharge features for the slope. Where these are present, much of the required depressurisation of the slope may potentially occur passively. The greater the extent that permeable features penetrate into the slope, the shorter will be the flow path between the poorly permeable rock mass and the high-permeability features, so the faster the slope will drain.

Some discharge can also occur through natural low-permeability discharge pathways, or by relatively slow passive drainage by *en masse* seepage to the pit face and pit floor. However, in many cases where there are no

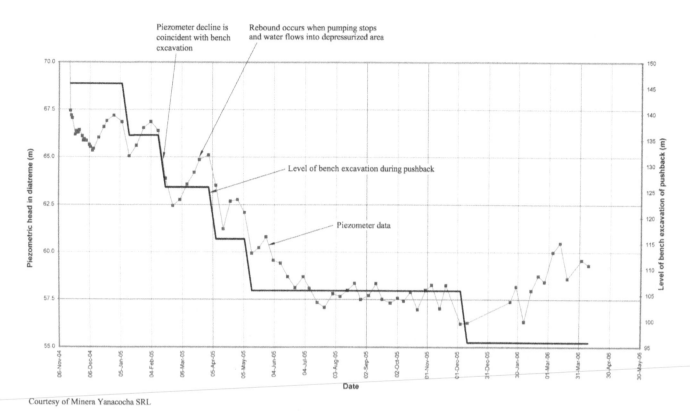

Figure 3.29: Illustration of a coupled response from Yanacocha Sur

Figure 3.30: Surface water control to reduce recharge and maintain the slope in good condition (an HDPE-lined surface water drainage channel is installed on every catch bench)

permeable discharge features which extend into the rock mass, artificial discharge points have to be created to allow the desired rate of depressurisation to be achieved. These usually take the form of pumping wells, or horizontal or angled drain holes, as discussed in Section 5.2.

Diavik has shown that blasting can potentially open up new discharge pathways from the slope materials. An example from the Sleeper Mine in Nevada is discussed in Section 5.2.7 and shows how blasting was used to create drainage across a 'range-front fault' barrier in the pit floor, located out from the toe of the slope. This allowed drainage of the material at the toe of the slope which consequently lowered the discharge point for the water in the slope. The resulting re-equilibration of the hydraulic gradient caused a reduction in pore pressures in the slope itself.

The following approximate rules of thumb may be applied.

1. Where the permeability is greater than about 10^{-8} m/s and the diffusivity is greater than about 10^{-4} m²/s, a significant amount of depressurisation can usually be achieved by drainage (Section 3.6.5.2).
2. Where the permeability and diffusivity are lower than this, depressurisation may potentially occur due to hydromechanical coupling effects (Section 3.6.6.2).
3. Materials with a permeability between about 10^{-7} and 10^{-8} m/s will often require closely spaced drainage measures.
4. If discrete permeable features are present to act as pathways, depressurisation will often occur passively; if they are absent, they must be provided in the form of wells or drain holes.
5. If water is entering the slope domain as recharge faster than it can be removed, depressurisation will not be possible.

The *rate* of depressurisation in a rock slope is governed by the balance between recharge and discharge and the ability of the fracture network to transmit this flux to the discharge point(s). Similarly the *influence* of active depressurisation measures is dependent on these measures being able to increase the rate of discharge, above that which can occur through passive drainage, by creating artificial discharge points.

Passive drainage occurs throughout the fracture network within which numerous flow pathways exist. Depressurisation is more rapid if there is a dominant flow conduit that provides the discharge 'sink' and to which the fracture network can drain. Active depressurisation measures rely on point sources of discharge (i.e. wells and drains) that intersect only a small part of the fracture network. These point-source depressurisation measures will be more effective when:

▪ they can intersect a zone of locally high permeability to extend their influence away from the point source;

▪ the interconnectivity of the fracture network is such that the depressurisation can propagate throughout the system.

In effect, this means that the influence of active depressurisation measures is dependent on high diffusivity in the fracture network.

The dewatering well GH03 at Jwaneng is a good example of a point source of discharge that satisfies the above conditions (Appendix 4). Although the fracture network appears well interconnected, most of the test holes drilled did not intersect fracture zones that could produce sufficient yield to cause widespread depressurisation. The yield in GH03 (7.5 l/s) is small compared with many other sites but is significant in comparison with the total groundwater flux moving through the South-east wall. The well intersected a relatively major fracture that could draw in water from an extensive part of the system. Therefore, the well has caused widespread depressurisation.

3.6.5 Hydraulics

3.6.5.1 *The fracture network and rock matrix*

The rocks that provide the settings for mineral deposits are typically hard, brittle and with low compressibility. Commonly they are igneous or metamorphic rocks, or well-indurated sediments that exhibit an element of dual-porosity behaviour. The matrix will usually possess very low drainable porosity and permeability, and the rock will be fractured. The permeability and porosity of these rocks are largely created by the presence of fractures (Section 3.6.2).

The matric pore space may be intercrystalline, occurring, for example, around the boundaries of crystals in granite, or it may be remnant intergranular porosity in a metasediment. The geological history of burial and compression will usually have decreased the porosity and permeability; or mineralisation and/or weathering may have increased it. Typically matric porosities will be less than 0.01 in granites and other volcanic rocks, and rarely more than a few per cent in rocks such as quartzites. Matric permeabilities will be correspondingly low, perhaps as low as 10^{-13} m/s in unaltered granite (Freeze & Cherry, 1979).

The effects of weathering in rocks such as granites rarely persist for more than a few tens of metres below ground surface (e.g. Trainer, 1988). However, this is often sufficient to cause healing of the fractures and reduce the permeability of the near-surface zone, or of the bedrock surface when this lies below permeable material such as recent alluvial deposits. In this way, the upper part of the bedrock may act as a partial flow barrier, impeding the rate at which groundwater can move downward into the bedrock from overlying saturated alluvium or other porous deposits. The effect of the poorly permeable nature of the bedrock surface is frequently shown around open

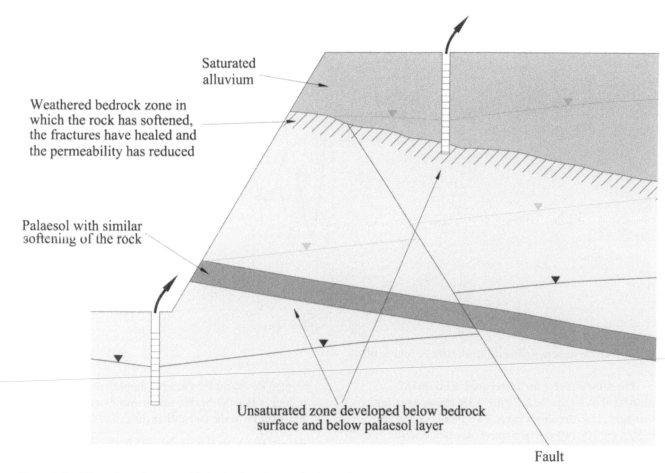

Weathered bedrock zone in which the rock has softened, the fractures have healed and the permeability has reduced

Saturated alluvium

Palaesol with similar softening of the rock

Unsaturated zone developed below bedrock surface and below palaesol layer

Fault

Figure 3.31: Effect of poorly permeable bedrock surface and palaeosols

pits (and underground mines) by a large downward hydraulic gradient that develops between the saturated alluvium and the underlying, fractured bedrock below the effects of weathering, which has been depressurised because of mining (Figure 3.31).

Below the near-surface effects of weathering, permeability in a fracture network is widely found to decrease with depth. Most experimental data indicate that the most pronounced depth-dependency occurs in the upper 100 to 300 m of bedrock. Higher permeability can exist at depth in fractures locked open by bridges of hard mineral filling or large shear dislocation, but these fractures are likely to be insensitive to stress (Rutqvist & Stephansson, 2003). At Diavik, for example, the packer-test data indicate a depth-permeability relationship that causes transmissivity to decrease by 1.5 orders of magnitude over 200 m.

The fracture network may impart significant heterogeneity to the distribution of storage and permeability. The permeability of any part of the fracture network is a function of:

- the permeability of the individual fracture flow pathways;

- the number of these flow pathways that occur in that part of the fracture network;
- the interconnectivity of the fracture network.

A critical component in the determination of fracture permeability is defining the scale at which a value of permeability is required. At the scale of individual fractures, permeability will be highly variable, depending on the specific properties of the individual fracture. As the volume of fractured rock under consideration is enlarged, the behaviour of the fracture network as a whole, rather than individual fracture sets, tends to become the primary control on groundwater flow. Thus, the representative elementary volume (REV) is much larger for fractured rock than for granular materials (Section 1.1.2.6).

As a general rule in fractured rock environments, the flow patterns will become increasingly more homogenous as the scale of measurement increases. In an intergranular medium such as an alluvial sand, for example, the flow through one pore channel will be different from that in another, and it would be impossible to measure all the flows through all the pore channels independently. Darcy's law in effect avoids this problem by treating a sufficiently large volume of the medium as a porous continuum.

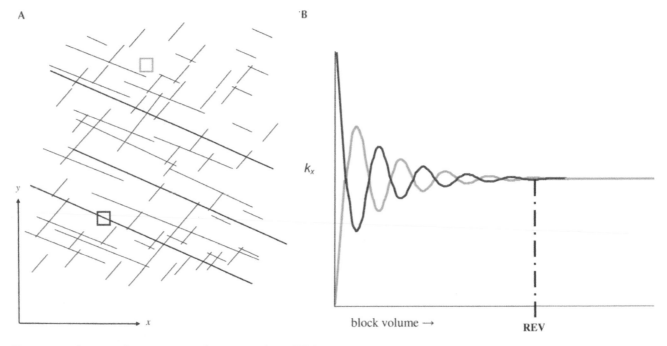

Figure 3.32: Concept of representative elementary volume (REV)

The same concept for a fractured medium is illustrated in Figure 3.32. Figure 3.32a shows a section through a fractured rock mass. Two small regions are represented by two initial squares, one blue and the other green. The permeabilities of the regions are plotted in Figure 3.32b. The squares are then assumed to expand to encompass increasingly larger regions. To begin with, the regions have different permeabilities but, eventually, as the regions become larger and include more fractures, increasing the size of the regions no longer makes a difference. The volume of rock under consideration when this happens is called the representative elementary volume (REV) (Bear, 1972) of the material. A material can be thought of as homogeneous for a property, such as porosity or permeability, when two volumes, each large enough to be a REV, have the same value of that property. The REV is typically larger for fractured rocks than for granular materials.

It should be noted, however, that:

1. the material can be homogeneous without being isotropic; in Figure 3.32a, for example, the value of K_x is likely to be larger than the value of K_y;
2. a series of rock units (such as a collection of beds), each of which may itself be *homogeneous and isotropic*, can in combination act in the same way as a single homogeneous, *anisotropic* unit (Section 1.1.2.3 and Figure 1.10) and such a series may divert flow in a direction other than that of the general hydraulic gradient. In this way sub-pathways parallel to the slope,

created for example by steeply dipping permeable beds or high-angle permeable faults, may divert water sub-parallel to the face rather than towards it.

Factors that impart homogeneity in the fracture network, and hence cause the fracture network to behave more like a porous medium, are:

■ tortuosity in the fracture network and within the individual fractures (from asperity);
■ variations in fracture apertures along their length (the tendency for fractures to pinch and swell);
■ alteration of the rock mass, mineralisation in fractures and healing of fractures;
■ infilling of fractures, and variations in fracture filling along their length, often at a scale of 1 m or less (Figure 3.33).

Of the above, the tendency for fractures to pinch and swell along their length (Figure 1.16a) is of great importance. This has the effect of limiting the maximum transmissivity of a fracture to the transmissivity at the point of minimum aperture. This factor is often overlooked by studies to characterise the hydrogeological regime.

Fracture networks that exhibit less difference in the permeability of different flow pathways will tend to approach the equivalent porous medium (EPM) type of flow at a smaller scale. Placing a conceptual model for a particular mine site within this framework (i.e. identifying the amount of heterogeneity in flow pathways)

Figure 3.33: Illustration of fracture variability and infilling seen in drill core (Photo: Geoff Beale)

is important for establishing how a fracture network should be modelled. Interpretation of the REV is clearly an important consideration for helping to determine what type of numerical model should be used and the grid (or mesh) size of the model (Section 4.2.1). The hydraulic response data from Diavik suggest that a fracture-network model may be necessary only at a scale less than about 40 m, and only for a time period of about a week after a new hydraulic stress has been applied; beyond that, an EPM model is valid.

3.6.5.2 Influence of diffusivity and response times on depressurisation of a fractured rock slope

An important implication of the A-B-C-D concept is that depressurisation caused by drainage will occur first in the largest fractures connected to the drainage system (e.g. wells or drains). Drainage will then occur from successively smaller fractures into larger ones in response to the successive declines in head caused by the drainage. Finally, depressurisation will occur in the blocks bounded by the least permeable fractures, as the matric porosity or micro-fractures depressurise.

Hard rocks usually possess some primary (matric) porosity, albeit small. When the rocks are fractured, both the matric and fracture components can contribute to storage as well as to groundwater movement. The time taken for the centre of a block to respond to the head change in the surrounding fractures will depend on the hydraulic diffusivity, D, of the matrix (Section 1.1.2.8). The hydraulic diffusivity is a function of the permeability or hydraulic conductivity divided by the storage property, and can be written variously as:

$$D = \frac{T}{S} \text{ or } D = \frac{K}{S_s} \qquad \text{(eqn 3.1)}$$

for confined conditions, where T is transmissivity, K is hydraulic conductivity, S is storage coefficient and S_s is specific storage. (For unconfined conditions, the storage parameter would be replaced by the specific yield, S_y.)

The specific storage of a rock is defined as the quantity of water that will be released from (or taken into) storage in a unit volume of rock for a unit decline (or increase) in hydraulic head. The determination of Specific storage (S_s) is described in Section 1.1.5. Generally, the compressibility of fractured rock is about one to two orders of magnitude greater than that of 'sound' rock and generally greater than the compressibility of water. It can therefore be assumed that:

1. the fracture component of compressibility is the same as that of the fractured rock mass;
2. since the porosity of either the matric or fracture component is unlikely to exceed 0.05 (5 per cent), the specific storage will usually be dominated by the compressibility of the fractures.

For a cubic system of three mutually orthogonal fracture sets, the characteristic time for a head change in fractures surrounding a block to reach the centre of the block is given (Barker, 1993) by:

$$t = \frac{x^2}{D} = \frac{x^2 K}{S_s} \qquad \text{(eqn 3.2)}$$

where x is the distance from the fracture to the centre of the block. The specific storage, S_s, in this case of the matrix, can be calculated as discussed in Section 1.1.5.

Figure 3.34 shows values of hydraulic diffusivity for a range of values of specific storage and hydraulic conductivity, and Figure 3.35 shows characteristic response times for various diffusivities and block sizes.

Where the matric porosity and storage are considered negligible, or for dealing with the response time of fractured blocks bounded by larger fractures, the conductivity, porosity, compressibility and storage properties needed will be those of the micro-fractured mass. The implication of these figures is that, in large blocks with low diffusivities, high residual pressures may remain for a long time after the depressurisation of the fracture network. Figure 3.36a shows how depressurisation progresses in a 2 m cube of rock with a hydraulic diffusivity of 10^{-4} m²/s (equivalent, for example, to a hydraulic conductivity of 10^{-8} m/s and specific storage of 10^{-4}/m). Figure 3.36b shows the equivalent depressurisation in adjacent factures.

As an example, when dealing with material with fracture spacings of around 10 m, if the acceptable time for pressures to stabilise in the matrix of the rock is taken as around 30 days (~2.5×10^6 seconds), then the diffusivity must not be less than 10^{-5} m²/s. For a totally unfractured material, the hydraulic conductivity may be as low as 10^{-12} m/s, but the specific storage will only rarely exceed 10^{-7}/m, so the criterion will usually be met. In a fractured material, the specific storage may be as high as 10^{-4}/m; this requires the hydraulic conductivity of the material between major fractures to be at least 10^{-8} m/s.

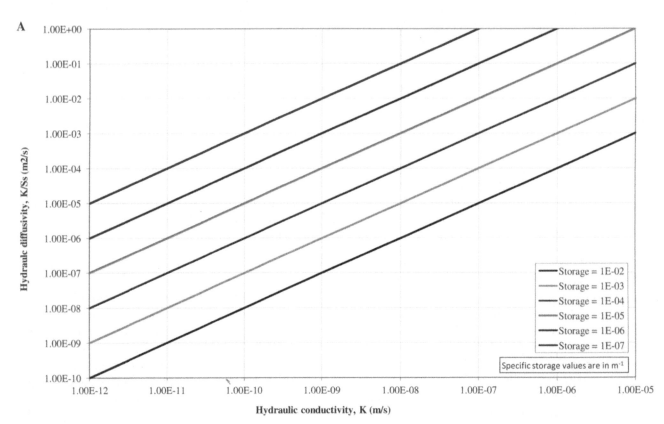

Figure 3.34: Values of hydraulic diffusivity for a range of values of specific storage and hydraulic conductivity

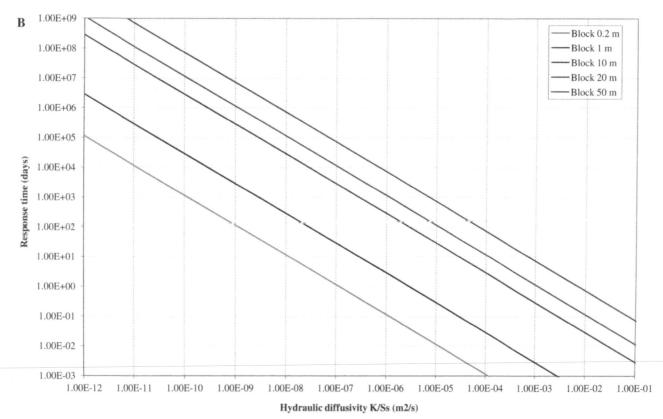

Figure 3.35: Characteristic response times for various diffusivities and block sizes

These times give the response of double-porosity systems to depressurisation by drainage. The response to depressurisation by deformation caused by unloading is considered in Section 3.6.6.2.

3.6.5.3 Controls on the nature and response of the fracture network to depressurisation

At the scale of specific hydrogeological units (i.e. units with similar hydraulic properties, which will typically be of a consistent lithology or alteration type), the important material properties in controlling how a fracture network depressurises over the scale of the pit slope are as follows.

- The bulk permeability of the fracture network.
- The storage and diffusivity of the fracture network.
- Anisotropy in the fracture network.
- The response of the fracture network to deformation.
- Interaction of the fracture network with the matrix.

The interaction of the various flow pathways with the recharge and discharge points is also a key component of evaluating pore pressure distributions within the slope. Differences in the permeability of individual flow pathways and variations in the permeability values of flow pathways over their length can complicate pore pressure distributions in the slope as follows.

- More permeable connections to the recharge source than to the discharge 'sink' will tend to increase pore pressures with respect to the average hydraulic gradient.
- More permeable connections to the discharge 'sink' than to the recharge source will tend to reduce pore pressures with respect to the average hydraulic gradient.

As described in Section 3.6.5.2, the hydraulic diffusivity of the system governs the rate of response of the fracture network to active depressurisation measures. A fracture system with a high diffusivity (i.e. high permeability and low storage) will respond rapidly to depressurisation. This was observed in the lower West wall at Chuquicamata, following the opening of underground drainage holes drilled from the hanging wall into the Granodiorite Fortuna in the footwall of the West fault (Appendix 4). Although the West wall as a whole has low permeability, these drains intercepted a fracture network with a significant local-scale permeability and very low storage. Drain holes with individual flow rates of about 0.5 l/s produced about 400 m of rapid depressurisation, with all installed piezometers responding in a similar manner. Some piezometers showed a slight lag in response; the probable explanation is that they were in blocks of lower permeability, more remote from

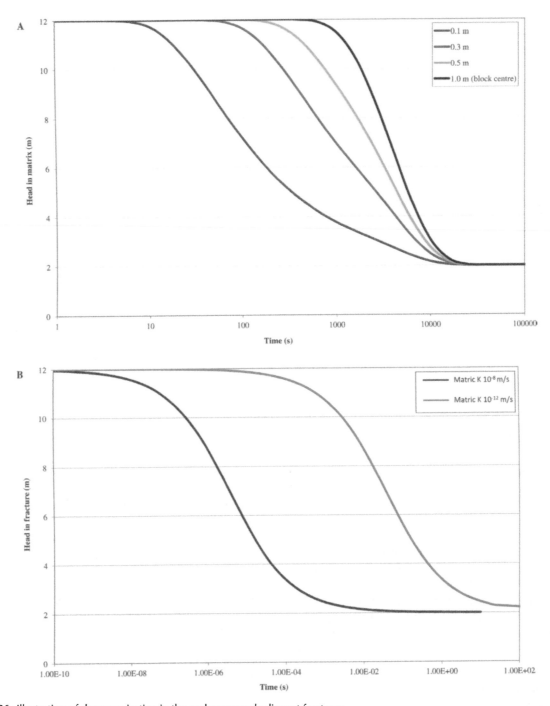

Figure 3.36: Illustration of depressurisation in the rock mass and adjacent fractures

major fractures. However, the difference in response time was only a few days.

Anisotropy in the fracture network controls the relative permeability of the rock mass in any given direction. The orientation of this anisotropy with respect to the recharge and discharge controls is important in governing the hydraulic gradients that can develop toward the pit face. Bulk anisotropy can result from the larger geological structures discussed above (Section 3.6.2.7) and/or can be present in the rock mass itself from the arrangement of the fractures in the fracture network (Section 3.6.5.1).

Prior to mining-induced deformation, the geological factors that exert the greatest influence on the properties of the fracture network are therefore lithology, alteration of the unit in question, and the structural evolution of the area (all of which may be co-dependent).

A key factor is the amount of breakdown of the rock structure that has occurred by weathering and/or alteration processes, and the extent to which the fractures have become healed because of either softening of the entire rock mass or filling and/or softening of the fractures themselves. As described in Section 3.5.3 (the Escondida Mine case history), some of the most difficult slopes to depressurise are in clay-altered rock.

3.6.6 Deformation

3.6.6.1 Pore pressure and effective stress
The relationship between fracture-normal aperture and stress field is particularly pertinent in open pit mining environments as stresses within the rock mass can change significantly during excavation.

The effective stress (σ_e) acting on a material is calculated from two parameters, total stress (σ) and pore pressure (p) according to:

$$\sigma_e = \sigma - p \qquad \text{(eqn 3.3)}$$

If pore pressure increases, this reduces the effective stress. If pore pressure decreases, the effective stress increases (Section 1.1.6.2).

Because the permeability of fractured rock depends in part on the apertures of the fractures and because the aperture of a fracture is partly dependent on the effective stress, it follows that permeability is related to effective stress; an increase in effective stress will usually lead to a decrease in permeability. Furthermore, because the effective stress is partly dependent on the pore pressure, it follows that there is a relationship between the pore pressure in the fractures and their permeability. The interdependence of rock properties and fluid behaviour is referred to as hydromechanical (HM) coupling.

Rutqvist and Stephansson (2003) and Wang (2000) identify two types of HM coupling: direct and indirect. Direct coupling occurs where a change in fracture aperture directly affects pore pressure (i.e. due to the change in storage). Indirect HM coupling occurs when the change in fracture aperture (from deformation) causes a change in transmissivity; pore pressures therefore change as a result of increased or decreased groundwater flow.

Direct HM coupling is termed solid-to-fluid if a change in stress produces a change in fluid pressure. In a pit slope, an example would be when a reduction in total stress caused by excavation of material leads to a reduction in pore pressure. Direct HM coupling can also be fluid-to-solid; this is where a change in fluid pressure causes a change in the volume of a porous or fractured material. An increase in pore pressure, for example, can lead to the dilation of fractures and an increase in permeability. Conversely, a decrease in pore pressure and corresponding increase in effective stress can lead to the closure of fractures and a corresponding decrease in permeability (Section 1.1.2.5).

3.6.6.2 Deformation resulting from excavation of the pit
Deformation of the fracture network occurs from a combination of lithostatic unloading and the tensile stresses that develop in the pit walls. Although relatively restricted to the pit area, in the case of large slopes the zone of deformation can extend 200 m to 300 m or more from the pit face (with a gradational boundary into the underlying undeformed rock). The lithostatic unloading causes a direct hydromechanical coupling response as the fractures dilate and an indirect hydromechanical-coupling response due to improved groundwater flow within these increased-aperture fractures. The magnitude of the deformation from lithostatic unloading depends on a large number of factors for any given rock or alteration type. The most important factors are:

- the amount of lithostatic unloading (i.e. the weight of material removed which, in turn, will depend on the position of a fracture within the pit slope and the depth of the pit);
- the 'stiffness' (compressibility) of fractures within the fracture network (i.e. the amount of potential deformation from stress relief);
- the orientation of fractures with respect to the changing stress field. Fractures orientated perpendicular to the changing stress field (e.g. slope parallel fractures or horizontal fractures on the pit floor) will show a greater propensity for deformation.

As the pit deepens, the amount of lithostatic unloading beneath the pit increases (i.e. very deep pits will typically show a large zone of deformation below the base of the pit). Tensile stresses are also greater closer to the toe of the slope, and extend further behind the pit slope and above the toe of the slope in deeper pits (Stacey *et al.*, 2003).

As transmissivity and storage are strongly depth-dependent, it is probable that the **relative importance** of lithostatic unloading to total depressurisation increases as the pit deepens. In the case studies, the only sites to exhibit the direct response to unloading are Chuquicamata and Radomiro Tomic. With a total depth of around 900 m, Chuquicamata is the deepest pit included in the case studies and also the one with the lowest permeability and the lowest recharge.

The direct response to unloading will depend on the hydraulic diffusivity of the system. The situation can be thought of in simple terms as the opposite of the response to drainage discussed in Section 3.6.5.2. With drainage depressurisation, head is reduced in the larger fractures and the response travels through successively smaller fractures to the matrix. In the case of unloading, the matric pore space and the finer fractures probably occupy a larger volume in total than the larger, more permeable,

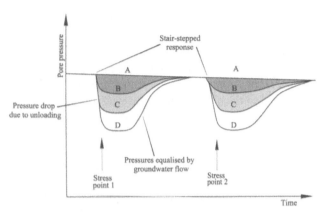

Figure 3.37: Unloading responses in a fractured medium illustrated for the 'A-B-C-D' fracture network

fractures and so will experience a greater increase in volume and therefore a larger initial drop in head. This effect will then be transmitted through successively larger parts of the fracture network.

The response time can be determined, but will be much shorter than was the case in Section 3.6.5.2 in response to drainage, as in this case the head change is moving from a block with relatively high storage to a fracture with low storage instead of the other way around. The compressibility of fractured rock is higher than that of unfractured rock (e.g. Freeze & Cherry, 1979); for granite it is typically around 10^{-10} to 10^{-8} Pa^{-1}. Taking the higher value and a fracture porosity of 1 per cent gives a specific storage value of around 10^{-4} m^{-1}, which is likely to be near the maximum for jointed rock. Figure 3.36 shows the time for the head in bounding fractures to respond to a head decrease of 10 m in a 2 m cube of rock with matric conductivities of 10^{-8} and 10^{-12} m/s. The response illustrated for the 'A-B-C-D' fracture network is shown in Figure 3.37. The time will increase as hydraulic diffusivity

decreases, but even so it is clear that equilibration of pressures will happen much more quickly in response to unloading than in response to drainage. This conclusion is supported by the conditions at Chuquicamata.

An understanding of the loading and unloading effects and the principles involved in the behaviour of clays is of value in understanding HM coupling in rock slopes. Direct hydromechanical coupling responses in clay units are regularly observed in the construction industry. Although not well documented in mining, they are likely to be significant in zones of clay-altered fractured rock in pit slopes.

The amount of deformation displayed by a volume of rock in response to a change in effective stress is governed by its compressibility. Compressibility, α, is the reciprocal of bulk modulus of elasticity, κ, so that elasticity is a measure of change of stress over resulting strain and compressibility is a measure of resulting strain over change of stress (eqn 3.4):

$$\alpha = \frac{1}{\kappa} - \frac{-\Delta V_B / V_B}{\Delta \sigma_e} \qquad \text{(eqn 3.4)}$$

where:
V = bulk volume (rock or soil)
ΔV = change in V
$\Delta \sigma_e$ = change in effective stress responsible for change in volume
The equivalent equation for water is:

$$\beta = -\frac{\Delta V_w / V_w}{\Delta p} \qquad \text{(eqn 3.5)}$$

The negative signs are necessary because volume reduces as effective stress increases.

The behaviour of a rock or soil during loading or unloading can be illustrated by the analogy of a 'spring in a cylinder', filled with water and with a piston that is free

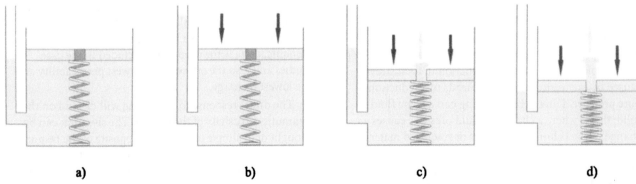

| a) | b) | c) | d) |

Figure 3.38: The 'spring in a container' analogy to explain the 'undrained' (b) and 'drained' (c and d) response to increasing total stress on a saturated material. (The manometer is shown only to represent the pressure of the water in the container – it does not represent a means of input or output for the water.)

to slide (Figure 3.38a). Here, the cylinder or container represents a volume of compressible formation. The spring represents the compressible mineral framework of the formation and the water that fills the cylinder represents the pore water. The lid or piston incorporates a plug or valve that can be open or closed; when it is closed, the arrangement is the equivalent of an 'undrained' system.

The weight of, or on, the lid represents the total stress and the compression of the spring represents the effective stress. A load is applied to the piston with the valve closed (Figure 3.38b). This represents an increase in total stress on a saturated rock or soil in the 'undrained' condition. If the spring is relatively weak the increased load will be resisted by the water, which is much less compressible, and the water pressure will rise. In the case of a rock or soil, the formation will frequently be more compressible ('weaker') than water but this compressibility comes about by deformation of the framework: it is generally assumed that the grains of the medium are barely compressible. The volume of the rock or soil can therefore decrease only because of a decrease in the pore volume. As the pore volume is filled with water, this requires the water to compress and its pressure to increase.

If the valve or plug is then opened, water will leave the container, the water pressure will drop (Figure 3.38c) and the load will be transferred to the spring, which will compress (Figure 3.38d), increasing the analogous 'effective stress' and giving the 'drained' condition.

Undrained behaviour in the field is generally the result of low permeability; in a permeable formation the increased pore pressure causes a rise in head which causes the water to migrate quickly away from the region that has experienced the increase in stress, so these materials typically show drained behaviour. In a poorly-permeable formation, the excess head may take years or longer to dissipate. As the head and pore pressure reduce, the load will transfer to the framework of the rock which, like the

analogous spring, will be compressed and reduced in volume – which usually means in vertical thickness.

A reduction in loading produces the opposite effect. In the 'spring in the cylinder' analogy, removing load from the lid of the cylinder allows the spring to expand (Figure 3.39a). This increase in volume means that the water contained in the cylinder will be at reduced pressure (Figure 3.39b): if the lid moves far enough that the volume of the cylinder is greater than that of the contained water, water will vaporise to fill the extra space. In the case of a rock mass, if the total stress is reduced quickly (as by removal of overlying material during excavation) the increase in volume will usually be accommodated by an increase in pore volume in the fracture network. This is because the compressibility (and therefore the scope for expansion) of the mineral grains or crystals is generally two or more orders of magnitude less than that of the framework, so it is the framework that expands and not the individual grains. This increased pore volume will lead to a reduction in pressure of the contained water; if the rock is highly permeable, then the corresponding decrease in head will cause water to move in quickly from surrounding material unaffected by the decrease in total stress, giving a drained response. If the rock is poorly permeable, the direct solid-to-fluid HM coupled response will be of the undrained type which, as with the loading case, may mean the reduced pore pressure persists for a very long time.

The way in which a change in total stress ($\Delta\sigma$) is distributed between a change in pore pressure (Δp) and a change in effective stress ($\Delta\sigma_e$), for an undrained situation in which all the volume change occurs as a change in pore volume, can be calculated from Skempton's coefficient, B, as follows:

$$\Delta p = B\Delta\sigma \qquad \text{(eqn 3.6)}$$

Skempton's coefficient, in a simple case where grain compressibility is negligible, is given by:

$$B = \frac{\alpha}{\alpha + n\beta} \qquad \text{(eqn 3.7)}$$

B is therefore equal to the tidal efficiency term of Jacob (1940) or the pore pressure coefficient (Domenico & Schwartz, 1997). An alternative version of Equation 3.7 using modulus (κ) in place of compressibility (α and β) is provided in Section 4.2.8.

Equations 3.6 and 3.7 are effectively saying that, as expected, for a rock that is relatively incompressible ($\alpha\rightarrow0$), B will tend to zero and for any change in total stress the change in pore pressure will also tend to zero, because the majority of the extra stress is taken by the rock framework. (In the spring analogy of Figures 3.37 and 3.38, the spring is so strong that the lid of the canister hardly moves in relation to the change in load.) For a weak rock, α, the

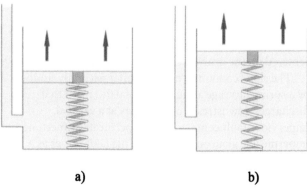

a)   **b)**

Figure 3.39: The 'spring in a container' analogy to explain the 'undrained' response to decreasing total stress (lithostatic unloading) on a saturated material

compressibility of the rock will be much larger than the term $n\beta$ in Equation 3.7, B will tend to unity and most of the extra load will be taken by the water unless and until drainage can occur. Domenico and Schwartz (1997, Table 8.4) suggest that Skempton's coefficient (which they have as R/H) varies from 0.99 for clay to 0.12 for basalt.

In fractured material, using the 'A-B-C-D' model, the bulk of the rock will have matric or 'D'-type fracture porosity and this is where the greatest volume change will occur. The effect will then spread through the larger ('C-B-A') fractures.

Rearranging Equation 3.4 gives:

$$\Delta V_B = -\alpha V_B \Delta \sigma_e \qquad \text{(eqn 3.8)}$$

Usually, it can be assumed that all rock expansion occurs by expansion of void space (intergranular or fracture) so that the change in bulk volume is almost exactly the same as the change in pore volume, i.e.:

$$\Delta V_B \cong \Delta V_P \qquad \text{(eqn 3.9)}$$

The change in pore pressure resulting from the change in volume of the water-filled pore space can be determined using Equation 3.5:

$$\Delta p = \rho g \Delta h = -\frac{\Delta V_P/V_P}{\beta} \, or -\frac{\Delta V_B/V_P}{\beta} \quad \text{(eqn 3.10)}$$

It follows from Equation 3.10 that the change in pore pressure as a result of unloading will be greater in a formation with a lower porosity than in a more porous unit (though it should be borne in mind that materials with lower porosity may also have lower compressibility and therefore expand less for a given reduction in total stress).

The reduction in pore pressure following unloading can have a significant depressurising effect in slopes of low porosity and permeability in deep pits. The drained response to unloading can be seen in the Yanacocha Sur example in Figure 3.29. Here the drop in pore pressure during or after the excavation of each bench is followed by a recovery of pore pressure, so that this is effectively a 'drained' condition.

In the case of Chuquicamata, there has been a significant and permanent reduction in pore pressure. Beneath the East wall at Chuquicamata there is a pervasive downward hydraulic gradient that can exceed 0.6. A downward gradient also occurs beneath the central and northern parts of the pit floor. With material of extremely low permeability, this gradient cannot be attributed to a normal groundwater flow system and is most probably the result of expansion of the rock caused by stress relief as a result of removal of the material excavated from the pit. In material of very low permeability, such as that at Chuquicamata, fluid migration will be so slow that the condition can probably be treated as effectively 'undrained' and any recovery will be very slow. As discussed in Section 3.5.4, the downward head gradient

that occurs beneath the pit walls and floor at Chuquicamata (and also at Radomiro Tomic) is likely to be the result of an increased relative importance of lithostatic unloading with increasing depth. The magnitude of deformation decreases with depth but, because the porosity also decreases with depth, the observed reduction in pore pressure is greater due to the 'undrained' conditions.

3.6.6.3 Potential for association of depressurisation and deformation

The most common consideration of hydromechanical coupling in mining environments is the response to lithostatic unloading, which is mostly observed to cause a dilation of fractures with the potential for a reduction in pore pressure and an increase in permeability (solid-to-fluid coupling). There is also the potential for poro-elastic contraction of fractures and consequently a reduction in permeability because of depressurisation (fluid-to-solid coupling) which could, in some circumstances, create a reduction in groundwater movement and could slow the rate of depressurisation. However, the effect of depressurisation on contraction of fracture aperture is thought to be small in comparison to the potential dilation response of a fracture to unloading, hence fluid-to-solid hydromechanical coupling is likely to be a consideration only if dewatering is initiated well in advance of excavation, or as a result of depressurisation after the pit has reached full depth. *In-situ* measurements of slope tilt and fracture deformation at the Coaraze test site, southern France, show field evidence for fluid-to-solid coupling (Guglielmi *et al.*, 2008).

The Coaraze test site corresponds to the lower 15 m section of a 40° to 60° dipping slope comprised of a thick sequence of fractured limestone (Figures 3.40A and 3.40B). The slope is cut by 12 parallel bedding planes, with a 40° strike dipping 45° SE, and two sets of approximately orthogonal near-vertical faults. There are 12 well-identified faults with a decametric continuity in the slope that form a fracture network with 0.5 m to 2 m spacing (Figure 3.40C). Discontinuities less than a metre continuity correspond to minor bedding planes and fractures, located within 200 mm to 500 mm thick bands centered on the near-vertical faults (Guglielmi *et al.*, 2008).

The slope contains a minor aquifer naturally drained by a spring (average annual yield of 0.012 m³/s) that discharges downstream of the valley at a vertical impervious fault contact between permeable limestone and impermeable glauconious marls. This fault serves as a natural dam for water stored upstream in the slope. The spring is artificially controlled with a water gate (Figure 3.40A). When the gate is completely closed, fluid pressures stabilise at about 8 m above the gate (Figure 3.40D).

The experimental approach consists of performing simultaneous and multi-frequency measurements of fluid

Figure 3.40A: Coaraze test site: The slope is naturally drained by a spring that can be artificially closed or opened by a water gate located at the shed to the lower right of the picture (Photo: John Read)

Figure 3.40B: Reverse shot of Figure 3.40A, looking downwards towards the water gate, showing the limestone bedding (Photo: John Read)

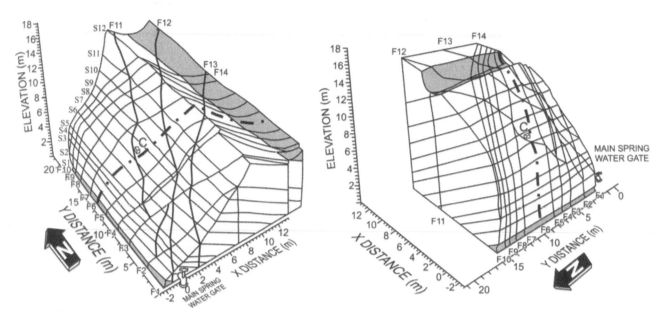

Figure 3.40C: Coaraze test site 3D geometrical model of rock mass fractures viewed from two angles (From Cappa *et al.*, 2005)

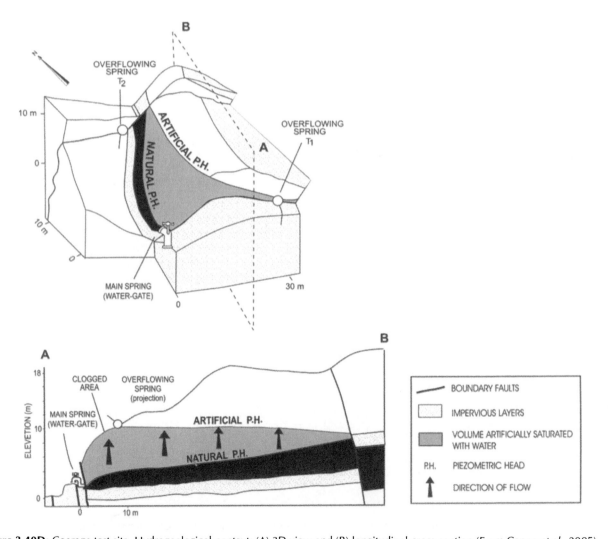

Figure 3.40D: Coaraze test site. Hydrogeological context: (A) 3D view and (B) longitudinal cross-section (From Cappa *et al.*, 2005)

pressures and displacements at different points and on different fracture types within the carbonate reservoir. Two kinds of experiments were conducted (Cappa, 2006).

1. At the fracture network scale, a global hydraulic loading by groundwater level change shows that the coupling between fluid flow and deformation is simultaneously governed by a dual-permeability hydraulic behaviour and a dual-stiffness mechanical behaviour.
2. At the single fracture scale, the hydromechanical behaviour was evaluated by performing hydraulic pulse injection testing. This test was monitored using high-frequency (f = 120 Hz) hydromechanical measurements conducted with innovative fibre-optic borehole equipment.

Closing and opening the gate valve produced very rapid pressure response in the vertical faults, with elastic

deformation shown by the extensometers in response to the rising and falling piezometer heads (Figure 3.40E). There was a lagged (but fast) piezometer response in bedding planes, with the response continuing after completion of the testing period, but with no associated movement on the extensometers. There was a poorly discernible piezometer response in the intact rock mass. The experiments show field evidence for fluid-to-solid coupling and confirm the potential for poro-elastic contraction of fractures and consequently a reduction in permeability.

The experiments also showed that discontinuities that are almost parallel to the direction and dip of the slope direction (vertical faults) are in general widely opened. The bedding planes which dip inwards from the face are more closed than slope-parallel faults, mainly because of the higher compressive stresses. This observation is particularly pertinent to mining applications, where

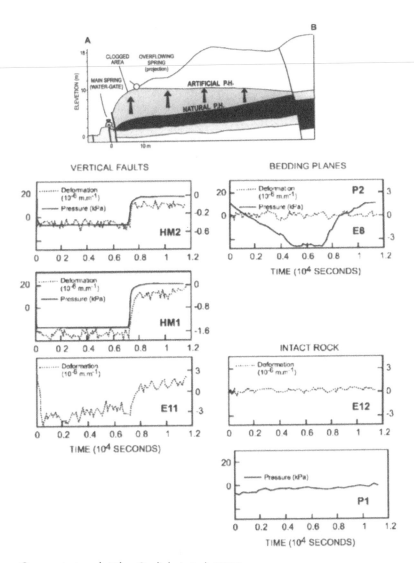

Figure 3.40E: Results from Coaraze test work (After Guglielmi *et al.,* 2008)

secondary structures perpendicular to bedding and parallel to slope could open preferentially. As a result they would be prone to enhanced recharge (or discharge), with the associated increases in pore pressure on the discontinuity.

Guglielmi et al. (2008) suggest that highly permeable elements of the slope will respond to short period oscillations, like those induced by daily infiltration, whereas long period oscillations, like those linked to seasonal recharge fluctuations, will affect a wider range of slope elements and even the poorly permeable parts of the slope will be affected. This observation compares well with sites such as Antamina, where major seasonal responses in pore pressure are observed, and further highlights the need for an evaluation of seasonal fluctuations in pore pressure that may be exacerbated by deformation processes. The freezeback which occurs at Diavik is another example, but the head fluctuations here are controlled by the availability of a route of discharge (analogous to the closing of the water gate at Coaraze).

When fluid pressure changes occur within the fracture network, the hydromechanical coupling is direct in the highly permeable faults where a pressure change induces a deformation change. No direct hydromechanical coupling occurs within the lower permeability zones where deformation is not directly correlated with pressure changes. This means that the mechanical deformation of the bedding planes and rock matrix is induced by the fault deformation (Cappa, 2006).

Moreover, from pulse tests it was found that the key parameters to coupled hydromechanical processes in such fracture systems are the initial hydraulic aperture and normal stiffness of the fracture, the stiffness of the rock matrix and the geometry of the surrounding fracture network (Cappa, 2006).

Subsidence in granite has been identified where depressurisation of the fracture network is not carried out alongside lithostatic unloading (as it is with an open pit mine). In an open pit, fracture dilation from lithostatic unloading tends to offset any potential aperture contraction that might occur from fracture depressurisation. Vertical settlements with magnitudes reaching 120 mm were measured in fractured crystalline rock several hundred metres above the Gotthard highway tunnel in central Switzerland, as a result of depressurisation around the tunnel (Zangerl et al., 2003). The magnitude of settlement could not be attributed to poro-elastic contraction alone, suggesting that some structural settlement has also occurred. It is conceivable that fractures that have previously dilated in response to lithostatic unloading could revert to their original condition when the pore pressure reduces.

3.6.6.4 Fluid-to-solid hydromechanical coupling from blasting

The Diavik case study shows how the pressure pulse caused by blasting can affect pore pressures within a fracture network, over both the short term and medium term. It is thought that gases released during blasting cause 'instantaneous' pressure increases in the interconnected fracture system. The pressure increases can result in fluid-to-solid hydromechanical deformation, potentially causing dilation of fractures. These effects may, in turn, reduce the frictional resistance to slip, and lead to small-scale movements in the fracture network in response to the stress fields present in the slope. The highly variable responses to blasting in the VWP datasets at Diavik support the idea of a fracture network that is evolving in response to the stress field developing behind the pit slope.

It is not known whether changes resulting from the pressure pulse entering the fracture network are significant enough to cause any major increases in the permeability of the rock mass as a whole. However, changes in connections to the recharge or discharge points between different parts of the pit face, and between single fractures, can cause changes to the **distribution** of pore pressures within the slope. In the Diavik dataset, these appear as deviations of pore pressure either side of the long-term trend. The predictability of the magnitude of these deviations may warrant further study.

3.6.6.5 'Over-break' zone (or blast damaged zone)

The 'over-break' zone (or blast damaged zone) represents the 'rind' where rock properties and hydrogeological conditions are altered as a result of the mechanical excavation processes. It is developed because of the shock wave of the blast, and potentially also because of fluid-to-solid hydromechanical coupling processes that result from pressure pulses being transmitted through the fracture network at the time of the blast (Section 3.6.6.4).

The 'over-break' zone should be considered differently from the zone of deformation ('zone of relaxation') created by the stress reduction that occurs due to mining of the overlying material in the slope (lithostatic unloading). However, the two are clearly inter-related, and the lithostatic unloading processes may start as soon as stress reduction occurs at the time of the blast.

Based on subjective observations at a number of mine sites, it is considered likely that changes in material properties resulting from the shock wave may extend from less than 5 m behind the slope (for a small pit with 6 m high benches and controlled blasting) to more than 40 m behind the pit face (for a large pit with 15 m high benches and large blast patterns). The thickness of the over-break zone will also be highly dependent on rock type. The case study at Diavik suggests that deformation because of fluid-to-solid coupling may also cause changes to fractures deeper into the face.

Many of the recent pore pressure models employed for pit slopes use permeability values that are about three orders of magnitude higher for the blast-damaged zone than for the undisturbed rock behind the face, but the increase may be higher than this in unaltered brittle rock types. As a result, there is a tendency for groundwater to move preferentially at shallow levels behind the pit slope. The porosity of the rock mass is also increased within this 'rind'.

3.6.7 Transient pore pressures

Transient pore pressures occur as a result of short-term changes to the hydrogeological system. They may be:

- seasonal as a result of climatic fluctuations (wet-season, dry season; frozen, unfrozen);
- short-term as a result of discrete rainfall events (or blasting events);
- short-term or progressive as a result of movement of the materials within the slope.

While transient pressure changes may occur at any depth or location within the slope, they are often of most concern at shallow levels. Many slope failures occur at a relatively shallow depth in the slope ('skin-failures') where confining stresses are low: it is clear that a rise in pore pressure at a depth of 30 m behind the slope would have a much greater **proportional** influence on the effective stress than a failure 300 m behind the slope, where the confining stresses are much greater.

Transient pressures may occur in slope sectors which have no pore pressure for the majority of the year. The movement of water in the over-break zone may be of particular relevance. In Figure 3.41, the foliation and joint orientation occurs sub-perpendicular to the strike of the slope, allowing shallow water within the slope to easily drain onto the bench. If the joint sets were to occur sub-parallel to the strike of the slope, drainage of the shallow water onto the bench would occur to a lesser extent. This situation may become compounded by the fact that joints that occur perpendicular to the slope have often become dilated, so a rise in pore pressure may aggravate

Figure 3.41: Example of foliation and joint orientation sub-perpendicular to the strike of the slope where drainage of shallow water can easily occur

Figure 3.42: Example of sub-horizontal clay bands that may cause a temporary build up of transient water pressure

the situation and create a condition where bench-scale toppling may occur. In Figure 3.42, sub-horizontal clay bands may cause a temporary build up of water pressure, reducing the strength of the weaker clay material and potentially creating a condition of instability.

Clearly, the potential for transient increases in pore pressure at shallow levels in the slope is strongly related to surface water events and the infiltration of in-pit runoff. Surface water management is discussed in Section 5.3 and is a major consideration for the slope design process in high rainfall (or seasonal) settings and/or in areas of weak rock. Many mining operations are becoming increasingly aware of the role of surface water in the performance of their pit slopes. Another potential consideration is the discharge from horizontal drains which may move down the over-break zone if not reticulated. Losses from within horizontal drains at shallow levels behind the slope may also cause water to enter the over-break zone. In some settings, the drains may require collar casing installed to a sufficient depth to prevent this from occurring (Section 5.2.3.5).

Transient pore pressures may also relate to seasonal fluctuations in groundwater conditions within saturated parts of the slope. In seasonal climates, there may be a large difference in piezometer readings between the dry season and the wet season. This needs to be carefully factored into the slope design analysis, considering that the potential for slope instability at many locations is greater during the wet season. The change in pore pressure that resulted from freeze-back at Diavik (Section 3.3.4) is another example of seasonally varying conditions.

Transient pore pressure changes related to blasting in the North-west wall of Diavik are described in detail in Section 3.3.3 and Appendix 3. While the magnitude of the pressure pulse may potentially be greater for Diavik than at other sites, the analysis suggests that increases in pore pressure during and immediately following a blast may warrant further consideration and study. At Diavik, the high pore pressure response was considered to be a combination of (i) the very high diffusivity of the granite wall rock, and (ii) the ongoing recharge moving downward behind the slope which created a form of hydraulic confinement and magnified the amount of pressure rise.

Examples of (i) how pore pressures above the back-scarp of an area of slope movement in a saprolite setting reduced rapidly in response to the slope movement, and (ii) how pore pressures gradually increased within an active failure plane and quickly dissipated at the time of accelerated slope movement are discussed further in Section 3.7.2.

Overall, the rapid rise in the general number of piezometer installations worldwide will undoubtedly lead to an increasing understanding of the relationship between movement of materials within the slope and transient pore pressure changes.

3.7 Conclusions

3.7.1 Key factors

There is no generic conceptual model for pit slope depressurisation. There are, however, some common factors and processes that must be evaluated for developing the conceptual model, and for making decisions regarding a slope depressurisation program.

The data that are currently available from the worldwide mining industry indicate that the following questions are most important and must be addressed.

1. Characterisation of recharge:
   - how much recharge is occurring to the slope?
   - where is it entering the slope?
   - is it possible to intercept it?
2. Evaluation of the water balance of the slope:
   - what is the balance between water entering the slope and water discharging from the slope?
   - how may the water balance change with time as the slope is progressively excavated?
   - will a pushback of the slope cause new hydrostratigraphical units to be encountered?
3. Understanding compartmentalisation:
   - is the structural regime causing lateral or vertical compartmentalisation of the slope?
   - do different parts of the slope need to be dealt with differently?
4. Evaluation of the discharge pathways:
   - are there natural permeable drainage pathways that provide the potential for water to discharge from the slope at a greater rate than it may enter the slope (resulting in lower heads)?
   - are these sufficient to create the desired level of depressurisation, or do additional discharge pathways need to be created by installing wells or drains?
5. Understanding the behaviour of the local-scale rock mass and fracture network:
   - how do point measurements from packer tests and other *in situ* measurements relate to the effective permeability at a scale greater than the REV?
   - is there sufficient permeability in the rock matrix or fracture network for the water to drain into the permeable discharge conduits?
   - if the permeability or connectivity is too low to achieve the desired drainage, will deformation of the fractures as a result of lithostatic unloading cause depressurisation at a sufficient rate?

6. Evaluation of any localised zones of low permeability:
   - are there zones of clay alteration, or zones of very low fracture permeability, where a significant lag in the drainage time can be expected?
   - does the surrounding rock mass need to be drained at an advanced rate to provide sufficient time to allow these less permeable zones to depressurise?
   - how can recharge entering these poorly permeable zones be minimised?
7. Understanding the importance of transient pore pressure:
   - is the climate such that seasonal variations in pore pressure need to be considered either farther behind the slope in the saturated zone or at shallow levels in the unsaturated zone?
   - what is the nature of the slope materials that may potentially be susceptible to seasonal or rapid increases in pore pressure at shallow levels in the slope where the proportional effect on effective stress may be greater?
   - is it necessary to implement a program of surface water or shallow groundwater control to reduce the potential for seasonal infiltration and therefore reduce the transient build up of water pressure in the shallow zone behind the pit face.

The case studies have demonstrated that, wherever it is practicable, the interception of recharge before it reaches the slope domain is often the most effective method of pore pressure control. Implementation of good in-pit surface water management may be important for reducing the potential for transient pore pressure increases.

3.7.2 Hydrogeological setting

3.7.2.1 General

A good understanding of the wider-scale hydrogeological setting is fundamental to the development of a slope-scale conceptual model. This incorporates recharge and interactions with the site-wide groundwater system, and how discharge may occur from the slope itself.

The hydrogeological setting can be thought of as comprising three regimes (sets of factors).

1. Geological: the framework in which the water occurs.
2. Hydrological: representing the input of water to and output of water from that framework.
3. Hydraulic: the factors that control the way that the framework constrains the water movement within the slope materials.

3.7.2.2 Geology

The geology is essentially immutable except for changes that may be caused by the activities of the mining operation itself. Given that the geology will have a major influence on the fundamental hydraulic properties (porosity, primary permeability, fracture frequency, and anisotropy) and may create boundaries to the groundwater inflow field, it is always of great significance.

3.7.2.3 Hydrology

Hydrological factors include recharge, surface and subsurface flow and discharge from the slope domain. The influence of the hydrology on the slope can usually be modified by the installation of interactive depressurisation measures.

For all case studies presented in this document with the exception of Diavik, it has been possible to implement an interactive program to enhance the depressurisation of the slopes.

3.7.2.4 Hydraulics

The hydraulic factors fall into the three major groups described below.

Presence and distribution of discrete permeable features

The density, orientation and interconnectivity of the discrete permeable features (fault zones, high-order fracture zones, or the permeable layers) within the slope control the nature and magnitude of flow (i) into the slope from the recharge zones, and (ii) out of the slope to the discharge point(s). Therefore, the presence of permeable features governs the resulting distribution of pore pressures within the slope.

The properties of the permeable fracture zones are a function of lithology, alteration and the structural history that has influenced the slope materials. An understanding of the site-scale structural geology and of the range of hydraulic properties attributable to the permeable fracture zones is therefore important for understanding and predicting the depressurisation of the slope.

The actual permeability of individual features that is required to achieve drainage is relative, and must be compared with the average permeability of the slope materials and with the water balance of the slope. For example, in the NW Wall of Diavik, fractures in individual holes were encountered with sufficient permeability to allow removal of >1 l/s from the pumping well, but this was not sufficient to overcome the *en masse* movement of water through the contributing lower order fracture network. The opposite example is the West wall of the Chuquicamata pit, where fractures in individual holes produced <1 l/s, but this was sufficient to allow complete drainage of the entire interconnected lower-order fracture network within about a week.

Distribution, orientation, permeability and interconnection of local-scale fracturing

Within the context of the total flow system in the slope domain, the local-scale distribution of lower order fracture and joint sets will control how quickly drainage can occur towards the more permeable features and discharge points.

The distribution, orientation, permeability and interconnection of local-scale fractures is controlled by the lithology, structural history, alteration and mineralisation, and degree of weathering, together with later events that may cause infilling or subsequent closing or opening of certain fracture sets.

When evaluating fracture hydraulics, it should be understood that the *contrast* in hydraulic properties *between* the different units often exerts a greater control on depressurisation patterns than fracture variability within a single unit.

Mining-induced changes

Mining may result in changes to the hydraulic properties of the slope materials in four ways:

1. **Changes induced by mining and mechanical excavation.** These predominantly result from the shock wave during blasting, and create the 'over-break' zone, which may typically be 5 m to 40 m in thickness. In addition to this, the pressure wave resulting from the blast may move through the interconnected fracture network and create fluid-to-solid hydromechanical changes, and these may propagate deeper into the slope than the mechanical effects of the shock wave.

2. **Deformation caused by removing the weight of the overlying materials (lithostatic unloading).** This may create solid-to-fluid changes ('direct coupling'), which most typically create an expansion of the fracture aperture and pore space, and consequently cause a reduction in pore pressure. Because of the low porosity of most fractured rock masses and the high bulk modulus of water, there is potential for significant drops in pore pressure to occur as a result of unloading. However, in reality, large pressure reductions are seldom observed in fractured rock because transient flow of water occurs towards the depressurised area to damp and equalise the pressure drop. The solid-to-fluid changes may also cause an increase in permeability if the deformation also affects the fracture interconnectivity, particularly if there are associated changes in the stress patterns, which lead to slippage on the fracture surfaces. As a result, the damping of pressure changes can occur quickly, depending on the availability of water in the system.

3. **Deformation caused by drainage and reduction of pore pressure.** This fluid-to-solid coupling may increase effective stress which, in turn, may lead to a reduction in fracture aperture and permeability. It may be permanent or temporary, depending on the nature of the materials. In unconsolidated materials, where the pore space often forms a large proportion of the rock volume, it may create a significant volume reduction, and lead to consolidation.

4. **Changes in the hydraulic properties of materials as a result of slope failure.** Failure of the slope materials causes changes to the mechanical properties of the intact rock. Although poorly documented in the field, failure is also expected to have a large influence on the hydrogeological properties of the materials. It is thought the response is often one of increasing porosity and permeability, leading to reduced pore pressure and improved drainage characteristics (Figure 3.43). However, failure may also result in compaction of poorly permeable materials within or below the failure plane, leading to increased pore pressure and reduced drainage characteristics (Figure 3.44).

The effects of lithostatic unloading and dissipation of pore pressure are primarily governed by the compressibility of the rock and its hydraulic diffusivity. Highly altered argillaceous rocks are typically prone to unloading affects, due to their combination of high compressibility and low hydraulic diffusivity. Fractured hard rocks also experience temporal variations in hydraulic properties during mining but, in more permeable formations, local-scale changes in pore pressure tend to create flow towards or away from the affected zone which, in turn, tends to attenuate the changes in pore pressure. Therefore, the effects are typically most notable in the less permeable parts of the flow system.

Experience from the case studies presented in this section indicates that effects of deformation on pore pressure start to become important in zones where the permeability is less than 10^{-8} m/s (Read & Stacey, 2009; Varona, 2009) and where the diffusivity is less than about 10^{-4} m²/s. In the case studies presented, only Chuquicamata and Radomiro Tomic show piezometer responses which can be attributed to deformation and these are the only sites where the permeability is generally less than 10^{-8} m/s.

3.7.3 Nature of the conceptual model

A conceptual groundwater model requires a series of assumptions which must produce a realistic, albeit simplified, representation of the hydrogeological system. The conceptual model must have a level of detail that is commensurate with both the available data and the objectives of the study.

Defining an appropriate scale of measurement for the study is therefore a critical step in the conceptual model development. Where spacing between permeable fractures is small in comparison with the volume of rock being studied, and where the fractures are reasonably well

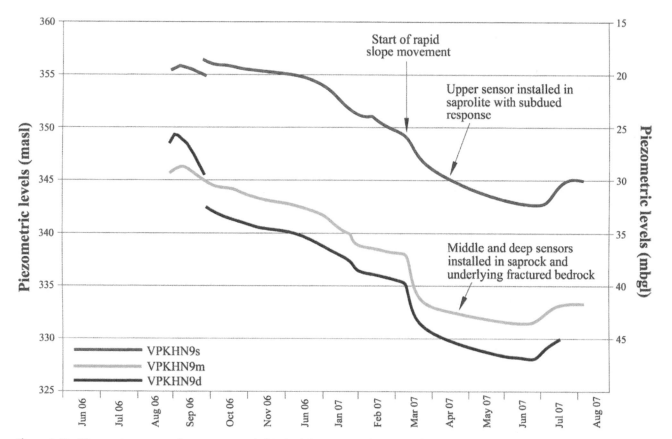

Figure 3.43: Observed response of pore pressures behind a failure in saprolite material

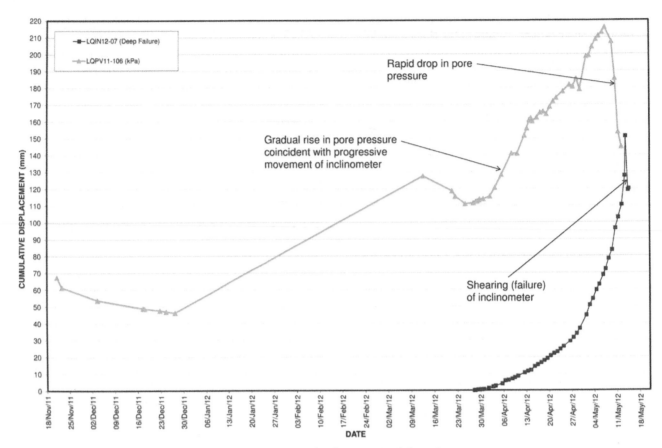

Figure 3.44: Observed pore pressures of piezometer positioned within an active failure plane

interconnected, the fractured rock usually behaves as an equivalent porous medium (EPM). In contrast, the characterisation of a rock volume in which the majority of flow is attributable to a relatively small number of widely spaced and perhaps poorly connected fractures will almost certainly benefit from a conceptual model that represents these 'flow arteries' discretely.

In the Diavik case study, it is postulated that flow through the fracture network can generally be considered to be EPM (except for the areas located close to the localised pumping influence for short time periods). Pumping from wells (or flow into drains) relies on the presence of higher-order fractures to transmit water into the hole at an adequate rate. Close to the point of a new or changed pumping stress, the pore pressure in high-order fractures may be significantly lower than in the adjacent lower order fractures.

The use of a discrete facture network (DFN) to support the conceptual model development can be considered for sites where sufficient field data can be obtained to constrain the uncertainties associated with the DFN. Where uncertainties in the DFN model output are large, a DFN-based conceptual flow model presents few, if any, benefits over an EPM model, although equivalent permeability distributions derived from the DFN have the potential to add value. The DFN approach is potentially more beneficial where flow is controlled by a fewer number of features that can be generated deterministically, provided that the hydraulic properties and interconnection of these features, and changes in property values along the length of the features, can be characterised.

The conceptual model should ideally be developed as early as possible in the planning process. It should be continually updated and refined, and ongoing field programs should be adjusted in line with the evolving conceptual model. The conceptual model forms the basis for underpinning and validating any numerical modelling work (Section 4).

4

NUMERICAL MODEL

Loren Lorig, Jeremy Dowling, Geoff Beale and Michael Royle

Section 4 deals with the set-up and operation of numerical models and the input of pore water pressures into the slope design studies (Figure 4.1). Section 4.1 outlines how numerical models are planned, Section 4.2 describes how they are developed and Section 4.3 outlines how they are used in slope stability analyses. Appendix 5 contains two case studies (Diavik and Cobre Las Cruces) that illustrate a range of modelling approaches.

As discussed in the Introductory section of this book, it is apparent from the case studies (Section 3) and from general worldwide experience that the wider site-scale hydrogeology (and particularly recharge) is the most important factor for the evaluation of water in pit slope stability. In many cases, it is not possible to achieve the required pit slope pore pressure goals without the operation of a general mine dewatering system. Therefore, the content of Section 4 includes some general principles of modelling and considers the modelling approach taken to simulate mine dewatering systems.

4.1 Planning a numerical model

4.1.1 Background

4.1.1.1 What is a numerical groundwater model?

A groundwater model is a means of representing a real groundwater system, allowing investigation of features and hydraulic behaviour of the system relevant to a particular study. The term 'model' could refer to a small-scale laboratory tank model of a field situation, or a conceptual model based on interpretation of field data.

A numerical groundwater model uses a computer program to apply the equations of groundwater flow (essentially a combination of Darcy's law and the principle of continuity or conservation of mass) to describe the behaviour of groundwater within a specified volume of geological materials called the model domain. The region to be modelled (the domain) is typically divided into a finite number of cells or elements, each of which can have its own distinct hydraulic properties (permeability, storage, specific yield) and boundary conditions (e.g. recharge, abstraction, or a specified groundwater head). Known values of head, groundwater flux or recharge must be defined at the model boundaries, along with the initial conditions; internal features can also be defined.

The computer then solves the governing equations for each element or node using an iterative process. Based on the parameters assigned within the domain, the model computes values of head at the centre of cell or the connecting corners of each element, depending on the code. If the groundwater heads and flows do not change with time, the model is steady-state, and will produce a single set of head and flow values. If the heads and flow vary with time, it is described as non-steady-state or transient, and the head values can be calculated for a chosen series of time increments.

Early models divided the model domain using a rectangular grid and calculated the results at the centre of the resulting blocks; these are finite-difference models. These models are still widely used (such as MODFLOW), but a later variation divides the domain into triangular or irregular quadrilateral elements and is known as a finite-element model (such as FEFLOW). This model type has the advantage that it is potentially able to cope better with complex geometry and anisotropy.

The value of any numerical model is heavily dependent on having sufficient reliable supporting data and having the correct conceptual model. The numerical model is usually calibrated to historical data on groundwater heads and/or flows to obtain an acceptable representation of the real system. The model can then be used to make forward predictions about the future behaviour of the groundwater (or surface water) system.

Groundwater flow models can be grouped broadly into three categories.

1. Equivalent Porous Media (EPM) models.
2. Discrete Fracture Flow (DFF) models.
3. Double-porosity models.

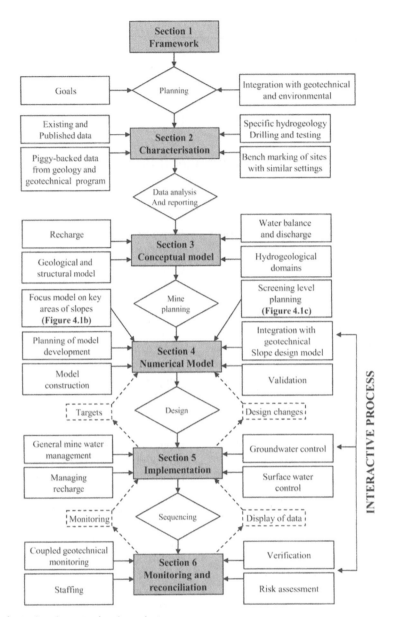

Figure 4.1A: The hydrogeological pathway in the slope design process

The majority of groundwater systems can be characterised and modelled using an EPM approach. The procedure when a discrete fracture flow model is needed instead is described in Section 4.2.6. Double-porosity models are seldom used in mining applications because their codes are typically at an early stage of development and require a level of supporting characterisation that is not practical to achieve for most mining situations. They are not discussed further in this book, but their application may become increasingly important for simulation of fluid flow and transport in heap leach operations.

Numerical models include analytical models. Examples include spreadsheet models that can be used, for example,

to estimate the inflow to an excavation based on radial flow equations and using broad simplifying assumptions about the system properties. When applying analytical models to mine settings, it is usually necessary to lump the input parameters into single terms with uniform hydraulic properties.

When using a model to support decisions, it is essential to have a good understanding of the uncertainties in both the supporting data and the model itself. Most numerical models of mine sites are much more sensitive to the broad scale assumptions that are used to define the conceptual model than they are to differences resulting from the choice of model code, numerical accuracy or the calibration statistics.

4.1.1.2 Basic questions to address prior to setting up a model

Numerical models can provide support for all phases of the mining cycle: from scoping; through pre-feasibility to feasibility studies; construction and start up; operation and expansion; to closure and post-closure. The most important hydrology issues commonly faced by mining operations are described in Section 1.3. The reasons for model development most often include (i) planning and managing the mine dewatering or pit slope depressurisation programs, (ii) defining potential impacts of the mining activities on the regional groundwater system, and (iii) demonstrating the environmental performance of a mine during operations and/or closure.

It would be unreasonable to expect that any single model could provide precise answers to such a broad spectrum of mining hydrogeology and environmental issues, and at scales ranging from the complete region where the mine is located down to a specific sector of the pit. Therefore, prior to planning any model, two basic issues need to be addressed.

1. What questions does the model need to answer and what decisions need to be made using the results?
2. What scale should be used for the model?

The conceptual and numerical model must be underpinned by adequate geological and hydrological data, the amount and quality of which may vary from project to project. There are inherent data gaps in any model, but two further issues need to be addressed prior to model construction.

1. Are the data set and the conceptual model adequate to reliably support the numerical model, and to back up the decisions that will be made using the model results?
2. What parts of the model are most critical for decision making and where is there a need to fill data gaps?

Every mine site is unique. However, most mines do have significant similarities to other sites. Therefore, the validation of the model by benchmarking is an essential step in model development, using the data and operating experience from as many different mine sites as possible. In some cases, the benchmarking process can be as important as the actual site-specific data.

4.1.2 Scale-specific application of the model

4.1.2.1 General considerations

Where numerous hydrogeological issues exist at a particular site (as is the case at most medium to large mines), it is often more efficient and realistic to construct individual models that focus on specific issues. That way, the hydrogeological detail and variability can be matched to the scale of the investigation. For example, the mine environmental department may develop regional-scale models that support permitting and closure, whereas the mine engineering or operations department may develop models related to mine dewatering and/or pit slope design. However, there should always be cross-referencing and general agreement between the models with respect to model boundaries, recharge and hydraulic property ranges.

A modelling limitation is the computational size of the model itself; this is normally defined by the number of cells or elements, the number of layers, and the degree of geological and hydrogeological complexity that is input to the model framework. A regional scale model, constructed to support environmental permitting of a mine, will usually contain insufficient hydrogeological detail and variability at the pit scale to make it useful for simulating details of the dewatering system or pore pressure fields in the pit slopes. However, the actual numerical size of a regional-scale and sector-scale model may potentially be similar and the number of cells or elements in each model may be approximately the same.

For current day software, the rule of thumb practical limit for a numerical model is roughly one million cells or elements. Therefore, a regional model, with a nominal domain size of 40 km × 30 km would have an average cell or element size of approximately 110 m × 110 m, assuming a 10 layer thickness. In contrast a mine scale model with dimensions of 5 km × 5 km, and centred on the open pit, could have an average cell or element size of 16 m × 16 m, which would be about the same as the mine geology block model although, in reality, the cell or element sizes would be varied to capture variable amounts of detail in the model domain. The amount of detail that could be incorporated at the mine scale would clearly be much greater but its numerical size and complexity may be the same as for the regional model.

4.1.2.2 Regional-scale models

A regional-scale model normally encompasses a wide area around the mine site. Depending on the setting and scale of the mining operation, it may extend for several tens of kilometres beyond the mine property boundaries, and may even extend across administrative boundaries or international borders.

A regional-scale model is most often required where there is a need to define:

- the hydrogeological interaction between the mine site and surrounding regional groundwater system, for example where there is potential for widespread regional groundwater flow to provide sustained recharge to the mine dewatering system;
- the potential impacts of a mine dewatering program on the regional hydrogeological system, including the

need to investigate sensitivity of the system to mining operations, quantify timing and location of impacts, demonstrate lack of impact, or to design mitigation systems;

■ the potential impacts of facility development (such as tailing, water storage dam, waste rock) on the regional hydrogeological system;

■ the potential hydrogeological impacts of developing groundwater resources for mine water supply;

■ the boundary conditions to constrain a more detailed site-scale or facility-specific model.

Many regional-scale models are built primarily to satisfy the demands of the regulatory environment. However, appropriately conceptualised and calibrated regional-scale models provide a good tool for examining the general interaction of a mining project with the regional system, including hydraulic gradients, flow paths and flux rates between the mine, its individual facilities and the surrounding regional system. The regional-scale model may investigate groundwater resource availability, mine dewatering flows and environmental performance at a gross scale, and may be important for demarcating the potential limits of mine impact.

A regional-scale model may not be justifiable in situations where:

■ there is demonstrated to be limited potential for regional groundwater flow to influence the mine dewatering system, and vice versa;

■ hydrological issues can be evaluated to the required level of detail with more simplistic calculations, and dewatering rates and potential impacts can be reliably predicted using empirical data or by analytical methods.

Regional-scale models are almost invariably constructed in 3D, but the scale is often such that the model represents multiple bedrock geology units as single 'lumped' hydrostratigraphical units with single bulk hydraulic parameters. For example, basin fill alluvial aquifers above bedrock are often represented as lumped hydrogeological units even though, in reality, the alluvial materials are highly variable and the basin is often layered. Mapped regional scale structures may be incorporated within the model, although there is often inadequate information to assign distinct hydraulic properties that are valid at a regional scale. Recharge may be distributed into sub-regions within the model, often based on elevation, surface water runoff and aspect. The reasons for lumping and generalising at the regional model scale are normally due to:

■ a need to manage the numerical size of the model;

■ the non-availability of data to differentiate distinct hydraulic properties at a regional scale. It is likely that

there will be large portions of the model domain where the distribution of data points is at a spacing of several kilometres.

The lumping of geology and hydraulic parameters often 'smears out' smaller scale (slope-scale) geological or hydrogeological details that may be important for the mine dewatering design, water-resource performance, and specific facility monitoring. Furthermore, a regional-scale model often contains relatively few layers because there are insufficient data to justify the vertical discretisation.

4.1.2.3 Mine-scale models

A mine-scale model normally includes part or all of the immediate mine site and the closest defined sub-basins and hydrogeological boundaries. Depending on the setting, a mine-scale model may focus on individual facilities (such as the pit) or the total area within the mine property boundary. The model may extend several kilometres from a facility. The purpose of the mine-scale model is to incorporate geological and hydrogeological details that influence the open pit (or underground mine) or facility performance, but cannot be captured adequately at the scale of the regional model.

The application of a mine-scale model can often include:

■ simulation of groundwater-head distributions and gradients in the vicinity of the mine, pertinent to environmental performance or mine design;

■ detailed prediction of the potential range of ground-water inflow rates to the open pit (or underground workings);

■ planning and design of the general mine dewatering system and/or the discharge system for the pumped water, and/or investigation of the sensitivity of alternative mine plans to dewatering requirements and sequencing;

■ assessment of the performance of a tailing facility and associated groundwater and environmental management;

■ understanding the uncertainty in the local hydrogeological system variables, including rock mass hydraulic parameters, the characteristics of known structures, or the variability in local recharge conditions;

■ prediction of the hydrogeological interaction between mine facilities, such as the pit and nearby heap leach or tailing facilities;

■ prediction of the post-mining pit hydrogeology, including groundwater recovery around the open pit.

Mines with moderate to high dewatering flows (roughly above 50 l/s) will often need to develop a mine-scale model to support the dewatering design to Pre-Feasibility or Feasibility level, and to provide a management tool for decision making during operations, including evaluation and testing of dewatering options, and the development of an overall dewatering strategy.

Mine facilities placed in an environmentally sensitive area may require a mine-scale model to examine facility performance (e.g. waste rock or tailing areas) in cases where the regional model results are too coarse. The purpose may be to identify potential flow paths between the facility and the down-gradient groundwater system (which may be the pit) and to provide a tool to optimise the placement of environmental controls or monitoring.

A mine-scale model normally requires local areas of increased data intensity, allowing for identification and characterisation of geological and hydrogeological features that are important to open pit or facility performance. For example, bedrock geological units surrounding an open pit or facility that are 'lumped' for the regional-scale model may need to be sub-divided based on lithology, alteration type or structural domain, and assigned distinct hydraulic properties. It may be necessary to have data points distributed at spacing intervals of hundreds of metres in order to support the conceptual model and to validate the numerical model. Facility or mine-scale structures or structural domains with known behaviour may be added as discrete features or as a broader 'fabric' represented with anisotropy in selected model zones. More localised sector-specific or facility-specific recharge regimes may be added, such as enhanced recharge caused by exposure of certain materials a pit slope, or due to losses from surface water diversions.

Addition of necessary detail requires that the model domain is limited in extent compared with the regional-scale model in order to manage the numerical size and complexity of the model. A mine-scale model usually incorporates a significant number of additional layers compared with a regional model, with its boundaries often being defined by the regional scale model by 'telescoping', as described in Section 4.2.3.7. For some mine sites, where the regional groundwater flow system is less important for mine development and operation, it is possible extend the mine-scale model to address the interaction with the regional conditions.

4.1.2.4 Pit slope (or sector-scale) models

A pit slope or sector-scale model is normally centred on one particular sector of the open pit. Pit slope models are often based on 2D geotechnical cross-section models. They are almost always done to support a geotechnical study, a slope design analysis, or to address an existing slope stability concern. The small physical scale allows a level of complexity that is normally not practicable to simulate with a mine-scale model.

The need for a slope-scale model generally falls into two categories.

1. Planning: where there is a need to support studies for the design of a new mine or for an existing mine expansion.
2. Operational: where there is a need to understand the role of pore pressure in a slope sector of concern or to determine the preferred way dissipate pore pressures in order to improve slope performance and/or economics.

The need for a sector-scale model is normally a decision made by the mining company rather than being driven by the regulator. Some mining houses and international financial institutions require that viable slope design angles are presented during declaration of reserves, and a robust pore pressure model that supports a viable slope design may therefore become mandatory during the Pre-Feasibility and Feasibility study process.

The applications of a sector-scale model most frequently include:

- prediction of pit slope pore pressures, and their changes through time, during the mining cycle of a particular pit slope, with a particular focus on the vertical hydraulic gradients;
- development of pore pressure profiles to feed into slope-stability models of the same sectors, and prediction of pore pressures impacts on slope performance for selected design alternatives;
- evaluation of alternative pit slope depressurisation options, and prediction of the impact of active pit slope depressurisation measures on slope performance;
- assessment of the pore pressure profiles during recovery of the groundwater system during closure, and evaluation of a range of options to manage long-term post-closure pit slope pore pressures. This particular application of pit slope modelling is becoming increasingly required by regulatory authorities.

A sector-scale model may either be a local 3D model of a particular slope sector, a 2D model or a hybrid model (Section 4.2.2). It normally incorporates inter-ramp scale detail in geology, structure, alteration and mine planning information. However, although the area of interest is limited to the pit slope itself, the model domain is normally extended a significant distance (hundreds of metres or more) beyond the pit crest, such that boundaries placed in the model do not influence the results in the areas of interest. Vertical hydraulic gradients are high in many pit slopes (sometimes greater than lateral gradients), so it is important to understand how the location of the model boundaries may influence the model development and resulting pore-pressure distributions.

A sector-scale pore pressure model must incorporate detailed mine planning information, including changes in slope position at small increments of time (minimum six months, sometimes less). For a 2D model, an additional challenge arises where future slope orientations change significantly relative to the 'current condition' and geotechnical sections that are aligned to examine future conditions may contradict the flow path requirements for the pore pressure models (Section 4.2.2.1).

4.1.3 Focussing the model on the slope design process

The slope design process (and/or back-analysis of slope movement) requires pore pressure as a standard input. Numerical simulation of pore pressures is typically required in situations where:

- the materials in the slope are of low or variable permeability, remain isolated and do not drain in response to general mine dewatering;
- the geological structure and/or hydrogeology of the slope materials creates compartments from which the groundwater does not freely drain;
- there is high precipitation to provide continual recharge to sustain pore pressure in the slope materials;
- pore pressure has the potential to significantly decrease the effective stress of the slope materials, and the slope design is determined by the rock mass or structural controls on an overall slope-scale or inter-ramp scale.

A specific pit slope numerical model may not be necessary in situations where:

- the materials in the slope are above the water table and contain no perched groundwater, and are not influenced by precipitation (rare);
- the permeability of the rock mass is pervasively high, and analysis with the district- or mine-scale model confidently demonstrates that a general dewatering program will successfully depressurise the slopes (either sectors or the entire pit) to the extent that 'dry' conditions prevail;
- the rock properties and strength of the material in the slope are such that pore pressures have little influence on the effective stress of the material (such as unaltered granite), and therefore pore pressure and seepage is not a major concern for the slope design;
- the slope performance is controlled by conditions at a bench scale and inter-ramp or overall design factors are not a consideration;
- the future mine plan remains coarse, the sector-specific hydrogeological data are sparse, and only rough estimates of future pit slope configurations and associated pore pressures are possible by any given means. In this case, the pit-scale model may provide adequate estimates, or pore pressure predictions may be based on analytical calculations or by forward extrapolation of piezometer data.

The results of sector-scale pore pressure models are always fed into the slope design or slope management process. It is therefore essential that the hydrogeological and geotechnical groups work interactively. It is often appropriate to focus the pore pressure analysis on specific issues that have been identified in the slope design process, rather than providing a generic pore pressure model. This applies mostly to operating mines, where the actual controls on slope performance are usually understood at a greater level of detail. However, it may also be applicable for the planning process. Specific areas of focus that frequently need to be considered are:

a. whether water is important for overall slope stability based on rock mass properties and/or lithological contacts (A on Figure 4.1B);
b. whether reducing the driving pressure gradient across identified structures should be a focus for minimising wedge or planar instability (B on Figure 4.1B);
c. whether issues caused by water may relate only to the presence of weak materials in the upper slope (C on Figure 4.1B);
d. whether the presence of isolated zones of low-permeability weak rock may be the most important factor (D on Figure 4.1B);
e. whether water may be important in the analysis of toppling in those pit sectors where jointing is sub-parallel to the slope (E on Figure 4.1B);
f. whether the issue of water pressure only relates to shallow conditions within the slope because of the lower confining stress, and may be influenced by transient conditions (F on Figure 4.1B).

It often advisable that the geotechnical model is first used for an initial screening analysis with the goal of:

- identifying the general sensitivity of the slope materials to water pressure;
- identifying the most important sensitive areas in the slope.

To support this screening analysis, the hydrogeologist may need to provide preliminary 'worst case' and 'best case' estimates of the phreatic surface based on the data to hand, prior to the development of any numerical pore pressure models.

4.1.4 General planning considerations

4.1.4.1 Planning of regional-scale models

In an increasing number of countries, a regional model is becoming a legal requirement for permitting a new mine or a mine expansion, even in circumstances where the mine and facilities are above regional groundwater levels or within very low-permeability crystalline rocks containing limited amounts of groundwater. It may also be required as part of the corporate governance policy at some mining

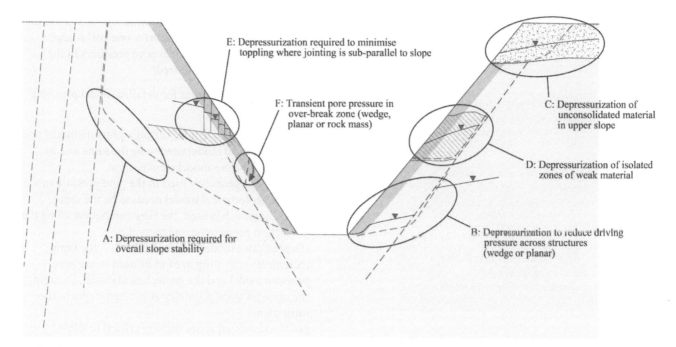

Figure 4.1B: Specific areas of focus that frequently need to be considered for sector-scale models (see Figure 4.1A)

houses for risk identification and environmental management. The model files and supporting data may need to be provided to a government agency or external reviewer for scrutiny. However, it is important that the model matches expectations and a common mistake is to assume that a regional model can precisely match heads or observed flows at the scale of the open pit.

Some important general considerations for planning a regional model are as follows.

- The model domain may encompass other sites with a hydrogeologic interest in the region, including active agricultural land, public water supply aquifers or industrial sites. This may create an elevated public profile or political sensitivity for the model (cumulative impacts).
- It is normally important to indentify all stakeholders that exist within the model domain, which may not be limited to the mining company and the regulator.
- If the model domain is extensive, it may overlap with other existing models, built for alternative purposes. For example, in the USA, the USGS has models for much of the Great Basin Province, which overlap mining districts. Therefore, there may be 'competing models' of the same general area.
- The model code will need to be acceptable to the regulatory community and a publically available valid code will need to be used. It is often advisable to agree the approximate domain and general conceptual model content with the regulators, before numerical model construction commences.

- Data beyond the immediate mine area and outside of the control of the mining company will be needed to develop the conceptual model and to calibrate the numerical model.
- It may be difficult to acquire new data for much of the model domain, since it is mostly outside of land controlled by the mining company.

4.1.4.2 Planning of mine-scale models

Although the drivers for a mine-scale model are often related to engineering feasibility studies or to support the design and management in an operating mine, they can also be required to support permitting or regulatory compliance. Some general planning considerations are as follows.

- A sophisticated level of geologic detail may be required including individual structures and the irregular geometry of the ore body. Alteration zones or country rock may need to be represented in the model, because the hydraulic contrasts between these materials may be important.
- The mine geology model (block model) is often a good platform on which to build the mine-scale groundwater flow model, particularly if the open pit or underground operation is the primary area of interest, although the groundwater model will need to be extended a considerable distance outward from the limits of the block model.
- The frequency and distribution of hydrogeologic data points need to be adequate to support the detail placed into the model domain. Geologic units, alteration types and structures can only be differentiated in the model

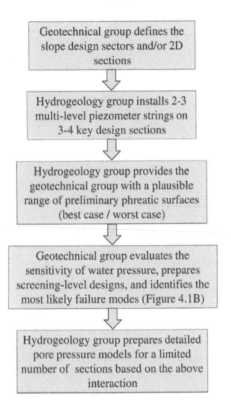

Figure 4.1C: Screening-level planning process for pore pressure models (see Figure 4.1A)

domain if there are data points or strong supporting evidence.

- Because the model likely resides within the limits of the mine property, it may be relatively easy to acquire additional field data to refine the model as it is being constructed and calibrated.

4.1.4.3 Planning of sector-scale models

A screening-level analysis using the geotechnical model as part of the early planning process is discussed in Section 4.1.3. The purpose of the screening analysis is to provide an initial understanding of the pit slopes prior to carrying out detailed pore pressure modelling. Potential steps in the screening-level process are outlined in Figure 4.1C.

Once the screening analysis is complete, the hydrogeology group can begin to evaluate the importance of the data obtained, consider potential data gaps, and plan for construction of pore pressure models. In all cases, the key questions to ask when determining the type of model are:

1. what type of geotechnical analysis will be conducted (2D or 3D; limit equilibrium or numerical)?
2. what inter-ramp scale geology and hydrogeology data are available?
3. what is the timeframe required to produce model output, and what are the client budget considerations for both data collection and for the modelling itself?

4. will the model be used on a continuing basis or are the results required only to support a 'one-off' study?
5. Will the analysis of transient pore pressures in the unsaturated zone be required?

The main considerations for detailed model planning are as follows.

- The coordinate system for the pore pressure model and the geotechnical model need to be the same and/or understood before model construction.
- The pit slope geometry used in the pore pressure model and the geotechnical model needs to be the same.
- If a 2D approach is used, the flow component along the 2D section needs to be understood.
- If a predictive model is being built, then the same increments of mining need to be used in the pore pressure model and the geotechnical model, including the same pit slope geometry at selected stages in the mine plan.
- For the identified zones that are critical to slope stability (Figure 4.1B), the amount of existing hydrogeologic data required to validate the pore pressure predictions needs to be defined.
- The expected level of confidence in the predicted pore pressures needs to be defined before modelling commences. The pore pressure model results are likely to have a significant impact on the predicted slope performance.
- If an active pit slope is being modelled, the amount of ongoing active drainage and the transient behaviour in the hydrogeologic system needs to be understood and incorporated into the model.

For a new mine, confidence in the model can usually be improved by:

- performing cross-hole testing of lower-permeability units (using low-flow injection or airlift tests);
- examining the behaviour of pit slopes in other operating mines with similar hydrogeology and in similar climatic settings;
- using the numerical model to assess a range of possible pore pressure responses and using judgement to select the most appropriate response on which to base the pore pressure input for the geotechnical model.

For a mature mine, confidence in the model can usually be improved by:

- comparing results to the current-day piezometer data and installing new piezometers in critical areas that can be accessed;
- comparing model predictions to historic depressurisation efforts, to gauge whether the predicted changes in pressure with time make sense compared to what has been observed in the past;

■ comparing the model fluxes with those from coarser/ larger modelling scales (3D mine scale) to ascertain that there is 'general agreement'.

As discussed in Section 3.6.7, it may be necessary to simulate transient pore pressures in the sector-scale numerical model. Where there are seasonal fluctuations in piezometric levels, it is often most appropriate to calibrate the pore pressure model to the high seasonal piezometric levels, as these are usually of most importance for slope stability. It may also be possible to calibrate the model to both dry and wet season conditions, depending on the time steps used.

However, many seasonal transient pore pressure responses are the result of infiltration of in pit precipitation into shallow 'dry' sectors of the slope where the confining stresses are less and where pore pressure can have a greater proportional influence on effective stress. Such transient pore pressures need to be dealt with on a case-by-case basis, often outside the main pore pressure modelling exercise. Some guidance on the input of transient pore pressures into geotechnical models is provided in Section 4.3.9.

4.1.5   Timeframe and budget considerations

4.1.5.1   Introduction

When developing numerical models, an important factor is the time required to produce the model results, and the budget that is available for the modelling study. Some guidance for time and budget considerations is provided in the following sections.

4.1.5.2   Regional-scale models

Regional-scale models are often fundamental to the mine permitting process. However, construction of the actual model itself may only be a small component of the overall permitting program. Many regulatory bodies require appropriate data collection and testing and one to two years monitoring of the hydrogeological system to support model construction and validation. Therefore, for a new project, it is normally beneficial to plan a regional model one to three years ahead of the permitting delivery schedule.

Many regulators have a general understanding that the model will not provide realistic results without the supporting field data. In some cases, the regulator will therefore allow the initial model to be constructed using available data interactively to define where data gaps exist. Once there is confidence in the model's ability to simulate the behaviour of the hydrogeological system, a series of forward runs can be performed to predict the potential impacts of the mine site on the region, and to define appropriate impact mitigation measures, as required. Obtaining confidence in the numerical model often requires multiple iterations of field work and model updates. The final model and predictions will often be incorporated into an Environmental Impact Statement (or similar) and permitting documents.

For a large mine, a regional model for regulatory compliance and permitting may cost in the range US$500 000 to US$1 million to construct, calibrate and use for the predictive analysis. As a general rule, the cost of the model should be no more than about 10 per cent of the cost of field testing and data collection.

Where a new regional-scale model is required to support an existing project expansion, the timeframe required for model development is typically less, probably of the order of three to 12 months. For a project expansion, there is usually a reasonable quantity of existing data available to support the numerical model, so it may be possible to focus the required test work and model development on certain parts of the known hydrological system.

4.1.5.3   Mine-scale models

The time needed to develop a mine-scale model is normally about three to six months, although this will depend on the availability of data to support the model, and which part of the system is the priority for model construction. Table 4.1 provides guidance for evaluating dewatering requirements for new projects. A dewatering model for an operating mine can normally draw on existing data and understanding, whereas one for a new mine will need to define areas of detail and data gaps that have to be filled in the field. For a large mining operation, a typical budget to construct a dewatering model will probably be within the range of US$200 000 to US$500 000, depending on the local detail and complexity.

4.1.5.4   Sector-scale models

Greenfield sites

Guidelines for the steps that may be followed to integrate hydrogeological and geotechnical models are outlined in Section 4.3.5. It is essential that the two models move forward interactively and that the pore pressure modelling is focused on the key identified areas for slope stability. The approximate time input required for pore pressure modelling will clearly depend on the scale of the project and the importance of water in the slope design process. Time requirements for modelling may be one to three weeks for Level 1 (Screening level), three to six weeks for Level 2 (Pre-Feasibility) and three to six months for Level 3 (Feasibility) and Level 4 (Detailed design). For a Level 3 (Feasibility), a guideline budget may be US$75 000 to US$250 000.

For new projects, it is important that the pore pressure modelling is focused on what is achievable within the time available at each stage of the investigation. Development of the geotechnical model is often carried out under major

Table 4.1: Typical numerical modelling sequence to evaluate dewatering requirements for a new pit or a major pit expansion

Stage	Type	Input	Calibration	Prediction	'Accuracy'
Conceptual (Level 1)	Analytical or simple numerical (axisymmetric)	Simplified (homogeneous & isotropic) geology Estimate of hydraulic properties using data from proxy sites Conceptual mine plan Screening-level geotechnical analysis	Preliminary or based on *benchmarking*	Preliminary estimates of inflow Screening analysis of dewatering and slope depressurisation options Identification of slower-draining pit sectors	Order of magnitude
Pre-Feasibility (Level 2)	Axisymmetric or preliminary mine-scale 3D	Layered geology with lateral-to-vertical anisotropy Preliminary field-derived values of hydraulic parameters Time-variable mine plan Geotechnical modelling and interaction	Based on initial site-specific data, but also relying on benchmarking	Estimates of inflow over time and phreatic surfaces for the pit slopes sufficient to allow estimate of costs Uncertainty analysis	Factor of 5
Feasibility (Level 3)	Regional or site-scale 3D and sector-scale 2D	Full 3D representation of geology Detailed field-derived values of hydraulic properties, including some cross-hole testing Actual mine plan and slope designs Data on recharge, pumping, and surface water Measured water levels and flows to allow calibration Interaction with geotechnical models	Good for overall site and for key defined pit sectors. Requires transient calibration	Amounts of water to be managed and effects of various dewatering schemes Sufficient to provide boundary conditions for pore pressure predictions with supporting data Impact analysis Uncertainty analysis	Factor of 2
Design & Construction (Level 4)	Regional or site-scale 3D and sector-scale 2D	Fully 3D representation of geology including structures Field-derived values of hydraulic properties and good understanding of range of values Detailed cross-hole testing and test pumping Pilot dewatering or slope depressurisation trials Detailed mine plan with applicable geotechnical input and interaction Data on recharge, pumping, and surface water Comprehensive water level and flow data (for calibration)	Good for all pit sectors, with specific calibration to pumping trials and/or cross-hole testing	Detailed design and optimisation (location and timing) of dewatering system Changes in water levels and impacts on water resources Close interaction with pore pressure models Input to water balance and detailed plan for discharge of excess water Uncertainty analysis Solute transport from tailing etc. Inflow to a 'pit lake'	±30% (better with time)

time constraints because the results need to be incorporated into the pit design. The pore pressure model will be of little value if the results are not available for timely input into the geotechnical model. It may therefore be of benefit to focus the pore pressure modelling on a limited number of design sections (potentially two to four, depending on the project) with a less rigorous analysis carried out for the remaining less critical sections, for which there will likely be limited pore pressure data available to constrain the model. This is discussed further in Section 4.3.5.

Thus, an in-depth analysis can be carried for the key design sections to determine the benefit of active slope depressurisation measures for increasing the slope angles or increasing the factors of safety, and the results can be applied to the other sections.

Brownfield sites

The amount and quality of available data to support pore pressure modelling for an existing pit expansion or for an update of the operating slope design is normally much greater than for a new mine. Therefore, it is usual that more robust models can be constructed for an existing operation. The time required for pore pressure modelling of any given slope sector may be within the range two to six months, depending on the importance of water. A typical budget for the work will probably be within the range US\$150 000 to US\$350 000 for a medium to large pit, although this will again depend on the availability of data.

Again, for an existing operation, the time available may be a limiting factor in the type of model that can be constructed, particularly if the model is being used to support a re-design in response to deterioration in the slope performance. In this situation, it may also be necessary to consider only one cross-section for detailed numerical analysis and to use the numerical results from that single cross-section to support a simpler analysis for the other sections. If this is the case, the mine operation needs to be made aware of any possible consequences.

For existing operations, it usually helps if the pore pressure model is initially set up so it can be rapidly adjusted to accommodate changes in the mine design and that the monitoring program provides sufficient data to allow rapid re-calibration of the model. An advantage of 2D and sector-scale 3D models over larger models is that they can be constructed and calibrated relatively quickly.

4.1.6 Modelling workflow

General guidelines detailing modelling workflows for study Levels 2–4 are provided in Table 4.2. The workflow for a 3D regional and a mine-scale model are generally similar. For sector-scale models, the integration with the geotechnical and mine design teams is also outlined in the

table. It is stressed that each mine site will have different objectives and requirements for modelling, so the actual work flow for each site must be tailored to the site specific considerations, the actual hydrogeology and the available data.

As mining proceeds, transient changes in rock hydraulic properties may be incorporated in the slope-scale model as new zones with distinct hydraulic properties are exposed (e.g. blast damaged rock or a dilating rock mass in poorly permeable materials).

4.1.7 Data requirements and sources

4.1.7.1 Introduction

The data sets required for model development are similar for all scales. However, the geographical extent of the data points needed for a slope-scale model is clearly much less than for a regional-scale model. Where there is good availability of data, the time required to calibrate and adjust the model is longer, but the model results are more credible.

The key data requirements for any model are as follows.

- Topography.
- Mine planning data.
- Climatology (meteorology).
- Hydrology, as it controls recharge and discharge to the groundwater system.
- Surface water drainage patterns and flows, including seeps and springs.
- Geology (stratigraphy, alteration and structure).
- Hydraulic parameters (conductivity and storage properties, ideally for each of the main defined geological units).
- Groundwater level data (static head, equivalent to elevation and pore pressure data).
- Important hydraulic stresses within the model domain (water balance data).
- Ideally, groundwater level monitoring data that show a transient response to the key hydraulic stresses.

4.1.7.2 Topography

For the vast majority of models, the ground surface topography forms the top of the upper model layer and is a fundamental input parameter. Virtually all new and existing mine sites now have (or can obtain) good digital topographical control, so definition of the upper surface of the model is relatively straightforward.

4.1.7.3 Mine planning data

The key mine planning information is defined in Table 4.3. Sector-scale models need to simulate changes to the pore pressures in the pit walls as mining proceeds, including those resulting from recharge and active pit slope depressurisation measures.

Table 4.2: Modelling work flows for study Levels 2–4

Regional and mine-scale models	Sector-scale models
1 Ensure the specific goals and required outputs of the model are clearly defined	1 Identify the importance of water and the key controls for geotechnical stability
2 Define the model domain, encompassing hydrogeological areas of interest, extended to natural hydrogeological boundaries	2 Define the pit slope sectors of prioritised concern and those of lesser concern, based on the initial screening-level geotechnical analysis (Section 4.1.3)
3 Develop a 3D geological framework for the model domain	3 Prepare hydrogeotechnical cross-sections through the prioritised pit slope sectors, to be utilised for both pit slope pore pressure modelling and slope stability modelling. An example of a hydrogeotechnical section is shown in Figure 2.4
4 Define hydraulic parameter ranges for the geological units within the framework, using field-testing data or established text book ranges if specific data are not available, and carry out benchmarking with other operations	4 Extend the model boundary from the pit slope area to a defensible groundwater boundary beyond the immediate pit area (Section 4.2.3)
5 Prepare a set of 3D hydrogeological units for the model domain. This normally involves 'lumping' and/or sub-division of geological units with distinct hydraulic properties	5 Populate the model mesh or grid with geological data to the level of detail present in the mine geological block model. If (as normal) the hydrogeotechnical section extends beyond the geological block model, then a coarser geological framework is used in the model in the area between the limits of the geological block model and the limit (boundary location) of the sector-scale model
6 Develop an analytical water balance for the model domain, encompassing recharge zones and amounts, groundwater flux amounts, and discharge zones and amounts	
7 Incorporate the future mine plan, including the open pit and other facilities, into the model grid design	6 Input each geological unit to the model grid or mesh. Each geological unit is normally assigned as a distinct hydrogeological unit with distinct hydraulic properties, based on either field testing or published parameters for similar rock types. Assign hydraulic properties [horizontal hydraulic conductivity (K_x, K_y, K_{xy}), vertical hydraulic conductivity (K_z), specific storage (S_s) and specific yield (S_y)]
8 Establish a set of steady-state calibration targets for the model domain. These would often include (i) 'reliable' stable groundwater levels within the established hydrogeological units, (ii) measured groundwater seeps (spring discharges), (iii) measured surface water flows, and (iv) groundwater evaporation rates (in defined zones where groundwater is at or near surface)	7 Incorporate specific geotechnical or structural geology features into the model as hydrogeological zones that may 'over-print' the geology or further sub-divide the geology
	8 Incorporate the future proposed mine shapes (annual pit shells) into the model mesh or grid during model construction, in preparation for predictive analysis
9 Establish a set of transient calibration targets for the model domain, if data are available. Transient calibration targets would normally involve groundwater pumping (including dewatering) and associated groundwater responses in observation wells and piezometers	9 Prepare an estimate of local-scale (pit slope sector) groundwater recharge and apply this to the model. Assign 'pumping' and drain nodes that coincide with existing hydraulic outflows (seepage faces or installed slope depressurisation measures)
10 Conduct a steady-state model calibration. The hydraulic variables in the model (hydraulic properties, recharge, boundary conditions) are adjusted within pre-determined bounds until a best attainable match between modelled heads and flows and observed heads and flows (the calibration targets) is attained	10 Develop a set of calibration targets for the model, normally involving groundwater pressure data from piezometers installed specific to each slope sector
11 Conduct a transient model calibration. This may be conducted using the steady-state model result as the 'initial condition'. Hydraulic stresses are input to the model to represent pumping for time periods at known locations (e.g. pumping wells, pit floor sumps) and the hydraulic variables within the model domain are adjusted until a best attainable match between modelled and observed (calibration data) transient groundwater responses and flow rates is achieved	11 Develop an initial conditions model ('calibration'). The level of sophistication varies considerably between Greenfield sites where no mining has occurred and mature pit slopes being considered for further development, where prolonged mining has occurred and strong variability may exist in 'current' groundwater head distributions
	12 Conduct sensitivity analysis to examine the level of confidence and to quantify uncertainty in the parameters utilised in model calibration
12 Carry out sensitivity analysis with the calibrated model in order to define the uniqueness of model results and to define the key areas of uncertainty	13 Perform a series of predictive model runs using the calibrated model as the start point or initial condition. The predictions may involve implementing annual mine pushbacks or other planned increments of mining into the model. This is normally done by migrating the seepage face to the location of the pit surface and 'removing' the mined material from the model (code dependent)
13 Use the model to define critical data gaps that may trigger additional field work and subsequent iterations of model calibration	
14 Carry out forward runs of the calibrated model to predict future conditions, based on the project-specific scenarios related to open pit mining and dewatering, facility development, mine closure or water-resource development	14 Generate a series of pore pressure profiles at selected increments of time and mine plan advance, ready for input to the geotechnical slope design model
15 Use the model to optimise the overall design of the dewatering system and the mine water discharge and mitigation plan (as appropriate). Note that most models can only be used to help specify general areas where dewatering wells are required rather than specific well locations (Section 5.2.5.2)	15 Use the model to evaluate and optimise various slope depressurisation measures (methods and timing). Work interactively with the geotechnical model to determine the cost–benefit of each option (Introduction Section 3.2)

Table 4.3: Key mine planning data required for model input

Regional- and mine-scale models	Sector-scale models
• Annual pit shells (if available at the time of modelling); annual facility footprints; annual development plans	• Monthly (near-time) and annual (far-time) cuts and pit shapes (common to the pore pressure model and the geotechnical model)
• The ultimate pit shell and final facility footprint	• Underground mine planning information where this may affect the hydrogeology of the open pit

4.1.7.4 Climatology (meteorology)

The patterns of precipitation and evaporation are clearly important for determining the amount and seasonal distribution of recharge that needs to be applied to the upper surface of the model, at any modelling scale. For a sector-scale model, it is important to assess the amount of recharge that may occur to the slope itself as a result of runoff accumulating on catch benches or other flat areas within the pit area. The assessment of in-pit recharge is particularly important where zones of weaker and deformable materials are exposed within or below the slope. Local-scale recharge may cause increases in pore pressure within these weaker materials.

Climate data are also used to help compute the evaporative losses from surface water bodies and from zones of groundwater discharge. In some arid low-permeability settings, these losses may balance the entire groundwater inflow rate to the pit.

At the regional scale in semi-arid areas, the use of annual or monthly data may lead to some misinterpretation. Recharge may only occur during and following occasional high-intensity rainfall events (which may be many years apart in some settings). Typical requirements of climatological data are shown in Table 4.4.

4.1.7.5 Hydrology and surface water drainage patterns

Knowledge of stream locations, springs and seeps, stream flows and flow variations is important for any regional-scale model. The surface water drainage pattern may represent significant recharge sources to the model (losing reaches) or discharge locations (gaining reaches). The drainage pattern may act as a strong control in the choice and location of model boundaries.

Where surface waters occur in the vicinity of the pit, they would form an essential part of any sector-scale model. In addition, the assessment of surface drainage and stormwater diversion designs can be a fundamental requirement for assessing potential recharge sources to the whole pit or to an individual slope. As described in Section 3.6.3, the presence of surface water and recharge associated with mine facilities is often an important consideration for pore pressures.

4.1.7.6 Geology

Any model requires a fundamental knowledge of the key geological units, lateral and vertical variations within those units, zones of weathering, alteration, and mineralisation, and geological structures (including major structures and joint orientations). As part of the model development, it is likely that several of the units will be lumped (or possibly divided) into distinct hydrogeological units. The geological input parameters are defined in Table 4.5.

4.1.7.7 Hydraulic parameters

There are only two basic hydraulic input parameters to any model: hydraulic conductivity (K) and storage (S and S_y). Each of the defined hydrological units in the model must have initial values for these parameters assigned, based on the field testing results and the conceptual model. Where testing data for a particular unit are not available, it is normal to use data from other similar locations where similar lithology is present, or to use published data (for regional-scale models). Benchmarking with other sites is also important.

It is normal to vary the hydraulic parameters laterally and vertically within each unit, if this is justified by the field testing results. Clearly, for defining hydraulic

Table 4.4: Key climate data required for model input

Regional- and mine-scale models	Sector-scale models
• Long-term average monthly precipitation; proportion of precipitation that occurs as snow; published regional isohyets	• Normally requires monthly average and annual average precipitation and evaporation data from at least one on-site climate station
• All available data on the intensity and frequency of precipitation events	• Any available data that may indicate local-scale variations in rainfall intensity
• Monthly pan evaporation and/or published regional potential evaporation isohyets	• Local climatological variables that may allow alternative calculations of actual evaporation
• Multiple stations, even if outside the model domain, are preferred	• Measurements of water pumped from sumps (that may help calibrate rainfall-runoff models for the pit)

Table 4.5: Key geological input parameters for model development

Regional- and mine-scale models	Sector-scale models
• Regional geological maps and cross-sections; including zones of alluvium, glacial or recent flood plain (or lacustrine) deposits	• Mine-property surface geological maps
• Maps showing the elevation of key geological contacts; and/or isopach maps showing the thickness of key geological units	• Mine geological block model and structural wireframe; in conjunction with maps showing the elevation of key geological contacts where these are continuous (such as the bedrock surface, the surface of the weathered zone or the surface of the leached zone
• In areas of sparse data, maps showing the elevation of the bedrock surface as a minimum starting point	• Structural geology model and an interpretation of the locations of the primary and secondary structures
• Regional structural geology maps; any information on the weathering profile	• Alteration model and/or mineralisation model
• Geophysical maps/interpretations	• Borehole logs (geological, geotechnical and geophysical)
• Borehole logs (geological, geotechnical and geophysical)	• Borehole geology logs for sterilisation/condemnation holes
• Remote sensing data (satellite imagery)	• Pit slope geology maps
• Existing public-domain hydrogeological studies, models or modelling reports	• Geotechnical domains

parameters, a larger number of measurements may reduce the uncertainty in parameter ranges.

In addition to the basic definition of hydraulic parameters, for a sector-scale model it will be necessary to define changes with time to the parameter values within the blast-damage zone, and potentially within zones of relaxation as the pit slope materials deform.

4.1.7.8   Groundwater level data

A knowledge of groundwater levels over as much of the model domain as possible is essential for achieving a credible model calibration, irrespective of scale. Ideally, there should be some groundwater level data for all hydrogeological units represented by the model. As a minimum for regional-scale models, basin fill alluvium and bedrock groundwater levels should be differentiated and, in most situations, additional sub-division of basin fill and bedrock groundwater levels is required. For sector-scale models, it is important that observations of seepage faces and water levels in old boreholes are included within the model. Vertical discretisation of groundwater

levels is also important. Typical sources of groundwater level data are outlined in Table 4.6.

4.1.7.9   Water balance data

It is necessary that a water balance is developed during the conceptual model development, and that the water balance information is used as a calibration target or for a check on the 'reasonableness' of the model results. The properties that are assigned to the hydrogeological units, and the calibration of the model to observed groundwater data, can only be considered valid if the main groundwater-related flows in the system are also reasonably matched by the model.

The principal water balance components, both 'natural' and those relevant during and post-development, are shown in Table 4.7.

4.1.7.10   Data to assist the transient calibration

At all scales, it is important that the model incorporates hydraulic stress to allow transient calibration. The hydraulic stress may take the form of recharge zones,

Table 4.6: Typical available sources of groundwater level data

Regional- and mine-scale models	Sector-scale models

• Published or government groundwater monitoring data and maps
• Monitoring data from other local landowners or third parties (which may require careful quality review)
• Water level measurements in mineral exploration holes, definition holes or condemnation (sterilisation) holes
• Mine site environmental monitoring wells (hydrographs over as long a time period as possible)
• Dedicated observation wells and piezometers from the hydrogeology program (hydrographs)
• Published regional aquifer maps
• Dedicated hydrogeological monitoring points that may include:
  – existing mine shafts
  – study monitoring wells
  – springs and seeps
  – base-flow sections of rivers and streams
	• Multiple-level vibrating wire piezometers completed within major lithological units, alteration types, structural compartments and potentially the blast damaged zone (where transient pore pressures are important)
	• Water levels in blast holes
	• Locations of water encountered in horizontal drain holes
	• Open standpipe piezometers (they are not ideal, but they provide useful data in some circumstances).
	• Observations in the pit, which may include old exploration drill holes, flowing drill holes, levels of seepage faces, sump levels

Table 4.7: Principal water balance components

Regional- and mine-scale models	Sector-scale models
• Measured flow in perennial reaches of rivers and streams and changes within the modelled area	• Pumping rates from individual mine dewatering wells
• Measured flow in significant springs or other identified groundwater discharge zones	• Discharge rates from individual pit slope horizontal drains or drain sets grouped by geology, domain and sector
• Evaporation rates from significant water bodies (lakes) in communication with the regional water table	• Discharge rates from underground drainage galleries and, where possible, with drain hole flows grouped by geological domain
• Bulk pumping rates from regional and third partly wells, wells associated with mine facilities other than the pit, existing underground mine operations, historical pits (sumps and/or wells)	• Discharge rates from seeps or flowing wells within the pit (estimates or measured)
• Pumping rates from the existing dewatering operation (wells, drains and/or sumps)	• Estimation of evaporation fluxes from wet zones within the pit walls or floors
• Sources of artificial groundwater recharge which may include infiltration from irrigation water, water supply collection reservoirs and other sources	• Local-scale groundwater recharge and resulting transient responses to the groundwater system, which may result from ponding of precipitation, runoff channels, leakage from mine facilities

previous pumping for regional water supply, historical mine dewatering, or passive groundwater inflow zones to an existing pit or nearby underground workings. Long-term test pumping is also highly valuable for model calibration. For the sector-scale model in zones of low permeability, this information may be obtained by means of an extended airlift or injection test (Section 2.2.8).

Transient calibration to hydraulic stresses and response data is normally the best way to address critical data gaps in the conceptual or numerical model. Any particular hydrogeological feature in the model can often be parameterised in many ways, without changing the quality of calibration, but profoundly changing the predictive results. The larger the scale of the stress testing, the more the model calibration can be refined to achieve a unique solution. The types of transient response data that are typically required are shown in Table 4.8.

4.2 Development of numerical groundwater flow models

4.2.1 Steps required for model development

The 'standard' approach for developing a general mine dewatering or sector-scale model is to:

▪ use an 'off the shelf' equivalent porous medium (EPM) flow code to construct the model and perform transient analyses;
▪ incorporate anisotropy and inhomogeneity into all or part of the model domain;
▪ include discrete features that may influence the groundwater flow system. The discrete features may include fault zones, more permeable stratigraphic horizons, less permeable zones or old mine workings.

The basis for the numerical model is the conceptual model (Section 3). Section 4.2 describes the construction of a numerical groundwater flow model for general mine dewatering and slope stability studies. Guidelines are provided for determining when the 'standard' approach is valid, and for deciding when an alternative methodology may be warranted.

The broad steps for model construction are as follows.

1. **Determining the model geometry** (i.e. whether to simulate flow behaviour using a three-dimensional (3D) model, a two-dimensional (2D) model, or a 'hybrid' model – which may be an axisymmetric model or a planar model (Section 4.2.2).
   ▪ This decision only really applies to pit slope models. For most situations, regional and mine-scale models are constructed in 3D.

Table 4.8: Transient response data

Regional- and mine-scale models	Sector-scale models
• Observation well response to seasonal climate fluctuations, surface water flows and/or recharge sources.	• Bulk groundwater system response to in-pit or near-pit pumping of single or multiple dewatering production wells, shafts or pit-floor sumps
• Observation well response to single-well or group pumping tests within distinct groundwater units.	• Bulk groundwater system response to pit slope depressurisation measures, including underground drain drilling or in-pit horizontal drain drilling
• Observation well or groundwater discharge zone response to known sources of artificial recharge	• Cross-hole or multi-hole testing using airlift, injection or other testing methods
• Observation well response to large precipitation events (when the recharge rate can be estimated)	• Observations of groundwater response to changing stress (such as mining) or to zones of slope movement

■ A 2D approach may allow more focused/detailed analysis at the sector scale, particularly to assess vertical hydraulic gradients that develop in pit slopes, and can be easily integrated with 2D geotechnical models and with detailed mine planning.

2. **Setting the model domain boundaries and boundary conditions**

■ The model domain includes the area covered by the model and the height of the model.

■ It is necessary to justify the location of model boundaries based on the conceptual model and the time-period of interest.

■ The lateral edges of the model are normally assigned with constant or general-head values based on measured piezometer data, or are no-flow boundaries based on measured or interpreted hydrogeological divides.

■ The pit topography in the model is usually a zero pressure boundary, but may have a specified infiltration rate to represent recharge into the materials exposed in the slopes.

3. **Defining the mesh or grid size**

■ The size of the grid or mesh is typically controlled by the amount and quality of supporting geological and hydrogeological information over the various parts of the model domain. It is normally very refined at the location of the current or future pit slope, so that mine cuts or slope drainage measures can be accurately included.

■ Mesh or grid density should consider hydraulic gradients, and it may be appropriate to use a fine grid or mesh in areas of large hydraulic gradients close to the pit, allowing for precision when matching and predicting pore pressure distributions.

4. **Determining whether to run steady-state, transient or undrained simulations**

■ At an active mine, the pit geometry and hydraulic stresses are constantly changing in response to mining or dewatering activity, so it is usually necessary to carry out a transient analysis.

■ There are certain (rare) conditions where a steady-state pit slope model may be appropriate. These are normally related to (i) permeable slopes with very rapid drainage, or (ii) permeable slopes with a very low excavation rate.

■ Steady-state analysis may also be applicable for predicting long-term post-closure conditions in pit slopes.

■ Undrained simulations may occasionally be appropriate in situations where very low-permeability settings and near-zero recharge conditions occur and the only significant control on water pressure is related to changes in the stress field (hydromechanical coupling).

5. **Determining whether the use of an EPM code is adequate**

■ At the scale and mesh size of most regional and mine-scale models, groundwater flow can usually be represented as EPM, with the addition of material property variation based on geological, alteration or geotechnical domains and with discrete features or hydraulic property anisotropy built into the EPM model to represent specific structures or structural fabric, as required, and where justified.

■ The decision on whether to use an EPM code (or whether the use of a fracture flow code is required across all or part of the model domain) normally only applies to sector-scale models. EPM models are usually satisfactory for most general mine hydrogeology settings.

6. **Selecting the appropriate stress periods**

■ The modelling process needs to specify the initial conditions (which, for a Greenfield site, may be pre-mining steady-state conditions).

■ The mine excavation sequence then needs to be incorporated into the model, with results obtained at times of interest.

■ The stress periods are the periods when the hydraulic stresses in the model are changed. New stresses may be applied at the start of each period and the model then calculates the conditions at the end of that period.

■ The number of stress periods used will depend on the goal of the model, the rate of mining and the time-frame for implementing active dewatering or slope depressurisation.

■ The frequency and length of the stress periods can also be varied over the entire simulation period. For many projects or studies, the level of precision in the long-range pit design is less than for the short-term mine plan. Therefore, coarser and less precise stress periods are often applied when predictions extend to more than about 5 years from current time.

When selecting the stress periods, the requirement of the model to simulate seasonal or sometimes monthly variations must also be considered. A discussion of the need to simulate transient pore pressure changes at time steps less than monthly is discussed in Section 4.3.9.

7. **Deciding whether a coupled geomechanical pore pressure modelling approach is required**

■ Field experience indicates that increased fracture permeability caused by unloading the slope results in a short-term pore pressure reduction that is usually followed by a pore pressure rebound as the rock becomes more open to recharge.

- The pressure rebound is typically very rapid so, as a general guideline, a coupled geomechanical pore pressure model for pit slopes should be considered in situations where the permeability of the rock is below approximately 10^{-8} m/s and recharge is very low.
- In theory, a coupled model may be applicable where continuing changes to the physical properties due to deformation of the slope materials, slope failure or blast damage, may have a significant effect on pore pressures, such that the hydraulic properties need to be progressively changed in the model. However, if a coupled model is applied without due consideration of recharge, it may have a tendency to over predict the rate at which depressurisation occurs.

8. **Incorporating active drainage measures into the model**
   - Mine-scale dewatering models and sector-scale pore pressure models are usually run to simulate a range of active dewatering and/or slope-drainage options. In lower-permeability settings, a dewatering model is often run to simulate a case where no active depressurisation is implemented and groundwater seeps into the pit passively, so that the cost–benefit of an active depressurisation system can later be quantified.
   - Active slope-drainage options include pumping wells, horizontal drains, vertical or angled drains and, potentially, drainage tunnels.
   - It is preferable to anticipate where future dewatering and slope depressurisation infrastructure may be placed in the pit slope during operations, and to design the grid or mesh to accommodate drains or wells to simulate these features.

9. **Calibrating the model**
   - The model first needs to be calibrated to an 'initial' observed condition which is defined with data. Sometimes it may be feasible to calibrate to a pre-mining condition. All models are required to reasonably match 'current day' conditions before they can be used for 'future time' predictions.
   - Transient calibration of the model is a strongly preferred step as this is the best way to determine whether the model provides a reasonable representation of the bulk-scale system response. The calibration should be with respect to the water level or pore pressure changes through time in response to changing hydraulic stresses through time.

10. **Interpreting and validating model results**
    The model results need to be benchmarked against the observed hydrogeological system including natural and induced responses in the system. For a Brownfield site, the model should compare reasonably with heads and flows observed during the operational history.

If the model identifies important areas in the system that are unsupported by data, then additional field installations may be necessary to fill data gaps. The model is typically validated as part of the transient-response calibration process and by ensuring that the model performs 'as expected' compared to observations at the site or known conditions at other similar mine sites. Peer review is an accepted way to validate a model.

When determining the modelling approach, it is also necessary to have an understanding of the needs of end-users of the model. Important factors to consider are:

- if the model will be used solely to support company decisions or whether it will be reviewed and used by third parties, such as review consultants, funding agencies, regulatory authorities, NGOs or the public at large;
- if it will be necessary to hand over the model for third parties to run. In this case, it will probably be necessary to use a publically available code;
- whether the staff of the mining company will use the model on a day-to-day basis. If so, it may be more appropriate to use a friendlier 'off-the-shelf' code to make it easier for staff at the mine site to operate, rather than a more complex specialist code.

4.2.2 Determining model geometry

For all models, it is necessary to use a simplified geometry to represent the hydrogeological system. All hydrogeological systems are three-dimensional. However, in many cases the system at the sector-scale can be adequately represented by using a 2D model that is constructed along a given cross-section line (mostly coincident with the geotechnical section). Other types of model include 'planar' or 'strip' models or 'axisymmetric' models. 'Planar' or 'strip' models are also constructed along a cross-section line, but include a thickness of the model perpendicular to the plane of analysis. They have the advantages of a 2D model but they also allow dewatering features (such as wells and horizontal drains) to be simulated either side of the analytical section.

The type of approach selected will depend on the nature of the hydrogeological setting, the available supporting data, the goals of the model and the time and budget constraints that are imposed by the mine operator. The following guidelines generally apply.

- Regional and mine-scale models are almost always constructed in 3D.
- Detailed sector-scale models can be constructed in 2D or 3D.
- A 'hybrid' approach may provide a good compromise for allowing a detailed analysis of vertical gradients

with the incorporation of wells and/or horizontal drains off the section. A 'hybrid' approach will also allow better representation of the groundwater flux moving down the slope than a 2D model, and will therefore allow the flow rates from the slope depressurisation measures to be more accurately determined.

4.2.2.1 Two-dimensional (2D) models

A 2D model is relatively easy to set up and is often not as computationally intensive as a 3D model. An advantage of a 2D model is the relative ease with which detailed geological information can be incorporated, and the ability to rapidly calibrate to vertically separated piezometers and simulate vertical heads and pore pressure gradients. A 2D approach is usually applicable for simulating pit slope pore pressures during the early stages of a project when there are limited site-specific data and the model can be used to evaluate 'what-if' scenarios. During operations, a 2D approach may also be more appropriate for providing rapid and practical answers for slope depressurisation programs. A 2D model has limited ability to simulate flow associated with dewatering at the pit-scale because it is only representative of a unit width.

The codes most often used for 2D pore pressure modelling are *SEEP/W* and *FEFLOW*. An example layout of a 2D model is shown in Figure 4.2.

A typical model contains a recharge boundary along the ground surface external to the pit and sometimes the upper part of the pit slope, a seepage face condition for the lower slope and pit floor, a general-head boundary to represent the far-field district groundwater system, no flow boundaries along the base of the model, and either a no flow or specified head condition along the inner pit boundary.

For many pit slopes, the general direction of groundwater flow is from the crest of the slope to the toe (downslope), in which case the interpretation of the model results may be relatively straightforward. However, it is perfectly valid, and often beneficial, to apply a 2D model in situations where the principal flow direction is sub-parallel to the strike of the slope (and therefore at right angles to the plane of the model), provided that the downslope flow component (along the plane of the model) can be reasonably estimated, and preferably with vertically discretised piezometers to calibrate the model. In the cross-section shown in Figure 2.56, the stratigraphical units are aligned sub-parallel to the strike of the slope and include a permeable unit (Bruno's band, as described in Appendix 4.7). The focus of the 2D model built on the same section was to fix the head in the permeable unit and study the response in the surrounding low-permeability units. By carrying out sensitivity runs, it was possible to use the model to determine what head needed to be maintained in the permeable unit to achieve the desired rate of depressurisation in the surrounding low-permeability units. The required pumping rate to achieve the necessary head in the permeable unit was predicted using a pumping test analysis outside of the model (not using the model).

Similarly, to simulate 'stair-stepped' head gradients across key structures (zone B in Figure 4.1B), it is possible

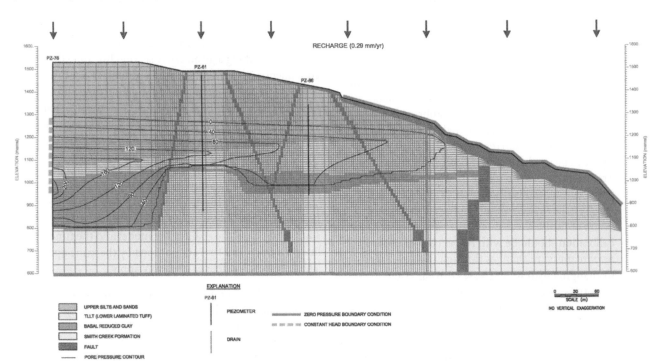

Figure 4.2: Finite-difference grid of 2D model

to use a 2D model to investigate the pressure gradients that may develop across the structures as a consequence of lowering heads on the 'pit side' of the structures. An advantage of using a 2D model is its ability to rapidly simulate the vertical head gradient component that develops either side of the structure, which is important for calculating the actual driving head. Again, the focus of the model is to study the heads, not to calculate flows.

Two important considerations for interpreting 2D model results are (i) that the groundwater flow component perpendicular (or oblique) to the cross-section is understood, and (ii) that variations in the 'off section' geological conditions are known. This second consideration also applies to 2D geotechnical models. For geological conditions where an important control on groundwater flow and water level distribution is the presence of geological structures or lithological contacts occurring oblique to the slope (Figure 4.3), the use of a 2D approach may require some significant interpretation. Enhanced permeability occurs along the strike of many structures, but the same structure may form a barrier to flow across its strike. In several 2D models, structures have been input as zones of lower permeability where they are oriented at an azimuth of less than about 35° relative to the orientation of the slope, and are input as zones of higher permeability where they are oriented with an azimuth of greater than about 35°. However, the exact model input details will require careful consideration of the local site conditions.

The typical benefits of using a 2D approach for pit slope modelling are as follows.

- Complex vertical pore pressure gradients can be simulated with a greater level of detail compared to a 3D approach. Calibration of the model is a simpler process because it only requires data points specific to the slope sector of concern and the model domain is smaller and more tightly controlled.
- The hydrogeological model domain can be made coincident with the 2D geotechnical model domain. Hydrogeological data collection can be focused around the key defined sections, guided by the geotechnical interpretation.
- The model can be set up quickly, often using the same information already available from the geotechnical model. Cross-sections of lithology, alteration and mineralisation can be used to conceptualise the hydrogeological units in the 2D model. In certain applications, it may be applicable to make the hydro-geological units consistent with the geotechnical units, but this needs to be considered on a case-by-case basis.
- The model can be easily managed and running times are normally less than a 3D model. This allows the model to more rapidly simulate different slope designs and alternative depressurisation methods.

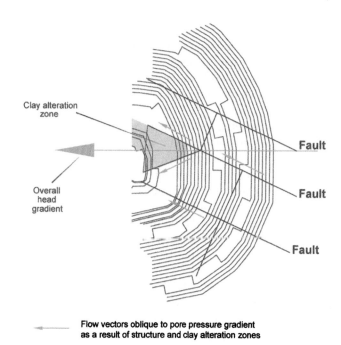

Figure 4.3: Flow vectors oblique to the pore pressure gradient (From Read & Stacey, 2009)

- Since the 2D model is easier to set up, calibrate and run, it is often more appropriate for the time and budget constraints that may be imposed by a mining study.
- Interpretation of the model results can be used to determine specific zones that may need to be studied in 3D.

4.2.2.2 Axisymmetric models

Because a 2D model has a unit thickness, it does not provide a good representation of the downslope flow into a pit. Radial flow may be simulated using an axisymmetric model, where the width of the model slice expands from the pit centre to the distal boundaries of the model. Figure 4.4 shows the configuration of the model using a finite-element approach. Three rays converge towards the pit from the boundary distal from the pit. The configuration of the grid represents the radial flow pattern towards the pit.

An axisymmetric model usually assumes no-flow boundary conditions along the lateral sides of the model, general-head conditions to represent the district system distal from the pit, and seepage face conditions along the surface of the lower pit slope and the pit floor. Figure 4.4 shows simulated pore pressures corresponding to different depths of the pit by simulating the progressive excavation of the pit benches.

By adding more rays to the axisymmetric model, the model contains slices that can simulate more detailed groundwater head distributions. If the impact of slope depressurisation does not induce groundwater flow through the lateral boundary (that is roughly at right angles to the direction of the slice or sub-parallel to the pit wall), the axisymmetric model can be used to predict the seepage inflow to the pit, or the inflow to the slope depressurisation measures that are incorporated into the model.

An axisymmetric model should be oriented parallel to the principle flow paths present within the pit slope. Figure 4.5 shows a slice from an axisymmetric model used for the prediction of pore pressure using the finite-difference approach. In contrast to the finite-element method shown in Figure 4.4, the grid in Figure 4.5 is rectangular in shape because of the orthogonal finite-difference grid. In order to simulate the radial-flow pattern, the model cells outside the slice boundary were simulated as inactive cells.

4.2.2.3 Planar or slice models

Limitations of working in 2D are the difficulty of simulating horizontal drains or vertical wells that are off the line of section of the model, and the inability of the model to simulate different drain or well spacings. Another

method that avoids the complexity of using a 3D model, but overcomes the difficulty of inputting drain spacing is a 'planar' or 'slice' model based on a rectangular grid, an example of which is shown in Figure 4.6. The rectangular grid can be thought of as representing a linear (flat) section of the pit wall, rather than the curved sector axisymmetric model described above. Rectangular grids are easier to conceptualise and set up than pie grids. In many large open pits, and at the scale of most pore pressure modelling, the curvature of the slope is small compared with the domain of the model. Therefore, a 'planar' or 'slice' model usually performs adequately well at the sector scale.

In Figure 4.6, the central axis of the 'planar' model is the 2D hydrogeotechnical section (Slice 6), and the figure shows how the hydrogeological units and model grid are extended in the third dimension: in this case five slices are used on each side of the central section. The model grid illustrated is symmetrical around the 2D axis. However, this is not always necessary, and Figure 4.7 shows how existing underground workings occurring off-section have been input to one part of a 'planar' model domain. The

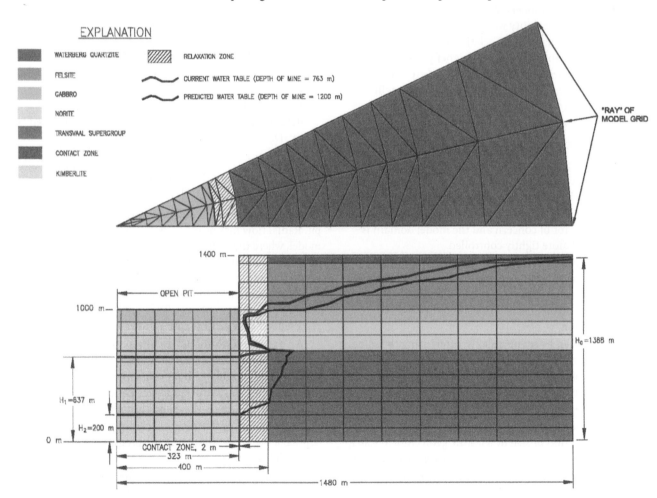

Figure 4.4: Plan view and cross-section of axisymmetric model using a finite-element method

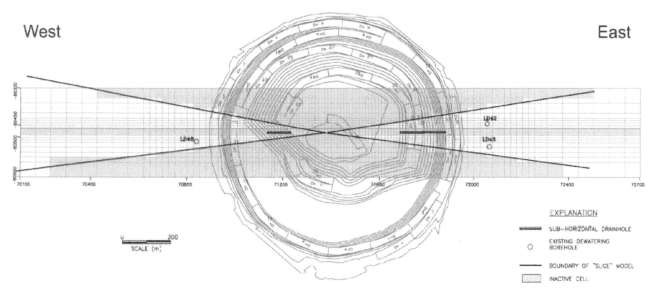

Figure 4.5: Plan view of slice model using a finite-difference method

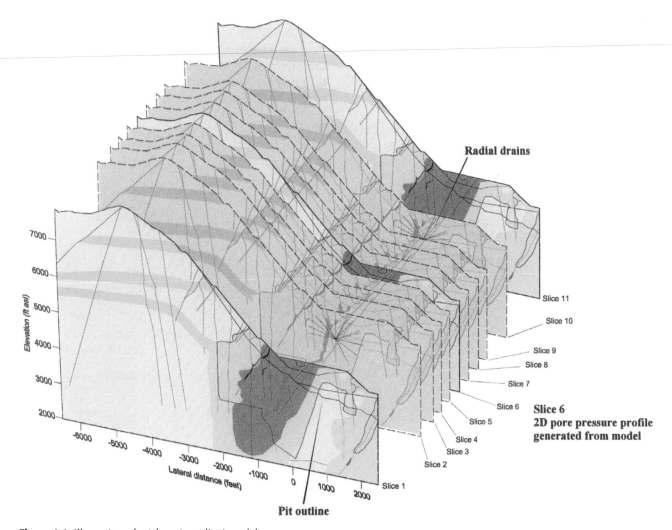

Figure 4.6: Illustration of a 'planar' or 'slice' model

model set-up allows the effect of different drain and well spacings surrounding the study section to be directly input and investigated by the model. The effects of different drain spacings and orientations can be directly simulated in terms of their influence on the pore pressure and slope performance, with the results brought back to the central hydrogeotechnical section (Slice 6 in Figure 4.6).

4.2.2.4 Three-dimensional models

A 3D groundwater flow model for a pit slope may be more appropriate than a 2D or a hybrid model in situations where:

- the pit slope hydrogeology is dominated by structures or other hydrogeological units that impart a strong oblique flow and groundwater head distribution that cannot be captured with a 2D approach;
- the dominant control on sector-scale pit slope pore pressures are large hydraulic stresses that are distal from the sector;
- the geology and structure is relatively simple and the necessary detail can be implemented to a 3D model without making it computationally complex;
- a 3D pore pressure grid is required for input to a 3D geotechnical model.

The mesh configuration of the finite-element method is theoretically more flexible for implementing the

irregular geometry of geology and hydrogeologic units around an ore deposit, when compared to the grid of a finite-difference method, although this is not always the case. Figure 4.8 and Figure 4.9 show the plan view and cross-section, respectively, of a 3D groundwater flow model using the finite-element approach. The complicated geological conditions are simulated in the model with different hydraulic parameters assigned to each geologic unit. In Figure 4.8, each different colour represents a different hydrogeological unit. However, when there are no data or justification to distinguish the hydraulic character of different geological units, the parameter zones may lump several geologic units and smear the same parameters across all of them. A finer mesh is assigned to areas where detailed simulation of vertical head gradients is needed and also where there are data available to support refinement of the model. Almost always, the finer mesh is centred on the open pit. As the mesh extends outwards from the pit area, it usually becomes much coarser.

The vertical layering of the model is often designed to coincide with the major stratigraphic divisions although these are often irregular in a mine setting. A benefit of finite element model is that layers can have irregular thickness as they extend through the model and can follow irregular stratigraphy. In sector-scale 3D models, several

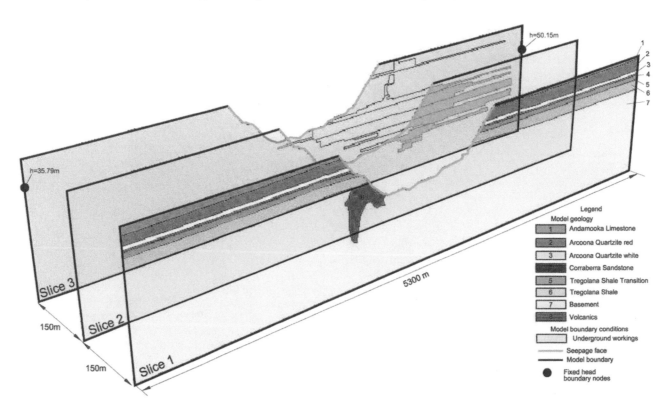

Figure 4.7: Illustration of a 'planar' model with underground workings simulated offset from the central hydrogeotechnical section (Figure 6.33 from Read & Stacey, 2009). The underground workings are shown in Slice 3.

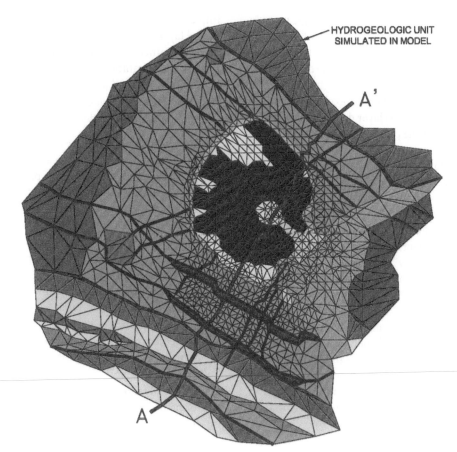

HYDROGEOLOGIC UNIT
SIMULATED IN MODEL

A'

A

Figure 4.8: Plan view of finite-element grid and simulated geological units

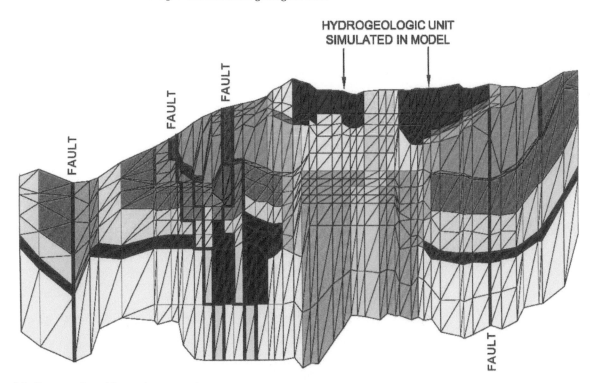

HYDROGEOLOGIC UNIT
SIMULATED IN MODEL

FAULT

FAULT

FAULT

FAULT

Figure 4.9: Cross-section of finite-element grid and simulated geological units (Section A–A' from Figure 4.8)

layers may be implemented within single geologic units in order to provide better vertical discretisation of groundwater head gradients in the vicinity of the pit slope. In the portion of the model that occupies the current or future pit shell, layers may be placed through the model at intervals of 15 to 30 m, even if there is no change in geology over this distance. This layer density is normally limited to the portion of the pit shell that is beneath the phreatic surface.

In order to maintain reasonable computational efficiency, the concept of 'layer pinch-out' can implemented in finite-element codes such as *FEFLOW* and *MINEDW*. The two main advantages of using the layer pinch-out method are (i) that it significantly reduces the total number of elements and nodes, and (ii) it essentially creates a 'window' model within a large regional groundwater flow model. Therefore, the layer pinch-out method may avoid the need to create a separate regional flow model and a 'window' model and any associated linkage between these two separate models. Figure 4.10 shows the layer pinch-out configuration of a model cross-section. In the pit area, 33 model layers were used to simulate the open pit, with a block cave operation simulated beneath the pit. Immediately outside the pit area, the number of model layers is 23. In the regional area, the number of model layers is seven. By

implementing the layer pinch-out approach, more vertical model layers can be assigned to the pit area, but these are gradually pinched out towards the boundaries of the models and common hydraulic parameters are smeared through the pinched layers.

The most important considerations for deciding whether to use a 3D model are:

- can the 3D model capture the sector-scale geological, hydrogeological and mine planning detail that is controlling the distribution of pore pressures in the current or future pit slope, whilst remaining at a manageable computational size (with a million cells or elements being a general rule of thumb for overall model size)?
- is the 3D model likely to be capable of matching the observed vertical groundwater gradients and head distributions with an acceptable level of accuracy during the calibration process?
- can the geometry of the future pit shells to be used in the geotechnical model be matched with the same accuracy (geometry, timing) in the 3D groundwater model?
- can the 3D model be constructed, calibrated and used for predictive analysis within a timeline that supports the overall project schedule?

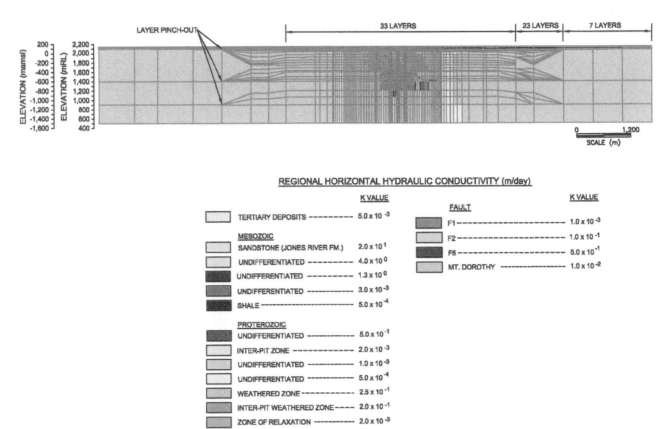

Figure 4.10: Configuration of 'Layer Pinch-Out' method of model layers

4.2.3 Setting the model domain and boundaries

4.2.3.1 Sizing the model domain

Construction of the model first involves determining the necessary lateral and vertical extent of the grid or mesh (the model domain). This is normally governed by the identification of appropriate hydrogeological boundary locations. An important part of model design is to make the domain and boundaries distant enough from the open pit or other mining feature to ensure that the effects of hydraulic stresses, such as dewatering pumping, do not arrive to the boundaries. If drawdown occurs at a model boundary, it will introduce an artificial gradient between the model boundary and the nodes within the model domain. This can force the model to produce an inappropriate result. For mine sites connected to a permeable regional groundwater flow system, the distance between the pit and the regional model boundary may range from a few kilometres to several tens of kilometres.

The upper surface of most model domains coincides with the topographic surface and may change with time due to the excavation of the pit. The base of the model domain is normally extended to a significant depth beneath the deepest future planned mining. The base of the model is almost always defined as a 'no flow' boundary and the intent is to locate this boundary deep enough beneath the base of the mine such that it cannot adversely influence the computation of groundwater levels in areas of interest. Also, it ideally should coincide with a low-permeability formation that is expected to contribute no upward flow towards the mine, though it may not always be possible to define this. The depth of the model base requires judgement and interpretation specific to each mine site. Typically, the depth of the no-flow boundary will vary between 500 m and 1000 m beneath the deepest planned elevation of mining. Where there are limited supporting data, the upper end of this depth range is normally used and a 'general assumption' of reducing permeability with depth is made.

For 2D models, the model domain often ends at the centre of the planned or proposed ultimate pit. Therefore, a boundary condition has to be placed at this location. A typical challenge is that the area of greatest interest in the model is often the pore pressure distribution in the lower slope and pit floor, which is close to this boundary (Section 4.2.3.4).

4.2.3.2 Specifying the boundary conditions

Figure 4.11 is a schematic diagram of a conceptualised groundwater system illustrating the simulation of boundary conditions. The groundwater system in Figure 4.11 is conceptualised in Figure 4.12. The most common boundary conditions are illustrated in Figures 4.11 and 4.12 and are discussed briefly below.

- The **no-flow boundary condition** is usually applied to the groundwater divides where there is little groundwater flux exchanged between the model domain and other basins. In this case, no flow is allowed to cross the boundary of the model. As shown in the figures, the boundary nodes in the mountains are no-flow nodes. The lateral sides of 'slice' models can also be likened to no-flow boundaries since they represent symmetry planes and are inherent in a 2D modelling approach.
- The **specified head** is the head value assigned to each boundary node at a specified time. The specified head can be constant (time-invariant) over the entire simulation period, or it can be time-variable.
- A **constant-head** boundary can be used where there is a constant recharge source at a given elevation (for example a river or lake) or a constant topographically or geologically controlled discharge point for the groundwater system (Figure 4.12). In this case, the head at the model boundary is fixed and the amount of water entering or leaving the model is controlled by the groundwater gradient inward from the boundary. In addition, specified-head conditions are often assigned to the model boundaries to maintain the regional groundwater flow direction and groundwater gradient.
- A **variable-head** boundary may be used where conditions at the margin of the model are allowed to be influenced by the spread of drawdown within the model and thus the amount of water entering the margin of the model may vary.
- **Recharge from precipitation** to the groundwater system is generally applied to the uppermost wet layer of the model. The amount of recharge (often in millimetres) is usually calculated separately from the main model. Attempting to compute the process of water movement through the unsaturated zone to the phreatic surface can add numerical complexity and instability to the model. The recharge rate over the model domain can differ significantly from one area to another because of:
  - → differences in the way water is concentrated at the surface (for example ephemeral stream channels, ponds or melting of accumulated snow banks);
  - → different permeability values of the hydrostratigraphical units that are simulated in the model;
  - → topographic elevation, causing different precipitation rates at different elevations.

It is normal to perform separate water balance modelling (separate from the groundwater flow model) to estimate recharge and its variation with time, considering precipitation, runoff, soil-moisture storage, evaporation and evapotranspiration, as described in Section 3.6.3. The estimated recharge rates are then applied to the groundwater model and may be adjusted during the process of model calibration. Recharge is strongly time dependent and it may be necessary to input averaged or smeared values into the model to avoid the requirement to use short model stress periods.

Streams can provide recharge to the groundwater system or can be receptors of groundwater discharge. Streams can be simulated with the stream routing package in all common 3D codes.

Groundwater discharge from 2D and 3D models is usually simulated using drains or seepage faces. Discharge may be input as discrete drain features at the locations of known springs or gaining streams and, in a 3D code, may also be simulated as an evaporation zone where the groundwater is close enough to the topographic surface to allow evapotranspiration (ET: the sum of evaporation and plant transpiration). In general, an 'extinction depth' and a maximum evapotranspiration rate should be assigned to the modelled ET area.

4.2.3.3 *Defining the external distant boundary*

The external boundaries of the model are normally defined by defensible physical features which may include the following.

- Natural boundaries of the groundwater system, for example a natural groundwater divide or a low-permeability formation that may transmit little water.
- Artificial boundaries, used to define some portion of a larger aquifer in the regional system that is unlikely to

change with time or be influenced by open pit mining or dewatering. A constant head condition or constant flux condition is often assigned to represent a 'distant' condition at the model boundary that is not expected to change through time.

Figure 4.13 compares the same model design in a Finite Difference and Finite Element format (based on the conceptual model shown in Figure 4.12). In the model, the upland areas in the basin are represented as recharge zones and a constant head is used to represent the downstream outflow from the basin. The sides of the basin represent hydrologic divides and are implemented in the model as no flow boundaries. The rivers are represented with drain cells and evaporation cells in lowland areas where evapotranspiration of groundwater is observed.

Although it is preferable to set the model domain large enough to avoid the propagation of mining-induced hydraulic stresses to the model boundaries, it is possible that a small drawdown (less than a few metres) can occur in the model boundary. In such cases, the model should simulate the boundary fluxes that would result were the modelled flow domain extended outwards. In *MODFLOW*, the regional boundary flux can be simulated as a general-head boundary condition. In *FEFLOW and MINEDW*, a variable-flux boundary condition can be applied.

Radial flow calculations can be used to provide an approximation of the area of drawdown ('radius of influence') surrounding the pit, based on the depth and timescale of mining beneath the pre-mining water level

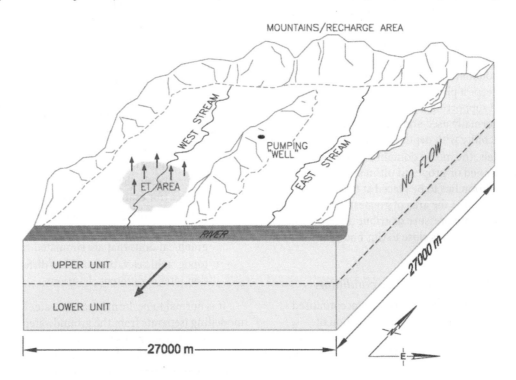

Figure 4.11: Schematic diagram of hydrogeological settings of a regional basin

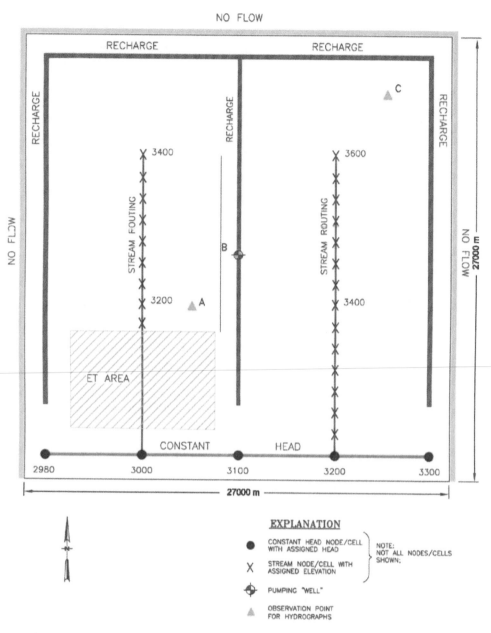

Figure 4.12: Conceptual model of hydrogeological settings (based on Figure 4.11)

and on bulk hydraulic parameter values for the rock units surrounding the pit. However, such analyses do not account for the geological and structural variability that normally occurs around an open pit and will often tend to over-estimate the radius of influence. Where there is doubt, the approach should be to extend the model domain and external boundary further from the pit, even if the amount of supporting geological and hydrogeological information is minimal for the distant portions of the model domain.

For a sector-scale model, the boundary is normally placed at an extended distance from the planned final pit boundaries (crest of the slope) such that it does not

artificially influence the pore pressure profile in the pit slope or the geotechnical domain being modelled. Ideally, there should be piezometers at or near the pit crest and extending 1 or 2 km beyond the crest. In this way, there will be control of groundwater heads in the ex-pit area and the distant boundary can be adjusted to ensure there is an acceptable match to both ex-pit groundwater heads and the heads within the pit slope itself. A boundary that is set too close is likely to result in artificially high heads, or fluxes being introduced to the pit slope area during the forward (predictive) analysis. It is sometimes possible to simulate the external boundary by using a time-variable head condition, usually with a reducing head through time to

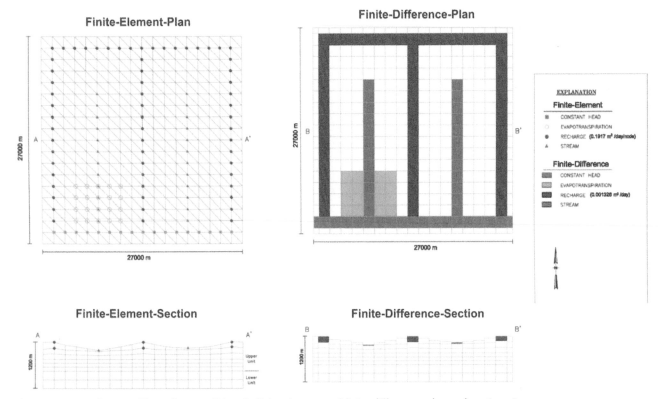

Figure 4.13: Simulations of boundary conditions in finite-element and finite-difference; plan and section views

represent the gradual lowering of groundwater levels surrounding the mine (Figure 4.14). The justification for this may be provided by the larger-scale model or by the overall groundwater interpretation.

4.2.3.4 The inner pit boundary

Most active pits are hydraulic sinks and so the central axis of the pit floor can usually be approximated as a no flow boundary. In general, groundwater converges radially towards the pit floor; thus there is often a vertical (upward) flow component in the lower pit walls and at the toe area. Exceptions to this condition may occur where the open pit is in an upland location and the regional groundwater flow-field is not convergent (meaning there is a gradient away from the pit floor towards the regional system in some sectors), or where underground mining is taking place below the open pit operation.

If the pit is a hydraulic sink, then it can usually be approximated that there is no component of groundwater flow that crosses the pit floor to the opposite slope, so a

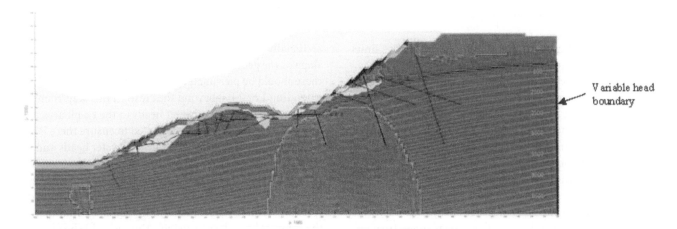

Figure 4.14: Use of a variable head boundary condition

no-flow boundary can be assigned. However, it is necessary to ensure that the model boundary coincides with the centre of the pit (or with the position of zero flow) for this condition to be applicable. For forward (predictive) modelling, if the pit floor is used to represent the model boundary, the location and potentially the features of the boundary condition may need to be adjusted to reflect the changing conditions caused by the deepening or changing nature of the pit in successive time steps. There is no need to include an inner pit boundary where the 2D (cross-section) model extends through the entire pit (crest-to-crest) or for 3D models that encompass the entire pit area within their domain. For 2D models, to better understand the changing conditions below the pit floor, it is sometimes prudent to develop one crest-to-crest model section with an 'unconstrained' inner pit boundary.

An alternative to a no-flow boundary is to use a prescribed head that may be transient where the lower slope and pit floor geometry are migrating with time. This condition can normally only be used at a mature mine where there are good data to support head values in the base of the pit sector being modelled, and where the rate of decline in head has been historically observed.

In situations where groundwater heads in the pit floor are influenced by active dewatering operations, or in some low-permeability settings by the 'zone of relaxation', the inner-pit boundary condition may be applied to the model below the level of the actual pit floor.

4.2.3.5   Pit slope boundary

The pit slope itself is an important model boundary. It is normally specified as a zero pressure condition, which may be achieved using drain nodes in the model. During the pit excavation, each model node or cell that represents the pit bench (or sometimes the base of the over-break zone behind the slope) is simulated as a drain node. The water that discharges from the drain nodes/cells represents seepage into the pit. During predictive analysis, as mine expansions and pushbacks occur, the position of the seepage face will need to be changed for each successive stress period where the pit slope geometry is changed, in accordance with the mine planned being analysed. These changes are normally implemented by modifying the pit slope boundary condition and by eliminating zones within the model to represent removal of material from the pit slope due to the progression of mining. When working in zones of high or seasonal rainfall, it may be necessary to add recharge to sections of the pit slope where there is no upward gradient or seepage into the slope. It can be important to simulate the influence of minor recharge to slope sectors that consist of weak materials.

Pore pressures in low-permeability materials exposed on the slope (or below the slope) are often very sensitive to recharge that may result from water accumulating on some

of all of the catch benches. The recharge is normally estimated outside of the model and added to the model as a specified flux (which is usually time-variant).

4.2.3.6   Coordination with the geotechnical model

It is usually a requirement, particularly for 2D models, that the hydrogeological model domain is coincident with the geotechnical model domain. However, it is often necessary that the hydrogeological model boundaries extend beyond the limits of the geotechnical model. In that way, the geotechnical model can be fully populated with predicted groundwater pressure values that are not influenced by artificial conditions that may caused by the model boundaries. Also, the model can better account for the wider groundwater flow field acting on the slope. In all cases, close coordination between the hydrogeology and geotechnical groups is needed.

4.2.3.7   Telescoping

It is possible to use 'telescoping' (sometimes called the 'window' technique) to set the boundaries of a smaller-scale model within a larger, coarser model. This method may allow construction of a detailed independent sector-scale model which has its boundaries linked to the mesh of a regional or site-scale model. This makes it possible to capture the detail required for the pit slope and the interaction with the regional flow field, while maintaining a manageable numerical model size. The physical parameter values and boundary conditions in the two models would be the same, although some value-averaging may be necessary where the layering or conceptualisation has been simplified in the larger-scale model. The outer boundary conditions in the sector-scale model (constant heads, for example) are derived directly from the wider-scale model.

Coupling of a regional model and window model can be implemented by extracting time-variable head values from selected locations in the regional model and then assigning them to boundary nodes at complementary locations in the window model. The position of the window model boundaries is normally determined using a combination of interpretive judgement of site data and/or the use of the regional model, which is run prior to window model construction to provide some coarse predictions of the timing and location of regional response to mine dewatering. Figure 4.15 shows the plan view of a regional model and two window models. In this example, the heads at the nodes in the regional model that correspond to the boundary nodes of the window models were extracted for each time step. These extracted heads were then assigned as time-variable specified-head boundary conditions for the window models.

An example of coupling a 2D pit slope model to a general mine dewatering model is shown in Figure 4.16.

Window Model of Pit Area

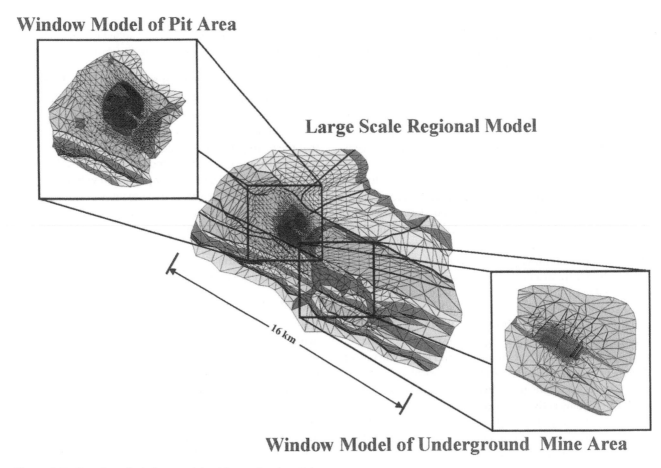

Large Scale Regional Model

Window Model of Underground Mine Area

Figure 4.15: Coupling of window models with a regional model

The dewatering model is used to determine the drawdown that may occur in the groundwater system related to an alluvial basin along the east side of the pit and how this may vary at any given time step. The pit slope model was then set up so that the heads at its external boundary are fixed for each time step based on the amount of drawdown (and the head field) predicted by the dewatering model. A progressively reducing head was used at the external boundary of the pit slope model to simulate the effects of general mine dewatering.

4.2.4 Defining the mesh or grid size

Once the boundaries of the model are determined, it is necessary to discretise the model domain by subdividing it into model grids. Depending on the numerical methods used, the grid may have rectangular or irregular polygonal subdivisions (Faust and Mercer, 1980). Figure 4.17 shows a comparison of typical two-dimensional (plan view) gridding for finite-difference and finite-element methods.

The finite-difference method (e.g. *MODFLOW*) requires that orthogonal cells be superimposed over the study area.

The outline of the cells may not coincide with features in the model (or the model boundaries) so that finer grid spacings (i.e. smaller spacings) are required. At each cell (the dot in the centre of the cell [block-centred]) a calculation of head is made. The more cells that are created, the more computational effort is required.

For the finite-element method, irregular elements can coincide more closely with features in the model and the model boundaries. Calculations of heads are made at nodes which form the corners or connection points between elements. With both the finite-difference and finite-element methods, the discretisation should be increased in areas of expected high hydraulic stress such as pumping features and high pore pressure gradients, in order to obtain accurate results. A comparison of key geometric features of the finite-difference method and the finite-element method is provided in Table 4.9.

Because of the grid requirements of a continuous orthogonal pattern, finite-difference methods are often used to simulate relatively simple geological conditions and geometries of mine workings, as shown in Figure 4.18. A typical approach is to centre an area of fine cell size over

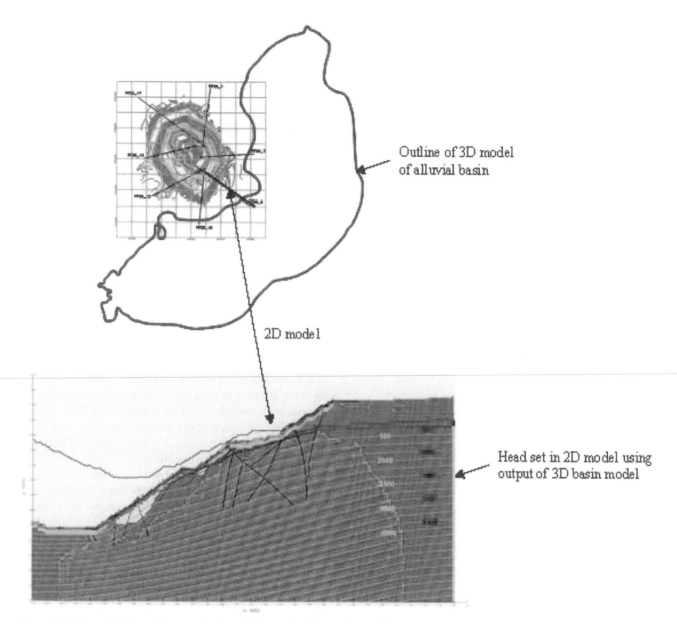

Outline of 3D model
of alluvial basin

2D model

Head set in 2D model using
output of 3D basin model

Figure 4.16: Example of coupling a 3D basin model to a 2D slope-scale model

the pit or a slope sector where the most detailed geologic knowledge will exist. Current code and computing capability allows for most *MODFLOW* models to run with cells implemented at the scale of the mine geologic block model provided there is significant coarsening of the grid outside of this area. In many mining applications, the irregular geological surfaces and the geometry of the open pits can be more closely represented with the irregular and more flexible mesh of the finite-element technique, as shown in Figure 4.19.

Table 4.9: Comparison of key geometric features of the finite-difference and finite-element methods

Feature	Finite-difference method	Finite-element method
Node representation	Block-centred nodes	At corners of elements
Calculation	Node-to-node calculation based on differentiation	Volumetric calculations within elements based on integration
Grid configuration	Some models require continuous orthogonal grids	Flexible grid

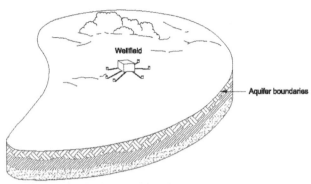

a) Map view of aquifer showing well field and boundaries.

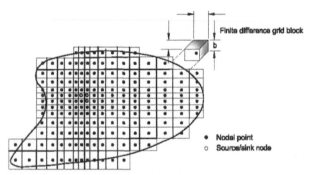

b) Finite-difference grid for aquifer study, where $\triangle$x is the
spacing in the x-direction, $\triangle$y is the spacing in the y-direction
and b is the aquifer thickness.

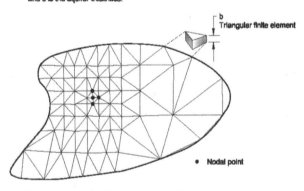

c) Finite-element configuration for aquifer study where
b is the aquifer thickness.

Adapted from Mercer and Faust, 1980a

Figure 4.17: Gridding for numerical models

For sector-scale models, the preferred practice is to
design the grid or mesh using the geological block model
as the basis. The mesh or grid is normally designed to
include (i) significant detail at the scale of the geological
block model for the entire slope domain inside the planned
pit crest-line, and (ii) coarsening of the mesh or grid in
areas extending outward from the crest line (or geological
block model limits) where the supporting geological and
hydrogeological information may be less. An example of a
finite element grid of a 2D pit slope model is shown in
Figure 4.20.

The number of elements in a mesh, or cells in a grid,
defines the numerical model size which, in conjunction

with assigned material properties, is closely related to
model run time and the numerical stability of the model.
The general objective is to incorporate the variability and
detail that are pertinent to the questions that need to be
answered by the model, at an appropriate model scale, and
commensurate with the amount of supporting data.

4.2.5 Determining whether to run steady-state, transient or undrained simulations

At an active mine, the pit geometry and hydraulic
stresses are constantly evolving, so it is usually necessary
to carry out a transient analysis. (In this context,
'transient analysis' means the response of the system to
changing hydraulic stresses, as opposed to the term
'transient pore pressure' which normally refers to
short-term local-scale variations in groundwater head or
saturation (as described in Section 3.6.7).) Time is an
important control on pit slope pore pressure. Events that
impose transient hydraulic stresses normally include the
excavation of the pit and the lowering of water levels due
to passive inflows, the lowering of groundwater levels as a
result of general mine dewatering, changes in recharge or
discharge with time and, in some cases, changes in
material properties with time. These cause a departure
from steady-state flow, so that prediction of the
subsequent evolution of the groundwater flow field
requires a transient analysis, both for model calibration
and for assessing future conditions.

There are, however, some cases where the analyses can
be simplified by ignoring time and performing either
steady-state or undrained simulations. Steady-state flow, as
the name implies, occurs when there is no (or minimal)
change in groundwater stress and resulting heads
throughout the model domain during the stress period (or
time step) being modelled. A pore pressure profile
generated from a steady-state model assumes that the
pressures have equilibrated to the particular slope design
and position and will remain constant until new hydraulic
stresses are applied (which may be due to changes in
seepage due to the changing shape of the pit).

The steady-state analysis assumes that infinite time is
available for pore pressures to equilibrate to the applied
stresses. In a steady-state analysis, the groundwater
inflows equal the groundwater outflows along the
boundaries of the model and there is no change in storage
within the model domain. Such conditions may occur, for
example, when predicting long-term post-closure
conditions in pit slopes, in pit slopes that are highly
permeable so that equilibration of pore pressures occurs
quickly, or in situations where changes in pore pressure
occur extremely slowly so that transient effects can be
ignored in favour of a simpler steady-state analysis.
Confirmation of such conditions is gained by study of
long-term site-monitoring data and demonstration of rapid

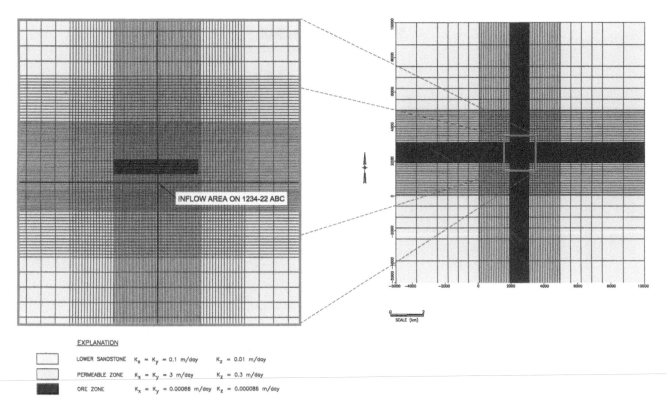

Figure 4.18: Plan view of a grid configuration of a finite-difference model

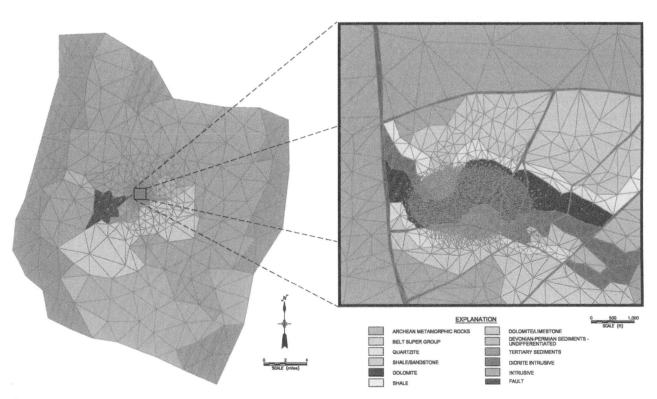

Figure 4.19: Finite-element grid of a complex geological condition commonly encountered in mining

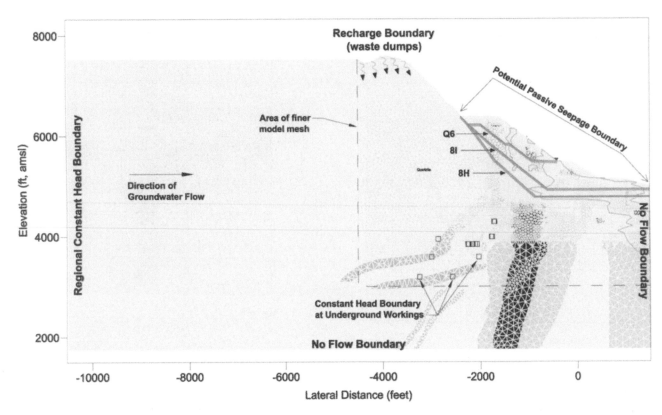

Figure 4.20: Example of a finite element grid for a pit slope model

pore pressure equilibration following application of new hydraulic stresses.

In situations where the rock mass permeability is low and hydromechanical coupling is not apparent, the rate of mining (and associated time requirement for pore pressures to respond) is a strong controlling factor in pore pressure dissipation and equilibration. Modelling of low diffusivity (Section 3.6.5.2) situations using a steady-state approach is not applicable except possibly when planning for long-term mine closure.

Analysis can be carried out to help guide whether steady-state or undrained simulations are applicable to the situation being studied. An undrained simulation is one which considers only the effects of hydromechanical coupling and where there is no component of groundwater flow in the analysis. In theory, two interacting factors influence pore pressure in a low-permeability setting. These are:

1. the speed with which excavation occurs (controlled by the rate of removal of rock from the slope area);
2. the speed with which head or pressure changes can dissipate through the rock mass to accommodate the change in total stress caused by the unloading; this is dependent on the hydraulic diffusivity.

An approach to assessing the combined effects of these two factors is to define a Dimensionless Excavation Rate (R), which takes into account the rock-mass diffusivity and the slope excavation rate. R is a function of the mining rate (M_r) and rock mass hydraulic diffusivity (D), as follows:

$$R = \frac{M_r}{D} \qquad \text{(eqn 4.1)}$$

The volumetric mining rate (per unit model thickness) is defined as:

$$M_r = L\frac{dH}{dt} \qquad \text{(eqn 4.2)}$$

where L is the excavation width, H is the slope vertical height and dH/dt is the rate of excavation (Figure 4.21). Separation of variables and integration (with boundary conditions of $H = 0$ at time $t = 0$) gives:

$$M_r t = LH \qquad \text{(eqn 4.3)}$$

If L is taken as the width of the pit expansion, then LH is the average cross-sectional area of the pit in the vertical plane which can be written as A_p, so that M_r can also be expressed as:

$$M_r = \frac{A_p}{t} \qquad \text{(eqn 4.4)}$$

The analysis suggests that if $R > 0.7$, no flow modelling is necessary because mechanical response is governed by

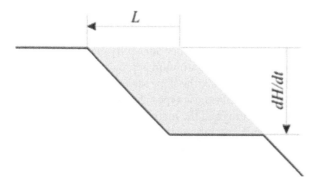

Figure 4.21: Definition of terms in Equation 4.2

undrained behaviour (Appendix 5). For $R < 0.7$, transient analysis is typically required but, if the recharge boundary is close, then steady-state analysis could potentially simplify the modelling work.

A rule of thumb can be derived to determine the dimensionless excavation rate that divides the steady-state and transient modelling regimes. The model can be assumed to have reached the steady-state if the distance, d_M, from the pit to the vertical model boundary (with fixed head) is a fraction of the radius of influence, or the distance to which the drawdown would have propagated in an infinite domain:

$$d_M \leq a\sqrt{Dt} \qquad \text{(eqn 4.5)}$$

where t is the time from the start of mining and a is a dimensionless factor less than 1. Substituting D from Equation 4.1 into Equation 4.5, and taking the expression for M_r from Equation 4.4, then the dimensionless excavation rate, R, for which the pore pressure field can be approximated with a steady-state approach can be expressed as:

$$R = a^2 \frac{A_p}{d_m^2} \qquad \text{(eqn 4.6)}$$

However, even though the pore pressure field will remain in equilibrium if the excavation rate is low, that equilibrium condition will change as excavation proceeds because – apart from anything else – the drainage levels will need to alter. Therefore, if steady-state modelling is carried out, the models will need updating to take account of this.

For models of typical open pit operations, a value of $a = 0.35$ was determined by Hazzard *et al.* (2011). Therefore, assuming that the pit is excavated at a relatively constant excavation rate and for rock mass porosity less than 0.01, the condition for validity of the steady-state pressure approximation can be expressed using the following relation:

$$R < 0.125 \frac{A_p}{d_B^2} \qquad \text{(eqn 4.7)}$$

where d_B is the distance from the recharge boundary.

The examples in Appendix 5 show that, for virtually all normal mining situations where there is continuing mine excavation, where there are low-permeability rocks in all or part of the model domain, and where there are recharge boundaries relatively distant from the pit, the use of a steady-state analysis underestimates the groundwater heads and pore pressures compared with a transient analysis, often by a large amount. The results predicted by a steady-state approach would therefore be non-conservative, meaning that the estimated pore pressures would be lower than the actual pore pressures, and their input for geotechnical modelling would suggest an unrealistically high factor of safety. Thus, in virtually all cases for simulating mine operations and active pit slopes, modelling should be carried out using a transient approach.

Calibration of numerical models is discussed in Section 4.2.10. For many models, it is advantageous to carry out both a steady-state and a transient model calibration. Again, this should not be confused with the general requirement for using a transient modelling approach.

4.2.6 Determining whether the use of an equivalent porous medium (EPM) code is adequate

Section 3.6.5.1 described the use of an EPM approach for groundwater analysis in a fractured rock medium. Most rock encountered at depths where instability may occur below pit slopes is pervasively fractured at the scale of normal field measurement. Therefore, unless there are significant localised varying stresses (for example, immediately after the start up of a nearby pumping well), the pore pressure field usually occurs as a continuum within individual structures and within the intervening subordinate joint and fracture sets. The Diavik case study (Appendix 3) indicates that discrete fracture flow conditions may become important close to points of new pumping stress for about a week following the application of the new stress. The numerical analysis (Appendix 5) supports the use of an EPM flow code to study the stability of closely jointed rock slopes when the spacing(s) of the most transmissive features is less than about 5 per cent of the slope height (H), as illustrated in Figure 4.22. This represents the situation for virtually all slopes greater in height than individual benches. For larger spacing values, there is a need to explicitly account for the transmissive features using a discontinuum approach, either by incorporating the discrete features into an EPM code or, occasionally, by using a discrete fracture flow (FF) code.

Fracture flow codes assume that the matrix (blocks between fractures) is impermeable. Thus, their potential requirement is restricted to situations in which fracture flow is the only mechanism for pore pressure dissipation. This situation will occur when the response time for flow from the centre of a matric block to the surrounding

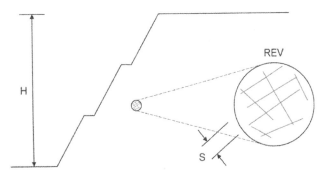

Figure 4.22: Relation of the Representative Elementary Volume (REV) to the fracture spacing and the slope height

fractures is long compared with the response time of the fracture system to a new hydraulic stress (e.g. drainage or unloading). The response time at a point increases with the square of the distance between the point in question and the origin of the head perturbation (Section 1.1.2.8); for the fracture system that distance can be taken as approximately equal to half the slope height, and for the matrix it can be taken as the distance from the centre of the block to the surrounding fractures, i.e. half the block size. The ratio of the response times will therefore be proportional to:

- the ratio of the squares of these distances (increasing with increasing in s^2/H^2);
- the ratio of the diffusivities (increasing with increasing in D_f/D_m), where D_f is the diffusivity of the fracture system and D_m is the diffusivity of the matrix.

As either or both of these ratios increases, the tendency will be for the EPM approach to become less valid.

An attempt has been made to characterise these situations by use of a chart in terms of values of two dimensionless parameters: the excavation rate, $R = M_f/c$, and the ratio $(s^2/H^2)/(D_f/D_m)$ (Figure 4.23). The chart was created by running numerical flow models in theoretical pit slopes. The fracture spacing, matrix properties and fracture properties were systematically perturbed and the resulting pressures were used to generate the approximate behavioural boundaries shown on the figure. The figure shows that the matrix can be considered to behave as impermeable in the case of large blocks, with a large ratio between fracture and matric diffusivities, and a fast excavation rate. At a slope scale, these conditions are not likely to occur. Thus, for the majority of pit slopes, it would be expected that the use of a continuum approach is appropriate.

By using Figure 4.23, practitioners with knowledge of fracture spacing, slope height, excavation rate and rock and fracture diffusivities can determine what type of analysis is required to simulate pore pressures in the slope. Rock slopes with closely spaced fractures and/or low-

permeability contrasts behave like an equivalent porous medium, whereas rock slopes with large fracture spacings and/or large differences in permeability may need to be modelled using a discontinuum approach. The numerical analyses carried out at the slope scale (Appendix 5) have also helped identify those types of conditions in which heads in fractures and the matrix may be significantly different, which represents the right hand sector of Figure 4.23. A discontinuum approach may need to be considered when those conditions are met.

For most practical mining situations, if the use of a discontinuum approach is warranted, the initial approach should be to apply an EPM code with the discrete transmissive fracture zones added. For situations where there are multiple transmissive zones that would make an EPM model too complex, and where these have a large permeability contrast with the subordinate fractures or with rock matrix, the use of a fracture flow code can be considered.

It should be noted that, at the current time, few, if any, practitioners have used a fracture flow code to reliably simulate pore pressures on an inter-ramp or slope scale. As described in Section 1.1.2.5, the transmissivity of any given fracture is determined by the transmissivity at the 'pinch points' which cannot be realistically determined by 'point' field tests (such as packer tests). The hydraulic behaviour of individual fractures can only be determined by monitoring of pore pressure responses due to changes in hydraulic stress. A very large number of closely spaced piezometers would therefore be necessary to justify and support the use of a fracture flow model.

For local areas affected by new pumping stress, where there is shown to be a difference between the head in the main fractures and the surrounding matrix, it would be prudent (conservative) to apply the pre-stress heads in any stability analysis until the pore pressures in the lower order fractures (and/or the matrix) reduce and the system equilibrates to the new (lower) level (Section 3.3.6). However, in situations where pumping wells are shut down,

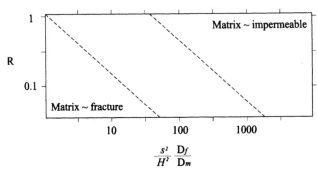

Figure 4.23: Flow regimes for different dimensionless excavation rates plotted against the ratio of fracture spacing to slope height squared times the ratio of fracture diffusivity to matrix diffusivity

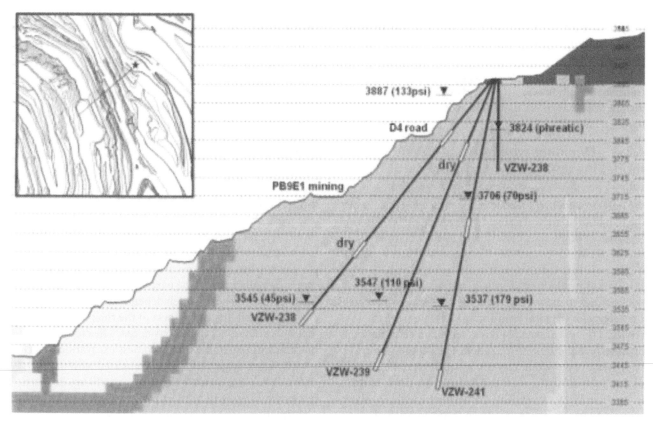

Figure 4.24: Example of head in a discrete fracture sustained by ongoing recharge where a discontinuum modelling approach needs to be considered. The discrete nature of the flow is illustrated by the anomalously high head in piezometer VZW-238 Shallow.

the resulting rise in head may occur first in the most transmissive fractures, so heads within those fractures may be temporarily higher than those within the surrounding matrix. In this case, it may be necessary for the stability analysis to consider the pore pressures in the more transmissive fractures. A similar situation may arise when ongoing recharge is transmitted to a pit slope along discrete fracture zones. Figure 4.24 shows piezometric levels in a fractured rock slope. Most of the piezometers show good levels of depressurisation. However, a single piezometer (VZW-238 Shallow) shows a high head that is sustained by a recharge source outside the pit and is being transmitted to the slope along a discrete fracture. In this case, it is necessary to use a discontinuum approach by adding the discrete fracture zone into the existing EPM model (such as the discrete features facility in *FEFLOW*).

4.2.7 Selecting the appropriate time steps (stress periods)

There are two types of time step in numerical flow simulations. The first type is the discrete time increments used by the numerical model 'engine' to compute flow. These numerical time steps are selected by the code to achieve convergence (or avoid numerical instability). The

use of small numerical time steps leads to longer computation times for the model. The second type corresponds to specific times of interest (stress periods) when changes to the flow or pore pressure regime might occur (e.g. pit excavation, drainage or well installation) or where some output of particular interest is required.

In many active mines, changes in hydraulic stress can occur very frequently. Therefore the modeller must select time steps that typically achieve a balance between accuracy (obtained by using smaller time increments) and practical considerations of the time required to construct and run the models.

Excavation of a mine is simulated differently by each numerical code. Codes that are able to simulate saturated-unsaturated conditions are normally best suited for mining applications (e.g. *MODFLOW-SURFACT*, *MINEDW*, and *FEFLOW*). Generally, the mine excavation is simulated using a series of mine plans, given as contour lines, on a monthly, quarterly, or annual basis. In many applications involving pit slope stability, the time periods of interest for generating pore pressure profiles are monthly or quarterly. For most models, the length of the time steps is varied within the overall simulation period, and it is normal practice to use small time steps (monthly or quarterly) for near-time (say < 5 years into the future) and longer time steps (annually or longer) for distant time, when conditions become less certain (say > 5 years into the future). For mine sites with seasonal climatic variations, the time steps used for near-time may have to consider transient conditions (Section 4.3.9) and time steps used for distant time may be six monthly to account for the seasonal changes.

4.2.8 Deciding whether a coupled modelling approach is required

When a pit slope is excavated within a saturated rock mass, the pore pressures will tend to reduce as a consequence of the reduction in total stress and consequent deformation (expansion) of the rock mass caused by the removal of overlying rock (Section 3.6.6). For most hard rock types, it is currently believed that most of the deformation is manifested as dilations (increases) in fracture aperture. As the fractures dilate in response to bench excavation, there is a tendency for the total head and the pore pressure to reduce. The area affected by stress relief and deformation is often termed the zone of relaxation. The depth to which pore pressure reduction may occur depends on the nature of the materials in the slope and the size and depth of the excavation. Data are available from Chuquicamata to show undrained pore pressure effects up to 450 m below the pit slope. At the present time, there are only a limited number of hard rock mine sites that have piezometer data to validate the pore pressure response due to unloading. There are mine sites with piezometer data that show a total head increase as a result of increasing total stress beneath a developing waste rock pile. There are also mine sites with piezometer data to show an increase in total head due to stress increases near the lower part of the slope.

The theoretical change in pore pressures during undrained deformation can be calculated from fully coupled hydromechanical simulation or, much more simply, from the mean stress change ($\Delta\sigma$), obtained from an uncoupled mechanical model:

$$\Delta p = B\Delta\sigma \qquad \text{(eqn 4.8)}$$

where B is the Skempton's coefficient and $\Delta\sigma = (\Delta\sigma_1 + \Delta\sigma_2 + \Delta\sigma_3)/3$.

Under undrained conditions, according to Biot's theory of consolidation, the change of pore pressure can be estimated from the change in mean total stress via Skempton's coefficient, B (Section 3.6.6.2). The general expression for Skempton's coefficient can be written as:

$$B = \frac{\alpha_B K_w}{[\alpha_B - n(1 - \alpha_B)]K_w + n\kappa} \qquad \text{(eqn 4.9)}$$

where α_B is Biot's coefficient, K_w is the bulk modulus of water, n is the porosity and κ is the drained rock mass bulk

modulus. The last three quantities are usually relatively well known. The information that is missing to evaluate Skempton's coefficient is an estimate of Biot's coefficient.

Lorig *et al.* (2011) describe a technique for calculating α_B based on the GSI (Geological Strength Index) of a rock mass. They show that for low porosity rock, α_B generally varies between 0.03 and 0.5. A realistic range in Skempton's coefficient is therefore $0.5 < B < 1$, where $B = 0.5$ represents a stiff, high-quality rock mass with a low α_B, and $B = 1$ represents a softer, lower-quality rock mass. This means that in a poor-quality rock, the instantaneous water pressure change will be equal to the mean stress change (the pore pressure decrease due to unloading is equal to the mean stress change). In a stiffer, high-quality rock, the pore pressure decrease due to unloading may be only half of the mean stress change.

During each bench excavation, the rate of pore pressure reduction will be partly controlled by the hydraulic diffusivity (which depends on permeability) and the compressibility or stiffness of the rock mass. Two extreme scenarios can be considered. The controlling parameter is again the dimensionless excavation rate (R).

In one extreme, if R is large compared with the rock mass diffusivity, then the pore pressure change between excavation steps will be mostly controlled by rock mass deformation. For this extreme within very low-permeability settings ($<10^{-8}$ m/s), the effect of fluid flow on pore pressure is relatively small so that, in theory, the system will behave in an undrained mode. Section 4.2.5 suggested that an *undrained-only analysis* may be considered when R > 0.7. However, the currently available field experience is that, for many low-permeability hard rock types, the fracture porosity is so small that even the smallest amount of recharge or groundwater flow will damp and offset the reduction in pore pressure caused by fracture dilation. It is therefore possible that the practical importance of an undrained response may be more relevant to situations where (i) the amount of recharge reaching the slope is very low, and/or (ii) the slope includes low-permeability silts, clays and marls, or rocks which contain a significant amount of clay alteration.

In the other extreme, when R is small compared with the rock mass diffusivity, then the undrained pore pressure change is negligible compared with the groundwater flow potential, so the system behaves in drained mode. This situation can be simulated by a conventional flow analysis (where R < 0.1) and where the permeability is > 10^{-8} m/s.

Coupled modelling can be considered in situations where *R* is between 0.1 and 0.7 (Appendix 5). However, in practice, a coupled approach may tend to under-estimate the amount of groundwater available to offset the undrained response and therefore may tend to over-estimate the reduction in pore pressures. Therefore, for any active mine site, it is advisable to collect piezometer data to validate the use of a coupled analysis.

It is likely that the understanding of coupled responses will gradually improve in the future as the depth of open pits continues to increase, and as an increasing number of piezometers are installed in rock with low permeability and low fracture porosity. The current evidence is that piezometers are much more likely to show a coupled response in situations where there is very low porosity and insufficient water available to damp or equalise the pressure change that is caused by the change in total stress. Figure 4.25 shows how a coupled response may be seen in piezometers. The figure has been prepared subjectively based on currently available piezometer responses for a number of mines in northern Chile. The figure shows that (i) deeper piezometers show a greater response – hence a downward pressure gradient develops, and (ii) piezometers show a stepped response to mining of the overlying benches. As a practical rule-of-thumb, if drawdown is evident in piezometers at depth beneath an active pit, and there is clearly no potential for discharge of groundwater at depth (considering that most pits are hydraulic sinks), then it is probable that the observed response may be due to deformation.

There are currently few software packages that offer the option to simulate coupled behaviour. Modules are becoming available for *MODFLOW* to incorporate stress changes into the analysis, and coupled behaviour can be simulated by *SIGMA/W*, *FLAC* and *FLAC3D*. *Slope Model* (Section 4.3.10) allows fully coupled simulations but there is currently little operating experience with this code for pore pressure modelling. Caution should be exercised in the use of an over-complicated approach until there are a

sufficient number of piezometers to illustrate the site-specific processes that control the pore pressure field in any given slope.

There are currently a number of Greenfield mine development projects in arid settings with limited recharge, and where the field testing results have indicated that the excavation will occur in rock with low permeability. These sites would fall under the right hand sector of Figure 4.23 and a conventional flow-only analysis would lead to an over-prediction of pore pressures and therefore an over-conservative (and potentially uneconomic) slope design. Some form of coupled modelling is required for these sites but, because no excavation has yet occurred, it is not possible to obtain any site-specific monitoring data to justify the approach or to validate the model. For such projects, it is necessary to provide an analogy with other mines in similar settings where a coupled pore pressure response has been demonstrated, and to use the analogous sites as the basis to support a coupled modelling approach. However, because of the currently paucity of active sites with data to illustrate coupled responses, it is likely that any modelling results will be poorly validated.

For practical applications, a simplified way of conducting a coupled analysis is to use a mechanical model (such as Equation 4.8) to compute undrained pressure changes after each bench excavation stage. The undrained pressure changes can then be imported into the flow model to modify the initial pore pressure state before subsequent transient analysis between periods of mine excavation. While this approach is not ideal, it allows the model user to interpret the results of each individual step and therefore to obtain a better 'feel' for the overall model output.

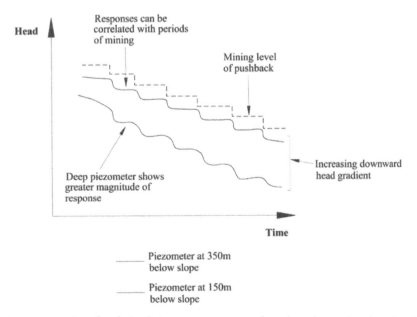

Figure 4.25: Diagrammatic representation of undrained piezometer responses (based on observations by mines in northern Chile)

Another situation where the use of a coupled analysis may be warranted is the back-analysis of a slope failure, where pore pressures may have changed rapidly as a result of the failure. Again, there is currently very limited field information documenting the actual behaviour of pore pressures during slope failures, so any modelling results will probably be poorly validated, and caution should also be exercised in the use of an over-complex coupled modelling approach.

4.2.9 Incorporating active drainage measures into the model

Groundwater removal within and around the pit can occur because of passive drainage to the pit slopes (seepage faces), horizontal or inclined drains which operate by gravity, or vertical wells which usually require active pumping. The implementation of dewatering and slope depressurisation measures is discussed in Section 5.2.

Model nodes along the pit wall should be simulated as drain nodes (i.e. fixed zero pressure nodes) and are assigned reference heads that are equal to the elevation of the pit wall and correspond to the elevations and timing of the mine excavation. The drain nodes are assigned to all model nodes which lie along the pit face and correspond to the elevation of the pit wall. If heads in the model are higher than the drain node at that location, water is removed from the model (the drain node acts as a seepage face), and the flux is computed to accommodate that boundary condition. If the heads are lower than the drain, the drain is inactive and no water is 'produced'. The assigned drain elevations and locations change with time as the mining progresses. For each new time step, the material that has been (or will be) mined during the time step is excluded from the model. Drain elevations are placed along the pit wall based on the changing shape of the pit for each model time step in order to simulate the transition from one point in time to the next. In some settings, it may be necessary to add seasonal recharge to the slope, as discussed in Section 4.2.3.5.

The removal of water through horizontal or angled drains drilled into the pit slope (or subsurface drains drilled from a tunnel) may be simulated using a combination of head dependent flux boundary conditions and/or enhanced permeability features. The *MODFLOW* drain module or *FEFLOW* discrete features combined with a seepage face boundary condition are commonly used to represent this process. When using the *MODFLOW* drain module, head-dependent flux out of the model is computed until head drops below a threshold value, defined as an elevation. Flux out of the model cell is zero when head is lower than the specified drain elevation. When using *FEFLOW* discrete features, head-dependent flow along a linear feature of enhanced permeability will simulate drain flow to a seepage face node at the point of discharge, where head dependent flux out of the model is

computed. After the head at that node has become reduced to the specified value (elevation), flux is defined to be zero. Alternatively, a *FEFLOW* discrete feature combined with the transfer boundary condition and specified transfer coefficient may be employed along a drain length to compute head-dependent flux out of the model.

For simulation of pumping wells, the modelling software simulates the active removal of water by the use of pumping or well nodes which remove a specified flux from the groundwater system. Depending upon the vertical discretisation of the model, and other factors such as the simulated horizontal and vertical hydraulic conductivity at the well site, the well can be simulated with:

1. a flux assigned to a single node when the well screen interval is limited to the vicinity of that node;
2. a distributed flux proportional to the transmissivity of a vertical column of nodes that correspond to the well screen interval.

FEFLOW allows definition of a multi-layer well that is conceptualised as a discrete feature and allows the user to set well radius, head constraints, etc. The total pumping rate is automatically proportioned between layers.

In some cases, model simulations can be carried out to determine the expected flow rate for maintaining a specific water level at the location of a given well. A specified head would be applied to the model to simulate such a pumping condition. The model would assume that the simulated well is efficient. Thus, a high hydraulic conductivity value should be assigned to elements associated with the well screen interval.

While models can be used to simulate the general effect of pumping at a given location within the model domain, caution is advised regarding the use of models to predict the yield of individual pumping wells, particularly in fractured rock settings where conditions can vary widely over very small distances (several metres; significantly less than the size of model cells or elements). Also, most models do not simulate the local head losses in the formation immediately surrounding the well or the well losses associated with the well itself.

4.2.10 Calibrating the model

4.2.10.1 The calibration process

Model calibration is the process of adjusting the model input parameters until a best match between field-observed values and model-simulated values is attained. Input parameters include rock properties (hydraulic conductivity and storage values), boundary conditions and specified hydraulic features and stresses. Monitoring data that form the basis for model calibration can include water levels in piezometers or monitoring wells, seepage face elevations, seepage rates, stream/river baseflow loss or gain, spring flow or other flow measurements (such as pumping wells, flowing artesian wells, recharge through irrigation, etc.). Before model calibration commences, the calibration data set is carefully selected, typically based on data location, quality, duration and overall reliability.

4.2.10.2 Steady-state calibration

Generally, a steady-state calibration is performed prior to calibrating to transient groundwater conditions. The steady-state calibration provides initial groundwater conditions for the subsequent transient model predictions and assures that groundwater heads and fluxes are reasonable.

A steady-state calibration assumes that, at a given point in time, the groundwater system is in equilibrium, or quasi-equilibrium, especially when compared with the transient conditions that occur during mining. It is preferable to match the steady-state calibration to pre-mining conditions. Although pre-mining conditions may show seasonal fluctuations, in general, the groundwater system is assumed to be in equilibrium or steady-state. However, for mature mines, a common issue is that there is a lack of pre-mining data against which to calibrate. In this case, a 'best estimate' of the pre-mining head and flow conditions can sometimes be made and a general steady-state calibration performed. Alternatively, a steady-state calibration could be matched to long-term stresses (such as long-term seepage in an open pit or long-term pumping) where the groundwater system has reached a new equilibrium. A common practice is to review the available data sets for the mine operating life and identify periods where groundwater conditions were relatively stable (e.g. relatively little change in piezometer levels or pit groundwater inflows) and configure the steady-state calibration to this time. Both approaches require judgement.

If at all possible, an estimate of flux through the groundwater system, or a portion of the system, should be made before starting a steady-state calibration to improve the likelihood of achieving a unique calibration. The sole use of measured heads as calibration targets during a steady-state groundwater model calibration, with no constraint of flux, could give a non-unique solution and erroneous results. For example, applying constant heads on the model boundaries with no restraint on the flux through the model could potentially result in fluxes predicted by the model being inaccurate by orders of magnitude if the assigned hydraulic conductivities are incorrect, even when predicted heads match those observed. For 2D pit slope models, it is necessary to estimate the component of groundwater flux moving downward through the pit slope (or into drains and wells), and to calibrate the model to both piezometer heads and the estimated flux along the model section. If a reasonable estimate of the downslope flux (along the model section) can be prepared, it is acceptable to use a 2D model to simulate conditions where the main flow direction occurs at right angles to the slope (i.e. perpendicular to the plane of the model (Section 4.2.2.1).

In hydrogeologically complex areas such as mines, most modellers use a trial-and-error approach to calibration. While automated parameter estimation (such as *PEST* coupled with *MODFLOW*) or inverse modelling can be used, each parameter should be constrained based on field observations in order for the model to produce reasonable predictions. Steady-state calibration results are generally reported using three different presentations: plots of head or flow, maps, or using a statistical analysis. Figure 4.26 shows graphical representations of the observed (field-measured) versus model-predicted heads for a 3D model (Figure 4.26, upper) and a 2D model (Figure 4.26, lower). A perfect match point, where the model exactly replicates a field value, falls on the X=Y line. The plot provides a quick way to evaluate the overall 'goodness of fit' between simulated and observed values. It shows where individual simulated steady-state heads are too high or too low compared to field observations. In combination with model statistics, the modeller can use the scatter plot to make decisions about how to adjust the model input parameters to provide a better overall fit or to improve the pit to specific points. A similar scatter plot is often used to present 2D calibration data. However, with 2D models, the data used in calibration often come from locations outside of the model domain (off section) and the calibration scatter plot should be viewed as a general guide. It is good practice to (i) group match points based on distance from the model domain and colour code them on the plot accordingly, so that additional context is available when reviewing the plot, and (ii) plot 'primary' calibration targets and 'secondary' calibration targets as separate series. Primary calibration targets represent piezometers falling close to the 2D section line (as a general guide within 50 m) and providing good-quality data. Secondary targets are further from the section but maybe used as a general reference.

A map showing the spatial distribution of the differences between measured and calculated heads (head residuals) is another way to evaluate 3D steady-state calibration (Figure 4.27). Colour-contouring of the results can enhance the understanding of areas where head residuals are high or low.

Statistical methods often are used to evaluate the average error of the steady-state calibration. Three standard methods often cited (Anderson & Woessner, 1992) to express the average difference between measured and calculated heads in the model are the mean error (ME), the mean-absolute error (MAE), and the root-mean-squared error (RMS). These three methods are ways of expressing the average difference between computed and measured heads. The ME is calculated simply by:

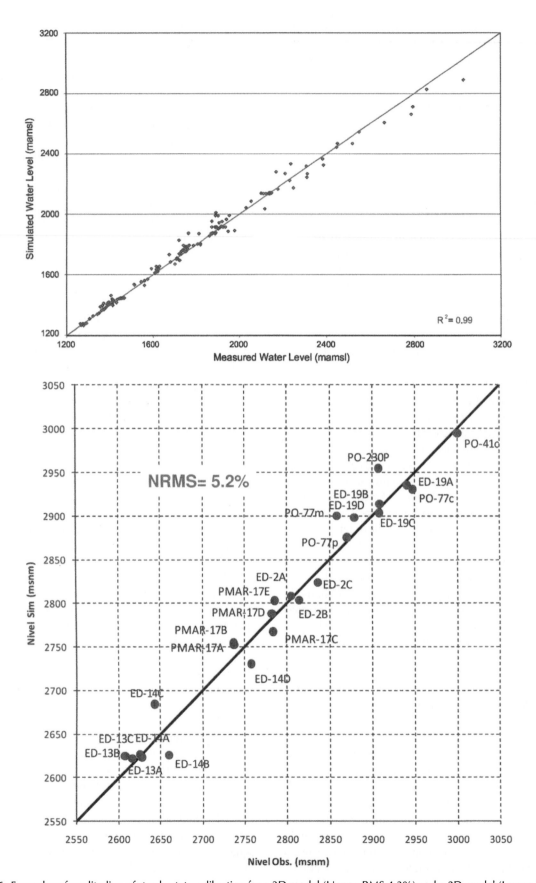

Figure 4.26: Examples of quality line of steady-state calibration for a 3D model (Upper, RMS 4.3%) and a 2D model (Lower, 5.2%)

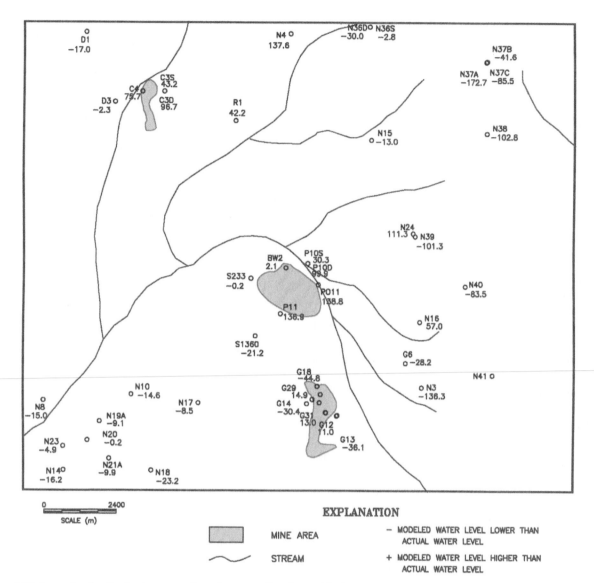

Figure 4.27: Example of a distribution map of steady-state calibration residuals

$$ME = \frac{1}{n}\sum_{i=1}^{n} E_i \qquad \text{(eqn 4.10)}$$

ME has the disadvantage that positive and negative errors may tend to cancel out, so that a model that produces significant errors, some of which are high and some low, may give an apparently small or zero ME. Both the MAE and RMS approaches avoid this by making all errors appear positive.

The equation for MAE is

$$MAE = \frac{1}{n}\sum_{i=1}^{n} |E_i| \qquad \text{(eqn 4.11)}$$

where the vertical lines indicate absolute value. RMS, perhaps the most commonly applied measurement of error, is calculated by

$$RMS = \sqrt{\frac{1}{n}\sum_{i=1}^{n} E_i^2} \qquad \text{(eqn 4.12)}$$

In each of the equations, $E = h_c - h_m$, E = error, h_c = calculated water level, h_m = measured water level, and i = calibration point ($i = 1$ to n).

4.2.10.3 Transient calibration

Transient calibration consists of applying historical (past) stresses on the groundwater system in the model and matching historical responses to those stresses. During transient calibration, the hydraulic conductivity, storage, or other input values may need to be adjusted from those used in the steady-state calibration. After each adjustment of the hydraulic parameters, a steady-state simulation should be re-run to ensure that the steady-state calibration

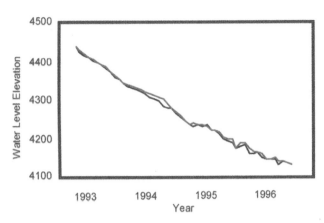

Figure 4.28: Example plot showing good transient calibration, showing measured versus model-predicted water levels

has not changed significantly, and to provide new computed steady-state heads to be used as the initial conditions for the subsequent transient simulation. A transient calibration is the only means to obtain modelled storage terms for the hydrogeologic system.

A transient calibration is highly recommended when the model will be used for predictions of future dewatering rates or pit slope pore pressures. A transient calibration builds significant confidence that the model is capable of simulating the correct ranges of response to active hydraulic stresses. One key objective of a transient calibration is to match the trend of measured water level over the calibration period. Figure 4.28 is an example of a good calibration at one point to long-term transient conditions. Figure 4.29 shows a calibration where the response to stress is mostly matched, but the initial head responses were not as closely matched. This calibration can be acceptable if it is not in a critical area. For pit slope models, it is normally of benefit to plot the actual and simulated data onto the model output as this allows the modeller to ensure the vertical head variations are being captured by the model (Figure 4.30).

It should be noted that in many geologically complex mine settings, it is nearly impossible to have all calibration

points match observed data. Professional judgement and a peer review process will help determine when the calibration is 'good enough.' Use of ME, MAE or RMS during the transient calibration can indicate the average difference between measured and observed heads at a given point in time, but history trend-matching is a better method to evaluate the transient calibration when transient data are available.

4.2.11 Interpreting model results

Numerical simulations are performed for both the model calibration period and the predictive (forward) analysis. Results must be interpreted to generate the desired final outputs: for example pore pressures at specific x-y-z-locations. The interpretation of model results typically involves the following.

1. Processing and presentation of the calibration statistics.
2. Processing groundwater head contours and phreatic surface contours for specific points of interest in time.
3. Preparation of cross-sections through the model showing total head or groundwater pressure, for selected points in time.
4. Processing of hydrographs of head versus time for specific points in the model, including calibration match points or other specific locations of interest.
5. Processing and reporting of selected flows (for example total pit inflow or flow to a set of drains) and the overall water balance for the model.
6. Assessment of the results for reasonableness including comparison to historic site operating experience and associated data sets, benchmarking against other site models and the peer review process.

All common software packages (e.g. Visual *MODFLOW*, *Groundwater Vistas*, *FEFLOW* and *MINEDW*) provide post-processing capabilities in addition to facilitating model inputs and executing the code. These packages typically allow for rapid calculation and export of the important results (e.g. calibration residuals/statistics, hydraulic heads at nodes, interpolated contours between nodes, global water balance, internal and boundary fluxes, velocities, etc.). There must always be a 'reasonableness' check on the model results. It is important to acknowledge and explain, to the extent possible, any notable deviations of the model results from expected conditions. If a satisfactory explanation cannot be provided, the model results should be questioned.

Model results are displayed in a variety of ways, including tables, graphs, contour or iso-surface plots, flow path lines and/or computer animations of spatial/temporal changes in a particular quantity. Figure 4.31 shows colour-enhanced contour plots of simulated pore pressures in a vertical cross-section through a pit at different points in time for two alternative final-pit configurations. Such

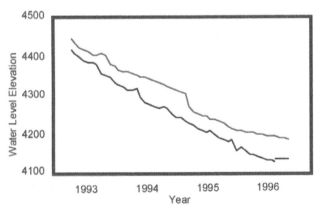

Figure 4.29: Example plot of acceptable transient calibration, showing measured versus model-predicted water levels

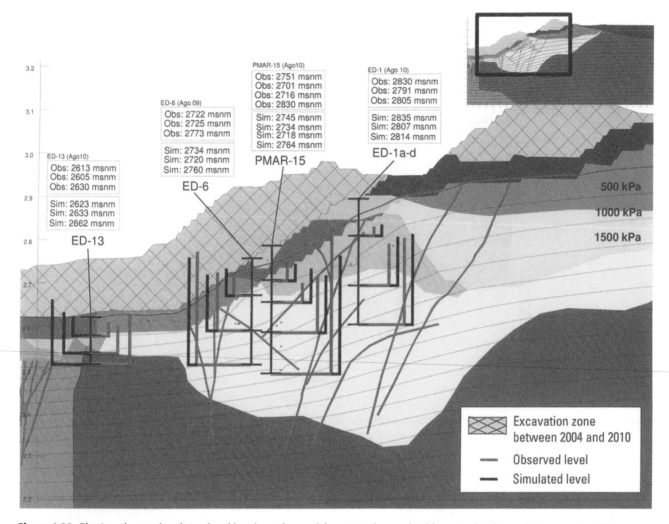

Figure 4.30: Plotting observed and simulated heads on the model output (observed in blue, simulated in red) (Source: Daniela Alejandra Abarca Argomedo)

plots immediately reveal areas of the pit wall where groundwater seepage and/or zones of elevated pore pressure may occur. Figure 4.31 facilitates side-by-side display of model results from different simulations. However, for evaluation of pore pressures, it is usually better to focus the area of model presentation rather than present the entire model (Figure 4.32).

In Figure 4.33, the results of several different simulations predicting the residual passive inflow to a mine are displayed as a time-series graph. An interpretation of these results might note the following.

▪ The residual passive inflow varies between approximately 7000 and 3000 m³/day for the base case (existing conditions) and is less for the other scenarios, all of which involved some form of active dewatering.

▪ The residual passive inflow is predicted to decrease with time under all scenarios, although there is an approximate two-year period of generally steady inflow between the initial decrease and the later reduction in residual passive inflow.

▪ The periods of declining and generally steady inflows are related to the mining schedule. (At the site in this example, the declining inflows correspond to periods of upper level mining, whereas the generally steady inflows correspond to periods of pit deepening.)

▪ Relative to the base case, active dewatering measures are predicted to reduce the residual passive inflow by up to approximately 50 per cent during the mid-years of mining, and by as much as 80 per cent during the final years of mining under the optimised plan.

▪ The use of horizontal drains is predicted to be ineffective in reducing residual passive inflows to the pit.

▪ The optimised dewatering plan is predicted to reduce the residual passive inflow by approximately 10 to 30 per cent compared with the current plan.

For pit slope models, results may be plotted as pore pressure (Figure 4.34, left) or total head (Figure 4.34, right). Pore pressures tend to increase with depth below the phreatic surface hence the contours are approximately horizontal. Flow lines tend to converge towards the pit so

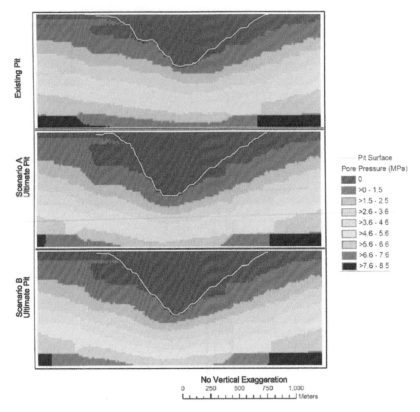

Figure 4.31: Pore pressures plotted 'side-by-side' from different simulations for comparative purposes

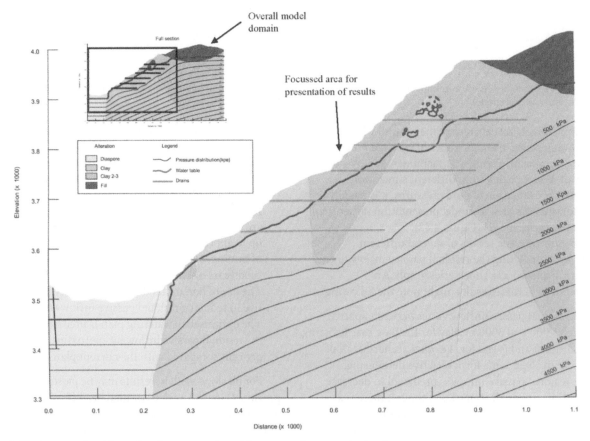

Figure 4.32: An example of focussing the presented model output on key areas in the slope

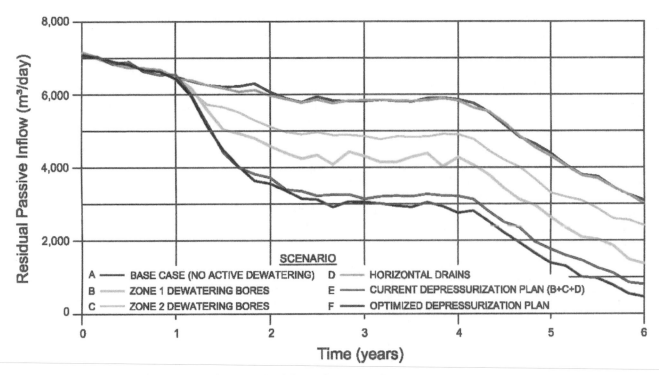

Figure 4.33: Simulated residual passive inflow to pit for different dewatering/depressurisation scenarios

total head contours are roughly vertical. In particular, the contour shape spacing should be examined to determine whether the model is indicating downward or upward hydraulic gradient. For pore pressure plots, downward gradients are reflected by a wide contour spacing while upward gradients are reflected by a close contour spacing. For total head plots, downward head gradients are reflected by outward sloping contours while upward head gradients are reflected by inward sloping contours.

4.2.12   Validating model results

Model validation is an attempt to demonstrate that the model can be used reliably to make future predictions. It has

been argued that verification and validation of numerical models of natural systems is impossible (Oreskes *et al.*, 1994) and that proclaiming that a groundwater model is validated carries with it an undeserved aura of correctness that is misleading to the general public (Bredehoeft & Konikow, 1993). This does not imply that model testing and evaluation should be abandoned or that the models are not useful; rather, it points to the inherent uncertainty in data and the modelling process and emphasises that caution should be exercised in describing any model or model result as being 'valid'. The comparison of the model results to actual observations at similar sites can provide confidence in the results, particularly if it is possible to understand the

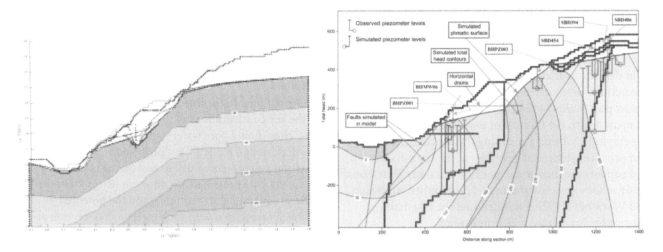

Figure 4.34: Illustration of pore pressures (left) and total heads (right) plotted onto a cross-section

differences in the conceptual model between the site under study and the comparative site.

At present, no standard criteria exist on how to demonstrate model accuracy. However, a common practice is to compare model data with a data set not used in the model calibration (Spitz & Moreno, 1996). This procedure is more useful when the simulated conditions differ significantly from the ones used in calibration (i.e. when predicting alternative conditions); for example, calibrating to steady-state conditions and using transient data to test the calibration, or reserving part of a transient data set (possibly subject to different boundary conditions) for 'validating' the model via history matching. If the calibrated model does not reproduce results accurately in such testing, the model is recalibrated to both data sets and other approaches to testing and evaluation must be used. Within the mining industry, peer review is often accepted as the best approach for validating the model. A seasoned mining hydrogeologist who has experience at many sites will be able to draw on a variety of knowledge for similar settings, and will be able to determine the factors of the model that are most important for the site under study, given the decisions that have to be made using the model results.

For pit slope models, it is obviously easier to validate the model for a Brownfield site with existing data, and particularly where the model is being use to simulate a mine expansion or the pushback of an individual slope for which there are previous monitoring data available. For a Greenfield site, the validation of an early stage pit slope model can be difficult, particularly if the rock units are poorly permeable and there may be a component of hydromechanical coupling to consider. While the confidence in Greenfield model results will be improved if there are cross-hole or multi-hole stress testing results available for model calibration, it is likely that early-stage modelling results for low-permeability slopes will be poorly validated (Section 4.2.8).

Post-audits are another means of demonstrating model accuracy. A post-audit compares model predictions with the actual outcome under field conditions. Typically, a post-audit is conducted several years after the modelling study is completed, when new field data are available to determine whether the model predictions were correct. Although many groundwater model post-audits have been made, their number is small compared with the number of model studies. In all cases, the emphasis falls on explaining the conceptual model and providing practical examples of site data that match and support the numerical model predictions. Based on a study of four groundwater model post-audits, Anderson and Woessner (1992) concluded that the two main reasons for inaccurate predictions were:

1. errors in the conceptual model;

2. the use of inappropriate values for assumed future stresses (e.g. the timing and magnitude of changes in recharge, pumping, and mine excavation rate).

It is also apparent that several different scenarios should be simulated using different assumed trends or shifts in the applied stresses in order to define a plausible range of predicted values.

4.2.13 Using the model for operational planning

The strength of a numerical flow model is its ability to predict potential future conditions and to underpin the design of future drainage programs. The use of pit slope models as input to the slope design process is described in Section 4.3. An important factor for pit slope depressurisation is time, and the numerical model is an important tool for estimating the required lead time for installation of drainage measures in each sector of the slope, to ensure the pore pressure targets are met.

The development of dewatering and slope depressurisation targets is discussed in Section 5.1.3.4. Figure 4.35 is a graph showing predicted groundwater pumping rates for a mine. The predictions can be used to design the dewatering system and to develop a plan for downstream management of the water. Figure 4.36 shows estimated pumping rates and residual passive inflow (the flow that would enter the pit without being captured by the dewatering system). This water would need to be pumped from sumps.

4.3 Use of pore pressures in numerical stability analyses

4.3.1 Background

4.3.1.1 Current issues surrounding pore pressure use in stability modelling

There is currently no industry standard or best practice regarding the method(s) of incorporating pore pressures into geotechnical slope stability analysis. It is, however, recognised that some of the methods commonly used today are often inadequate for the level of detailed slope design that is required for the deeper and larger pits being proposed.

As outlined in the introduction to these guidelines, one of the key considerations of the LOP hydrogeological study was 'How should pore water pressures be incorporated into the slope design model?' Section 4.3 addresses this question and presents a discussion on the methods of inputting pore pressure data into stability models, the differences among the methods, and the potential problems associated with each method. The section also discusses what are currently considered the best practices for the various stages of the slope design process.

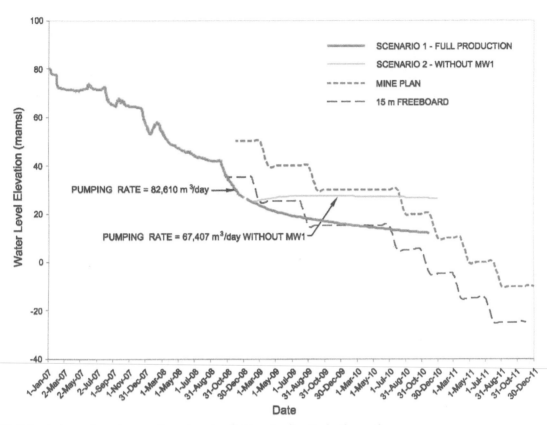

Figure 4.35: Estimated pumping rates over time at a mine that is extending its depth over five years

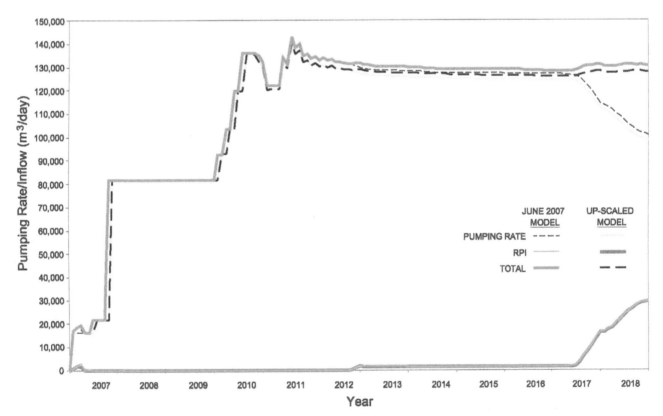

Figure 4.36: Predicted dewatering rates and residual passive inflows over time

4.3.1.2 *The role of pore pressure in slope stability analysis*

The key impact of water in the slope is the potential reduction of effective stress acting on the soil or rock mass or the structural discontinuities and hence the shear strength of the materials in the slope, which results in reduced stability. Therefore, geotechnical stability models for pit slopes almost always require groundwater pressures as an input to ascertain whether they have a significant impact on slope performance. The only exception would be in a situation where the real-life pit slope is dry and devoid of a groundwater flow system. This exception is a rarity when the potential for transient recharge events is considered.

Numerical models for slope stability analysis are used to supplement the simpler limit-equilibrium analysis when there is a need to determine not just the Factor of Safety (FoS) for a given failure mode, but also the more complex conditions such as non-linear stress-strain behaviour, anisotropy, and changes in geometry (deformation and/or displacement) that may occur in the slope.

For slopes that include materials with significant porosity (e.g. alluvium), the degree of saturation above the phreatic surface (in the 'unsaturated zone') can also be important because it will affect the soil density and hence weight. If required, the percentage of drainable and retained porosity and the weight of the material are normally calculated separately using either published estimates or using an unsaturated flow analysis.

4.3.2 How pore pressure modelling differs from stability analysis

In general, stability analyses are carried out on static problems using deterministic methods while hydrogeological modelling must consider transient conditions (Section 4.2.5). To integrate water pressures effectively into the stability analysis, the geotechnical engineer must take into account a number of factors that differentiate the behaviour of water from that of the slope materials themselves. The key difference is that, until mining/excavation begins, the rock mass will not experience any significant changes in material properties or parameter distribution, whereas the hydrogeological system is continually influenced by external sources or boundary effects that are typically controlled by the wider-scale groundwater system. Because of this, the scale of the hydrogeological characterisation invariably extends beyond the spatial and temporal limits of the pit slope (and beyond the geotechnical model domain).

The primary differences between the geotechnical and hydrogeological modelling processes are as follows.

- Geotechnical stability modelling is (usually):
  - → static (no time component);
  - → not impacted by significant changes in material properties (unless in regions of blast damage or unloading);
  - → difficult to calibrate.
- Hydrogeological modelling of pore pressures is (usually):
  - → transient (strongly dependant on time);
  - → subject to changes with time as the pit is excavated;
  - → influenced by active depressurisation measures to improve stability;
  - → sensitive to changes in boundary conditions;
  - → dependant on a good transient calibration.

Historically, both geotechnical and hydrogeological analyses have used 'point' samples or measurements as input to models that simulate a 'gross system' behaviour. The hydrogeologist has the advantage of being able to add cross-hole or multi-hole testing that encompasses a wider (and therefore more representative) volume of rock, and the importance of these tests for pore pressure studies is discussed throughout Section 2.2.8. For the geotechnical engineer, it is necessary to wait for some deformation or movement in the slope to occur before any studies can be carried out using back analysis methods to simulate a wider volume of rock. For Greenfield sites, reliance must be placed on a benchmarking analysis.

Hydrogeological modelling requires calibration to current conditions (transient calibration) and, at the slope-scale, must often allow for changing material characteristics with time (i.e. changes in hydraulic conductivity in regions of blast damage or unloading). It is often necessary to update the calibration as a result of changing boundary conditions and material properties.

4.3.3 Methods for inputting pore water pressure

The pore pressures incorporated into either the limit-equilibrium or numerical stability analysis have usually taken the form of:

- **The use of an R_u value.** Pore pressure for the analysis can be defined by the pore pressure ratio, R_u, which is defined as the ratio between the total upward force due to water pressure and the total downward force due to the weight or overburden pressure. The R_u value is a simple method of assigning pore pressure for each material in the geotechnical model, although complications may arise in a layered system. This method is the least accurate because it ignores any water flow.
- **The use of a phreatic surface (water table).** Use of a phreatic surface as the indicator of pore pressure requires the assumption of hydrostatic conditions beneath the phreatic surface. This can be appropriate for early stage assessments where vertically discretised piezometers have not yet been installed and for

conditions where significant vertical gradients do not develop. In most hydrogeological settings, however, the assumption of hydrostatic conditions will lead to an over-simplistic analysis when applied to a detailed slope design.

- **The use of a pore pressure field that includes the horizontal and vertical distribution of pore pressures**. Most analyses for large pit slopes require this method and it is considered to represent current best practice. The pore pressure field can be obtained by contouring of data if a 2D analysis is being used, but would typically require the output from a 2D or 3D numerical model, as described in Section 4.2. This is becoming standard practice for most pit slope studies, and it is normal to consider the effect of vertical pressure gradients from Level 2 (Pre-Feasibility) forward.

4.3.4 Pore pressure profiles versus phreatic surface (water table) assumptions

The phreatic surface (water table) is the surface of zero pore pressure (relative to the atmosphere) and therefore represents the free groundwater level within a slope. When a simple hydrostatic pressure distribution is used, the pressure at any point below the water table is simply $\rho_w gd$ where d is the depth (the vertical distance between the point and the water table). This means that all equipotentials are assumed to be vertical, so that there is no vertical component of hydraulic gradient (Figure 4.37a), any flow is horizontal, and at every depth the potentiometric surface is the same as the water table. Strictly speaking, this condition is impossible where any groundwater flow is occurring because:

- if any flow is occurring there must be a hydraulic gradient;
- if there is a hydraulic gradient the water table will be sloping; and
- if the water table is sloping, there must be a vertical component of flow, hence a vertical gradient.

Prior to the start of mining, natural flow may be negligible and the assumption of hydrostatic conditions may be approximately valid. Vertical gradients will develop with flow into the pit and there will be vertical

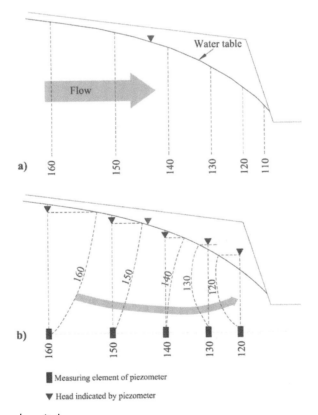

Figure 4.37: Groundwater flow towards a pit slope
Notes: a) Hydrostatic assumption. Head does not change with depth so equipotentials are vertical and a piezometer at any depth will register the same head as the water table. The shortcoming is that the flow clearly cannot be purely horizontal as it would have to cross the water table. This assumption is therefore valid only for static groundwater with a horizontal water table. b) Dynamic assumption with vertical flow components (curved equipotentials). At any point head varies with depth. Piezometers installed at the depth of the green dashed line describe a potentiometric surface for that depth that will cross the water table as flow moves from downward (recharge zone) to upward (discharge zone) near the toe of the slope and beneath the pit floor.

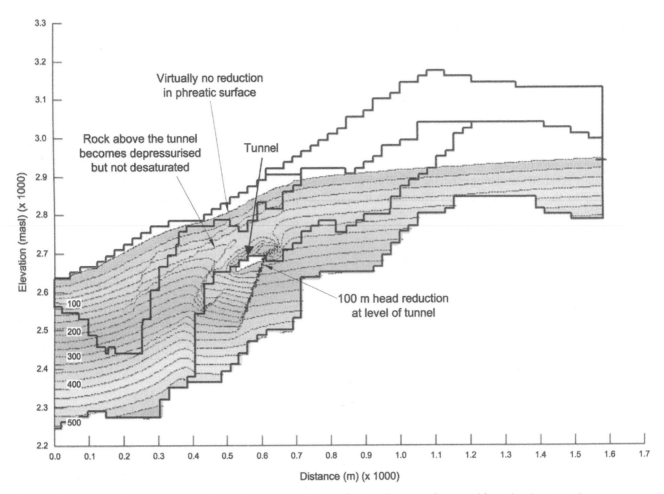

Figure 4.38: Pressure dissipation surrounding a drainage tunnel located below the general water table in the slope (Read & Stacey, 2009)

components of seepage force. As shown in Figure 4.37b, this means that equipotentials will be curved and the potentiometric surface at depth will be different from the water table, being deeper where recharge is occurring, crossing the water table where flow locally becomes approximately horizontal, and being above the water table in the discharge area. This means that pore pressures in the lower pit slope and beneath the pit floor will usually be higher than predicted by the hydrostatic assumption (Figures 4.37b and 1.22).

In the vast majority of active pit slopes, the distribution of pore pressure in the pit slope is non-hydrostatic. In other words, the first occurrence of groundwater below ground level cannot be used as a reliable means to estimate groundwater head distributions at greater depth. Furthermore, in an active mine setting, pore pressure distributions will normally vary with time. Therefore, the development of a pore pressure profile for the pit slope, with transient changes, is usually the most reliable approach.

In terms of pit slope geotechnical modelling, the outcome of predicted pit slope performance can be

significantly different depending on whether the model uses a phreatic surface and hydrostatic condition or a pore pressure profile distribution. The position of the water table is the same in both cases, but the hydraulic gradient and pore pressure distribution beneath the water table may be entirely different. Some examples are as follows.

- In situations where groundwater pressures in the pit slope are elevated above a hydrostatic condition, the normal result would be a reduction in slope performance compared with the hydrostatic condition.
  - → An example would be where a conductive district-scale groundwater unit contacts a poorly conductive intrusive rock unit in the lower part of the pit, and the driving head and rate of flux from the district-scale unit sustain excess pressure in the intrusive rock, due to the contrast in head and relative hydraulic conductance of the units.
- In situations where groundwater pressures in the pit slope are reduced below a hydrostatic condition, the result would be a relative improvement in slope

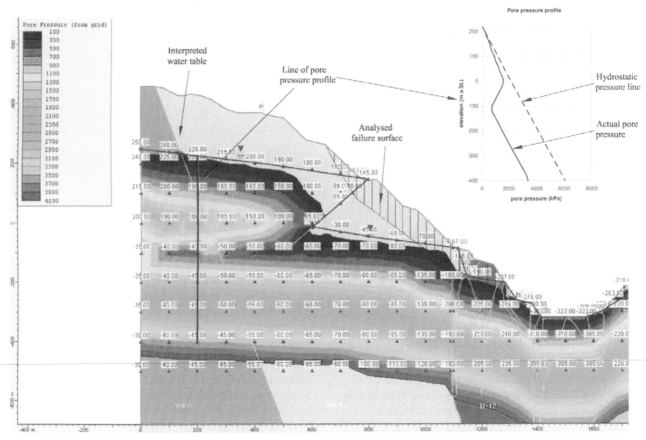

The colouring represents pore pressure in kPa. The figures on the section are
values of static head (relative to sea level) in metres

Figure 4.39: Total head distribution in a pit slope undergoing significant depressurisation (Source: Meutia Alfri)

performance, compared with a hydrostatic condition.
Examples include:
→ where bedding or faulting high in the pit slope
creates horizontal impediments to vertical groundwater movement and the groundwater system
becomes vertically 'de-coupled' such that the
phreatic surface does not impart a hydraulic load on
the deeper system, or the lower part of the pit slope.
An example would be where the geological unit
high in the pit slope has low vertical hydraulic
conductivity and is underlain by a geological unit
with high hydraulic conductivity. In essence, the pit
wall is naturally drained;
→ in low hydraulic conductivity settings, where pore
pressures can dissipate due to geomechanical
unloading and dilation of fractures. An example of
this is Cobre Las Cruces Mine (Appendix 5);
→ Around a drainage tunnel, where the pressures have
been reduced immediately around the tunnel, but
this effect is reduced moving upward from the
tunnel and closer to the pit wall. Assumption of
hydrostatic conditions would not take into account

the potentially significant depressurisation around
the tunnel illustrated in Figure 4.38;
→ Where other drainage or depressurisation
measures have been installed behind the pit slope
to depressurise a unit at some depth below the
water table. Figure 4.39 shows an example from a
large pit that extends below sea level; drainage has
caused a major reduction in head and pore pressure
behind a section of the slope. The water table is
sustained by shallow recharge. Below the water
table down to about –100 m elevation, the rock is
saturated but the pore pressure is low. This
situation would be difficult to incorporate into a
simple hydrostatic model.

Figure 4.40 shows a pit slope where a large downward
pressure gradient existed and where a 2D geotechnical
analysis of the current slope conditions was required
within a time frame that was too short to allow any
numerical modelling of pore pressures. In this case, a set
of 2D pore pressure contours of total head was prepared
by forward-projecting actual data from vibrating wire

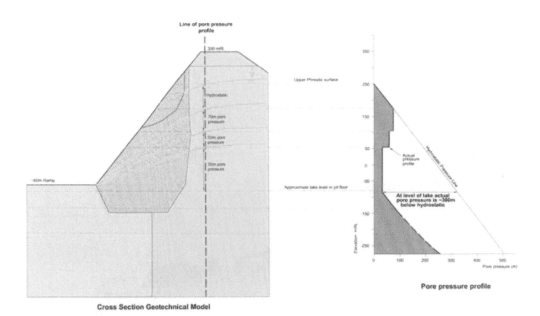

Figure 4.40: Illustration of rapid approximation of pore pressure for input to a geotechnical model (Source: Raimundo Almenara)

piezometers using a trend analysis, using the known geology and historical pore pressure responses to constrain the interpretation. A vertical profile of pressures was developed as shown on the right hand side of the diagram. At about −80 m elevation, the pore pressure was about 300 m (3000 kPa) lower that the hydrostatic condition so the use a phreatic surface in the analysis was not realistic. In order to achieve rapid results, the observed piezometer data were used to develop a number of pore pressure zones, which were then input to the geotechnical analysis shown on the left hand side of the figure.

In most pit slopes, there is a distribution of different hydraulic conditions (often termed hydrogeological domains), and there is often a combination of pore pressure regimes, the relative importance of which may vary with

time. The net effect of the pore pressure profile on slope stability is best assessed through a process of pit slope pore pressure modelling that is completed in collaboration with pit slope geotechnical modelling, as part of a dedicated hydrogeotechnical modelling program.

Figure 4.41 shows the difference in pore pressure contours calculated (i) assuming a hydrostatic analysis below the phreatic surface, and (ii) using a transient groundwater flow model. Both cases assume a homogenous hydraulic conductivity distribution. The position of the water table is the same in both cases. However, hydraulic gradient and pore pressure distribution beneath the water table is different.

Figure 4.42 shows the percentage pressure difference between the two approaches. It shows that, if only hydrostatic conditions are considered, pressures in the

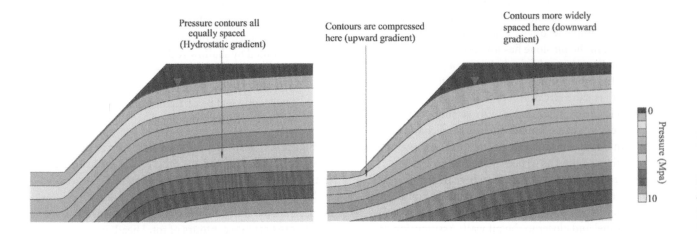

Figure 4.41: Comparison of pore pressures calculated using (a) a hydrostatic analysis and (b) a transient groundwater flow model

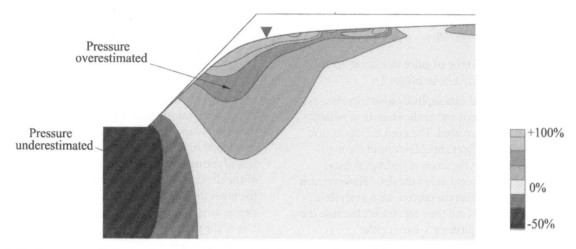

Figure 4.42: Comparison of calculated pressures between the hydrostatic analysis and the pore pressure model from Figure 4.41

upper slope are overestimated while pressures in the lower slope and toe area are underestimated.

The theoretical pit slope factor of safety (FoS) was calculated for these two models by importing the pore pressures into a finite difference mechanical model. Typical rock properties were assigned and effective stresses were used to evaluate failure states. It was found that, when using pressures derived from the hydrostatic condition (phreatic surface), the FoS was 15 per cent lower than when using a calculated pressure profile. This is because the pressures are overestimated in the region of expected failure in the lower part of the slope. In other cases, depending on the location and mechanism of failure, and the distribution of geomechanical properties within the pit slope, the opposite result may occur.

4.3.5 Integration of the hydrogeology and geotechnical models

4.3.5.1 Discussion of the two models

Generally, a two-stage process is used to bring the pore pressure and slope stability models together. The first stage involves developing a predicted pore pressure distribution for the slope using a groundwater modelling package. The second involves importing the pore pressure profile into the geotechnical model. The first stage is almost always best performed by the hydrogeological group, with the second stage being performed by the geotechnical team.

In order to accomplish this process in a way that produces meaningful results, the hydrogeology and geotechnical models require identical pit slope geometry and an identical sequence of mine cuts. The main geology and structure is normally included in each model, with hydrogeological or geotechnical domains normally assigned collaboratively between the hydrogeology and geotechnical study team. A fine grid is normally used, and a 10 to 50 m cell or element size is typical. This is usually sufficient to accommodate the detail of vertical head gradients, pit slope surface position, and geology and structural features at an appropriate scale to see their influence on the pore pressure distribution within the slope.

In most cases, the hydrogeology and geotechnical models are constructed and run separately. This is because the pore water conditions in the slope are most often influenced by the wider-scale regional hydrogeology and, in particular, by recharge from outside the slope domain. As a result, most groundwater models require their domains to be extended beyond the geotechnical model domain. In addition, groundwater modelling packages are designed to handle a more comprehensive range of hydrogeological conditions (such as evaporation, routed streams and processes) than geotechnical codes, hence more complete and accurate pore pressure profiles can be generated from dedicated groundwater flow models in most situations. Close collaboration between the two models is, however, essential.

Pore pressures that are generated from the groundwater model must be exported and provided as an input to the geotechnical model. In most situations where a sequence of mine cuts is being studied, the groundwater model will need to be run in transient mode, since time is a key factor in pore pressure distribution. Multiple pore pressure 'snap-shots' should be extracted from the transient groundwater model, representative of the slope geometry at selected stages in the mine plan.

In situations where rock of low permeability occupies the pit slope and recharge to the slope materials is low, the effects of hydromechanical unloading may exert a significant transient influence on the pore pressure profile in the slope (Section 4.2.8). In this case, it may be possible to implement a pore pressure analysis that is integral to the geotechnical model (a 'coupled model'). When considering the application of a coupled model, it should be noted that significant effects of hydromechanical coupling on pore pressure are observed in a relatively small number of situations and, because it tends to reduce the predicted pore pressures and provide a less conservative analysis, the general rule should be to include its effects only where they can be confirmed by monitoring.

4.3.5.2 *Analysis and input of pore pressure to the geotechnical analysis for a new project*

The level of modelling and data sophistication increases as a new mining project develops and as the amount of reliable hydrogeological data is increased. The need for inputs into geotechnical stability analyses should be one of the main considerations in defining the scope of hydrogeological investigations, and subsequent use of the data. However, it is important to understand that the geotechnical analysis is often carried out under critical time constraints because the results are needed for the pit design. Because the hydrogeology results need to feed into the geotechnical analysis, there is often insufficient time to carry out detailed numerical modelling in the early stages of an investigation. Furthermore, there is seldom enough data to justify detailed numerical modelling during the early project stages.

Broad guidelines for pore pressure inputs to a geotechnical analysis are as follows.

Initial screening (Conceptual) analysis
- Development of a basic hydrogeological model of the pit area.
- Identification of the key hydrogeological units that may affect the slope design.
- Preparation of preliminary best case and worse case estimates of the phreatic surface for anticipated critical and representative cross-sections (identified by the geotechnical group).
- Geotechnical screening analysis based on the sections selected, and using the early phreatic surfaces to investigate (i) the general sensitivity of the slope materials to water pressure, (ii) the key sectors for which the slope design is likely to be most sensitive to water pressure, and (iii) an initial assessment of potential failure modes.

The screening analysis may occur at Level 1 (Conceptual level) or Level 2 (Pre-Feasibility) in the overall project development, depending on the expected importance of water in the slope design process.

Level 2 (Pre-Feasibility) analysis
- Identification of a small number of important (and representative) geotechnical sections (the number is site-specific will depend on the scale of the project and the importance of water; one to three sections may be reasonable for a mid-sized project; two to five for a larger project).
- Installation of multi-level piezometers based on the results of the screening analysis and focused on the key identified sections and hydrogeological units.

- Addition of piezometer results and other hydrogeological data to the cross-sections.
- Preparation of first-generation 2D numerical pore pressure models of the key sections. It is usually better to analyse a few sections thoroughly than a larger number at a more superficial level. The models should be prepared jointly with the geotechnical group and the results input to the geotechnical models being developed in parallel.
- Use of the models to further investigate the sensitivity of the slope materials to water pressure and to identify the potential requirement, and targets, for active depressurisation.
- Use of the models to identify options for active slope depressurisation.
- Input of modelled pore pressures into the stability analyses for the most critical geotechnical sections. Simpler analysis of the remaining geotechnical model sections as justified on the basis of modelling results.

Level 3 (Feasibility) analysis
- Installation of additional piezometers based on the Level 2 results (Section 2.1.5.5), focused on the important hydrogeotechnical sections.
- Implementation of a cross-hole or multi-holes testing program to provide bulk-scale responses (Section 2.2.8). Depending on the importance of water for the slope design, this program may be commenced during the Level 2 studies.
- Preparation of updated and additional hydrogeotechnical cross-sections.
- Identification of the important (and representative) geotechnical sections (potentially selecting two to four for mid-sized projects and three to seven for larger projects, depending on the variability in geotechnical properties).
- Preparation of detailed numerical pore pressure models of those sections. Where possible, these should be calibrated using cross-hole testing results or other hydraulic stresses.
- Working closely with the geotechnical group to focus the models on the identified potential failure modes and achieving an identified level of drawdown for stability.
- On the basis of the target levels of drawdown, use of the models to design the active slope depressurisation program.

Analysis for Level 4 and Level 5
- Compilation of a robust conceptual model for all pit slope sectors.
- Monitoring of the piezometric response to mining on critical geotechnical design sections.

■ Implementation of a pilot drainage trial (where possible).

■ Recalibration of the models.

■ Ongoing verification of rates and magnitudes of drawdown against model predictions required for stability.

The numerical analysis should be focused and resources should be placed on those sectors of the pit that are considered most important for pore pressure in the slope designs. For a large operating pit, up to ten 2D numerical models and/or two sector-scale 3D models may be manageable for the detailed design stage and during operations.

In all cases care must be taken when using a phreatic surface and assuming hydrostatic conditions. It is necessary to evaluate the potential for, and consequences of vertical gradients which may impact slope stability.

4.3.5.3   Work flow for the development of pore pressure models at an operating mine

Guidance on the steps required for the development of numerical pore pressure models at an operating mine are as follows.

1. The mine plan being studied is reviewed by the hydrogeology, geotechnical and mine engineering groups. Based on conceptual understanding of geology, geotechnical properties, hydrogeology, mine design and infrastructure, slope sectors are prioritised in sequence through time.

2. The location of 2D analysis sections, and/or a 3D sector-scale model domain, is agreed by the group, for use in groundwater and geotechnical models.

3. The domain for each model is agreed, to include the 'current condition' at the mine and also the location and geometry of future mine cuts and future slope geometry.

4. The pore pressure model grid or mesh is constructed, incorporating key features of the hydrogeological system plus the current site topography and the selected stages of the future mine plan and slope geometry. As noted in Section 4.3.5.1, the groundwater flow model grid will generally be much larger that the geomechanical grid because the groundwater flow model must extend to clearly definable boundaries.

5. A similar process is completed for the geotechnical model.

6. The pore pressure model is first calibrated in a steady-state mode (as possible) and then to 'current conditions' in transient mode. The model calibration is validated by comparison with site data.

7. The calibrated pore pressure model is forward run in transient mode. The changing pit slope geometry is implemented into the model run, using a time schedule that matches the proposed mine plan.

8. Pit slope pore pressure distributions are generated from the model, for selected increments of the mine plan and slope geometry.

9. The pit slope pore pressures are exported from the model and are post-processed in collaboration with the geotechnical group, so that they can be imported to the coincident geotechnical model to predict the performance of the pit slope for the same time-dependent slope geometry.

10. The results of the geotechnical model are used to determine whether slope performance is acceptable given the predicted pore pressure regime.

11. Further iterations of the modelling process may be run to assess a range of active pit slope dewatering/depressurisation approaches and their relative benefit to slope stability. To speed up the iterative process, several dewatering/depressurisation options can be simulated initially and all options can be evaluated by the geotechnical model.

12. A sensitivity analysis is normally performed to assess how much difference in modelled slope stability occurs if pore pressures are varied within the potential ranges of uncertainty for either (i) key hydrogeological parameters, (ii) the construction schedule and engineering performance for active dewatering/depressurisation measures, or (iii) the geotechnical properties of the rock mass and structures in the pit slope.

The interactive feedback between Steps 7–12 may go through several cycles before an optimised design is achieved. Note that it is often beneficial to consider additional piezometer installations following Step 10 in order to provide better data control on the parts in the slope that are identified as being most sensitive to pore pressure.

In some cases, a numerical slope stability model can be used to predict zones of increased permeability that may develop through time due to slope deformation. If a reasonable site-specific empirical relationship between deformation and change in hydraulic properties can be developed, it may provide a basis for defining zones of increased hydraulic conductivity in the pore pressure model that are specifically tied to geotechnical domains.

4.3.6   Model codes

Table 4.10 lists the groundwater and geotechnical model codes most frequently used in slope stability analysis. For most pore pressure modelling, the commonly used codes are the *SEEP/W, FEFLOW* and *MINEDW* finite element codes and the *MODFLOW-SURFACT* finite difference code. However, these are not exclusively utilised, and the

Table 4.10: Common groundwater and geomechanical model codes for pit slope modelling

Interactions (couplings)	Continuum (implicit representation of discontinuities)		Hybrid (explicit representation of major discontinuities)	Discontinuum (explicit representation of many discontinuities)	
	Flow	Coupled		Flow	Coupled
One-way	FEFLOW MODFLOW-SURFACT SEEP/W MINEDW		UDEC 3DEC Phase2**	FRACMAN* 3FLO	
Two-way		SIGMA/W Plaxis FLAC/FLAC3D Eclipse/Visage	UDEC* 3DEC* ABAQUS		Slope Model

* Fluid flow in joints only, not matrix ** Steady-state flow in matrix (no flow in joints)

development of new and updated codes is a continuing process. The actual code selection should be based on the merits of each individual study.

In low-permeability settings, where a single code can be used to simulate both pore pressure and mechanical behaviour, the degree of interaction can be greater because of the inherent ability of the model to couple the mechanical behaviour with rock mass hydraulic properties. However, the geotechnical codes that can be applied to model this coupled process do not have the full ability to simulate important hydrogeological process such as recharge or regional fluxes or active slope drainage, and they are not commonly used for assessing the interaction between the materials in the slope and the wider-scale groundwater flow system. Hence, the application of a coupled mechanical and hydraulic code requires careful consideration and is normally only carried out at mature mines where high confidence in the conceptual model exists and model results can be validated with site monitoring data.

There are a number of actual examples where the excavation of the pit slope (i.e. unloading) can be directly related to changes in pore pressure. One of the best case studies of coupled hydromechanical analysis is from the Cobre Las Cruces open pit in Spain (see Appendix 5 and Galera *et al.*, 2009). In this example, the coupled analysis using the *FLAC* code showed a significant pore pressure drop due to the volumetric expansion associated with the excavation of the pit, and was supported by monitoring data collected from piezometers. The lower pore pressure distributions predicted by the analysis, and verified in the field, allowed for a more aggressive and economical slope design. However, for the majority of pit slopes, the potential for reduced pore pressures due to increasing fracture aperture (dilation) is offset by groundwater flow and/or surface recharge into the zone of dilation. Therefore, a groundwater flow model is normally required as the basis for developing pit slope pore pressure profiles.

4.3.7 Requirements for groundwater input to the slope design

4.3.7.1 General guidelines

Currently, there are no universal guidelines for the required accuracy for predicting pit slope pore pressures as this is a risk-based decision by the mine owner. Most mine studies involve completing a sensitivity analysis to express the relative change in slope performance that may be expected given a range of simulated pore pressure conditions, within a defensible range of hydrogeological parameter uncertainty. Most major mine development projects will require a review process and many will adopt a technical review board to 'reality truth' the results based on benchmarking with other mine sites and global experience (Section 6.2).

4.3.7.2 Input of pore pressure to the rock mass and discontinuities

The research work carried out in Section 3 has largely confirmed that, at the scale of most practical field measurement and modelling, pore pressures are pervasively distributed throughout the rock mass, including the micro-fractures within the rock mass and the major fractures and discontinuities (considering that most slope instability occurs within about 300 m of the slope). Therefore, for a given pit slope pore pressure profile, it is normally appropriate that 100 per cent of the predicted pore pressures are universally applied and input to all geotechnical units and structures within the geotechnical model domain (for example within both the blocks and the joints in *UDEC*). In all cases, effective stress (Section 1.1.6.2) is used to account for water pressure within the geotechnical model.

4.3.7.3 Time steps (stress periods)

Time is a fundamental factor in pit slope pore pressure distribution and rate of pore pressure change. Generally, the evolving pit slope geometry is communicated by the mine planning or engineering group on a monthly, quarterly or annual basis. It is essential, at the initiation of a project, that the mine engineering, geotechnical, and hydrogeology groups agree to the phases of the mine plan and slope configuration through time that are important to the mine design and performance, and which will form the basis for modelling. A fundamental requirement is that the pit slope geometry is consistent between hydrogeological and geotechnical models.

The selected increments from one phase of mining to the next are then input into the pore pressure model both in terms of the physical change in the pit slope geometry and the time taken to progress from one selected phase of mining to the next. Models may use monthly or quarterly time steps for the near term (< 5 years) increasing to annual time steps for the far term (> 5 years). The choice of time steps is not simply a function of model performance, but is also a function of practical factors such as planned mining versus actual mining or proposed (modelled) slope depressurisation measures versus those actually implemented by the operation.

4.3.8 Transferring output from the hydrogeological model to the geotechnical model

4.3.8.1 Output from the hydrogeological model

The only hydrogeological input required for most slope design models is the pit slope pore pressure distribution, which is then used in the calculation of effective stresses. The approach of exporting pore pressure grids from a pore pressure model and directly importing the grids to a geotechnical model is valid and represents current best practice for most slope design studies. The approach relies on the following key assumptions.

- Pore pressures are pervasively distributed throughout the rock mass and fractures.
- The grid (mesh) size of the models is large relative to the fracture spacing.
- The time intervals at which pore pressures are predicted are greater than about one month.

Currently, the most detailed geomechanical modelling for larger slopes typically uses a minimum 10 m grid or mesh size. It is generally not practical to resolve hydrogeological or geomechanical conditions or pore pressures within the rock mass at a scale less than this. Where discrete features are known, they can be input directly into the hydrogeological and geomechanical models, ideally in a coincident manner.

In an EPM model, pore pressures are calculated at discrete locations (finite-element nodes or finite-difference grid points), and the pressures are assumed to vary linearly, or sometimes quadratically, between the calculated values. A challenge is therefore to export the pore pressures calculated at nodes in the hydrogeological model and import these into the stability model in a manner that applies pore pressures at appropriate locations in the geotechnical model. It is almost impossible to have identical grid cell or element distribution and geometry between the groundwater and geotechnical models. Therefore, a degree of interpolation is normally required and a rigorous quality assurance (QA) process is necessary to ensure that the pore pressure profile imported to the geotechnical model is 'lined up' correctly relative to the pit slope surface being analysed.

Typically, the following steps need to be performed.

1. From the groundwater flow model, export to an external post-processing file (often a simple text file or Microsoft XLS) the results of the pit slope pore pressure prediction. The export should involve each model node or grid coordinate plus predicted pressures or head at each coordinate. Normally an X, Y coordinate plus the head or pressure value (for 2D models) or X, Y, Z plus the head or pressure value (3D) format is used. A sequence of predictions is usually run, involving predicted pore pressure distributions for several increments of a mine plan, so it is necessary to create separate files for each increment. They are normally given a file name that identifies the stage of the mine plan. This is a simple QA step that helps to ensure that the correct pore pressure profiles are linked into the correct stages of the geotechnical model.

2. The X,Y or X,Y,Z coordinates from the groundwater model are converted to the coordinate system for the geotechnical model, with extreme care taken to ensure that coordinates place the pore pressure values in their correct location in the geotechnical model, relative to the pit slope surface being analysed. The conversion requires individual thought for each project and normally involves coordination between the hydrogeological and geotechnical modeller. Normally, a conversion algorithm that translates the groundwater model coordinates to geotechnical model coordinates is developed and then implemented within the post-processing file.

Each pressure or head value in the post-processing file is converted to the default working units of the geotechnical model, most commonly psi or Pa. Again, the actual conversion is specific to each project and requires coordination between the hydrogeological and geotechnical modellers to ensure the correct conversions are implemented.

3. Once the coordinates of the groundwater model have been converted to the coordinate system of the geotechnical model, and the pressures have been converted to the working units of the geotechnical model, a further step is required to assign the pressure values to the exact grid or node locations within the

geotechnical model. This is called 'interpolation' and is further described in Section 4.3.8.3.

4. The pressure values are imported from the post-processing file into the geotechnical model. Interpolation of the pressure values to the exact grid or node locations within the geotechnical model is completed.

5. A graphical representation of the pressure contours within the geotechnical model is normally produced. The position of the pore pressure profile is qualitatively reviewed, relative to the pit slope position, and to key geological units and structures, to ensure that there is a very closely match to the groundwater model. If there is not, the trouble-shooting must be continued until a good match is obtained.

The most difficult step in the above process is the interpolation. Some programs (e.g. *Phase2* for steady-state flow models) automatically perform Steps 3 and 4 when importing the pore pressure. Other programs (e.g. *UDEC*, *3DEC*) require that the user perform the conversion and interpolation via a macro.

4.3.8.2   Negative pore pressures

Some codes used for groundwater modelling can produce values of negative pore pressure in the zone above the water table. For the majority of situations where this occurs, the negative pressures are an artefact of the model code and should not be exported to the geotechnical model. Careful review of the results is required. Negative pore pressures have been rarely observed in bedrock associated with mining, but they may occur in some mudstones, shales or zones of clay alteration. However, expected negative values would likely be low (of the order of a few metres of water head) and would probably not occur if there were any continuous fractures or joints that provided continuous groundwater flow paths. It is fairly common to observe negative pore pressures in plastic clays and other low-permeability materials around excavations in the construction industry. While these may be significant for relatively shallow excavations, it is less likely they would be important for improving the stability of large pit slopes.

A rule-of-thumb is to only include negative pore pressures in the geotechnical analysis if there are piezometer observations to support them at the operation in question or at other strongly analogous sites, and only to use the negative pressures as part of a sensitivity analysis. Care must be exercised to properly interpret piezometer readings. There are many examples of vibrating wire sensors recording negative pressures because they are poorly calibrated or, for example, because the calibration was carried out at sea level and the sensors were installed at altitude.

4.3.8.3   Importation of pore pressures to a regular grid or mesh

No matter what the process, it is important to avoid using a scheme that involves any kind of search algorithm to find the appropriate nodes for interpolation, especially when performing 3D computations. Search algorithms are notoriously slow because they must query each node in the flow model to determine which are the closest. A more efficient way is to convert groundwater model output (total heads or pore pressures) to a regular 2D or 3D grid or mesh. This eliminates the need for any search algorithm, since any point in the stability model can be immediately mapped into X, Y space in a 2D model or X, Y, Z space in a 3D model. Thus, the surrounding points are immediately known. *Surfer* is often used to do this for 2D models and has several interpolation algorithms available, the most popular of which is Kriging (see Jounel & Huijbregts, 1978). The output is a regular mesh with interpolated pressures.

4.3.8.4   Determining actual pressures in the stability model

The next step involves determining the actual pressure at each point in the stability model. The manner in which this is accomplished is dependent on the software programs being used. A good method that is frequently used is based on an inverse distance-weighted interpolation, sometimes called 'Shepard's method' (Shepard, 1968). Essentially the pore pressure at surrounding grid points is 'weighted' according to their distance from the point in the stability model:

$$pp(x,y,z) = \sum w_i \, pp_i \qquad \text{(eqn 4.13)}$$

where $w_i = 1$, $\Sigma \, w_i \, \text{dist}^2$

If a distinct-element program is being used for the mechanical simulation (e.g. *UDEC* or *3DEC*), then the user must decide what pressures should be applied to the joints and what pressures should be applied to the rock blocks (matrix). As described in Section 4.3.7.2, for most practical modelling scales, it is usually necessary to apply the same pore pressure distribution to the model joints and the model blocks. However, this may need to be considered if there are large areas of low hydraulic conductivity (and low diffusivity) rock and the excavation rate is rapid, in which case either the pore pressure dissipation response in the blocks may lag behind the pore pressure responses in the joints (the blocks drain to the joints), or the blocks may demonstrate an unloading response which is slowly equalised by flow in the joints (the joints recharge the blocks).

Several programs can perform pore pressure and mechanical analyses within the same program (e.g. *FLAC*, *Phase2*) or within the same program suite (e.g. *SEEP/W* with *SLOPE/W*). The pore pressures from the grid points in the flow model are then simply applied at the grid points of the mechanical model.

4.3.9 Input of transient pore pressures to the slope design model

Transient pore pressure responses may occur below the water table or in zones of otherwise 'dry' rock above the water table (Section 3.6.7). Where transient pressures relate to seasonal fluctuations in groundwater conditions within saturated parts of the slope, it is usually appropriate to apply the higher pore pressure in the geotechnical model, as this usually represents the most conservative condition for the slope design, and may reflect conditions in the wet season when instability of the slope may be more likely. In some settings, there may be a large difference in piezometer readings between the dry season and the wet season (50 to 100 m in some cases). This needs to be carefully considered when selecting input pressures for the geotechnical model.

It is more difficult to simulate transient pore pressures where they involve a seasonal or short-term change to the saturation state of the slope materials. The Batu Hijau Mine in Indonesia has a number of piezometers that have become 'dry' as the pit has been progressively expanded and dewatered. Following recent pushbacks, several of these piezometers now occur within a distance of 50 m of the slope, in the zone where confining stresses are lower. 'Dry' piezometers installed deeper in the slope (below wide pushbacks where standing water may accumulate) have shown pressure increases of up to 100 m during the wet season, and have recorded a seasonal 'pulse' of water moving downwards from the slope to the underlying phreatic surface (about 250 to 300 m below the slope). 'Dry' piezometers closer to the slope have shown a seasonal pore pressure rise of between 15 m and 40 m. Figure 4.43 shows an example of a dual vibrating piezometer installation where both sensors record zero pore pressure in the dry season. In the wet season, the shallower sensor shows pore pressure rises of 10 m to 15 m, and the piezometer responds to discrete rainfall events. The deeper piezometer has shown a more consistent seasonal transient pressure rise of 30 m to 40 m (it should be noted that the piezometers are installed in an angled drill hole and are both located at a relatively shallow distance behind the active pit slope).

There are no prescribed methods for inputting transient pore pressures to a geotechnical analysis. Procedures should generally follow those described in Section 4.3.8 for a saturated slope. A method to simulate

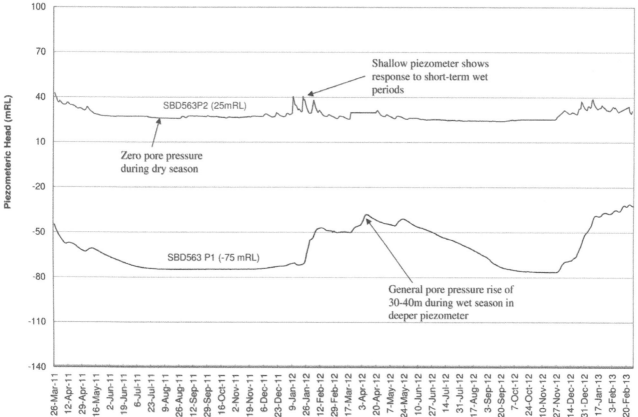

Figure 4.43: Transient response of 'dry' piezometers at the Batu Hijau Mine (Source: Ida Bagus Donni Viriyatha)

zones of positive pore pressure that represent the transient passage of water through the pit slope materials that lie above the phreatic surface is described as follows.

- Installation of piezometers to characterise the site-specific conditions at shallow levels in the slope where transient pressures are expected (to characterise, for example, whether the majority of recharge would move down over-break zone or would percolate deeper into the slope materials).
- Use of a small and highly focused model of the local area to be studied, rather than use of a wider-scale model of the entire slope.
- Evaluation of the amount of geological information available for the materials immediately behind the slope, whether it is realistic for a numerical model to produce absolute answers, or whether it is better to use the model to produce comparative results for a range of 'what if' conditions for risk-based decisions to be made.
- Selection of the model time steps – which may be short (possibly hourly) if the localised transient pore pressure response to an extreme high-intensity precipitation event is to be studied (clearly such short time steps would require a very focused model of a small area).

Similarly, to numerically simulate a rapid transient response to a pore pressure pulse due to a blast would require piezometer control because the site-specific response is likely to be widely variable.

The advantage of a numerical analysis to simulate transient pore pressures is that it would be a means of providing a pore pressure grid as input to the geotechnical model. It is likely that the transient grid of the highest simulated pressures would be most appropriate for the stability analysis (considering that the highest pressures at different locations may occur in different time steps if there is a downward pressure pulse).

Unless there is good site-specific data control, it is likely that the broad assumptions required for a numerical model would be similar to those for a more simplistic analysis. The staff at Batu Hijau carried out a screening analysis using R_u values to investigate the effect of transient pore pressures on shallow zones in their slopes. The R_u values were selected based on the observed rises in shallow piezometers as a proportion of the calculated confining stress at the position of the sensor. The R_u values were varied within a plausible range to investigate the sensitivity of transient water pressure at shallow levels in the slope. In a separate analysis, a saturated horizontal

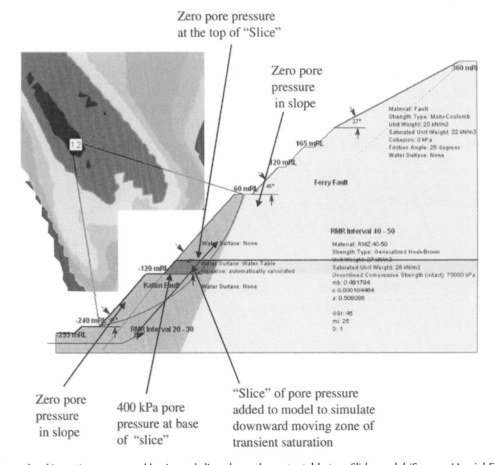

Figure 4.44: Example of inputting a saturated horizontal slice above the water table to a *Slide* model (Source: Hemiel Dayan Lelono)

slice of 40 m thickness was input to a *Slide* model analysis to simulate a transient pore pressure 'pulse' moving downward in the slope above the water table (Figure 4.44).

The behaviour of transient pore pressures at shallow levels behind pit slopes is an area where further monitoring and research is required, particularly given the high number of shallow slope failures that occur in pits with high rainfall. Future work may focus on how the high permeability in the over-break zone may influence transient pore pressures below this.

4.3.10   Introducing *Slope Model*

Slope Model is a new generation rock slope modelling code which was developed as part of the LOP research program. The intention of the code is to reproduce and predict the behaviour of a rock slope based on the interaction of both the rock mass and fractures, in contrast to the continuum (rock mass) or discontinuum (fractures or joints) approaches that are commonly used at present. In addition, the code was developed to incorporate pore pressure predictions, coupled with the geo-mechanical processes.

Numerical modelling in geomechanics is performed with two main classes of computer codes, continuum and discontinuum. Continuum codes such as *FLAC* and *FLAC3D* require constitutive models to represent the behaviour of soil, rock and concrete. These models can become quite complicated if they are to represent a wide spectrum of nonlinear responses, especially for cyclic loading and complex strain paths such as during dynamic liquefaction. Continuum codes are also not well suited to representing discontinuities, such as joint and fractures, which are included either in a smeared fashion or as interfaces. At the other extreme the distinct element method (DEM), embodied in codes such as *PFC3D* (Itasca, 2008), treats the material as an assembly of discrete particles that may or may not be bonded together. Simple rules at particle contacts lead to complex emergent behaviour at the level of the ensemble. Such behaviour is found to reproduce most of the rich response of soils and rocks (Cundall, 2000). Although complex constitutive models are avoided, calibration processes are necessary in order to select micro-properties that will lead to quantitative matches of the ensemble response to results from laboratory tests.

An active field of application of the DEM is in the modelling of rock fracturing (e.g. Potyondy & Cundall, 2004). It is possible to include existing discontinuities (here, collectively termed 'joints') with a scheme that treats the sliding and opening behaviour by reference to the direction of the joint plane rather than the orientation of local particle contacts. The overall model, which includes fracturing of intact rock as well as slip and opening on joints, is called the Synthetic Rock Mass (SRM) (Pierce *et al.*, 2007; Read & Stacey, 2009, Section 7.3.1.2).

Most applications of the SRM do not exercise the intrinsic ability of the DEM to handle large deformations, because failure initiation in brittle rock commonly occurs at small strains (e.g. much less than one per cent). Thus, a simplification is possible whereby finite-sized particles are replaced by point masses and the contacts between particles are replaced by springs that may break. This 'lattice' model achieves high computational efficiency (e.g. five to 10 times the execution speed of *PFC3D*) because the interaction geometry (location and apparent stiffness of springs) can be pre-computed, eliminating contact detection as an overhead. Further, the equations of motion often can be simplified by the elimination of the rotational degrees-of-freedom and the associated cross-products (for example) needed for interaction.

A version of the lattice code designed to simulate quasi-static problems in slope stability is called *Slope Model* (Cundall & Damjanac, 2009). The code was written by Itasca as part of the research studies for the LOP project and incorporated validation and test simulations with both steady and non-steady flow, and in both coupled and uncoupled modes.

The feature of the code is that it embodies the SRM concept so that a DFN may be imported into the model, with failure allowed both by movement on the faults and joints that intersect the rock mass and by breakage through the intervening rock bridges. The code also has the ability to couple fluid flow, pore pressure distribution and rock deformation. The hydrogeological computation has (i) a flow-only version (quasi-static flow within joint segments) and (ii) a fully coupled version to model transient flow within a heterogeneous environment, including transient evolution of pore pressures as the slope is excavated.

At the time of writing, *Slope Model* includes most of the hydrogeological processes pertinent to pit slopes, such as active depressurisation (e.g. pumping wells, drains, tunnels). *Slope Model* also has standard boundary conditions (e.g. impermeable or fixed head). It does not yet have specialised boundary conditions to simulate regional boundary conditions and/or surface infiltration.

The lattice formulation and the *Slope Model* code are presented in Appendix 6.

5

IMPLEMENTATION OF SLOPE DEPRESSURISATION SYSTEMS

Geoff Beale, John De Souza, Rod Smith and Bob St Louis

The purpose of this section of the book is to outline the methods that can be used to dissipate pore water pressures in the pit slopes as part of the mining process (Figure 5.1). Section 5.1 outlines general site water management procedures, how to integrate them with the mine plan, how to manage recharge, and how to develop general dewatering and slope depressurisation targets. Section 5.2 describes the various types of depressurisation systems and the physical processes of installing and operating them. Section 5.3 outlines the factors that must be considered for integrating surface water control plans into the pit slope design.

5.1 Planning slope depressurisation systems

5.1.1 General factors for planning

5.1.1.1 Site water management
A mine water management (or water control) program must incorporate planning for both groundwater and surface water that may impact open pit mining operations, including the performance of excavated slopes. Due to the increasing size of open pits, and the increasing regulatory constraints imposed on mining operations, planning of the water-management program must be an integral part of the initial Conceptual studies (Level 1, Introduction, Table 1) for any new project and for all subsequent stages of study, as described in Section 2.1.5. By the Feasibility stage (Level 3), the detailed water management plan should include an appropriate level of engineering design, capital and operating cost estimates and risk assessment that is commensurate with the overall project investment. At Level 4, detailed design is normally carried out for the coming five-year period, but the design must also be consistent with the overall dewatering and water management strategy for the life of mine (LOM) plan (which is normally developed at the Feasibility stage).

During Project Operations (Level 5), the water management plan must be adaptable based on interactive monitoring results.

5.1.1.2 Predicting the pit inflows and the general dewatering rate
A key part in any system design is the prediction of the potential rate of inflow to the pit and hence the amount of water that will need to be pumped to achieve general mine dewatering. Prediction of pumping rates is often less important for slope depressurisation programs because the flow rates are usually lower. Development of a detailed conceptual model (Section 3) is essential for providing a reliable prediction of the future groundwater inflows and mine dewatering rate. Without a good conceptual understanding, any method used to estimate the dewatering rate is unlikely to produce reliable results. Drawing upon the experience of other operators in similar groundwater settings is an essential part of the evaluation, particularly during the early stages of project development.

Estimating inflows of surface water can potentially be more straightforward than groundwater but can be complicated by the fact that, in many regions, the period of climatological record is too short to correctly represent the variability of the rainfall so there can be large uncertainty in the prediction of extreme events. This uncertainty may need to be factored into the planning and design process. Prediction of surface water flows is discussed further in Section 5.3.4.

The majority of mining operations that require high dewatering rates are either (i) hydraulically connected to an alluvial or sedimentary groundwater basin, (ii) connected to a regional carbonate aquifer system, (iii) located close to a surface water body that provides continuing recharge, or (iv) located in an area of high rainfall. The key factors for the prediction of the groundwater inflow rate are therefore the regional hydrogeology, the nature and extent of any alluvial basins

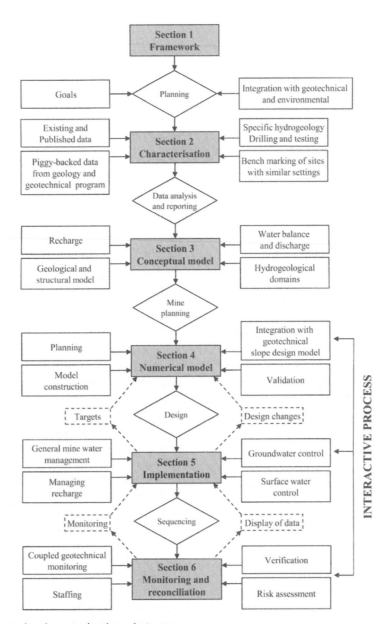

Figure 5.1: The hydrogeological pathway in the slope design process

and the nature and extent of any sedimentary groundwater units. Clearly, mines with high dewatering rates have the potential to create regional-scale drawdown which may, in itself, lead to uncertainties because of the difficulty in characterising a larger area.

Of particular importance for any dewatering prediction is the identification of regional or site-scale flow barriers (such as regional faults or impermeable groundwater units) and the use of the available data to attempt to assess the nature of these boundaries. Many of the mine sites that require large dewatering rates are located in areas of lower-lying topography, although this is not always the case in areas of high rainfall (e.g. Grasberg).

Identification of potential sources of surface water or aerial recharge sources throughout the region is an important consideration. In many cases, recharge will account for a large part of the total dewatering flow rate following the initial removal of groundwater storage from the immediate area of the mine. The spatial and temporal variations in the recharge patterns that occur in most natural systems must also be incorporated into the analysis.

There are three basic approaches for predicting the future mine dewatering rate.

1. **A water balance approach.** When using a water balance approach, the important regional and site-scale groundwater units are identified and estimates are

made for (i) the release of storage from each unit as the pit is progressively deepened, (ii) the continuing recharge to each unit, and (iii) the potential for flow between the units. Estimated storage and recharge components for each unit are summed to produce an estimate of the dewatering rate. As for all methods, an estimate of the drainable porosity for each unit is required. The use of a water-balance approach often provides the most reliable method for estimating the dewatering rate, particularly in the early stages of a project when detailed pumping trials have not yet occurred.

2. **An analytical approach**, using radial flow equations that can be adapted to simulate inflow to large excavations or the adaptation of Darcy's law to estimate flow to the various sectors of the excavation through the defined hydrogeological units, and summing the results. When using an analytical approach it is invariably necessary to 'lump' parameters together and careful judgement is required as to how this may influence the sensitivity of the predictions. In addition, it is often necessary to add the effects of continuing recharge into the analysis as well as adjusting for unconfined conditions.

A method of estimating flow into a pit by treating it as equivalent to a large-diameter well in an unbounded groundwater system is described by Singh *et al.* (1985). It is based on the method described by Jacob and Lohman (1952) for determining T and S from tests in which the drawdown is known and constant and the measured discharge varies with time, such as when a well in a confined aquifer is opened and allowed to flow freely. To use it to estimate the inflow to a well or equivalent discharge structure such as a pit it is necessary to know the 'well' diameter and the hydraulic properties of the groundwater system.

The method is also described in detail in Lohman (1972) which is readily available from the USGS. The equation is based on the assumption that the formation is homogeneous, isotropic and laterally extensive and that confined groundwater conditions are present. For these conditions, the radial inflow to the pit, Q, is given by:

$$Q = 2\pi TG(\alpha)s_w \qquad \text{(eqn 5.1)}$$

where:

T = formation transmissivity

r_w = radius of pit or well

s_w = required (constant) drawdown in discharging well or pit

$G(\alpha)$ = Jacob-Lohman well function of α, where α (not to be confused with rock compressibility) is given by:

$$\alpha = \frac{Tt}{Sr_w^2} \qquad \text{(eqn 5.2)}$$

Lohman (1972) provides a table of values of $G(\alpha)$ for values of α together with a set of type curves but comments that for all but very small values of t (which will be irrelevant to pit-dewatering calculations) the function $G(\alpha)$ can be approximated very closely by $2/W(u)$, where $W(u)$ is the Theis Well Function (Appendix 2) and

$$u = \frac{r^2 S}{4Tt} \qquad \text{(eqn 5.3)}$$

Values of $W(u)$ are more widely available, e.g. in Kruseman and de Ridder (1990), than values of $G(\alpha)$. As discussed in Appendix 2, for small values of u (i.e. at long times or small values of r), $W(u)$ can be closely approximated by:

$$W(u) = (-0.5772 - \ln u) = 2.30\log_{10}\frac{2.25Tt}{r^2 S} \qquad \text{(eqn 5.4)}$$

Using the approximation in Equation 5.4, and the substitution of $2/W(u)$ for $G(\alpha)$ in Equation 5.1, gives the approximate inflow as:

$$Q = \frac{4\pi Ts_w}{2.30\log_{10}\left(\dfrac{2.25Tt}{r_w^2 S}\right)} \qquad \text{(eqn 5.5)}$$

Equation 5.5 is identical to the Cooper-Jacob 'universal equation' (Appendix 2, Equation A2.18) with the modifying subscripts.

a. It should be noted that the Jacob-Lohman method is derived for confined conditions. Under unconfined conditions: the equation is not strictly valid, as the transmissivity will reduce with declining phreatic surface (this will probably lead to an erroneously high predicted value of Q, so in this sense will be conservative);

b. the effective storage coefficient will be the 'long-term' value of the specific yield.

To account for the presence of observed groundwater flow boundaries, a factor may be applied to reduce the potential radial flow, based on the position of the boundary relative to the pit.

It may also be possible to estimate the dewatering rate by conceptually replacing the mine with a simulated wellfield that would produce the required drawdown over the pit area. If unknown barrier boundaries are present, the simulation will give an overestimate of the required production rate, potentially with large errors. Alternatively, boundaries can be simulated using image wells (situated as far away on the other side of the boundary as the 'pumping' wells are on the pit side and pumping at the same rate), recharging image wells (used to simulate recharge boundaries) and abstracting image wells (used to simulate barrier boundaries). The process can be carried out graphically (Heath (1983) gives instructions for simple graphical wellfield calculations) or by use of spreadsheets into which the Cooper-Jacob approximation (Appendix 2) has been entered to calculate drawdown values after known times and at known distances. For a first estimate, the error in using the Cooper-Jacob approximation is unlikely to be significant but, for greater accuracy, the Theis equation can be entered into the spreadsheet and up to the fifth or sixth power of u (Appendix 2) can be used. Many of the proprietary aquifer-test analysis packages incorporate simple radial-flow models or, even more usefully, wellfield simulation packages that greatly simplify and speed up the process.

For all analytical calculations, it is necessary to add the continuing recharge component (which may include recharge from local surface water sources). A process of trial-and-error is usually needed to achieve the required drawdown patterns.

3. **A numerical approach,** initially using simple models (during the early stages of project development) and progressively increasing the complexity of the model as more information for the groundwater system is gathered (Table 4.1). For larger dewatering models, it is normal that a large part of the model domain would not be fully characterised, so the assumptions and uncertainty of the model need to be carefully understood when using the model for dewatering predictions. It should also be realised that, if the broad-scale assumptions on which the numerical model is based are the same as those applied to a water balance or analytical approach, the predictions of the numerical model may be no more reliable than predictions made using a simpler method.

Whichever method is used to predict the future dewatering rate, a 'reality check' on the dewatering predictions, using the experience gained at as many sites with similar settings as possible, is essential. Defining the parts of the hydrogeological system which are uncertain, and assessing the extent to which those uncertainties may affect the predictions, is also necessary. It may be appropriate to develop a 'base case' dewatering estimate and then to quantify the extent to which each area of uncertainty may influence that estimate.

An example may be the prediction of the dewatering rate for a pit developed in regional limestones. A 'base case' inflow prediction may be made using general hydraulic parameters for the limestone units with no geological boundaries. 'Case 1' may provide an assessment of the extent to which identified regional structures may form flow barriers and reduce the base case estimate. 'Case 2' may determine the extent to which overlying saturated alluvial deposits may produce sustained leakage into the limestone and increase the base case estimate. Thus, the importance of each cause of uncertainty can be assessed, and the field investigation can be focused accordingly.

At the Conceptual and Pre-Feasibility levels, the most important information is usually the head (groundwater level) measurements made from as many drill holes as possible. Similar head values in several holes will often indicate a good groundwater connection; varying head values will often indicate a poor connection. Where stair-stepping of water levels can be correlated with geological features, it can provide a good basis for characterising groundwater barriers. Where measured head values in the bedrock are different to those in overlying saturated alluvial deposits, it often indicates a stratified system with the potential for interconnected flow in the bedrock. Where measured groundwater levels show a strong correlation with topography, it is often an indicator of a poorly connected groundwater system and that flow is localised.

There is no substitute for pumping trials (Section 2.2.10) for helping to provide a reliable prediction of future dewatering rates, and prolonged pumping trials should form part of the Feasibility study for any project where high dewatering inflows are anticipated. Within the constraints of the investigation, it is normally good practice to pump at as high a rate as possible for as long a time period as possible (unless the system goes quickly into steady-state). The goal is to induce a response in as many surrounding piezometers as possible. A poor response of piezometers across geological features often provides a reliable basis for characterising important flow barriers. In addition to underpinning the engineering design and cost estimate, an extended pumping trial can also provide the regulator with confidence that potential regional impacts are properly understood. It is also much more difficult for an NGO group to question the results of pumping trials than it is to dispute the results of a numerical model.

5.1.1.3 The importance of time for pit slope depressurisation

A key factor for designing a slope depressurisation program in poorly permeable materials is the lead time required to achieve the target pore pressure profile for the critical sectors of the pit. The lead time must be factored into both planning and implementing the program, and must also be considered carefully when determining what level of depressurisation is actually achievable. Slopes excavated in higher permeability environments typically require less time for the depressurisation process to be effective than do those in poorly permeable rocks.

Equations 1.25 and 1.26 (Section 1.1.2.8) describe the significance of hydraulic diffusivity in controlling the speed of response of a formation to a perturbation in head, such as the onset of pumping from a well. Appendix 2 explains the use of the Cooper-Jacob approximation ('universal equation') in calculating the expected head change at a given time and distance in response to pumping from a line source (such as a drain or well) at a known rate. These equations give some help in understanding the effects of pumping, but it must be remembered that they are based on various assumptions (e.g. homogeneity, isotropy) that are unlikely to be met in practice given the complex geology around many mine sites. Therefore, they should always be treated as approximations.

In poorly permeable, weak rock environments, such as in operations with deep weathered zones or thick zones of argillic (clay) altered rock, it can take months to years to depressurise slopes to the desired targets. Depending on the time available, it may be necessary to consider advancing the general dewatering program to lower the water table and so provide additional time for drainage of the less permeable sectors. In the late 1990s at the Goldstrike Betze-Post pit in Nevada, four years of lead time were required to achieve partial depressurisation of the low-permeability East wall Carlin Formation in advance of final high wall development. Where such advanced planning is not possible, it may be necessary to accept more modest levels of depressurisation that can be practically achieved within the time frame of mine development.

5.1.1.4 Reduction of recharge

For many mine sites, ongoing recharge into the slope domain is the most important factor to be accounted for when designing a slope depressurisation program. For low-permeability (non-pumpable) groundwater units that occur within pit slopes, the key to effective slope depressurisation is the reduction of recharge to the slope domain.

The more that recharge can be intercepted before it enters the slope domain, the more efficient the slope depressurisation program will be. An assessment of how recharge may be minimised or prevented should therefore form part of any planning exercise, including:

- general reduction of infiltration at the surface, for example by improving surface water drainage and by avoiding surface water depressions (hollows) and other areas of ponding water;
- lining of surface water systems and/or key mine facilities;
- interception of shallow groundwater that may move towards the pit in near-surface hydrogeological units;
- interception of deeper groundwater before it reaches the mine area; to reduce the overall pumping head and

prevent contact of the water with the ore body, so that the pumped water chemistry can be optimised;
- working with mine and process plant operations to minimise the recharge that may result from leaky site facilities;
- use of positive drainage or lined ditches on catch benches and haul ramps to minimise the infiltration of incident precipitation to the slope domain.

For many mine sites, particularly those in drier environments, artificial recharge can be an important and sometimes the major factor influencing the slope depressurisation program. The planning process must therefore account for potential recharge that may occur from on-site facilities. Recharge may occur from the tailing impoundment, the base of waste rock facilities (in wetter climates), other pits which may accumulate water, historic underground workings, and ditches and ponds that may have been constructed as part of the surface water management or stormwater control system. In addition, many pit slopes are locally affected by leakage losses from tanks and pipes associated with the process plant, or as a result of concentrated runoff from areas of sealed hard-standing.

5.1.2 Integration with mine planning

5.1.2.1 Overall considerations

It is one thing to design a slope depressurisation program for a given mine design, but it is another to ensure that it is sufficiently practical for implementation by the mine operations group.

The ability to integrate the required pore pressure controls with mine planning and operations while recognising what is practical to achieve is always a major consideration, and therefore is a key issue for design. The mine operations group must appreciate the need for and benefit of dewatering and/or slope depressurisation.

5.1.2.2 Providing sufficient lead time

In addition to the hydrogeological characteristics described in Section 5.1.1, the time required to achieve the general dewatering and pore pressure profile targets depends on a number of other factors, including:

- factors external to mine operations, such as the remoteness of the site, the rate at which the required contracts can be secured with drilling contractors and other third parties, and the supply chain and required lead time for shipping and delivery of materials;
- factors relating to the mine plan and mine operation such as the rate of mining, the timing and nature of mine expansions and pushbacks, and the extent to which the mine plan is flexible.

Within the constraints of the mine plan, the following issues need to be considered.

- Provision of working areas within the mine plan for the drilling and installation of any wells and drain holes, given the likely heavy production vehicle traffic.
- Sequencing of the installation of the wells and drain holes with the active mine operations.
- Avoidance of safety concerns due to rock fall, or due to overspill as a result of mining of the benches above the sites that are selected for in-pit drilling and/or monitoring installations.
- Provision of sufficient lead time for the installed system to achieve the desired level of depressurisation.
- Provision of ongoing access for maintenance of both the installations themselves and any associated ancillary equipment such as pumps, reticulation pipelines or drainage ditches.
- Location of the installations so they will survive for the required period of time, considering the ongoing mining operation and the timing of future mine expansions and slope pushbacks.
- Design of the slope depressurisation measures with sufficient flexibility to accommodate changes to the mine plan.

The required access can normally be provided by advance planning and sequencing the drainage measures, or sometimes by making minor adjustments to the mine plan. Slope depressurisation often results in the opportunity to steepen slope angles, thus reducing stripping and waste rock handling and increasing productivity. Minor adjustments to the mine plan to accommodate slope drainage measures may therefore have a large payback.

5.1.2.3 Providing work areas within the pit

A number of operational considerations can be useful when planning the location of in-pit drainage measures.

- The potential to use semi-permanent or permanent upper benches for production wells (and the need to provide long-term access to the benches).
- Creation of slightly wider catch benches at certain elevations to provide access for well or drain drilling (Figure 5.2).
- Installation of drainage measures on catch benches close to the point where the benches intersect the haul ramps.
- Installation of drainage measures on wider sections of the haul ramps, such as the 'dead' areas on the inside or outside of bends (provided visibility is maintained).

In many cases, slope depressurisation measures can also be installed along haul ramps, particularly where the ramp is not currently being utilised for haul traffic. A suitable situation for production wells is often the junction between catch benches and the haul ramp, as shown in Figure 5.3, where suitable access can often be provided with a minimal amount of dozer work.

Where a large amount of infrastructure is required to meet targets, the mine plan may need to incorporate dedicated dewatering benches to support long-term access, for example, for drain holes, wells, piping, and pump stations. It is often possible to combine dewatering benches with step-ins that are required for the geotechnical design. With appropriate planning, the design and location of the geotechnical step-ins can often be modified to suit the dewatering or slope depressurisation program.

A challenge that frequently occurs is when the slope is rapidly excavated in weak, low-permeability materials that require an array of closely spaced drain holes to achieve the required pore pressure targets. In this case, the program of drain hole drilling may require integration with the drill and blast operation. Modifying the drill and blast sequence to allow installation of the drain holes may also be necessary.

Figure 5.2: Examples of in-pit drainage installations

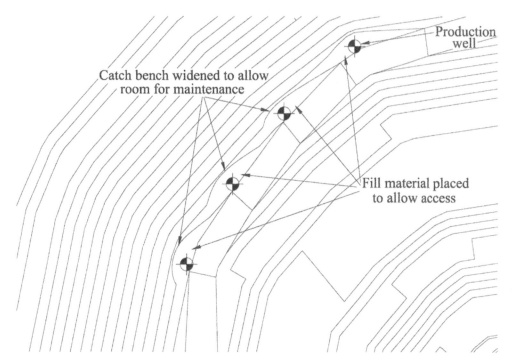

Figure 5.3: Location of a production well at the intersection between a catch bench and a haul ramp

The reality of slope depressurisation programs is that a compromise must often be achieved between the ideal locations to install the wells, piezometers and drain holes and the actual locations that can be accessed within the pit. For installations that require ongoing maintenance, the access must be permanent, or they may need to be replaced.

5.1.2.4   Medium- to long-range planning

Development of a practical dewatering and slope depressurisation system that satisfies requirements of the life of mine (LOM) plan is an integrated task for the mining operation. The process requires the hydrogeologist and the geotechnical engineer to have a comprehensive understanding of the medium- and long-range development sequence and schedule; and the mine planning engineer must fully understand the nature of mine dewatering and related logistical requirements. The process will carry on throughout the life cycle of the mining operation. Figure 5.4 shows a typical interactive sequence for optimising individual in-pit hydrogeology installations.

System maintenance and continual improvement can often be challenging in the active mining environment, where priorities and cost-pressures for management and operations personnel are continually changing. Therefore, a process of interactive review and periodic design updates is usually of benefit, based on the changing goals. This would include risk assessments that address the consequences if drainage targets are not achieved (Section 6.3). Adjustments to the program can then be made in time to ensure that the dewatering system achieves the objectives of the mine plan.

5.1.2.5   Creating permanent access for ongoing maintenance

In many cases, drain hole arrays and associated reticulation systems require no further maintenance after installation. Maintenance-free gravity drainage can be a significant benefit for horizontal drain programs. Figure 5.5 shows an example of a horizontal drain array high in the slope. The drains continue to work by gravity even though ongoing access to their collars is no longer possible. Where the continued operation of the drains is necessary to achieve the required depressurisation targets, mine operations should be aware that measures to reduce the risk of damage to reticulation lines due to rock fall from above or fly rock from blasting below may need to be employed.

In some situations, installation of an angled drain hole array may reduce the required access and maintenance. Figure 5.6a shows a zone of low-permeability material behind a pit wall that requires depressurisation. In this case, it was not possible for the operator to provide access for drilling of horizontal drains, but it was possible to take advantage of a wide step-out and drill angled drains in the slope to cross-connect the low-permeability material behind the slope with the underlying permeable and dewatered rock. A similar approach was taken in Figure 5.6b, but in this case there was only a limited step-out width, so the drains were drilled into the slope, and were consequently farther behind the face.

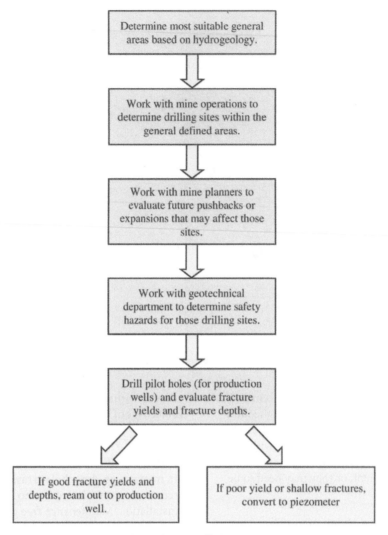

Figure 5.4: Interactive process for planning of mine hydrogeology installations

For pumping wells and certain high-volume gravity drain holes, there is a requirement to conduct ongoing maintenance, so there is a need to provide continued access to the wellhead or collar location. Installations located on catch benches adjacent to haul ramps are therefore often preferred. However, where the well installation is critical to the success of the program, access between benches can sometimes be created, as shown in Figure 5.7.

It is usually desirable to continue monitoring of drain holes and possibly shut-in pressures for as long as access is possible after the drains have been installed. Where there has been good scaling of the wall above the drain collars, access by foot, or by light vehicle is often possible, even though access for heavy maintenance equipment may have been lost.

In many cases, the creation of long-term access simply is not possible within the constraints of the mine plan, and the wells need to be plugged and abandoned and

replacements installed at alternative locations. In other cases, operation of some wells or drains becomes redundant because of the ongoing lowering of the water table due to new or deeper installations. Where it is known in advance that wells and drains will have a relatively short life, installation costs should obviously be reduced to the extent possible. Pumping of in-pit wells into adjacent staging sumps will lead to a minimal pumping head so that a small pump diameter, and therefore a smaller well diameter, is often possible. The installation of lower-specification casing (or line pipe) and perforated pipe may also be possible.

5.1.3 Development of targets

5.1.3.1 Overview

For new mining operations, the mine plan often involves multiple phases of mining early in the mine life, with relatively flat 'de-coupled' slopes and small overall slope heights. At this stage in the mining cycle, pore pressure

Figure 5.5: Example of operational horizontal drains where access to the collars is no longer possible

and slope depressurisation may be less of a priority. Rapid deepening of the pit floor often occurs early in mine life to access ore and recover capital, thus general dewatering and removal of groundwater volume may be more important than depressurising the slopes. For a more mature operation, later in the mine life, the overall slope heights typically increase, and final slope angles become important to later-stage economic performance. Therefore, pit slope depressurisation measures may become a critical element in the later stages of the mine plan.

However, in some cases, early slope depressurisation may be important, for example for those start-up operations that require a large amount of pre-stripping to access the ore. Steeper slopes in the starter pit will often mean less pre-stripping and earlier ore production, but the consequences of a slope failure could have a major impact on early cash flow. Depressurisation of the starter-pit slopes may therefore be important for reducing the risk of failure and loss of early ore production.

5.1.3.2 General mine dewatering targets

For general mine dewatering, the development of water level targets is based on the mine plan. The targets are usually based on lowering the water table below the lowest part of the main pit floor (or below the pit floor in a given sector) ahead of the mine excavation, and are often specified on a quarterly, six monthly or annual basis, depending on the stage of mining. In Figure 5.8, the mine plan calls for a rapid lowering of the pit floor, so it is necessary to gradually lower the water table in advance of this. The figure shows actual water levels achieved to date and required future (target) water levels, plus the total pumping rate to date and the predicted (modelled) pumping rate necessary to achieve the future target levels.

Typically, the dewatering target is specified for lowering the water table a given distance below the advancing pit floor by a given date; for example, 20 m below the pit floor by year-end. However, in some instances, in order to provide adequate time to achieve the required drainage for lower-permeability units, the

(a)

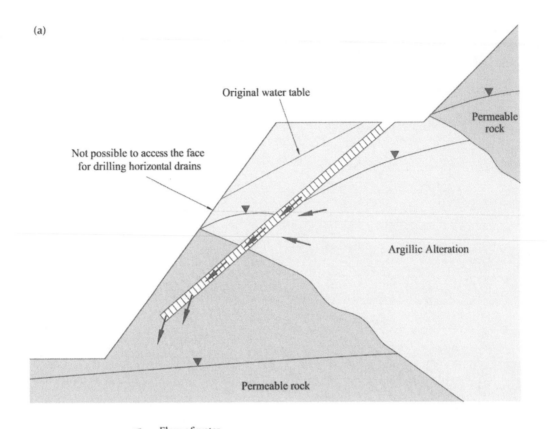

Original water table

Not possible to access the face
for drilling horizontal drains

Permeable
rock

Argillic Alteration

Permeable rock

← Flow of water

(b)

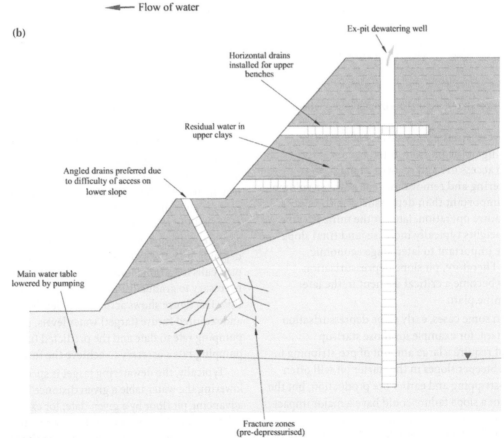

Ex-pit dewatering well

Horizontal drains
installed for upper
benches

Residual water in
upper clays

Angled drains preferred due
to difficulty of access on
lower slope

Main water table
lowered by pumping

Fracture zones
(pre-depressurised)

Figure 5.6: (a) Angled drain holes into face being used when access for drilling horizontal drains is not possible. (b) Angled drain holes away from face used in preference to horizontal holes because of access constraints

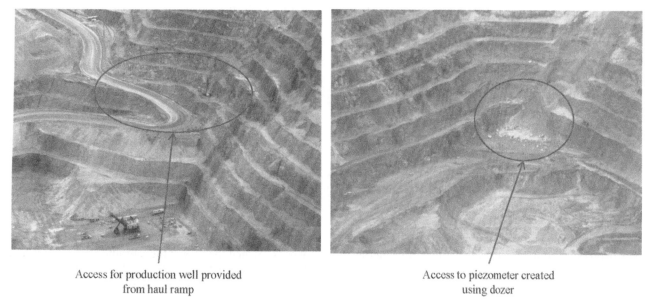

Access for production well provided
from haul ramp

Access to piezometer created
using dozer

Figure 5.7: Maintaining access to an important well location

dewatering target may be specified to lower the water table to a certain elevation by a given time (say 12 months) ahead of the mine plan.

5.1.3.3 Targets for slope depressurisation

For pit slope depressurisation, the development of pore pressure targets is also important, particularly if the slope design has been predicated on achieving a given pore pressure profile by a specific date based on the mine plan. The targets can sometimes be set based on achieving a simple phreatic level by a given point in time. However, it is more often prudent to provide purpose-placed piezometers on each key geotechnical design section, and to specify the targets based on achieving

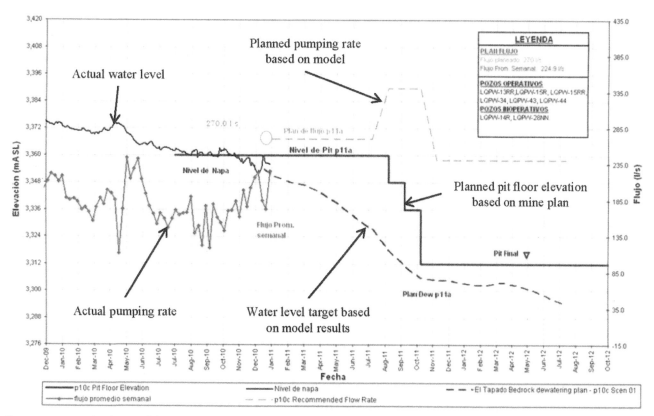

Figure 5.8: Example of general dewatering targets (Source: David Rios, MYSRL)

given head values in some or all of the piezometers by a given time (often annually). Performance monitoring using the piezometer arrays will be essential to ensure that the program stays on track.

The development of realistic targets for slope depressurisation must consider the practicality of achieving the target levels. In addition to the hydrogeological characteristics, the assessment must also consider the support and logistics of the particular mining operation and its ability to install the 'ideal' system within the prescribed time. Given the production constraints of an operating mine, it is easy to under-estimate the time required to install in-pit drainage measures.

For the North-west wall of the Diavik Mine (Section 3.3), depressurisation targets were defined by using the geotechnical model to determine the necessary pore pressures to allow the stability acceptance criteria to be achieved for each mining bench. Piezometer hydrographs were forward-predicted to determine the time when pore pressures would fall to the target levels to allow each new bench to be safely excavated. The use of piezometric

profiles to predict pore pressure targets is illustrated (for a different site) in Figure 5.9.

The need to carry out a trade-off study to define the cost of the depressurisation measures versus the savings in waste rock stripping is described in Section 3.2 in the Introduction to the book. At the Cowal Mine in New South Wales (Appendix 4.7), initial targets were set assuming that maintenance of a dewatering system continued to achieve the same rate of depressurisation in the future that was experienced in the past (Case B). However, it was necessary to confirm that the operation could achieve the target slope depressurisation using the Case B dewatering system in an acceptable time frame. Targets were set on a year-by-year basis, in accordance with the mine (slope) development milestones. Key milestones were when the layback reached the base of the saprolite zone, and when the layback reached a saturated weak rock interval. In the event that monitoring indicated the depressurisation targets would not be achieved, additional (contingency) pumping wells and/or horizontal drains were planned to increase the rate of pressure dissipation to the required level. The alternative was to slow the mining rate.

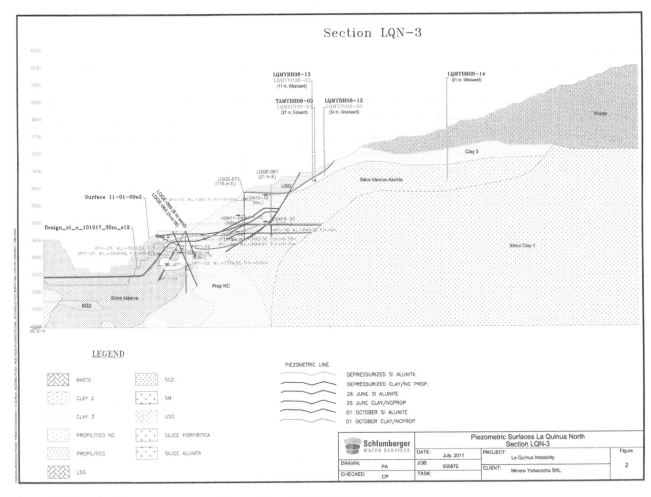

Figure 5.9: Example of pore pressure targets (Source: Franz Soto, MYSRL)

5.1.3.4 *Determining the required number of installations*

Where general mine dewatering is the main objective, the required number of dewatering wells is best determined by site-specific testing and by gaining experience in the short- and long-term performance of wells in various sectors of the pit. In permeable settings, where a high overall dewatering rate is required, it is likely that individual well yields will also be relatively high, so there will be a manageable number of operating wells (typically five to 25 for a moderate size pit). However, where the mine is surrounded by high permeability formations (or recharge sources) but the mining area itself is of lower permeability, the yield of individual wells may be small compared with the overall required dewatering rate, in which case a large number of wells may be necessary. There are many mines that operate in excess of 50 general dewatering wells and some operate over 100. If the pit scale hydrogeology is relatively complex (as at many mines), the yield and performance of dewatering wells may show a large variation between sectors.

Typically, when general mine dewatering is required, production dewatering wells and associated infrastructure are constructed six months to two years in advance of planned mining activity below the water table. The general location and approximate number of dewatering wells can be predicted using the numerical model, but the exact locations and yields are often dependant on local-scale conditions, including variations in the detailed hydrogeology and structural geology in the pit area, and in-pit access constraints. In many hard-rock settings, the yield of a dewatering well can vary greatly if the collar of the hole is moved by just a few metres, as described in Section 5.2.5.2.

Planning of a slope depressurisation program and the required number of installations usually carries more uncertainty than planning of general mine dewatering systems, particularly in the early stages of Greenfield projects before the pit has been open up and before any pilot-scale testing can be carried out. The best-form of early stage planning is often benchmarking with other similar sites. It is also possible to use a slice or planar model to investigate the required lateral and vertical spacing, and the length of horizontal drains for each material type that will be exposed in the slope. The 'wall area' of each material type that will be exposed every year can be obtained from the mine plan, such that the total number of required drain holes each year can be calculated for each material type. For operating mines, the number of drain hole installations required each year for each material type is best determined through the use of pilot-scale trials and by calibrating the numerical model to the pilot results. Again, the logistics of installing the planned slope depressurisation measures at an operating mine must be carefully considered.

5.2 Implementing a groundwater-control program

5.2.1 Types of control systems

The following types of groundwater control are commonly used for open pit mining operations.

- Horizontal (or sub-horizontal) drains.
- Gravity flowing steep angle or vertical drains.
- Pumping wells.
- Drainage tunnels.
- Passive groundwater cut-off systems.

The application and construction of **horizontal drains** is described in Section 5.2.3. They are commonly used worldwide to relieve groundwater pressure behind the pit slopes in rock units that are not amenable or accessible to pumping via vertical wells. There are many construction methods for horizontal drains, but a typical construction will involve 100 mm to 150 mm diameter holes drilled to depths of 150 m to 350 m, with 25 mm to 50 mm diameter slotted pipe installed in the drain.

Prior to starting a horizontal drain program, it is important to determine the objective of the drains, and to design the program to accomplish the specific objectives and targets. The installation of drains at an azimuth such that they intersect the maximum number of joints, fractures and individual hydrogeological compartments may be advantageous.

The application of **vertical or angled drains** is discussed in Section 5.2.4. They can be considered where a large vertical hydraulic gradient is developed within the slope, or where groundwater is perched above a less permeable unit that is underlain by more permeable units at depth. The goal is to cross-connect the low-permeability unit that requires drainage to a more permeable zone which is already depressurised.

There are many applications for **pumping wells** (Section 5.2.5) for both general mine dewatering and pit slope depressurisation. In permeable settings where a general lowering of the groundwater table around the pit area is required, most open pit mine dewatering systems use some form of vertical pumping wells, either external to the pit (outside the crest; ex-pit) or within the pit (in-pit). Many dewatering projects use a combination of interceptor wells to remove active recharge to the groundwater system at shallower levels, ex-pit wells to remove groundwater flow that would otherwise enter the pit, and in-pit dewatering wells to remove groundwater storage and accelerate drawdown inside the pit shell. Figure 5.10 shows a permeable oxide zone exposed in a pit slope overlying lower-permeability fresh rock. In this case, shallow wells could be used to remove water within the oxide zone where most of the permeability and storage occurs. The

Most flow towards pit occurs in oxide zone

Base of oxide zone

Water runs from oxide zone into underlying fracture zones that have become dilated in the slope

Figure 5.10: Example of shallow flow in high permeability oxide zone

underlying fresh rock does not have sufficient open fracturing to sustain pumped yields.

Drainage tunnels installed behind the pit slope or underneath the pit are being increasingly considered by some of the larger open pit mining operators, both for dewatering and to achieve pit slope depressurisation (Section 5.2.6). A significant operational advantage of a drainage tunnel is that the drains and other slope depressurisation measures can be installed and operated from within the tunnel itself, without interfering with mining operations (once the portal is established). An obvious potential downside of a tunnel is the up-front cost and time required for planning although, for larger pits, the cost of a tunnel is often comparable with the overall cost of drilling a large number of drains from within the pit for a number of sequential pushbacks. As for all of the depressurisation options, the overall cost must be viewed in terms of the potential benefit of achieving steeper slope angles.

Passive **groundwater cut-off systems** applied to open pits include (i) slurry walls, (ii) grouting programs focused on permeable fracture systems, and (iii) ground freezing. Slurry walls (and other types of low-permeability barriers) are commonly used in the construction industry and around smaller or shallower open pits, particularly in the coal mining sector. Grouting of permeable fracture zones is used at some time or another in the majority of underground mines for

reducing the permeability and inflow rate around development tunnels and, in some cases, around the production workings themselves. Ground freezing is commonly carried out in shafts and tunnels to increase the strength of the wall materials and to reduce inflows. Although freezing is more commonly applied to softer or sedimentary materials, a number of underground mine operators use widespread freezing programs to seal the connections from permeable units over wide areas.

For the majority of large open pit applications, passive cut-off systems are most commonly applied to specific zones within key sectors of the pit, rather than globally around the pit perimeter. However, there are a number of new mine projects that are currently considering a full low-permeability 'curtain' around the entire pit perimeter to minimise the amount of water that needs to be pumped. A major driver for this is to reduce potential impacts and meet regulatory constraints.

The primary objective for placing a low-permeability barrier around an open pit is to reduce the permeability of a particular formation or zone along the flow path, with the goal of reducing the amount of groundwater reaching the pit. A common application for **slurry walls** is within shallow permeable saturated alluvial deposits that occur in the upper pit walls (Figure 5.11) or adjacent to surface water bodies (Figure 5.12).

The normal maximum depth of a long slurry wall is 25 to 30 m although much greater depths can be achieved

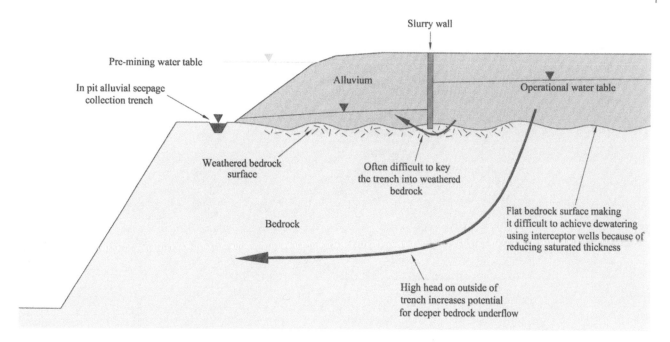

Figure 5.11: Example of slurry wall being used to cut off groundwater flow in alluvium

with 'soil saw' type excavation equipment and shorter diaphragm walls have been constructed to more than 50 m deep. Figure 5.13 shows an example of a slurry wall being constructed within saturated alluvial materials around the margin of the pit using a long-boom excavator. During construction, the stability of the trench is maintained by continually topping up with a 'heavy' bentonite-based slurry.

When considering the application of a passive cut-off program, the operator must appreciate that the system will create a 'groundwater dam', such that the total head will not reduce (and may increase) on the outside of the wall. As a result, there could be a tendency for groundwater to find alternative pathways towards the pit as groundwater heads on the inside of the wall become progressively lowered as a result of mine dewatering, and as the hydraulic gradient across the low-permeability zone steepens. A common issue with the construction of both slurry walls and freeze walls is the ability to key the structure into the underlying bedrock surface where this is weathered and/or broken, and a great deal of care and time is frequently required to ensure a good seal along the base of the wall. Even so, the

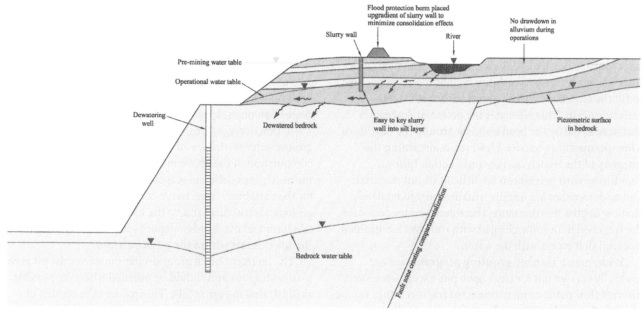

Figure 5.12: Example of slurry wall installed adjacent to a river

Figure 5.13: Illustration of slurry wall construction

groundwater head build up along the outside of the wall will increase the potential for leakage at the alluvial-bedrock interface (at the base of the wall) or for enhanced groundwater underflow in the deeper permeable units beneath the wall. As with the application of any water control system, the geology needs to be carefully investigated prior to specifying and designing the system.

For certain applications, rather than the creation of an impermeable barrier by filling the trench with bentonite slurry or other low-permeability membranes, it is possible to fill the trench with clean gravel such that it acts as a collector drain. This alleviates the potential downsides that are caused by the head build up around the outside of a low-permeability barrier. However, maintaining the integrity of the trench during construction (prior to backfilling with gravel) can be difficult in soft materials and such trenches are usually installed only to relatively shallow depths. Furthermore, there may be a tendency for the trenched to become clogged with time by fine-grained material that enters with the water.

Widespread 'blanket' **grouting programs** are not normally carried out for large open pits except where very discrete flow paths or interconnected fracture zones can be identified and where the sealing of the zone will not cause

the groundwater to find an alternative pathway into the pit wall. The application of widespread grout curtains around open pits has been likened to sealing the holes in a colander; when one area is plugged the water merely enters the pit slope by another route. Grouting programs are more normally used in underground mines where a common application is to exclude short-term inflows to development headings in order to allow the heading to be advanced past a specific inflow zone. Clearly, the surface area of the rock exposed in an underground development heading is vastly smaller than the rock exposed in the slope of an open slope.

In practice, most mine sites are located in complex groundwater settings, and the project must utilise a combination of the above measures in different sectors of the mine, or at different stages of the mine life to account for the variability in recharge conditions, bedrock types, geological structure, and/or the nature and extent of the alluvium and overburden deposits that may occur at shallower levels within the pit slope areas.

The interception of groundwater outside of the pit area is advantageous and should be pursued wherever possible, as illustrated in Figure 5.14. This reduces the burden of constructing, maintaining and replacing production wells

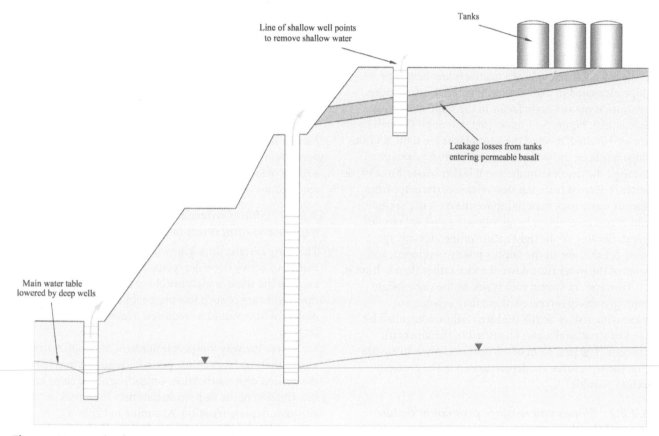

Figure 5.14: Example of intercepting water at shallow levels

and/or drains inside the mine operating area, and also minimises nuisance water inside the pit. Interception of groundwater away from the immediate pit area may also lead to an overall decrease in pumping head and pumping costs, even if it is necessary to produce a higher volume. In many cases, the operation of ex-pit wells, and particularly alluvial interceptor wells, leads to an overall improvement of the pumped water chemistry, since groundwater is intercepted and pumped before it contacts the mineralised ore deposit. An associated reduction in downstream water management costs may be realised.

Depressurisation of pit slopes using horizontal or angled drains can be equally (or more) effective than vertical wells where the required depressurisation targets can be reached using gravity flow. Perched and remnant water might be accessed with vertical drains where under-drainage conditions exist. Horizontal and vertical drains are typically easier and less expensive to install and maintain than pumping wells.

5.2.2 Passive drainage into the pit

In Section 4.2 of the Introduction, passive drainage was identified as allowing pore pressures to dissipate as a result of seepage to the pit slope, with removal of the water using in-pit sumps. This 'do nothing' approach may be

applicable in situations where the rock mass in the pit slope is strong, there is a high tolerance to groundwater pressure in the slope design, and there are relatively low amounts of groundwater entering the pit, with little or no impact on mine operations.

There are generally four settings where the required depressurisation may be achieved solely as a result of passive drainage into the pit.

1. Slopes excavated in high-strength materials.
2. Slopes where other prominent factors control the overall design, such as the bench face angle or the foliation of the rock mass.
3. Slopes excavated in permeable materials.
4. Slopes where the installation of active depressurisation is not practicable within the available time constraints.

5.2.2.1 Slopes excavated in high-strength materials

For those pit slopes excavated in relatively strong rock types, the main slope design control is often the bench or multi-bench-scale kinematic analysis, or the achievable geometry of the bench face angle (which may in itself be controlled by the dip of the joint sets) and catch berms. In some cases, a reduction of the pore pressure in the wall rocks would provide little added benefit. A seepage face in the slope may therefore be tolerated, and there may be no requirement to install horizontal drains or other slope depressurisation systems.

The Hypogene pit in at the Andacollo Mine in Chile is a good example of a relatively low-height pit slope excavated in high-strength materials where pore pressure is not the main factor in the slope-stability assessment. Figure 5.15 shows that, although the wall rocks are well jointed, most of the joint planes are tight, and the slope has been the well protected from blast damage through the use of trim shots and buffer blasts. Most of the water is derived from the slow release of drainage from porous waste rock material above the crest of the slope which had received recharge following a recent rainfall event. Because of the tight nature of the jointing, the 'over-break' zone in the slope is poorly developed, and most of the water runs down the face rather than behind it.

However, in strong rock types, as the slope height is progressively increased, there may become an increasing risk of brittle (sudden) failure controlled by the key structural zones identified in the kinematic analysis. The best form of mitigation is often to ensure that the structures are depressurised to the greatest extent possible.

5.2.2.2 Slopes where other prominent features control the design

Where other prominent geological controls are apparent in the wall rocks, such as joint foliation or bedding, the design of the overall slope may not be sensitive to pore pressure. Examples of strong foliation controlling the slope design are shown in Figures 5.16 and 5.17. In these examples, the slope angle is controlled by the foliated bedding, regardless of pore pressure conditions. Again, as the slope heights increase, there will likely be a need for depressurisation to reduce the risk of larger scale failures controlled by structures.

In some settings, the nature or angle of the bedding or foliation is such that the slope design would not be possible without depressurisation. In such cases, although the bedding and/or foliation is the main control on the overall design and slope angle, it would not be possible to achieve the design without dissipating the pore pressure to increase the shear strength of the structured rock mass.

5.2.2.3 Slopes excavated in permeable materials

Where the wall rocks are excavated in well-jointed materials, and where the joints are open, interconnected and permeable, passive seepage from the pit slope may itself provide an adequate level of pressure dissipation to achieve the desired slope performance. An example is the North wall of the Round Mountain pit (Nevada) where the well-jointed metasediment unit drains readily into the pit and/or into the adjacent tuff unit where the water table is being lowered using dewatering wells. The passive seepage

from the metasediments is sufficient to dissipate any elevated pore pressure in the North wall of the pit.

Figure 5.18 shows the North-east wall of the Gold Quarry pit in Nevada. The lower slopes have been excavated into strong siltstone and limestone which fully drains and depressurises as a result of the general mine dewatering system (deep, high-capacity dewatering wells). The upper slopes are excavated in weaker materials which remain saturated and decoupled after the water table has been lowered in the underlying siltstone and limestone and a range of local slope depressurisation measures is required for the upper slope.

5.2.2.4 Slopes where the installation of active depressurisation is impracticable

There are certain situations where it is not practicable to install an active slope depressurisation program, either because the slope is inaccessible or because the materials themselves are of such low permeability that a very long period of time would be required to achieve the desired targets.

Where the very low permeability of the slope materials means that too large a time period is required to achieve the desired depressurisation, unloading of the slope and deformation of the slope materials may itself lead to adequate depressurisation. At a mine in Chile, a permeability of 10^{-8} m/s coupled with a hydraulic diffusivity of 10^{-4} was identified as a threshold below which it is not practical to drain the slope, but where deformation of the slope materials may start to become an important contributor to pore pressure dissipation.

The Mag pit at the Pinson Mine in Nevada is a good example of achieving drawdown in the slope materials using passive drainage. In the East wall of the pit, it was necessary to mine through about 25 m of saturated, poorly permeable alluvium. The alluvium extended a considerable distance into the pit and, without pre-drainage, it was not possible to run heavy equipment over the bench being mined. In addition, there was a seepage face in the alluvium that caused piping and instability in the lower part of the slope, above the toe. Figure 5.19 shows that a 6 m deep trench was cut at the toe of the slope prior to mining each new bench. Pumping from the trench allowed the water level to be lowered sufficiently in the alluvium to allow mining. The trench also increased the drainage rate of the alluvial material in the slope. Figure 5.20 shows a photograph of the operations. The seepage water was collected and managed using a series of sumps located below prominent zones of seepage.

5.2.3 Horizontal drain holes
5.2.3.1 General
The use of horizontal or sub-horizontal drain holes is one of the most commonly applied methods worldwide for locally reducing pore pressures behind open pit slopes.

Figure 5.15: Example where seepage from the benches is of little concern because of the high-strength materials in the slope (Andacollo Hypogene pit, Chile; source: Mauricio Tapia)

Figure 5.16: Example of Western Australian iron ore pit where the slope angle is controlled by foliation regardless of the pore pressure conditions

Figure 5.17: Strong foliation in the West wall of the Piment pit at Tasiast, Mauritania (Courtesy Kinross Gold Corporation)

Figure 5.18: Example of weak materials and water pressure in the upper slope overlying the strong rocks in the main slope below (Courtesy Newmont Mining Corporation)

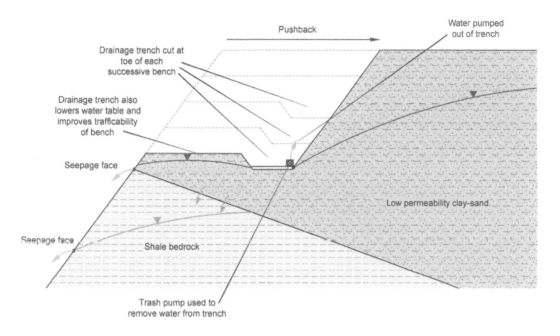

Figure 5.19: Use of a drainage trench to improve access and reduce pore pressures in the slope (East wall of the Pinson Mag pit, Nevada, USA)

Figure 5.20: Photograph showing the construction of an in-pit drainage trench to help depressurise the slope (East wall of the Pinson Mag pit, Nevada, USA)

When planning and implementing a horizontal drain program, it is important that sufficient information is available to allow the hydrogeological conditions behind the slope to be characterised. This allows the drain program to be properly focused.

Compared to other commonly used drainage methods, some typical advantages of horizontal drains are as follows.

- They can often be conveniently installed from inside the pit to target zones of elevated pore pressure behind the slope.
- They can be installed concurrently with mining, as the benches are cut, and as the pit is progressively deepened.
- They flow passively and do not require pumping. Therefore, they often require minimal maintenance.
- They are the most efficient way to dissipate pore pressures in a rock mass that is compartmentalised by steeply dipping structures.
- They are the most efficient way to depressurise poorly permeable clay-alteration zones that occur behind the pit wall.
- They are relatively low cost to install (compared with vertical dewatering wells).
- They have minimal impact on mining (compared with in-pit vertical dewatering wells).

However, the use of horizontal drains may have some disadvantages, the most common of which are as follows.

- The drains are often inefficient, because the first portion of the drain is usually drilled through unsaturated material.
- The drains cannot lower the potentiometric surface below their collar elevation. Therefore, they cannot be used to achieve advanced depressurisation prior to excavation. They can only be installed after the slope has been cut.
- If the water flowing from the drains is not captured, piped and reticulated, it may infiltrate into the catch bench below the drain collar and enter the over-break zone, cascading downward above or behind the slope.
- Horizontal drains typically become cut and lost during a pushback of the slope. If this occurs while they are still flowing, they may subsequently feed water into the rock near the slope face.
- Installation can be difficult, particularly in soft material where the borehole caves or plugs during drilling. As a result, the planned hole length may not be achieved, which reduces the effective drainage distance behind the slope.

5.2.3.2 Determining targets for horizontal drains

It is important to develop specific targets for the horizontal drain program. Typically, there are four situations where horizontal drains may be required.

1. To remove water trapped in permeable but compartmentalised fractures behind the pit slope (the 'A' drains in Figure 5.21).
2. To depressurise water contained in low-permeability material. This may be rock that has become altered to a clay, such that it exhibits dominantly porous medium flow characteristics (the 'B' drains in Figure 5.21).
3. To remove water 'dammed' behind geological structures (the 'C' drains in Figure 5.22).
4. To dewater new geological units that will be encountered in pushbacks as the pit is expanded (the 'D' drains in Figure 5.23).

Once the hydrogeological conditions behind the slope are understood, the targets for the drain program can be determined, and the design of the drains can be optimised.

Water trapped in permeable but compartmentalised fractures ('A' drains)

Figure 5.21 shows that a relatively small number of long drains are needed to dewater permeable fractures that contain compartmentalised water in unaltered rock greater than 175 m behind the slope. For these drains, intersection of permeable fractures is more important than the actual drain spacing.

In fractured rock settings, the same principles apply as for any other type of drilling, and drain yields may be 'hit and miss' because of the inherent variability in fracture locations. If open fractures are encountered, the yield of the drains is good. If the adjacent hole misses the fractures, there may be virtually no yield. It is common to have a high-yielding drain adjacent to a low-yielding drain.

For fracture-controlled groundwater, prediction of the drain yield in advance of drilling is often not possible. Therefore, when planning the program, there is a need to allow for redundancy in the system. For 'A' drains, installing the drains at an azimuth may be advantageous to intersect the maximum number of open joints and fractures, based on the structural geology model.

Water contained in low-permeability clay-altered material ('B' drains)

Figure 5.21 shows that a greater number of shorter drains are needed to depressurise the low-permeability clay-altered material close to the slope. More consistent (but low) drain yields are to be expected because of the porous-medium nature of the flow system. The drain spacing is important and is dependent on the permeability of the material.

Water trapped behind compartmentalising structures ('C' drains)

The purpose of the 'C' drains on Figure 5.22 is to release water that is 'dammed' by structures that are oriented sub-parallel to the slope. The number of drains required

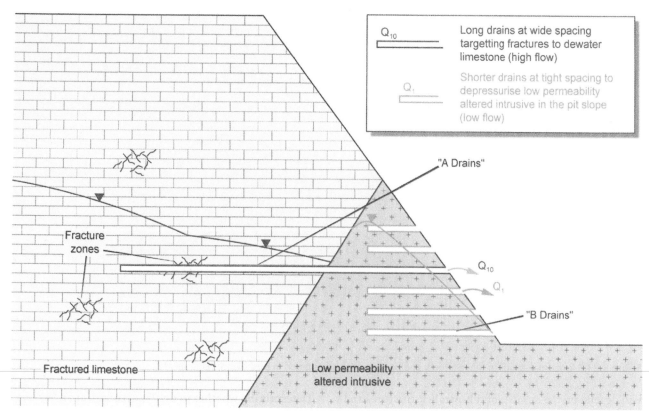

Figure 5.21: Horizontal drains to remove water in compartmentalised fracture zones ('A drains') and to dissipate pressure in poorly permeable rock ('B drains')

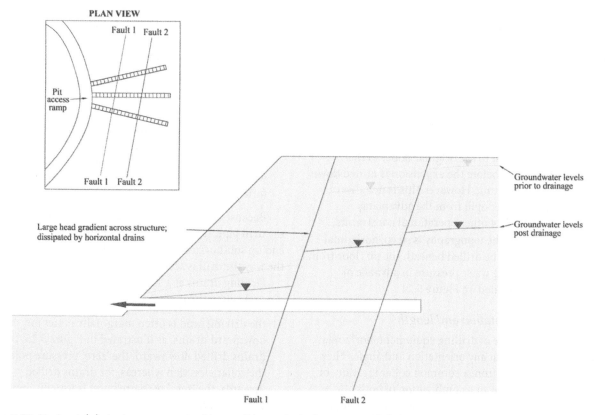

Figure 5.22: Horizontal drains to remove water 'dammed' by geological structures ('C drains')

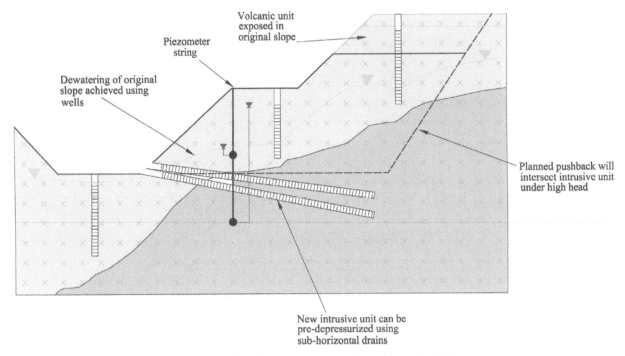

Figure 5.23: Horizontal drains to dewater new geological units encountered in pushbacks ('D drains')

will depend on the number of sub-parallel structures and the nature of the fracturing associated with those structures. Again, the intersection of permeable fractures is more important than the actual drain spacing, and significant variability in yields is to be expected because of the nature of the fracture-controlled groundwater system.

Water within new geological units that will be encountered in pushbacks ('D' drains)

Figure 5.23 shows the situation where new geological units are encountered because of a pit expansion. The purpose of the 'D' drains is to provide advanced dewatering of these units. It is often best if the drains can be installed from the lowest possible location in the original slope, so that dewatering can occur before the expansion is mined down to the level of the new unit. However, this is not always possible because of over-spill from the advancing pushback, or because of other operational constraints.

In settings where the topography is steep, horizontal drains can sometimes be drilled beneath the pit floor from downslope to lower the water pressure in advance of mining. This is illustrated on Figure 5.24.

5.2.3.3 Drain orientation and length

Depending on the type of drilling equipment, horizontal drains may be drilled at any orientation and angle. They may be drilled in fans from a common collar location, or linearly along the slope, or a combination of both.

Once the target for the drains has been identified, the following decisions may have to be made regarding the layout of the drain program.

Determining whether to drill horizontal, at an 'up' angle, or a 'down' angle

The primary factor for the selection of drain angle should be how best to intersect the defined target zones. Once the target has been defined, there may be advantages for varying the drain angle, as illustrated in Figure 5.25.

Some common advantages of drilling at a slight upward angle (say 5° to 10° up) are:

- for drains drilled upwards, gravity may facilitate clearance of any collapse or blockage along the drain length (note that this is not applicable if a perforated liner is installed along the length of the drain);
- because there is usually 'non-full' flow within the drain, there may be marginally less tendency for the water to flow out of the drain into shallow fractures (often in the over-break zone).

Because the drains often 'droop' due to the weight of the drill pipe, drains drilled at an upward angle will often end up sub-horizontal or slightly downward, regardless of the way the drill is set up.

Drilling drains at a slight down angle (say 5° to 10° down) is often the preferred approach because:

- the driving head is often marginally better for downward drains, as illustrated in Figure 5.25. For drains drilled downward, the 'zero' pressure point is the collar elevation whereas, for drains drilled upwards, the 'zero' pressure point is usually above the collar elevation;
- drains drilled at a down angle will remain full of water. Therefore, chemical oxidation and 'clogging' of the

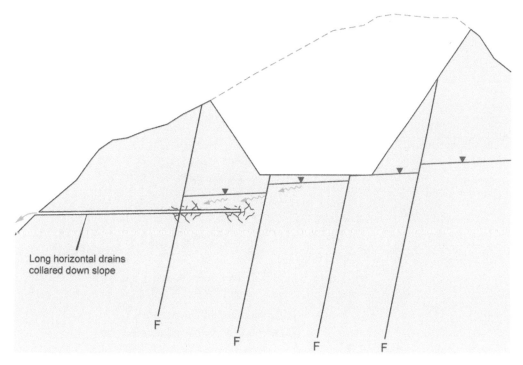

Figure 5.24: Horizontal drains drilled in a topographically steep area

inflow zones or perforations is often less of a concern than for drains drilled upwards;

- it is usually easier to pressure cement the collar casing into competent bedrock, or to the base of the over-break zone, when drilling at a slight downward angle;
- it is easier to drill at a downward angle and also install the drain pipe within the borehole compared with upward drilled holes;
- drains drilled at a down angle facilitate the running of a temperature probe which may have the benefit of allowing groundwater ingress zones in the hole to be identified, if this hasn't been identified during the drilling process.

Note that the discussion in this section only pertains to 'horizontal' drains. A discussion of vertical and angle drain drilling is contained in Section 5.2.4.

Determining the required length of the drains
There are no specific rules governing the length of the drains. Again, the most important consideration is the knowledge of the hydrogeology behind the slope and the target zone where pressure relief is required. This is illustrated in Figure 5.21 where, for the clay-altered zone immediately behind the slope, a drain length of 150 m to 175 m was sufficient, whereas the fractured rock zone deeper in the slope required drains of at least 300 m in length.

If there are no specific target zones identified for the program, it is usually better to drill the drains as long as possible because:

- longer drains typically have a lower proportion of unsaturated or 'dead' wellbore;
- longer drains typically encounter a greater driving head and a higher number of fracture intersections, particularly where the fracture zones are sub-vertical.

However, in situations where groundwater is compartmentalised in pit slopes, care must be taken when planning drains if the deeper compartmentalised water is not adversely impacting slope stability. It may be best to avoid bringing this water close to the high wall via drain holes.

A common issue with the drilling of horizontal drains is that the first portion of the wellbore is often unsaturated, as illustrated in Figure 5.25. The greater the unsaturated zone behind the collar, the greater the 'wasted' drilling meterage. In addition, there may be a tendency for water to flow out of the drain into fractures within the unsaturated portion of the bore.

Furthermore, as the phreatic surface becomes lowered by the drains, the length of the unsaturated or 'dead' portion of the wellbore increases and the overall drain efficiency becomes further reduced. If drains are drilled longer, the proportion of unsaturated or 'dead' wellbore is less. All other things being equal, a good rule-of-thumb guideline is to keep initial unsaturated or 'dead' zone less than 25 per cent of the overall drain length.

Figure 5.25 also illustrates that longer drains may typically have a greater driving head than short drains. As the drain is drilled deeper behind the slope, it will normally encounter a progressively increasing head.

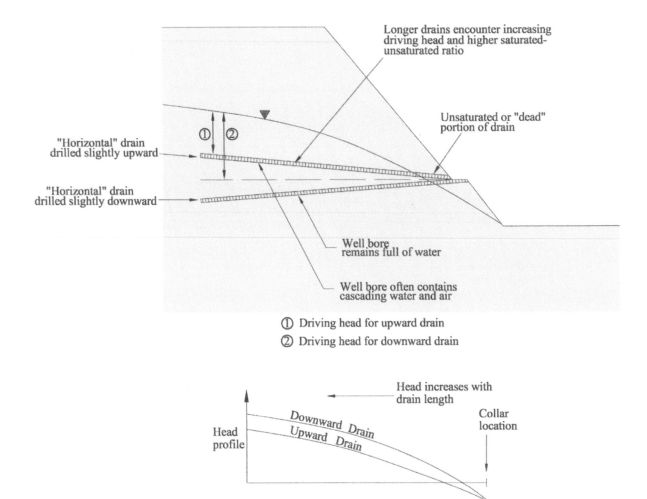

Figure 5.25: Comparison of upward and downward 'horizontal' drains

Therefore, the efficiency of the drain usually increases with greater drain length.

The depth capability of most commonly used horizontal drills is within the range 300 m to 500 m, depending on the actual drilling conditions, which primarily relate to the competency of the materials encountered by drilling and the pressures encountered in the hole.

Selecting the drain orientation

Again, the main guideline for drain orientation is the knowledge of the hydrogeology behind the slope and the target zone where pressure relief is required. However, there are often two key factors to consider:

- drains drilled perpendicular to the main structural orientation often have the most potential for intersecting the largest numbers of water-bearing fractures;
- drains drilled directly into the slope (perpendicular to the strike of the slope) may penetrate less of an unsaturated zone and may encounter a higher overall 'driving' head.

Orienting the initial drains sub-perpendicular to the main water-bearing structures is often beneficial. Once a number of drains have been installed at differing azimuths, the results can be evaluated to determine if there is a relationship between drain orientation and flow rate.

Figure 5.26 shows that drains drilled from individual collar locations spaced laterally along the slope may be more efficient than fan drilling from a single collar location. However, where access to the slope is constrained, the locations for drill collars can be limited, so a fan configuration may become necessary. Collection and piping of the produced water can also sometimes be easier for fans.

5.2.3.4 Sequencing the drain drilling program

Often the most challenging part of the horizontal drilling program is sequencing the drain installation within normal mine operations. Drains can often be installed on step-out benches or inactive haul ramps. With some advance planning, it may be possible to design dedicated dewatering benches which can be built into the slope

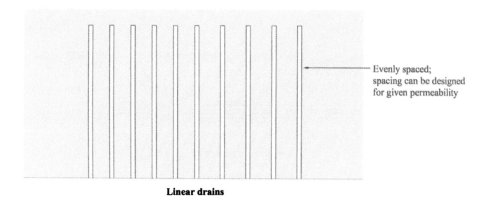

Linear drains

Evenly spaced; spacing can be designed for given permeability

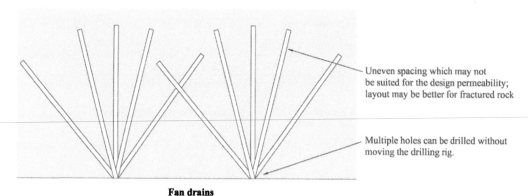

Uneven spacing which may not be suited for the design permeability; layout may be better for fractured rock

Multiple holes can be drilled without moving the drilling rig.

Fan drains

Figure 5.26: Comparison of 'fan' drilling and 'longitudinal' drilling

design. These allow the safe installation of the drains. They also facilitate maintenance of the drain reticulation systems and continued monitoring of the flow rates after the drains have been completed.

For slope sectors where depressurisation of poorly permeable clay-alteration zones is necessary, and where more drains at a closer spacing are required, installation from the pit can be difficult to manage. It may become necessary to sequence the horizontal drain program within the drill and blast operations, and to adopt similar steps to those outlined in the following sequence.

Step 1: Take the production blast on the new bench.
Step 2: Muck out the material from the blast.
Step 3: Scale the newly excavated face.
Step 4: Install the drain holes in the newly excavated face.
Step 5: Install reticulation lines for the drain holes.
Step 6: Carry out production drilling for the next bench.

Steps 4, 5 and 6 can often be carried out simultaneously. For slopes where a trim blast is required, it may be possible to continue drain hole drilling during and after the trim blast for the next bench.

Access for drain installation may also be constrained by the poor condition of the bench faces, the slope as a whole, or by other safety concerns. This may require drain designs, angles, orientations and lengths to become modified from the ideal design.

5.2.3.5 Design of the drain hole
Hole diameter
Many construction methods are successfully used for horizontal drains, but a typical drilling diameter is 75 mm to 125 mm. This usually allows for 25 mm to 50 mm diameter perforated liner pipe to be installed in the drains, if required. The drains may be installed using a diamond drill by coring methods, but are more typically installed using conventional tricone or down-the-hole hammer drilling methods with air or water flush.

Collar casing
Figure 5.27 shows a typical simple design for a horizontal drain. Collar casing is frequently installed to depths ranging from 2 m up to about 6 m. In some cases, it may be advisable to install the collar casing deep enough to penetrate the over-break zone in the slope. This will minimise the risk of water seeping from the completed drain into the over-break zone. If the collar casing is pressure cemented, it allows the drain to be 'shut in' if desired.

a) Simple design

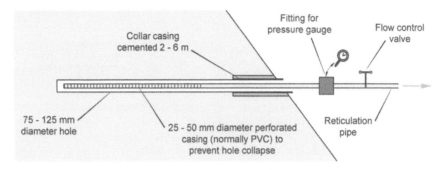

b) With packer installed to isolate flow

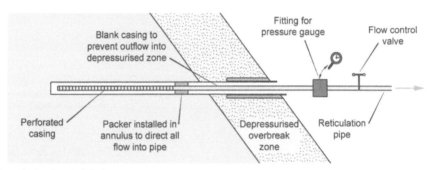

Figure 5.27: Typical design of a horizontal drain

Liner pipe
It is often advisable to install perforated liner pipe within the drain hole in order to minimise the potential for subsequent blockage due to collapse of the hole. The installation of liner pipe is particularly important where discrete collapse zones were identified during drilling, where clays or other squeezing materials are encountered, or if ongoing blasting will be required in the vicinity of the drain collar. The purpose of the perforated liner pipe is to prevent blockage of the hole. The installation of any gravel pack or formation stabiliser in the annulus is not practical in virtually all 'everyday' applications.

For drains that are drilled in weak ground, it may become necessary to adapt the drilling method to allow the perforated liner pipe to be installed inside the drill rods, and to remove the drill pipe while leaving the perforated liner in place.

Use of packers
As shown in Figure 5.28, a packer can be used to limit the amount of water flowing in the annulus and to prevent it from recharging depressurised fractures at a shallower depth in the hole. In these cases, a packer is set around the liner pipe in front of the target water-bearing zone. The packer forces the water to enter the screened section of liner pipe, and therefore prevents the water from flowing along the annular space around the casing and back into the formation at a shallower depth in the hole.

Similar packer systems can be used for horizontal drains installed in cold weather climates to prevent water pressure building up in the drain annulus immediately behind the drain collar. If the drain collar freezes, the pressurised water is retained within the liner pipe. The water in the annulus remains at low pressure. As a result, there is no pressure transferred to the materials immediately behind the slope.

5.2.3.6 Drilling procedures
Equipment used
Historically, horizontal drain holes have been drilled using wireline core equipment, underground drilling machines, rotary drilling machines with the mast in the down position and small 'utility' blast hole drills. However, drilling with 'modified' equipment may lead to a program which is depth-limited or too slow. Safety procedures also need to be determined for the type of equipment being used and location of drilling platform with respect to identified areas of high geotechnical risk (rock fall). It often takes time for the driller to adapt to site-specific conditions, and there is a learning process at most sites. As a result, the unit costs for small horizontal drain programs are typically much higher than those for an 'established' contractor at a particular site.

When it becomes clear that a significant drain drilling program will be required, obtaining a purpose-built or purpose-adapted horizontal drilling machine is

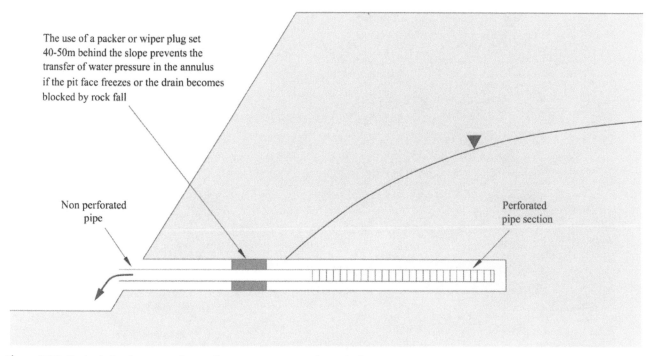

The use of a packer or wiper plug set 40-50m behind the slope prevents the transfer of water pressure in the annulus if the pit face freezes or the drain becomes blocked by rock fall

Non perforated pipe

Perforated pipe section

Figure 5.28: Drain design for preventing outflow or pressure transfer to shallower depths

normally of benefit. Many drilling companies operate dedicated horizontal drills, such as those illustrated in Figures 5.29 and 5.30. Most dedicated machines are capable of installing drains to depths of 300 m or more and often up to 500 m. Once the 'learning-process' has been completed and site-specific procedures have been developed, a dedicated machine with an experienced crew is often able to install a 300+ m long drain hole in a single shift.

Measurements while drilling
The three key pieces of information to obtain during drilling of horizontal drains are:

1. the geological log of the drill hole;
2. the penetration rate for each drill pipe;
3. the flow rate from the hole, measured at the end of each drill pipe.

The flow rate can be measured simply by recording the time taken to fill a container placed below the collar casing. In situations where access to the face is more limited, the flow can often be channelled through a weir system away from the face.

As for any hydrogeology field program, the measurements taken while drilling will form a necessary part of the overall investigation. The ongoing program should be interactively modified, based on the results obtained. Many operations now import their drain hole results into the 3D geology block model so that 'jumps' in flow rate during drilling of the drain can be correlated with geological structures in the block model.

Deviation surveys
Deviation surveys of horizontal drains are normal practice to allow the actual alignment of the drain holes to be determined and to facilitate the calculation of hydrostatic heads and correlation of water strikes between holes and with known geological structures.

Deviation of horizontal drain holes is often quite severe; one degree of deviation per 30 m to 40 m of hole may occur with wireline equipment, and even larger deviations can occur if down-the-hole hammers are used. The amount of deviation will clearly depend on the dip of the formations penetrated and the orientation of the prominent geological structures. Deviation is generally more exaggerated for holes drilled upward from the horizontal.

Survey tools determine borehole inclination and azimuth using accelerometers, gravimetric strain gauges or gyroscopes. Survey instrument diameters range from 30 mm to 40 mm to fit most drill rods. Directional or deviation surveys can be run using conventional survey tools conveyed through the drill rods using 13 mm continuous fibreglass duct rod (Figure 5.31).

Inside the drill rods, the survey tools can be gravimetric (using accelerometers and/or strain gauges to determine inclination and azimuth) or gyroscopic, for example Devico-Devitool or Deviflex Icefield tools M13 multishot or Gyro shot. In non-magnetic formations, open boreholes can be surveyed with magnetic tools. In water-filled holes, the vertical deviation can be approximated by measuring water pressure with a small memory pressure sensor conveyed to total depth with 6.7 mm diameter fibreglass duct rod.

Figure 5.29: Example horizontal drilling operations (Source: Read & Stacey, 2009)

Figure 5.30: Horizontal drilling operations with smaller equipment (Courtesy Aitik Mine, Sweden)

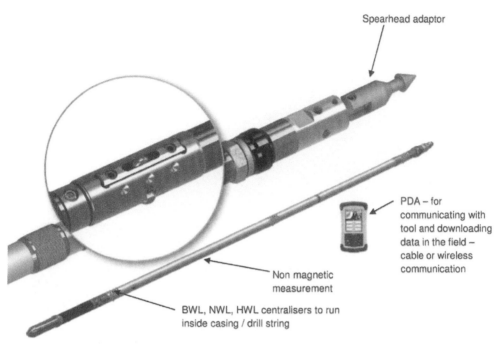

Spearhead adaptor

PDA – for
communicating with
tool and downloading
data in the field –
cable or wireless
communication

Non magnetic
measurement

BWL, NWL, HWL centralisers to run
inside casing / drill string

Figure 5.31: Deviation survey equipment typically used for horizontal drain holes (Courtesy Devico)

The most efficient method is to trip the survey tool to total depth inside the drill rods. Tools can be conveyed through the drill rods using PVC or steel pipe, or more quickly using continuous fibreglass duct rod. There are continuous fibreglass rods (6.7, 10 or 13 mm diameter) that are made for pushing a cable or rope into electrical ducts or pulling wires. Hydraulic-powered duct rod pushers are also available for the 13 mm rods. Vertical borehole rodding machines have recently been tested it at the Cameco Cigar Lake Mine for underground logging of vertical-up freeze holes, up to 78 m straight up. Spools of 6.7 mm flex-rod are also available for tripping either deviation tools or vibrating wire piezometers into NQ exploration boreholes. The rods are continuous and a rapid installation can be achieved. This methodology has advantages over a sacrificial tremie pipe because holes smaller than HQ can be instrumented.

Single shot surveys can also be taken say at 30 m to 50 m intervals, either during drilling or continuously on completion of the hole. The normal procedure is to pump the tool into the hole using a sub that is manufactured to fit inside the particular drill pipe being used. Difficulties can occur pumping the deviation survey tools in the hole when high flows and heads are experienced in slim-hole conditions and, in many cases, single-shot survey tools have now been replaced by multi-shot or continuous electronic instruments (which typically record borehole data in five second intervals). The data are directly uploaded to a field computer after the tool is removed from the hole and are subsequently processed to determine borehole trajectory.

Safety considerations

The safety procedures employed for any given horizontal drain program will depend on the requirements of the mining operator, and also those of the drilling contractor. Important considerations include the following.

- **Keeping the drill at a safe distance from the pit face.** While the 'allowed distance' will vary with site conditions, it is normally good practice to employ a minimum distance of at least 3 m. Figures 5.29 and 5.32 show examples of the drilling operation located remotely from the slope. Figure 5.29 shows the safety berm at the toe of the slope and the protective mesh cover placed above the collar casing.

- **Optimising the condition of the benches above the drill collar.** It is good practice to clean and scale the benches above known future locations for horizontal drilling, and to place rock-fall control berms on the outside crest of the benches above the drill area. Field crews should routinely examine the bench faces above the drilling location for unstable rock. If a night shift is operated, the benches above the working area should be illuminated. If the overlying bench faces are clearly unstable, the installation of mesh on the faces above the drill area and/or the use a spotter may be required. Such onerous procedures may make fan drilling from a limited number of collar locations more attractive than drilling holes linearly along the face.

- **Ensuring that collar casing is correctly installed.** It may be necessary to ensure the surface collar casing is installed down to the depth of competent rock and

Figure 5.32: Safety 'gap' between the pit face and the horizontal drill (Courtesy Yan Adriansyah)

grouted so that no washouts occur. If washouts were to occur, the pit wall may become destabilised, and control of the drilling operation may be lost. Grouting of the collar casing is particularly important if high flow and high head conditions are anticipated. Cementing the surface casing in the hole is easier if the hole-angle is below horizontal, as this negates the need for a packer in the annulus for cementing. Shut-in tests can be used to provide supplementary monitoring information in holes where collar casing is installed.

- **Blow out prevention when drilling into zones of high pressure.** If zones of high pressure during drilling are anticipated, special procedures may be necessary to ensure blow-back of the drill pipe does not occur. These may include a pressure-controlled blow out system around the drill pipe, or a packer system around the drill pipe that can be chained to the collar casing to prevent blow-back. Under extreme cases, where high flows in conjunction with high pressures are possible, consideration may need to be given to securing the drilling machine to anchors set in the face.

5.2.3.7 Monitoring of horizontal drains

Monitoring of the pore pressures during drain installation

The best way to monitor the effectiveness of the horizontal drain program is an array of piezometers installed behind the slope. The piezometer array should ideally be installed prior to the drain drilling program. This will enable the magnitude and distribution of the drain responses to be quantified throughout the slope domain and will allow the resulting pore pressure profile to be quantified.

Horizontal drains that are creating depressurisation should also exhibit a steady decline in their individual and combined flow rates. This occurs because the drains lower the water pressure, and therefore also lower the driving head that forces the water into the hole. Where horizontal drains exhibit consistent or steady-state flows, it usually means there is recharge occurring within the area of influence of the drains. If this occurs, the amount of depressurisation that can be achieved is often reduced. Figure 5.33 shows some typical response curves.

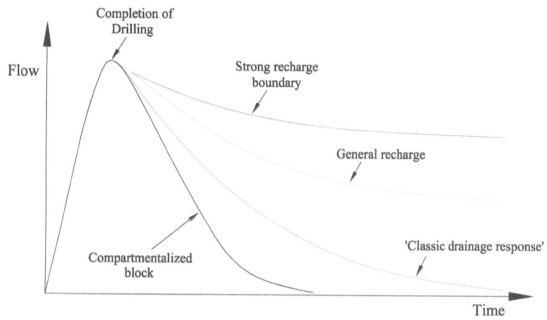

Figure 5.33: Common types of horizontal drain responses

For horizontal drains that can be periodically shut in, there will also be progressively declining 'static' pressure measurements at the collar, as water pressures behind the slope are lowered. However, the 'static' shut-in pressure measured at the drain collar is a reflection of average pressure along the drain trace. The shut-in pressure may not be a true reflection of conditions behind the slope if there is some leakage out of the drain at shallow levels in the slope (or at some other location along the drain trace).

The best way to test the effectiveness of horizontal drains is to carry out a pilot field trial (Section 2.2.10). The procedure is to install one to two strings of multi-level vibrating wire piezometers at a convenient location behind the slope, and to target an array of about four to six trial drains towards the lowest sensors of the piezometer string. The results of the pilot test can be used to help design of the full-scale drain program.

A useful procedure to assess the performance of individual drain arrays as they are drilled is as follows.

- Use the first-drilled drain to make pressure observations during the drilling of subsequent drains by carrying out the following procedure.
  a. Install and grout a suitable length of collar casing in all drains.
  b. Install a pressure sensor in the initial drain and shut in the drain at the collar.
  c. Drill the other drains in the array, monitoring the pressure in the first drain.
  d. After five days monitoring following completion of the array, remove the sensor and open the first drain.

- During the drilling of each drain, the flow should be measured regularly with depth during drilling. The goal would be to correlate 'jumps' in the flow rate with known structural zones or lithological changes.
- Ideally, the flow rate from each drain should be monitored for as long as access is available after the drain has been drilled. Ideally, the schedule for measuring the drain hole flows should be every second day for seven days, twice-weekly for three weeks, then monthly for as long as possible.

Figure 5.34 shows an example of a pit slope where the installation of drains would have been possible but where the high strength of the rock provided an acceptable slope performance without their use. The permeable nature of the over-break zone is illustrated in this photograph with the seepage infiltrating within about a metre or so of the toe of each bench and moving downward to the bench face below. The photograph illustrates that, in strong materials, the bench faces and the benches themselves can be maintained in good condition despite the seepage, with minimal crest loss and very little loss of catchment on the benches.

Target flow rates from the drains

The amount of water removed by a drain array will depend on:

1. the permeability and degree of compartmentalisation of the groundwater behind the slope;
2. the drainable porosity of the materials encountered;
3. the amount of water head above the trace of the drains;
4. the amount of recharge entering the slope domain within the area of influence of the drains.

Little crest
damage

Good catchment
on benches

Figure 5.34: Example of seepage in strong rock with little impact on slope performance

Successful depressurisation results have been achieved with flow rates ranging from 0.1 to 10 l/s in individual drains and 0.5 to 100 l/s for the drainage program as a whole.

Most hard rock settings exhibit a low drainable porosity; both in the fresh rock itself, and in zones of clay alteration and weathering. In weak materials and clay-alteration zones, the total porosity may be significantly increased, but the drainable component of the porosity is often still relatively small (Section 1.1.1). Individual drain flows in these poorly permeable environments are typically low (less than 0.2 l/s). Provided there is no significant recharge reaching the slope domain, total combined flows of about 1 to 2 l/s magnitude can create a significant amount of depressurisation within a period of several months.

The most important factor controlling the required drain flows is the amount of recharge entering the area of drain influence. A good example in the case studies of the importance of recharge is the comparison between the granite in the North-west wall of the Diavik A154 pit and the granodiorite in the West wall of the Chuquicamata pit in Chile. Both lithologies are similar and have comparable hydraulic properties, but vastly different results were observed, as follows:

- For the North-west wall of the Diavik A154 pit, Lac de Gras provides a constant-head source of water at the crest of the wall. Water from the lake can readily enter the fracture systems outside the crest of the slope. The horizontal drain program was not able to intersect fractures behind the wall that would allow water to be removed at a faster rate than recharge could enter the slope domain. Combined drain flows were about 5 to 10 l/s and they produced 5 m to 10 m of local-scale depressurisation, following which steady-state conditions became re-established after a period of seven days or less. Following an initial drop in the flow rate by about 20 per cent to 30 per cent, the drain flows exhibited steady-state conditions.

- In the West wall of the Chuquicamata pit, recharge to the granodiorite that comprises much of the slope is minimal. Drain holes with initial individual flow rates an order of magnitude lower than those at Diavik (about 0.5 l/s) produced about 400 m of depressurisation within a time period of a few days. The drain flows declined rapidly to near-zero as the depressurisation occurred.

5.2.4 Vertical and steep-angled drains

5.2.4.1 General discussion

Vertical or steep-angled gravity (non-pumping) drain holes are also commonly installed around mine sites in situations where vertical hydraulic gradients and/or 'perched' groundwater zones occur. For these systems to function, there needs to be a zone of already depressurised and usually permeable materials into which the lower part of the drains can discharge. In some cases, underground workings or tunnels can be used as the lower, pre-depressurised zone.

The purpose of the vertical drains is to cross-connect the upper high-head formation with underlying low-head formation. Water will flow into the drains from the upper zone, flow down the drain-bore, and will flow out of the drains into the underlying depressurised, more permeable zone below, which is typically being (or has been) dewatered by another method. The angle and orientation at which the drains are installed depends on the nature and geometry of the hydrogeological units that make up the pit slope (Figure 5.6). The drains may be installed at an angle such as to increase the probability of encountering principal structures or permeable fracture zones, or to provide better cross-connection of perched groundwater zones. As for any type of depressurisation system, the target zone for the drains should first be defined.

Figure 5.35 shows a field situation where the low-permeability materials in the slope are being depressurised using a combination of horizontal and vertical drains and low-yielding pumping wells. The upper part of the slope is excavated in tailing that are partially depressurised using

vertical drains but require excavation at a flat angle because of the weak material and residual water pressure.

5.2.4.2   Examples of the use of vertical and steep angle drains

Figure 5.36 illustrates the use of vertical drains to prevent a zone of perched water from causing saturation in the slope. The use of angled drains drilled inward to the slope is illustrated in Figure 5.6. The drain holes were installed at steep angles to cross-connect the upper low-permeability zone and the adjacent pre-depressurised permeable zones. The target zone for the drains was first defined, following which the appropriate drilling angle was then determined.

Vertical or steep angle drains can be installed with any type of conventional drilling equipment. A typical construction method involves holes drilled at about 100 mm to 150 mm diameter. The RC drilling method is particularly suitable for installation of vertical drains, firstly because of its speed and secondly because it allows airlift flow and water level measurements to be made while drilling the hole.

The drains can have 25 mm to 50 mm diameter perforated liner pipe installed, as required. Occasionally, when dealing with hard rock formations that do not contain any significant clay-altered zones, it is possible to

Figure 5.35: Illustration of weak tailing materials cut at low angle in the upper slope with water intercepted by vertical drains (Courtesy Newmont Mining Corporation)

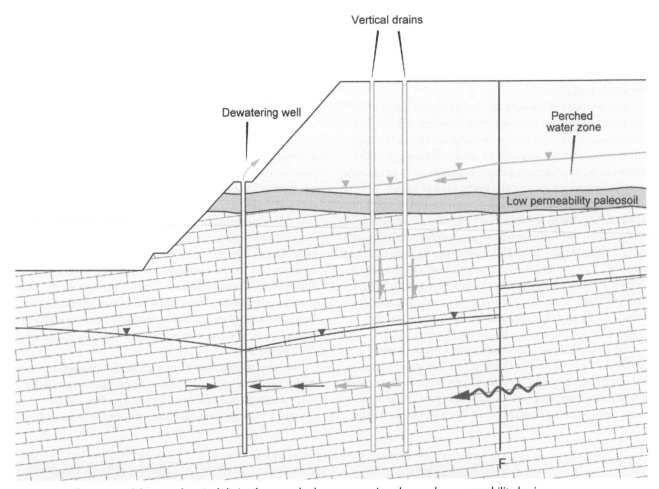

Figure 5.36: Illustration of the use of vertical drains for a perched zone occurring above a low-permeability horizon

leave the drains as 'open-hole' completions with no liner pipe. At many mine sites, holes drilled for the mineral exploration program remain open and have allowed improved drainage of the ore body (although there are several case histories of old exploration holes cross-connecting permeable features and creating undesirable inflows).

In unconsolidated materials, a sand or gravel pack is typically installed in the annulus between the pipe and the borehole wall. In some instances, where flow rates are relatively small, the hole can be backfilled with permeable gravel to provide a permeable conduit, with no casing installed (Figure 5.37). However, some operators have experienced a declining efficiency due to plugging of the gravel by fines. In rare cases, when depressurising unconsolidated sedimentary formations, it may be necessary to consider centralising the pipe and installing a gravel pack.

Where the upper part of the drain may become destroyed in the future by mining of a pushback, it is possible to 'back-off' the pipe at some point down the hole, to allow the upper section of pipe within the zone of future

mining to be removed from the hole. This is illustrated in Figure 5.38. The 'backing-off' of the pipe is normally achieved using a left hand coupling or by simply leaving a loose joint and unscrewing it.

The best way to test the effectiveness of angled or vertical drains is to carry out a pilot field trial (Section 2.2.10). The procedure is to install one to two strings of multi-level vibrating wire piezometers within the upper and lower units and to target an array of about four to six

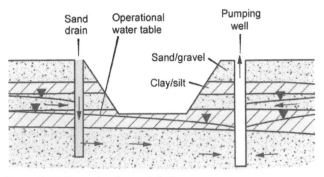

Figure 5.37: Example of a sand drain

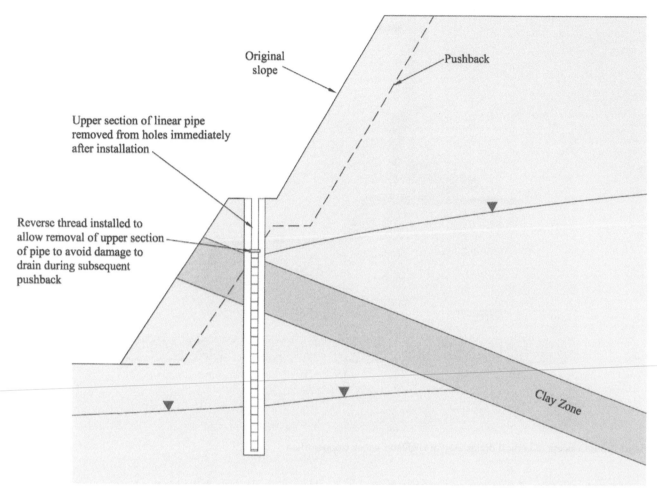

Figure 5.38: Example of removing the upper section of liner in a vertical drain hole

trial vertical drains across the intervening low-permeability unit. The piezometric response of the drain drilling can be readily monitored, and the results of the pilot test can used to help plan and design the full-scale drain program.

5.2.4.3 Example of an integrated program illustrating the use of vertical drains and vacuum wells

A good example of the integration of vertical drains into the dewatering and slope depressurisation program can be found at the Highland Valley Copper Mine (HVC) in British Colombia, Canada. The East wall of the Valley pit has been excavated through approximately 200 m of overburden as part of a pushback. The overburden includes three aquifers: the Upper Aquifer; the Main Aquifer; and the Basal Aquifer. The Upper Aquifer and the Main Aquifer are separated by a glacial till sequence and the Main Aquifer is separated from the Basal Aquifer by a lacustrine sequence with sand seams and layers within a silt and clay deposit. The three aquifers had been pumped

for an extended time which had substantially reduced piezometric pressures. However, it became necessary to further depressurise the lacustrine deposits so that the overburden slopes could be safely steepened in the pushback, thereby reducing the overall mining cost. Depressurisation of the lacustrine deposits required multiple methods.

1. Additional pumping wells installed in fan material adjacent to/and co-depositional with the lacustrine silts and clays as well as in the Main and Basal Aquifers.

2. Wells with a vacuum assist were selected as a primary means of achieving adequate and timely pore pressure reduction within the lacustrine silts and clays. These wells were constructed in 300 mm diameter boreholes with 150 mm diameter stainless steel continuous slot screen paired with a 50 mm slotted PVC screen. Most of the hole was backfilled with a filter pack, but the top of the hole was sealed with cement plug. Testing of the wells showed that steady-state drawdown was achieved in the horizons of concern in a relatively short period

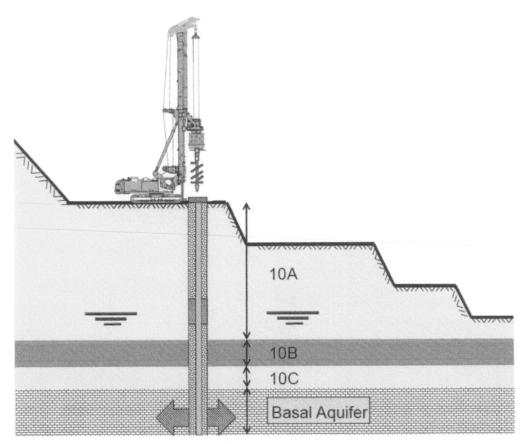

Figure 5.39: Concept of vertical drains used at Highland Valley Copper Mine

of time but the use of a vacuum system effectively increased the drawdown achieved with each well, both above and below the water level in the well. The vacuum was applied to the 50 mm PVC riser and a pump was installed in the hole to pump the collected water. Following excavation, the well materials were recovered.

3. Passive vertical drains (PVDs) were included near the toe of the excavation to help drain the weaker silts and clays into the underlying Basal Aquifer that, under the pumped condition, was not saturated at its top. The concept of the drains is shown in Figure 5.39, the design in Figure 5.40 and their installation in Figure 5.41.

4. Horizontal drains were installed into the overburden slopes from upper bedrock benches to supplement other methods, where required.

Although the vacuum assisted wells were able to provide adequate depressurisation, they would not be accessible at the toe of the excavation during the critical period. The PVDs were installed to drain the weaker silts and clays into the Basal Aquifer during this period. In addition, the vacuum assisted wells were susceptible to ground movement, especially near the toe of the overburden slope. The PVDs were constructed within a

large-diameter hole drilled with a bucket auger (Figure 5.41), making them less susceptible to ground movement. The completed wells were airlifted, pumped and flushed out, and treated with sodium hypochlorite to improve their performance. The PVDs were generally successful. However, their performance varied depending on the transmissivity of the draining soils and the Basal Aquifer as well as the effectiveness of the post installation development activities.

Successful completion of the project required careful planning so that wells were re-commissioned as soon as possible after each bench excavation was complete.

5.2.5 Design and installation of pumping wells

5.2.5.1 Introduction

Groundwater pumping wells are installed all over the world for many purposes. There are many reference texts providing guidelines and procedures for drilling, installation and development of pumping wells. For practical mining applications, the most common references are as follows.

- *Groundwater and Wells* (3rd edition) Johnson Screens (2008).

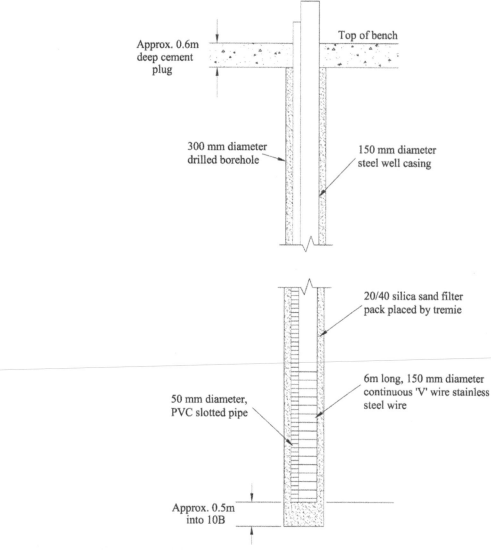

Figure 5.40: Detailed design of Highland Valley Copper vertical drains

- *Handbook of Ground Water Development* published by the Roscoe Moss Company (1990).
- *Drilling* by the Australian Drilling Industry Training Committee Ltd (1997).

Most general mine dewatering systems employ some type of dewatering wells. These may be interceptor wells, ex-pit wells or in-pit wells. The aim of dewatering wells is usually to achieve the maximum possible drawdown in the surrounding formation for a minimum pumping rate, which is typically the opposite of water supply applications.

There are many situations were production wells are installed specifically for slope depressurisation purposes, usually in lower-permeability or isolated sectors of the pit where the dewatering influence is not sufficiently rapid. Production wells installed specifically for slope

depressurisation often have lower target yields than wells installed for general mine dewatering.

The discussion in the following section applies mostly to wells drilled in fractured rock settings. For wells drilled in intergranular flow settings, the selection processes for screen and filter-pack selection are generally more straightforward, and are well documented in the above texts.

5.2.5.2 Constraints for production wells in fractured rock settings

In many fractured rock settings, much of the groundwater inflow to a production well can occur in a limited number of discrete fracture zones (sometimes just one zone). Therefore, in order to provide a sustained pumping rate, the water must flow through the fracture zone into the wellbore at a high rate. For that to occur easily, the aperture

Figure 5.41: Drilling of vertical drains at Highland Valley Copper Mine

of the fracture must be large, and the fracture must be connected with a network of other open fractures that can collect water from the largest possible volume of rock.

For this reason, the broken rock zones (with high fracture aperture) associated with main faults often make good targets for production wells. It is generally not possible to resolve these discrete zones using geophysics so, while known fault zones can be targeted, it is not possible to predict the exact yield of the well in advance of drilling. Therefore, the normal approach for dewatering well installation in fractured rock formations is to first drill a pilot hole (Section 2.2.4.2). If the yield of the pilot hole is good, the hole can be reamed and made into a production dewatering well. If the yield of the pilot hole is poor, the hole can be converted into a monitoring well or piezometer, and another pilot hole drilled.

It is important to understand the requirement to achieve high groundwater flows when evaluating production wells in fractured rock. An understanding of this can be gained from the 'A-B-C-D' fracture model

(Figure 3.2). In any given fractured rock setting, there may be only a very limited number of 'A' fractures with a sufficiently wide aperture (and continuity of aperture) to provide sustained high flows. If one of these discrete fractures is encountered by a drill hole and pumped, groundwater will flow *en masse* through the subordinate interconnected 'B-C-D' fracture network to feed that 'A' fracture. However, if the drill hole intersected only parts of the 'B-C-D' fracture network, those fractures might not have a sufficiently wide aperture to feed an adequate flow of water into the hole directly.

Fractures can change in aperture or roughness over short distances along their dip and the strike, and it is normal for fractures to pinch and swell along dip or strike. In addition, some fractures may be completely 'healed' with fillings such as clay or calcite, and the degree of filling may also change along the dip or strike of the fracture. Therefore, the water-transmitting characteristics of any given fracture may increase or decrease within a short distance, as illustrated in Figure 5.42.

Figure 5.42: Illustration of the variability of fracture aperture over a short distance (Photo: Geoff Beale)

As water converges on a production well, the same volume of water must flow through a progressively smaller area (Figure 5.43); this means that the specific discharge, q, must increase. When flow is occurring through a few very narrow fractures, this can require a significant increase in speed which, in turn, means that the flow may:

1. become non-linear (non-Darcian), and the head loss will increase significantly;
2. reach the speed where kinetic energy becomes significant, causing more loss of static head.

The end result is that there will be a *local* increase in drawdown around the well which will not spread to any distance through the formation; this will effectively increase the drawdown in the well so that the pump will use more energy without achieving the aim of dewatering or depressurising a large volume of rock (Figure 5.44).

This situation will be much less likely to occur if the well intersects a fracture of large aperture ('A') that is well-connected to the surrounding network of smaller fractures (the 'B-C-D' network) because the larger fracture will, in effect, be collecting water from the surrounding formation and conveying it with minimal head loss to the well. The overall goal of the dewatering well is to achieve widespread drawdown in the formation (maximise 'aquifer loss') with minimal *additional* local drawdown in the formation immediately surrounding the well and within the well itself (minimise 'well loss'). If the dewatering well encounters a large aperture ('A') fracture, it will allow this goal to be achieved. If the dewatering well only encounters subordinate ('B-C-D') fractures, then the yield will be low and most of the drawdown will be localised in the immediately vicinity of the well, with little widespread drawdown. Thus, depending on whether high aperture ('A') fractures are encountered, the yield and specific capacity of dewatering wells may vary dramatically between wells located only metres apart.

The depth of the fractures that are encountered by the wellbore will also affect both the well yield and the general

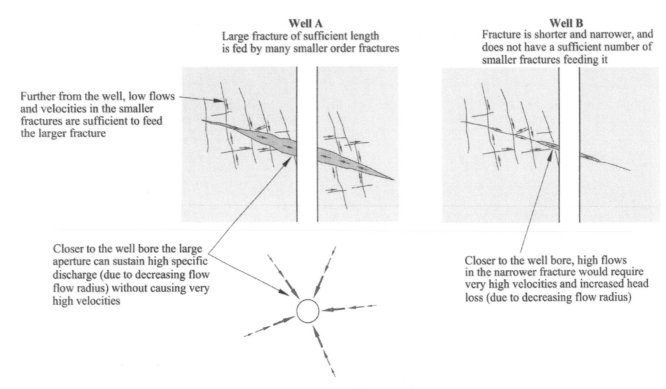

Well A
Large fracture of sufficient length
is fed by many smaller order fractures

Well B
Fracture is shorter and narrower, and
does not have a sufficient number of
smaller fractures feeding it

Further from the well, low flows
and velocities in the smaller
fractures are sufficient to feed
the larger fracture

Closer to the well bore the large
aperture can sustain high specific
discharge (due to decreasing flow
flow radius) without causing very
high velocities

Closer to the well bore, high flows
in the narrower fracture would require
very high velocities and increased head
loss (due to decreasing flow radius)

Figure 5.43: Illustration of increasing water velocities surrounding a production well

effectiveness and operational life of a dewatering well. A well that encounters a fracture at great depth can be made more productive than a well that encounters a fracture of similar hydraulic conductivity at shallower depth, as the well with the deeper fracture has more available drawdown. In Figure 5.45, Well A encounters a fracture zone at a depth of about 100 m below the initial water table, while well B encounters the same fracture zone at a depth of about 200 m below the water table. Initially, in Well A, only 100 m of drawdown can be applied, whereas 200 m of drawdown can be applied to Well B. Therefore, the initial yield of well B is greater. After two years, when 50 m of general dewatering has occurred, the available drawdown that can be applied to Well A has reduced by 50 per cent, whereas the available drawdown that can be applied to Well B has only reduced by 25 per cent. After four years, when 100 m of general dewatering has occurred, the fracture zone in Well A is dewatered, so the well has no yield, whereas there is still 100 m of available drawdown that can be applied to Well B.

Many ore bodies are associated with large geological structures and fault systems. Fault planes can vary from clay-filled aquitards to large fractured rock zones that may extend for a significant distance each side of the strike of the fault. Large faults and associated conjugate or anastomosing fractures from the fault are often the main targets for siting pilot holes for dewatering wells as they can provide a major collector zone for the water in the surrounding lower aperture fracture network. The key is

to intercept these open or fractured faults at a depth lower than the required dewatering target elevation. To be fully effective, dewatering wells must derive at least some production from below the targeted dewatering elevation, or full dewatering of the area cannot be achieved. A productive fracture near the bottom of the well will, in the end, be much more effective than several similar fractures located much higher in the wellbore.

The above discussion illustrates the discrete nature of groundwater production zones in fracture flow settings. It is important that the hydrogeologist or mining engineer understands that:

- dewatering wells need to be located where the fractures will provide yield, which is not always the best location for the mine plan;
- if the mine plan changes and access to a production well is lost, it is not always easy or possible to replace that well.

5.2.5.3 Pilot holes

The most cost-effective approach to ensure that a high cost production well encounters large aperture fractures that are sufficiently open and interconnected to transmit appreciable quantities of water into the drill hole is to first install narrow diameter (low-cost) test holes (pilot holes). The goal of the pilot holes is to confirm the presence of large aperture fractures prior to production well drilling.

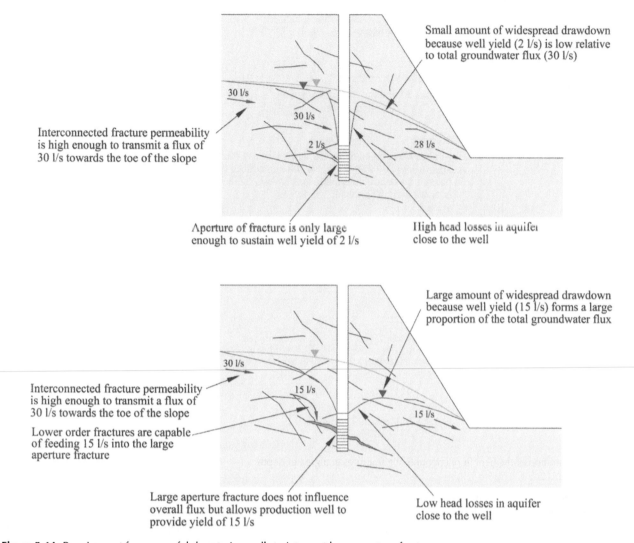

Figure 5.44: Requirement for successful dewatering wells to intersect large-aperture fractures

There are many examples from the worldwide mining industry where costly pumping wells have produced too little water because fracture zones have not been intersected, and where considerable savings could have been made by first proving ground conditions by drilling narrow diameter pilot holes.

The main goals of pilot holes are as follows.

■ Determining the presence and assessing the nature of the producing fracture zones, and their depth. Locating the productive fracture zones by pilot hole drilling will usually help maximise the long-term yield of subsequent production wells, which will ultimately lead to fewer wells being required for any given application.

Where several productive pilot holes are drilled, airlift flow and injection test results can be quantitatively compared to assess which of the various pilot hole sites will make the best location for a production well.

■ Defining the depth of the groundwater producing zone and how this may relate to the long-term well yield.

A dewatering well with producing fractures too shallow in the hole may suffer a rapid loss of yield, as described above.

In addition to testing of pilot holes, a good way to determine the depth of water-producing fractured intervals is to run fluid conductivity and temperature logs or carry out a program of hydrologging (Section 2.2.7.5). Downhole fluid velocity ('spinner') logs can be run to quantify the depth and amount of ingress in the hole although, because they usually require pumping stress, they are normally run in the completed well rather than in the test hole.

Pilot holes (Section 2.2.4.2) can be drilled by any method but best results are typically obtained using the RC method with air. There are no set rules for how many pilot holes will be required to site a production well in any

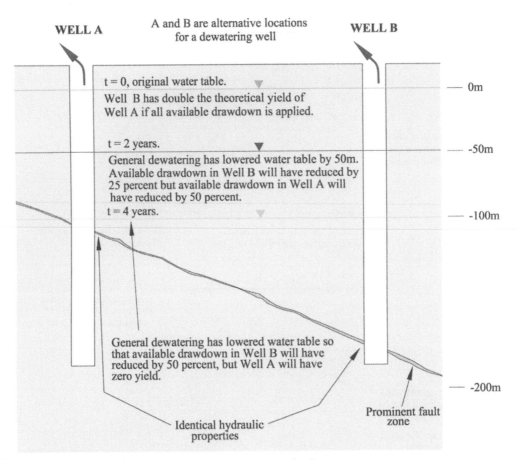

WELL A   A and B are alternative locations for a dewatering well   **WELL B**

t = 0, original water table. — 0m

Well B has double the theoretical yield of Well A if all available drawdown is applied.

t = 2 years. — -50m

General dewatering has lowered water table by 50m. Available drawdown in Well B will have reduced by 25 percent but available drawdown in Well A will have reduced by 50 percent.

t = 4 years. — -100m

General dewatering has lowered water table so that available drawdown in Well B will have reduced by 50 percent, but Well A will have zero yield.

— -200m

Identical hydraulic properties

Prominent fault zone

Figure 5.45: Illustration of the benefit of encountering fractures at a greater depth

given fractured rock formation. This will depend on the complexity and the level of understanding of the site hydrogeology, and may therefore change with time as a particular project develops. The minimum number should be one pilot hole, the normal is two to four per production dewatering well, but more than five could be required in some settings where there are only a limited number of higher aperture fracture zones within the formation.

The hydrogeology of the site, the mine plan, and purpose of the pilot holes will dictate how deep the holes should be drilled. A general rule-of-thumb is to drill the pilot holes at least 100 m deeper than the ultimate dewatering target (which is often the lowest planned elevation of the pit floor). Productive fractures must be intercepted well below the target future dewatering elevation in order to achieve fully the dewatering goals; how much deeper depends on the hydraulic conductivity and degree of interconnection of the fracture zones.

If the yield of a pilot hole is good, and the fractures are at a sufficient depth, the hole can be directionally surveyed to determine the true location of the fractures. If the hole is determined to be vertical enough, it can be reamed into a test production well. If the hole is

significantly off vertical, the collar location of the well can be adjusted to line up vertically over the fractures and a new well drilled. If the yield of the pilot hole is low, the hole can be converted to an observation hole or piezometer.

5.2.5.4 Considerations for production well design in fractured rock settings

There are a number of important guidelines to consider when installing production wells for general mine dewatering or for slope depressurisation in fractured rock systems, the most important of which are as follows.

1. The depth of the groundwater producing zone and the **changing well hydraulics** and yield as dewatering progressively occurs.

   - As dewatering or slope depressurisation progresses, there will be an increasing amount of groundwater drawdown in the formations surrounding (and contributing to) the well. Therefore, there will be ongoing changes to the hydraulics of the producing fracture zones, which will typical lead to a progressive decline in well yield. The well will have a tendency to 'dewater itself'.

- As a result of this, the yield in the well may progressively decline and the pumping head may progressively increase. Knowledge of the level of the groundwater producing horizons will allow the future well dynamics to be predicted so that the pump selection can be optimised.
- The changing well hydraulics is an important factor for planning the pump assembly, including diameter, length and NPSH (Net Positive Suction Head) required.

2. The potential for **cascading water zones** to develop within the well and the implications of this for oxidation, chemical precipitation and clogging, and pump performance and cavitation.

- As dewatering proceeds the pumping level within the wellbore may drawdown below productive fractures and/or perched groundwater zones. As a result, water may cascade from high-level fractures down the wellbore.
- The cascading water may cause a change in oxidation of the inflowing water and therefore changes in the pumping water chemistry and/or rates of chemical precipitation within the wellbore and within the pumping system. Note that this is usually more of a factor for in-pit wells and is generally of less concern for interceptor wells.

3. **The need to minimise the use of drilling mud** and, if the use of mud is unavoidable, adapt the well design accordingly.

- In intergranular flow settings, when drilling mud is more often used, a mud cake (or filter cake) will build up on the wall of the drill hole. However, when drilling in fractured rock settings, it is important to realise that most of the mud losses occur into discrete fracture zones and, depending on the nature of the rock, much of the loss may occur into as few as one or two zones. In horizons where the wall of the hole has low porosity and permeability, development of a filter cake may be poor to non-existent.
- Depending on the fracture aperture and connection, the mud may travel a considerable distance into the fractures, and may be difficult to remove by traditional techniques. If cuttings are carried into the fractures along with the drilling mud, extensive well development may be required to remove the debris from the fracture zones.

4. **The use of an appropriate well screen and filter-pack design** for the application of the well.

- From the point of view of minimising head loss on water entry, the most efficient possible well design is to install no casing and screen. In this case, there is nothing to impair the flow of groundwater into the well. However, there are very few formations that will stand up sufficiently (and indefinitely) to allow this, and most wells in hard rock require some type of lining.
- For fracture flow settings, because much of the water enters the well over discrete intervals and therefore well entrance velocities are typically high, conventional gravel pack design is often not appropriate.
- If the well screen and gravel pack is installed prior to full development, debris may flow from the fracture zone into the gravel pack, permanently clogging the gravel pack and reducing the well yield. Therefore, under certain circumstances, installation of a well screen with no gravel pack produces the best results.
- It is therefore important to monitor the mud level for evidence of circulation loss and to take early remedial action. However, in wells with small aperture fractures, it is not always possible to identify and correct a loss of drilling fluid as the hole is being drilled. The potential for clogging of wells can often be reduced by using a very coarse formation stabiliser in place of a gravel pack, or by not installing a gravel pack at all.
- In fractured rock production wells that are generally competent, the advantages of using any kind of gravel pack or formation stabiliser material within the annulus may be outweighed by the risk of clogging the material (though this may depend on local regulations for well construction). However, in many settings, a very coarse formation stabiliser (say 5 mm to 20 mm diameter grains) can provide a good compromise where:
  → intervals of the hole are liable subsequently to squeeze around the casing and cause damage or collapse, so that some kind of hole stabilisation is required;
  → it will subsequently be necessary to mine around the well, so that a formation stabiliser will dampen the effects of blasting on the well casing;
  → local regulations state that some form of filter medium is required on the outside of the well screen.
- For intergranular flow settings, selection processes for screen and filter-pack selection are well documented in the texts cited in Section 5.2.5.1.

5. **The potential short life of the well** within an active and changing mine environment.

- Many in-pit wells have a short operating life compared with typical water wells. Wells may either 'dewater themselves' as a result of ongoing drawdown, or they may be in the way or destroyed by the mining operation.
- The materials should be selected considering the active life of the well. In many cases, there is a need

to minimise drilling and well construction costs because the well life may be short.

- Some mining operations may also prefer to avoid the use of steel well casing for in-pit wells as the steel may eventually create difficulties in the crushing process when the well is mined out.

- Where the well life is predicted to be short, it may also be advisable to minimise the diameter of the downhole pump unit in order to minimise the required well diameter and installation costs. This can often be accomplished by pumping from the downhole pumps to staging sumps or transfer systems, rather than using high-lift downhole pumps to remove the water directly from the well to the pit rim.

6. The possible need for **future blasting around the well**.
   - Another consideration is that wells may subsequently be mined around. Mining around wells is a common procedure when in-pit wells cannot be avoided.
   - Guidelines for well construction and protection for in-pit wells are addressed in Section 5.2.8.

7. **Possible shearing of the wellbore** if it is within a zone of active slope movement.
   - In some instances, it may be necessary (unavoidable) to place pit slope depressurisation wells so that the wellbore passes through potential future shear or failure zones. A good example of this is discussed in the La Quinua case study (Appendix 4).
   - If this is the case, it is possible the well casing may shear down the hole. Figure 5.46 provides an example of this, as viewed through a downhole camera. A consequence of this is that the pump and riser pipe may become lost in the hole.
   - If it becomes necessary to install slope depressurisation wells through a failure plane, continuous pumping from the wells will minimise the effective stress acting on the failure surface itself.

In several countries, prescriptive standards are in place for water well drilling. Any deviation from the local regulations in order to meet the specific needs of a dewatering or slope depressurisation well may therefore require a waiver. It should also be noted that some countries require that all wells are eventually plugged and abandoned after use, even if they are inside the pit. Again, it may be necessary to obtain a waiver to do otherwise.

5.2.5.5 Interceptor wells

Figure 5.10 shows a pit slope where most of the inflow is occurring at shallow levels within the oxide zone. In this case, it would be possible to consider interception of the water using shallow wells targeting the permeable flow horizon. Interceptor wells are often employed away from the immediate pit area to prevent water in a relatively

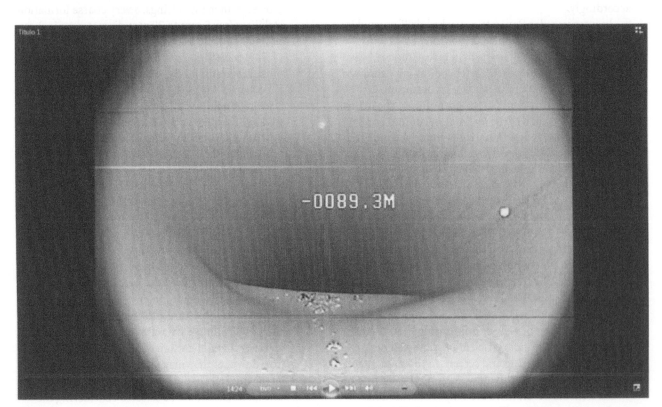

Figure 5.46: Example of video inspection, showing damage to well casing caused by shearing (Source: Victor Perez)

shallow permeable unit (often either weathered bedrock or alluvial sand or gravel) from moving towards the pit, with the goal of cutting off recharge to the slope materials.

The use of interceptor wells as part of general dewatering or slope depressurisation systems may potentially lead to a higher overall pumping rate than would otherwise be the case if all the water was pumped from the immediate pit area. However, interceptor wells offer a number of important advantages.

- They prevent water from moving towards the pit area and frequently reduce the amount of recharge that is able to enter the slope materials.
- They frequently lead to a reduced overall pumping head because they remove water from shallower horizons, rather than from the lower sectors of the pit.
- They minimise the contact between groundwater and the excavation, and therefore lead to an overall improved pumped water chemistry, which can in turn lead to a significant reduction in water management and discharge costs.
- They generally lead to less water within the excavation itself and therefore lead to an overall reduction in general mining costs.

In most situations, where the geometry is suitable, the upsides associated with the use of interceptor wells greatly offset the downside of having to handle an overall greater volume of water, particularly if the water produced by the interceptor wells is of good quality.

Figure 5.47 shows an example of pumping wells used to depressurise an isolated steeply dipping permeable unit at an iron ore mine. Pumping from the wells also allows depressurisation to occur across strike in the surrounding steeply dipping lower-permeability units. This situation is also common in some open pit coal mines where the side walls may consist of steeply dipping strata, some of which are permeable. It would be possible to achieve depressurisation using horizontal drains, but the use of pumping wells allows a greater amount of drawdown to be achieved in the permeable unit and therefore a more rapid rate of depressurisation in the adjacent lower-permeability units.

In the event that a groundwater recharge source in the vicinity of an excavated slope is significant (e.g. from a nearby lake or river), localised pumping wells or well-points can be installed specifically to intercept recharge flow. Such wells could target specific lithological or alteration units through which water may move from the recharge source into the pit slope (Figure 5.48).

5.2.5.6 Ex-pit wells

Dewatering wells that are installed around the crest of the pit, or in the upper few benches of the pit, are typically referred to as ex-pit wells. The usual advantage

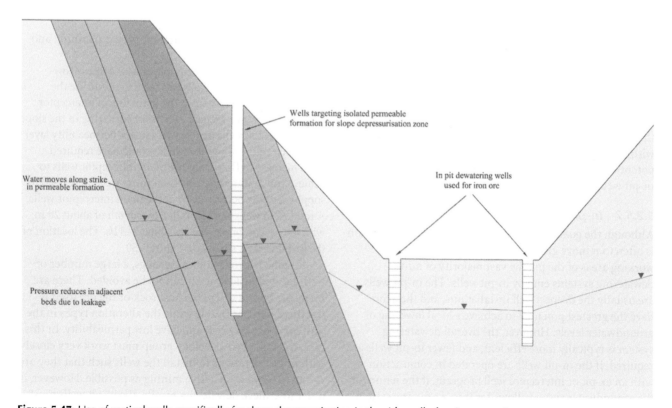

Water moves along strike in permeable formation

Pressure reduces in adjacent beds due to leakage

Wells targeting isolated permeable formation for slope depressurisation zone

In pit dewatering wells used for iron ore

Figure 5.47: Use of vertical wells specifically for slope depressurisation in the side wall of an iron ore mine

Figure 5.48: Example of a well-point system to cut off recharge to the slope at shallow levels

of ex-pit wells is that they are out of the way of the mining operation and therefore easier to operate with minimal disruption to mining activity. They also tend to remove water prior to it contacting the mining operation and therefore may have a better water chemistry than in-pit wells. A frequent downside of ex-pit wells is that they become under-drained because of pumping from within the pit, or from in-pit wells, and may therefore potentially suffer from more rapidly declining yields than in-pit wells.

5.2.5.7 In-pit wells

Although the goal of the general mine dewatering program is often to remove groundwater before it reaches the working areas of the pit, the vast majority of mine dewatering systems employ in-pit wells. The in-pit wells are usually the deepest well installations, and therefore have the greatest potential to achieve a local lowering of groundwater levels. However, the overall dewatering system is typically more efficient, and fewer in-pit wells are required, if the in-pit wells are operated in conjunction with an ex-pit or interceptor well program, if the nature of the groundwater system allows for this. As for any system design, a good understanding of the district and pit-scale

groundwater system is essential for proper planning and long-term design.

Figure 5.49 shows an example of how a line of low-yielding in-pit alluvial wells was used to minimise the required pumping rate from the main line of interceptor wells and prevent a seepage face from occurring in the slope. In this case, the wells penetrated a lower-permeability layer at the base of the alluvium which would have required a large increase in flow from the main interceptor wells to achieve full dewatering. The local in-pit wells were used to supplement the performance of the main interceptor wells. A total of 16 wells were installed to a depth of about 20 m, with a total combined yield of about 2–3 l/s. The location of the in-pit wells is shown on Figure 5.50.

In some hydrogeological settings, a large number of wells located inside the pit cannot be avoided. There are many ore bodies where the host rock or the ore itself is fractured and permeable while the alteration types in the wall rocks are clay rich and have low permeability. In this case, the mine hydrogeology group must work very closely with mine operations to install the wells such that they are as 'out of the way' of active mining as possible. However, in order to achieve dewatering results, the in-pit wells must be placed according to the locations of productive fracture

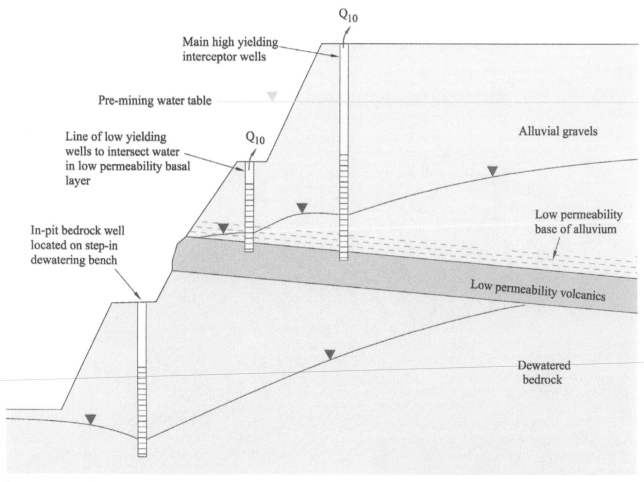

Figure 5.49: Use of in-pit alluvial wells to supplement the performance of the main interceptor wells (Courtesy Round Mountain Gold Corporation)

zones, which are often discrete and structurally controlled. Individual wells must also be placed at a particular location governed by the extent of hydrogeological boundaries and compartments.

At the El Tapado pit in Peru, the required general mine dewatering rate was about 225 l/s and the yield of individual wells was about 40 l/s, so six operating wells were required. However, the nature of the fracturing in the silica-altered ore body was such that, although there were no individual well locations that could sustain flows of more than about 40 l/s, the wells could be placed within about 10 m of each other with minimal interference. Thus it was possible to cluster all the in-pit dewatering wells in one location where lower grade silica occurred 'out of the way' of mining. The fractures within the silica exhibited such a good interconnection that the 'cone' of drawdown was virtually flat across the entire pit area and general dewatering of the entire pit was possible by operating the six wells in one location (Figure 5.51).

For situations where all the dewatering wells must be inside the pit, it is normal for the pit area to be less

interconnected and/or compartmentalised, such that it is necessary to distribute the wells around the entire pit area. The need to provide power to the wells and run discharge lines from the wells compounds the logistical difficulties. Some operators choose robust well installations where it is possible to reliably mine around the wells (Section 5.2.8). Other operators elect to install lower cost 'disposable' wells where pump sizes and completion diameters can be minimised by pumping to in-pit sumps (minimal head) and where wells are designed only to achieve a relatively small amount of drawdown (say sufficient for the coming two years) such that the well depths can be minimised. In the future, it is possible that directional drilling technology may be used to help reduce the number of in-pit wells by allowing the wells to be collared at the pit rim with the production section of the well within (below) the pit footprint.

While horizontal drains are the most commonly used technique for slope depressurisation, it is estimated that about 30 to 50 per cent of all active slope depressurisation programs use pumping wells installed specifically for the

Figure 5.50: Location of the in-pit wells shown in Figure 5.49 (Courtesy Round Mountain Gold Corporation)

purpose. Examples of where pumping wells have been specifically applied to slope depressurisation are shown on Figures 5.52a and b. Figure 5.53 illustrates a line of low-yielding wells being used to pump water from highly-compartmentalised intrusive rock, to depressurise a veneer of highly clay-altered volcanics within the slope In this particular case, it was necessary to lower the water level below the working bench, so it was not possible to use horizontal drains.

The use of **a vacuum system** to increase the amount of effective drawdown in wells drilled in low-permeability materials is more commonly used in the construction industry than in mining. With an efficient vacuum system at sea level, the maximum amount of additional drawdown that can be applied is about 8 m. The incremental effect of this is often minimal for the deeper wells drilled for mine dewatering, but it can make a major difference to the well performance in shallow excavations and can be effectively used to help stabilise silts and clays. Several large open pits are currently looking at expansions that will expose tailing

materials in their upper slopes and it may be possible to use vacuum wells to help stabilise the tailing deposits that become exposed. The design of the vacuum wells used for the Highland Valley Copper Mine is shown in Figure 5.54.

Low-yielding pumping wells for slope depressurisation can often be installed at relatively low cost using an RC drill. Holes can be drilled at 150 mm to 300 mm diameter, to allow installation of 100 mm to 200 mm diameter casing and screen, depending on the anticipated yield and the pumping head. Maintenance requirements are often small for low-yielding wells, and small-diameter well pumps (30 kW or less; 100 mm diameter or less) can often operate with minimal maintenance, provided that the dynamic water level in the well is not allowed to drop to the level of the pump. Where groundwater storage becomes progressively reduced and the wells show declining yields with time as dewatering of the surrounding formations occurs, an automated system of low-level power cut-off and high-level re-start can be operated. The cut-off/re-start system allows the pumping

Figure 5.51: Operation of six in-pit wells in a small, 'out of the way' area at El Tapado, Peru (Courtesy Minera Yanacocha SRL)

time of the wells to cycle. Alternatively, variable speed drive pumps may be used to maintain continuous pumping from the wells at a progressively reducing rate.

In-pit wells may pump to a staging sump, or may pump directly out of the pit if conditions are suitable. However, pumping directly out of the pit may increase the power requirements and hence the diameter of the pump to overcome the head which may, in turn, lead to larger diameter, higher cost well installations. In some cases, it has been possible to pump the water from low-yielding slope depressurisation wells into larger dewatering wells for removal using the main dewatering well pumps.

To decrease the cost of multiple well installations, it is also possible to consider **airlift pumping for wells or steep angle drains**. In the case of the low-yielding angled drains shown in Figure 5.55, 15 mm to 25 mm diameter airlines were installed in each drain, and a single small compressor was used to remove the water from all drains in the array. Typical flow rates from individual drains were 0.02 l/s to 0.2 l/s, and 30 drains were used in the array. For this

application, because of the geometry of the slope and access constraints, the use of angled drains and a low-cost airlift pumping system was more efficient than using horizontal gravity drains.

When planning a dewatering program using in-pit wells, it is necessary to account for the lower **operational availability** of each well that is typically achievable in an active mining area. Wells located outside of the pit and active mining areas may achieve an availability of 75 to 95 per cent, whereas the overall average availability for an in-pit well within an active mining area may be 55 to 70 per cent and even less in some instances. Maintaining maximum pumping availability is critical to the efficiency of any dewatering system (particularly for in-pit wells) and requires careful planning and cooperation between the hydrogeology and mining operations groups. Nevertheless, it is necessary to plan for the installation of contingency wells to achieve the specified dewatering rate. For example, if the required pumping rate is 240 l/s, and the average well yield is 40 l/s, then six wells would be required with 100

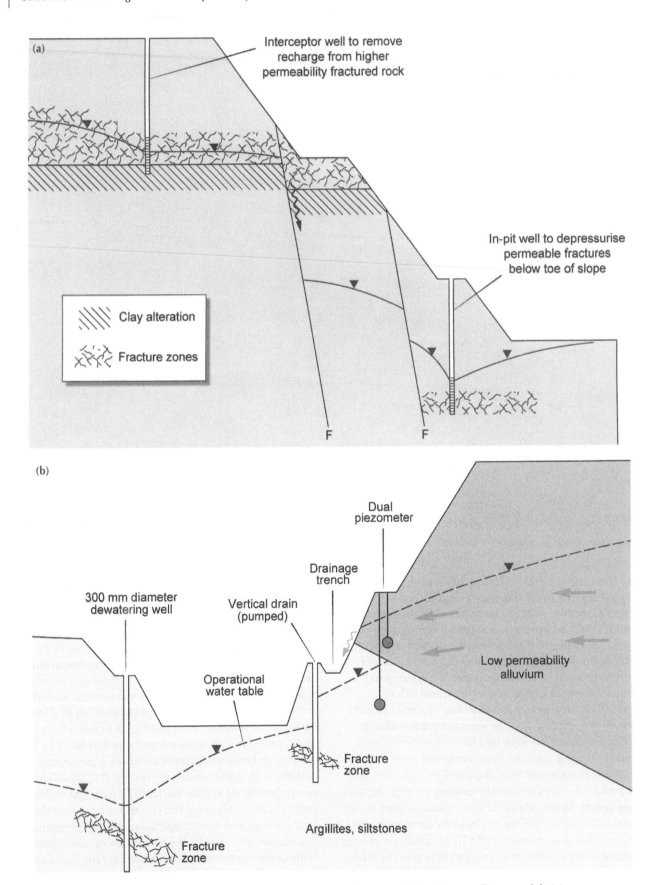

Figure 5.52: (a) Pumping well installation for pit slope depressurisation; (b) Low-yielding ejector well (pumped drain)

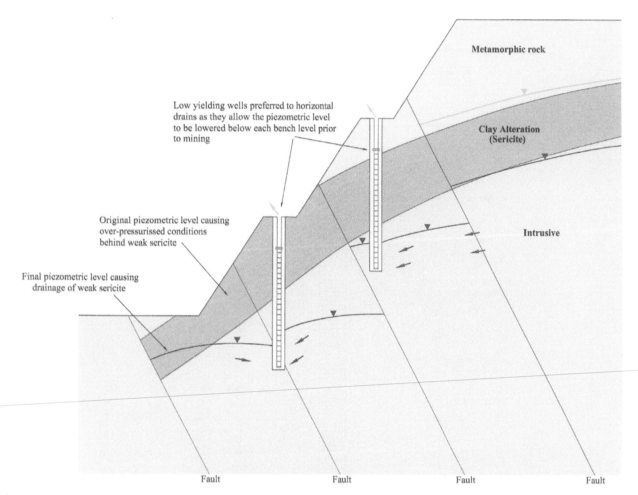

Figure 5.53: Low-yielding wells being used to depressurise compartmentalised intrusive rock and overlying clay-altered material in the slope

per cent operating availability, eight wells would be required at 75 per cent operating availability and 10 wells would be required at 60 per cent operating availability. Clearly, the higher operating availability that can be achieved, the less the drilling requirements and costs, and the less the overall disruption to the mining operation.

Case study from Nevada, USA

A good example of using vertical wells to depressurise low-permeability materials comes from the Gold Quarry pit in Nevada, USA. The upper part of the South wall is excavated in the Carlin Formation; a Miocene-age assemblage of fine sands, silts, tuffs (generally altered), and clays with minor gravel. Lateral facies changes, reworking, and discontinuities resulting from both syn- and post-depositional faulting result in highly compartmentalised, frequently perched, saturated zones throughout the sequence (Figure 5.56). The fairly thick saturated zone on the right of the photograph is truncated by a structure, whereas the thinner upper saturated zone crosses the bounding structure.

The predominantly fine grain size of these units results in permeability values on the order of 10^{-6} to 10^{-8} m/sec. The low permeability, combined with the lateral and vertical anisotropy, creates a tremendous challenge for dewatering. This issue is compounded by the fact that the Carlin Formation is waste rock, so the density of geologic drilling is insufficient to provide detailed resolution within the geological model. In other words, the hydrogeologist does not have prior knowledge of where the compartmentalising structures are located when planning for dewatering efforts.

Dewatering of the Paleozoic carbonate (limestone) bedrock underlying the Carlin Formation has been highly successful using deep high-yielding wells. The contact between the Carlin Formation and underlying Paleozoic rocks is usually marked by a 20–40 m thick zone of altered tuff. This basal tuff is dominated by fat clay, and possesses very low geomechanical strength (less than 12° internal friction angle at residual strength). As a result, the degree of success in dewatering the strata above the basal tuff is a major factor for slope stability.

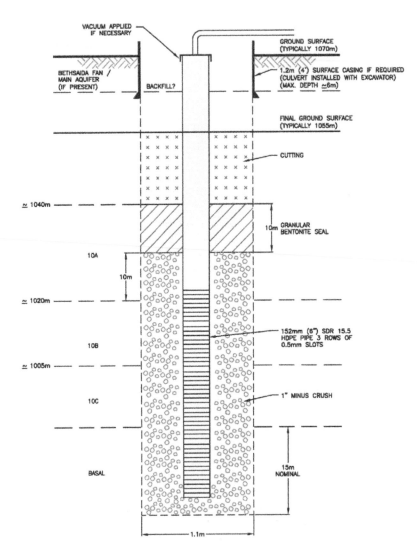

Figure 5.54: Vacuum well design used by Highland Valley Copper Mine

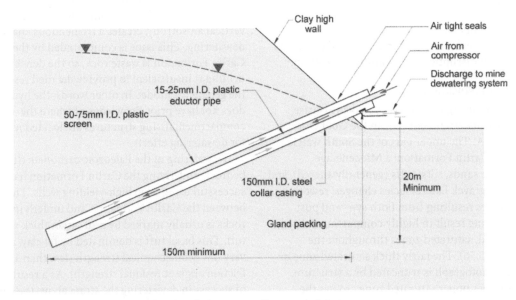

Figure 5.55: Use of an airlift system to pump from small diameter wells or angled drains

Figure 5.56: Vertical and lateral anisotropy within the Carlin Formation (Courtesy Newmont Mining Corporation)

Pumping wells installed to dewater the Carlin Formation generally produce 0.3–2 l/s (about 5 to 30 gallons per minute), and frequently exhibit extreme well losses (upwards of 30 to 50 m). Efforts to overcome the high well losses include application of variable frequency drive pumps to the control head in the wells, and a large number of closely spaced wells.

Compartments that do not communicate with the wells must be dewatered using drains (vertical, angled, and horizontal). The drains provide a conduit through the basal tuff into the underlying dewatered bedrock. Passive dewatering through drains is generally not as effective as active dewatering by wells, but it is a necessity in many areas because of high wall geometry. It is common for drains to be drilled on 15 m centers in order to achieve effective dewatering. However, over time most drains lose effectiveness, both as a result of near-drain head reduction and silting of the fractures within the bedrock (the silt/fine sand is transported from the Carlin Formation in suspension as water flows down the drains).

5.2.6 Drainage tunnels

5.2.6.1 Tunnels for general mine dewatering
There have been many applications of dewatering tunnels for general mine dewatering. The general goal of virtually all dewatering tunnels was to install the portal at a lower elevation than the final pit bottom, and to excavate beneath and underdrain the mining area using a gravity system. The 9.6 km long Sutro tunnel was completed in 1878 to dewater the mines of the Comstock Load (Virginia City) in Nevada. The 12.8 km long Carlton tunnel was completed in 1941 from a portal to dewater the Cripple Creek Mine in Colorado.

More recent examples of gravity dewatering tunnels are the Amole tunnel beneath the Grasberg pit (Indonesia), the tunnel driven beneath the Boinas and El Valle pits for the Rio Narcea project in Spain, the San Pablo tunnel driven beneath the El Indio Mine in Chile, and the Socavon tunnel driven beneath the Mina Sur pit in Chile. Each of these examples is gravity draining. In each case, the tunnel was driven below the lowest ore zone from a suitable portal location downslope of the pit, and relied on good vertical connection by gravity drainage through the ore body to achieve the required dewatering. In most of the above cases, angled drain holes were drilled upward or laterally from the tunnel to improve the vertical hydraulic connection within the ore body, or to allow the drainage of fault controlled hydrogeological compartments. Figure 5.57 shows a general setting for a gravity drainage tunnel.

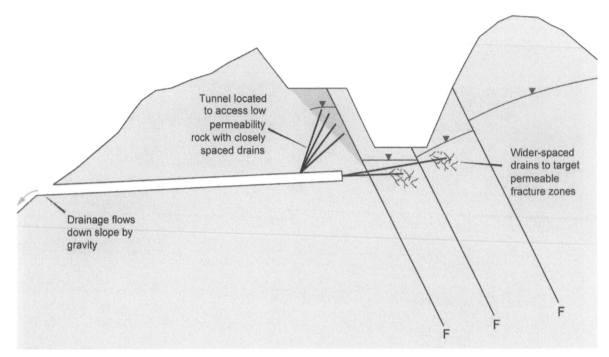

Figure 5.57: Gravity drainage tunnel

5.2.6.2 Utilisation of exploration or underground mine tunnels

There are also many examples of drainage tunnels driven from existing underground mine operations to provide dewatering of the open pit above. The installation of exploration tunnels, or underground mining operations beneath the open pit or behind the pit slope, provides an opportunity to enhance slope depressurisation or mine dewatering. Bedrock dewatering at the Nchanga pit in Zambia is carried out from the underground workings. At the Cortez Hills pit in Nevada, the concurrent mining of open pit and underground has provided the opportunity to install an underground drainage gallery with the goal of increasing the rate of pore pressure dissipation in the South wall of the pit. Dewatering from the underground drains provided the most effective means of depressurising the North-west wall of the Diavik A154 pit (Section 3.3.5.3).

Several mines around the world are currently considering the transition from open pit to underground operations (specifically block caves). For several of these, the underground mineral exploration tunnels may provide access behind the slope for the installation of slope depressurisation measures for the later stages of the open pit. It may be possible to apportion the up-front cost of the tunnel to a number of different cost centres.

5.2.6.3 Purpose-built tunnels for slope depressurisation

Within the last 10 to 15 years, the increasing size of open pits has lead mining operators to consider the installation of drainage tunnels specifically for the purpose of depressurising the pit slopes. In each case, the main purpose of the tunnel is to provide access for drilling drain holes from behind the pit slope to depressurise the slope materials.

The distinction between the recent tunnels constructed solely for the purpose of depressurising pit slopes and most previous dewatering tunnels is that, in all cases, the portal of the depressurisation tunnels is located within the pit, and the tunnel is driven locally behind a pre-designated sector of the slope.

To date, the principal examples of drainage tunnels driven solely for the purpose of slope depressurisation are as follows.

- Chuquicamata South wall K-1 tunnel (Chile), constructed in 1997.
- Escondida North-east wall 2660 L tunnel (Chile), constructed in 2001.
- Chuquicamata East wall drainage tunnel (Chile), constructed between 2004 and 2007.
- Bingham Canyon Highland Boy Corner drainage tunnel (Utah, USA), constructed between 2008 and 2009.

Each of these tunnels provided access for the drilling of drains from behind the slope to allow depressurisation of the slope, and to allow installation of piezometers from the tunnel to monitor the progress of the depressurisation. Figure 5.58 shows the layout of drains and piezometers drilled from the two phases of the Escondida North-east wall tunnel.

The key factors for each of the tunnels were as follows.

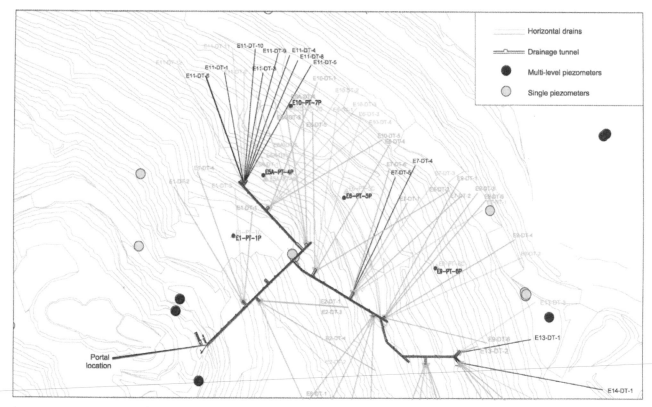

Figure 5.58: Layout of pit wall drainage tunnel (Source: Escondida Mine, Chile)

- The tunnel was constructed specifically to depressurise the slope and therefore allow ore recovery that would otherwise not have been possible.
- The tunnel was constructed following an economic analysis similar to that described in Section 3.2 of the Introduction section to this book.
- The portal of the tunnel was within the pit and located close to the target area of the pit slope.
- The main purpose of the tunnel was to provide access for drilling of drains. Any inflow to the tunnel itself was considered to be a 'bonus'.
- The tunnel orientation was selected such to minimise the amount of bad ground and water that was intersected by the tunnel itself.

When selecting the layout of a drainage tunnel for a large open pit operation, the inclination of the tunnel, and its ability to achieve positive (gravity) drainage, is an important consideration. However, for drainage tunnels installed behind a pit slope, it may not be possible to achieve a positive inclination, if deepening of the tunnel is subsequently required to depressurise the toe of the slope. If a drainage tunnel is designed that requires pumping water at high rates, interruptions in power, or maintenance of pipeline or pumps may need to be catered for by back-up equipment.

An increasingly important consideration for the design and layout of a drainage tunnel is the ability to close the tunnel within the objectives defined for the overall mine closure plan, and the potential need to install permanent bulkheads to seal the tunnel upon completion of mining. This may be less relevant for tunnels installed behind the walls from within the pit, but is nonetheless an important factor to consider when designing a drainage or depressurisation system.

5.2.6.4 Installation of depressurisation holes from a tunnel

The layout of the Bingham Canyon Highland Boy tunnel is described in Section 3.4 of the Introduction section to this book. Drains drilled from a tunnel behind the slope are more efficient in that they fully penetrate saturated rock, whereas drains drilled from within the pit are typically drilled through dry rock for the first part of their length (Figure 5.59).

Drilling drains from a tunnel behind the slope offers the following advantages.

- All drain collars and collection pipes are outside the active open pit mining operation.
- The drains are not affected by subsequent slope pushbacks (the end of the drain may be cut, but the drains are still able to function as designed).
- The drains are under a higher driving head and are more efficient than drains drilled from within the pit.

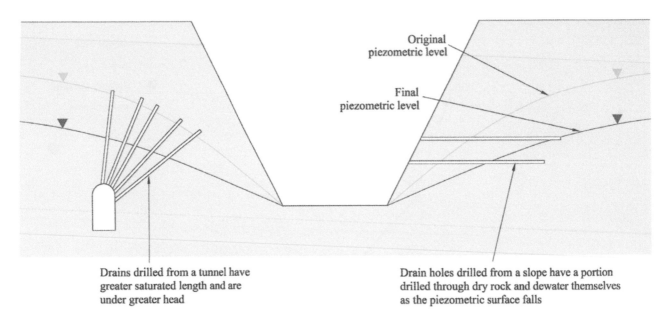

Drains drilled from a tunnel have greater saturated length and are under greater head

Drain holes drilled from a slope have a portion drilled through dry rock and dewater themselves as the piezometric surface falls

Figure 5.59: Advantages of drilling drains from within a drainage tunnel

- The drains are saturated throughout their entire length, again leading to greater efficiency.

However, there are a number of potential downsides for installing underground drainage holes, which are:

- pressures during drilling may be greater so there may possibly be safety considerations;
- the drainage tunnel may create humid conditions that can lead to corrosion of installations at the collar;
- because the tunnel is not typically active, routine maintenance is not always considered to be high priority, and this may ultimately lead to a loss of access to part of the tunnel.

The main design factors for underground drain holes are:

- holes can be drilled at any angle and any orientation, to suit the ambient geological conditions and prevailing structural orientations (although installation is typically easier for down-drilled holes);
- holes can be drilled with any underground equipment. Depending on the equipment, typical hole diameters may be 50 mm to 125 mm;
- the length of the holes should be determined based on the target zones to be depressurised. Hole lengths of between 125 m and 300 m are common;
- providing at least six metres of cemented collar casing allows a flange plate to be welded onto the collar which allows the hole to be shut in, the water to be controlled, and the shut-in pressure to be measured using a purpose sensor or a pressure gauge fitted to the collar (but note that, in higher pressure settings, and depend-

ing on the nature of the rock around the 'skin' of the tunnel, it is often advisable to install more than six metres of collar casing);

- if unstable ground conditions are encountered, perforated liner pipe can be installed to minimise the potential for subsequent blockage due to collapse of the hole. The installation of liner pipe is particularly important where discrete collapse zones were identified during drilling, or where clays or other squeezing materials are encountered.

Design of the tunnel location and alignment should depend on the conceptual hydrogeological model, the target zones requiring slope depressurisation, and the lead time required to achieve the depressurisation goals. For example, if the goal of the tunnel is to allow drain drilling into permeable compartmentalised fracture zones, fewer drains will be needed, the tunnel alignment can be further from the target zone, and the required amount of depressurisation can be achieved relatively quickly. However, if the goal of the tunnel is to allow closely spaced drain drilling into low-permeability clay zones, more drains will be required, the tunnel alignment must therefore be closer to the walls of the pit, and planning and installation must allow for a greater lead time for depressurisation to occur once the drainage holes are in place (potentially more than a year).

5.2.6.5 Monitoring from within a tunnel

The installation of vibrating wire piezometers within drainage tunnels is now standard procedure for monitoring the progress of the drainage system. The readings of sensors installed underground should be consistent with those installed in surface holes. Because the sensor is recording pressure above the instrument, it is very important to know the exact location of the sensor, so that the pressure measurement can be converted to total head. It is therefore advisable to carry out a directional survey of holes drilled for the installation of underground piezometers.

Drilling techniques for underground piezometer holes are identical to those used for underground drain holes. It is usually advisable to install some or all of the piezometers prior to installing the drain array so that the progress of the drainage program can be monitored. In addition, to provide supplementary information, consideration should be given to converting any selected low-yielding drain holes into a piezometer hole. Again, the installation of underground piezometers is typically easier for down-drilled holes, but the vibrating wire sensors can be installed within holes drilled at any angle and any orientation (Appendix 2).

In situations of low-permeability or compartmentalised wall rocks, the drainage tunnel may achieve the required amount of depressurisation of the material behind the slope, but may not actually dewater the material above the tunnel. This is shown in Figure 4.38, which represents the actual pore pressure profile following installation of the Escondida drainage tunnel. The rocks immediately surrounding the tunnel are depressurised, and the pressure head was successfully reduced by more than 100 m (~1 MPa) over much of the critical area. This led to a reduced amount of movement in the slope above. However, the level of the phreatic surface above the tunnel was affected to a much lesser extent. This example provides a good illustration of the difference between 'drainage' (usually removing the drainable porosity) and 'depressurisation' (reducing the pore pressure, but the materials remain fully saturated). Although the pore pressure targets were achieved, most of the material within the slope above the tunnel has remained saturated as a result of minor recharge moving down the slope at shallow levels.

5.2.7 Opening up drainage pathways by blasting

In some cases, it is possible to use controlled blasting to increase the permeability of tight ground and open up

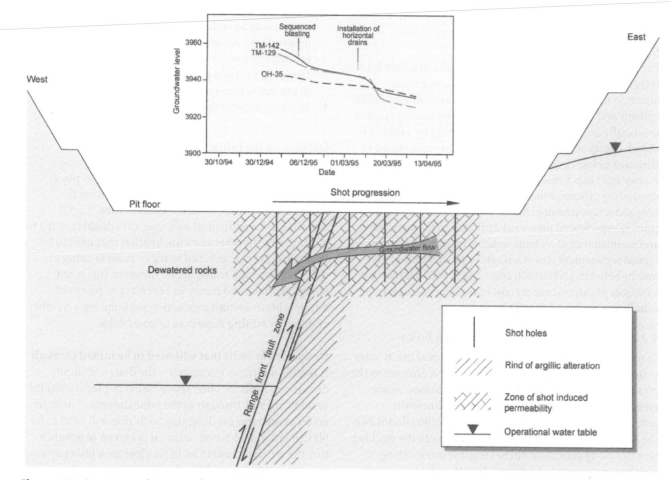

Figure 5.60: Opening up drainage pathways by blasting

drainage pathways. An example is shown in Figure 5.60, where a low-permeability clay gouge zone was impairing drainage of rocks on the footwall side of the structure which, in turn, was causing elevated pore pressures in the pit slope above the bench.

Three lines of blast holes to a depth of 45 m were installed across the 'tight' zone. The lines of holes were about 20 m apart and the aim was to create three drainage pathways from the saturated rocks in the footwall into the dewatered rocks on the hanging wall side of the structure. The shot was progressed from the hanging wall to the footwall with long delays. The resulting drainage caused pore pressures in the pit wall above the bench to reduce, as shown by the inset on Figure 5.60.

At the Liberty pit in Nevada, where development of new drop cuts was difficult due to the presence of water in a low-permeability rock mass, a practice was established to drill 30 m deep blast holes to open up fractures below the pit floor and to act as sumps, and subsequently to install temporary well casing and submersible pumping equipment to dry out localised areas ahead of mining.

5.2.8 Protection of in-pit dewatering installations

5.2.8.1 Introduction

For many operations, in-pit dewatering infrastructure is an integral part of the mining cycle and requires similar attention to protection and maintenance as all other mining systems and equipment. Although most dewatering systems are usually designed with 25 to 35 per cent spare capacity to absorb effects of dewatering downtime, the replacement of damaged or lost dewatering wells and equipment costs money and takes time, and often leads to setbacks in the dewatering process, which may ultimately cause a delay in ore production schedule. It is therefore important to have a team of experienced personnel dedicated to the installation and maintenance of all mine water-management systems. A typical replacement cost of a single in-pit dewatering well may be between US$150 000 and US$450 000, but there are a number of cases where the cost of replacing a single in-pit installation is over US$1 million.

5.2.8.2 Protection of horizontal drain holes

In many operations, the water from horizontal drain holes is collected within a reticulation system and directed to the pit sumps (Figure 5.61). In hard rock formations, where erosion of the slope is not of concern and where the potential for (or consequences of) bench infiltration is low, the water may be allowed to flow directly onto the working benches along collection ditches (e.g. toe drains along benches or ramps) to the sumps. In soft rock formations, where there is a risk of infiltration, all ditches/trenches

should ideally be lined to prevent infiltration (recharge) and erosion.

The main cause of damage to horizontal drain systems is fly rock during blasting activities or over-spill from benches that are being worked above the installations. Where a blasting pattern is within range, the design of the pattern should consider the fly rock damage potential for the drain collars and reticulation lines. Where possible, the blast heave should be directed away from the installations. Protection against fly rock can also be taken by covering the collector pipes and HDPE lined trenches using old conveyor belts. These are commonly available at most mines sites and can be cut into manageable sizes for easy handling. In active mining areas, the conveyor belt protection covers may be temporary or permanent installations.

5.2.8.3 Protection of dewatering wells

Given the cost and time to install dewatering wells, particular care is required to protect these systems from damage. There are four aspects to consider.

1. Optimising the initial location of in-pit wells.
2. Design of any wells that will need to be mined through, or planning to replace the well (twinning) as mining progresses to achieve dewatering to progressively greater depths.
3. Design of the blast pattern and preparation of the wells to minimise damage.
4. Mucking around the wells following blasting.

Optimising the initial location of in-pit wells

For many operations, the nature of the hydrogeology means that there is no alternative other than to place individual wells within the pit. Some guidelines for locating in-pit wells are described in Section 5.2.5.7. Clearly, for any individual well site, it is advisable to try to minimise the number of future benches that need to be mined around the well, and to try to avoid locating the well in future high traffic areas. However, this is not always possible, and it may be necessary to carry out multiple blasts around production wells during a number of different mining sequences or pushbacks.

Design of any wells that will need to be mined through

It is often advisable to customise the design of in-pit dewatering wells (or observation wells) in preparation for periodic mining through as the mine deepens. There are no set procedures for designing wells that will need to be blasted around. However, where it is known in advance that the well will need to be in (or close to) a blast pattern, there are a number of design modifications that have been successfully used by mining operators, as follows.

Figure 5.61: Reticulated horizontal drains that require protection during nearby blasting (Courtesy AngloGold Ashanti)

- Use of steel rather than PVC casing and well screen, in all cases (or constructing the well with double casing – steel outer casing; PVC inner casing).
- Installing thicker-wall casing over the interval that will be mined around.
- Using centralisers to separate the production casing (or outer casing) from the borehole wall to protect against the shock wave.
- Using a high-strength well screen (such as louvered steel) that will not pull apart if the well casing is lifted by the blast.
- Placement of a cement plug in the annulus above the screened section of the well to minimise the potential for lift and breakage of the underlying well screen.
- Welding straps on the casing and screen joints to minimise the potential for joints to pull apart if the casing is lifted during the blast.
- Use of casing materials that allow some vertical flexibility at their joints (Figure 5.62).
- Placement of a second (outer) protective casing string over the interval to be blasted.
- Placing a coarse formation stabiliser in the annulus outside of the production casing to help protect against the shock wave created by the blast.
- Installing production casing larger than is required (typically 50 mm larger) so that a smaller sleeve can be

inserted inside the production casing if this becomes damaged.

In some of the West African goldfields, double casing design has been used in soft rock environments for protection and also to allow the placement of a gravel pack to help absorb part of the shock wave. The operators often use a 305 mm steel outer protective casing, and a

Figure 5.62: Example of louvered well screen designed with vertical play in the threads (Photo: Geoff Beale)

203 mm internal PVC casing, with gravel pack in the annulus. In hard rock environments, a single steel casing may be sufficient, providing the casing is centralised and the gravel pack (or formation stabiliser) is placed successfully.

Design of the blast pattern to minimise damage to wells
Prior to mining through the well, the pump and all accessories would normally be removed from the well and stored in a safe area. However, some operators that use smaller equipment prefer to keep the equipment in the well and may lower it to the bottom prior to blasting. When blasting nearby but not directly around the well, it is good practice to switch off the pumping system for the duration of the blasting to avoid possible damage to pumps by earth tremors or electrical shorts. When a blast is considered to be within range of dewatering equipment, old conveyor belts can be used to cover the discharge lines and other items such as electrical switch boxes, flow meters and valves.

The design of a blast pattern around production or observation wells needs to consider the type, hardness and variability of the rock, the amount of water that may accumulate in the blast holes, the type of explosive used, and the height of the mining bench (and therefore the depth of the blast). Many operations have successfully developed their own procedures for blasting around wells, and there are no set rules. However, some of the following guidelines are regularly adopted.

- Keep the blast pattern as small as possible.
- Blast to as many free faces as possible so that the blast energy can dissipate outwards.
- Sequence the blast from the margins of the pattern inward to the well, using a relatively long delay.
- Use a row of buffer holes around the wellbore, potentially with a second row of lightly loaded holes, with minimal stemming.

Figure 5.63 shows an example blast from a gold mine with 10 m bench heights (Round Mountain Mine, Nevada). The blast contained two production dewatering wells. The photograph shows how the design of the previous blast patterns was carried out in order to

Free face from
Previous blast

Figure 5.63: Example blast around in-pit wells (Courtesy Round Mountain Gold Corporation)

maximise the number of free faces in the blast around the wells.

There are a number of operators who follow the procedure of removing the pump and riser pipe from the well, inserting a packer in the well three to five metres below the base of the blast, and using down-hole cutting equipment to pre-cut the casing above the installed packer prior to the blast. The wellbore above the packer is cleaned following the blast, the packer is removed, and the pump and riser are re-installed. The advantage of using this procedure is that the casing below the blast will not be subject to movement (particularly lifting) during the blast.

Mucking around the wells following blasting

Generally, damage caused by heavy equipment to the dewatering or monitoring wells and related infrastructure (e.g. discharge pipelines) most often occurs during night shift work. This is mainly due to poor visibility and often the lack of supervision as well as proper demarcation and protection around the installations. Flagging the well locations is important and can be achieved by placing an earth berm of approximately 1.5 m height with either a flashing light signal or a painted used truck tyre for protection (Figure 5.64). Each dewatering installation

should be mapped on mine plans and made available to shift supervisors for discussion and awareness during shift safety meetings. For blast protection, old conveyor belts can be placed over sensitive equipment for protection against fly rock near active mining areas. A housing structure may also be placed over the wellhead or over pumping equipment to protect against blasting in an adjacent area (Figure 5.65).

Mining through the well should ideally be carried out during day shift and supervised by a dewatering/ engineering foreman. The use of smaller equipment for mining around the well is normally good practice, ideally a loader, or a small shovel, as shown in Figure 5.66. Once mining through is complete and the well is cut to the new collar position, the pumps can be re-installed and the well re-commissioned. If the casing has not been cut down the hole prior to blasting, a crane may be required to hold and stabilise the casing when cutting it down following the blast. For some operations, electrical cables and discharge pipelines leading to and from the borehole are buried within trenches at least 0.5 m deep and demarcated on surface with a mound of earth delineating the trench path. Following re-installation and commissioning of the dewatering system it is critical to

Figure 5.64: Example of delineating and protecting a wellhead from traffic (Courtesy Siguiri Gold Mine, Guinea)

Figure 5.65: Example of a protective housing to protect a production well from blast damage (Courtesy Minera Antamina SA)

update the new borehole collar elevation in the dewatering monitoring database.

Specific considerations in soft rock environments
In soft rock environments, blasting near a dewatering well is usually discouraged and should only be allowed if there are no other means to mine the area. Normally a 10 m radius blast exclusion zone is sufficient to guard against blast damage. It is a good practice to safeguard the dewatering well by placing a 1.5 m high sand or gravel berm demarcating the blast exclusion zone clearly. The 'dressing' down of the pillar can be carried out using a small excavator under the supervision of a trained person that will direct the machine operator at all times during the removal of the pillar.

Special considerations for hard rock environments
A geotechnical assessment of the ground conditions is important to provide the blast engineer with the rock mass characteristics such as internal rock strength, fracture

frequency and RQD. The blast design must minimise lateral ground shifts that may shear the casing whilst at the same time achieve suitable loading fragmentation. The design of the wells should allow for a minimum of 100 mm annular spacing for the installation of a gravel pack (formation stabiliser) around the centralised casing. The gravel will assist in 'cushioning' the blast vibrations as well as filter out the fine materials entering the hole from the fractured rock mass.

Experience at the Cerro Vanguardia Mine
At the Cerro Vanguardia Mine in Argentina, the methodology for mining through and cutting down the well casing is to leave a protection pillar of approximately 20 m diameter during normal production blasting. The protection pillar is then removed by initially carrying out a pre-split blast at approximately 0.8 m beyond the steel casing position. This is followed by low intensity trim blasts designed to heave material away from the well. Once the steel casing is exposed, a small excavator is used to

Figure 5.66: Example of careful mucking around in-pit dewatering wells (Courtesy Minera Yanacocha SRL)

clean around the well casing. The casing is then cut down leaving a 0.8 m stickup. When mining advances approximately 30 m past the well, the pump and accessories are re-installed. This process usually will take about one week to complete. The experience has shown that blasting design configuration, timing, hole spacing and correct powder factors are critical to a successful outcome. Figure 5.67 shows an example of a well that has been blasted around at Cerro Vanguardia.

A good practice when setting up the dewatering system is to prepare by trial blasting in 'dummy' holes designed identical to the real dewatering wells. Various experimental blasts can be done, while documenting the results and making adjustments to the blast design until a suitable design is found. The site-specific nature of blast designs is emphasised.

5.2.9 Organisational structure

For many mine operations, dewatering and pit slope depressurisation are not currently regarded as a key part of the mining cycle. In some cases, this causes an increasing risk for future ore production, operating costs and mine safety. For larger projects, there is often a need for organisational restructuring to allow direct reporting of hydrogeology staff to a higher management level. In addition, there is always unquestionable benefit provided by the close coordination and integration between the hydrogeology, mining and geotechnical groups.

Operations that fail to implement good planning and staff recruitment programs may find themselves in 'fire fighting' mode, not only for current operations, but also for studies related to future projects. There is often a need to start dewatering operations early in the mining cycle to allow sufficient lead time ahead of the mine plan, particularly for pit slopes that are cut in weak, low-permeability materials. Given the current worldwide shortage of experienced technical staff, it may be beneficial for mining operators to employ more technicians, and to carry out more 'on-the-job' training. This also helps 'free up' senior technical staff for essential work that is currently not carried out, or is carried out retrospectively.

Figure 5.67: Example of blasted well in hard rock at the Cerro Vanguardia Mine, Argentina

Worldwide experience shows that a 'Hydrology Group' and a 'Water Services Maintenance Group' is often an effective operational structure for implementing and maintaining dewatering or slope depressurisation programs. Typically, the 'Hydrology Group' would be responsible for the planning, design and engineering of the systems. The 'Water Services Maintenance Group' would be responsible for implementing and maintaining them.

It is common that the 'Hydrology Group' reports through mine planning and/or mine engineering to mine management, whilst the 'Water Services Maintenance Group' reports through mine maintenance to mine management. A frequently used reporting structure for mine hydrology programs is shown on Figure 5.68. The two groups clearly need to work closely together and there also needs to be a reporting function between the head of the 'Hydrology Group' and the head of the 'Water Services Maintenance Group'. The two groups need to work closely with purchasing and warehouse departments to provide appropriate inventories of equipment, pumps, pipe and other spares. Where operators have more than one mine project in the region, the working inventory for all projects could be partly maintained at a centralised location.

The following division of responsibilities between the two groups can be considered.

- **The 'Hydrology Group'** is responsible for groundwater and surface water studies, dewatering design, engineering design (pump and pipeline engineering), with appropriate engineering skills provided. The performance indicators of the group may relate to system design and engineering to determine the required water level and pore pressure targets. The following are often important considerations:
  - → for many operations, the technical staff may spend a large amount of their time on administration-related matters. Aside from not allowing the necessary technical work to be undertaken, this may result in staff dissatisfaction and higher rates of turnover than desirable. Hydrological skill development is therefore important;
  - → key motivators for hydrological staff are often cited as a) good mentoring, and b) demanding projects, so that staff members feel their careers are progressing. The ability to provide good mentoring may help staff retention;
  - → the 'Hydrology Group' normally reports through mine engineering or planning to mine management. For mining operations with large hydrology requirements, employing a hydrology superintendent who reports directly to mine management is often desirable;
  - → a close relationship with geotechnical and mining personnel at the mine site is essential for the Hydrology Group. The right co-ordination is often best achieved by having a common reporting structure for the 'Hydrology Group' and the 'Geotechnical Group'.

- **The 'Water Services Maintenance Group'** is focused on implementation of the designs provided by the 'Hydrology Group' and mainly carries out the operation of the wellfield and maintenance. The performance indicators of the 'Water Services Maintenance Group' can be based on an agreed target monthly utilisation and/or pumping volume for each well or pump installation. Important considerations are as follows:
  - → the 'Water Services Maintenance Group' normally reports through mine operations to mine management;
  - → for larger operations, a 'dewatering superintendent' who reports directly to the head of mine operations

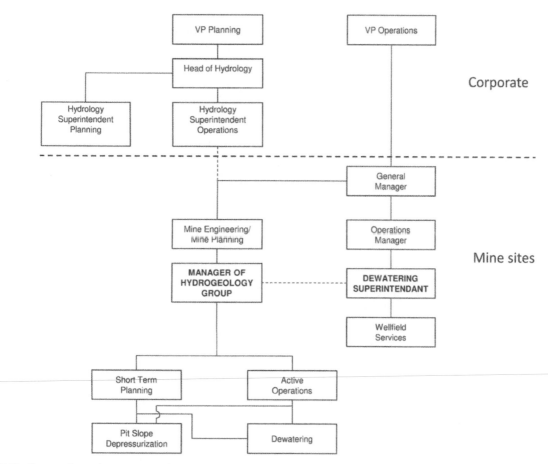

Figure 5.68: Frequently used organisational structure for larger mine hydrology programs

is desirable. There needs to be close coordination between the 'dewatering superintendent' and the head of the 'Hydrology Group';

→ the group should optimise the coordination and timing of blasting and mucking around in-pit dewatering wells with pump removal and installation, so that the wells have the highest possible availability.

Many operators are currently experiencing delays in getting drill rigs to meet short-term requirements. Operations are currently considering factors such as reviewing the basic organisational structure for drilling operations, diversifying the drilling contractor base, and procuring more water well rigs to meet the ongoing expansion and operational needs. For drilling rig requirements, the 'Hydrology Group' may need to work closely with geology and other planning groups to ensure the efficient use of all drilling equipment at the site. In addition, careful medium to long-term planning is required to obtain all necessary dewatering equipment on site and to maintain sufficient equipment inventory. This is particularly relevant at remote sites with poor transport infrastructure where typically a lead time of three to four

months may be required to get, for example, a well casing order to the site.

5.3 Control of surface water

5.3.1 Goals of the surface water management program

In-pit surface water runoff is an important factor for slope stability, particularly in wetter climates. A large number of slope failures occur at a relatively shallow depth in the slope. Increases in pore pressure at shallow depths can have a greater *proportionate* influence on effective stress than an equivalent pore pressure rise deeper in the slope.

Uncontrolled surface water flow may lead to (i) erosion and back-cutting of the weaker slope materials, (ii) shallow recharge causing a transient pore pressure rise in the over-break zone creating bench-scale instability which, in turn, may lead to a mass movement of the slope materials, and (iii) deeper recharge into the slope creating inter-ramp-scale instability that is often associated with a transient pore pressure rise within larger structures (Section 3.6.7).

As a general rule, a detailed surface water management plan is required for all operations where the mean annual rainfall is greater than about 250 mm year. A stormwater management plan is required for virtually every major mining operation. It is clearly beneficial to intercept surface water runoff externally to the pit (or on the upper benches), and surface water management plans must consider the construction and maintenance of diversion or drainage ditches which have the dual function of (i) preventing run-on to the pit slopes, and (ii) minimising the potential for surface water ponding and therefore infiltration and recharge.

Development of surface water controls is often neglected during mine construction because of the need to focus on other priorities. Even though site planning may include the development of surface topographic gradients to manage stormwater runoff, large flat surfaces in the mine area may be left by construction activities that sourced borrow materials from the vicinity of the open pit. These often cause ponding, infiltration and recharge.

There are two principal goals of a surface water management program related to pit slopes.

1. To protect the slope from erosion damage caused by surface water.
2. To shed the water from the slope as quickly as possible in order to minimise the amount of recharge to the slope materials. This is particularly important if low-permeability poorly drainable materials occur within the slope.

In addition, a well-managed surface water control program will reduce the potential for transport of eroded materials into the lower part of the pit, which may itself lead to damage of dewatering infrastructure or loss of access.

5.3.2 Sources of surface water

Surface water may enter the slope domain from five possible sources.

1. Runoff from incident precipitation onto the slope itself.
2. Surface water channels or runoff that may flow onto the slope from the upgradient catchment area above the crest of the wall. This includes shallow sub-surface gravity drainage, for example from rivers or dams close to the pit crest.
3. Uncontrolled groundwater seepage that may flow into the slope domain, for example from alluvial seepage faces higher up the slope, or perched water bodies.
4. Controlled groundwater that may enter the slope from wells and horizontal drains that have no reticulation system.
5. Leakage losses from pipes, tanks or other mine facilities.

Of these five sources, the first two are generally most important for the design of an in-pit surface water management program. However, there are also cases where uncontrolled groundwater seepage may potentially occur at a high rate. All five sources may contribute to recharge and sustained pore pressure in the underlying slope materials.

5.3.3 Control of surface water

In wet and seasonal climates, slope damage due to surface water can be difficult to control, and there is often no easy solution. The following surface water control measures are typically considered for large open pits.

- Diversion of any streams that may affect the pit area or may create recharge to alluvial materials that are exposed in the upper pit walls.
- Diversion of runoff from the upgradient catchment area around the crest of the slope.
- Diversion of runoff from the upper benches along drainage channels (toe drains) constructed on catch benches or step-out benches that allow gravity drainage to low points on the pit crest.
- Placement of surface water collection facilities on individual catch benches to shed localised runoff water as quickly as possible.
- Placement of surface water collection facilities below prominent zones of groundwater seepage in the slope.
- Installation of interception trenches or slurry walls (Section 5.2.1) to prevent shallow sub-surface seepage into the pit from sources of water near to the pit (e.g. rivers or dams).

5.3.3.1 Stream diversions

Many mine operators are faced with the need to construct substantial diversion channels for surface water streams that affect the mining area. Where necessary, diversions may have to start a long way behind the pit crest. For larger streams, the scale of construction may be significant, and may involve constructed dikes, culverts or pipelines, staging ponds and constructed channels, complete re-routing of streams, and potentially diversion tunnels.

For a major surface water diversion, the mine operator will generally carry out a cost–benefit analysis and risk assessment to select an appropriate return period for the design event. For some situations, the diversion can be a major construction exercise and may significantly affect the economics of a new project or Brownfield expansion. Potential trends in climate may also influence the accuracy of predictions for the magnitude and frequency of high rainfall events or cycles. In many areas, the downstream and riparian consequences of the diversion need to be understood, and community engagement as well as environmental permitting can be significant. In addition to the diversion itself, the effects of recharge from the diversion to shallow groundwater may need to be considered.

Where failures of surface water diversions occur, it is often because of either an extreme flow event or poor maintenance, and these factors may combine in arid climatic settings where flow events are infrequent and poorly understood, typically occurring as flash-floods. In arid or seasonal climates, it is often essential for extensive maintenance of diversion structures prior to and during the wet season. There have been several instances of tension cracks related to ground movement influencing surface water diversions and increasing recharge at the crest or in the upper slope (Figure 5.69).

The consequences of a failure of the surface diversion will obviously need to be factored into the design and planning process. Many of the large mining operations in Nevada USA are located along mountain range fronts meaning that ephemeral drainages from the upgradient

mountain ranges require diversion. Large localised thunderstorms that occur in the mountains immediately upslope of mine facilities may cause flash flooding that exceeds the design capacity of the diversion system. There have been several cases where intense rainfalls have produced extreme flows in the ephemeral drainages creating a wash out and failure of the diversion system. Such failures may cause the flow of water and transport of sediment down the pit high wall onto the lower benches. Most operators are aware of the risk and, because of the large size of many of the Nevada pits, the operators retain the ability to use multiple production dig faces to avoid impacts of flash flooding to mine production. Nonetheless, the costs related to removing the water and in-wash sediment from the pit can be high.

The perennial Granite Creek flows from the Osgood Mountains and intersects the high wall of the CX pit at the Pinson Mine in Nevada. The management plan for the creek has three components.

Dam and pipeline installed to bypass tension crack

Prominent tension crack

Figure 5.69: Example of tension cracks in surface water diversion

- Normal high flows enter an intake structure above the high wall from where they are piped around the pit and are allowed to re-enter the creek immediately downslope from the pit.
- Flows in excess of the normal high flows up to the 1 in 50 year, 24-hour event are allowed to overflow the intake structure and enter a diversion channel which conveys the water around the high wall into the nearby mined-out A pit, from where the water is allowed to evaporate from the pit floor.
- For extreme flows in excess of the capacity of the diversion channel, the excess water would be allowed to flow down the high wall in an area of strong wall rock (limestone) and enter a sump in the lowest part of the pit floor.

5.3.3.2 Diversion of upgradient overland runoff

If there is a reasonably sized upgradient catchment area, it may also be necessary to consider construction of a gravity (free draining) diversion channel around the pit crest to minimise the amount of water reaching the slope (Figure 5.70). Construction of a pit crest diversion can itself be a major undertaking in areas of steep topography where there is potential for intense localised runoff, even if there is normally no stream channel. The following design parameters need to be considered for a surface water diversion where flows are expected to be occasional and high.

- The potential magnitude of peak runoff flows, and the extent to which these events are locally understood. In remote drier climates, there are often limited data to assess the magnitude of peak rainfall events. It is often prudent to incorporate substantial contingency into the design.
- The ability to provide ongoing access to allow routine cleaning and maintenance of the diversion system. Any diversion will require regular maintenance such as clearing the diversion of in-wash material to prevent blocking of the system. If the maintenance isn't carried out on a routine basis, it may compromise the entire structure.
- The nature of the materials that will form the bed of the diversion structure and the need to line the structure. As a general rule of thumb, diversion ditches should be lined if there is a risk of infiltration into the materials above the crest. Lining also minimises erosion scouring thus maintaining the integrity of the structure.

Figure 5.70: Example of gravity diversion channel around pit crest (Photo: John Read)

- The nature and strength of the materials immediately underlying the diversion and the risk of slope movement undermining the diversion. The potential for localised ground movement to create cracking from stress release following excavation may compromise the integrity of the diversion. In this regard, it is often beneficial to place permanent ditches far enough away from the pit crest to minimise the potential for damage of the diversion due to the development of tension cracks and thus minimise the potential for water to enter these tension cracks.

- The nature and strength of the materials in the pit slope immediately below the diversion and the risk of damage to the slope in the event of a failure of the diversion. In some cases, it is possible to use lesser design criteria for the diversion design if there are strong rocks exposed in the slope below the diversion and the risk of damage to the slope is lower.

- The extent to which sediment may be washed into the diversion and reduce its capacity. Sediment transport during extreme events may be so severe in some situations that it severely affects the design or, in some cases, precludes the construction of a diversion. Transport of large rocks and boulders may create damage to the channel, liner or structures placed within the diversion.

- The extent to which snow may accumulate and become compacted within the diversion structure. The accumulation of windblown snow may render the diversion channel ineffective to carry a large runoff event, which may occur during the time of spring breakup. In some cases, it may be more appropriate to use runoff diversion berms, which may be less prone to over-topping due to accumulating compacted snow than excavated channels.

Maintenance is a critical aspect of any site water-management system and the ability of the operation to implement a maintenance program is an important decision factor. Extensive cleaning is often required immediately prior to, and regularly throughout, the wet season. Because diversions concentrate runoff flow within a channel, the potential damage to the pit slope caused by a poorly maintained diversion may be greater than the potential damage created by diffuse 'natural' runoff with no diversion structure in place.

Surface water diversions that carry only occasional flows are notoriously prone to failure. It is therefore necessary to specify a regular cleaning and maintenance program for the diversion structures. In drier climates, simple runoff control berms around the crest of the pit may be adequate for preventing water from entering the slope.

Downstream management and potential treatment of 'contact water' flows is becoming an increasing consideration for mine sites, and there is a need to minimise contamination of clean water. Surface water designs may need to ensure that mine water and clean (upgradient) runoff water (i.e. clean and dirty water) are kept separate, and managed separately to minimise water quality deterioration. Sediment control ponds (reservoirs) may be required depending on the topography and climatic conditions.

5.3.3.3 Diversions on upper-level catch benches

In areas of steeper topography, diversion ditches may be placed on high catch benches in the upper slope rather than at the pit crest. These may be routed to low points on the pit crest, allowing the water to drain out of the pit by gravity. Such diversion benches usually need to be designed into the mine plan in advance of operations. If in-pit diversions are implemented, they will need to be regularly inspected for cracking at their base. A robust cleaning and maintenance program is essential. In tropical areas, where the upper slope materials are weathered, removal of sediment from ditches and sumps needs to be factored into the maintenance plan. In cold climates, late winter clearing of snow drifts may be required.

5.3.3.4 Collection of runoff water on catch benches

The installation of collection ditches around prominent catch benches, or along the inside of haul roads, is commonly used to help shed surface water runoff as rapidly as possible from the slope. The ability to collect and store runoff from high-intensity storm events must be included in the mine design. Ideally, the diversion ditches should be designed to minimise the potential for ponding and infiltration into the slope materials. In arid settings, grader-cut ditches around the inside of the haul roads may be all that is required.

For wet and seasonal climates, it is often necessary to allow for the temporary storage of large volumes of in-pit water in the mine design. Most operations have the ability to store water inside the pit and do not attempt to remove extreme runoff events in short time periods. However, there is an increasing number of large operations where short-term flooding of the pit floor during stripping or early in the mine life would negatively impact the operation and the operator may need to consider a system to remove peak events early in the mine life (when the pit footprint area is smaller).

A cost–benefit analysis may be warranted to provide a balanced design between minimising the capacity of the installed (nameplate) pumping equipment and minimising the time that the working sumps or lower pit levels are flooded, and the situation may change as the mining cycle progresses. Mine operators in wet climates typically ensure that several alternate dig faces are accessible during the wet season to allow for flexibility in the event that the lowest working benches become flooded. Some operators elect to close the bottom of the pit and work only at higher levels during the wet season.

Snow accumulation inside the pit has been an issue for some operations with higher snowfall. Most operators try to minimise the amount of snow that gets trucked out of the pit. Sump pumping systems are often designed for peak capacity during the spring break up (snow and ice melting). However, the peak runoff may occur as a result of a summer storm.

5.3.3.5 Protection of weak materials in the slope

Surface water management is a critical aspect for pit slopes that are cut in weak materials, particularly if those materials are of low permeability (as is frequently the case). In addition to erosion and back-cutting, the need to minimise recharge is critical, and any ponding of water on the weak materials should be avoided. Frequently, in wetter or seasonal climates, the surface water management plan must include:

- wider benches every 40 to 60 m vertical height in areas of moderate to strong rock, with the bench cut at a positive grade (typically 0.5 per cent) to allow the installation of surface water control ditches, and minimise the potential for ponding on the benches;
- protection of the slope above the drainage or diversion with mesh if there is potential for rock fall that may block the ditches;
- construction of a berm on the outside of the bench to minimise the potential for flow onto the slope below;
- installation of lined ditches on each catch bench in areas of weak wall rock.

The installation of lined ditches on each catch bench is an onerous undertaking, but may be the difference between an intact slope and a failure, with associated sterilisation of resources. A good example is the North wall of the Yanacocha Sur pit in northern Peru, where infiltration during the wet season would otherwise create recharge and sustain high pore pressures in the weak clay-altered materials that occur throughout the slope domain. Key features of the drainage system are as follows.

- HDPE-lined drainage trenches are installed on each catch bench. The benches are designed with an adequate grade to allow them to flow to central locations where downpipes are installed to convey the runoff down the slope.

- The purpose of the drains is to prevent erosion and cutting back of the slope, and also to shed water as quickly as possible and reduce the amount of recharge.
- Because recharge has been largely prevented, the piezometers have demonstrated good depressurisation as a result of deformation of the underlying slope materials.
- Cleaning and maintenance of the drains is carried out on a routine basis, and at a higher frequency during the wet season. Maintenance is carried out using light vehicles and foot crews. Protection of the slopes above the drains is therefore critical.
- Repair of damaged sections of drains and downpipes is carried out rapidly to prevent the onset of erosion and damage to the system.

Figure 5.71 illustrates the lined ditches to protect weak materials in the slopes. Figure 5.72 illustrates the use of downpipes to transfer water between catch benches, and Figure 5.73 illustrates the use of down pipes on a permanent slope.

5.3.3.6 Collection of uncontrolled groundwater seepage

Uncontrolled groundwater seepage into the slope should be managed in the same way as surface water; the difference being that the groundwater flow rates are usually lower, more sustained, less seasonal, and not as prone to peak events. Many operators route the inflows to lined ditches on catch benches and into sumps, from where they are pumped out of the pit. Ponding of the water on the catch benches should be avoided if possible, particularly above areas of weak and low-permeability materials.

An extreme example illustrating the difficulties of managing groundwater seepage is the South wall of the Kumtor Central Pit in Kyrgystan, where the ice tongue of the active Davidov glacier is exposed in the high wall for over 1400 metres. The challenge at Kumtor has not only been the continual movement of ice into the high wall, and the high hydraulic pressure in the till layer (head of over 140 m) but the high surface water flows that occur beneath the ice above the underlying glacial till material. It has been possible to install pumping wells at the pit crest to intercept some of the sub-glacial flow, but a large amount of seepage enters the slope at the base of the ice tongue. The majority of this water is stored (or was) in discrete compartments mainly in ice. The release of this water during mining operations comes in short, sudden large flows of over 500 l/s that last a maximum of 48–72 hours and then recedes. Nonetheless, movement of the slope has been reduced in recent years by (i) implementing a program of dewatering wells to depressurise the thick glacial till materials that occur beneath the ice flows, (ii) reducing the water stored behind the ice and waste dump mass, and (iii)

Strong rocks
with no surface
water control

Weak rock with
surface water
control

Figure 5.71: Illustration of ditches only in weaker slope materials, with downpipes between benches

Figure 5.72: Down pipes being used on an active slope

Down pipes to pass water between benches

Lined channels on each bench

Figure 5.73: Illustration of fully engineered downpipes on a permanent slope (Photo: Geoff Beale)

intercepting surface runoff high above the crest of the pit. Depressurisation of the till has been successfully achieved by installing an array of pumping wells located on catch benches and also above the crest of the pit. In order to minimise the possibility of shearing and to maintain as long an operating life as possible, most of the wells have been installed lateral to the main ice tongue and also just below the ice mass to intercept runoff and/or stored water.

Another example illustrating the difficulties of uncontrolled seepage comes from the Andacollo pit in Chile. In this case, the occasional intense (seasonal) rainfall runs into waste rock material that has been placed above the crest of the slope. The waste rock acts as a storage unit (a 'sponge'), soaking up the precipitation and upgradient run-on, and allowing the water to be released gradually over a period of several weeks after the rainfall event. The water runs from the toe of the waste rock and onto the slope below. However, in the case of Andacollo, the high strength of the material in the slope, and the high quality of blasting and the poorly developed over-break zone, combine to prevent any erosion of the slope below, or any significant recharge to the materials in the slope. The seasonal water can therefore be allowed to flow down the bench faces and into a sump, from where it is pumped out of the pit.

5.3.4 Estimating flow rates

Conventional surface water modelling methods, such as HEC-HMS (Hydrological Model System) using the NRCS (US Natural Resources Conservation Service) Curve Number (CN) is usually applied to small catchment areas to help predict the overland flows, and flows from small drainage basins that need to be diverted around the pit and above its crest.

The CN approach is an empirical method that combines infiltration losses with surface storage and antecedent soil moisture conditions to estimate the rainfall excess or storm runoff volume. These variables are a function of a single parameter, the CN.

Therefore, the CN of a catchment indicates the runoff potential of the catchment and, as a guide to the hydrologist, CN values are tabulated for various land uses and for four hydrological soil groups which can be modified to represent wetter or drier than normal antecedent soil moisture conditions.

The CN ranges from 0 to 100, however, practical values usually range from over 50 to 100. Lower values denote a low potential for runoff and higher values a higher potential for runoff. The hyetograph could be either a recorded or a synthetic storm, for instance one of the synthetic storms proposed by the NRCS (Type I, Ia, II or III).

The HEC-HMS software has several options to transform storms into runoff and obtain peak flows for channel design. Among the options is the NRCS unit hydrograph which is essentially dependent on the lag time of the catchment. It is common practice to adopt as a design criteria for a pit diversion channel a 24-hour storm event with a return period of 100 years. However, the design return period should be chosen considering the following.

- The amount of data and knowledge that exists on the ambient rainfall patterns in the area.
- The nature of the drainage area above the crest of the pit and its applicability for modelling.
- The nature of the materials in the slope below the diversion and the risk of erosion of the slope.
- The ability to clean and maintain the diversion, and the degree to which it is necessary to over-design the system.
- The consequences of failure (overtop) of the diversion channel.
- The planned life of the mine.
- Applicable local regulations.

Alternatively, a risk-based approach can be used to size the diversion channel.

For assessing the magnitude of runoff from the slopes themselves, a frequently used technique is to apply the CN approach. If this approach is used, caution should be taken in selecting the CN value as this technique was originally developed for agricultural lands. Due to the unique nature and geometry of pit slopes, the applicability of any modelling technique must be questioned and the degree of uncertainty recognised. It is often best to apply a runoff coefficient based on empirical evidence or based on experience at mine sites in similar settings. Again, the following factors need to be considered.

- The nature of the materials that form the slope, whether they consist of porous alluvium that may absorb some of the incident precipitation, or hard bedrock when most of the water will tend to run off.
- The extent of fracture damage of the rock areas and the ability of water to infiltrate and therefore reduce the runoff coefficient.
- The overall slope design and particularly the bench face angle and the width and separation of the catch benches.
- The extent to which rock fall and sloughing of the material from above has created a porous cover that may attenuate runoff on the benches.
- Whether there are any retaining berms on the outside of the catch benches that may reduce the amount of water moving downslope and increase infiltration on the benches.

Thus, runoff coefficients for pit slopes are extremely variable and may range from less than 0.1 for an occasional storm event in an arid climate to over 0.9 during an intense rainfall period in a wet climate. A good rule of thumb has been developed for the Batu Hijau pit in Indonesia, where typically about 70 per cent to 80 per cent of the mean annual rainfall of about 2700 mm generally occurs within a five month wet season. The operating experience at Batu Hijau has indicated that, during the wet season, about a third of the incident rainfall on the pit slope runs off and appears within the pit floor sump within three hours, about a third is retained on the catch benches and within the over-break zone and appears within the sump within three days, and about a third recharges the groundwater system and is removed by 300 m long horizontal drains installed from the slope, contributing to groundwater flows of the following three-month period.

5.3.5   Control of recharge

The way in which local-scale recharge may influence the pore pressures in pit slopes is described in Section 3.6.3. When assessing the importance of recharge, the following factors need to be considered.

- The potential for any surface water to enter tension cracks above the crest or within the slope.
- The nature of the materials that comprise the slope; whether they are strong and highly permeable, or weak, deformable and poorly permeable
- Minimising the extent to which ponding of the water can occur on the catch benches or above the crest of the slope.
- Minimising the extent to which infiltration into the over-break zone can create high transient pore pressures in the near-surface materials (even if the rocks deeper in the slope are depressurised). Shallow transient pore pressures can often lead to bench-scale failures which may propagate to larger scale mass movement.

5.3.6   In-pit stormwater management and maintenance

Stormwater management design for runoff generated within the pit footprint must consider the flow paths and volumes to be diverted, using trenches and culverts, to intermediate and pit bottom sumps. It is normal to cut drainage ditches around the inside of haul ramps (Figure 5.74).

In circumstances of high-intensity and short-duration events, the objective is to provide sufficient containment capacity to ensure that, in the worst case, discharge arising from stormwater runoff does not cause significant adverse effects on mining operation and pumping equipment. Regular upkeep of stormwater trenches, pit sumps and pumping equipment is required by dedicated teams (Figure 5.75).

Drainage channel cut
on inside of haul road

Figure 5.74: Drainage ditch around outside of haul ramp (Courtesy Mustakim Hamnur, PT Newmont Nusa Tenggara)

Careful consideration to design and planning is required to determine where best to locate and manage dewatering equipment such as electrical cables, pipelines, pumps, flow meters, and telemetry systems. For example, flow meters and control valves and transfer tanks can be positioned some distance from the active mining area in

Figure 5.75: Examples of easily maintained runoff control ditches (Courtesy AngloGold Ashanti)

protected locations. All installations require highly visible demarcation and protection against heavy mine equipment. Electrical cables and water pipelines located within the mine area, where possible, should be buried in temporary shallow trenches.

The in-pit surface water management plan requires consideration of current conditions and potential changes that may occur to the mine plan. Planning for the program will need to consider:

- whether a single stage (single lift) system is appropriate;
- whether a multi-stage system is required using booster pumps, either from the outset or becoming phased in as the pit gets deeper;
- whether it is better to integrate the pumping system for groundwater and surface water, considering that groundwater flows tend to be more sustained and less seasonal, and groundwater pumped from wells or collected from drain systems tends to have a low suspended solids content.

In high rainfall areas, barge-mounted electrical pumps triggered by high and low level switches are often the preferred method for pumping from pit bottom sumps (Figure 5.76). These can be powered remotely via the mine's electrical grid systems or by generators positioned along higher-lying areas which are easily accessible for maintenance and diesel re-filling (pumping can continue uninterrupted during storm events). Temporary pit sumps in active mining areas will often require wheel mounted diesel driven pumps which can be moved from sump to sump or removed to safe areas during blasting. Routine preventative maintenance and servicing of the generators and pumps is essential.

The approach for designing the pit floor pumping capacity is variable throughout the industry, depending on:

- the nature of the mining operation, and the ability to create alternate dig faces or stockpile ore if the pit floor is flooded;
- the magnitude and return period of peak events (Figure 5.77), and the potential time when access to the pit floor may be lost if the pumping design capacity is exceeded.

Most mines normally design the system assuming some loss of access during the wet season. For example, within the AngloGold Ashanti mines, the approach is to design the pit bottom sump to cater for 1:100 flood surges and to have the pumping capacity to remove the surge within 30 days. Analyses carried out at Iduapriem mine after a flood event in Ghana central region indicated a rainfall event estimated at 1:50 intensity (two per cent probability of occurrence) generated peak flows of 8000 m^3/hr (Figures 5.77b and 5.78). With these high volumes, provision for flood surge using pumping capacity only is not practical. The approach adopted following the event was to increase the pit bottom sump to cater for the water generated during a 1:100 storm and have sufficient pumping capacity to remove the flood surge within a given time period. It is planned that mining during wet periods is also possible at higher elevations in the pit, rather than at pit bottom.

5.3.7 Maintenance of surface water control systems

Ongoing maintenance is a key factor for surface water management, and one that is often overlooked. The surface water management system is an integral part of the mining cycle and requires similar attention to protection and maintenance as all other mining systems and equipment.

Figure 5.76: Examples of barge pumping in an active sump with high-solids water

Figure 5.77: Examples of extreme in-pit runoff: (a) Western Australia and (b) West Africa

Surface water maintenance programs can be difficult and cumbersome. However, if maintenance is not carried out on a systematic basis, the control structures and/or pumping systems will have a high risk of failure. Experience at many mine sites has shown that a maintenance program works best if there is a designated individual or team of people that is specifically responsible for surface water maintenance, rather than placing it under the general mine maintenance program, where it may not be prioritised because it is 'too difficult'. Figure 5.79 shows an example of severely cracked shotcrete in a surface water control ditch causing recharge

Figure 5.78: Main pit at the Iduapriem Mine (Ghana) following an extreme rainfall event (Courtesy AngloGold Ashanti)

Figure 5.79: Severe cracking of shotcrete liner in surface water control ditch allowing recharge to the underlying slope materials

to the slope materials below. A good maintenance program should therefore start with a robust initial system design, using durable materials, considering as long a return period as possible for 'design events' and ensuring there is ample room for ongoing access.

5.3.8 Integration of in-pit groundwater and surface water management

5.3.8.1 General

Many open pits are designed with individual collection systems for groundwater and surface water. The most common situation is for groundwater from the general dewatering wells to be piped out of the pit while groundwater from horizontal drains and low capacity wells is routed along surface water ditches to sumps, and integrated with the surface water management system.

Normally, in pit runoff would be routed to the pit floor or other storage areas as rapidly as possible to minimise the potential for infiltration into the slope. It is often necessary to provide one or more runoff water storage

areas inside the pit (Section 5.3.3.4). Typically, the main storage area is the pit floor. However, the system must sometimes be designed to minimise resident (storage) time of water in the pit-bottom sump to reduce infiltration and recharge.

It is often beneficial to integrate the pumping system for surface water and groundwater. The potential advantages of an integrated water removal system are:

- groundwater flows are more sustained and so the system will operate on a continuous basis. This often makes maintenance easier than running a surface water only system where flows are seasonal or occasional;
- if the water from in-pit dewatering wells is routed to a sump it means the downhole pumping head is less. This, in turn, may reduce the required diameter and hence the installation cost of dewatering wells. This may be an important factor where the location of wells interferes with mining operations and there is thus a need to regularly replace them.

Potential downsides of running an integrated pumping system for surface water and groundwater are:

- there is often a difference in water quality, and therefore the two water sources may need to be managed differently in the overall mine water-management system;
- surface water typically has a high total suspended solids (TSS) loading whereas groundwater has a low TSS loading. This may have implications for pump design. Again, it may also have implications for the downstream management of the water with respect to discharge compliance;
- it may not be desirable to operate sumps on a year-round basis.

For seasonal surface water runoff, the pump design for removal of the water may need to achieve a balance between the use of higher capacity (and higher cost) pumping equipment and the length of time that water is allowed to remain inside the pit. A rule of thumb for a large pit is to design the surface water sump pumping system to remove a 1 in 100 year runoff event in 30 days, but this will depend on many factors, including the intensity of the runoff events, the flexibility of the mine plan to provide alternate dig faces away from the pit floor, the ability of the mine to carry a stockpile of ore and withstand periods of low ore production, and the frequency of use of the pumping equipment. Some mines in tropical regions operate a split bench at the lowest level of the pit so that only half the pit floor is inundated should there be a large runoff event. Other mines plan their operations such that the lower levels of the pit are completely closed during the wet season and all mining is carried out on higher benches above the water. Several operations attempt to carry out sequential sump pumping to help ensure a workable block of ground on the pit floor during part of the wet season.

5.3.8.2 Removal of water from sensitive areas of the slopes

The consequences of any localised infiltration to the slope materials need to be determined when designing the in-pit water-management program. As any excessive infiltration may defeat the purpose of the slope depressurisation program, the water produced by in-pit dewatering methods should be removed from the vicinity of the slope as quickly as possible. In particular, it is necessary to keep water away from materials that are sensitive to re-wetting, such as argillic materials, so that when these materials relax and new fracture surfaces develop, their strength is not reduced when they react with water.

Figure 5.79 illustrates severe cracking of shotcrete liner in surface water control ditch allowing recharge to the underlying slope materials. Figure 5.80 shows an example where the surface water control ditch along the inside of the haul ramp had failed, leading to uncontrolled water flow down the ramp. The infiltrating water penetrated joints below the ramp that were sub-parallel to the slope and had become dilated due to the loss of confinement. The infiltration caused a rise in transient pore pressure within the joints and created outward movement of the slope. This resulted in the cracks observed on the ramp.

Management options that have been employed for removing water from the pit slopes themselves include:

- collection of water on the benches and using down-pipes to remove the water from the slope (Figures 5.72 and 5.73);
- routing of the water from the slope depressurisation system and/or surface water from the benches into unlined drainage ditches constructed on adjacent haul ramps to convey the water to intermediate or pit floor sumps;
- piping of the water from horizontal drains or low volume wells by gravity into higher-volume dewatering wells for removal from the pit (if the incremental pumping rate is relatively low).

5.3.8.3 Reticulation of flows from drain holes

It is general practice for many operations to allow water from horizontal drains or low volume pumping wells to flow on the catch bench and either evaporate, infiltrate into the bench and move downslope through the over-break zone. The acceptability of allowing the water to flow onto the bench must be judged for the operation. Two factors must be considered. First, the flow of water down the over-break may lead to bench-scale failures or compromise the benches below. Second, the water may infiltrate and sustain pore pressures deeper in the slope. The presence of low-permeability clay-altered materials may increase the potential for, and consequences of, infiltration and recharge.

Ideally, the water produced from horizontal drains would be collected within reticulation piping systems that are routed directly to sumps, with no contact with the wall rocks or catch benches (Figure 5.81). However, allowing the water from drain holes to flow directly onto the working benches may be acceptable where the water can be routed directly into collection ditches (such as for drains drilled along the inside of a haul ramp). Allowing the water to infiltrate into the catch benches below, where it may join the 'invisible' water moving down the slope within the over-break zone, may help sustain pore pressures in the wall rocks below (Figure 5.82). Maintaining access to catch benches to manage sustained horizontal drain flows can be difficult and there is often no easy operational solution.

Uncontrolled water on outside of ramp

Infiltrating water causing transient pore pressure in slope parallel joints

Cracking due to outward movement of slope

Figure 5.80: Example of surface water infiltration creating transient pore pressures in joint sets underlying a ramp

Figure 5.81: Example of reticulation of horizontal drain flows

Horizontal drain flow infiltrating into the
bench below

Seepage from base of leached zone
providing recharge to entire wall

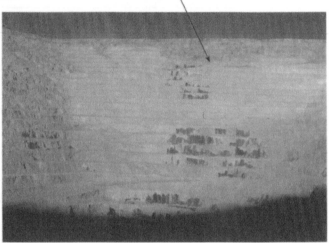

Figure 5.82: Example of horizontal drain flows entering the over-break zone in the slope below

5.3.9   Protection of the slope from erosion

In high rainfall areas it is sometimes necessary to prevent erosion of the weaker materials in the pit slope, even if a robust surface water control program is employed. The goal is usually to minimise the extent to which erosion and back-cutting into the slope occurs due to surface water erosion. Figure 5.83 shows an example of severe surface water back-cutting leading to instability.

In Figure 5.84, coalescing of the back-cuts into a cemented gravel has the potential to create bench-scale failures and a general deterioration of the slope, potentially creating larger-scale failures.

Measures that have been employed to protect bench faces are as follows.

- Mesh or matting draped over vulnerable areas (Figure 5.85).
- Crest berms on individual catch benches to prevent run-on to the bench face below.
- Lined drains on the catch benches, feeding into downpipes, as described in Section 5.3.3.4.
- Rip-rap protection of the bench toes (Figure 5.86).

Figure 5.83: Illustration of severe back-cutting in a slope due to surface water runoff

Figure 5.84: Bench-scale erosion due to runoff

Figure 5.85: Example of mesh use to protect a pit slope from water erosion

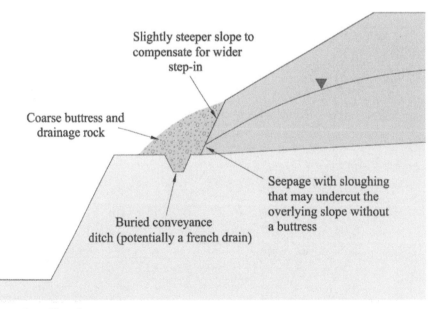

Figure 5.86: Rip-rap protection of bench toes

6

MONITORING AND DESIGN RECONCILIATION

Chris Lomberg, Ian Ream, Rory O'Rourke and John Read

If a given slope design is predicated on achieving a certain pore pressure distribution, then it is important that a monitoring program is implemented to ensure the required pore pressures are being achieved. Section 6 describes the elements of the performance monitoring system together with the periodic reviews and updates of the design criteria that are carried out as part of the monitoring and reconciliation process. Figure 6.1 illustrates the role of monitoring and reconciliation within the overall hydrogeology program.

Section 6.1 describes the procedures used to monitor the dewatering and slope depressurisation program and Appendix 7 provides a case study for monitoring the dewatering systems at a number of West African gold operations. Section 6.2 describes how the results of the monitoring program form the basis for performance assessment studies, and Section 6.3 summarises the management process for identifying and mitigating risk associated with water.

6.1 Monitoring

6.1.1 Overview

Verification of the actual performance of the dewatering and slope depressurisation systems is a key component of the mining cycle. It requires a robust monitoring system that (i) allows responses to the dewatering and depressurisation programs to be observed, (ii) provides a comparison between predicted and actual conditions, and (iii) is sufficiently flexible to respond to new hydrogeological and geotechnical data, revised geology interpretations, and changing mine plans.

The monitoring systems for specific pit slopes are typically integrated with the site-wide water management monitoring system to track the effective area of influence of the mine dewatering network and to evaluate the influence of the site-wide hydrogeology on the pit slopes. The components of the monitoring system are described below.

6.1.2 Components of the monitoring system

6.1.2.1 Groundwater monitoring

A dewatering and depressurisation monitoring network will typically consist of a combination of the following.

- Single or multiple **pumping wells or well point systems**, for which monitoring is normally carried out as follows.
  - → Flow rate, cumulative and instantaneous, recorded at minimum during a daily inspection using an in-line flow meter installed at the wellhead (at the pressure side of the pump).
  - → Downhole water level recorded at minimum daily, using a sounding tube installed inside the well (normally installed at the same time as the pump installation).
  - → Wellhead pressure during the daily inspection, using an in-line pressure gauge installed immediately before the flow meter.
  - → Wellhead power (line-to-line voltage and amperage) typically recorded during the daily inspection.
  - → Water chemistry field parameters (TDS, pH), usually carried out monthly.
  - → A full water chemistry sample, usually obtained quarterly.

Many production well systems are equipped with automated logic controls. These systems can be set up to monitor downhole water levels in wells, water levels in staging tanks, system pressures, flow rates and other operational

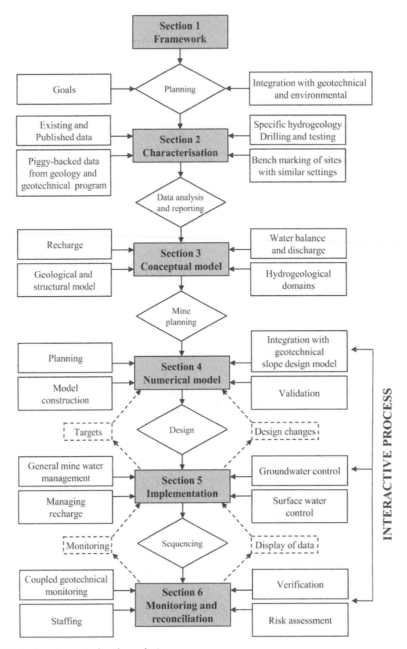

Figure 6.1: The hydrogeological pathway in the slope design process

parameters. They can be used to control wells or wellfields automatically based on demand, and it is becoming increasingly common for monitoring to be carried out using these systems, with the greater density of data providing opportunities for analysis of pumping effects over time in addition to planning for dewatering (or water supply). Some automated systems offer monitoring of intake pressures as well as temperature and vibration, and can alert operators to potential problems, or can be set up to shut down a well automatically. In systems where dewatering wells cycle on and off frequently, the instrumentation of the wells with pressure transducers provides crucial data on the duty cycle of the pumps that cannot be collected with daily visits.

- **Horizontal drain holes,** for which the following monitoring program would allow a good understanding of their behaviour.
  - → Flow rate at the end of each rod during drilling.
  - → Flow rate at the end of drilling.
  - → Water chemistry sample collected about 24 hours after completion of drilling (as required).
  - → Flow rate daily for the first seven days after drilling.
  - → Flow rate weekly for two to three months after completion.
  - → Flow rate monthly for as long as access can be retained.

→ Shut-in pressures, as needed, where shut-in valves are installed.

- **Vertical or down-angle drain holes**, for which monitoring is normally limited to downhole water level on an as-needed basis.
- **Individual sumps**, for which the following parameters are normally monitored.
  → Flow rate, cumulative and instantaneous, at minimum during the daily inspection, using an in-line flow meter on the pressure side of the pump system.
  → Water level in the sump, daily or more frequently in the case of smaller sumps.
  → Pumping pressure, during the daily inspection.
  → Power supply, during the daily inspection.
  → Water chemistry field parameters (TDS, pH), usually carried out monthly.
  → Suspended solids content, carried out as required based on the type of pumping system and down-stream management of the water
  → Full water chemistry sample, usually obtained quarterly.

Automated logic controls are also frequently installed on sump pumping systems, particularly where the sump pumps cycle on and off frequently and where it is necessary to control the pump running times based on the fluctuating water level in the sump.

- **Staged pumping systems**, where the following parameters are recorded at each pumping station.
  → Flow rate, cumulative and instantaneous, at minimum during the daily inspection, using an in-line flow meter on the pressure side of the pump station.
  → Pumping pressure, during the daily inspection.
  → Power supply, during the daily inspection.

Systems to automatically monitor, shut down and re-start pumps are commonly installed on multi-stage systems where the failure of another pumping stage may require a rapid shut down of the entire system.

- **In-pit seeps**, for which the following monitoring program is common.
  → Mapping and flow measurement (where possible) on a quarterly basis.
  → Monthly inspection of seeps observed on important parts of the slope, or more frequently depending on nearby activities
- **Water truck supply stands**, which are normally monitored for cumulative flow, using an in-line flow meter, reported weekly.
- **Observation wells and piezometers**, for which monitoring can be carried out automatically with data logging equipment and telemetry, or manually by the mine hydrology or geotechnical staff.

Downhole pressure transducers are being increasingly used for automated monitoring, with a shorter measurement frequency, particularly when nearby pumping wells or drain holes are brought on line. The use of data loggers greatly reduces field time and allows for flexibility in the resolution of data. For example, a data logger collecting daily readings can easily be reconfigured to take rapid readings to monitor a pumping test, and then set back to the original data collection rate. The data recorded by the instruments can be downloaded at the wellhead or transmitted automatically using telemetry systems. Where piezometer monitoring is carried out manually, the frequency is usually bi-monthly or monthly for regional installations, weekly for in-pit installations and daily (or more frequently) for any installations close to points of new pumping stress.

There is often a need to integrate the pore pressure monitoring program with the geotechnical monitoring program for extensometers, inclinometers or TDR systems. The frequency of pore pressure readings is often modified to match the monitoring frequency of geotechnical instruments, and there is an increasing tendency to integrate plots of geotechnical parameters (radar, prisms, inclinometers or extensometers) and pore pressure hydrographs.

6.1.2.2 Surface water monitoring

The nature of the surface water monitoring program will clearly depend on the climatic setting of the site. Some level surface water monitoring is usually required for any operation where the precipitation is greater than about 100 mm per year, and potentially less than that in regions where intense rainfall and flash flooding may occur or where there are regulatory requirements. The surface water monitoring program for the pit may have the following components.

- Scheduled or automated flow and/or field water quality parameter collection of features such as seeps, springs, creeks and diversion channels.
- Regular inspections of diversion channels, detention basins or structures for condition, freeboard capacity, sediment deposition and erosion.
- Scheduled or event-based water quality sampling.

Monitoring systems can involve any combination of the following, based on the objectives of the field monitoring program.

- Flumes or weirs with or without instrument wells.
- Pressure transducers and data loggers to track water flow at a flume or weir.
- Water quality parameter probes (typically pH, EC and temperature, and potentially other parameters).

Regular inspection of all surface water ditches and diversion systems is commonly required, with the findings documented in a field inspection report. This should include any erosion, damage, blockage or cracking. If structures are designed to contain specific events, capacities and freeboard must be maintained through regular maintenance. When the functionality of the surface water system is important for the slope design, then it is normally beneficial that the inspections are carried out by geotechnical staff.

Inspections would typically be monthly during the dry season, and weekly (or more frequently) during the wet season (wet climates) and for winter snow accumulation (cold climates). In cold climates, the extent to which compacted snow accumulates in surface water channels may need to be carefully monitored as this could reduce the channel capacity at the time of snowmelt. In arid climates, surface water flow and water quality monitoring is often event-based, with data collected during seasonal or occasional runoff. Erosion of the upslope drainage area and the transport of sediment into surface water channels during runoff events can be a key factor that requires careful monitoring and maintenance.

If significant surface water control structures exist on site which are critical to the safe mining operation, these structures will usually be inspected daily and in some cases monitored continuously. In particular, streamflow diversion channels above the pit crest or in the upper part of the slope need to be carefully monitored for general condition, blockage, siltation and cracking.

Tracking of surface water flows and water quality parameters may be mandated by regulators to record the effects of mining or dewatering, or to better assess the site-wide water balance. Surface water datalogging and telemetry systems can be integrated into other monitoring systems for automated data collection. In addition to surface water stations, many mines will install one or more weather stations (along with air quality stations) that provide detailed site-specific climate data. These facilities can be very important for interpreting the observed surface water, groundwater and pore pressure responses.

6.1.3 Setting up monitoring programs

6.1.3.1 Distribution of monitoring points

How the monitoring points are distributed is a key consideration for any mining operation. In most mine site settings, the hydraulic conductivity of the site materials varies widely with depth and spatially across the site, so the distribution of the monitoring points will depend on local conditions and specific requirements.

Important factors to consider when setting out the monitoring network include the following.

- The overall groundwater flow system and the nature and location of the groundwater recharge sources.
- The need for monitoring in each of the defined hydrostratigraphic units.

- The key areas of active mining that need to be dewatered or depressurised in the short term (over the coming two to five year period).
- The need to place observation points to ensure general dewatering targets are achieved.
- The need to monitor depressurisation in the various slope sectors and the need to meet any defined targets for slope depressurisation.
- Important areas of weaker materials and sectors that are of greater risk in the slope design.
- The ability to protect the monitoring installation from damage by the mining operation.
- The ability to retain access for monitoring.

For a general open pit dewatering system, it is common that between 10 and 30 active monitoring points are operated within and adjacent to the pit to monitor the overall groundwater distribution. An example of a dewatering well and monitoring system in an open pit operation is shown in Figure 6.2. The exact number of monitoring points will depend on the scale of the mine, the scale and required timing of the dewatering and depressurisation operations, and the variability of the geological materials.

For specific pore pressure monitoring, and if the achievement of pore pressure targets is critical for a given slope sector, it is not uncommon to maintain between five and 10 multi-level piezometer installations for a single sector. As outlined in Appendix 3, there were a total of 12 active multi-level piezometer installations during the critical periods of mining of the North-west wall of the Diavik pit.

Observation wells and piezometers can often be placed on ramp switchbacks, wider parts of the haul ramps, specifically designed wide dewatering benches, or at the junction between catch benches and the haul ramp, as illustrated in Figures 6.3A and B. Disused haul ramps can make good monitoring locations if access to these can be retained.

Where appropriate, both angled and vertical piezometers can be installed at multiple orientations from a common drilling platform to provide pore pressure information at various distances and depths behind the slope, as illustrated in Figure 3.7. For active pushbacks, it is obviously better to install the new piezometers above the advancing pushback, rather than below it.

Many mining operations successfully mine around and preserve monitoring points, particularly where smaller mining equipment is being used. An example is shown in Figure 6.4. Mining around observation wells and piezometers can use the same procedures as mining around production dewatering wells.

Many open pit hydrogeology programs have evolved over decades and include a combination of standpipes, grouted-in VWPs, environmental monitoring wells, and disused production wells, where the yield has previously

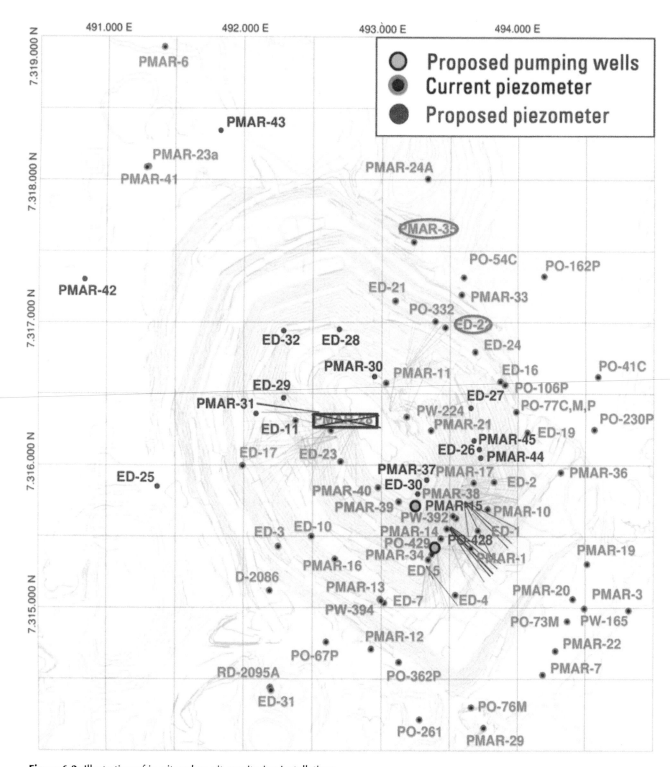

Figure 6.2: Illustration of in-pit and ex-pit monitoring installations

declined and the well is retained solely for the purposes of water level monitoring. However, it should be appreciated that production wells are commonly screened across much wider zones than monitoring wells.

In addition, a typical open pit monitoring system will need to be adapted to the changing monitoring needs and priorities of the operation. Where there is a need to obtain critical pore pressure monitoring information and there are no other alternatives, it is sometimes necessary to install an observation well or piezometer that may only last for several weeks before being destroyed by the mining operation.

Figure 6.3: (A) In-pit production well on widened catch bench adjacent to haul ramp. (B) In-pit piezometer being drilled on widened catch bench with long-term ramp access

Figure 6.4: Mining around production wells and piezometers

6.1.3.2 *Monitoring frequency*

Typical monitoring frequencies for a dewatering and/or depressurisation system are listed in Section 6.1.2. More frequent readings are often warranted at certain locations, for example after a new dewatering well has been brought on line. Conversely, less monitoring may be acceptable for zones where hydraulic stresses have been constant for many months.

Inside the pit, the frequency of automated monitoring can range from seconds to weekly, or more, depending on the resolution required. If the measurement of short-term responses to pumping is important, for example for calculating hydraulic parameters, or to provide data where pumping wells cycle on and off frequently, an appropriate data collection period would within the range 15 seconds to several minutes. Changes in pumping stress often produce a rapid response, which may be less than a minute at nearby observation points. If short-term response data are not required or water levels are stable, data collection can be set to hourly or daily to keep the size of datasets down. It is common to pre-set data collection intervals to a short time period initially, and then lengthen the time period after the behaviour of water levels has been established. Manual

monitoring with a dip sounder or hand-held readout in the mine area can be done as often as needed.

Outside the pit, the monitoring system needs to be integrated with the site-wide, district or regional monitoring program, and with any monitoring that may be carried out by the mine environmental department. Groundwater level measurements outside of the immediate pit area are frequently carried out on a bi-monthly or monthly (occasionally quarterly) time interval, with the exceptions being those points close to other mine site facilities or close to other points of hydraulic stress, where more frequent monitoring may be required. As discussed in Section 2.1, the environmental monitoring data are often invaluable for helping to support the site-wide and regional-scale models and for helping to characterise the recharge sources to the pit as a whole, or to a specific slope domain.

6.1.3.3 *Monitoring of short-term pore pressure changes*

In addition to the groundwater response caused by new pumping stresses (new production wells or drains), short-term changes to the pore pressure distribution may occur due to a number of other factors, including:

- changes in rainfall patterns (often causing an hourly or daily response);
- seasonal climatic changes (often causing a daily to monthly response);
- blasting (very rapid response, potentially milliseconds);
- slope failure (often a rapid response, potentially minutes, or sometimes more gradual associated with stick-slip response to regressive slope movement);
- mining-induced deformation response of the slope materials (potentially on the scale of days in areas of rapid mining, but generally longer term).

Pressure transducer installations, whether in a standpipe or grouted-in, can be connected to a data logger with flexible data collection rates to effectively monitor short-term changes in pore pressures. Data collection can be set to milliseconds if monitoring around a blast is required. Conversely, if the measurement of trends is the main objective and immediate responses are not important, daily or weekly readings are generally adequate.

Figure 6.5A shows an example of pore pressure responses in a pit slope influenced by seasonal recharge occurring directly to the slope. Piezometers at higher elevations (typically at a small vertical distance below the phreatic surface) show a strong seasonal response whereas piezometers at deeper elevations (typically at a large vertical distance below the phreatic surface) show a more subdued response to seasonal changes. In this case, the addition of monthly rainfall totals to the plot is essential.

Piezometers that go 'dry' should be retained if there is a potential for pore pressures to re-establish, or if there may be seasonal re-saturation and therefore a transient pore pressure response due to seasonal recharge. 'Dry' piezometers can provide valuable information for the slope performance monitoring. In certain instances, it may be appropriate to install a piezometer above the inferred phreatic surface to confirm there is zero pore pressure at a particular location, or to measure the shallow transient response to recharge events.

Figure 6.5B plots data from a dual-level VWP installation in a single borehole where the phreatic surface was below the shallow installation at the time of installation, then fell below the deep installation in January 2008. The shallow piezometer shows transient positive pore pressures for up to three months each year, with the pore pressure rising from zero to over 40 m head (~400 kPa). The deeper piezometer typically shows a greater transient response for up to four months each year, with the pore pressure rising from zero to over 100 m head (1000 kPa). There is typically more seasonal response in the deeper piezometer than in the shallower piezometer, and a time lag can be observed between the shallow piezometer and the deep piezometer as the seasonal recharge front moves downward through the slope materials. The data were collected using grouted-in vibrating wire piezometers with a manual readout box.

6.1.4 Water level monitoring

Monitoring of water levels is typically carried out using one of two basic methods.

1. Manual water level sounders.
2. Vibrating wire or other types of pressure transducers.

Water level sounders measure the water level below a fixed reference point at the surface. Therefore, it is essential to know the location and elevation of the reference point at the surface. On the other hand, pressure transducers measure the water pressure or head above the sensor. In this case it is essential to know the exact position and elevation of the sensor down the hole. Some field hydrologists prefer to 'zero' the transducer at the time of installation, so that all subsequent readings are added or subtracted from the known starting point. In this case, it is necessary to know the exact depth to water at the time of the first data point.

Inadequate procedures for recording of the sensor position and elevation are a common cause of measurement error. Errors may result simply from not recording the proper length of cable down the hole, or they may result from deviations in the drill hole. If the drill hole is deviated, the inferred elevation of the sensor may not be accurate so there may be errors when converting the measured pressures above the sensor into total head values. Where it is known that the drill hole has deviated, it may be necessary to check the total head values obtained from a downhole sensor with those obtained using a manual water level sounder.

Pressure transducers come in a range of accuracies and types depending upon the desired use. Self-contained transducers and data loggers can offer the simplest option for monitoring water levels in standpipe piezometers. Some types of system comprise a water pressure transducer and a small scale data logger to measure the water level, and an air pressure transducer and data logger to compensate for barometric fluctuations. If there are numerous wells in the area (within 10 km or so), only one station to compensate for barometric fluctuations is required for the whole network. Other self-contained transducer and data logger systems use a vent tube to self-compensate water level measurements for barometric pressure changes. There are advantages and disadvantages to both approaches. The need for venting of VWPs for barometric pressure compensation is discussed in Section 2.2.6.5.

Positive and negative aspects for manual water level sounders are summarised in Table 6.1 and for transducers in Table 6.2.

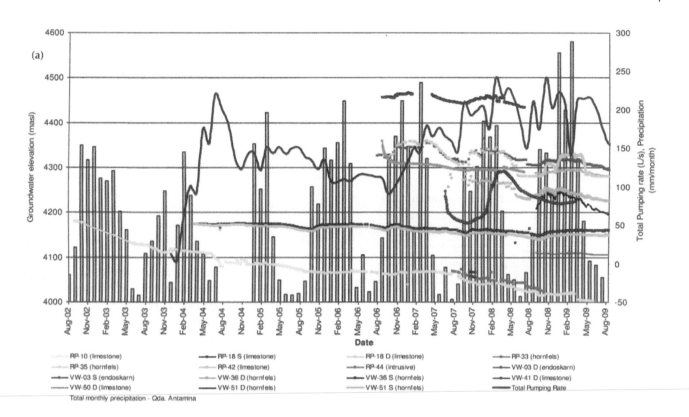

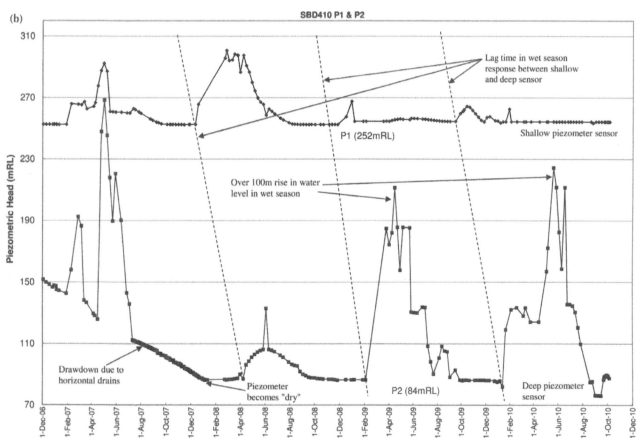

Figure 6.5: (A) Example of transient seasonal responses in piezometers effected by direct seasonal recharge to a pit slope. (B) Example of transient seasonal pore pressure responses in 'dry' piezometers

Table 6.1: Positive and negative aspects of water level sounders

Positives	Negatives
Low initial cost.	Operational costs can be high because personnel will be required to take regular measurements.
One can be used for all wells.	Can only identify long-term changes in water level.
Accurate provided data collection protocols are followed.	Affected by contamination within the borehole, especially LNAPLs, which affect the sensitivity.

6.1.5 Telemetry

Data logging and data transfer systems come in many forms, all with the same objective of recording and transmitting data from the field to the office. Real time data availability via the internet or other forms of telemetry is now standard practice in open pit mining where monitoring is a prerequisite tool for assessing design performance and failure risk, and/or for aiding risk mitigation. Other drivers of telemetry systems include the increased safety of personnel, by reducing the need to manually collect the data, and the reduced time spent in the field collecting the data, especially if the instruments are widely separated.

Data sent via a telemetry system are in the form of digital numbers, representing the output of various sensors. Typically these numbers will relate to the actual output of the sensor measured for example in frequency, current, voltage or resistance.

Telemetry systems must have some form of physical link between each end, as for example between the piezometers and computers illustrated in Figure 6.6. The most common methods are a hardwired solution, or a radio link. There are also optical systems available, using modulated laser beams to transmit information, but these are currently not widely in use.

Hardwired systems require the running of a cable between the two ends of the system. This cable can be copper carrying an electrical signal, fibre carrying an optical signal, or even tube carrying a pneumatic signal.

The most common of these is the copper cable, but whatever the medium, the hardwired system requires the cable to be installed at some point. Use can be made of existing cabling, such as an existing local area network, but typically some additional cable installation will be inevitable.

An example of remote telemetry using existing cabled infrastructure is the modem to modem link over the public telephone system. Conventional telephone calls are made over an existing network by the telemetry system, using modems to convert the values the system wishes to transmit into a form that can be sent over a telephone connection. Whilst generally reliable and easy to implement, this approach can run into significant installation cost issues if there is no existing telephone line at one or both ends of the link.

With the emergence of cellular mobile phone systems, an alternative is to use a cellular modem, a device which performs exactly the same function as a hardwired telephone line modem, but over a cellular telephone network. A dial-up method as used with landlines can still be adopted, but with the spread of 'mobile internet', the always-connected GPRS/3G/4G method is now far more common. This provides a telemetry system with a connection to the internet at the remote site. Information can then be sent over the internet to the other end of the system, which can be located wherever internet connectivity is available.

There are data-only networks available which are geared to high reliability data transmission and have a good degree of redundancy built into them. The cost of the specialised modems needed to connect to these systems, and the higher data tariffs may limit their use to applications which are critical to operations.

In most cases, the monitoring data are not critical in terms of the requirement for uninterrupted transmission, such that the loss of the telemetry link for a period of time causes major concern. However, operations

Table 6.2: Positives and negatives of pressure transducers

Positives	Negatives
Lower operational costs because fewer visits to each well are required.	Higher initial costs.
Can identify very small scale trends in water level as well as long-term changes.	Require a water level sounder when installing and downloading.
Accurate to around 0.05%.	Affected by density of water within the borehole so electrical conductivity sensors are required if water is brackish or saline.
If the well is backfilled after installation, multiple sensors can be installed at a variety of depths.	

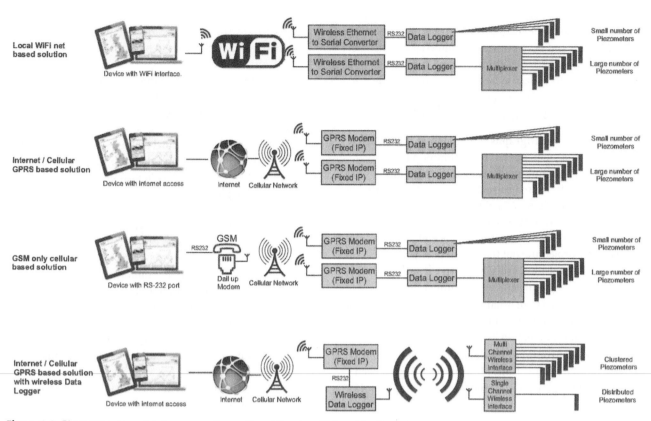

Figure 6.6: Piezometer to computer connection options (Courtesy Datum Monitoring)

integrating radar-based slope performance monitoring with their telemetry system for data communications, transfer and storage will need to consider this potential limitation.

In the case where there is no available existing communications system, be it hard wired or cellular based, then the installation of a dedicated radio-based system may be unavoidable. Alternatively, satellite modems provide internet connectivity for use over longer distances, although an issue with satellite systems is that they generally require more power than a cellular modem.

Ultimately the choice of a telemetry system has to be a balance of availability, reliability, criticality of data, quantity of data, speed of data transmission, and cost. The integration of a system is dependent on how the data are used and applied. The speed that data are retrieved from site should be related to the level of risk, how quickly the current status can change, and how quickly actions need to be taken based on the monitoring results.

6.1.6 Display of monitoring results

The objectives of the monitoring program will also dictate the required frequency, type of data analysis and display of the monitoring data. If the data collected are important for day-to-day operations, then an automated data collection system can be incorporated into a SCADA (Supervisory Control and Data Acquisition) type system, allowing operators to see the data in real time. Diagnostic data can also be displayed, such as signal strengths and battery voltages. While these types of systems are more common for permanent facilities (such as municipal water supplies) than for mining applications (where operating conditions may change rapidly with time), many packaged telemetry options now include automated data collection and report generation software. As these applications become globally widespread, they will become more commonly utilised by the mining industry.

Where data are collected for the purposes of long-term mine planning, weekly or monthly cursory system status and performance updates may be sufficient, with quarterly detailed reports prepared to show graphical representation of the data and other pertinent analysis. However, where the data are critical for short-term dewatering or for slope stability concerns, it is usually appropriate to provide real-time monitoring and daily reports.

It is often good practice to post print outs of key piezometer hydrographs and required targets (e.g. Figure 2.53) on notice boards within the mine operations and/or mine engineering departments so that crews are aware of the importance of the hydrogeology program. The hydrogeology and pore pressure monitoring can also be integrated with the geotechnical hazard map that is becoming commonplace at large mining operations.

Multiple formats can be used to display the data, as follows (Section 12.2.2.5, Read & Stacey, 2009).

- **Piezometer hydrographs** showing total head plotted against time, which is the most common format. Records from multiple piezometers, particularly vertically discretised piezometers, may be included on a single plot (Figure 6.7), illustrating vertical gradients between the transducers. Future mining elevations or pressure targets may be included as end points on the plots. By projecting out the rate of pore pressure declines, these plots will show if the drawdown rate is sufficient to stay ahead of the mining rate.
- Piezometer plots are usually grouped by their physical location within the pit but may be grouped by hydro-geological unit (Section 2.3.2.7).
- **Plots of horizontal drain data**, usually showing flow rate plotted against time. Multiple drains from each sector of the slope may be included on a single plot, with the total flow rate from all the drains in the sector shown on the plot. Drain flow plots are usually grouped by their physical location within each pit sector but may be grouped by hydrogeological unit (Section 2.3.2.8).
- **Plot of flow rates from pumping wells**, showing flow rate plotted against time. Wells from each sector of the slope may be included on a single plot, with the total flow rate from all wells in each sector also shown. The flow rate plots are grouped by their physical location within each pit sector and the wells may also be grouped by their hydrogeological unit (Section 2.3.2.8).
- **Composite plots**, which may show:
  → changes in pore pressure shown with changes in the flow rate from drains or wells;
  → changes in horizontal drain flows or pore pressure shown with changes in the rate of movement of prisms or other slope monitoring instruments (Figure 6.8). It is often useful to annotate the plots with dates where new wells came on line, where a pushback was mined in a particular slope sector, or precipitation events and other factors that may have influenced the observed hydrograph response (Section 2.3.2.11).
- **Water level maps**, which show total head or pore pressure at a given moment or which show change in

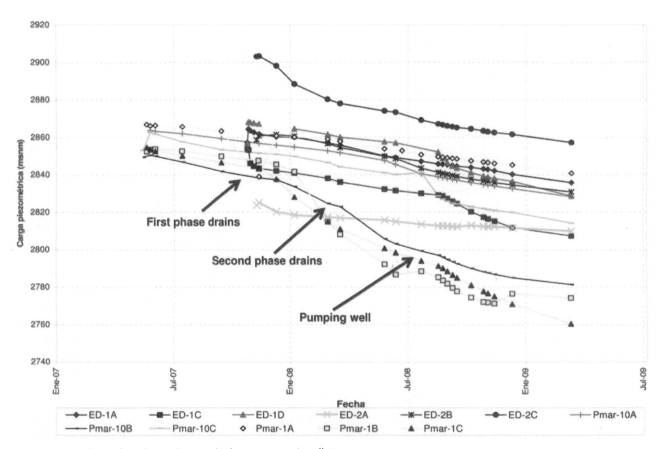

Figure 6.7: Hydrograph with data from multiple piezometer installations

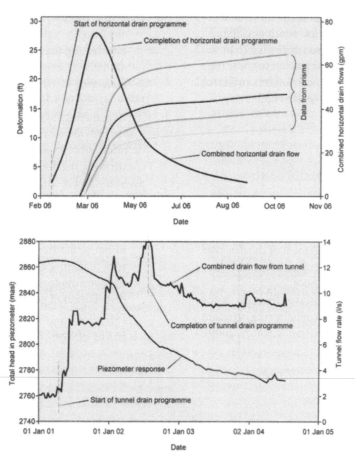

Figure 6.8: (a) Plot of prism data vs horizontal drain flow, for the SE wall, Chino Mine, New Mexico, USA; (b) Plot of piezometer response vs drain flows from tunnel, for the NE wall, Escondida Mine, Chile (Read & Stacey, 2009)

total head or pore pressure over a given time period (Section 2.3.2.3). To assist with data interpretation, it is normal that all piezometer and pore pressure data are plotted as pore pressure elevations (total head) onto a base map showing the key geological structures, and possibly lithology and/or alteration. Vertical differences in total head values can be presented by colour-coding the data from vertically discretised piezometers. When preparing water level maps, it should be noted that, because of local aquifer losses and well losses (Appendix 2, Figure A2.19), dynamic water levels in pumping wells are usually much lower than the water levels in the adjacent formation and therefore need to be distinguished. If the water level data are contoured, it is often good practice to ignore the dynamic water levels in pumping wells, particularly when working in a fractured rock setting where local aquifer losses can be high.

- **Distribution maps** showing, for example, the distribution of drain hole and production well flows, which can include:
  - → drain hole locations, with drain holes colour-coded by their initial flow rate (Figure 2.47);

  - → pumping well locations with initial and current flow rates. Depending on the amount of data available, the flow rate map can be integrated with one or more water level maps.

- Hydrogeological cross-sections which show:
  - → vertically discretised piezometers plotted along the section (Section 2.3.2.10);
  - → changes of the water table or pore pressure distribution with time;
  - → numerical model cross-sections that show actual monitored pressures compared with model output at a given point in time; an example is shown in Figure 6.9.

Figure 6.10 is presented as a unique example of pore pressure responses to seismic events in Chile in 2010. The record is from the Chuquicamata Mine, Calama, northern Chile, and the particular event is the February 2010 Magnitude 8.9 Concepción earthquake that resulted in considerable damage and loss of life in the Concepción and Santiago regions. Calama is located about 1650 km north of Concepción and the earthquake was not felt in the surrounding region. However, despite the geographical distance, piezometers at the mine, and also at the Escondida Mine about 200 km south of Calama, reacted to the earthquake. It is also interesting to note that piezometers at Chuquicamata reacted to a less than Magnitude 7 earthquake located near Calama in March, after the Concepción earthquake (end of upper hydrograph in Figure 6.10).

6.2 Performance assessment

6.2.1 Overview

Pit slope design is an iterative process that relies to varying degrees on empirical science, the design philosophy implemented at the mine, the judgement of the designer, experience gained at other similar operations, and feedback based on actual performance.

Actual slope performance is assessed from the output of a visual and instrumented slope monitoring program that has the following basic goals (Read & Stacey, 2009).

- Detecting and recording slope movement as a basis for:
  → ensuring the safety of the operation;
  → establishing the basis for the movement (failure mode);
  → managing instability.
- Investigating failures and ongoing instability. Monitoring of instability assists in identifying failure mechanisms, providing crucial data for back analysis and defining appropriate remedial work.
- Confirming the design model or providing a basis for assessing and modifying the designs, including:
  → geology (rock type distribution, alteration and mineralisation);
  → structural model, with consideration of major and minor structures;
  → rock mass properties;
  → the hydrogeological model, in particular the distribution of groundwater and the resulting pore pressures;
  → *in situ* stress levels, particularly for high slopes.

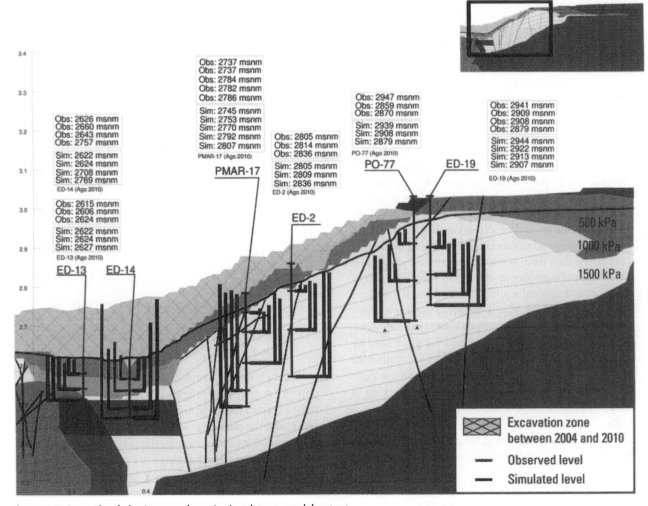

Figure 6.9: Example of plotting actual monitoring data on model output

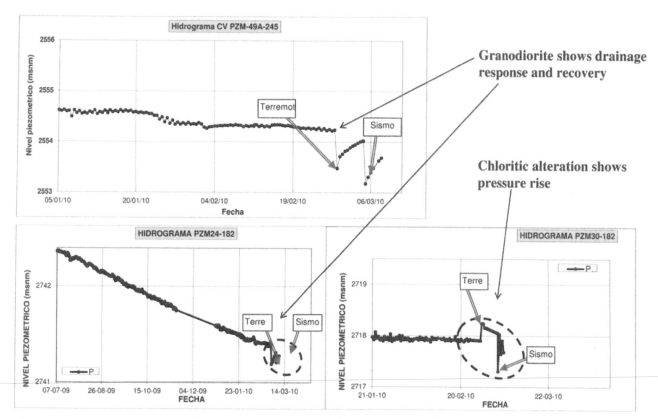

Figure 6.10: Pore pressure responses due to 2010 seismic events in Chile (Source: Luis Olivares)

- Ensuring that the slope design criteria and assumptions are being achieved in terms of operating procedures.

Of these goals, groundwater monitoring is key to producing an optimised design in the context of safety, ore recovery and financial return. Slope deformation can occur independently of pore pressures, but more commonly groundwater has a detrimental effect on stability. Hence, a critical element in the assessment process is the tracking of changes in the distribution of the groundwater and the resulting pore pressures as mining proceeds.

Essentially, the assessment process uses results of the monitoring described in Section 6.1 in three ways:

1. To verify the pore pressure model used in the slope design studies.
2. To verify that the slope depressurisation targets have been met (or otherwise).
3. To make iterative and ongoing changes to the slope depressurisation program going forward.

6.2.2   Operational groundwater flow model

A numerical groundwater flow model is a means of representing a real groundwater system. As outlined in Section 4.1.1.1, it uses a computer program to apply the equations of groundwater flow (essentially a combination

of Darcy's law and the principle of continuity or conservation of mass) to describe the behaviour of groundwater within a specified volume of formation called the model domain. The region to be modelled (the domain) is divided into a finite number of cells or elements, each of which can have its own distinct hydraulic properties (permeability, storage, specific yield) and boundary conditions (e.g. recharge, abstraction, or a specified groundwater head). Known or inferred values of head, groundwater flux or recharge must be defined at the model boundaries, along with the starting conditions.

When all input data are in place, the computer code solves the governing equations for each cell (element) or node using an iterative process. For the vast majority of pit slope applications, the groundwater heads and flows are time variant, so the model is described as non-steady-state or transient, and the head values are calculated for a chosen series of time steps.

The accuracy and reliability of the numerical model used for operations is heavily dependent on having sufficient reliable supporting data; both geology and hydrogeology. In order to obtain an acceptable representation of the real groundwater system, the model is usually calibrated to historical data on groundwater heads and/or flow and the data available from the early stage planning and design studies (Levels 1 through 4).

When calibrated, the model can then be used to make forward predictions about the future behaviour of the groundwater system, how the future pore pressures will change with time as mining is advanced, and the required design of the slope depressurisation system.

It is normal that the models which are developed during the planning and design stages of the project require significant re-calibration during mine operations (Level 5). In the early stages of active mining, as the pit is initially advanced below the water table, there is an increasing amount of hydraulic stress in all sectors of the pit. This represents a major opportunity to understand the actual hydrogeological conditions and to validate the predictions made during the study and design stages. Clearly, it is essential that the designated groundwater monitoring network is in place before mining advances below the water table as this will provide information both to verify and advance the model. In addition to the piezometer network, the opening of the pit provides forensic information about the newly exposed geology (lithology, alteration, mineralisation and structure) and any groundwater seeps or other inflows to the pit walls or floor.

As mining proceeds, and the pit floor is mined deeper and the slopes are pushed outward, so the magnitude of the inward hydraulic gradients increases. An increasing amount of drawdown occurs in the piezometer network. The footprint area of geological and hydrogeological understanding expands. As more information is gathered, there will almost certainly be modifications to the conceptual model which may result in changes to the approach adopted for the numerical model. Thus it is normal that the operational models evolve with time as mining progresses and as more data become available to support and validate the modelling process.

Provided that sufficient supporting data were collected during the planning and design stages, it is usually possible to take the numerical models that were progressively developed during the Level 1 through 4 studies and to convert them into operational models (Level 5). However, it would be normal to go through a major update for both the conceptual and numerical models within the time period of about six to 12 months following first excavation below the water table. Once the operational models are developed, a reasonable update and re-calibration period is yearly for the first three to five years and two-yearly after that once the mine has become established. More frequent calibration is often warranted where major changes occur to the mine plan or where pushbacks expose new geological units that are not fully characterised.

6.2.3 Process for ongoing assessment

While the monitoring data are important for validating the model and improving the forward predictions, the 'bottom line' for the performance assessment is to confirm that the actual pore pressure values which have been used as the basis for the slope design are being achieved in the field.

It is usually of benefit for the mining operation to implement an organised process for performance assessment and for interactively modifying the action plan going forward as dictated by the monitoring results. The process is typically carried out as follows.

- **Weekly** updates of pore pressure hydrographs that compare the actual pore pressure values (or total head values) to the required targets. These are normally discussed during a weekly site meeting to review progress during the previous week and to determine the action plan for the coming week. The monitoring data may indicate that is it necessary to modify the weekly drilling and/or maintenance schedule to ensure that flow rates for critical wells and/or drains are maximised. The weekly meetings are typically attended by mine planning, mine engineering, geotechnical, hydrogeology and mine operations staff.
- **Monthly** reporting to review progress towards the required targets and to recommend modifications (as required) to the detailed short-term action and mitigation plan (three to six months). The monthly report may form part of a monthly management meeting and, depending on the importance of the program, may be attended by corporate technical staff, mine management, mine planning, mine engineering, geotechnical, hydrogeology and mine operations staff.
- **Quarterly** data update, which would typically include:
  → plots of pumping well flow rates for each sector of the pit and the total pumping rate;
  → plots of horizontal drain flow rates for each sector of the pit;
  → piezometer hydrographs for each sector of the pit annotated with flow rates and rainfall (as relevant);
  → maps of total head and drawdown achieved during the preceding period;
  → updated piezometer levels plotted onto hydrogeotechnical cross-sections;
  → pertinent chemistry data plots or maps;
  → a listing of new installations during the previous quarter with relevant field data and well completion logs;
  → modifications to the action and mitigation plan that are required for the coming quarter. The quarterly data update is normally prepared as a factual report by mine hydrogeology staff, often with the help of external consultants. The quarterly report may provide the basis for making major modifications to the short-term action plan (three to six months) or for changing the overall drainage strategy for certain sectors of the pit.

■ **Six-monthly or annual review** of the results by mine staff and external experts to confirm that targets are being achieved or to evaluate the reasons and consequences of not achieving the targets. The review will normally include the verification (or modification) of the overall dewatering and slope depressurisation strategy and the development (or conformation) of a rolling action plan for the coming 12 months. At many mines, the hydrogeology review process is now combined with the geotechnical review and may form part of the review by the Geotechnical Review Board. The consequences of not achieving the pore pressure targets may be a change in the dewatering and/or slope depressurisation plan, a change in the slope design or a change in the mine plan. It is important that the review also identifies the upsides in sectors where pore pressures may be dissipating faster than expected and where it may be possible to consider a partial or full steepening of the slope.

The time interval for the meetings, data reports, reviews and program modifications will depend on the scale and urgency of the hydrogeology program and may be more or less frequent than indicated above.

6.3 Water risk management

6.3.1 Overview

Mining companies are, by definition, engaged in activities that carry significant risk at many different levels within the organisation. At the highest level, risk analysis and management is used to satisfy external compliance with international standards of corporate governance, for example, the Higgs Combined Code and Turnbull in the UK, the Sarbanes-Oxley Act in the USA, the ASX Guidelines in Australia, and the King Codes in South Africa. Internally, it is used by most mining companies as an integral part of proactively managing their business at all levels.

The principles and procedures of risk analysis and management are widely applied by mining companies to an ever increasing range of activities and situations to improve their understanding and management of a variety of hazards at all stages in the mining cycle. For specific applications, the general procedure of risk identification, evaluation, mitigation response and reporting is adapted to account for the factors and features particular to that situation. In most cases, the process retains a structured approach and methodology to compare the assessed levels of risk with predetermined standards, target levels or other criteria. A uniform process ensures that consistency and quality is maintained, and risk levels can be compared as the process is repeated at regular intervals.

This section complements the comprehensive description and treatment of risk management applied to the geotechnical risks associated with each stage of the open pit slope design process as presented in Chapter 13 of the LOP project's book, *Guidelines for Open Pit Slope Design* (Read & Stacey, 2009). It provides an overview of the risk management process, accompanied by a more detailed discussion of how this process can be applied to describe and evaluate risks associated with water in open pit mine operations. Whether the water acts to elevate pore pressures and reduce effective stresses in the walls of the pit, contributes to deformation response during mine development, or acts as a medium to promote block instability during blasts, it can be a significant contributor to the risk of both major and minor slope failures. Similarly, uncontrolled inflows, pit flooding and wet blasting can impact access, schedules, production rates and mining costs.

6.3.2 Process of risk analysis

Effective risk management relies on a competent definition and analysis of the specific risks for any given situation. There are many frameworks for assessing risk. Many of these are similar and generally follow clearly defined and inter-related steps. However, there are differences in application between mining operators and the type of risk being evaluated. For purposes of illustration, the generic process described in AS/NZ 4360:2004 is presented in Figure 6.11.

This process can be broadly applied and contains six basic components.

1. **Initiate the risk process** and establish the context – establish the external, internal and risk management contexts in which the rest of the process will take place. Establish the criteria against which risk will be evaluated and define the structure of the analysis.
2. **Risk identification** – identify where, when, why and how events could prevent, degrade, delay or enhance achievement of the objectives.
3. **Risk analysis** – identify and evaluate the existing controls. Determine the consequences and likelihoods of particular occurrences and therefore the associated levels of risk, considering the range of potential consequences and how these could occur. Generally, the risk is quantified as the product of the likelihood and consequence of the particular occurrence.
4. **Risk evaluation** – compare estimated levels of risk against the pre-established criteria and consider the balance between potential benefits and adverse outcomes. This enables decisions to be made about the extent and nature of treatments required and their priorities.
5. **Treat the risks** – develop and implement specific cost-effective strategies and action plans for increasing

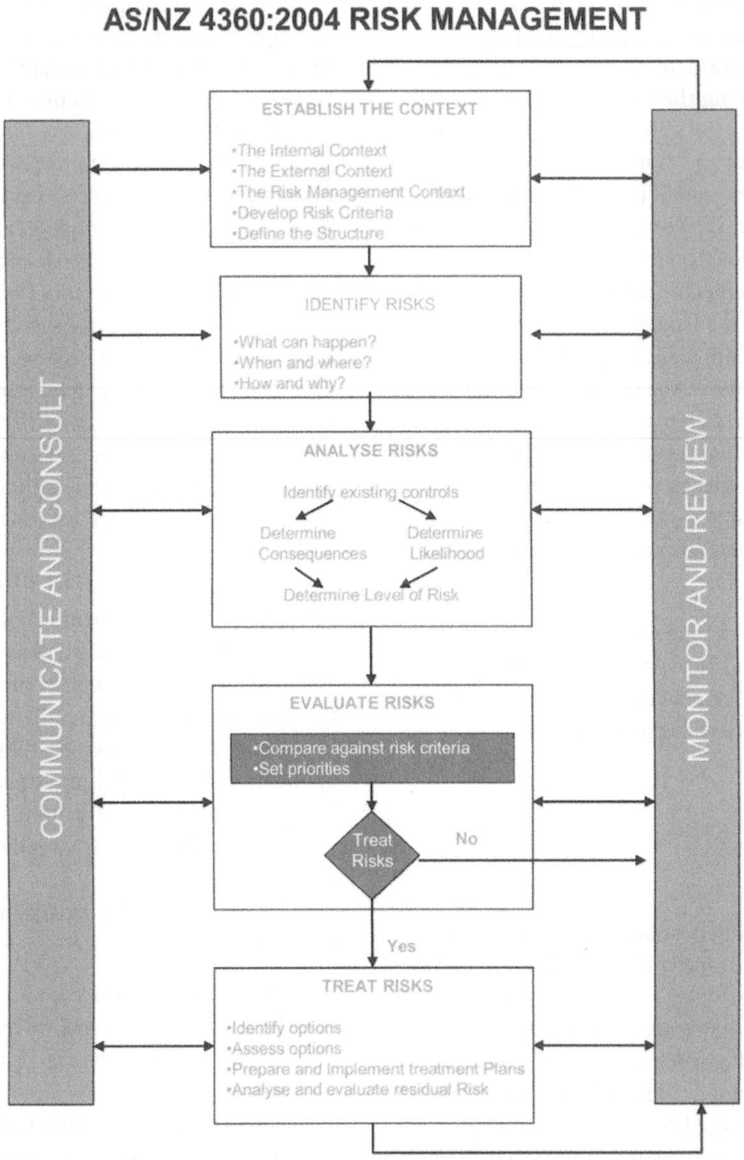

Figure 6.11: Risk management process (Read & Stacey, 2009)

potential benefits and reducing potential costs or adverse effects.

6. **Monitor and review** – monitor and review progress and the effectiveness of all steps in the risk management process to ensure continuous improvement, and that the risk management plan is implemented effectively and remains relevant.

Figure 6.11 shows that communication and consultation is required at every stage in the process, and that monitoring and review create feedback loops that may require modifications to earlier results.

It is important that the process is effectively documented. A risk analysis will generate a risk register that documents identified risks along with the analysis,

evaluation, and agreed responses. This forms the basis for implementation of mitigation strategies and monitoring of progress against agreed targets or milestones. Generally, risk owners are nominated for each treatment strategy and are accountable for ensuring that appropriate resources are allocated, and for maintaining progress toward targets.

6.3.3 Risk assessment methodology

6.3.3.1 Types of risk assessment
Risk assessment can be qualitative, semi-quantitative or quantitative, depending on the business level of assessment (e.g. corporate, management or operational), experience and expertise of the evaluators, the stage of the project or operation, and the availability of experience-based and factual data.

- Qualitative methods use verbal and subjective or judgemental descriptions of likelihood and consequence where data or resources are insufficient to support quantitative analyses. A qualitative approach is often effective at a high level to identify those risks which require more detailed analysis.
- Semi-quantitative methods apply weightings or scales to qualitative descriptions to generate a more detailed ranking. The weighting process carries a risk of subjective evaluation that may lead to inconsistencies in relative risk. Quantitative likelihood information is either subjective or judgement-based using the input from experts, or calculated; and is most often expressed as either a probability for a single event or a frequency of occurrence for repeat events. This can be useful if the requirement is to rank risks in order to prioritise the mitigation responses or to allocate funds or resources. Semi-quantitative methods are most applicable at an operational level when relevant expertise is present and probability or weighting can be meaningfully assigned.
- Quantitative methods are the most objective form of risk analysis. The approach uses calculated probabilities for the likelihood and consequence of an occurrence. This implies that sufficient data are available to quantify the consequences in terms of environmental,

economic, operational, technical or safety impacts. Quantitative methods apply at a discipline level, where values or data can be compared against relevant acceptance thresholds or trigger levels from other industries, or society in general.

Examples of the way these approaches can be applied in assessing water risk at a mining operation or at the study stage are illustrated in Figure 6.12. Each element of the water-management system can be subjected to increasingly rigorous levels of risk analysis to improve the reliability of any aspect of water management where the risk is judged to be unacceptable, given the prevailing data or knowledge. It is usual that some elements are common to more than one aspect of water management. For example, water chemistry may pose a specific risk to the pumping and conveyance equipment used for pit drainage, but also an environmental water management risk for water treatment and discharge. Invariably the consequences in each area are different, and the risk should be covered in each area, particularly as different departments within the mining organisation may be accountable for managing the associated risks.

Once mitigation and action plans are formulated, risk ratings can be revised in higher level assessments, while residual risks are reassessed once data on the performance

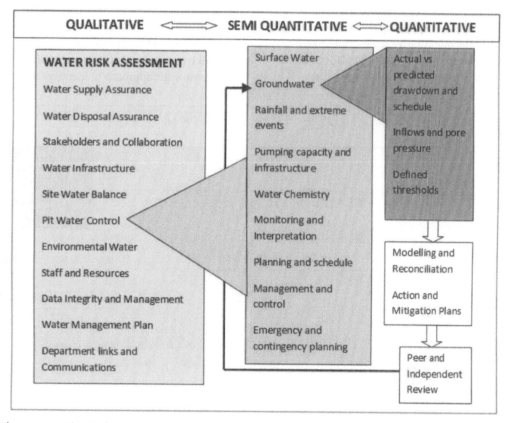

Figure 6.12: Risk assessment for pit slope water management

of mitigating actions becomes available. Peer or independent review for high and extreme risk elements is important, as is the iterative nature of the process.

In the example shown in Figure 6.12, pit water control is identified as an area of high risk in the qualitative assessment. Subsequently, the elements of pit water control are subjected to a more rigorous semi-quantitative risk assessment that exposes groundwater factors as a key issue. A full quantitative risk analysis may then be initiated to understand the full business or economic risk associated with pore pressures or groundwater inflows. In a similar way, any element of the site water-management system (e.g. water supply assurance or water-management infrastructure) may be subjected to progressively more rigorous risk assessment approaches.

6.3.3.2 Risk matrix

Qualitative or semi-quantitative likelihood and consequence scales are commonly defined by 4 or 5 points or ranges providing a 4×4 or 5×5 matrix. The scales should be sufficient to allow adequate discrimination between different classes of risk, for example low, medium, high, or extreme. Whichever the choice, it is essential that the scale points are carefully defined by specific risk acceptance thresholds during the planning phase when the context is being developed. An example 5×5 risk matrix is presented in Figure 6.13 to demonstrate combinations of scales that may be used for qualitative, semi-quantitative and quantitative risk matrices depending on the combination of scales used. For the quantitative scales, time-based impacts can be replaced by costs (Opex or Capex), annual production, resource or reserve, or NPV to assess

economic consequences. Non-economic consequences such as safety, environment, compliance and reputation generally use qualitative or semi-quantitative classifications. In the same way, the scale of likelihood can use statistical probability to create a semi-quantitative or quantitative risk matrix.

Likelihood and consequence scales, and risk classification thresholds, are selected and tailored to suit specific purposes. Different criteria, values, and ways of combining likelihood and consequence can be used depending on the amount of data available, the stage of the project, or the purpose of the risk assessment. During early study or project stages, before verified data or experience is available, qualitative techniques are appropriate.

For addressing the risks associated with water, at least a semi-quantitative analysis should be possible during mine operations, although a quantitative analysis of economic risks is preferred as a management decision tool. There is often a significant amount of factual data and experience that supports the development of a quantitative risk matrix. Developing a matrix to support decision making with respect to pore water pressures in pit slopes may require:

- several seasons of groundwater and climate data monitoring;
- groundwater model development, calibration and verification for events that may lead to significant or rapid changes in pore water pressure, and the possible magnitude of these pressure changes;
- a failure modes and effects analysis (FMEA), using a deterministic approach to identify and rank all potential

			CONSEQUENCE or IMPACT				
LIKELIHOOD	90%	Likely	M	H	H	E	E
	50% --	Probable	M	H	H	E	E
	25% --	Possible	L	M	M	H	E
	10% --	Unlikely	L	L	M	H	E
	5% --	Rare	L	L	M	H	H
Semi-Quantitative		Qualitative	Insignificant	Minor	Moderate	Major	Catastrophic
Quantitative			< 1 shift	< 1 day	< 1 week	< 1 month	> 1 month

Figure 6.13: Example 5x5 risk matrix

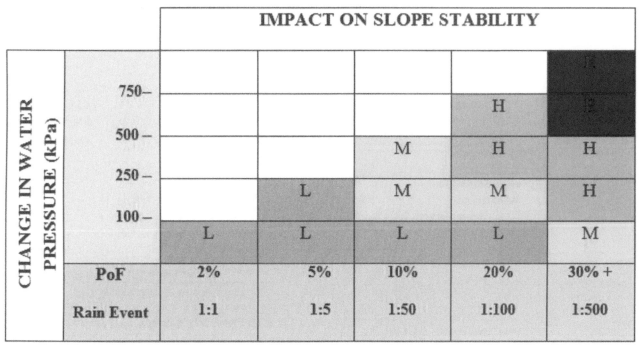

Figure 6.14: Use of 5x5 risk matrix to quantify the impact of water pressure on slope stability

risk elements, which may be performed to identify those risks for which more detailed quantitative risk analysis may be justified (e.g. fault tree, or event tree analysis);

- fault tree analysis to establish a Probability of Failure (PoF) due to changes in water levels;
- event tree analysis to determine appropriate risk classes;
- an understanding of the slope design Factors of Safety (FoS), and how variations in water pressure may influence the PoF for various sectors of the open pit.

All this information is site specific and can change as the mine develops or expansion projects are initiated.

Figure 6.14 illustrates a quantitative type evaluation of the impact of a variety of head conditions on slope stability by customising the scales for likelihood and consequence. This example uses the PoF for slopes, or typical rainfall return periods that may result in an increase in pore water pressure, the magnitude depending on the types and efficiency of dewatering methods employed.

A matrix such as this is developed for a specific set of conditions affecting a specific site over a given time period. It allows changes in slope pore water in different pit sectors to be understood in terms of risk, and appropriate actions to be initiated without delay. Calibration to rainfall return events or open pit depth can add a practical element that allows the change in risk to be predicted and suitable mitigation actions to be planned in advance.

Ultimately, the objectives are to identify and manage threats to the mining operation, and to maximise opportunities through the evaluation process inherent in a risk assessment. Risk mitigation and management plans are the communication tools to ensure that risks which are identified become actions in geotechnical or mine water management plans.

6.3.4 Identifying the risks

Risk identification is a subjective view of reality and cannot be accomplished by a single person. As water-related risks impact performance at many mining operations, risk identification normally involves a facilitated workshop that includes key stakeholders and experts. This ensures that all possible hazards are considered from a variety of perspectives and the associated risks are determined. These can be threats or opportunities. There are a number of tools and techniques available to define risks and opportunities including the following.

- Interviews (individual/group/expert).
- Brainstorming.
- Reports and reviews.
- Questionnaires or checklists.
- SWOT (Strengths, Weaknesses/Limitations, Opportunities, and Threats) or similar analysis.

It is important that a structured approach is used to ensure that potentially critical risk items are not overlooked and the full spectrum of relevant impacts is explored.

Water management is often split between several departments. Mining Planning and Mine Operations Departments may be responsible for water in the pit; Dewatering and Geotechnical Departments for monitoring and instrumentation; Processing Departments for water

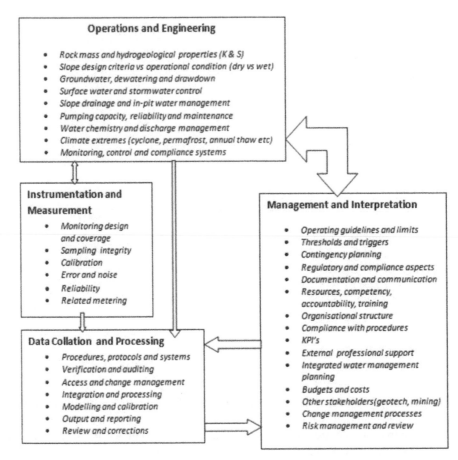

Figure 6.15: Systems approach for characterising water-associated risk

supply, reticulation and tailing; and the Environment Department for discharge, water quality monitoring and compliance. Water-related risks are not only present within each of these control areas, but also may present operational and communication gaps that exist between departments.

Water risks at each mine are unique and a range of approaches can be considered to identify key risks.

6.3.4.1 Systems approach

A systems type approach is one way of defining risks associated with control of water in pit slopes. Figure 6.15 illustrates in a simple form the inter-relationships and inter-dependencies of operations and engineering, instrumentation, data interpretation and the management functions. Each point may represent an area of concern or opportunity that may define some risk or a series of risks. For example, a lack of technician training may lead to issues with sampling accuracy or data collection, or sulphide rich formations may initiate Acid Mine Drainage (AMD) and issues with compliance for discharge criteria. Arrows represent data flow and, more importantly, the communication that is required between the system functions which, in itself, is often an area of significant risk.

A possible issue with using a systems type approach to defining risks is that individual parts of the system may be the responsibility of different mine operational areas. This may lead to difficulties in implementing risk mitigation measures. For example, pumping capacity may be compromised by maintenance issues because maintaining haul trucks takes precedence over maintaining pumping systems with the resources that are available. In addition, the approach may not be consistent with the level, scope or objective of the risk assessment, or the organisational structure at a given site (Section 5.2.9).

6.3.4.2 Management area approach

Another approach for identifying risks may be based on the assignment of management responsibilities or areas. Such areas of water management related responsibility may include: groundwater management (which includes geotechnical factors); surface water management; water monitoring; water chemistry; water-related targets (e.g. mining schedule, environmental compliance); infrastructure and capacity; emergency and contingency planning; and resources and skills. An example list to demonstrate the management area type approach is shown in Table 6.3. It should be appreciated that, for any given

Table 6.3: Management area approach for characterising water-associated risk

Management area	Hazard/Risks	Consequence
Groundwater	Uncontrolled high inflow	Production impacts/pumping costs
	Ineffective dewatering	High pore water pressures/geotechnical issues Pit flooding
	Rapid/seasonal recharge	Transient water pressures/geotechnical issues
	Water chemistry and AMD	Environmental and discharge compliance, treatment
	Insufficient drawdown vs mine schedule	Wet blasting Geotechnical issues Pit flooding
	Spatial drawdown impacts	Environmental, social impacts
	Dewatering system maintenance and operation	Pit flooding Geotechnical issues Cost Reliability
	Offsite hydrogeologic influences e.g. neighbouring pit closure and flooding	Increased inflows High pore water pressures/geotechnical issues Pumping capacity
Surface water	Extreme climate events	Recharge – transient pore pressures Pit flooding Lack of pit access Lost production Safety
	Haul road damage	Loss of access Tyre wear
	Pumping capacity	Lost production
	Silt loading and discharge compliance	High sediment load entering pit floor and impairing access Suspended solids in off-site discharge water Fines or censure
	Offsite hydrologic influences e.g. changes in catchment land use, river diversion or new dams, distant rainfall events	Supply security Change in discharge impacts Flash flooding
Water chemistry	High-volume AMD or ARD	Potential changes to the downgradient water chemistry Treatment costs
	High corrosion potential	Reduced equipment life – increased costs Protection of equipment and maintenance personnel
	Blasting agents contaminate pit groundwater	Potential changes to pumped water chemistry
Water monitoring and interpretation	Inaccurate sampling or data collection	Poor control of risk (geotechnical, production, environmental) Invalid interpretation
	Insufficient monitoring coverage	Compromised interpretation Conservative design
	Insufficient modelling experience	Results and predictions of groundwater, surface water and water balance models are incorrect leading to safety, cost, design and operational issues
Infrastructure and capacity	Poor design and insufficient pumping capacity	Reliability Pit flooding Lost production Higher capital and operating costs
	Horizontal drain discharge control	Recharge and lubricate bench scale jointing in overbreak zone Nuisance water
	Operability of in pit dewatering bores	Wet blasting Pit flooding Higher capital and operating costs Lost production
Management and control systems	No SOPs, standards or guidelines	Poor control Lack of consistent responses No KPIs or targets

Table 6.3: Continued

Management area	Hazard/Risks	Consequence
	Poor communications and integration	No business-wide planning or co-ordination
	Poor data management and documentation	Poor interpretation and understanding of risk Succession issues Compliance and safety issues
Resources and skills	Operator developed software for data management	Succession planning Loss of monitoring data Poor interface and compatibility
	Inadequate maintenance of monitoring equipment	Incomplete data record Inaccurate measurements Data reliability
	Insufficient competency and training	Poor or incorrect interpretation High levels of external support which may themselves be of poor standard
Emergency and contingency planning	Drought events	Water supply assurance Licence allocations Lost production due to mill downtime
	Flood events	Access Health and safety of personnel Stability of slopes/waste dumps/tailing facilities Loss of equipment
	Large retention impoundment failure	Downstream social and environmental impacts Lost production Contaminant clean up Reputational damage
	Unplanned discharge	Breach of licence with financial penalties Downstream impacts Reputational damage

site, the actual list developed in a risk identification process must consider site-specific conditions, and will therefore be more detailed and thorough.

6.3.4.3 *Operational management approach*

Risks may also be identified according to the specific criteria stipulated by the mine's operational management plans. Water management requirements are documented in parts of the mine plan and schedule, the ground control or geotechnical management plan, the water management plan, the environmental management plan, and the waste (rock and tailing) management plan. All these plans lay out details of the specific water management components for a mining operation. Each plan may contain target KPIs (e.g. water levels, water quality, pumped volumes), monitoring and sampling procedures, contingency and emergency planning, management responsibilities, and accountabilities. As a result, they generally cover all the areas and disciplines where technical and business risks can be readily identified. Whichever approach is taken to identify the water-related risks relevant to a particular mining operation, it should be practical and avoid excessive duplication.

It is best to avoid evaluating risks during the risk identification stage so as to minimise bias. The risk identification process should only aim to generate the definition and listing of the risks. The evaluation of the identified risks should be carried out as a separate part of the exercise.

6.3.5 Defining the consequences

The consequences of any identified risk can either be economic or non-economic. An experienced risk facilitator may be able to manage the evaluation of economic and non-economic risks during the same process; otherwise these can be completed separately.

Economic consequences are scaled relative to the risk tolerance of the mine business, operation or study. They can be either negative (threats) or positive (opportunities). The evaluation of economic consequences will be different for a study than for an operation. In general terms, economic consequences that may impact the NPV can be evaluated as follows.

- Capital expenditure ($).
- Project or mining schedule (months/weeks).
- Operating cost ($).
- Production volumes (Mtpa).
- Revenue ($/a).

These types of economic consequences each require different scales and risk thresholds to be defined proportional to the total project investment or operational

Table 6.4: Example risk acceptance thresholds for a project with a capital cost of about a $1 billion

Consequence type	Consequences			
	V. Low	Low	Moderate	High
Capital expenditure	<1.6% <$16 M	1.6–5% $16–50 M	5–10% $50–100 M	>10% >$100 M
Project schedule	<2.5% <1 Months	2.5–7.5% 1–2.5 Months	7.5–15% 2.5–5 Months	>15% >5 Months
Operating costs	<0.6%	0.6–2.5%	2.5–7.5%	>7.5%
Annual production	<0.6% <0.1 Mtpa	0.6–2.5% 0.1–0.4 Mtpa	2.5–7.5% 0.4–1.1 Mtpa	>7.5% >1.1 Mtpa
Investment (NPV)	<1.6% <$14.4M	1.6–5% $14.4–45 M	5–10% $45–90 M	>10% >$90 M
Resource/Reserve	<0.6% <1.5 Mt	0.6–2.5% 1.5–6 Mt	2.5–7.5% 6–18 Mt	>7.5% >18 Mt

budget. For a project with a nominal capital cost of a billion dollars, the risk acceptance thresholds that are selected may be along the lines of those shown in Table 6.4.

Given the potential consequences of poor slope performance that may result from water pressures, or the potential consequence of pit flooding on mine operations, the associated economic risks are generally high and it is normally necessary to ensure there is a program of ongoing active management to minimise the likelihood of occurrence. Even then, the risks that result from in-pit water may plot as 'high' or 'extreme' in many risk matrices, and will therefore require further detailed evaluation and pro-active management.

The most important non-economic consequences may include:

- health and safety;
- environment;
- social;
- compliance and regulatory issues;
- reputation and workforce relations.

In addition, there are many other potential non-economic consequences and risk acceptance thresholds that will define what is and what is not acceptable to any organisation. Scales for non-economic consequence are normally qualitative, although they often have collateral economic consequences, and quantitative techniques can also be used.

6.3.6 Implementing a water risk management program

6.3.6.1 Risk management and the water management plan

Water management performance in operations is guided by the site Water Management Plan. The Water Management Plan details the business and operational strategy for water management and is also the primary governance mechanism for management of water-related risks at a mine site. This document typically describes:

1. the physiographic and climatic setting of the operation;
2. the interaction of mine water with the external environment and visa versa;
3. the identified water-related risks at a mine site with the defined the approach to managing these risks;
4. the approach to water management on the site from interception, supply to discharge or disposal.

Responsibilities and accountabilities for water in each area of the operation, monitoring and control, and water risk management are integral to the site Water Management Plan.

The plan essentially summarises a number of supporting operating strategies or manuals that apply to each area of water management on the mine. These may include for example a Borefield Operations Strategy, a Tailing Operating Manual, a Stormwater Management Plan, a Mine Dewatering Systems and Procedures Manual, and a Dams, Pipelines and Pumps Maintenance Manual.

Of most relevance to pit slope water management, the Mine Dewatering Systems and Procedures Manual defines the aims and objectives of the pit slope depressurisation program. Key details may include the following.

- Slope depressurisation approach and KPIs to achieve or maintain slope stability.
- Monitoring approach, density, positions and other metered data.
- Frequency of measurement or sampling.
- Data recording and database management.
- Processing and comparison with thresholds and triggers.
- Reporting channels, internally and inter-departmental.
- Responsibilities and accountabilities for design, implementation and verification.

It is important to recognise that implementing risk management programs and maintaining an iterative and interactive process to ensure these programs are effective requires good governance, typically in the form of Corporate Standards, supported by an auditing protocol.

6.3.6.2 Staffing and responsibilities

The site hydrogeologist and/or geotechnical engineer/slope practitioner normally identify risks associated with water issues in pit slopes or water impacting open pit operations. In the same way, the hydrologist or process engineer may assess and advise on surface water or tailing management strategies, and the hydrogeologist and geotechnical staff are key to completing semi-quantitative and quantitative risk assessments. They are also responsible for:

- verifying conditions and reconciling monitoring data and modelling results;
- ensuring that identified risks are effectively communicated to other departments that may need to manage consequential impacts;
- implementation of risk mitigation strategies;
- ensuring mitigation strategies or action plans developed do not have unintended impacts in other parts of the operation;
- responding to concerns from other departments and modifying mitigation strategies to minimise overall operational and business impacts;
- updating the site Water Management Plan to reflect the risks and revised strategies.

It is common for regular technical risk reviews to be carried out by corporate staff, staff from other operations within the corporations and/or external consultants. For many mid-sized projects, the review can be carried out by one or two experienced individuals; usually a combination of in-house or corporate staff and an external consultant. For larger projects, an independent Technical Review Board may be warranted, and it may be necessary that the review board be given the responsibility of updating the risk register (or providing the basis for updating the risk register).

6.3.6.3 Developing an iterative program

Completing the initial risk evaluation is the start of a management process that continues throughout the mining life cycle and, in many cases, beyond closure. It focuses on any areas of high or extreme risk and tends toward a quantitative approach as more factual data and experience become available. As suggested by the elements presented in Figure 6.13, progressive iterations may require further data collection and interpretation, calibration of predicted versus observed conditions and further modelling before risks are suitably quantified. Generally, the identification of high or extreme risk issues should also

be subjected to peer or independent technical review before proposed mitigation measures progress to an agreed action plan. The iterative risk management program should encompass the following.

- The regular review and evaluation of monitoring data (Section 6.2.3).
- The re-evaluation of the risks based on interpretation of monitoring and modelling results.
- Peer or independent technical review to validate risk status and mitigation action plans.
- The preparation of an updated action plan going forward, including provision for all staff, equipment, management support and financing.
- Update of the risk register and Water Management Plan.

Risk mitigation and management responses developed by experts in their relevant fields allow residual risks to be gauged and the value of mitigation responses to be established prior to implementation.

It is common that a regular review of interactive monitoring results will cause some of the previously identified risks to be downgraded, and will cause other risks to become elevated. If a risk cannot be avoided, transferred or minimised in some other way, an owner or responsible person is appointed and resources allocated to address each active risk response.

6.3.6.4 Risk register

The results of the risk analysis are documented in a 'Risk Register'. This includes identified risks together with notes on their evaluation (likelihood and consequences), and also responses developed along with the risk owner, and progress toward completion of the agreed risk response. A variety of software is commercially available to assist the production of a risk register (e.g. Riskman, Riskeasy and Dyadem enterprise software, or a range of Excel-based spreadsheet templates), but care should be taken to avoid overly complex systems. These programs often also support the ongoing risk management process with reminders and prompts for update or review.

Risk registers are created for different disciplines (e.g. mining, water, health and safety), and at a variety of levels within an operation and organisation. Key risks from lower level registers can be promoted into higher level risk registers that summarise high and critical risks from all disciplines to the project, mining operation or company organisation.

While a risk register contains the facts and details of risk elements, analysis and grouping of risks in a variety of ways is often highly informative and can identify systemic weaknesses in the operation. Completing an analysis and report on the risks and risk process communicates the outcomes to a broader audience (senior management) without them having to sift through the detail of the risk register looking for particular areas of interest.

The development of the initial risk register is often best carried out as part of the internal peer review (IPR) process using senior operational and corporate staff together with senior experts for certain key technical aspects of the program. Once the risk register has been developed, appropriate milestones can be set, and the register can be rolled into the iterative review process, as described in Section 6.3.6.3.

6.3.6.5 Risk owners

Risk owners need to consider risk management measures that:

- prevent occurrence of high-risk events;
- detect triggers or realisation of risk and assist with response measures;
- protect the operation from the immediate consequences of the risk;
- mitigate the effects of the risk through prompt recovery to an 'acceptable' state.

Risk owners are responsible for identifying and validating risk management actions, and for assigning Action Owners and due dates for completion of actions. They also need to monitor progress and report on completed actions for re-assessment of risk levels.

6.3.6.6 Review and updates

Identified risks (including and new or emerging risks) can be reviewed at different stages of the project. In general, the project process focuses on eliminating or reducing key project risks as it progresses from Pre-Feasibility (Level 2) to Feasibility (Level 3), and on to the final design and engineering phases (Level 4). Within each phase, the risks should be closed out or mitigated as far as possible with planning or design. For any given phase of the project, this residual risk register becomes the starting point for the subsequent phase.

For active mining operations, the risk assessment should be updated periodically, or when a major change takes place. The trigger could be an increase in production rates, a change in the mine plan or a change in licensing or compliance conditions. Sometimes operational improvements reduce or eliminate a risk. Sometimes the change causes new risks to emerge and these changes need to be addressed. The risk register should be updated with changes such as new or emerging risks, mitigation progress, and closure of risk actions. These closed risks should be retained in the register (perhaps in a separate section) as an audit trail.

If a risk mitigation strategy is not effectively progressing or achieving the intended objectives, the strategy or response may need to be changed or revised. Sometimes it is necessary to change the risk owner or set up KPIs directly related to effective management of particular risks. A particular risk or risks may trigger a technical review. It is common for reviews to be carried out quarterly during critical development phases, and then backed-off to six-monthly or annually.

6.3.7   Value of water risk management

Developing a risk evaluation program and formally defining consequence and likelihood generates an understanding and ranking of risks facing a project or operation. It makes it possible to balance different types and sources of risk in terms of relative importance and value to the business. It forces attention and resources on managing critical and high risks, and generating programs and actions to mitigate these risks to acceptable tolerance levels. In the same way, it defers resources from lower priority risk areas and focuses attention where the business can minimise risk and maximise opportunity.

As a consequence of the water-management program being fragmented across several departments and/or operational functions, it is often necessary for the risk assessment to consider an integrated flow sheet, incorporating operational mine water management (dewatering and slope stability), process water infrastructure, and environmental water management. Focussing on water-related risks through the entire project or operation tends to highlight upstream and downstream dependencies on effective water management in each operational area. It may become apparent that local risk can potentially become aggregated and consequences can be elevated through the system until a major event leads to a serious interruption to the business.

Nowhere is this more apparent than for an operation mining below the water table, where general mine dewatering, slope depressurisation and surface water management combine with related infrastructure, monitoring systems and organisational structures to provide both threats and opportunities. Getting these elements right provides the environment and conditions necessary for an efficient and cost effective mining operation. A formal integrated risk evaluation provides a framework for optimising the interactive management of the entire mining project.

Appendix 1

Hydrogeological background to pit slope depressurisation

Michael Price and Geoff Beale

1 Darcy's law

The flow of groundwater through rock or soil is governed by Darcy's law. Derived experimentally by Henry Darcy in 1855–56 (Darcy, 1856, Appendix D; Bobeck, 2004, Appendix D) it describes the relationship between the rate of flow and the loss of head or energy. Although derived to quantify the rate of flow through filter beds and based on flow through a cylinder filled with sand, it describes the flow of any liquid through any porous medium and the concept of flow through a sand-filled cylinder is a useful starting point to illustrate the principles and terminology involved.

Figure A1.1 represents a simple laboratory demonstration of Darcy's law in which water fully saturates the sand in a cylinder and flows through the cylinder at a steady rate. The cylinder is inclined to make the demonstration representative of flow in any direction. The cross-sectional area, measured at right angles to the direction of flow, is A. The head in the sand is measured in two manometers, separated by a distance, l, measured along the flow path. In Figure A1.1, the hydraulic gradient, i, is given by:

$$i = \Delta h/l \qquad \text{(eqn A1.1)}$$

where Δh is the change in head between the two manometers.

In simple terms, Darcy's law states that the volume rate of saturated flow, Q, of groundwater will be directly proportional to the cross-sectional area, A, through which flow is occurring and directly proportional to the hydraulic gradient, i. The hydraulic gradient is the difference in head between two points on the flow path divided by the distance (measured along the flow direction) between them. Thus Darcy's law can be written

$$Q = KAi = KA\Delta h/l \qquad \text{(eqn A1.2)}$$

where the constant of proportionality, K, is termed the **hydraulic conductivity** and depends on the pore geometry of the rock and on the properties – principally viscosity – of the water or other fluid filling the pores (Section 1.1.2.1). K has dimensions of L/T and in mining literature is usually reported in units of metres per second (m/s).

2 Head and pressure

Head is the mechanical energy of the water per unit weight. Head loss occurs because mechanical energy is converted into heat in overcoming friction between the water molecules and between the molecules and the sand grains as the water flows through the sand. Water can possess mechanical energy in three ways: by virtue of height (elevation potential energy); by virtue of being under pressure (pressure potential energy); and by virtue of movement (kinetic energy). Groundwater usually flows so slowly that its kinetic energy can be ignored, and the remaining head (referred to as **static head**) therefore has two components:

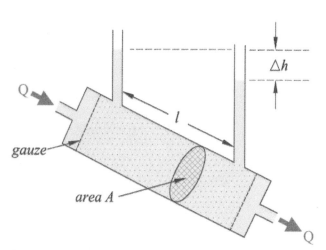

Figure A1.1: Laboratory demonstration of Darcy's law

1. **pressure head**, contributed by the pressure of the fluid in the pore space of the rock or soil.
2. **elevation head**, contributed by the height of the water above the measurement datum level.

Static head is the head measured by the manometers and can alternatively be defined as the height to which the water can raise itself above some arbitrary **datum level**. In a laboratory this datum level is usually the floor or bench; in a mine situation it is almost universally taken to be the national datum level or mean sea level and the static head is often referred to as the groundwater elevation or pore pressure elevation.

In Figure A1.2, for example, the head, h_1, at point P_1 consists of:

- elevation head, z_1 (which is simply the height of P_1 above the datum level) and
- pressure head, $(h_1 - z_1)$.

The head, h_2, at point P_2 consists of:

- elevation head, z_2 (which is simply the height of P_2 above the datum level) and
- pressure head, $(h_2 - z_2)$.

Pressure head is sometimes given the symbol h_p and sometimes the symbol ψ. Because the pressure at any depth, d, in a liquid is given by $\rho g d$, where ρ is density and g is acceleration due to gravity, pressure head is given by:

$$h_p = p/\rho g \qquad \text{(eqn A1.3)}$$

where p is the pressure and ρ is the density of the water. The static head is therefore given by:

$$h = z + \frac{p}{\rho g} \qquad \text{(eqn A1.4)}$$

The head h_2 at P_2 is less than the head h_1 at P_1 by an amount Δh but note that the pore pressure at P_2 is higher than that at P_1. Therefore water does not necessarily flow from high pressure to low pressure.

3 Darcy's law in field situations

The simplest field situation to consider is that of a horizontal permeable unit (an aquifer), of constant thickness, underlain and overlain by impermeable beds and containing water of which the pressure is everywhere greater than atmospheric pressure. When a borehole is drilled into the aquifer, the water therefore rises above the top of the aquifer, which is said to be **confined** (Figure A1.3).

The rate of flow through the aquifer can be found by simple application of Darcy's law. To determine the head, the manometers are replaced by boreholes at X and Y. The head in each borehole is simply the height to which the water stands above some common datum level, which is usually taken for convenience to be mean sea level or the datum level used by the topographical survey organisation of the country in question. Note that because it is the difference in heads between the boreholes that is required, the choice of datum level is unimportant so long as it is used consistently.

In the example of Figure A1.3, the flow rate, Q, is simply

$$Q = Kai = K \times b \times w \times (\Delta h/l) \qquad \text{(eqn A1.5)}$$

In this example, boreholes X and Y lie along the direction of flow, so the hydraulic gradient is easily determined, but this occurrence is unusual. If more boreholes were drilled into the aquifer, the levels at which the water stood in those wells would define an imaginary surface called the **potentiometric surface**, which represents the head at any point in the aquifer. The hydraulic gradient then lies in the direction in which the

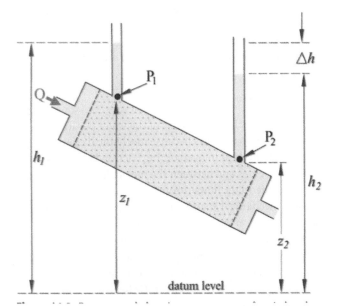

Figure A1.2: Pressure and elevation components of static head

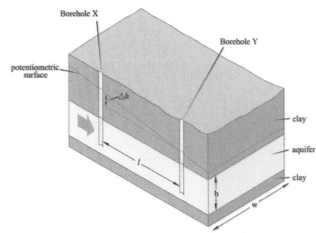

Figure A1.3: Flow through a confined aquifer

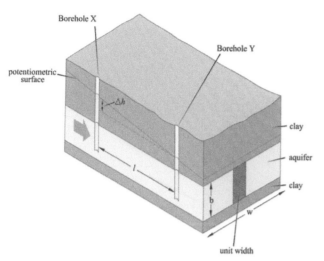

Figure A1.4: The meaning of transmissivity

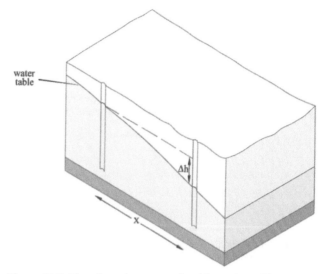

Figure A1.5: Flow through an unconfined formation with a water table

slope of the potentiometric surface is a maximum and its magnitude is the slope of the potentiometric surface in that direction.

In dealing with flow through an aquifer or to a well, the product Kb occurs frequently. This product is called the **transmissivity (T)** of the aquifer, defined as the rate at which water at the prevailing kinematic viscosity will move under a unit hydraulic gradient through a unit width of aquifer (measured at right angles to the direction of flow) (Figure A1.4). Then Darcy's law can be written as:

$$Q = Tw\Delta h/l \qquad \text{(eqn A1.6)}$$

If the aquifer is homogeneous, the saturated thickness is constant, and the flow is steady, then the potentiometric surface will have uniform slope and the hydraulic gradient will be constant. In a confined aquifer, this is theoretically possible, though rare in practice.

In an unconfined aquifer, the potentiometric surface lies within the permeable material, so in this case is the water table (Figure A1.5). The water table is the level at which the pressure of the pore water is exactly equal to atmospheric pressure; below the water table the pore pressure increases with depth and above the water table it decreases with height (though not necessarily in a uniform way). Because the water table is also the level at which water normally stands in shallow wells, the water table is also called the **phreatic surface** (from the Greek *phrear* for well).

The slope of the water table brings three complications:

- the saturated thickness cannot be constant;
- because the water table is usually curved, the slope of the potentiometric surface (and therefore the hydraulic gradient) is not linear;
- there must be vertical components of flow – it cannot be purely horizontal.

The non-linearity of the hydraulic gradient is solved by writing it in differential form, so that instead of $\Delta h/l$ it is written as

$$\left(-\frac{dh}{dl}\right)$$

the rate of change of head with distance. (The minus sign occurs because the flow is occurring in the opposite direction to the increase in head.) Also, to deal with the fact that the saturated thickness (and therefore the cross-sectional area of flow) is changing, Darcy's law can be written to describe the flow through a unit cross-section of the rock, so that:

$$q = \frac{Q}{A} = Ki = -K\frac{dh}{dl} \qquad \text{(eqn A1.7)}$$

where q, the flow through a unit area, is called the **specific discharge**.

The presence of a vertical component of flow means that there must also be a vertical hydraulic gradient. These vertical head differences and gradients can be measured using **piezometers** (Figure A1.6). Vertical gradients become more obvious and important when considering flow systems between recharge and discharge areas and flow towards pit slopes or around wells.

In most real-life situations, piezometers or observation wells do not lie conveniently along the direction of groundwater flow as they do in Figures A1.3, A1.4 and A1.5. In most cases, measurements in many piezometers are used to derive the shape of the potentiometric surface, from which the magnitude and direction of the hydraulic gradient (the maximum slope) can be derived. If the potentiometric surface is a plane, at least three values are needed to derive its slope.

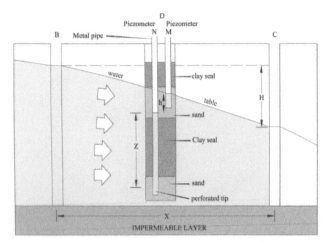

Figure A1.6: Vertical components of flow and hydraulic gradient. H is the head difference for horizontal flow; h is the vertical head difference.

The discussion so far has dealt with materials that are uniform or homogeneous (i.e. the properties are the same everywhere) and isotropic (i.e. the properties at any point do not vary with direction). Most rocks and soils are to varying degrees both heterogeneous (inhomogeneous) and anisotropic. The heterogeneity and anisotropy can arise in several ways at different scales, as discussed in Section 1.1.

4   Flow in three dimensions

In a laboratory demonstration of Darcy's law like that shown in Figure A1.1 or Figure A1.7, the flow is occurring parallel to the axis of the cylinder and the hydraulic gradient is parallel to that axis. There is no flow across the cylinder and therefore no component of hydraulic gradient across the cylinder. It follows that the manometer at A could be tapped into any point around the cylinder and it would measure the same head; the same is true of the manometer at B; there are planes of equal head running across the cylinder from L to L′ and from M to M′ (Figure A1.7).

Head is mechanical energy of fluid per unit weight; fluid potential is energy per unit mass, so that potential (ϕ) is directly proportional to head (*h*), the relationship being simply

$$\phi = gh \qquad \text{(eqn A1.8)}$$

It follows that a plane or surface of equal head is also a plane or surface of equal potential and such surfaces are called **equipotentials**. Extending the idea of the flow through a cylinder to flow through a homogeneous, isotropic rock formation and incorporating it into a set of Cartesian coordinate axes gives a situation like that shown in Figure A1.8. In this figure there are two equipotentials, with the potential (and therefore the head) at A greater

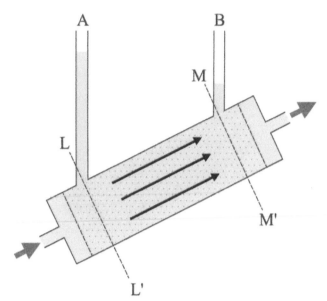

Figure A1.7: Planes of equal head (equipotentials) in a laboratory demonstration

than that at B. Equipotential A has been extended to cut the coordinate axes.

When groundwater is flowing under saturated conditions through an isotropic rock, specific discharge, *q*, is a vector pointing in the direction of flow. The magnitude of *q* is given by

$$q = -K\frac{dh}{dl} \qquad \text{(eqn A1.9)}$$

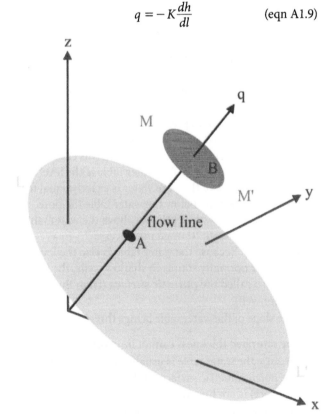

Figure A1.8: Equipotential planes in a geological formation

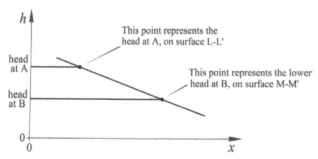

Figure A1.9: Component of hydraulic gradient in the x-direction

where l is also measured in the direction of flow, along the flow line. The hydraulic gradient is also a vector and in an isotropic medium defines the direction of the specific discharge, q.

Because in general the direction of groundwater flow will vary from place to place within a system, it is more useful to refer flow to a conventional Cartesian coordinate system. This is done by resolving the hydraulic gradient into components in the x, y and z directions. From Figure A1.8 it can be seen, for example, that both equipotentials A and B will intersect the x-axis if they are extended. Figure A1.9 shows the resulting component of hydraulic gradient in the x-direction.

For the component of flow in the x-direction the specific discharge, q_x, can be written as:

$$q_x = -K\frac{\partial h}{\partial x} \qquad \text{(eqn A1.10)}$$

where $\partial h/\partial x$ is the partial derivative of h with respect to x. It represents the component of hydraulic gradient in the x-direction.

Similarly, for flow in the y and z directions:

$$q_y = -K\frac{\partial h}{\partial y} \qquad \text{(eqn A1.11)}$$

$$q_z = -K\frac{\partial h}{\partial z} \qquad \text{(eqn A1.12)}$$

The specific discharge, q, is the resultant of these three components. In terms of magnitude, the relationship is given by:

$$q^2 = q_x^2 + q_y^2 + q_z^2 \qquad \text{(eqn A1.13)}$$

but this gives only the magnitude of the specific discharge, not its direction. Specific discharge is a vector quantity, which needs direction as well as magnitude to specify it fully. To obtain the direction, the components (including the components of hydraulic gradient) also have to be treated as vectors. Head is a scalar quantity (it is specified fully if its magnitude is known) but its gradient (the hydraulic gradient) is a vector. Vector quantities are usually represented in print by underscores or by bold type.

To make clear the directions of the components of hydraulic gradient, the convention of unit vectors is used; these possess the property of direction but a magnitude of one, so when multiplied by a scalar quantity they define its direction without altering its size. By convention, unit vectors are usually assigned as follows:

- $\mathbf{i}$ is a unit vector in the x-direction
- $\mathbf{j}$ is a unit vector in the y-direction
- $\mathbf{k}$ is a unit vector in the z-direction.

Expressed as vectors, the components of $\mathbf{q}$ become:

$$\mathbf{q_x} = -\mathbf{i}K\frac{\partial h}{\partial x} \qquad \text{(eqn A1.14)}$$

$$\mathbf{q_y} = -\mathbf{j}K\frac{\partial h}{\partial y} \qquad \text{(eqn A1.15)}$$

$$\mathbf{q_z} = -\mathbf{k}K\frac{\partial h}{\partial z} \qquad \text{(eqn A1.16)}$$

so that

$$\mathbf{q} = -K\left(\mathbf{i}\frac{\partial h}{\partial x} + \mathbf{j}\frac{\partial h}{\partial y} + \mathbf{k}\frac{\partial h}{\partial z}\right) \qquad \text{(eqn A1.17)}$$

The quantity

$$\left(\mathbf{i}\frac{\partial h}{\partial x} + \mathbf{j}\frac{\partial h}{\partial y} + \mathbf{k}\frac{\partial h}{\partial z}\right)$$

adds up to a vector that points in the direction at which the rate of change of head is a maximum, i.e. a direction at right angles to the equipotentials, and therefore represents the direction of the hydraulic gradient. It is written variously as 'grad h', 'del h' or ∇h.

The del operator, ∇, can be written as

$$\left(\mathbf{i}\frac{\partial}{\partial x} + \mathbf{j}\frac{\partial}{\partial y} + \mathbf{k}\frac{\partial}{\partial z}\right)$$

With this notation, Darcy's law in three dimensions *in an isotropic medium* can be written as:

$$q = -K\nabla h \qquad \text{(eqn A1.18)}$$

This is essentially saying that, in an isotropic medium, the specific discharge is proportional to the hydraulic gradient and in the same direction, which will be at right angles to the equipotential surfaces (or, in two dimensions, at right angles to the head contours). This is because in this case, where K has no directional property, it behaves as a scalar and simply multiplies the hydraulic gradient to affect the magnitude but not the direction of flow.

Very few geological formations are isotropic. In general, rocks have a maximum permeability k_{max} in one direction and a minimum permeability k_{min} in a direction that will be at right angles to the maximum. These are two of the three principal directions of anisotropy. The third principal direction is at right angles to both the first two.

In an anisotropic system the direction of flow is not usually in the same direction as the hydraulic gradient. The gradient is trying to send the water in its direction but the medium is trying to divert it towards the direction of maximum permeability. The hydraulic conductivity no longer behaves as a scalar quantity; instead it behaves as a **tensor**. When a vector is multiplied by a tensor, not only does this affect the magnitude of the product but it will generally also affect the direction. In sedimentary rocks it is common to find the direction of minimum permeability at right angles to the bedding; the other two directions are parallel to the bedding and are often approximately equal to each other. In metamorphic rocks with a prominent foliation, the permeability is usually at a maximum parallel to the foliation, but in many igneous and metamorphic rocks the permeability is due mainly to the presence of fractures.

Where the principal directions of anisotropy correspond with the x, y and z directions of the coordinate system, Darcy's law in three dimensions can be written as:

$$\mathbf{q}_x = -K_x \frac{\partial h}{\partial x} \quad \mathbf{q}_y = -K_y \frac{\partial h}{\partial y} \quad \mathbf{q}_z = -K_z \frac{\partial h}{\partial z} \quad \text{(eqn A1.19)}$$

and

$$\mathbf{q} = \mathbf{q}_x + \mathbf{q}_y + \mathbf{q}_z \quad \text{(eqn A1.20)}$$

so that

$$\mathbf{q} = -K_x \frac{\partial h}{\partial x} - K_y \frac{\partial h}{\partial y} - K_z \frac{\partial h}{\partial z} \quad \text{(eqn A1.21)}$$

If the bedding or foliation is horizontal, it is relatively simple to set up a coordinate system for modelling such that the principal directions of anisotropy correspond to the Cartesian coordinate axes of the model. Even if the directions of anisotropy do not line up with horizontal and vertical directions (if, for example, a sequence of beds has been tilted by tectonic action), the axes of the model can still be aligned with the directions of anisotropy. In the most general case, however, the directions of anisotropy will not correspond with the coordinate axes (Figure A1.10). In this case, both the hydraulic gradient and the principal permeability values have to be resolved into components along the coordinate axes. This leads to a situation where, for example, the component of the gradient in the x-direction can induce flow in the permeability component in the y-direction (see, e.g. Deming, 2002, p.34) so that to consider flow in the x-direction it is necessary to write:

$$\mathbf{q}_x = -K_{xx} \frac{\partial h}{\partial x} - K_{xy} \frac{\partial h}{\partial y} - K_{xz} \frac{\partial h}{\partial z} \quad \text{(eqn A1.22)}$$

Here, K_{xx} represents the hydraulic conductivity that controls flow in the x-direction in response to the component of hydraulic gradient in the x-direction, K_{xy} represents the hydraulic conductivity that controls flow in

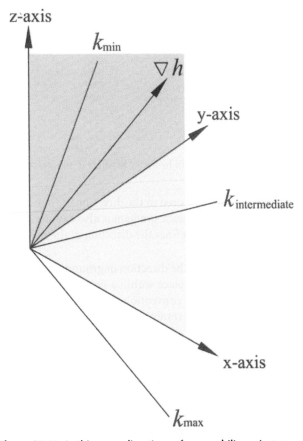

Figure A1.10: In this case, directions of permeability anisotropy (in blue) are not coincident with the coordinate axes (in black). The x–y plane is coloured yellow: the y–z plane is coloured green. The x-, y- and z-axes are mutually at right angles, as are the k_{max}, k_{min} and $k_{intermediate}$ axes, but the two sets are rotated relative to each other.

the x-direction in response to the component of hydraulic gradient in the y-direction, and K_{xz} represents the hydraulic conductivity controlling flow in the x-direction in response to the component of hydraulic gradient in the z-direction.

In total, the components of specific discharge are:

$$\mathbf{q}_x = -K_{xx} \frac{\partial h}{\partial x} - K_{xy} \frac{\partial h}{\partial y} - K_{xz} \frac{\partial h}{\partial z}$$

$$\mathbf{q}_y = -K_{yx} \frac{\partial h}{\partial x} - K_{yy} \frac{\partial h}{\partial y} - K_{yz} \frac{\partial h}{\partial z} \quad \text{(eqn A1.23)}$$

$$\mathbf{q}_z = -K_{zx} \frac{\partial h}{\partial x} - K_{zy} \frac{\partial h}{\partial y} - K_{zz} \frac{\partial h}{\partial z}$$

The nine components of K in three-dimensional space define the hydraulic conductivity tensor (Bear, 1972). They can be written in matrix form:

$$\mathbf{K} = \begin{bmatrix} K_{xx} & K_{xy} & K_{xz} \\ K_{yx} & K_{yy} & K_{yz} \\ K_{zx} & K_{zy} & K_{zz} \end{bmatrix} \quad \text{(eqn A1.24)}$$

and the three equations of (A1.23) can be written as a single matrix equation:

$$\begin{array}{c} q_x \\ q_y = \\ q_z \end{array} \begin{bmatrix} K_{xx} & K_{xy} & K_{xz} \\ K_{yx} & K_{yy} & K_{yz} \\ K_{zx} & K_{zy} & K_{zz} \end{bmatrix} \times \begin{bmatrix} \left(\dfrac{\partial h}{\partial x}\right) \\ \left(\dfrac{\partial h}{\partial x}\right) \\ \left(\dfrac{\partial h}{\partial x}\right) \end{bmatrix} \quad \text{(eqn A1.25)}$$

For the special case where the principal directions of anisotropy coincide with the x, y and z directions of the coordinate system, $K_{xy} = K_{xz} = K_{yx} = K_{yz} = K_{zx} = K_{zy} = 0$ so that the nine components of specific discharge reduce to the three of equation A1.21.

When using finite-difference numerical models it is necessary that the principal directions of anisotropy correspond to the coordinate axes (Freeze & Cherry, 1979). This can usually be arranged, provided that all the formations represented by the model have the same principal directions of anisotropy. When two or more formations with different directions of anisotropy are involved, this is impossible. In these situations – which are probably the most common in mine settings – the more flexible arrangement of the finite-element method will usually need to be considered.

Appendix 2

Guidelines for field data collection and interpretation

Greg Doubek, Rowan McKittrick, Michael Price and Simon Sholl

1 Summary of drilling methods commonly used in mine hydrogeology investigations

1.1 Direct push method

The direct push method can be used to perform cone penetration tests (CPT) to characterise soft soils (usually overburden materials). The method also works well for the installation of piezometers in tailing deposits. It can provide soil, water, and gas samples. Data obtained during drilling on the rise and decay of pore pressure can be used to derive estimates of permeability.

Piezometers can be constructed in the resulting hole using pre-packed screen and bentonite seals that are installed inside the 'drill pipe'. The pipe is then removed exposing the screen and seals to the formation. The bentonite seals expand due to the presence of water. The resulting standpipe piezometers are usually small diameter. Vibrating wire piezometers can also be installed and, under the right conditions, the seals may be provided by allowing the formation to collapse around the sensors.

1.2 Auger drilling

Hollow-stem auger drilling is a rotary drilling technique mainly used for the testing and sampling of soft soils and unconsolidated ground. It may be a useful technique for installing relatively shallow piezometers in wet or swampy soils as the piezometer can be constructed inside the hollow-stem auger before, and concurrently during the removal of, the auger pipe and tools. This can be very useful in ground that would otherwise cave before the piezometer could be constructed. Large-diameter (1 m+) non-hollow-stem auger bits are sometimes used to start large-diameter water wells in order to set surface conductor casing in unconsolidated ground.

1.3 Sonic drilling

Sonic or 'vibratory' drilling has gained increased popularity for working in unconsolidated formations, including tailing deposits, leach pads and waste rock dumps. The hole is advanced by sending high frequency resonant vibrations down the drill pipe. The resonance 'fluidises' the soil or formation particles at the bit. Samples are obtained by tripping out the drill pipe. Again, piezometers may be installed inside the drill pipe before its removal, with sand pack and seals being emplaced concurrently with the removal of the drill pipe.

1.4 Cable tool drilling

Cable tool is a traditional drilling method that is suited to shallow production wells in somewhat unstable 'overburden' containing cobbles and boulders (colluvium). The hole is advanced by raising and dropping a heavy drill cable with a carbide chisel-type bit on the end. The drill string is raised and quickly dropped by way of a cable operating off a 'walking beam' or 'spudding arm', which is driven by an eccentric wheel.

Surface casing is usually installed semi-concurrently with the drilling process, by driving it with the weighted spudding bar. The cuttings are removed from the bore using a bailer which is lowered and raised in the wellbore by way of a cable and a winch. The production casing is either perforated *in situ* at the end of the drilling process, or a string of casing and well screen is 'telescoped' into the water production zone at the bottom of the well. The method is much slower than rotary methods but is often economic in situations where it is necessary to construct a limited number of holes in a non-critical time frame.

1.5 Rotary core drilling

Core drilling is a rotary method also known as diamond drilling (due to the diamonds that are usually impregnated in the bit). It is usually the preferred method of drilling for geologists and geotechnical engineers because a core of the relatively undisturbed rock is brought to the surface to be examined, logged, tested, and/or assayed. Geological, geotechnical and geochemical testing results from the core are considered second to none when compared to other drilling methods in hard rock. The drill cuttings are cleaned from the hole using conventional circulation of the drilling fluid (usually a bentonite or polymer-based drill mud), so only limited hydrogeological information may be obtained. However, records of fluid losses or gains that may be associated with groundwater producing zones can be important in the early stages of a hydrogeological investigation.

As with all drilling methods that use fluid-based conventional circulation, the hole is normally in a condition that is known as 'overbalanced' during drilling, i.e. the pressure of the column of drilling mud in the wellbore is kept higher than the fluid pressure in the formation that is being drilled. While this helps maintain borehole stability, it also tends to drive the drilling mud and the water into the formation, so rarely will any formation water be produced.

1.6   Conventional mud rotary drilling

Conventional mud rotary drilling is one of the oldest forms of drilling and is still widely used, particularly by the petroleum industry. For this method, drilling mud (which is often bentonite clay or polymer-based) is pumped down the drill pipe where it exits the bit and is returned to the surface in the annular space between the borehole wall and the drill pipe. A mud cake develops on the wall of the hole because the pressure of mud in the hole is greater than the water pressure in the adjacent formation, so the mud liquid filters out of the hole, and the mud forms a fine-grained deposit ('cake') on the wall of the hole. Mud cake formation can be minimised through the use of polymer drilling fluids or thin muds (or water). For porous medium settings, the more the filter cake can be cleaned subsequent to drilling, the better the yield of the hole. Procedures for removing the mud cake are well documented in Driscoll (1986) and other texts.

A combination of sufficient viscosity and up-hole velocity of the mud is required to bring the drill cuttings to the surface where the cuttings are separated and the mud is re-circulated down the hole. Similar to core drilling, the hole is usually kept in an 'overbalanced' state. This is often considered desirable in the petroleum or geothermal industries to prevent 'blowouts' that can occur when formations holding oil/gas or hot water under high pressure are encountered.

When drilling with this method, if mud circulation is lost, drilling progress usually comes to a halt until the zone of fluid loss (often a fracture zone) can be plugged, usually by lost circulation material (LCM) which is added to the mud. LCM may consist of a combination of materials that may include cedar fibre, paper, mica, cottonseed hulls or nut shells. Polymer-based granules that expand several hundred times their original volume when hydrated are also commonly used. Acid soluble LCM has also been

developed so that, if it is used in a production zone, it can be acidised and developed out.

1.7   Conventional air/foam drilling

Conventional air/foam rotary drilling is widely used to install piezometers, monitoring wells and production dewatering wells. The method is similar to conventional mud drilling but, instead of using a liquid to circulate out the drill cuttings, air or air with a mixture of water, a foaming agent, and often a polymer is used to lift the drill cuttings out of the hole. The air/water/surfactant/polymer mixture is 'pumped' down the centre of the drill pipe and the resulting foamy mud carrying the drill cuttings comes back to the surface through the annular space between the borehole wall and the drilling pipe. The mixture of foam and cuttings exits from the casing by way of a diverter and a 'blooey line'; usually at a high velocity, into a collection pit.

Although the compressed air may be used to try to keep formation water from entering the borehole, new water strikes are usually evident by an increase in water produced at the surface. However, it is usually somewhat difficult to quantify increases in water using this method because of the rate that the discharge comes to the surface and exits into the collection pit. Because this is essentially an 'underbalanced' drilling method, borehole stability can sometimes be problematic. Additionally, the returns coming up the annular space can be rapid, causing borehole instability and erosion issues.

1.8   Flooded reverse-circulation drilling

Flooded reverse-circulation is often the rotary drilling method of choice for drilling large-diameter production water wells (Figure A2.1). The method combines the use of drilling mud and air. The drill pipe may vary, depending on the drilling company performing the work, but usually differs from reverse circulation because of the use of dual-wall drill pipe.

For this drilling technique, mud is introduced from the surface down the annular space between the surface casing and the drill pipe. Air is introduced down the annular space between the two pipes in the dual-wall pipe and the mud with the drill cuttings is returned to the surface by airlifting from the centre tube of the dual-wall pipe. The air is introduced into the centre tube normally fairly close to the drill bit. The air is compressed at the surface (and remains so) until it reaches its 'turn-around' point near the bottom of the hole. Once the air 'turns-around' into the centre tube of the drill pipe, the air begins to rise up the pipe (which is flooded with drilling mud). At this point the air breaks into bubbles and they expand in size thus lightening the mud column inside the centre tube. This causes a U-tube effect and the inertia of the expanding/rising air brings the mud and cuttings to the surface.

The mud, cuttings and air are normally discharged into a cyclone which dissipates the energy and allows the discharged mud and cuttings to drop out the bottom of the cyclone. Usually the mud and cuttings feed a 'shale shaker' that is located above a skid-mounted tank which removes the larger drill cuttings. The mud drops into the tank where it may be further conditioned by pumping it through de-sanders before being returned back down the drill hole.

The flooded reverse method has several advantages, particularly for drilling at larger diameters. Firstly, it is more difficult (but not impossible) to lose circulation. If a large fracture is encountered in the borehole, the fluid will drop to the water table but, provided there is still enough submergence, it will still circulate. Secondly, because the drillers do not have to keep the hole completely full of mud, if the mud level drops down to static, all the driller has to do is adjust the amount of mud going in the hole to equal the amount of mud coming out of the hole. The driller can monitor this by several different ways but chiefly it is done by gauging the fluid level gaining or losing in the mud tank, and by monitoring the air pressure. In this way, the hole is advanced where the mud column in the hole is in balance with the formation pressure. This is desirable when drilling most water wells

(in situations where high pressure gas or very hot water that could convert to steam is not expected) because there is less potential for building a 'wall-cake' on the borehole wall or causing mud to infiltrate into the water-bearing formation. Lost circulation material (LCM) is typically not required so there is less chance of plugging up potential production zones. After building the well, it is easy to develop the hole clean because the rig is already set up with the required compressed air. At the end of the well development period, it is also easy to perform an extended airlift test of the well using the dual-wall pipe.

1.9 Dual-tube reverse-circulation (RC) drilling with air

Dual-tube RC drilling with air is a widely used drilling method for geology exploration drilling (Figure A2.2). It is also an ideal method for water exploration and characterisation, particularly in fractured rock. Compressed air is pumped down the annular space between the outer pipe and the centre tube of the dual-wall pipe from where it is conveyed to the drill bit. It then picks up drill cuttings and formation water, and returns back to the surface in the centre tube of the dual-wall pipe.

Figure A2.1: Flooded reverse drilling rig during construction of in-pit dewatering well

Figure A2.2: Photograph of dual-tube reverse circulation drilling rig (Source: Jose Valenzuela)

and diamond) are easily adaptable. Inclined geology exploration holes are commonly used to intersect dipping ore zones or structures to provide better characterisation of the thickness of structures by drilling perpendicular to their strike. Angled drilling can also be used to minimise drill pad construction and surface disturbance by allowing several holes to be drilled from one platform.

Angled holes are routine in underground mines to characterise as much ground as possible from few drill stations. Underground drilling bays are expensive to construct and so are usually limited in number. These holes are often drilled in 'fans' and may be drilled at an angle up, down or horizontally.

Directional drilling is commonly used for utility corridor crossings of rivers or underneath major transportation infrastructure such as highways. The petroleum industry commonly employs directional drilling in order to produce oil or gas from thin or tight formations by drilling wells that parallel the production zones.

Applications for horizontal and angled drilling for pit slope depressurisation are described in Section 5. Directional drilling technology is being increasingly applied to mineral exploration as the depth of ore targets increases. In the future, it may be possible to use directional drilling technology to help reduce the number of in-pit dewatering installations by allowing dewatering wells to be collared outside the pit and steered beneath the pit footprint.

If the hole is dry, the rock chips are essentially 'blown' back to the surface (by the compressed air). Once significant formation water has been encountered, the system begins to airlift, when the rising and expanding air bubbles cause the rock cuttings and water to rise to the surface with significant velocity. At the surface, the fluid returns are directed into a cyclone to dissipate the energy. The water and cuttings drop out the bottom of the cyclone where the drill cuttings can be sampled and the volume of the water can be measured. Usually the water is directed away from the drilling rig in a ditch to a sump and a sample of the rock cuttings is bagged for assay and geologic logging.

In addition to providing a good method for hydrogeological data collection, the RC method typically provides good support for the drill hole because the diameter of the drill pipe is generally only just smaller than the diameter of the bit (and therefore the hole). The typical set-up uses a 140 mm (5½ inch) drill bit with 112 mm (4½ inch) drill pipe. The risk of caving and collapse of weaker materials into the hole is therefore reduced.

1.10 Horizontal, angled and directional drilling

Inclined, angled or directional holes are commonly drilled in the mining industry and the 'standard techniques' (RC

2 Standardised hydrogeological logging form for use with RC drilling

A typical form for recording hydrogeological data during reverse circulation drilling is shown in Table A2.1. Its application is discussed in Section 2.2.4.2.

3 Interpretation of data collected while RC drilling

The following notes are intended to assist with the interpretation of the data collected from an RC drilling program. As part of the data interpretation, it is important to understand the principles of airlift pumping.

3.1 Airlift pumping

The concept of airlift pumping has been credited to a German engineer by the name of Carl E. Loescher who, in 1797, used it for pumping wells associated with the mining industry. Airlift pumping was once widely used in deep water wells before deep well turbine pumps had been developed. Today it is often utilised in wastewater treatment plants, aquaculture, and drilling. The method

Table A2.1: Standardised hydrogeological logging form for use with RC drilling

Hole #: _____ Compressor CFM: _____ Driller: _____

Azimuth: _____ Inclination: _____ Bit Diameter: _____ Driller: _____

Contractor / Rig: _____ Pipe Diameter: _____ Driller: _____

Start Date: _____ Date Complete: _____ Driller: _____

Depth	Data	Penetration Rate (military time)		Rod Time	Bit Hours	Unloading Pressure	Bit Type	Degree of Fracture	Clay Content	Flow Rate	Temp	Air Pressure	Geology	Bridging	Comments
	Drillers initial at Shift Change	Start Time	End Time			PSI High Point	T = tricone H=Hammer	0=none 5=extreme	0=none 5=100%	(gpm)	°F	PSI end of rod	Major Contacts	Indicate Bridging With X	Compressor/ CFM Change Describe Hole Stability Issues/ Abnormal Torque/ Contamination Static Water Should be Checked Daily Drilling Method

has several disadvantages when compared to other pumping technologies, but it is ideal when coupled with drilling. Airlift pumping is one of the best ways to pump very dirty water with a high suspended solids content, as required during drilling. The principle behind airlifting is simple, but the actual design theory is somewhat more complex. The basic configuration of an airlifting pumping system is illustrated in Figure A2.3.

The concept of airlift pumping may also be illustrated by way of a U-tube submerged in a body of water with a way to introduce air into one leg of the tube at a point below the water level as shown in Figure A2.4. On the left side Figure A2.4 shows the U-tube equipment in a static condition where compressed air is not being introduced into the system. The right side of the figure represents the system in a dynamic state where compressed air is introduced. For the dynamic state, the compressed air is aerating the right hand water column thus changing the specific gravity of the column of water above. If enough air is introduced, the *two-phase* mixture will overflow the elevated right hand side of the U-tube as shown. The concept also applies to a straight pipe in a body of water.

Figure 2.8 in the main text shows three common configurations for airlifting equipment. Figure 2.8A and 2.8B are common equipment configurations for water well pumping. Figure 2.8C shows a typical RC drilling set-up.

The principle and the theory of airlifting have been described in detail in several technical papers, some of these include Brown (1967), Gibbs *et al.* (1971), Doubek *et al.* (1992) and Driscoll (1986). Brown (1967) is considered the definitive reference on air or gas lifting which was developed for applications pertaining mainly to artificial lift in the petroleum industry.

Airlift systems are affected by a number of factors, the most significant of which are:

- the volume of air (or gas) flow;
- the area of the riser pipe;
- lift height; and
- submergence.

During the drilling of a pilot hole, or several holes located in the same proximity where roughly the same equipment is used, the first two major factors remain roughly the same. While it is possible to vary the

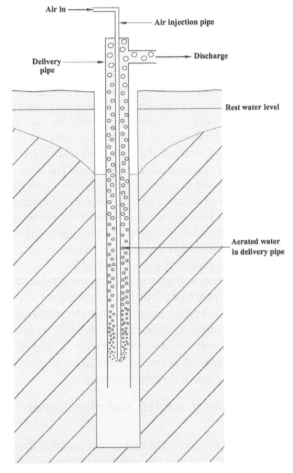

Figure A2.3: Usual configuration of an airlift pumping system

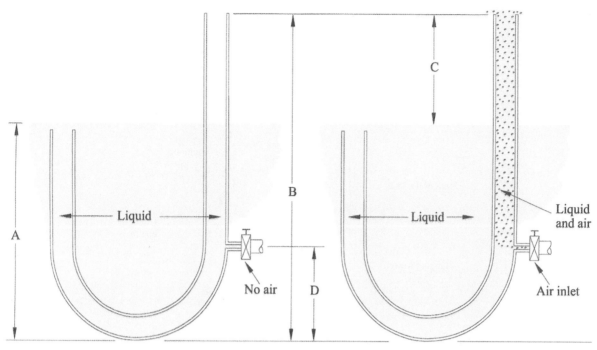

Figure A2.4: Principles of airlift pumping system (Source: Greg Doubek)

volumetric air flow rate produced from air compressors installed on some drill rigs, this is not normally done, so the air flow (in cubic feet per minute) will remain constant. The cross-sectional area of the riser pipe (the centre tube of the dual-wall drilling pipe) remains constant during the drilling so this also is not a variable. The lift height during the drilling of a pilot hole is roughly the depth to the water table plus the distance to the highest point the water must rise up the drill mast and, additionally, the amount of drawdown of the water in the borehole. The first two factors remain relatively constant during the drilling of any given hole and may also be similar to other pilot holes nearby. The most significant *variable* during the drilling of any one hole is submergence which is in part a product of the drawdown of the water level in the hole while drilling.

3.2 Submergence

It is important to understand the effects of submergence when interpreting the results of an RC drilling program. The amount of water that is airlifted and measured at the surface while drilling is not directly correlatable with how much water a well will subsequently produce at the same location. As described below, a pilot hole (pilot hole 'A') drilled at one location may make 6.3 l/s (100 gpm) at its total depth and a production well drilled subsequently at the site may initially only be capable of producing 6.3 l/s (100 gpm). But another pilot hole (pilot hole 'B') located 30 metres away from the first test hole may also make 6.3 l/s (100 gpm) at its total depth, but a water well later drilled at this second location may initially produce 63 l/s (1000 gpm). Such a situation is commonly observed in mine

hydrogeology settings. A clear understanding of main controls on airlift yields requires close analysis of both conditions of submergence and borehole specific geological conditions.

The typical terminology used in an airlift pumping system is shown in Figure A2.5 and summarised as follows.

- Lift above ground (LAGL). This is the distance from the ground to the highest point in the airlift system which is usually about 7 metres (25 ft) up in the drilling mast.
- Depth to water (SWL). The depth to water from the ground surface. In a truly static condition, the depth to water in the well or borehole is the same as the depth to water in the aquifer.
- Total starting lift (LS). The sum of the lift above ground (LAGL) and the depth to water (SWL).
- Starting submergence (Ss). This is the length of the airlift pipe to the air inlet (defined below) that is below the water table in a static condition. This measurement is mostly used to calculate the pressure that will be required to start the airlifting process (in order to properly size the air compressor) and is usually expressed as a percentage of the total length of the airlift system.
- Drawdown (d.d.). The distance the surface of the water drops in the well or borehole due to pumping. The static water level in the formation will usually be higher than this.
- Total pumping lift (LT). The sum of the lift above ground (LAGL), the depth to water (SWL), and drawdown (d.d.).

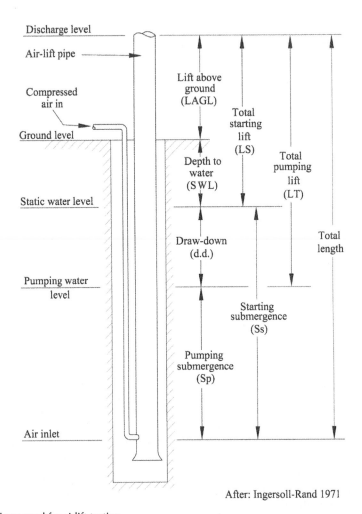

After: Ingersoll-Rand 1971

Figure A2.5: Typical terminology used for airlift testing

■ Air inlet. The depth from the ground to where the air is allowed to discharge into the airlift system. While drilling, the air is often discharged immediately above the bit so the depth the bit is located below the ground is the measurement commonly used.

■ Pumping submergence (Sp). This is the length of the airlift pipe (technically just to the air inlet) below the water level in the well or the borehole during pumping. It is usually expressed as a percentage of the total length of the airlift system (see below).

■ Total length. The total distance from the air inlet to the highest point in the system.

Submergence is simply the length of the airlift pipe (or drill pipe) that extends below the water table in the well or borehole either at static conditions or during stress (pumping or drilling). Submergence is most meaningful when expressed as a percentage of the total length of the airlift system. Thus:

$$\% \, Ss = (Ss/total \, length) \times 100 \quad (eqn \, A2.1)$$

$$\% \, Sp = (Sp/total \, length) \times 100 \quad (eqn \, A2.2)$$

The most important factor for analysing pilot hole data is the pumping submergence (Sp) expressed as a per cent. When the air volume and the cross-sectional area of the airlift pipe are fixed (as in drilling), then the airlifted flows produced at the surface will increase as the per cent of pumping submergence increases, up to the point where the airlift system is 'maxed out' and no more water can physically be produced by the system. A longer length of pipe will also produce more water to the surface, assuming the same percentage submergence. In the case of a 300 m (1044 ft) deep borehole, if the airlift pipe is 50 per cent submerged, it would make more water than a 100 m (350 ft) borehole with an airlift pipe at 50 per cent submergence. Figure A2.6 shows the results of field testing carried out at the Sleeper mine (Nevada, USA) in 1992 to illustrate these two key points.

Figure A2.7 illustrates the two pilot holes that were considered at the beginning of this section. They were located 30 m apart and were drilled to the same depth. Because they are so close to each other, they have similar depths to water (75 m), and they both produced similar flows at their total depth. However, a water-well located

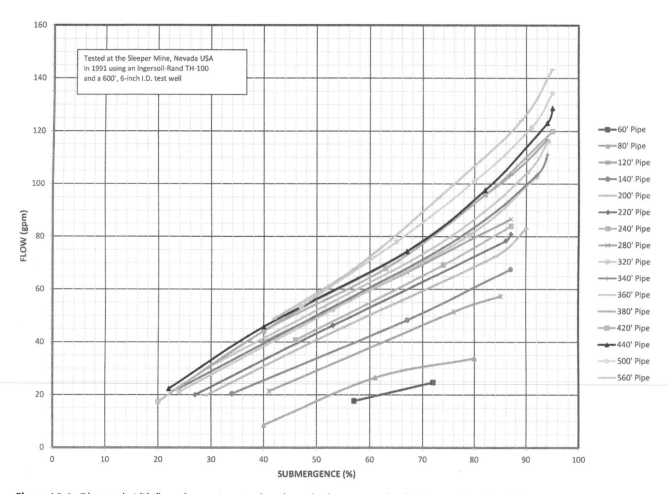

Figure A2.6: Observed airlift flows for varying pipe lengths and submergence depths (Source: Greg Doubek)

at pilot hole 'B' will out produce a well at the site of pilot hole 'A' by a factor of ten. Figure A2.7 illustrates how this can happen assuming the depth to water is 75 m below ground level.

The results from the two holes differ as follows.

- Pilot hole 'A' produces no water at all until it is 40 m below the water table where it encounters a large fault. The fault at this elevation is only capable of producing a maximum of about 6.3 l/s but, because of the low submergence, the airlift system cannot produce more than about 2.5 l/s (40 gpm). After this, even though the borehole encounters a few more minor fractures, the airlift flow just increases slowly due to the increasing submergence as the borehole is advanced toward its total depth. If the borehole was extended deeper, the airlift flow graph would flatten out because the airlift flow system would not be capable of lifting any more water.

- Pilot hole 'B' shows a slightly increasing airlift flow rate as small fractures are encountered but the flows remain well below the capacity of the airlift system until it encounters the large fault. The airlifted flows increase almost immediately and get back on trend with the submergence line.

3.3 Examples of pilot hole comparison and data interpretation

Figure A2.8 shows a comparison of several holes drilled in the same hydrogeological regime in an area of a mine where a dewatering well was needed. The pilot holes were drilled in a granodiorite intrusion adjacent to a major range front fault.

The figure illustrates several points. The bottom group of holes (PH-20, 21, 22, 22A, 23, 24, and 37A) were drilled with nominal 4 ½-inch RC drill rods, whereas the top group of holes (PH-38 and 41) were drilled with nominal 5 ½-inch RC rods and a drill with a larger air compressor. The bottom group of holes shows a distinct stair-stepped pattern because they were drilled by a driller who did not need to time a calibrated container to measure flow because 'he had a trained eye'. This illustrates one reason why airlifted flows need to be graphed. This figure also shows that the airlift system for the upper group of holes eventually was 'maxed out' indicating a maximum airlift capacity of the system of approximately 16.5 l/s (260 gpm). It should be noted that once an airlift system has reached its maximum capacity, additional groundwater 'strikes' cannot be identified by increases in airlifted flow. In the

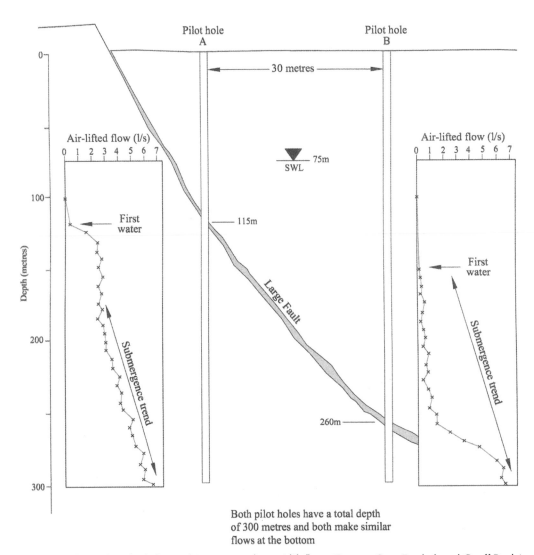

Figure A2.7: Importance of specific pilot hole conditions on resultant airlift flows (Source: Greg Doubek and Geoff Beale)

end, a production dewatering well was sited at the location of PH-38 which was capable of producing over 50 l/s (800 gpm). It was chosen because of the deep and higher yielding fracture zone that can be identified between 396 to 427 m (1300 and 1400 ft) on this figure.

Figure A2.9 shows another example of airlift data from drilling of a pilot hole. In this example, the area under investigation was located at depth, below an upper aquifer unit. The upper aquifer had a relatively shallow water level of about 115 m (400 ft) bgl and the impermeable base of the aquifer comprises a shallow dipping sill with extensive argillic alteration. Below this aquitard unit, the water table dropped to about 400 m (1400 ft) bgl. Intermediate casing was installed in most of these pilot holes to this depth, which effectively isolated the upper aquifer from the lower aquifer. This can be seen in the figure where the airlifted flows drop off significantly. Ultimately a production dewatering well was drilled at the site of

PH-93A. The well yielded a flow of over 190 l/s (3000 gpm). This site was chosen based on the relatively deep production zone shown by the airlifted flows at about 720 m (2500 ft) bgl and from *in situ* testing of the borehole.

Figure A2.10 shows airlift data collected while drilling a pilot hole plus the spinner log from the subsequently drilled water well. The figure shows the airlift flow while drilling, the air pressures associated with the drilling, and the raw spinner logging data with the locations of the main water production zones identified (and quantified). The fractured zones (as noted by the driller) correspond reasonably well to rises in the airlift flow, the air pressures, and the ultimate production zones verified by the spinner log. The production well was ultimately only drilled to a depth of 400 m (1400 ft) bgl because no significant fracture zones were identified below this depth. The well initially made 25 l/s (400 gpm).

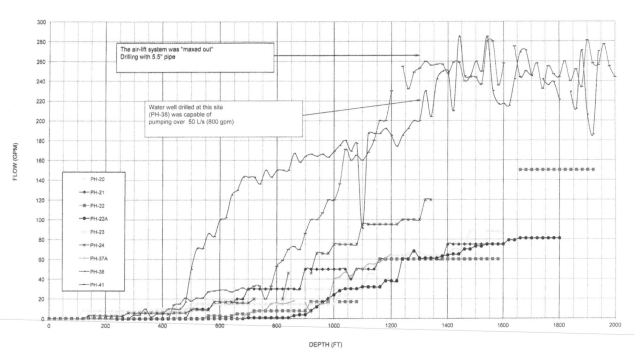

Figure A2.8: Comparison of airlift flows from several pilot holes of varying diameter and airlift pumping capacity

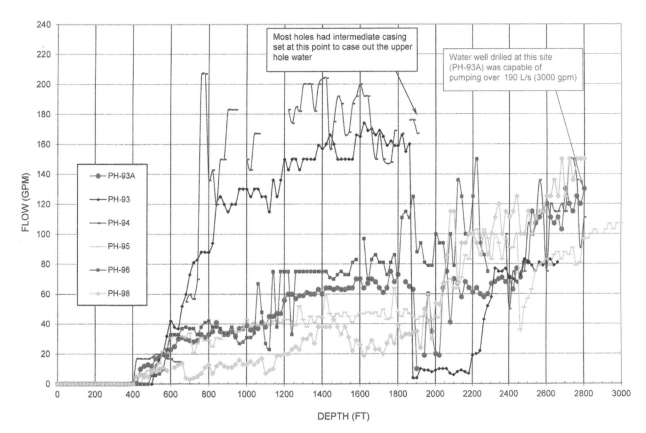

Figure A2.9: Comparison of airlift flows from multi-layered aquifer system

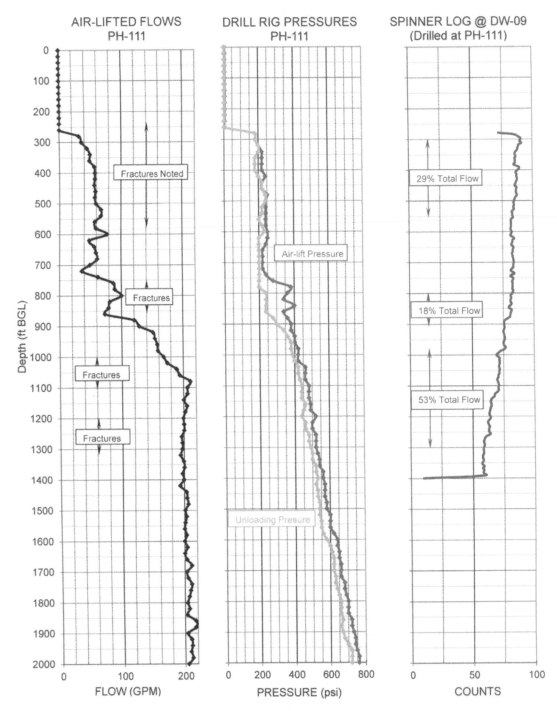

Figure A2.10: Comparison of airlift flows from pilot hole and spinner flow logs from production water well

These examples have shown actual data collected from typical programs. However, every project is different, and the data that may be collected will vary from site to site, so the site hydrogeologist must take into account the site specific conditions for their particular project. If water temperature is collected, a sudden change in the slope of temperatures might be important. Field water chemistry may show similar important changes.

4   Guidelines for drill-stem injection tests

Drill-stem injection tests can be carried out in RC holes to help characterise the fracture zones penetrated during drilling. It can provide an easy, low-cost method of investigating the hydraulic response of fracture zones during RC drilling.

The purpose of the injection test is to provide quantitative specific capacity information for the fracture zones that are being drilled. In holes that encounter large fracture zones, the water level may recover very rapidly and hence an injection test rather than slug test may be more beneficial for characterisation of fractures.

Ideally, multiple tests should be carried out to assess the main fracture zones. The number of tests performed will depend on the depth of the hole and the characteristics of the formations penetrated during drilling.

The test takes advantage of the design of the RC drill pipe. Water is injected down the annulus of the drill pipe and the resultant rise in head in the drill hole is measured in the inner tube.

The test requires that a special sub-adapter be fabricated to connect a water supply to the RC drill rods. This can be easily made by the drilling contractor and should introduce the water supply into the annular space between the inner and outer tubes of the drill string. In addition, a 50 mm in-line flow meter, a 3–5 kW trash pump, a water source, a water level sounder and a stopwatch will be required. The typical set-up for injection testing is shown in Section 2, Figure 2.10.

The general procedure for injection testing is as follows.

- Stop drilling when the drill bit passes through a significant fracture zone (this may be indicated by an increase in water airlifted to the surface or jerky motion of the drill rods). Lift the drill bit off the bottom of the hole, to just above the fracture zone of interest. Circulate air to clean the borehole and clear cuttings from the drill pipe.

- Allow the water level to stabilise for a short period (usually less than 10 minutes), then measure and record the static water level inside the RC drill pipe using a water level sounder.
- While waiting for water level to stabilise, set up the equipment for the test.
- Pump water into the hole at a constant rate via the sub-adapter.
- Record the change in water level using a water level sounder, and flow rate using an in-line flow meter or by recording the time taken to empty a tank of known volume.
- Pumping should be continued until the water level stabilises. Note that in lower-permeability formations, water may rise to the top of the hole without stabilising. If this occurs, discontinue the test when water reaches the surface.
- At the end of the injection period, continue to monitor and record the fall (recovery) in water level after the pump is turned off.

Figure A2.11 shows an example of the data obtained from an injection test. In this case the response to injection is superimposed on a background rise in water levels due to recovery after completion of drilling.

Injection tests can be run for as long as required, but valuable data on the hole and hydraulic characteristics of fracture zone can be obtained from a test lasting 20 to 30 minutes.

It is important that the hole is airlifted prior to testing to make sure that the bit and the inner tube are clean. Air in the formation subsequent to drilling can produce spurious results. Better results are usually obtained if the

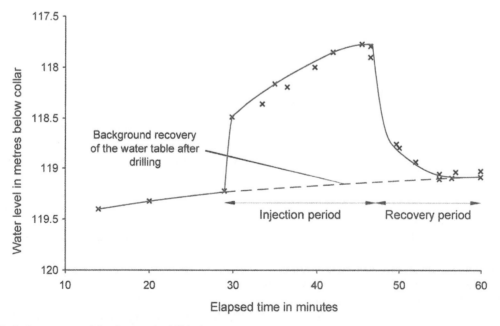

Figure A2.11: Typical response to injection test in drill hole

bit is pulled back to above the fracture zone being tested, as this allows the injected water to freely enter the fractures.

Testing at frequent depth intervals, as the hole is being drilled, allows the depth of groundwater inflow zones to be determined. A rise in specific capacity is observed after each zone is penetrated.

The test procedure is better suited to fractured rock units and is usually not applicable in unconsolidated formations. As with any injection test, the results may be invalid if there are 'dry' permeable fracture zones within the interval of water level rise.

Injection testing generally involves a small amount of rig standing time. It can be carried out with minimal cost on as many exploration or test holes as required. It allows a wide areal spread of groundwater data to be built up about the properties of individual permeable fracture zones across the site.

A limitation of the injection test is that it cannot be conducted through a down-the-hole hammer as the check valve located within the hammer prevents water being pumped down the annulus at low pressure. Clay zones and boots around the drill pipe may also affect interpretation of the data.

5 Guidelines for running and interpreting hydraulic tests

5.1 Single-hole variable-head tests

Variable-head tests are also known as rising- or falling-head tests, or slug tests. They offer the simplest and lowest cost method of *in situ* measurement of hydraulic conductivity, either in an existing hole, or in a hole that is actively being drilled. The procedures for carrying out rising- and falling-head tests are described in Section 2.2 and further information on test procedures is available in numerous technical papers and publications, including Freeze and Cherry (1979), Kruseman and de Ridder (1990), Brassington (1998), and Domenico and Schwartz (1998). Many of these refer to the detailed descriptions provided in the standard methods publications by the British Standards Institution BS 5930 (1999) and the American Standard and Testing Methods ASTM D 4044-91 (1991).

Numerous methods for analysing slug tests have been used and are described by various authors. Probably the simplest and most robust analytical method is that devised by Hvorslev (1951) and described by Freeze and Cherry (1979). This method was apparently devised for testing sections of hole that are relatively short compared with the thickness of the formation; it assumes formation homogeneity and that both formation and water are incompressible so it cannot be used to determine storage properties. Other authors, including Cooper *et al.* (1967),

have described methods of slug-test analysis suitable for the full thickness of an aquifer and which take into account compressibility of the formation and water. They use a curve-matching technique that theoretically will provide a value for storage coefficient (*S*) in addition to hydraulic conductivity. In practice, the families of curves to be matched are so similar in shape that the value of *S* obtained is unlikely to be reliable. For analysis of slug tests in unconfined conditions, the Hvorslev (1951) and Bouwer and Rice (1976) methods may be applied. Many of the methods are also summarised in Kruseman and de Ridder (1990) and British Standards Institution (1999). Probably the most important advice when using one of these methods is to consider carefully the conditions for which it was intended and how closely these conditions are being met at the test site in question.

The Hvorslev method of analysis uses a simple graphical technique, based on the concept of 'shape factors'. The shape factor considers the radius and length of the interval being tested and its position in relation to bed boundaries. The most generally useful 'shapes' are those involving a cylindrical test section entirely within a permeable uniform medium (Hvorslev's Figure 12, Case 8) or cylindrical test section in a permeable medium immediately overlain by a unit of low permeability (Hvorslev's Figure 12, Case 7).

The approach for both falling-head and rising-head tests is identical except that the signs for drawdown and flow direction will be reversed. Taking the rising-head test of Section 2, Figure 2.14 as the example (because positive drawdown and inflow are generally more familiar concepts), Hvorslev argued that the inflow at any time will depend on the hydraulic conductivity of the formation, the head difference between the water in the formation and that in the well or piezometer, and the geometry of the piezometer intake section which he summarised as the 'shape factor', *F*. The rate of inflow, *Q*, at any time, *t*, can therefore be written as:

$$Q = FKh_t \qquad \text{(eqn A2.3)}$$

where h_t is the remaining drawdown at that time.

This flow, *Q*, must be equal to the rate of change of water volume in the cased or standpipe portion of the piezometer, so that also

$$Q = \pi r_c^2 \frac{dh}{dt} \qquad \text{(eqn A2.4)}$$

where r_c is the radius of the casing and therefore

$$FKh_t = \pi r_c^2 \frac{dh}{dt} \qquad \text{(eqn A2.5)}$$

Hvorslev introduced a quantity called the **basic hydrostatic time lag**, which is here given the symbol t_L. This is the time that would be required for the head in the piezometer to return to its original level if the inflow *Q* were to be maintained at its initial value (the value immediately after the slug is removed). This value of Q would be FKh_0 while the volume of water that has to enter the pipe is $\pi r_c^2 h_0$, so the time lag is given by:

$$t_L = \frac{\pi r_c^2 h_0}{FKh_0} = \frac{\pi r_c^2}{FK} \qquad \text{(eqn A2.6)}$$

Manipulation of the equations and solving the resulting differential equation gives (Freeze & Cherry, 1979):

$$\frac{h_t}{h_0} = \exp -\frac{t}{t_L} \quad \text{or} \quad \ln \frac{h_t}{h_0} = -\frac{t}{t_L} \qquad \text{(eqn A2.7)}$$

A plot of h_t/h_0 on a logarithmic scale against time on a linear scale should therefore yield a straight line, as in Figure A2.12.

At the instant the slug is withdrawn, $h_t = h_0$, so that $h_t/h_0 = 1$; when the water level has returned to its original position, $h_t/h_0 = 0$. Inspection of Equation A2.7 shows that when $\ln (h_t/h_0) = -1$, $t = t_L$; this occurs when $h_t/h_0 = 0.37$.

The procedure for Hvorlsev analysis is therefore as follows.

1. Prepare a semi-log plot of h_t/h_0 against time. Note that unlike most semi-log methods of analysis, in this case time is on the linear scale.

2. Draw a straight line of best fit through the points.
3. Determine the value of time at which $h_t/h_0 = 0.37$; this time is t_L, the basic hydrostatic time lag.
4. The value of K can then be found from Equation A2.8 by using this value of t_L and substituting the appropriate value of the shape factor F. For Hvorslev's (1951) Case 8 in his Figure 12, where the length of the test interval is l and its radius is r_w, and the ratio l/r_w is greater than about 8, the equation becomes:

$$K = \frac{r_c^2 \ln\left(\dfrac{l}{r_w}\right)}{2l t_L} \qquad \text{(eqn A2.8)}$$

where r_c is the radius of the cased section or standpipe in which the changes in water level are being observed.

Whichever analytical method is used, and particularly when using automated software packages, it is necessary to screen the data carefully and to consider that the observed water level decay may be the result of several contributing factors (for example multiple ingress or outflow zones).

When analysing the data from tests in open-hole standpipes or monitoring wells, it should be understood that the head change is being applied to the formation around the circumference of the drilled hole and not around the smaller circumference of the monitoring pipe.

Slug tests affect only a limited volume of material around the test hole. The measured permeability is also likely to be affected by damage caused by drilling, especially drilling mud cake or smeared walls caused by natural clays or finely ground rock. While these 'point' measurements are useful for basic characterisation of the groundwater system, it should be understood that the 'bulk-scale' system behaviour is normally controlled by discrete features (conduits or barriers) rather than by the general hydraulic conductivity of the formation (Section 1.3.2). Therefore, for many mining applications, the 'point' test results are often most useful for characterising the least permeable units.

Slug tests can also be used to check for good hydraulic connection between observation wells or piezometers and the formation; Black and Kipp (1977) provide guidance on the technique.

It is important to note that although fairly simple, variable-head tests need to be carried out with care as small errors in measurement can lead to significant errors in estimated hydraulic conductivities. Repeating the tests can improve the confidence assigned to the final estimates of hydraulic conductivity.

During interpretation of the results it is also important to note that, in many hydrogeological settings where stratified sediments or fractured/fissured rock may have a wide range of hydraulic conductivities, the bulk hydraulic conductivity derived from a variable- or constant-head test may largely be a result of a single thin sedimentary layer or single fracture or fissure of very high hydraulic

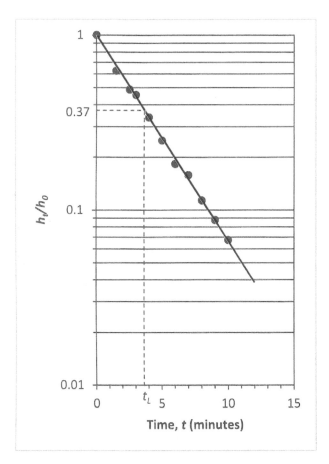

Figure A2.12: Hvorslev method for slug test analysis

conductivity. Where more reliable estimates of *in situ* hydraulic conductivity are required it may be necessary to consider implementation of pumping tests and/or packer tests.

5.2 Packer tests

5.2.1 General

The principles of packer testing are described in Section 2.2. Packer tests involve isolating a length of a borehole between two packers or between one packer and the base of the borehole and then injecting water into or pumping water out of the interval.

Packer tests are sometimes referred to as Lugeon tests, after Maurice Lugeon (Lugeon, 1933). The test described by Lugeon involves pumping water into the isolated interval at a pressure in the test interval of 10 kgf/cm^2 (approximately 1 MPa) above the natural formation pressure. The rate of water flow in litres per minute into each 1 m length of the interval is the *Lugeon coefficient* or permeability in Lugeon units. The method was apparently derived for assessing the permeability of rock at dam sites before and after grouting, where the holes were of narrow diameter and an index value would be useful. For general hydrogeological application the Lugeon approach suffers from three disadvantages.

1. There is no allowance for the diameter of the borehole, which will have some effect on its ability to accept water.
2. The Lugeon unit is not directly comparable with other units of hydraulic conductivity.
3. The suggested pressure may be high enough, especially at shallow depths, to result in artificial fracturing of the formation.

Although (3) could theoretically be overcome simply by reducing the pressure and adjusting the Lugeon value accordingly, this assumes a linear response between flow rate and applied pressure or head which, as discussed below, may not be true, especially in fractured materials (Lancaster-Jones, 1975).

5.2.2 Field preparation

In principle, a packer assembly needs one packer with an open bottom for a single packer test or two packers joined by a perforated middle pipe and a closed cap on the bottom for a straddle packer test. Mining investigations commonly use a wireline system where the assembly is deployed though drill pipe and that pipe is often used, via a system of valves, both to carry the fluid for packer inflation and the water for the injection test. As shown in Figure A2.13, this wireline system has the advantages of being compact enough to be used in small-diameter holes, of not needing a separate downhole inflation line and of

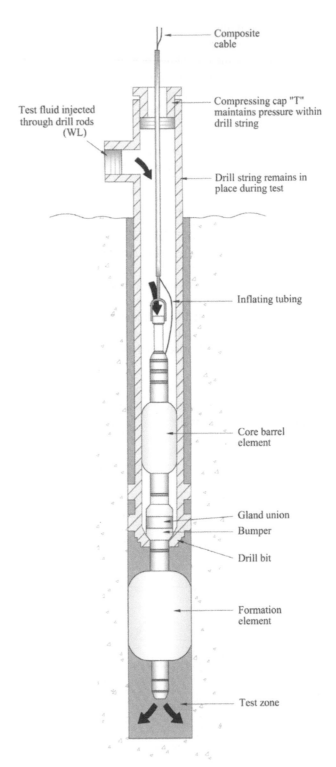

Figure A2.13: Typical configuration for wireline packer system

being rapid in operation. It suffers from the disadvantage that the pump used for inflation and injection is usually a piston pump, which imparts a severe pulse to the injection flow and makes it difficult to determine the injection pressure accurately.

Figure A2.14: Installation of wireline packer string and running a test

The following guidelines refer to the most commonly applied wireline packer injection test using a drilling rig, for which field preparation is often carried out as follows (Figure A2.14).

1. Prepare packer assembly: one packer with open bottom for single test or two packers with perforated middle pipe section and closed cap on the bottom for a straddle packer test.
2. Review and check all equipment:
   i. packer assembly for leakage by inflating to maximum working pressure in an appropriate length and diameter of drill casing at ground surface;
   ii. overshot and wireline connectors;
   iii. water (and gas if being used) supply system: tank, supply, pump, connection hoses, pressure gauges, valves and flow meter.
3. Define: depth and length of test zone, required depth of drill bit, depth position of packers, inflation pressure and water pressure for ascending (minimum of 3 incremental steps) and recovery pressure steps (minimum of 2 steps).
4. Drill hole preparation: removal of drilling mud and cuttings, flush with clean water and hypochlorite if drilling muds need to be removed from borehole walls.
5. Locate drill bit at selected depth and prepare wireline winch.

6. Measure groundwater level or artesian pressure prior to installing packer system.
7. Lower the packer assembly to the landing ring at the drill bit, using the overshot and wireline – check if it seats on landing ring by 'listening' to rods using, for example, a wrench.
8. If possible, check the depth marking on the wireline, if this has been marked for the expected depth, before disconnecting the overshot and tripping it out (this is not required in some wireline systems).
9. Fill the rods with water. Remove the water supply and check the assembly has sealed correctly – the water level should be static (not rising or falling).
10. Attach the loading chamber or stuffing box (a VWP can be installed at this stage to monitor the water pressure inside the rods) and connect the water supply via the flow skid (or nitrogen gas supply).
11. If using water to inflate the packer(s), do it slowly, stopping at the predefined testing pressures (e.g. 15, 30 and 45 psi) to record the stable flow rate (this provides an estimate for rod leakage).
12. Inflate the packer(s) slowly (in 50 psi steps) until they are at inflation pressure. To ensure correct inflation and an effective seal with the borehole walls, this pressure can be calculated, although allowances for the hydrostatic pressure (pressure from overlying water column) must be accounted for. Where the hole is overflowing (i.e. there is an 'artesian' pressure), this

excess pressure at the well head must be added to the pressure calculated from the length of the water column, as described in Equation A2.9.

If the water column is assumed to be approximately 1.4 psi/m of water (for fresh water), then the total working inflation pressure (gauge pressure at surface) required may be estimated using:

$$p_i = p_w + \left(h_{wc} \times 1.4 \text{ psi/m}\right) \qquad \text{(eqn A2.9)}$$

where:

p_i = indicated pressure on the surface inflation gauge

p_w = packer working pressure

h_{wc} = head of water above the higher packer.

5.2.3 Testing sequence

The general sequence for conducting the packer injection test is as follows.

1. Using the flow skid to control the injection pressure, maintain constant initial pressure for step A until it stabilises (possibly 10 to 15 minutes), recording the volume of water injected every minute for a minimum of three minutes under stable conditions.
2. Increase the injection pressure to step B and record the time and corresponding flow rates as for step A.
3. Repeat point 2, for every increase in injection pressure required (a minimum of three steps should be used), taking into account the pressure injection capacity of the water supply pump, ensuring that the maximum injection pressure step does not exceed 80 per cent of the maximum injection pressure capacity of the pump.
4. Following stabilisation of the maximum pressure injection step, repeat the measurement for each step in a sequence of reducing injection pressures, recording the corresponding flow rates every minute until a stable flow rate is seen for at least three minutes. If the formation is tight, release the injection pressure using the bypass valve on the water supply system to decrease pressure from C to B and B to A quickly.
5. After stabilisation of the final step in the sequence, step A, the packer test is completed.
6. To record the static water level of the test section, remove the loading chamber and monitor the water level until it recovers. If the head is artesian, leave the loading chamber in place, close any taps and perform a shut-in test (a form of recovery test recording pressure rather than water level).
7. If required, trip the overshot back into the rods. Deflate the packer assembly, waiting at least five minutes for full deflation.
8. Pull the assembly carefully to top of drill rods, checking the wireline pressure to prevent damage to the assembly.

5.2.4 Analysis and interpretation

Analysis and interpretation of packer tests has been described in detail in a large number of technical publications; U.S. Bureau of Reclamation (1951), Moye (1967), Muir-Wood and Caste (1970), Lancaster-Jones (1975), Houlsby (1976), Ziegler (1976), Pearson and Money (1977), Price et al. (1977 and 1982), Bliss and Rushton (1984), Barker (1988), BSI (1999) and S-Y Hamm et al. (2007). Many mining companies and regulatory authorities have a preferred method. A careful review of the raw data and an assessment of the integrity of the packer seals is an essential part of the interpretation.

For general applications, the basic formulae derived by Hvorslev (1951) are generally satisfactory. Hvorslev considered various configurations of permeable and impermeable material and of cased and open hole, giving varying 'shape factors'. The most useful of these arrangements for packer testing are Hvorslev's Case 8 in his Figure 12 (Hvorslev, 1951); the essential dimensions are shown in Figure A2.15 (a) and (b) below. For these arrangements, Hvorslev used an equation based on flow from a line source for which the equipotentials are ellipsoids, so that when applied to a cylindrical source the solution is an approximation.

For an **isotropic** medium, the relevant equation for hydraulic conductivity K is:

$$K = \frac{Q}{2\pi\Delta hl} \ln\left\{ \frac{l}{2r_w} + \sqrt{1 + \left(\frac{l}{2r_w}\right)^2} \right\} \qquad \text{(eqn A2.10)}$$

where l is the length of the test interval, r_w is the radius, Q is the flow rate and Δh is the applied head difference.

If the test interval is long in relation to its radius ($l \geq 10r_w$), this can be simplified to:

$$K = \frac{Q\ln\left(\frac{l}{r_w}\right)}{2\pi\Delta hl} \qquad \text{(eqn A2.11)}$$

For an **anisotropic** medium:

$$K_h = \frac{Q}{2\pi\Delta hl} \ln\left\{ \frac{ml}{2r_w} + \sqrt{1 + \left(\frac{ml}{2r_w}\right)^2} \right\} \qquad \text{(eqn A2.12)}$$

where:

$$m = \sqrt{\frac{K_h}{K_v}} \qquad \text{(eqn A2.13)}$$

K_h is the horizontal hydraulic conductivity and K_v is the vertical hydraulic conductivity.

Again, if the test interval is long in relation to its radius ($l \geq 10r$), this can be simplified to:

$$K_h = \frac{Q\ln\left(\frac{ml}{r_w}\right)}{2\pi\Delta hl} = \frac{Q\ln\left(\frac{1}{r_w}\sqrt{\frac{K_h}{K_v}}\right)}{2\pi\Delta hl} \qquad \text{(eqn A2.14)}$$

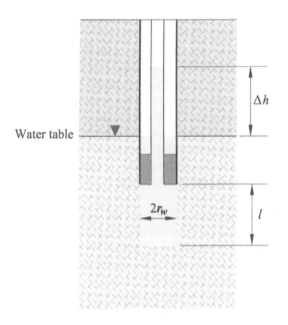

a) Single packer test

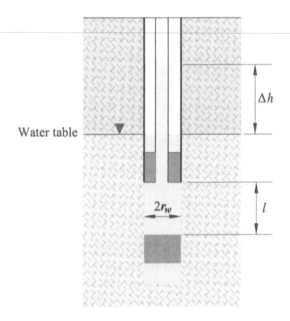

b) Double packer test

Figure A2.15: Explanation of dimensions used in packer-test formulae: (a) Single packer test; (b) Double packer test

Generally, the value of the anisotropy ratio m is unknown and K_h occurs on both sides of equations A2.12 and A2.14, so a synergetic approach is used – laboratory values can often be used to restrict the range of values of K_v. Two points are worth noting:

■ because the anisotropy ratio occurs as the natural log of the square root, even significant errors in m do not have a great effect;

■ in longer test intervals, the anisotropy ratio is less significant and the result is dominated by the K_h value.

The value of Δh is the excess head (or drawdown) applied to the test interval. For systems using a pressure transducer or VWP sensor within the test interval, it is the difference between the head measured during pumping and that measured with the test zone isolated (packers inflated) under static conditions (this last measurement is also useful for checking for vertical hydraulic gradients). If no transducer is available, the static head must be measured with a water-level sounding probe and the injection head is obtained by using a pressure gauge at the surface. In this case, the effective applied head Δh is given by:

$$\Delta h = h_g + h_s \ h_l \qquad \text{(eqn A2.15)}$$

where:

h_g = indicated gauge pressure
h_s = vertical height of gauge above **static** level of water in the **isolated** test interval
h_l = head loss in pipework between gauge and test interval.

This arrangement is unsuitable where the water table is deep, because h_s becomes so large that it may be impossible to inject water fast enough to keep the pipework full to the surface, or it may result in too high an injection head. It is also problematic when using high flow rates through narrow tubing or pipework, especially with internal-upset joints, when the head loss component h_l may become very large and difficult to determine. At relatively low flow rates, h_l is often ignored. At higher rates it can be found from tables or charts such as those produced by the U.S. Bureau of Reclamation (1951, 2005) or by applying the Darcy-Weisbach equation; either approach requires knowledge of some form of friction factor which will vary with the age and internal condition of the tubing.

The Hvorslev equation is valid for anisotropic materials, but is not strictly valid for fissured materials. However, for minor fissuring it seems to work reasonably well (Bliss & Rushton, 1984). As shown in Figure A2.15, Hvorslev made the assumption that the hole above the test interval is lined with solid casing, so that there is no possibility of flow occurring to or from the test section through the surrounding rock and effectively by-passing the packer. Both Bliss and Rushton (1984) and Braester and Thunvik (1984) examined this possibility and concluded that the presence of the uncased portion of the hole would not in general lead to significant errors in measuring permeability. In the case of material with significant vertical fracturing this flow must be a possibility but it would always be unwise to place packers adjacent to vertical fractures; geophysical logging or borehole television or televiewer inspections should be undertaken wherever possible to avoid such cases.

Early practice in so-called Lugeon tests appears to have been to test at only one flow rate and test pressure; this involves the tacit assumption that flow is conforming to Darcy's law and that the test procedure does not alter the natural conditions. A better approach is to follow the multi-step procedure outlined in Section 2.2.5.6, giving at least three values of flow and injection pressure with the two lower ones repeated in the sequence A-B-C-B-A.

The data are then plotted as a graph of flow vs applied head. In an ideal situation of Darcian flow behaviour, with no change to the test conditions (e.g. no fracturing caused by over-pressuring, no scouring or clogging) all five points should lie on a straight line through the origin. Several authors, notably Houlsby (1976), have discussed the implications and interpretation of non-linear behaviour. Figure A2.16 shows examples of the responses that might be encountered; parts (a) to (e) of represent five possible conditions.

The left-hand diagram for each case shows the plot of flow rate (or injection rate), Q, against the applied head, Δh, required to cause that flow rate and the right-hand diagram represents the values of hydraulic conductivity that would be calculated from the flow-head relationship for each of the five points in the left-hand graph, using the simplified Hvorslev equation for an isotropic formation. The conditions shown in Figure A2.16 are:

a) linear relationship – a near-ideal example in which all the points on the flow-head curve lie approximately on a straight line, with no hysteresis. The conditions conform to Darcy's law and all five values of K (right-hand diagram) are equal;

b) non-Darcian relationship – at higher flow rates, the head increases faster than the flow rate. This means that the flow, while not necessarily truly turbulent, is exceeding the point where Darcy's law is valid. There is no hysteresis. The calculated values of K are highest at the low flow rates and lowest at the high flow rate;

c) dilation – the flow rate increases disproportionately to the increase in applied head, so that the calculated K value is much higher for the high head value. There is no hysteresis, so the K value reduces again at the lower applied heads. The most likely cause is dilation of fractures caused by the injection pressure exceeding the effective stress and causing temporary enlargement of fractures;

d) clogging/void filling – the flow rate decreases disproportionately to the increase in head throughout most of the test, with marked hysteresis. There are two possible causes for this;

 i) reduction in permeability caused by clogging of pores or fractures. This is most likely to occur in injection tests if the water being injected contains suspended sediment or if the borehole walls are lined with mud which is moved during the test, for example into fine fractures.

 ii) filling of empty 'blind' voids. Houlsby (1976) describes this behaviour as 'an indication that the test is gradually filling empty voids, joints, etc., which are semi-blind (i.e. water cannot easily escape from them)'. This behaviour is likely to occur only in tests at or above the water table, where the ground is not fully saturated. If such conditions are suspected it is preferable to run the test until a stable permeability value is reached at several flow rates.

e) enhancement (scouring) – this behaviour is the converse of (d), in that the measured permeability progressively increases during the test; again, there is marked hysteresis. The probable cause is material being washed out of fractures but it may be the result of the test pressure being too high and causing the creation of new fractures or a permanent or semi-permanent opening of existing ones (this is in contrast to the dilation described under (c), where the increase in permeability was sustained only while the high test pressure was effective).

Where the response is approximately linear, the best interpretation is obtained by drawing a line of best fit through the points and the origin and using the gradient of $Q/\Delta h$ to calculate the value of K. In interpreting results with non-linear behaviour, it is important to select a conservative value of K. Houlsby generally takes this to be the highest 'Lugeon' or K value. This is generally true when planning dewatering or grouting programs but, in cases where depressurisation is relying on passive drainage and therefore a high value of hydraulic diffusivity, it may be more appropriate to take the *lowest* permeability value as being conservative.

Packer tests typically affect a larger volume of rock than slug tests but their effects are still limited. As Bliss and Rushton (1984) point out, although derived on the basis of equipotentials that are semi-ellipsoidal in form, the Hvorslev equations (2.11 and 2.14) are identical to the equation for radial flow through a confined aquifer of constant thickness l (the Dupuit equation), but with an outer fixed-head boundary at a distance l (or ml for the anisotropic case) from the centreline of the well. This means, in theory, that when a packer test in a homogeneous formation of constant thickness has reached equilibrium, the radius of influence of the test is the same as the length of the test interval. The difficulty with this is that few packer tests will be in a confined homogeneous unit of constant thickness the same as the length of the test interval, and that even fewer will probably achieve a true equilibrium. However, Bliss and Rushton used a numerical model to compare the theoretical response to packer

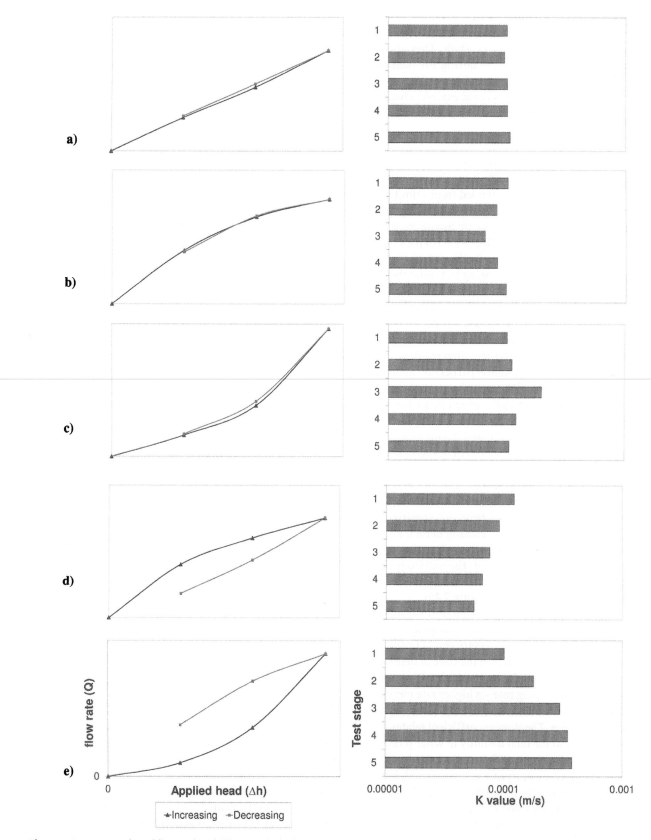

Figure A2.16: Examples of flow vs head relationships (left-hand diagram) and calculated values of hydraulic conductivity (for each of the measured points) (right-hand diagram) for five situations: a) linear relationship; b) non-Darcian ('turbulent') relationship; c) dilation; d) clogging/void filling; e) enhancement (Modified after Houlsby, 1976)

testing of various configurations with the results of actual packer tests, and concluded that their effect typically extended to around 10 m from the borehole. They also concluded that effective equilibrium during any stage of a test should be achieved within a few minutes, although for materials of low permeability this could extend to hours.

5.3 Pumping tests

5.3.1 Basic principles

Pumping tests offer a means to characterise the mass behaviour of a large volume of rock. It should, however, be understood that results are still obtained from a limited number of monitoring points, so there is still a sampling problem. Furthermore, mine settings are usually very different from the theoretical configurations of rock units on which many of the analytical methods are based. Nevertheless, applied sensibly, pumping tests offer the best approach to understanding the bulk behaviour in many settings. Brief guidelines on the more basic methods of analysis of tests are given in this section but it must be understood that the correct interpretation of test data, as opposed to their mechanical analysis, often requires specialist knowledge and experience; particular care must be exercised when using proprietary software packages that can impart a false confidence to the inexperienced user.

When water flows to a production or drainage well, the difference between the head in the well during pumping (the pumping water level or PWL) and that when the well was at rest (the 'static' or rest water level or RWL) is called the drawdown (s or s_w). The presence of the drawdown creates a head difference between the water in the well and that in the formation. Water flows from the formation to the well to replace that abstracted, causing in turn a drawdown in the formation around the well so that the potentiometric surface slopes towards the well from all directions, creating a cone of depression of the potentiometric surface (Section 2, Figure 2.49).

The steepness of the cone depends on the hydraulic gradient, which in turn depends on the pumping rate and on the transmissivity and storage coefficient of the aquifer. It should be noted that the radius of the cone of depression depends on time and hydraulic properties but is independent of the pumping rate – a higher rate merely results in increased drawdown and not in a faster radial expansion of the cone. The storage coefficient relates the volume of the cone to the total quantity of water pumped out; the smaller the storage coefficient, the larger the cone must be for any given quantity of water abstracted. It follows that by knowing the pumping rate and observing the shape and rate of expansion of the cone, it should be possible to deduce the properties of the formation; this is the principle of the pumping test.

Ideally, a test is carried out using a pumping well and several observation wells or piezometers; these allow the drawdown (and hence the shape of the cone of depression) to be monitored over time at known distances from the pumping well. If no observation points are available, data from the pumping well can still be used to derive a value of transmissivity but, for the reasons given in 5.3.4 of this appendix, a value for the storage coefficient cannot be derived.

5.3.2 Analytical methods for aquifer tests: the Theis method

The most common pumping-test methods use a constant pumping rate and a non-steady-state analysis. The simplest is based on the Theis equation (Theis, 1935) for confined aquifers. The method is based on a number of assumptions including homogeneity, isotropy and an infinite horizontal aquifer; in practice, minor departures from the ideal are usually tolerated. The Theis equation relates drawdown (s), pumping rate (Q), distance from centre of the pumping well (r or r_w), time since pumping started (t) and the formation properties transmissivity (T) and storage coefficient (S):

$$s = \frac{Q}{4\pi T} \int_u^\infty \frac{e^{-y}}{y} dy \qquad \text{(eqn A2.16)}$$

where

$$u = \frac{r^2 S}{4Tt} \qquad \text{(eqn A2.17)}$$

The integral term in equation A2.16 is termed the exponential integral: in the context of aquifer-test analysis it is often written as $W(u)$ and termed the 'well function of u' or the 'Theis Well Function', so that equation A2.16 can be written as:

$$s = \frac{QW(u)}{4\pi T} \qquad \text{(eqn A2.18)}$$

The solution of the integral is an infinite exponential series:

$$W(u) = -0.5772 - \ln u + u - \frac{u^2}{2 \times 2!} + \frac{u^3}{3 \times 3!} - \frac{u^4}{4 \times 4!} + \cdots$$
$$\text{(eqn A2.19)}$$

There are published tables of $W(u)$ for different values of u (see, for example, Kruseman & de Ridder, 1990). Knowing S and T, the user can calculate u for any distance r from the pumping well and at any time t from the start of pumping, then look up $W(u)$ and calculate the drawdown s.

The form of the Theis equation is such that it cannot readily be used to derive T or S directly from values of drawdown, distance and time. Instead a graphical, curve-matching method was developed. Values of log $W(u)$ are plotted against values of log u or log ($1/u$) to form a type curve.

Equation A2.17 can be rearranged to give:

$$t = \frac{r^2 S}{4T} \times \frac{1}{u} \qquad \text{(eqn A2.20)}$$

For a given aquifer (with constant T and S) and pumping rate and at a given distance from the pumping well, equations A2.18 and A2.20 can be summarised as saying:

$$s = [\text{constant term}] \times W(u)$$

$$t = [\text{constant term}] \times \frac{1}{u}$$

therefore

$$\log s = [\text{constant}] + \log W(u)$$

$$\log t = [\text{constant}] + \log \frac{1}{u}$$

so a data plot of log s against log t has the same shape as a type curve plot of log $W(u)$ against log $1/u$, but is displaced by constant amounts that depend on T and S. This principle underlies most methods of aquifer-test analysis, whether done manually or using a computer-based method. There are later variations to deal with unconfined (water-table) conditions (the Boulton or Neuman methods) or leaky conditions (the Hantush-Walton methods). More advanced methods deal with double-porosity or heterogeneous aquifers. The principal methods of analysis are summarised by Kruseman and de Ridder (1990). All of these methods make use of the fact that when a well is pumped at constant rate, drawdown increases with time until some boundary condition, or leakage or drainage effect, slows down the increase or causes the drawdown to stabilise.

The procedure for manual analysis using the Theis method is:

1. Obtain or prepare a plot of $W(u)$ versus $1/u$ on log-log paper (Figure A2.17). This plot is called a **type curve**.
2. Plot the measured values of drawdown against time on tracing paper using logarithmic scales of the same size as those on the type curve.
3. Superimpose the data plot on the type curve and move it around, keeping the coordinate axes parallel, until as many as possible of the data points fall on the type curve.
4. Select an arbitrary **match point**. This can be anywhere on the data plot – it does not need to lie on the curve – but it is convenient to use the point where $1/u = 1$ and $W(u) = 1$. Read off the values of s and t at the match point.
5. Substitute the match-point values in equations A2.21 and A2.22 to derive T and S. **Remember to use consistent units**.

Re-arranging equations A2.18 and A2.17 gives:

$$T = \frac{QW(u)}{4\pi s} \qquad \text{(eqn A2.21)}$$

and

$$S = \frac{4Tut}{r^2} \qquad \text{(eqn A2.22)}$$

Pumping-test analysis using curve matching is now normally done using proprietary software. The data are entered by copying from spreadsheets or data loggers and plotted in the correct form by the software, which then presents the chosen type-curve against which the data points can be matched. The analytical packages usually offer the choice of an 'autofit' or a manual curve-matching option. It should be noted that the autofit options are normally unreliable except where the data conform precisely to the theoretical curve, which is very rare especially in complex, fractured or mixed-rock settings. It is also important to ensure that the measurement units are entered correctly.

5.3.3 Analytical methods for aquifer tests: the Cooper-Jacob ('straight-line') method

Cooper and Jacob (1946) provided a simpler graphical approximation to the Theis method for long times and/or observation points close to the pumping well. This is based on the same assumptions as the Theis method and on the fact that for small values of u (i.e. after long pumping times or for small values of r) the terms in equation A2.19 after ln u become insignificant, so that for values of u less than 0.01, $W(u)$ can be replaced by $(-0.5772 - \ln u)$ and equation A2.18 can be written as:

$$s = \frac{Q}{4\pi T}(-0.5772 - \ln u) \qquad \text{(eqn A2.23)}$$

Recognising that $-\ln u = \ln 1/u$ and that $0.5772 = \ln 1.78$ and substituting for u leads to:

$$s = \frac{Q}{4\pi T}\ln\left(\frac{2.25Tt}{r^2 S}\right) \qquad \text{(eqn A2.24)}$$

and

$$s = \frac{2.30Q}{4\pi T}\log_{10}\left(\frac{2.25Tt}{r^2 S}\right) \qquad \text{(eqn A2.25)}$$

Equation A2.25, the Cooper-Jacob equation, is a very useful 'universal equation' for estimating the effects of pumping, for example in deciding the distances to observation wells when planning a pumping test. When using it, it is important to remember the time and distance limitations to its validity.

The procedure is as follows.

1. Plot drawdown (linear scale) against time (logarithmic scale) (Figure A2.18).
2. Draw a straight line of best fit through the points; early data may not fit because the time is too short (and therefore u is too large) for the Cooper-Jacob approximation to be valid.
3. Measure the rate of change of drawdown over one log cycle; this value is termed Δs.

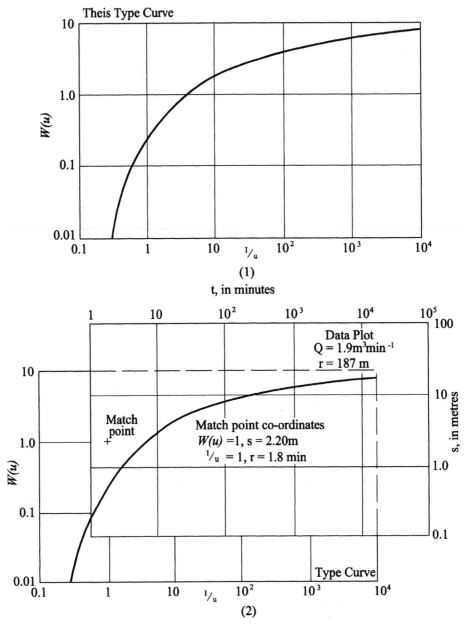

Figure A2.17: The graphical procedure for analysis of pumping test data by the Theis curve-matching technique (from Heath, 1983 and 2004)

4. Project the straight line to reach the time axis; the value of time at this point is termed t_0.

5. Substitute the values of Δs and t_0 into the following equations, which are derived from equation A2.25:

$$T = \frac{2.30Q}{4\pi\Delta s} \qquad \text{(eqn A2.26)}$$

and

$$S = \frac{2.25Tt_0}{r^2} \qquad \text{(eqn A2.27)}$$

Where more than one observation well is available, the Cooper-Jacob method can also be used for distance-drawdown analysis, with distance plotted on the log scale. The procedure and limitations are similar to those for time-drawdown analysis and the simplified equations are:

$$T = \frac{2.30Q}{2\pi\Delta s} \qquad \text{(eqn A2.28)}$$

and

$$S = \frac{2.25Tt}{r_0^2} \qquad \text{(eqn A2.29)}$$

where r_0 is the intercept of the straight line on the distance axis. Note the difference in denominators between equations A2.26 and A2.28.

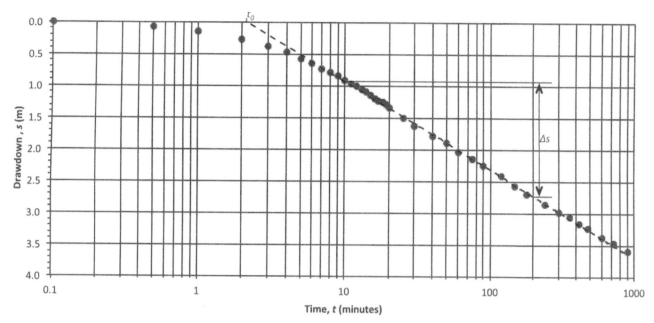

Figure A2.18: The Cooper-Jacob or 'straight-line' method of test analysis

5.3.4 Well losses and their effect on dewatering performance and test analysis

Part of the drawdown in a production well represents the energy loss necessary to make the water flow through the formation to the outer boundary of the well (outside any filter pack or zone of disturbance caused by drilling); this is termed the aquifer loss or formation loss. In many fractured rock and lower permeability settings, the local formation loss immediately surrounding the well can be a large proportion of the total formation loss. The remainder of the drawdown represents the energy loss incurred as the water moves through the zone of drilling disturbance, filter pack, and well lining or screen (if any of these zones or items are present), then changes direction towards the pump (and therefore undergoes a change of momentum); these components are referred to as well losses or collectively as the well loss. Figure A2.19 illustrates them diagrammatically in a confined fractured rock aquifer unit; the same principles apply in an unconfined formation but with the added complication that, as saturated thickness is reduced as a result of drawdown, the transmissivity of the formation is also reduced.

From the point of view of dewatering, what matters is the drawdown in the formation away from the well; large additional drawdown in the local formation immediately surrounding the well or inside the well contributes nothing to dewatering or slope drainage, but merely adds to the pumping costs.

s_{w1} = true aquifer formation loss

s_{w2} = localised formation loss

s_{w3} = linear flow well loss

s_{w4} = turbulent flow well loss

When calculating the pumping rate required to achieve a given level of dewatering, the variable of interest is the true aquifer or formation loss (s_1 in Figure A2.19). This is the drawdown present in the formation away from the immediately vicinity of the well. The true aquifer loss is proportional to the pumping rate, Q, and increases with time. The localised formation loss (s_2) occurs in the formation immediately surrounding the well as a result of increasing groundwater velocities within the fracture system because of radial flow to the well. The well losses (s_3) are also dependent on Q (though the proportionality is not necessarily linear) but, once established at a given value of Q, do not increase with time. When analysing a pumping test to determine transmissivity, it is the rate of change of the drawdown that is important; it follows that the presence of well losses does not prevent drawdown data from a pumping well being used to derive an estimate of transmissivity. Where there are significant well losses, the large early drawdown can make matching to type curves on a log-log scale difficult, so the Cooper-Jacob semi-logarithmic 'straight-line' approach is more useful; the small value of r (in this case the radius of the well) means the method is valid after a very short time.

To derive a value for storage coefficient or specific yield, however, it is necessary to know the drawdown at a specific time and distance from the centre line of the pumping well. When only data from the pumping well are available the drawdown is that at the well face, distance r_w. Unfortunately, the drawdown at this distance is affected by well losses and is therefore not the true drawdown in the

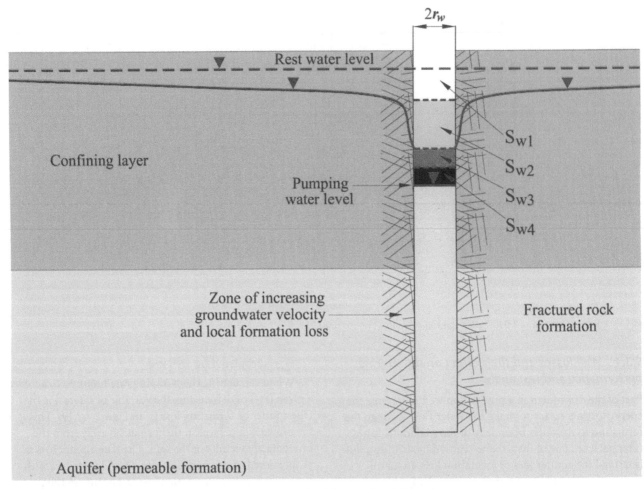

Figure A2.19: Components of drawdown in a well within a confined fractured rock formation

formation, so that single-well tests are unable to yield a reliable value for storage coefficient.

The difference between actual drawdown (s_{w1}) and theoretical drawdown ($s_{w1} + s_{w2} + s_{w3} + s_{w4}$) is sometimes known as the 'skin effect'. Jacob (1946) showed that, for a homogenous aquifer, the well loss due to pumping can be regarded as the difference between the drawdown measured at the actual well radius, r_w, and the theoretical drawdown at that radius computed from the properties of the aquifer and the production rate. However, for fractured rock and low-permeability settings, the local formation loss (s_{w2}) also requires consideration. The concepts of skin factor and effective well radius provide a relatively simple way of understanding losses around the well. The distance from the centre line of the well at which the measured drawdown is equal to the theoretical drawdown is the effective well radius, r_{ew}. In almost all cases, the actual drawdown in the well face will be significantly greater than the drawdown in formation surrounding the well.

Figure A2.20 shows a theoretical distance-drawdown plot in a confined aquifer of $T = 500$ m²/day, $S = 0.001$, after pumping from a well of radius (r_w) of 0.15 m at 50 l/s for 1000 minutes. Observation wells at 30 m and 300 m

define the slope of the semi-log plot. The theoretical drawdown s_w (resulting from true aquifer loss alone) is approximately 12 m. The difference between actual and theoretical drawdown is the skin effect which can be written (Kruseman & de Ridder 1990):

$$\text{skin effect} = \text{skin factor} \times Q/2\pi T \quad \text{(eqn A2.30)}$$

The skin factor or 'skin' is a dimensionless quantity which in the oil industry is usually given the symbol S. Here, to avoid confusion with storage coefficient, it is written out in full. It can be related to effective well radius by the equation (Matthews & Russell 1967):

$$r_{ew} = r_w e^{-skin} \quad \text{(eqn A2.31)}$$

Taking natural logs gives:

$$skin = \ln r_w - \ln r_{ew} \quad \text{(eqn A2.32)}$$

The exponential nature of equation A2.31 means that relatively small skin effects can greatly influence the effective well radius, and this is also illustrated by Figure A2.20. For example, if a well of radius 0.15 m were to intersect a highly-transmissive fracture extending for 2.5 m from the well centre line, then the effective radius of the well could be as high as 2.5 m and the skin factor (from equation A2.32) would be –2.8. Typical skin factors range from as high as 100 for a badly gravel-packed well to as low as –6 for a well that has been hydraulically fractured ('fracked') to produce a major fracture (Schlumberger, 2011).

Many mine settings are in rock where the principal flow mechanism is through fractures, with relatively impermeable material between fractures. If a dewatering well intersects a well-connected fracture (Case A in Figure A2.20), that well can become an efficient structure for drainage or depressurisation; it may require a high pumping rate to lower the water level significantly but the lowering of water level or pore pressure will be transferred to the surrounding formation for some distance. If the well fails to intersect such a fracture (Case B in Figure A2.20), the drawdown in the well may be very high for a low production rate but will have little effect on lowering the water level in the surrounding rock.

There are several methods of quantifying well efficiency by means of step-drawdown tests, in which the well is pumped for short periods at increasing rates. If the conditions in the aquifer and well system are obeying Darcy's law then the specific capacity of the well (Q/s_w) should be the same at all rates, but if non-Darcian flow (loosely called 'turbulent' flow) is occurring, the specific capacity will reduce at higher rates. The principle was first set out by Jacob (1947) and developed by, among others, Rorabaugh (1953), Bierschenk (1963) and Hantush (1964); one commonly used method, attributed jointly to Hantush and Bierschenk, is described in detail by Kruseman and de Ridder (1990). The method essentially separates linear and non-linear components of drawdown and ascribes the linear component to aquifer loss and the non-linear component to well loss; its shortcoming is that it therefore ignores the fact that some (often most) of the well loss is linear in nature.

Local formation and well losses can be especially significant in fractured rock settings where much of the water may enter the well at a small number of discrete and relatively thin fractured intervals, so groundwater velocities in the fractures immediately around the wellbore can become very high. The presence of well losses should always be borne in mind when designing or reporting on dewatering systems; in particular it must be remembered that a large drawdown in a pumping well does not necessarily mean a large drawdown in the formation (Section 2.2.8). For this reason, it is usually not appropriate to include dynamic water levels in pumping wells on contoured groundwater maps.

5.3.5 'Real life' departures from the ideal case

Characterisation of the large-scale properties and behaviour of the groundwater system requires some form of multiple-well testing. As described above, this is usually

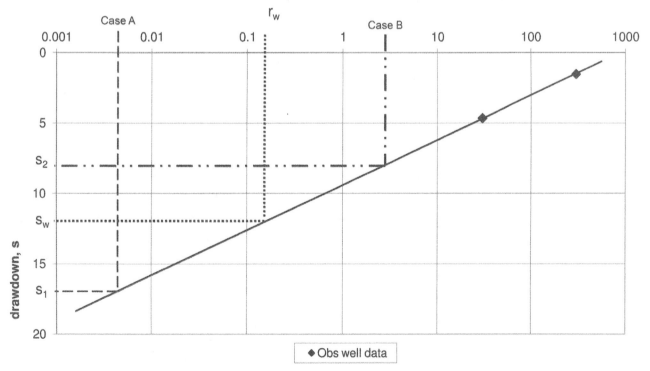

Figure A2.20: Semi-logarithmic distance-drawdown plot showing concept of effective well radius: Case A – Inefficient well. In a fractured formation, this could also be one that intersects lower permeability fractures; Case B – Efficient well. This could be a well whose effective radius is increased significantly by intersecting a major permeable fracture.

carried out by pumping from a single well and making additional measurements in nearby observation wells or piezometers. In permeable formations, where the pumping rate is likely to be relatively high, the distances from the pumping well to the observation wells can be relatively large, so that a large volume of the formation can be characterised. For slope depressurisation studies in less permeable formations, it is often necessary to carry out the testing program at low pumping rates, sometimes less than 1 l/s. In this case, the surrounding observation points will need to be closer to the pumping well, thereby reducing the volume being tested, and the design of the test program may also need to account for the presence of strong anisotropy and preferential flow paths.

It should also be noted that, in a mine dewatering operation, the saturated thickness of the groundwater units, and therefore their transmissivity, will typically decrease with time so that values of transmissivity derived from initial pumping-test data may not be applicable to longer-term analysis.

Most pumping-test analysis is now carried out using software packages that permit the use of a much larger number of theoretical type curves to cover a wide variety of conditions, such as anisotropic conditions with piezometers that are screened over different intervals from the pumping well. It can often (perhaps usually) be very difficult to achieve a unique interpretation of the results from such tests. Around many hard rock mine sites, the complex geology may lead to a range of interacting responses in the pumping-test data, incorporating anisotropy, horizontal and vertical preferential flow paths, and groundwater barriers. These factors introduce complications to conventional analysis, but the difficulty is compounded by the fact that different factors may produce similar-looking responses; it is important to realise that type curves (and data plots) for different formation configurations can be very similar.

This emphasises the need for a sound conceptual understanding of the hydrogeological situation and of a willingness to vary the conceptual model if increased data prove it to be incorrect; it also emphasises the need to obtain specialist advice where necessary. The presence of boundaries – either barrier or recharge – can be one of the most critical factors in the success or failure of dewatering and depressurisation operations and the full significance of boundaries may become apparent only after fairly prolonged pumping, perhaps from several wells, with careful monitoring at multiple points and depths.

In areas of complex geology around mine settings, it is often of more benefit to take a semi-quantitative approach when interpreting the results of pumping tests, using judgement as to how the results may reflect the geological setting.

5.3.6 *Value of recovery data*

The measurement of water level data at the end of a pumping test should be a normal part of the data collection process. These recovery data are unaffected by the minor variations in pumping rate and may therefore be easier to analyse; methods of analysis are described in standard texts.

Another point to note is that recovery data are not, as is often assumed, a 'mirror image' of drawdown data. Full recovery from a period of pumping often takes much longer than the length of the pumping period. Also, recovery occurs not from the edge of the cone of depression inwards, but from the centre of the cone outwards; in poorly permeable material, it is possible for the drawdown near the edge of the cone still to be increasing while the water level in and near the production well is recovering.

It is often useful to assess the amount of residual drawdown in the formation that remains following a pumping test. If the recovery occurs quickly, is it usually indicative of a large and potentially near recharge source. If the system does not fully recover, it may indicate that storage has been 'permanently' removed from the groundwater units that have been tested, indicating the lack of a recharge source.

6 Guidelines for the installation of grouted-in vibrating wire piezometer strings

6.1 Drilling methods

A string of grouted-in vibrating wire piezometer sensors may be installed in boreholes drilled using a variety of drilling methods including conventional mud rotary, RC, simultaneous casing advance systems (Odex, Tubex, Barber) and diamond core drilling. Diamond core and RC methods are the most commonly used around mine sites. Where reliable data on anticipated geological and geotechnical conditions are already available, then RC may be the preferred method given its lower cost. Diamond core drilling at standard diameters of HQ (96 mm) or NQ (76 mm) is more regularly deployed where better control on specific geological and geotechnical conditions is required for specification of the depth setting of each sensor.

6.2 Depth setting for vibrating wire sensors

The installation of multiple vibrating wire sensors, whether in a vertical or angled borehole, requires the suspension or fixing of the individual vibrating wire sensors at predefined depths in the borehole. Criteria normally considered when specifying installation depths for vibrating wire piezometers in pit slopes, include:

- depth to groundwater, and the presence of any perched water conditions;
- geological conditions - lithology, alteration and mineralogy;
- structural conditions – joints, fractures, fissures and faults;
- geotechnical parameters – fracture frequency and RQD;
- pit floor elevations (current, staged and ultimate);
- elevation of nearby dewatering/slope depressurisation infrastructure (dewatering wells, drains and tunnels), and the extent to which water levels are expected to decrease as mining progresses.

It is important to note that vibrating wire sensors measure pressure in height above the instrument. Therefore, to enable an accurate determination of groundwater head and to provide a meaningful groundwater analysis, it is essential that the exact position of the sensor in the hole is known with confidence.

6.3 Installation of multi-level VWPs using the guide-tremie pipe method

The guide-tremie pipe method is currently the most frequently applied for installation of multi-level, grouted-in VWPs, particularly where there are a larger number of installations programmed and ground conditions are competent such that there is limited risk of borehole collapse during installation. The method comprises the installation of the VWPs down the inside of the open borehole with each of the VWP sensors strapped to the outside of a guide pipe which also serves as a grout injection or tremie pipe. Figure A2.21 illustrates a typical design of a multi-level VWP installation completed using this method.

There are many site-specific procedures developed for this method, but general steps for vertical and negative inclined boreholes are summarised as follows. A similar approach for horizontal and positive inclined boreholes is described in Section 6.5 of this appendix.

1. Select appropriate depth settings for the installation of each VWP and ensure adequate annular space within the open hole, normally PQ, HQ or NQ, for installation of the guide-tremie pipe, each of the VWP sensors and the communication cable(s).
2. Select an appropriate guide-tremie pipe for grout injection to total depth of borehole (or a suitable depth below the lowest planned VWP position) and for fixing VWP sensors. Typically either 26 mm to 33 mm outside diameter (Sch 80) PVC or 35 mm diameter galvanised steel pipe are commonly used.

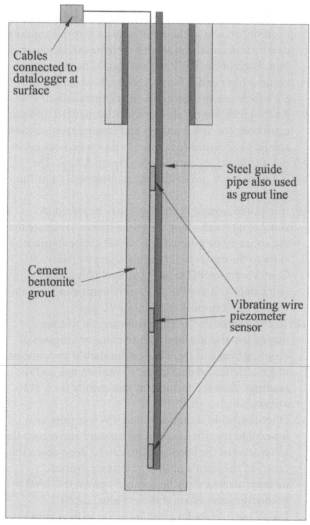

Figure A2.21: Typical configuration of guide-tremie pipe method for grouted in VWPs

3. Measure and mark corresponding depth settings on the guide pipe and on communication cables for each VWP.
4. Record the zero readings (temperature and pressure) according to the manufacturer's instructions (usually in a bucket of water out of the sun). Check each sensor is working correctly by comparing the readings against the factory reading specified on its accompanying calibration sheet.
5. Slot or drill holes in the lower 15 to 24 m length of guide-tremie pipe to allow release of grout to the bottom of the borehole during grout injection.
6. Prior to running the guide-tremie pipe with the VWPs into the open borehole, number sequentially all lengths of the pipe and ensure that all VWP cables are clearly marked with corresponding sensor and depth setting at ground surface.
7. Measure and record the static water level in the borehole using a dip meter or sounding probe.

8. Run the guide-tremie pipe to the total depth of open borehole (or a selected depth below the lowest VWP position) attaching the VWP sensors at corresponding depth marks on the pipe with cable ties and industrial tape. Each of the VWP sensors should be inverted so that the filter end is facing upward, to avoid trapping air bubbles on the filter end during installation.

9. As the guide-tremie pipe is slowly introduced into the open hole, the VWP sensors and communication cable should be strapped tightly to the pipe with both cable ties and heavy industrial tape (Figure A2.26). The industrial tape can be applied continuously or at 1 m intervals.

10. If required, simple centralisers may be placed at suitable intervals along the guide-tremie to ensure that the entire guide and string of VWP sensors remain as close to the centre of the borehole as possible. Centralisers for PQ to NQ size boreholes may be fabricated with 'Jubilee' or 'Hose' clamps and 3 × 10 to 15 mm sections of 19 to 38 mm PVC pipe.

11. To ensure that the VWP sensors are not damaged during installation (and later to monitor progress of grouting), each of the VWP cables should be connected to a multichannel data logger or readout box and the readings observed to confirm that each of the VWPs are working.

12. Once the guide-tremie pipe and VWP sensors are lowered to specified depths, re-measure and record the water level in the borehole (this can be done down the centre of the tremie pipe). The current pressure measurements should be checked against this level to ensure the sensors are at the expected depth.

13. Install the T-piece for grout injection to the top of the guide-tremie pipe and prepare the grout mix according to the guidelines indicated in Section 6.8 of this appendix.

14. Prior to grout injection, adjust injection rate to less that 0.5 l/s by adjusting pumping rate of mud pump during an injection trial using a container at the ground surface.

15. Commence the grout injection until the grout mix discharges at ground surface from the annular space between the open borehole and the guide-tremie pipe.

16. Continue to monitor the VWP sensors throughout the grouting process, and particularly the uppermost VWP sensor to enable estimates of additional grouting requirement to be made.

17. After a day or two, top up the grout (from surface), as the level will go down as grout is lost to fractures in the formation and due to the curing shrinkage of the cement.

18. Prepare a protective cover for the communication cables and an enclosure for multichannel data logger, and connect each of the communication cables to the data logger.

19. Check that all VWPs are functioning correctly.

20. Program the data logger to record pore pressure readings from each VWP at required frequency of measurement.

Although the above procedure has been well tried and tested, a number of risks may remain, as follows.

- If there are permeable fractures in the hole, a considerable loss of grout can occur into the fracture systems, which may greatly increase the amount of grout required, and may require a multi-stage grouting program.
- Where the depth of the hole is significantly greater than the planned depth of the VWP installations, it may be necessary to partially backfill the hole prior to installation of the guide-tremie pipe and piezometers. Alternatively, the guide-tremie pipe and piezometers can be 'hung' in the hole from the collar and the grout can be pumped in the usual way, but with an additional volume of grout to fill the lower part of the hole.
- Collapse of the hole may make it difficult to install the guide-tremie pipe and piezometers without causing damage, in which case it may be necessary to install the guide-tremie and VWP sensors down the inside temporary casing or the drill rods and to progressively remove the casing during the grout injection, being careful to remove the casing ahead of rising grout level.
- For boreholes where the static water level is deep (over 50 m), the zero readings must be taken inside the borehole approximately five metres above the water level. This can be done during the installation process. The manufacturer's instructions should always be consulted to ensure the correct calibration method is used.

6.4 Installation of multi-level VWPs using the wireline method

Where collapsing borehole conditions are expected, an alternative method is to line the borehole with casing, perforated over the interval where the VWPs will be located. The VWP sensors and cables are then strapped to a sacrificial wireline which is then run inside the casing to the predefined depths (Figure A2.22). A number of materials may be used to lower and fix the VWP sensor at specific depths for the wireline method, these include 5 mm steel wire and 13 mm fibreglass rod. Grouting is then facilitated via the casing with an appropriate grout injection T-piece connected to the top, as shown in Figure A2.23.

It is generally considered that this method is less reliable than the guide-tremie method because there is less control on the grout placement. Furthermore, it cannot be used in horizontal or positive inclined boreholes and can be unreliable in shallow, negative inclined boreholes (0 to –50°). However, the method is beneficial where there are collapsing conditions or where the static water level is very

Figure A2.22: Installation of multiple VWP cables and sensors on sacrificial wireline

deep (over 100 m). The latter may be important because the pressures produced by grouting may be higher than the specified pressure rating on the VWP sensor. The wireline method allows the grouting to be carried out in stages, reducing the pressure on the sensor.

Figure A2.23: Grout injection T-piece

A number of VWP installations have been made using spun-out (old or damaged) diamond drill casing (usually PQ or HQ), perforated at the specific depths defined for installation of each VWP sensor, and installed to total depth of the borehole. Alternatively, PVC or polyethylene casing can be used. Figure A2.24 illustrates a typical design of a multi-level vibrating wire installation completed within perforated drill casing. The general guidelines for this method are summarised in the following steps.

1. Select the appropriate casing material and diameter to fit into the borehole and to ensure adequate annular space within the pipe for wireline cable, each VWP sensor and all of the communication cables.

2. Select appropriate depth settings for the installation of each VWP.

3. Measure and mark corresponding depth settings on communication cables for each VWP. Number sequentially all lengths of the casing (solid and slotted/perforated) to ensure that open sections correspond with the VWP sensor depths. Typically, the lowest 6 to 12 m of casing should be slotted and an interval of 6 to 9 m should be slotted at the selected depths for VWPs.

4. Run in the casing to total depth of the borehole and cut at ground surface to allow a stickup of approximately 0.3 m. Alternatively, the casing can be 'hung' in the hole.

5. Attach a sacrificial plumbing weight (e.g. 25 mm rebar) and run the wireline down the inside of the pipe to the base of the borehole. This ensures that the pipe has remained clear and straight during installation. Trip the wireline out of the borehole once this has been checked.

6. Measure and record the static water level in the borehole using a dip meter or sounding probe.

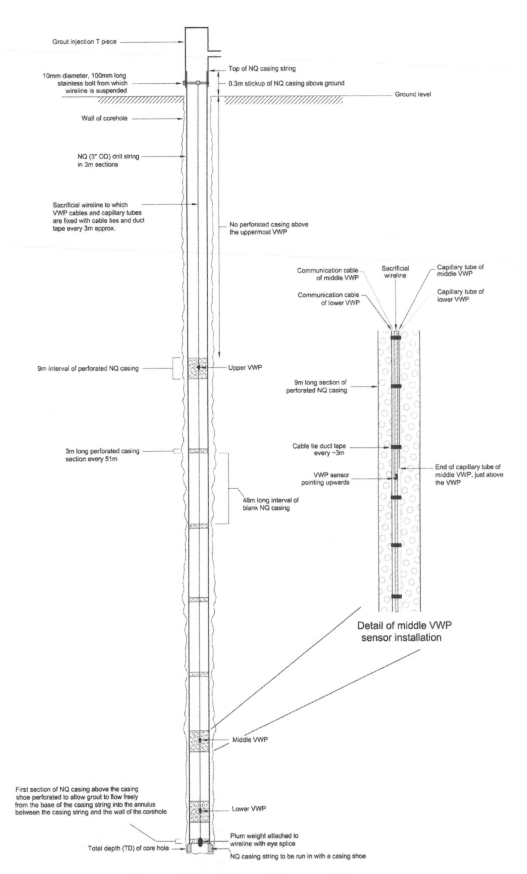

Figure A2.24: Slotted drill casing method for unstable borehole conditions

7. Record the zero readings (temperature and pressure) according to the manufacturer's instructions (usually with a bucket of water out of the sun). Check each sensor is working correctly by comparing the readings against the factory reading specified on its accompanying calibration sheet.

8. Commence running the sacrificial wireline inside the casing attaching each of the VWPs, pointing upwards, at the predetermined locations, and binding the sensors and communication cables tightly to the wireline with cable ties and heavy industrial tape at 1 m intervals (Figure A2.22). The wireline or fibreglass rod may be run over a measuring wheel during installation to verify exact depth location for fixing of VWP sensors.

9. To ensure that the VWP sensors are not damaged during installation (and later to monitor progress of grouting), each of the VWP cables should be connected to a multichannel data logger or readout box and the readings observed to confirm that each of the VWPs are working.

10. Once the sacrificial wireline and VWP sensors are lowered to specified depths, re-measure and record the water level in the borehole taking care not to snag the cables and wireline. The current pressure measurements should be checked against this level to ensure the sensors are at the expected depth.

11. Install a T-piece for grout injection to the top of the casing using a modified loading chamber to allow the VWP cables to pass through without grout-leakage. Prepare grout mix as described in Section 6.8 below.

12. The remaining steps are the same as numbers 14 to 19 in Section 6.3 of this appendix.

6.5 Installation of VWP sensors in horizontal or positive inclined drill holes

The installation of VWP sensors is frequently carried out in horizontal holes (as part of the horizontal drain program) or in positive inclined drill holes. Installation is comparable to the guide-tremie pipe method outlined in Section 6.3 with modifications for grouting due to the angle of the borehole. These additional procedures need to be adapted to suit site specific conditions, but the general approach is typically as follows.

1. Before the main borehole is drilled, drill a larger diameter (over 50 mm larger) borehole to between three and six metres depth.

2. Install conductor casing to the end of the borehole with a 12 mm (1/2") steel tremie pipe welded to the top. Weld on a flange plate with a hole for the tremie pipe and an injection point inside the conductor casing.

3. Grout the entire borehole (inside and outside the conductor casing) by injecting through the conductor casing, monitoring the returns through the steel tremie

pipe. A guideline for the grout is that it should comprise cement:expander:Sika brand rapid 251 accelerant in the weight ratio of 50:1:24 respectively with enough water to produce a pumpable grout (consistency of heavy cream).

4. The grout must be left to set or harden for a minimum 10 hours, after which the flange plate can be removed and the borehole drilled to the planned depth at the planned diameter, drilling out the grout inside the conductor casing. Remove the drill rods upon completion.

5. After the installation depths for the VWP sensors are defined, prepare 20 mm (3/4") diameter galvanised steel guide pipes and couplings. The end of the first section of steel guide pipe, to be installed at the deepest point in the borehole, should be cut at an angle of approximately 30° and a one metre length slotted to allow venting of trapped air from the base of the borehole during grout injection. Similarly this will allow grout to return to the borehole collar down the inside of the steel guide, confirming that the borehole is fully grouted to total depth.

6. Run the steel guide pipe to the total depth of open borehole (or a selected depth below the lowest VWP position) attaching the VWP sensors at corresponding depth marks on the pipe with cable ties and industrial tape. Both the VWP sensors and the communication cables should be attached to the upper side of the steel guide and centre of the borehole to avoid abrasion and damage as the guide pipe is inserted into the borehole. Centralisers may be used as required.

7. Install the grout injection pipes (PVC of 12 mm (1/2")) in diameter or steel in some particular cases) naming the shortest section 'primary' followed by 'secondary' and potentially 'tertiary' in the case of the longest (Figure A2.25). These vary in length for each borehole, depending on the depths proposed for the installation of the VWP sensors, and to ensure that the grout is injected over the corresponding instruments.

8. Attach and weld the flange plate to the tremie pipes and conductor casing at the collar of the borehole.

9. Grout injection may be carried out in two or more stages, starting at the lower end (collar end) of the borehole.
   - Inject the grout via the primary tremie until a return is seen from the secondary tremie.
   - Seal off the primary tremie and begin injection through the secondary tremie until a return is seen from the tertiary tremie (or steel guide pipe depending upon number of stages required).
   - Continue injections until a return is seen from the steel guide pipe.

For most VWP applications, the best seal is often achieved by grouting from the bottom of the hole up. Thus,

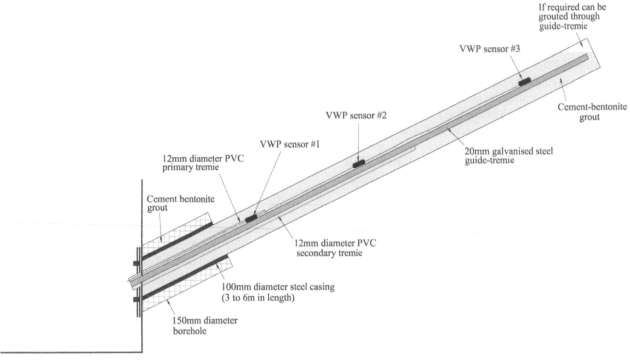

Figure A2.25: Positive inclined boreholes in open pit slopes or underground

for 'horizontal' holes drilled at a slight upward angle, grouting is carried out through a short tremie pipe installed at the collar flange and grouting is complete when grout returns are observed from the longer guide pipe to which the VWP sensors are attached. For 'horizontal' holes drilled at a slight down angle, grouting is usually best carried out from the end of the hole through the guide-tremie pipe to which the VWP sensors are attached, with the grout returns being monitored in a short pipe installed at the collar flange. For most of these 'horizontal' holes, it is normal that grouting can be carried out in a single stage because the grout pressures are low.

6.6 Installation of multi-level VWPs in underground boreholes

The installation of underground multi-level VWPs (Figure A2.26) does not differ significantly from surface boreholes with the exception that, due to the limited working area, modifications may have to be made to the installation procedures. These may include different casing lengths, the marking up of cables completed in advance at the surface, and the use of smaller grout pumps.

6.7 Prefabricated multi-level VWP installations

As well as single VWP sensors on individual cables, prefabricated multi-level sensors are also commonly used (Figure A2.27). The 32 mm (1.25") diameter Schedule 40 PVC pipe and 71 mm (2.8") diameter Schedule 40 PVC

housing allows installation of up 6 separate VWP sensors at pre-defined depths.

Advantages of this method include reduced handling and preparation time required at surface prior to and during installation in the borehole, and added protection to cables and sensors facilitated by PVC pipe and housing for VWPs. Also, the PVC pipe used to house and protect the VWP communication cables can be used as the grout delivery pipe during grout injection, provided that the maximum cable capacity (six) of the 32 mm (1.25") diameter Schedule 40 PVC pipe is not occupied. The main disadvantages of the system are associated with the limited collapse strength and tensile strength of Schedule 40 pipe and couplings, which limits the maximum depth of installation to about 200 to 250 m. In addition, many VWPs are installed immediately after the borehole is drilled. Therefore, when the VWPs are ordered, it is not known precisely what depth the sensors will be installed at. If a prefabricated VWP is used, a sensor may not be in a suitable position to monitor a zone of interest (such as a fracture) identified during drilling.

A similar installation procedure may be followed as described above for the guide-tremie VWP installations. The main difference is that the prefabricated VWP assembly should be suspended from the top of the borehole and should not rest on the base of the borehole. In the case that there is limited space for grout injection through the 32 mm PVC pipe, then an independent grout

Figure A2.26: Photos illustrating VWP installation underground using the guide-tremie method

injection or tremie pipe should be installed to ensure grout reaches the base of (and completely fills) the borehole.

6.8   Commonly used grout mix

To achieve a grout mix with an *in situ* hydraulic conductivity of less than 10^{-8} m/s, it is common that a mix of relative proportions by weight of cement:water:bentonite is 1:2.5:0.3 is used (Mikkelsen, 2002). This is achieved by mixing cement and water proportions initially and then adding the bentonite, whilst maintaining the 'pumpability' of the mixture. The result should be a grout consistency similar to heavy cream. Only pure sodium bentonite powder, Type 1 Portland cement and fresh, clean water should be used in preparing the grout.

The following grouting mixture preparation procedure is based on the assumption that mixing will be done using 2 × 2000 litre tanks and a hydraulic mixer. Two tanks are used to allow a fresh batch of grout to be mixed while the preceding batch is being injected.

A typical guideline for the grout mixing procedure is as follows.

1. The first mixing tank is filled with approximately 1000 litres of water.
2. 400 kg of cement is slowly added to the tank, the cement being slowly poured through the hydraulic mixer into the tank.
3. Once all the cement slurry is properly mixed, 120 kg of bentonite powder is very slowly added to the tank via the hydraulic mixer.
4. The bottom of the tank should be regularly checked for lumps.
5. Once the bentonite has been added, the grout is mixed for a further 15 to 20 minutes.
6. The grout mix should slowly drip off the hand with the consistency of heavy cream.
7. Once the mix is lump-free, it is ready to be injected.
8. As grout mix from the first tank is being injected, mixing begins in the second tank.

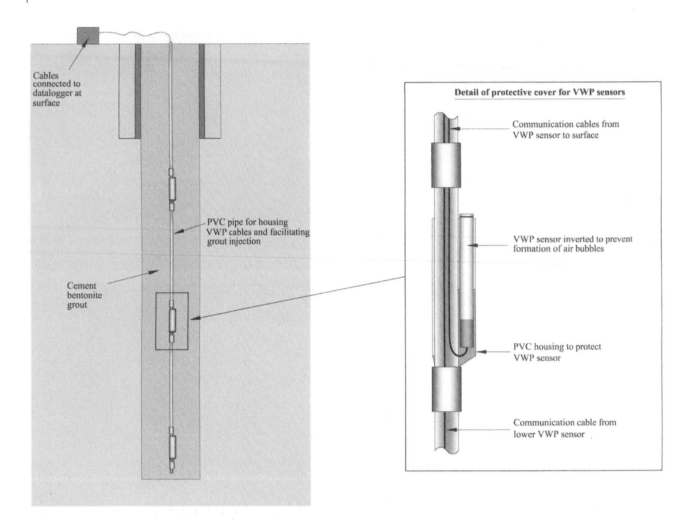

Figure A2.27: Typical configuration of prefabricated VWP installations

7 Westbay multi-level system

7.1 Installation

Installation of the Westbay system is a specialist procedure that involves prior consultation with the manufacturer. For most geotechnical applications in competent bedrock, the Westbay system is installed directly into an open drill hole. After logging the hole to determine the intended measurement depths, a single Westbay casing assembly is installed. The assembly consists of a single PVC casing string with valved 'ports' located at each selected measurement depth, and hydraulic packers to isolate and separate each measurement zone (Figure A2.28). The system is permanent, and some installations have remained in operation for over 25 years. Some installations include monitoring of up to 50 or more discrete zones and are therefore ideal for detail pore pressure assessments. They have deployed to depths of up to 2200 m (7200 ft).

The Westbay MP38 casing is compatible with HQ wireline drilling equipment, which allows the system to be installed through the drill casing (with bit removed) in holes which may have stability issues. The system can therefore be installed to the full depth of any drill hole. The drill rods or casing is then removed from the drill hole in conjunction with the inflation of the Westbay packers. The installation is complete when the packers are inflated and the measurement zones have been isolated.

7.2 Operation

Once installed, the Westbay system is operated using wireline tools that are lowered inside the Westbay casing to operate the valved ports. Pore pressure is measured in situ at each measurement zone as the wireline tools make a hydraulically isolated connection to the formation fluid through the valved port. In low-permeability environments, each measurement port provides a similarly quick response time to a vibrating wire piezometer and can therefore measure rapid responses to drainage.

A set of pore pressure response readings from a multi-level Westbay during a fractured rock pumping test is shown in Figure A2.29.

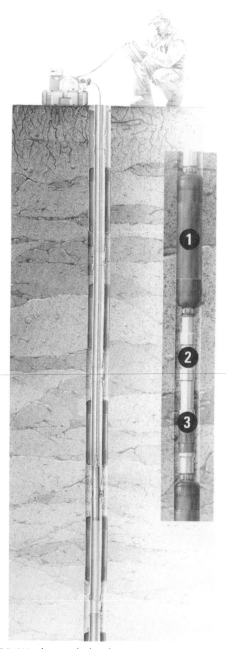

Figure A2.28: Westbay multi-level piezometer system

❶ Packer
- Engineered seal in a range of borehole sizes
- No dedicated inflation lines
- Controlled hydraulic inflation with record of pressure and volume
- Quality control tests to confirm performance at any time after installation

❷ Measurement Port
- For fluid pressure measurements, fluid sampling and low-k testing

❸ Pumping Port
- For purging, hydraulic testing, and quality control testing

The level of data density offered by the Westbay system has lead to an improved understanding of the pore pressures and hydraulic gradients in natural slopes, landslide areas and pit slopes. A greater number of measurement ports can be placed in a single hole than is possible using vibrating wire piezometers, although it is possible to use adjacent holes to increase the vertical discretisation of vibrating wire piezometer arrays.

The major advantage of the Westbay system over vibrating wire piezometers is that it allows hydraulic testing and water sampling from each measurement zone.

A variety of hydraulic testing methods are available, including:

- pulse tests (for low-permeability environments);
- single-zone rising-head, falling-head or slug tests;
- single-zone constant-head tests;
- cross-zone tests (within an individual Westbay completion);
- cross-hole testing (using multiple wells).

Such detailed testing can be valuable in identifying flow zones, flow barriers, lateral and vertical and

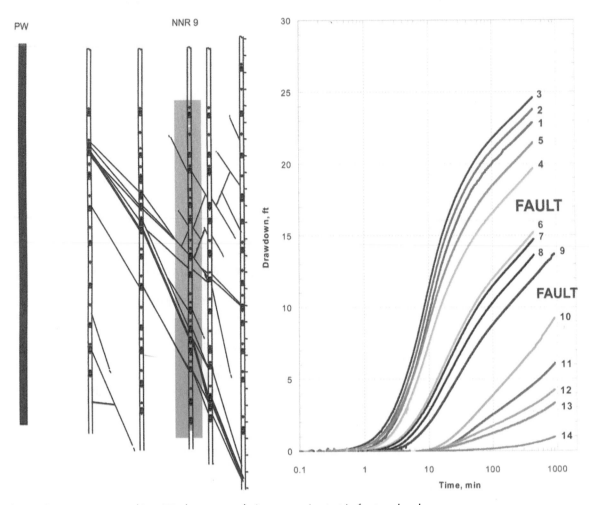

Figure A2.29: Response measured in a Westbay system during a pumping test in fractured rock

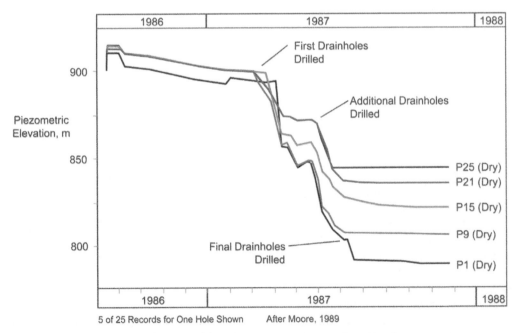

5 of 25 Records for One Hole Shown   After Moore, 1989

Figure A2.30: Real time monitoring during installation of depressurisation measures using a Westbay system

compartmentalisation in complex fractured rock masses. Perhaps as importantly, such testing ability also allows the testing and verification of 'anomalous' pressure measurements, and allows hydraulic testing of the 'anomalous' zones. Water samples can also be collected from each zone. Thus, drill holes equipped with the Westbay system can be used for supplementary data collection beyond geotechnical purposes.

Figure A2.30 illustrates the measurement of drainage vs time in a natural rock slope, and illustrates how the increased monitoring provided real-time feedback of the effectiveness of the depressurisation measures being installed in the slope.

Appendix 3

Case study: Diavik North-west wall

Pete Milmo, Steve Rogers, Mark Raynor and Geoff Beale

1 Background

1.1 Location

Diavik is located adjacent to Lac de Gras, 300 km north-east of Yellowknife, in the Northwest Territories of Canada. The diamond operation is a joint venture between DDMI (60%) and Harry Winston Diamond Limited Partnership (40%). Presently, DDMI is mining the A154 South and A154 North kimberlite pipes through the A154 open pit and associated underground workings. A second pit to the south has been excavated for access to the A418 pipe, together with initial underground development headings.

The mine site is located in the Canadian Shield within the region of continuous permafrost; however, mining operations are taking place in unfrozen ground (talik) within the confines of a diked portion of what was formerly Lac de Gras lake bottom. The kimberlite pipes, A154N and A154S, are hosted within a strong (>100 MPa), moderately fractured rock mass comprised of granite with smaller proportions of pegmatite and metasedimentary rafts in the granite intrusives. The A154S pipe is centrally located at the bottom of the pit while the A154N pipe is exposed over several of the upper benches in the north-east wall of the pit.

The granite in the pit area is affected by the presence of Dewey's Fault zone. This regional feature trends in a NE–SW direction through the kimberlite pipes and extends northwards beneath the A154 dike and under Lac de Gras, and to the south onto the island.

1.2 The North-west wall, A154 Pit

The North-west wall was identified early in the mining of the A154 pipes as an area where additional slope stability modelling was required as it represents the steepest wall on the lake side of the open pit (Figure A3.1). Because of the sensitivity of the slope stability to groundwater pressures (particularly for shallow, bench-scale movements), and the uncertainties inherent in predictions of these pressures, including potential effects of pit wall freezing, it was considered prudent to further investigate groundwater conditions. Consequently the North-west wall has become well instrumented.

1.3 Climate

The average annual temperature at Diavik is –12°C, and permafrost is widespread under landmasses (extending to a depth of 380 m under the East Island). 'Taliks' (thicknesses of unfrozen ground) are present under the lake, as it provides an insulation effect which prevents freezing of the lake bed.

The A154 and A418 show permafrost conditions in their South wall. The northern half of the pits are outside the influence of permafrost, but seasonal freezing does occur at the surface because of the exposure of the mining area. Mean annual precipitation at Diavik is less than 300 mm and does not provide any significant recharge to the pit slopes.

1.4 Hydrogeological setting

The regional geological setting of the Lac de Gras area is within syntectonic and post-tectonic intrusive rocks of the Slave province. The main Archean rock units consist of sedimentary greywacke, metaturbidites, biotite-tonalite, 2-mica granite, and granodiorite. The open pits are developed in competent granitic country rocks with Rock Quality Designation (RQD) of 95 per cent (Kutchling *et al.*, 2000). Sub-horizontal metasedimentary rafts occur through the granite rock mass. These may vary from a few centimetres to several metres in thickness.

Dewey's Fault zone is a zone of enhanced permeability. The A154N and A154S pipes occur along this feature. Dewatering at the pit floor is mostly focused on pumping from sumps excavated close to this higher-permeability

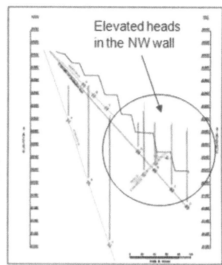

Figure A3.1: View from the north of the A154 pit and the NW wall (left) and profile of heads in the NW wall showing elevated heads in lower section (right)

structure. The toe of the NW pit wall lies about 70 m to the west of the expression of Dewey's Fault zone in the pit floor.

Hydrogeological investigations at the site have been extensive and have taken the form of packer testing, borehole flow meter testing, borehole temperature logging, borehole video imaging and well testing. The packer test data show a strong depth-dependency in estimated transmissivity, with average transmissivity values reducing by 1.5 orders of magnitude over 200 m (Golder, 2010). Packer testing results influencing the current North-west wall study are shown in Table A3.1. Test pumping carried out in a well located close to the crest of the North-west wall shows anisotropic drawdown extending in a NE–SW direction.

Major groundwater inflows into the underground declines are typically associated with single open fractures rather than zones of increased fracture density. During early hydrogeological investigations, the presence of

broken rock zones in the core was not necessarily indicative of significantly water-bearing intervals (Kutchling *et al.*, 2000).

1.5 Depressurisation of the North-west wall

Groundwater levels in the A154 pit were controlled primarily through sump pumping in the lower pit benches and at the pit floor. Since late 2009, with the development of the decline beneath the A154 pit, an increasing amount of the water has been intersected underground. Figure A3.2 shows the total dewatering rate for the A154 open pit and the underground workings.

As the A154 pit was progressively deepened, it was noted that pore pressures in parts of the North-west wall were decreasing at a somewhat lesser rate than the mine sinking rate. Consequently, total heads in the middle to lower slopes became artesian, with up to 80 m of head above the level of

Table A3.1: Transmissivity measurements from Packer testing (Golder, 2010)

Test domain	Geometric mean of transmissivity estimates (m²/s)
Country rock	2.E-05
Dewey's Fault zone	2.E-04
All test data	9.E-07
100–150 mamsl	7.E-08
150–200 mamsl	4.E-07
200–250 mamsl	9.E-07
250–300 mamsl	3.E-06
250–350 mamsl	1.E-06
300–350 mamsl	3.E-07

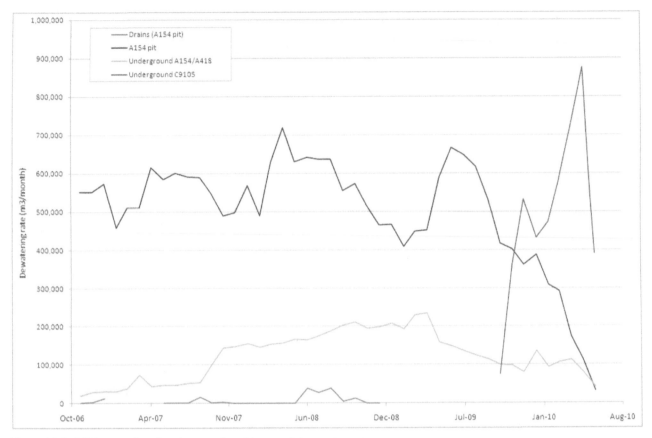

Figure A3.2: A154 open pit and underground workings total dewatering rate

the slope. As a result, a program of piezometer installation was carried out in mid-to-late 2006 with active depressurisation measures installed between 2006 and 2009. A timeline for the installation of active depressurisation measures is shown in Table A3.2, and the positions of the various wells and drains are shown in Figure A3.3.

Compared to the total groundwater flux entering the A154 pit and underground workings, the drains and wells installed within the NW wall were all low yielding. The reported flow in many of the drains and wells was zero.

The total dewatering rate for the in-pit wells collared on the 280 m bench (DW01-05, DW07, DW09, DW10, DW29 & DW30) was between 5 and 12 l/s in 2008, which was less than three per cent of the total dewatering rate shown for the A154 pit in 2008. The total combined flow rate from the 280 m bench wells was steady at around 10 l/s during 2009.

An important feature of Figure A3.2 is the initiation of dewatering from the C9105 underground drains. From October 2008, when these drains came on line, the

Table A3.2: Timeline for the installation of dewatering measures in the NW wall

Year	Activity
2005	Initial pumping test carried out.
2006	Drilled holes MS12-15 from the 280 bench at 70 degrees into the slope.
	Drilled 12-inch vertical wells PW-1 to 5 drilled from the 280 bench, of which PW2 and 5 were completed as wells, but did not have sufficient flow to install pumps.
2007	Drilled more than 20 drainage holes (54 m long) drilled with the L8 drill (low yields).
2008	Drilled 5 x 150–170 m long HQ holes (DW1-5) on the 280 bench, inclined 70 degree to the NW (intersected water in all holes, ranging from 0.6 to 1.3 l/s), installed downhole pumps and surface pipes, system brought on line on March 8, combined flow was 4–5 l/s.
	Drilled 4 x 114 m long, 6-inch diameter holes with L8, parallel to the HQ holes. Three holes intersected water, one zero flow. Infill wells installed on the 280-bench in May and June (DW7-10).
	Drilled A-154-MS-29, 30 and 31. A154-MS-29 and 30 drilled at the end of 2Q08 and were converted to wells, MS 31 to a piezometer. MS-29 and MS-30 came on line on 15 and 20 August 2008, respectively.
2009	Drilled 6 vertical and angled drainage holes on 160 m bench at the toe of the NW wall.

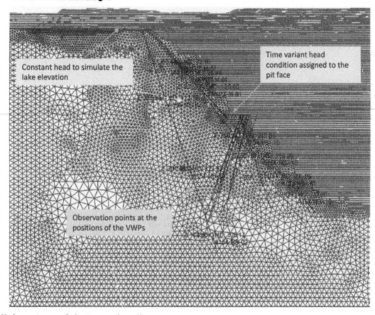

Figure A3.3: Diavik NW wall, locations of drains and wells

dewatering rate from the A154 open pit showed a sharp decline. Figure 3.5 (Section 3) shows the position and layout of the underground workings and drainage galleries. Drains extend towards the NW wall from two positions at either side (i.e. to the south and south-east of the wall), and from two different levels; at elevations 120 masl, and 20 masl. The pit sump for the A154 pit is currently (June 2010) dry, and all the dewatering for the pit is now through underground drainage.

1.6 Piezometer installations in the North-west wall

The vibrating wire piezometer (VWP) array behind the North-west wall consists of five deep pit-crest piezometer strings, drilled slope-parallel from the pit crest (A154-MS-08, 9, 10 & 11), and two shallower piezometer strings of a similar orientation (NWPT-M-01 & 02). A154-MS-09,

10 and 11, and NWPT-M-01 and 02 dip toward the pit at angles of 45 to 50°, whilst A154-MS-02 is inclined at 70°, resulting in VWP positions at depth which are further behind the face (160 m behind the face for A154-MS-02C). All completions are fully grouted and all are connected to data loggers which record measurements typically at 10 minute intervals (although the measurement interval has changed over time).

In addition to the pit-crest piezometers, there are two shallow in-pit piezometer holes collared on the 250 m bench: A154-MS-23 and A154-MS-24. Both holes have two VWP sensors installed at elevations of 202 masl and 227 masl.

The VWP hole locations are shown in Section 3, Figure 3.6 and the specific piezometer details are shown in Table A3.3. The Diavik VWP sensors are assigned the letter A, B, C or D, depending on their position within each

Table A3.3: Vibrating wire piezometer details

Piezometer	VWP sensor	VWP elevation (mamsl)	Approximate distance behind current pit face (m)
A154-MS-02	A	305.6	85
	B	201.5	125
	C	104.5	160
A154-MS-08	A	249.3	35
	B	219.1	40
	C	188.9	45
	D	158.7	45
A154-MS-09	A	250.5	18
	B	221.0	18
	C	191.4	24
	D	161.8	27
A154-MS-10	A	250.4	34
	B	221.1	34
	C	191.1	35
	D	164.0	39
A154-MS-11	A	248.6	25
	B	218.6	25
	C	188.6	26
	D	158.7	30
NWPT-M-01	A	358.4	14
	B	342.0	18
	C	321.8	19
NWPT-M-02	A	362.7	14
	B	340.4	19
	C	308.7	16
A154-MS-23	A	202.2	23
	B	226.5	40
A154-MS-24	A	202.2	23
	B	226.5	40

piezometer string (increasing in depth from A to D). For the deep near-face piezometers (A154-MS-08 to 11), VWPs of the same assignation have the same depths and are in the same position relative to the pit benches. This is shown in Figure 3.7 (Section 3), with a cross section through A-154-MS-10 and A154-MS-02. Also shown is the in-pit piezometer A154-MS-24.

For brevity, VWPs are hereafter also referred to with a shortened ID consisting of the number and VWP letter (e.g. 10A for A154-MS-10A).

2 Hydrograph analysis

2.1 Overview

2.1.1 Long-term hydrographs

The VWP dataset for the North-west wall is extensive, and has therefore required processing to create a dataset of daily mean hydraulic head, in order that the complete data period can be plotted. The average water level elevation in Lac de Gras is 416 masl. Therefore, prior to mining, it is expected that the total head at the location of each of the installed VWP sensors was also 416 masl. Any downward departure from 416 masl total head represents a change caused by the mining operation.

Hydrographs of daily mean head are shown in Figure 3.8 (Section 3) for piezometers A154-MS-08, A154-MS-11 and NWPT-M-02 as examples. Also shown are certain 'timeline events' such as the installation of drains, activation of pumping wells, and the changing elevation at the toe of NW wall to demonstrate the rate of pit advance.

2.1.2 Reproducing the long-term trends with a 2D numerical model

Section 3.3.2.4 describes that the long-term depressurisation of the North-west wall, particularly in the deeper VWPs, is

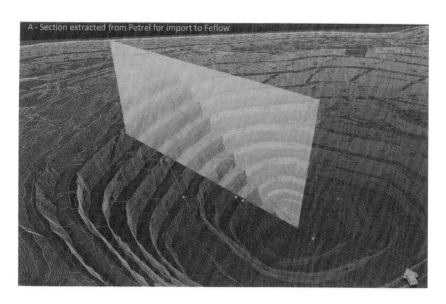

A - Section extracted from Petrel for import to Feflow

B - Feflow finite element grid

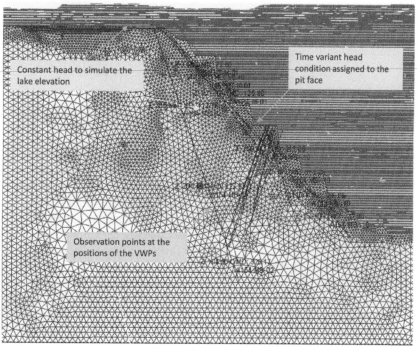

Constant head to simulate the lake elevation

Time variant head condition assigned to the pit face

Observation points at the positions of the VWPs

Figure A3.4: Diavik NW wall, numerical 2D model for the A154-MS-10 section

broadly similar to the rate of pit advance, suggesting that depressurisation is occurring primarily through a process of passive drainage towards the pit floor. In order to test the validity of this conceptual model, a simple two-dimensional numerical flow model was constructed.

2.1.2.1 Model set-up
The numerical flow model uses the *Feflow* code. It consists of a vertical 'slice' section through the pit at full depth, with a no-flow boundary along the majority of the outer edge, and a constant head at an elevation of 380 masl to

simulate the influence of seepage from Lac de Gras. A time-variant head condition was assigned to the pit face to simulate deepening of the pit so that, at any time, the hydraulic head is equivalent to the elevation of the pit base. The numerical model section is shown in Figure A3.4.

For simplicity, homogeneous material properties were assigned to the entire section. Whilst a depth dependent function for hydraulic conductivity would be more appropriate, homogeneous properties are more helpful for identifying the sensitivity of the system to the head boundary conditions. Observation points were assigned to

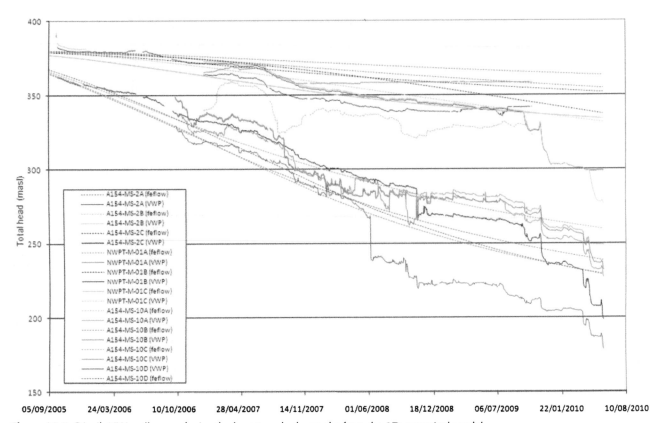

Figure A3.5: Diavik NW wall, reproducing the long-term hydrographs from the 2D numerical model

nodes at the positions of the various VWPs along the A-154-MS-10 section.

A basic model calibration was carried out by making changes to the assigned hydraulic conductivity and anisotropy. The best match to the VWP data was achieved with a hydraulic conductivity of 1×10^{-8} m/s, and a K_h/K_v anisotropy factor of 0.5 (giving a horizontal K of 5×10^{-9} m/s). It should be noted, however, that these values are representative of the 'overall' flow system. They incorporate the effects of barriers along the flow path. In the North-west wall, barriers can be expected where discrete structures occur at an orientation perpendicular to the overall hydraulic gradient. The overall hydraulic gradient is from NW to SE, but a dominant structural direction is perpendicular to this (NE–SW). At any location within the flow system, the actual 'in situ' hydraulic conductivity may be significantly higher than these values.

2.1.2.2 Model results

The model heads are shown against the long-term VWP data in Figure A3.5. Given the simplicity of the model set-up, the simulated heads are a reasonably good match to the actual data. The model has been able to reproduce the rates of depressurisation for the various VWPs, except for A154-MS-02C, which is positioned farther behind the face. The simulation of 10D is also poor because the VWP experienced

a large step drop in pressure in mid-2008 as a result of a discrete blasting event (discussed in Section 3.3.3.2).

As discussed above, the assigned hydraulic conductivity values are lower than field data suggest. It has been necessary to incorporate the lower values in the model to keep the pressures in the upper VWPs high, to limit the influence of the head boundary assigned to the pit face and to reproduce groundwater flow oriented off-section (i.e. due to x-y anisotropy). The low hydraulic conductivity has contributed to modelled heads in A154-MS-02C being much higher than observed at this location.

Overall, the main conclusions are as follows.

- The distribution of hydraulic heads in the data can be reproduced in a 2D simulation by assigning constant-head boundary conditions to simulate Lac de Gras (recharge boundary) and the changing pit floor elevation (discharge boundary). This provides evidence that the long-term trend of depressurisation is the mostly result of passive drainage towards the pit floor, with the rate of depressurisation controlled by the overall hydraulic gradient between the recharge and discharge boundaries.

- The best fit to the field data was achieved with $K_h:K_v$ model anisotropy of 0.5. The model is simplified, but an horizontal–vertical anisotropy in the fracture system is likely to be present in the wall rocks.

2.1.3 Localised influences

Within the overall 'long-term' piezometric response caused by progressive mining of the pit and pumping of water from the pit floor, there are many 'medium-term' (on the scale of days to weeks), and 'short-term' (responding in minutes to hours) fluctuations in pore pressure, which are typically the result of localised influences. These 'events' can be due to either hydraulic or mechanical stresses.

The following types of 'short-term' event can be commonly seen in the Diavik dataset.

- Waste, trim or ore blasting of the North-west wall.
- Drilling for the installation of drains or wells.
- Hydraulic testing of wells, via airlifting, or longer-term pumping tests.
- Power outages, resulting in the short-term cessation of pumping.
- Opening of relatively low-flow drains in the pit face.
- Opening of higher-flow drains in the underground workings.

In the 'medium-term', the above events can cause deviation from the long-term trend of depressurisation. 'Medium-term' pressure changes are brought about through:

- large magnitude responses with slow (or no) recovery;
- compound effect of successive 'events';
- seasonal freezing and thawing of the slope.

2.2 Analysis of specific events

2.2.1 Deviations from the long-term trend

The 'long-term' depressurisation of the NW wall fits an overall linear trend (see Figure A3.5). However, deviations from this trend occur in the 'medium-term' to 'short-term', caused by events which range in timescale from seasonal to 'instantaneous'.

In this section, the mean daily pore pressure data are examined in closer detail to identify the following.

- The nature and cause of seasonal trends in pore pressure.
- Major 'instantaneous' influences on depressurisation (e.g. hydraulic or mechanical 'events').
- Changes in the long-term rate of depressurisation.
- Changes in the magnitude or nature of responses to 'events' over time.
- The importance of the pit floor position with respect to the VWP position on the rate of depressurisation.

In total, 37 events have been examined. Hydrographs are plotted to show the 'event response' in terms of pressure head, and the spatial patterns of these responses have been analysed. Given the generally similar elevations of pit face VWPs that have the same letter assignation (e.g. A, B, C or D), it was decided to analyse the data using these groups.

The following sections present a detailed description of depressurisation in the piezometer data.

2.2.2 VWPs – A Group (250 m elevation)

The 'A' VWPS are positioned at an elevation of about 250 m. The data are plotted in Figures A3.6 and A3.7, which also shows the pit advance (displayed in benches). Piezometer 10A is not shown as there is inferred to be a downhole hydraulic connection with 10B and the pressure response recorded in 10A is the same as that of the lower elevation 10B.

The hydraulic head in VWP 08A is much higher than that of 11A and 09A. This is likely due to its greater proximity to the Lac de Gras recharge area and perhaps also its position set back slightly from the pit face. The magnitude of blasting responses in 08A is also generally smaller and this could also be due to a combination of these influences (i.e. 'swamped' responses because of the constant recharge influx, and/or buffering of the response to blasting because of its greater distance behind the pit face).

The following points refer to the numbers marked on Figure A3.6:

1 – Major depressurisation (60 m head drop) during excavation to the 250 m bench (September–November 2006)

Much of the depressurisation in 09A and 11A occurs between September and mid-November 2006. This period of rapid depressurisation coincides with deepening of the toe of the slope below the 280 m level.

To identify the specific events leading to this period of rapid depressurisation, the relevant data period was expanded in scale (see Figure A3.7). Both 09A and 11A exhibit relatively rapid 'passive' depressurisation, which is compounded by large drops in pressure associated with blasting of the bench above (280 m) and the bench face adjacent to the VWP positions (250 m).

The effects for 09A are most obvious in the pre-shear blast of 25 October 2006, causing around 20 m of depressurisation. A similar drop in pressure in 11A is evident a day later on 26 October 2006, from a second pre-shear blast in the vicinity of the piezometer. VWP 11A depressurises further on 10 November 2006, with 10 m pressure reduction from a pre-shear blast (250 to 280 m level). Following the pressure drop in 11A from the 26 October blast, pressures quickly show a degree of recovery. However no such recovery is evident following the 25 October blast which affects 09A.

The following points can be drawn from these blast responses.

- The near-instantaneous responses to blasting, and the large magnitude pressure reductions that occur, indicate that these responses are the direct result of a rapid change in porosity in the fracture network. This may be comparable to the 'direct' hydromechanical response that can be caused by unloading. However, in

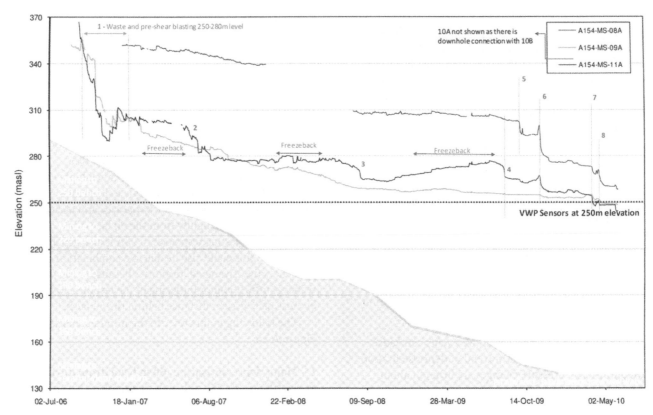

Figure A3.6: NW wall, hydrographs for 'A' group VWPs

this case, the depressurisation is likely also associated with blast damage causing additional fracturing and potential dilation of existing fractures through the intrusion of high pressure blast gases.

- The degree to which pore pressures recover following depressurisation is dependent on the permeability of flow pathways connecting the affected rock mass to the upgradient fracture network (and ultimately to the recharge source).

Later events in Figure A3.7 show the influence of blasting increasing the hydraulic head in 09A and 11A. A series of successive blasts between December 2006 and January 2007 caused a significant increase in pore pressures. It is thought that gases released during the blast created a pressure pulse through the fracture network, without causing significant fracturing and fracture aperture dilation that would result in pore pressure drops. These short-term increases in pore pressure can take some time to dissipate, particularly in winter when drainage to the pit face is reduced. Hence, a series of consecutive blasts can cause a 'hydraulic jacking' effect on pore pressures.

2, 3 & 4 – Freeze-back and thawing

Freezing of the pit faces typically occurs at Diavik between September/October and May/June. The freezing affects

near-face fractures (probably to a depth of about five to six metres at the end of the winter), blocking flow pathways and seeps. Freezing, in turn, causes a reduction in the rate of depressurisation or, in the presence of sustained recharge, an increase in pressure behind the face as heads 'back up'.

Following the 'hydraulic jacking' of pressures from blast gas intrusion in 11A between December 2006 and January 2007, further depressurisation was slowed by the freeze-back (Figure A3.7). In May 2007, when temperatures rose above freezing, the rate of depressurisation increased as the discharge points for some of the drainage pathways for 11A opened up. The influence of freezing and thawing is not similarly evident in 08A and 09A, which depressurised at a relatively constant rate during this time, and also during the following 2008 and 2009 winters. This indicates that 11A is in better hydraulic connection with the near-face over-break zone undergoing freezing.

As the pit advanced to greater depths below 11A, the onset of depressurisation during thawing undergoes a time lag associated with the increasing distance from the pit floor.

Point '4' in Figure A3.6 shows depressurisation associated with a blasting event during the thawing period of summer 2009. It is possible that blasting during this thawing period assisted in breaking down the remains of the ice in the near-face fracture network (over-break zone).

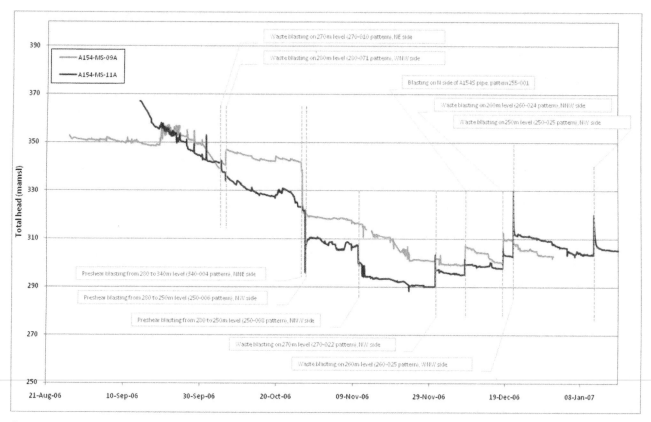

Figure A3.7: Diavik NW wall, detailed hydrographs of depressurisation events for the 'A' group VWPs (September 2006 to January 2007)

5, 6, 7 & 8 – Opening of the underground dewatering network

In late 2009 and early 2010, VWPs 08A and 11A experienced a large drop in pore pressure in response to opening of various parts of the underground drainage network (Figure A3.5). There are four periods of depressurisation associated with events on 27 September and 16 November 2009, and 29 March and 15 April 2010. Events 5 and 6 are described as the general opening of underground dewatering holes, whilst events 7 and 8 are caused specifically by the opening of valves U40/58 and the 9020 level dewatering holes. The greatest amount of depressurisation in response to these events is in 08A, which shows a 40 m drop in pressure head. VWP 11A and 09A show 20 m and 5 m of pressure head drop respectively, and the phreatic level falls below the elevation of these sensors as a result.

The main points which can be drawn from the responses exhibited in Figure A3.6 are as follows.

- The response to dewatering of the underground drains is a series of 'recession curves', which show hydraulic equilibration in response to each successive phase of depressurisation. The tendency for pore pressures to re-equilibrate, rather than drain freely, is characteristic of a system controlled by its hydrology (i.e. the recharge and discharge 'boundaries').

- The fast equilibration, and the general absence of lag times in any of the responses, indicates high hydraulic diffusivity in the direction of the underground drains (high transmissivity and low storage).

- The underground drainage galleries yield flow rates which are able to match the groundwater flux which was previously occurring through passive drainage toward the pit floor. As such, a widespread depressurisation occurs in response to this new lower discharge 'boundary'. Replacing the pit floor discharge point with a deeper underground discharge point eventually causes the pit floor sumps to completely dry out by mid-2010.

- The magnitude of the response is greatest in 08A, which is closest to the recharge source of Lac de Gras and showed the least amount of depressurisation prior to opening of the underground drains. Conversely VWP 09A, which showed the greatest amount of depressurisation prior to these events, showed the smallest response when the underground drains were opened.

- The difference in the magnitude of these responses is indicative of the high level of heterogeneity and anisotropy in the fracture network. VWP 08A, which is very close to the Lac de Gras recharge area, does not show depressurisation through flow toward the slope. However, when a vertical component of flow is initiated towards the underground drains, the pressures are relieved as the vertical flow pathways are sufficiently permeable to overcome the incoming recharge flux. The anisotropy that exists is one of greater permeability in the vertical, with reduced lateral permeability toward the slope. This is in agreement with the anisotropy necessary for calibration of the 2D numerical model (Section A2.1.2 of this appendix).

2.2.3 VWPs – B Group (220 m elevation)

The 'B' VWPS are positioned at an elevation of 220 m (Figure A3.8). VWP A154-MS-24A is included in this group as it is at a similar elevation. VWP 24A is positioned just 10 m from 11A.

As was the case with the 'A' group piezometers, the influence of Lac de Gras is evident in the distribution of heads in the 'B' group VWPs. The following points refer to the event numbers marked on Figure A3.8.

1 – Depressurisation during excavation to 250 m bench

Pore pressures reduced by 20 to 30 m between September and mid-November 2006 in 09B, 10B and 11B (the record for 08B is discontinuous over this period). Depressurisation in the 'B' VWPs occurs at the same time as the period of large magnitude depressurisation in the 'A' group VWPs and, like the 'A' group sensors, the 'B' VWPs exhibit stepped depressurisation in response to pre-shear blasting.

During this period, the toe of the North-west wall was advanced through the 280 m bench to the 250 m bench level. The 'B' VWPs are positioned beneath the 250 m bench, but not directly, and it is unlikely that depressurisation is the result of unloading. Drops in pressure that accompany pre-shear blasting suggest that the fracture network at the level of the 'B' VWPs is responding to the large depressurisation occurring in the fracture network of the bench above (i.e. there is good hydraulic connection in the fracture network between 'A' and 'B' group VWPs).

2 – Responses to blasting and freeze-back (November–December 2006)

The increase in heads between late November 2006 and early January 2007 occurs in a similar trend to 09A and 11A, with successive blasts causing 'hydraulic jacking' of pore pressures. Freezing of the pit face causes these higher pore pressures to persist until spring breakup starts in May 2007.

A similar period of blast-induced 'hydraulic jacking' occurs in March 2007 in 09B and 10B.

3 – Pre-shear blasting from 250 to 220 m level (12 February 2007)

VWP 11B shows a 12 m drop in hydraulic head following a pre-shear blast at the 250–220 m level. The position of this blast is in the vicinity of A154-MS-11. The effects of this

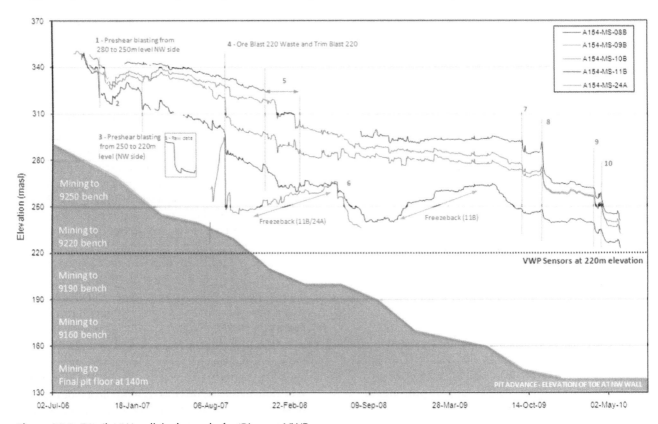

Figure A3.8: Diavik NW wall, hydrographs for 'B' group VWPs

blast are very localised, with little or no response in VWPs of the adjacent piezometer A154-MS-10.

The 'recession curve' type response to the blast in 11B, is typical of 'blast-induced drainage'. Blasting causes deformation of existing fractures and/or opening of new fractures which improves the permeability of flow pathways. For 11B, the depressurisation occurs as the pre-shear blast opens new flow pathways and creates an improved connection to the pit floor discharge area.

4 – Ore Blast 220 Waste and Trim Blast 220 (13 September 2007)

A large drop in hydraulic head in VWPs 10B, 11B and 24A is caused by a blast on the 220 m level on 13 September 2007. The position of the blast is in the bench face directly in front of these piezometers. VWP 08B and 09B, which are further around the pit face, are largely unaffected by the blast. The blast does, however, create a small response in most piezometers.

VWPs 10B and 11B show blast induced drainage in a similar way to that described above for 'Point 3'. The very large pressure response in VWP 24A (45 m head reduction) is thought to be caused primarily by blast-induced fracturing and/or fracture deformation resulting from the blast. Piezometer A154-MS-24 is an in-pit piezometer, which is inclined into the slope from the 250 m bench (Section 3, Figure 3.7). With this orientation, the piezometer is likely to intersect several slope-parallel structures, which may show the most deformation in response to blasting and material removal.

5 – Blasting events (January to mid-March 2008)

A series of blast events associated with pit advance below 220 masl (i.e. below the 'B' group VWP elevations) causes several large drops in hydraulic head recorded at VWP 9B, 10B and 11B. The raw data plotted in Figure A3.9 show that depressurisation occurs as a result of blast-induced drainage and/or fracture deformation. The response of 10B to the late December waste and pre-shear blasting exhibits the recovery usually associated with an unloading response. Deformation of fractures in this instance is not due to unloading as the pit has advanced below the VWP elevation. Fracture dilation might, however, occur under conditions of tensile stress.

6 – Freeze-back

Freeze-back is evident in VWP 11B and 24A during the winters of 2007 and 2008 (Figure A3.8), whilst 08B, 09B and 10B show no response to freezing. As discussed for 11A, 11B is likely to be connected to a near-face drainage

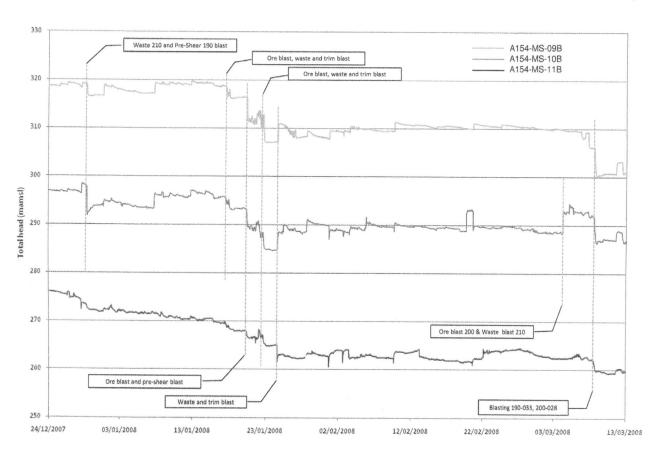

Figure A3.9: Diavik NW wall, detailed hydrographs of depressurisation events for the 'B' group VWPs (January 2007 to March 2008)

pathway, or to the over-break zone, which shows greater influence from freeze-back. Piezometer 24A shows a greater response during both the freeze-back and break-up phases of this cycle, and is clearly measuring pressure in fractures which respond more rapidly to the changing conditions at the face.

In 11B during the winter of 2007/8, freeze-back is evident as a temporary reduction in the rate of depressurisation and any pressure in excess of the long-term trend rapidly dissipates during the spring break-up (as was the case for the 2006/7 winter). Depressurisation during the break-up period is typically through 'passive drainage', accompanied by larger drops in pressure as blasting breaks up the residual ice.

7, 8, 9 & 10 – Opening of the underground dewatering network

In response to the opening of the drains in the underground decline, a very similar type and magnitude of depressurisation is evident in the 'B' group VWPs as in the 'A' group VWPs. There are four separate phases of depressurisation marked on Figure A3.8, caused by opening of drains on 27 September 2009, 16 November 2009, 29 March 2010 and 15 April 2010.

2.2.4   VWPs – C Group (190 m elevation)

The 'C' VWPs are positioned at an elevation of 190 m. Hydrographs are shown in Figure A3.10.

The downhole cross-connection between 10A and 10B may also extend to 10C. The rate of depressurisation in 10C is much slower than 8C and 9C, with a long-term trend which is closer to that of VWPs 10A and 10B.

The following points refer to the numbers annotated on Figure A3.10.

1 – Depressurisation during excavation to 250 m bench

The 'C' group VWPs responded to mining of the 250 m level over the same period in October 2006 in which 'A' group piezometers showed much of their depressurisation. The magnitude of this response was less in the 'C' group piezometers. Piezometer 11C showed the greatest response following the pre-shear blast marked as '1' in Figure A3.10, with a 9 m drop in head.

2 & 3 – Long-term drainage response in 11C

VWP 11C showed two periods of significant depressurisation starting at the beginning of October 2007 and at the end of May 2008 (marked as '2' and '3' respectively in Figure 3.10). These periods of depressurisation conform to a drainage type of response. The depressurisation in 11C amounted to over 80 m of head between October 2007 and October 2008. During this time, depressurisation in 11C occurred primarily from 'passive drainage' initiated by single blast events, rather than stepped depressurisation from successive blasting events.

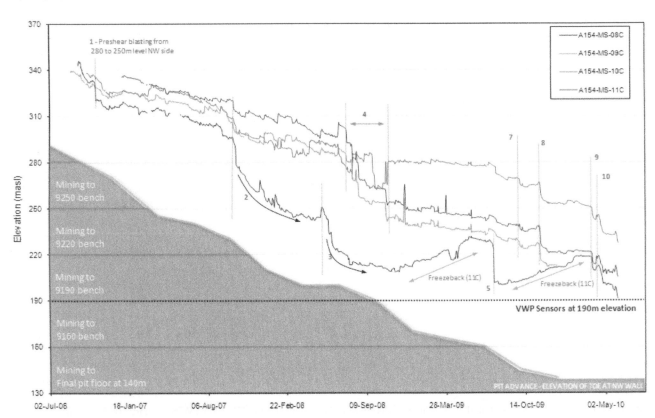

Figure A3.10: Diavik NW wall, hydrographs for 'C' group VWPs

The blast that caused initiation of the first phase of drainage ('2') is described as a pre-shear blast on the 220 to 190 m level, occurring on the 17 October 2007. The actual position of the blast is on the face directly in front of the 11C/10C position. The second phase of drainage ('3') appears to have been caused by an ore blast on the 21 May 2008. VWPs 08C, 09C and 10C showed relatively minor depressurisation in response to these events.

The large magnitude of the depressurisation responses in 11C suggests that blasting caused the opening of flow pathways, which affected a large volume of fractured rock. As identified previously in the large response of 11C to freeze-back, the fracture network monitored in 11C shows good hydraulic connection with the near face fracture network.

4 – Depressurisation in 08C and 09C (July–October 2008)

In VWPs 08C and 09C, the greatest amount of depressurisation occurred in the period July–October 2008 as mining exposed the bench face adjacent to, and then below, the elevation of these VWPs. This period is shown in more detail in Figure A3.11, which is plotted with the raw VWP data.

Figure A3.11 shows large drops in pressure associated with blasting events. The blasts caused depressurisation with very little, if any, recovery between events. These responses are likely to reflect deformation in the fracture network from the development of tensile stress normal to the pit face (i.e. dilation in fractures oriented parallel to the

pit face). At this point in time, excavation was occurring at the face in front of these VWPs.

Dewatering well MS-29 also came on line during this period. The position of MS-29 is directly between A154-MS-08 and A154-MS-09. Theoretically, it would be expected to cause an increase in the rate of depressurisation in these piezometers. The activation of MS-29 is marked on Figure A3.11. In reality, the activation of the well caused no obvious response in the VWPs.

5 – Freeze-back

Freeze-back is evident in 11C (Figure A3.10), with increasing heads occurring between late October and early July. As described for the freeze back response in 11B, the greater influence of freezing in 11C is probably due to good hydraulic connection with the near-face over-break zone.

Similarly to the freeze-back response of VWPs at higher elevation, a large freeze-back response is only evident once the pit face has advanced below the VWP elevation. For 11C, freeze-back occurs in response to the 2008/9 and 2009/10 winters, once the bench face adjacent to the 'C' VWPs is exposed.

For 11C, depressurisation during the spring break-up period commenced in late May 2009. The rate of depressurisation was increased by a blast on 23 July, which caused pressure in 11C to return abruptly to the long-term trend. The blast could have assisted in the removal of ice during break-up, or may have resulted in unfrozen rock in the newly exposed face.

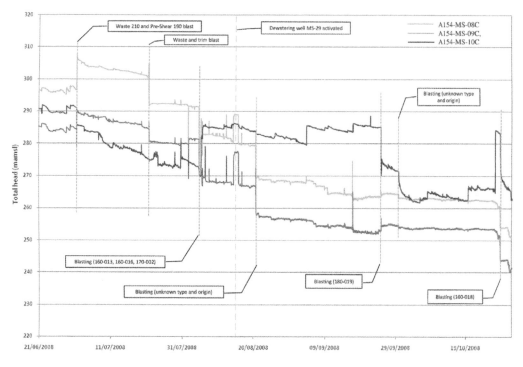

Figure A3.11: Diavik NW wall, detailed hydrographs of depressurisation events for the 'C' group VWPs (July to October 2008)

7, 8, 9 & 10 – Opening of the underground drainage network

Depressurisation of 'C' group VWPs from the opening of the underground drainage network was similar to the pattern displayed in the shallower VWPs. VWP 11C did not show depressurisation to events 7 and 8. During this time, the influence of freeze-back appeared to be the dominant control on the localised pore pressure at 11C.

2.2.5 VWPs – D Group (160 m elevation)

The 'D' group VWPs are positioned at an elevation of 160 m. Unlike the VWP groups at shallower elevations, distance from Lac de Gras does not appear to influence the hydrographs of the deeper sensors. In Figure A3.12, 'D' VWPs of Piezometers A154-MS-08 and A154-MS-09 show the most depressurisation, whilst 11D shows the least. This is at odds with the shallower VWPs, in which A154-MS-10 and A154-MS-11 typically show the most depressurisation.

Freeze-back effects are not as evident in the 'D' group piezometers. Freeze-back only becomes evident in VWPs once the pit face is exposed at, or below, their elevation. As in the shallower VWPs, the overriding control on pore pressure during the winter of 2010 (which was the first winter with an exposed face at the elevation of 'D' group VWPs) was the opening of valves in the underground drainage system, hence freeze-back effects are not seen in the data.

The following points refer to the numbers marked on Figure A3.12.

1 – Large increase in pore pressure in 11D (24 January to 3 February 2008)

At the end of January 2008, 11D showed the largest increase in pore pressure measured in any of the VWPs monitoring the North-west wall. The onset of this rise, which amounted to around 30 m of hydraulic head, did not coincide with the onset of freeze-back. These pressure increases did, however, follow three consecutive blasts at the face.

The causes of the pressure increases are not obvious, as the position of 11D at this time was below the base of the pit and likely beyond the direct influence of blasting. The response was an instantaneous increase in pore pressure, which is typical of an instantaneous change in porosity from fracture deformation. A reduction in fracture aperture is not usually associated with the advance of the pit floor, as unloading would tend to cause dilation of fractures. A reduction in fracture aperture can, however, occur where developing tensile stresses at the foot of the wall cause small-scale slip along fracture planes.

2 & 3 – Blast responses (12 April 2008 & 7 June 2008)

Mining of the face below the 220 m bench caused two significant blast responses in 'D' group VWPs. The first in

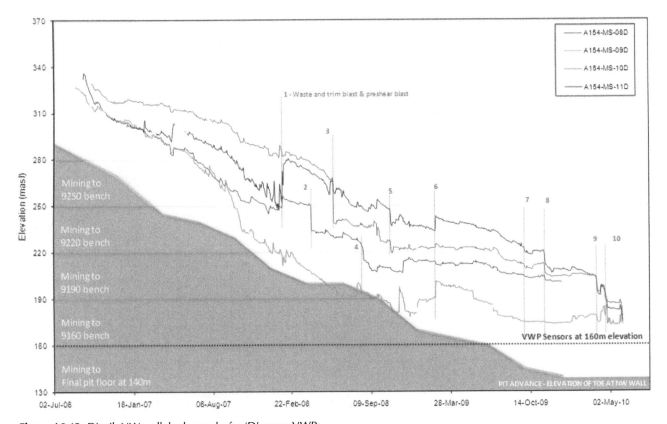

Figure A3.12: Diavik NW wall, hydrographs for 'D' group VWPs

08D was 20 m of depressurisation which occurred in response to a waste and trim blast (12 April 2008) and the second was a similar response in 10D to an ore blast, which caused 30 m of depressurisation (7 June 2008). Both of these events show drainage type 'recession curves', with the rate of depressurisation tailing off as pressures within the fractures re-equilibrate.

Despite the positions of the 08D and 10D VWPs beneath the pit floor at the time of these events, blasting causes a significant improvement in the permeability of flow pathways connecting them to the pit floor discharge area. Creation of the near face over-break zone at positions above 08D and 10D allows a pathway of groundwater flow within the sub-vertical fracture sets. The very fast initial depressurisation suggests that a component of fracture deformation may also be contributing to these responses. Unloading of the bench above is a potential explanation, however, if this was occurring, some recovery would be expected because of the relatively permeable fracture network and high recharge potential.

4 – Activation of MS-29 (15 August 2008)
On 15 August 2008, the MS-29 dewatering well, which is positioned between A154-MS-08 and A154-MS-09, was brought on line. VWP 08D responded with a fall in pore pressure equivalent to eight metres of hydraulic head, whilst 09D showed no response. There was also no perceivable change in the long-term depressurisation rate, despite pumping from MS-29. This event is included in the 'event response analysis' in Section 2.2.8.

5 – Depressurisation following blasting 160-018 (28 October 2008)
VWP 11D and 10D showed a drop in pressure associated with a blasting event on 28 October 2008 (which was also apparent in 8C and 9C, see the end of event group 4, Figure A3.10).

This blast response is particularly interesting as it was positioned across the pit on the 155 m level on the southern face.

6 – Pressure increase following blasting 160-033 (18 February 2009)
A blast on the 160 m level caused an instantaneous increase in pressure in 09D and 11D. This event may have been expected to cause further depressurisation in 'D' group piezometers, through development of the over-break zone and the exposure of the bench face adjacent to the VWP positions.

In this case, it appears that the blast was located in a position in which residue gases caused a pressure pulse to rapidly propagate through the fracture network, and may also have been in a position that opened up a connection with an upslope recharge source.

7, 8, 9 & 10 – Opening of the underground drainage system
The responses in the deep 'D' group VWPs to underground drainage were similar to the shallower VWPs. VWP 09D showed very little further depressurisation from the opening of underground drains as, by this time, 9D had shown the greatest amount of depressurisation and was probably at, or close to, hydraulic equilibrium.

2.2.6 VWPs farther behind the face – A154-MS-02

The VWPs installed in A154-MS-02 are further behind the pit face and so provide useful information about how pressure variations propagate from the near-face fracture network (as recorded in VWPs of A154-MS-08, 09, 10 & 11). Hydrographs for the VWPs of A154-MS-02 are shown in Figure A3.13

The most obvious difference between the VWP responses in A154-MS-02 and those closer to the pit face is the magnitude of response to blasting events. Hydrographs from VWPs 02A, B and C appear relatively smooth in comparison to those discussed above, as the extra distance from the pit face presumably allowed the buffering of pore-pressure changes as they moved through the fracture network.

As a consequence of the smoother hydrographs, it is easier to pick out subtle changes in the gradient of the depressurisation trend. The similarity in the hydrographs of 02A and 02B might indicate downhole hydraulic connection between these two VWPs.

1 & 2 – Freeze-back cycles
Full freeze-back cycles can be identified for the winters of 2005/6 and 2006/7 in 02A, 02B and 02C. As identified in shallower hydrographs, freeze-back is evident as a slowing in the rate of depressurisation rather than as an increase in pressure during the winter.

In the near-face piezometers, freeze-back can only be identified once the pit has advanced to below the level of the VWP. This was attributed to the need for an exposed bench face within which freeze-back can occur. The A154-MS-02 VWPs were at elevations below the base of the pit when these effects were evident in the hydrographs. This indicates the VWPs still fall within the overall 'flow net' of the slope, and therefore respond to reduced discharge at the pit face during the winter period.

3 – Depressurisation following blasting 160-018 (28 October 2008)
There is only one blast which caused a large, rapid pressure response in A154-MS-02 VWPs. The blast on 28 October 2008 (which also affected 'C' and 'D' group sensors), showed a large (10 m) drop in pressure in 02C. This is a larger response than was evident in the near-face piezometers.

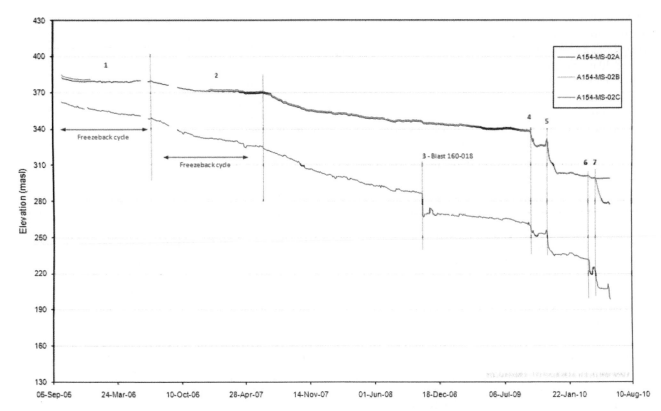

Figure A3.13: Diavik NW wall, hydrographs for A154-MS-02

4, 5, 6 & 7 – Opening of the underground drainage system

Drainage into the underground workings caused large drops in pressure in all VWPs of A154-MS-02. The last increment of this drainage ('7') is not recorded in 02A as the VWP reached zero pressure (i.e. the phreatic surface had dropped below the VWP position).

Although the VWPs of A154-MS-02 showed little response to any of the blasting or hydraulic events (which caused pressure displacements near the face), all of the VWPs showed a response when the underground drainage galleries come on line between September 2009 and April 2010.

2.2.7 Freeze-back

Figure A3.14 (from Golder, 2007) shows ambient air temperatures at Diavik. Figure A3.15 shows temperatures recorded in VWPs of varying depths (selected as showing the least interference). The thermographs show the following.

- VWPs at depth and further behind the pit face are largely insulated against seasonal variations in temperature.
- Shallower VWPs show larger seasonal fluctuations in temperature.
- A long-term trend of decreasing temperature is evident, which is strongest in the shallowest VWPs.

Correlation of mean temperatures recorded at each of these VWPs with their depths reveals a linear trend of increasing temperature with depth. The temperature gradient of around 0.13°C/m describes the geothermal gradient of the granite country rock. The thermographs show the equilibration of thermal gradients from the mining void, and appear to show the development of permafrost around the shallow near-face VWPs. For example, NWPT-M-02C shows temperatures decreasing at a rate of around 0.5°C every six months.

Consideration needs to be given to the seasonal freeze-thaw effect on the hydraulic properties of the near-face fracture network. The freezing of pore water will cause fracture apertures behind the zone of freezing to increase and they may not elastically recover to their previous positions, depending on the fracture orientation, and on the type of infilling on the fracture surfaces. Therefore, it is probable that repeated seasonal freeze-thaw will enhance the permeability of near-face fractures and potentially also fractures further behind the face, depending on the extent of the seasonal 'backing up' of heads.

In addition, if the VWP temperature data are reliable in indicating gradually reducing temperatures in the rock mass, a pit-face permafrost layer may develop which will permanently alter the hydrogeology of the NW Wall.

2.2.8 Discussion of key events

Many of the 37 events analysed are related to blasting. However, five events are described below where hydraulic stresses have produced a significant response in the

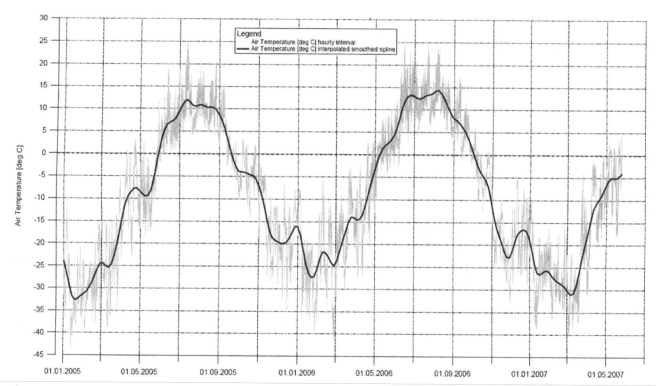

Figure A3.14: Air temperature data for Diavik (from Golder, 2007)

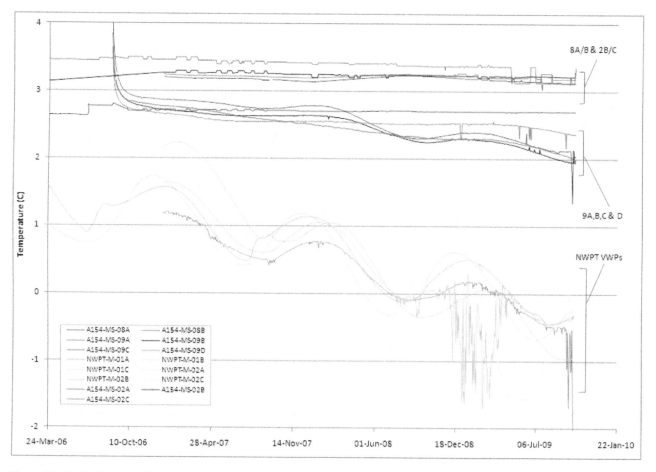

Figure A3.15: Diavik NW wall, temperature data for selected VWPs showing trend variation with depth

piezometers. The focus of this discussion is to interpret hydraulic connections and particularly the presence of permeable flow pathways, rather than connections apparent during blasting.

Responses from blasting, though possibly similar in appearance, do not necessarily imply hydraulic connection, but may be partly the result of deformation resulting from the shock wave. A single blasting event is discussed, blast 160-018 of 28 October 2008, which caused widespread depressurisation and drainage responses, apparently because of its influence on Dewey's Fault zone.

Response to blasting event 160-018, 28/10/08 (Figure 3.13, Section 3)

Blast 160-018 caused significant depressurisation in several deep VWPs. The blast was located in the South wall of the pit, at an elevation of 155 mamsl. This position, away from the North-west wall, makes this event particularly interesting, as it appears to show the influence of a well-connected 'plumbing' system of permeable fractures at the base of the pit, through which a pressure response can propagate into the NW wall.

The cause of depressurisation is not known with certainty. The most plausible explanation is that either the shock wave or the pressure pulse created by the blast caused an enhancement of one or more discrete fracture connections between the already depressurised fracture zones surrounding Dewey's Fault zone and the lower pit floor, and the location of the deep VWPs in the North-west wall.

Response to the initiation of pumping from the 280 m bench, 8 March 2008 (Figure 3.18, Section 3)

On 8 March 2008, wells DW1-5 on the 280 m bench were brought on line with an initial combined flow rate of 5.9 l/s. These wells are inclined into the face, dipping at 70 degrees, and are evenly spaced along the face behind the slope VWPs. On 14 March 2008, the average pumping rate in individual wells was 1.9 l/s (DW-4), 1.4 l/s (DW-1), 1 l/s (DW-3 & 5) and 0.6 l/s (DW-2).

As the pumping was occurring in several different wells, it is not possible to infer hydraulic connections from a specific well to an individual VWP. The main depressurisation occurred in the A and B group piezometers above an elevation of 200 m, which was around the elevation of the pit floor at the time. The only 'C' group piezometer to show a response was 10C, although this may have been due to the downhole inter-connection between the higher piezometers in the same hole (10A and 10B; previously discussed in Section A2.2.4). The VWPs of A154-MS-02 showed very little measurable response, being further away from the pumping stress.

The overall pattern of responses show the following.

- The magnitude of response in the individual piezometers appears to be governed by general proximity to the pumping stress and also by the amount of previous depressurisation that has occurred through passive drainage. VWPs which had shown more prior depressurisation, show less incremental depressurisation because of the pumping.
- The rapid re-equilibration of pressures, and the similar lengths of time required in different VWPs to reach hydraulic equilibrium, is indicative of high fracture permeability and good interconnectivity in the fracture network.
- The response in all the VWPs to pumping from the dewatering wells is the re-equilibration of pressures and this happens very quickly (within 1 day), with a relatively small drop in pressure head.
- This fast re-equilibration of pressures also highlights the overriding control of the hydrology (recharge and discharge) on the system. The inability of the dewatering wells to match the diffuse groundwater flux moving from the Lac de Gras recharge area towards the pit floor makes them largely ineffective, despite producing well yields which would otherwise (i.e. in the absence of recharge) cause large-scale depressurisation within the low storage fracture network.
- The absence of VWPs showing a lag time in the rate of equilibration suggests an absence of less permeable lower order fractures in the network. The system is able to equilibrate rapidly to recharge, so only the VWPs with a direct connection to the pumping centres show a response, and that response is rapid. Because of the dominance of recharge moving diffusely through the fracture network, it is not possible to observe a progressive 'A-B-C-D' lag-type response in the VWPs (Section 3.2.2, main text).

Response to activation of pumping from MS-29, 15 August 2008 (Figure 3.19, Section 3)

Pumping from dewatering well MS-29, located on the 280 m bench, started on 15 August 2008. The response to pumping was localised. As the hydrographs in Figure 3.19, Section 3 show, most of the response occurred in 8C and 9C (9–10 m decrease in head), with a smaller response in 8D (2.5 m decrease in head). Many of the other near-face VWPs showed responses of less than 1 m.

This event stressed the fracture network from a single 'point source', so the magnitude of responses in all VWPs has been looked at in detail with respect to distance from the pumping source as discussed in Section 3.3.5.2.

Response to activation of pumping from MS-30, 20 August 2008 (Figure 3.20, Section 3)

MS-30 is also positioned on the 280 m bench, in the northern corner, around the pit slightly from the North-west wall (see

Table A3.4: Properties assigned to the theoretical (Theis) fit for the distance-drawdown relationships

Event	Time (days)	Transmissivity (m²/day)	Storage coefficient (dimensionless)	Discharge (m³/day)
Well MS-29	0.25	10	10^{-4}	70
Well MS-30	0.25	20	10^{-4}	120

Figure A3.3). Pumping from MS-30 commenced on 20 August 2008 and, as with MS-29, the depressurisation was localised in the fracture network close to the dewatering well.

Depressurisation was greatest in 8C (10 m), 9C (9 m), 8D (6.5 m) and 8B (3.5 m). A154-MS-08 is the closest piezometer in the North-west wall to MS-30, and the greatest amount of depressurisation occurred in VWP 08C, which is closest in elevation to the base of well MS-30.

The larger than anticipated response in A154-MS-09C may be due to its position closer to the pit face than A154-MS-08. Groundwater may move preferentially through the more permeable near-face fracture network (over-break zone) towards the producing well. VWPs in piezometer A154-MS-10 were mostly unaffected by pumping.

Depressurisation caused by pumping from MS-29 and MS-30 is localised around the pumping well. The magnitude of responses conforms to a distance-drawdown relationship, as discussed in Section 3.3.5.2. For the Theis analysis described, the inputs (T and S) were adjusted to fit the data, and are not representative of any actual hydraulic properties of the fracture network. The Theis curve input values which result in the best fit to the distance-drawdown data are shown in Table A3.4.

Response to opening of the underground drainage galleries (Figure 3.22 and 3.23, Section 3)

Figure 3.22 shows the pore pressure responses to a short-term underground flow test (28/09/09). Figure 3.23 shows the longer-term depressurisation brought about through opening of the underground drains (September 2009 to June 2010). The main observations made through analysis of the hydrographs were as follows.

- A rapid pressure response occurs following opening of the underground drainage galleries. The 'recession curve' response, which is characteristic of drainage, is actually the re-equilibration of pressures to the new, lower elevation discharge point.

- The rapid equilibration of pressures is testament to the relatively high permeability and interconnectivity in the fracture network. The immediate onset of depressurisation following the opening of underground drains indicates the very high diffusivity of the fracture network. Reasonably high fracture transmissivity coupled with low storage enables pressure responses to propagate very quickly throughout the fracture network.

- The main controlling factor in the magnitude and time of the response is the distance of the piezometer from the discharge zone. There is a progressively smaller and marginally slower response going from deeper to shallower VWP installations.

The short-term underground flow test shows the above trends in the majority of the VWPs. The rapid response time is measured as a lag time of approximately 1 hour between the onset of the hydraulic response in VWPs of A154-MS-08 (closest to the nearest underground drain) and A154-MS-11 (the furthest away).

Several VWPs also exhibit damping of the response. VWPs 02A and 02B show a slower rate of depressurisation, and also a dampened short-term recovery from closing of the drain valves on 5 October 2009. In addition, the shallow sensor NWPT-M-02B shows gradual depressurisation that does not conform to the 'recession curve' type response and shows no recovery in response to the short-term closure of the underground drains on 5 October 2009. The deeper 'C' VWP sensors in A154-MS-02 and NWPT-M-02C do, however, share the typical depressurisation response exhibited in other near-slope VWPs.

In the longer-term response hydrographs, sensors 2A and 2B show the greatest amount of overall depressurisation. NWPT-M-02B would also likely have shown a large magnitude of depressurisation, however, the phreatic level dropped below the elevation of the sensor in early November, following the first phase of depressurisation.

The behaviour of A154-MS-02A & 02B and NWPT-M-02B are explained as follows.

- A154-MS-02A and NWPT-M-02B show a slow rate of response to depressurisation originating from drainage into the underground workings, but show the largest amount of overall depressurisation from the underground drains.

- These VWPs are shallow in comparison to the deep near-face VWPs and, as such, are closer to the recharge source of the Lac de Gras and farther from the point of discharge at the pit floor. Depressurisation via passive drainage is slow in the shallow fracture network as recharge is the overriding control.

- The combined yield of the underground drainage galleries is high relative to the rate of recharge and is sufficient to cause widespread depressurisation. In the shallower fracture network, the resulting drainage flux exceeds the rate of recharge influx.

■ Prior to drainage to the underground, heads in the shallow fracture network were much higher than those of the fracture network monitored in the deep near-face VWPs, which have already undergone some depressurisation. As a result, VWPs in the shallow fracture network, or those closer to the Lac de Gras, show the largest amounts of depressurisation when the underground drains are opened.

■ The event responses in the shallower VWPs still show the effects of constant recharge influx. The proximity of the recharge source causes a slower, 'damped' response as the shallow fracture network equilibrates with the rapidly reducing pore pressures in the deeper fracture network.

The relatively large magnitude depressurisation that occurs in 02A, 02B, 08A and 08B from underground drainage is also due to greater proximity to Lac de Gras which, prior to opening of the underground drains, had resulted in slow depressurisation in the fracture network monitored by these sensors.

2.2.9 Key observations

Key observations from the event-scale analysis can summarised as follows.

■ The overall depressurisation in the North-west wall occurs through the re-equilibration of pressures to new, lower discharge conditions. 'Free drainage' of the fractures would only occur if the recharge sources were removed.

■ Responses to hydraulic stresses are generally very rapid. There is little noticeable time lag in the responses of the different VWPs to dewatering pumping or the activation of the underground drains. The combined effect of very low storage and reasonably high trans-missivity in the fracture network is that even small pressure changes propagate rapidly through the system.

■ Although the magnitude of the pressure response to a new hydraulic stress is generally controlled by distance from the point of stress, parts of the fracture network with good connection to Lac de Gras show larger magnitude depressurisation responses from opening of the underground drainage galleries. Depressurisation of parts of the fracture network which are close to the Lac de Gras can only occur when a new discharge condition creates a flux that is in excess of the recharge flux (which is the overriding control in proximity to the lake). The underground drainage galleries yield the necessary flow rate to achieve this.

■ A 'plumbing' system of interconnected fractures appears to connect Dewey's Fault zone to the deeper rock mass and to the pit face. This is evident in the propagation through the fracture network of the pressure response caused by blast 160-018 on 28 October 2008.

As discussed in Section 3.3.6, it is postulated the Diavik data set supports that EPM flow conditions are valid at the 'slope-scale' and away from the immediate locations of discrete pumping stress, but that fracture flow conditions occur close to discrete pumping locations for limited periods of time. The transition is gradual and, in the case of the North-west wall at Diavik, occurs typically about 20 to 40 m from the point of new hydraulic stress.

3 DFN modelling

3.1 DFN-based data analysis

3.1.1 DFN modelling

DFN modelling attempts to simulate the rock mass fabric by describing the fracture system in realistic ways, allowing a description of the fracture geometry that is driven by verifiable data. DFN models seek to describe the heterogeneous nature of fractured rock masses by explicitly representing key elements of the fracture system as discrete objects in space with appropriately defined geometries and properties. By building geologically realistic models that combine the larger observed deterministic structures with smaller stochastically inferred fractures, DFN models capture both the geometry and connectivity of the fracture network as well as the geometry of the associated intact rock blocks.

The aim of the DFN modelling is to condition the fracture model as much as possible to available data and then use Monte Carlo simulations to quantify the uncertainty of extrapolation of the fracture pattern throughout the mine volume. It is a stochastic process allowing multiple but equi-probable realisations to be created (Figure A3.16).

Depending on the nature and application of a DFN model, a number of differing fracture properties need to be defined. The primary properties that are needed for all models in order to represent the fracture geometry reasonably are:

■ fracture orientation distribution;
■ fracture size distribution;
■ fracture intensity;
■ spatial variations.

When there are hydraulic and geomechanical applications of the DFN model, a number of other properties need to be defined including:

■ fracture aperture;
■ fracture transmissivity;
■ fracture storativity;
■ fracture shear strength properties and stiffnesses.

The primary data used for the Diavik study and their application are summarised below in Table A3.5. A

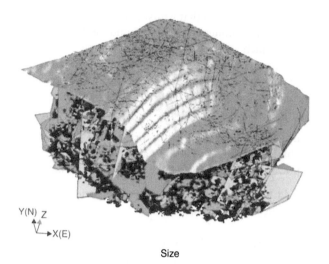

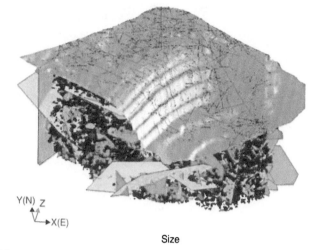

Size

Size

Figure A3.16: Equi-probable realisations of Diavik A154 NW wall model

photograph of the North-west wall during the winter freeze-back period is shown in Figure A3.17. The broad workflow involved with the DFN process is summarised in Figure A3.18.

3.1.2 Fracture orientation

Fracture orientation can be defined from either borehole imaging data, orientated core or trace mapping. Where fracture orientation data are highly systematic and organised into distinctive fracture sets, the statistical properties of these sets can be defined and used as a key stochastic input into the DFN model. However, in the case of Diavik, where the orientation data have a more dispersed nature, an alternative method of 'bootstrapping' can be used. This is a statistical method based upon multiple random sampling with replacement from an original sample to create a pseudo-replicate sample of fracture orientations. This allows the original data to be reproduced both accurately and efficiently.

The use of both borehole and trace mapping data for conditioning the DFN orientation distribution were investigated, with ultimately bench mapping data from the North-west wall being selected. The primary justification for this is the strongly biased nature of the borehole data relative to the bench mapping that has

resulted in a very high proportion of low-angle structures being observed in the borehole data. The bench mapping data was considered more representative of the overall population. Figure A3.19 shows a stereoplot of all features mapped on the North-west wall comprising three broad fracture sets: one sub-horizontal set and two sub-vertical sets (ENE–WSW and WNW–ESE striking).

3.1.3 Fracture size analysis

3.1.3.1 Introduction

The derivation of the fracture size distribution is critical to any DFN modelling campaign yet is generally among the most difficult parameters to constrain. The primary difficulty in determining fracture size is that it cannot be measured directly as any measurements relating to fracture size are actually measurements of the trace that a fracture or fault make with a geological surface or mining exposure (see Figure A3.20).

The sensitivity of a DFN model to fracture size can easily be understood when one considers two simple DFN scenarios (Figure A3.21).

Both models have been constrained by the same fracture frequency of 0.5 #/m yet the fractures in the left hand model have a two metre radius and the right hand model fractures have a four metre radius. The

Table A3.5: Main fracture data sources used for DFN analysis, their location and application

Data source	Location	Application
Geotechnical boreholes	A154 area & A21 area	• Constraining fracture intensity • Testing fracture apertures
Borehole televiewer data	A154 area & A21 area	• Testing fracture orientation
Bench mapping	A154 pit	• Constraining fracture size distribution and fracture orientations
Underground mapping	A21 underground	• Constraining & testing fracture size distribution
Hydraulic testing data	A154 pit – NW wall primarily	• Constraining fracture transmissivity and aperture

Figure A3.17: View of the North-west wall in winter

consequence of this is that whilst both models have the same fracture frequency, the right hand model has four times as much available fracture area. Clearly constraining this property is important if reasonable estimates of permeability and connectivity are to be made. To examine the underlying distribution of fracture sizes, two methods have been used:

- derivation of fracture radius distribution that produces obtained trace distributions; and
- power law scaling from trace data and major faults.

Additionally any dependency between orientation and size has been investigated.

3.1.3.2   Fracture radius distribution

Simulation of all trace length data shows that it can be best described using a log-normal distribution of radius with a mean of 1.8 m and a standard deviation of 2.2 m as shown in Figure A3.22.

However, it is clear that much of this trace length data is censored as it fails to show the longer structures that are known to exist. The maximum mapped length is approximately 60 m but features >100 m are observed. Figure A3.23 shows the North-west wall during mining, showing the broad-scale fracture distribution and the influence of the metasedimentary rafts on seepage to the slope.

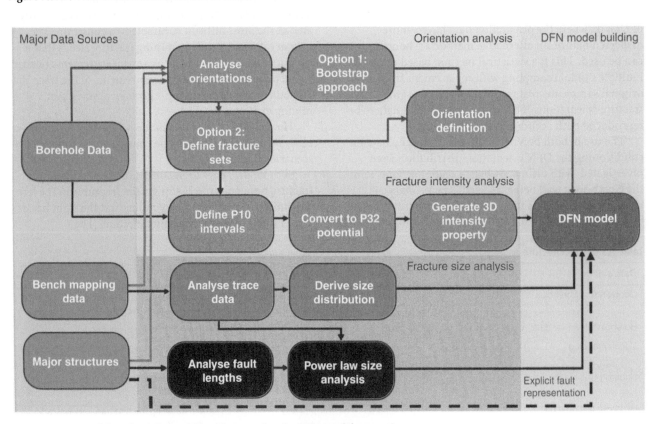

Figure A3.18: Workflow for DFN analysis of fracture data for DFN model generation

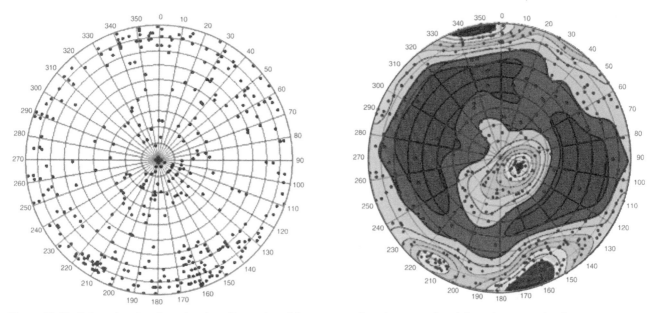

Figure A3.19: Orientation data from the pit wall mapping of the A154 pit plotted as point data (left) and contoured (right)

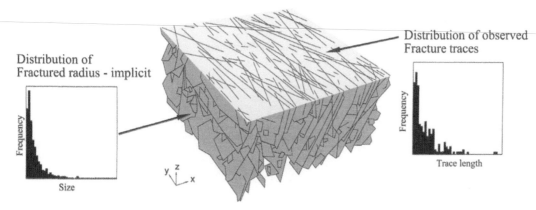

Figure A3.20: The problem in determining fracture size (radius) from observed fracture trace lengths

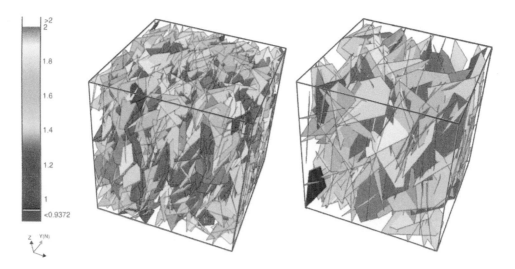

Figure A3.21: Two DFN scenarios having the same fracture frequency (0.5 #/m) but differing fracture size descriptions. The left hand model has a constant radius of 2 m and the right hand model has a constant fracture radius of 4 m.

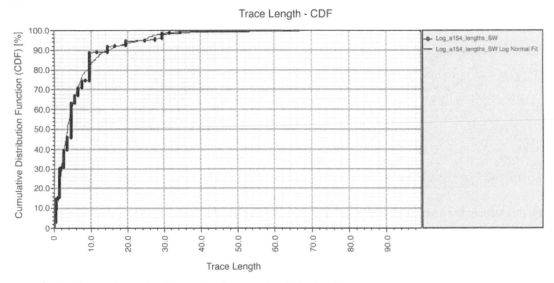

Figure A3.22: Trace length cumulative distribution function actual and simulated

Figure A3.23: View of North-west wall showing the metasedimentary rafts and their influence on seepage to the slope

3.1.3.3 Power law scaling

Field studies have shown that in many rock masses, fractures and faults scale according to scale invariant power laws (within certain size thresholds). This means that observations at the 100s–1000s metre scale can be proportionally scaled to say the 1s to 10s metre scale. By taking fault/fracture length data taken at different scales (e.g. regional lineaments, mine scale or district faults and fracture traces in surface or underground exposure), a power law function can often be fitted to these data, allowing the prediction of fracture size and frequency at scales not directly measured.

Data are available at two slightly different scales for the power law analysis.

- Bench mapping data from the A154 pit.
- Underground drift mapping from the A21 underground.

The methodology for the analysis is as follows.

- Take raw trace length data (see Figure A3.24, top right, column labelled 1).

- Sort into order from smallest to biggest (see Figure A3.24, column labelled 2);
- Calculate the cumulative number greater than or equal to the trace (see Figure A3.24, column labelled 3).
- Normalise this cumulative number by the area of the outcrop or map (see Figure A3.24, column labelled 4).
- Repeat for larger-scale data and plot normalised number (y axis) against trace length (x axis) for both trace data and map data.

The graph of data from A21 underground and A154 wall mapping used for power law mapping is shown in Figure A3.25.

The log-log linear fits (dashed lines) on Figure A3.25 to both wall mapping and drift mapping show very similar trends and consistent short and long data bias. Given that there are no mapped major structures at Diavik it was felt that the use of a Pareto distribution ('Power Law') for fracture radius was better as it reproduced a fuller range of known trace lengths in the pit, particularly for the long metasediment rafts. The slope of the line for trace length

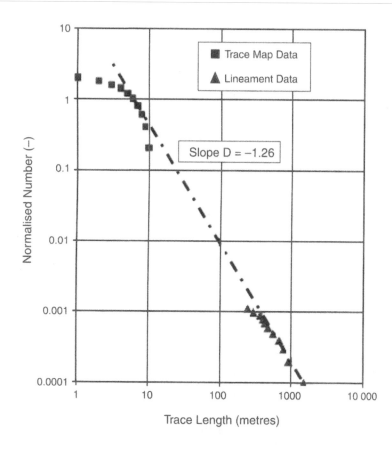

Trace Map Data

1 Trace Length	Sorted 2 Trace Length	3 Cumulative Number	Normalised 4 Cumulative Number
1.2	0.6	10	2
2.1	1.1	9	1.8
3.5	1.2	8	1.6
0.6	1.2	7	1.4
1.1	2.1	6	1.2
4.5	2.1	5	1
2.1	3.2	4	0.8
3.2	3.5	3	0.6
3.6	3.6	2	0.4
1.2	4.5	1	0.2

Area = 5 m^2

Lineament Data

Trace Length	Sorted Trace Length	Cumulative Number	Normalised Cumulative Number
250	250	11	0.0011
1500	300	10	0.0010
920	370	9	0.0009
300	410	8	0.0008
410	440	7	0.0007
480	480	6	0.0006
690	560	5	0.0005
560	690	4	0.0004
780	780	3	0.0003
440	920	2	0.0002
370	1500	1	0.0001

Area = 10 000 m^2

Figure A3.24: Graph showing an example of normalised number (cumulative frequency/sample area) plotted against trace length showing how the power law graph describes well, both trace map (1–10 m) and lineament data (100–1000 m) (right); and raw data used to generate the graph from lineament and trace mapping data (left). See text for explanation.

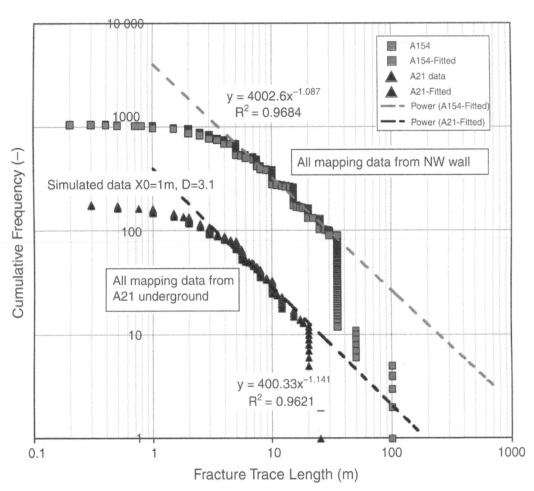

Figure A3.25: Fracture trace length distribution curves for data from the A21 underground and A154 pit showing the best fit power law curves. These curves are not area normalised which is why they are off set.

distribution is 1.1, yielding a slope for fracture radius of −3.1 (Gradient + 2).

3.1.3.4 Orientation dependence and size

To investigate whether there is a significant orientation dependence on fracture trace length (and therefore size), the trace data have been split into three orientation sets using FracMan's ISIS module. This found that a Bivariate Bingham distribution could be usefully used to describe the 3 fracture sets (Figure A3.26).

The trace length data were then plotted on a set by set basis as shown in Figure A3.27. Over the portion of the curve interpreted as linear, the three independent curves show a high degree of agreement, albeit with a slightly higher slope then the traces as a whole.

A slight degree of separation is noted between curves for shorter trace lengths but it is not excessive. There is a larger separation at longer fracture trace lengths between the sub-horizontal set and the sub-vertical sets owing to bench-height restrictions on the measurement of sub-vertical sets.

Given these observations, it is reasonable to model the fracture population as a single size population.

3.1.4 Fracture transmissivity analysis

3.1.4.1 Introduction

Hydraulic testing derives the transmissivity of the rock mass over selected intervals, but this needs to be converted to a fracture transmissivity distribution for inclusion in the DFN. Data for constraining fracture transmissivity are provided from an extensive hydraulic testing program with the test results shown in Figure A3.28. Grouping the data into 50 m intervals (Figure A3.29) reveals a strong depth dependency with a reduction in transmissivity of approximately 1.5 orders of magnitude over 200 m. Note the anomalous low T values at higher elevations.

3.1.4.2 Converting interval transmissivity to fracture transmissivity

Since packer tests only measure interval transmissivity, a methodology is needed to take these data and convert them to a fracture transmissivity distribution. This is done

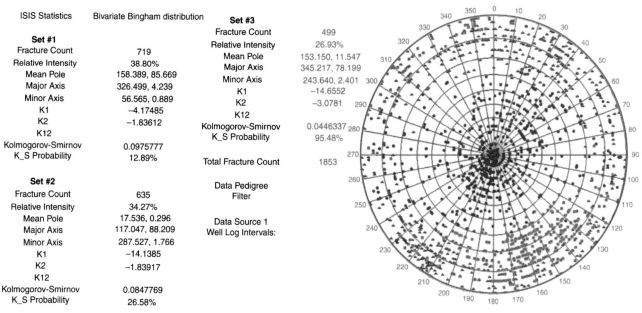

ISIS Statistics	Bivariate Bingham distribution

Set #1	
Fracture Count	719
Relative Intensity	38.80%
Mean Pole	158.389, 85.669
Major Axis	326.499, 4.239
Minor Axis	56.565, 0.889
K1	−4.17485
K2	−1.83612
K12	
Kolmogorov-Smirnov	0.0975777
K_S Probability	12.89%

Set #2	
Fracture Count	635
Relative Intensity	34.27%
Mean Pole	17.536, 0.296
Major Axis	117.047, 88.209
Minor Axis	287.527, 1.766
K1	−14.1385
K2	−1.83917
K12	
Kolmogorov-Smirnov	0.0847769
K_S Probability	26.58%

Set #3	
Fracture Count	499
Relative Intensity	26.93%
Mean Pole	153.150, 11.547
Major Axis	345.217, 78.199
Minor Axis	243.640, 2.401
K1	−14.6552
K2	−3.0781
K12	
Kolmogorov-Smirnov	0.0446337
K_S Probability	95.48%
Total Fracture Count	1853
Data Pedigree	
Filter	
Data Source 1	
Well Log Intervals:	

Figure A3.26: Distinct fracture sets as determined by a Bivariate Bingham distribution

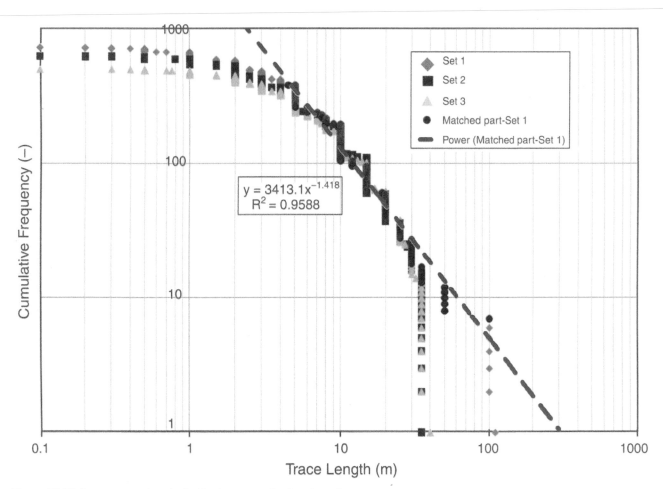

$$y = 3413.1x^{-1.418}$$
$$R^2 = 0.9588$$

Legend:
- Set 1
- Set 2
- Set 3
- Matched part-Set 1
- Power (Matched part-Set 1)

Figure A3.27: Fracture trace length distribution curves for data from the orientation three sets

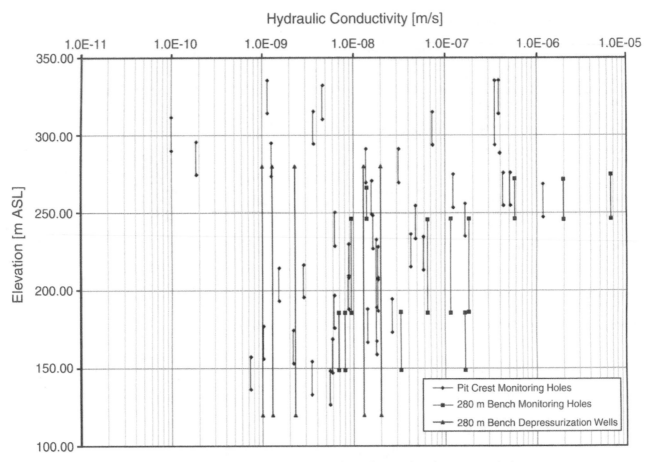

Figure A3.28: Hydraulic conductivity data from pit crest monitoring holes and 280 m bench monitoring holes

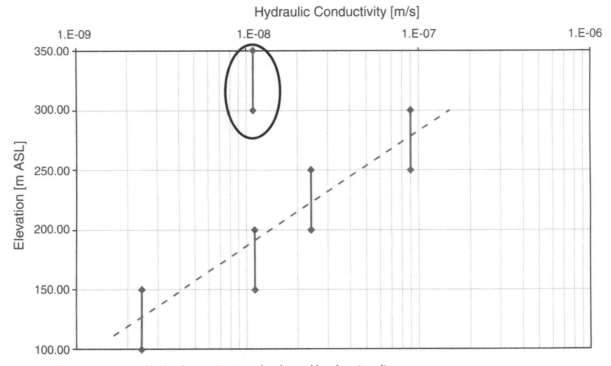

Figure A3.29: Geometric mean of hydraulic conductivity data binned by elevation slices

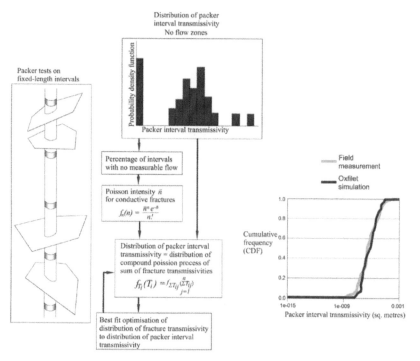

Figure A3.30: Basic workflow used by Oxfilet for deriving the fracture transmissivity distribution (left); and OXFILET simulated packer test results vs measured packer test results (right)

using Golder's OXFILET software that derives the underlying fracture transmissivity distribution that results in the sampled hydraulic test transmissivity distribution using the work flow shown in Figure A3.30.

An initial fracture transmissivity distribution is estimated and the resultant simulated packer test distribution determined. This is an iterative process using an optimisation function in order to obtain the best possible fracture transmissivity distribution. The results of these simulations are shown in Table A3.6.

The tabulated results for the NW wall have been examined in order to derive a relationship between elevation and mean fracture transmissivity to ensure that

the model reproduces as accurately as possible the distribution of data observed in Figure A3.29. The relationship implemented within the DFN model is shown in Figure A3.31. Note the clipping of the maximum and minimum fracture transmissivity.

3.1.5   Fracture intensity analysis

3.1.5.1   Definitions

Defining fracture intensity is somewhat problematic as there are a wide range of possible measures, often with ambiguous definitions. In order solve this problem, the DFN community developed a unified system of fracture intensity measures (known as the P_{ij} intensity system)

Table A3.6: OXFILET simulated packer test results

	Actual transmissivity [m²/s]	Std Deviation measurements	Simulated transmissivity [m²/s]	Std Dev simulation	Fractures per m	Significance
Country rock	2.E-05	4.E-05	1.E-06	1.E-05	1	90.5
Dewey's Fault zone	2.E-04	5.E-04	2.E-05	8.E-04	0.8	99.8
All testing data	9.E-07	3.E-05	1.E-07	3.E-04	2.17	97
100 to 150 masl	7.E-08	7.E-08	1.E-08	2.E-07	0.93	99.6
150 to 200 masl	4.E-07	2.E-06	4.E-08	1.E-06	0.9	95.5
200 to 250 masl	9.E-07	3.E-06	3.E-08	9.E-07	1.14	90.5
250 to 300 masl	3.E-06	5.E-05	5.E-06	1.E-01	2.15	90.5
300 to 350 masl	3.E-07	5.E-06	6.E-07	2.E-05	0.93	30.8
250 to 350 masl	1.E-06	4.E-05	5.E-07	3.E-03	2.5	35.8

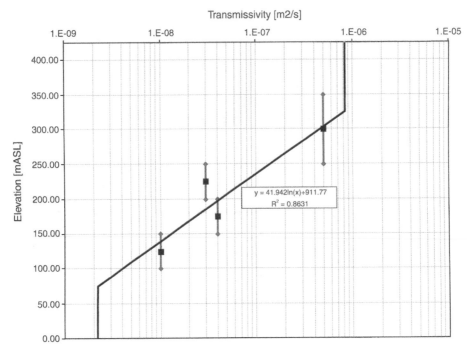

Figure A3.31: Elevation vs fracture transmissivity as calculated by OXFILET

that provides an easy framework to move between differing scales and dimensions.

Given the P_{ij} intensity system described in Table A3.7, fracture intensity for DFN modelling is typically defined either from borehole data (fracture logging or borehole imaging tools) as fracture frequency (P10) or from trace mapping upon surfaces such as benches or tunnel walls (P21). Care needs to be taken in the use of both of these data sets as they are highly directionally biased. The preferred measure of fracture intensity for a DFN model is known as P32 (fracture area/unit volume) which represents an intrinsic rock mass property. Whilst it cannot be directly measured, it can be inferred from the 1D and 2D data above using a simulated sampling methodology.

Through sampling of simulated fracture networks of differing intensity, a relationship can be developed between observed fracture intensity and its associated P32 that allows the volume of the model to be populated with the appropriate fracture intensity.

3.1.5.2 Derivation of P10 interval data

In order to derive a spatially varying description of P32, borehole fracture logs need to be processed in order to identify zones of the rock mass where the average degree of fracturing remains constant over borehole lengths in the order of 10s to 100s of metres. To achieve this, cumulative fracture intensity (CFI) plots have been generated for all boreholes in and around the A154 pit, with these

Table A3.7: Fracture intensity measures based upon the dimension of the sample and the dimension of the fracture measure

		Dimension of measurement				
		0	1	2	3	
Dimension of sample	1	P10 *No of fractures per unit length of borehole*	P11 *Length of fractures per unit length*			Linear measures
	2	P20 *No of fractures per unit area*	P21 *Length of fractures per unit area*	P22 *Area of fractures per area*		Areal measures
	3	P30 *No of fractures per unit volume*		P32 *Area of fractures per unit volume*	P33 *Volume of fractures per unit volume*	Volumetric measures
		Density		Intensity	Porosity	

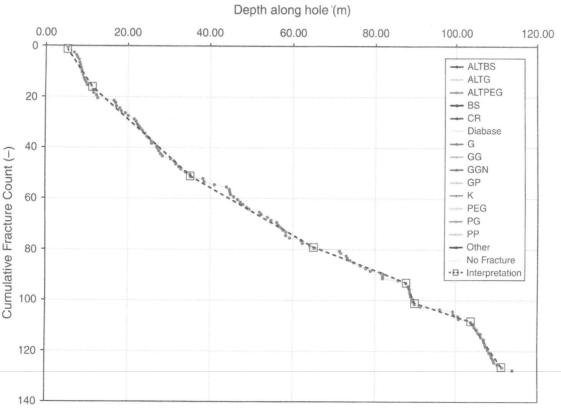

Figure A3.32: Example cumulative frequency intensity plot for A154-U05 showing the simplified interpreted line overlaid on top of the raw data. The raw curve is coloured by lithology.

displaying depths on the horizontal axis and cumulative fracture number on the vertical axis (Figure A3.32). They are interpreted as follows.

- Where the slope (gradient) of the CFI curve is constant, the fracture frequency over that interval is constant. The measured gradient is the fracture frequency in fractures per metre (#/m).
- Where the gradient of the curve is flattening, the fracture frequency is decreasing.
- Where the gradient of the curve is steepening, the fracture frequency is increasing such as through a fault zone or other highly fractured interval or unit.

The advantage of CFI plots over depth histograms is that they emphasise common fracture frequency rather than the variation, and represent a practical way to approximate the variation of fracture frequency along the length of the boreholes.

All available boreholes were reviewed and zones of constant fracture frequency defined for those boreholes where there were sufficient data. Several of the boreholes have zones where there are missing or non recorded fracture data and these holes were ignored to avoid the possibility of introducing zero intensity artefacts into the data.

3.1.5.3 Conversion of P10 to P32 Potential

The previously described derivation of P10 values from the borehole logs provides a local measure of fracture intensity that is highly dependent upon the orientation of the sample (the boreholes) and the orientation of the predominant fracture sets. To account for this, a technique has been devised (Wang, 2006) that allows the conversion of the P10 values to a property known as P32 Potential. P32 Potential (P32P) is a proxy for P32 but contains no fracture size component. P32 Potential is calculated using the equation:

$$P32 = C31 \times P10 \qquad \text{(eqn A3.1)}$$

C31 is a function of the orientation of the borehole and the orientation of the mean fracture sets and has been derived for every borehole from the mean borehole orientation using the table and methodology shown in Figure A3.33.

As the P10 data obtained from the CFI analysis have been derived from unclassified orientation data (i.e. not split into different sets) some pre-processing is required before the C31 values can be derived. The relative proportion of each of the three broad fracture sets was determined and the derived P10 data split into equivalent P10_set1, P10_set2 and P10_set3, see Table A3.8.

ρ \ κ	0.1	1	2	5	10	50	100	200	500	∞
0	0.50	0.53	0.62	0.79	0.90	0.98	0.99	0.99	1.00	
5	0.50	0.53	0.62	0.79	0.89	0.97	0.98	0.99	0.99	
10	0.50	0.53	0.61	0.78	0.88	0.96	0.97	0.98	0.98	
20	0.50	0.53	0.59	0.75	0.84	0.91	0.92	0.93	0.93	
30	0.50	0.52	0.56	0.68	0.77	0.84	0.85	0.85	0.85	
40	0.50	0.51	0.54	0.62	0.67	0.74	0.75	0.75	0.75	Cos ρ
50	0.50	0.51	0.51	0.54	0.57	0.62	0.62	0.63	0.63	
60	0.50	0.49	0.48	0.47	0.45	0.47	0.48	0.48	0.48	
70	0.50	0.48	0.45	0.39	0.34	0.32	0.32	0.32	0.32	
80	0.50	0.48	0.44	0.34	0.26	0.18	0.16	0.16	0.16	
90	0.50	0.48	0.44	0.33	0.24	0.11	0.08	0.06	0.04	

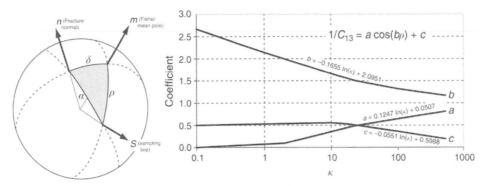

Figure A3.33: Methodology for deriving C31 values for P10-P32 conversion, from Wang (2006)

The methodology for converting the borehole P10 to P32 potential using Wang's method is as follows (Figure A3.34).

- Take undifferentiated borehole P10 *(column H)* and split into P10_set1, _set2 and _set3 *(columns J, K & L)* according to approximate proportions of data observed from previous orientation analysis (set 1 = 60.5%, set 2 = 25.8% and set 3 = 13.7%).
- Derive C31 values for each borehole and set *(columns N, O & P)* assuming the polar orientation of sets from Table A3.8.
- Derive P32P for that borehole interval from the sum of P10_set$_n$ x C31_set$_n$ *(column R)*.

It is this property P32 Potential *(column R)* that has been used in the geostatistical modelling of fracture intensity through the Diavik A154 volume.

3.1.6   Geostatistical modelling of fracture intensity

Having created a data set of orientation-corrected P32 Potential (P32P) values, geostatistical methods were used to interpolate these values through the Diavik A154 volume.

Table A3.8: Fracture sets and their relative proportion

	Pole_Dip	Pole_Trend	Dispersion K	Proportion
Set 1	169.3	86.5	11.5	60.5
Set 2	311.6	0.583	6.3	25.8
Set 3	34.8	0.76	4.7	13.7

This 3D intensity volume is critical to the DFN modelling but can also be used as a valuable input to engineering design or hydrogeological analysis where it can be used to predict zones of poor rock quality.

All analysis was undertaken using ISATIS, a geostatistical package developed by the Paris School of Mines, Fontainebleau, France, and marketed globally by Geovariances.

Figure A3.35 shows the distribution of Diavik boreholes used for the geostatistical analysis. The underlying data were highly irregular in sample length as a result of the interpretation method (i.e. the picking of zones of constant fracture frequency rather than conventional point samples). Therefore, to aid modelling and interpolation, it was first regularised by converting it into standardised sample lengths of five metres. From these data, histograms of the information were derived and the data are clearly identified as having a log-normal distribution, see Figure A3.36. Additionally all fracture intensity data from within the two kimberlite pipe volumes was removed to ensure that only granitic country rock was used in the analysis.

Using an understanding of the main geological structures of the deposit, variographic analysis of the data was undertaken using the orientations shown in Figure A3.37.

This analysis used the trends of the three main fracture sets with semi-variograms being calculated in these three principle directions on the P32P values which exhibited appropriate continuity in 3D space. This was carried out

	Initial P10		P10 split into 3 sets from ISIS analysis				Calculated C31 results			Calculated P32P From Σ (P10_set$_n$ × C31_set$_n$)

C	D	E	H	I	J	K	L	M	N	O	P	Q	R
HoleID	DepthFrom	DepthTo	P10		P10_Set1	P10_Set2	P10_Set3		C31_Set1	C31_Set2	C31_Set3		P32P
A154-GT-04	00.00	5.37											
A154-GT-04	5.37	27.19	1.88		1.137	0.485	0.258		1.2317	3.30189	2.19139		3.57
A154-GT-04	27.19	39.55	5.26		3.182	1.357	0.721		1.2317	3.30189	2.19139		9.98
A154-GT-04	39.55	39.56	100.00		60.500	25.800	13.700		1.2317	3.30189	2.19139		189.73
A154-GT-04	39.56	52.63	2.75		1.664	0.710	0.377		1.2317	3.30189	2.19139		5.22
A154-GT-04	52.63	88.18	3.80		2.299	0.980	0.521		1.2317	3.30189	2.19139		7.21
A154-GT-04	88.18	200.70	1.86		1.125	0.480	0.255		1.2317	3.30189	2.19139		3.53
A154-GT-05	0.00	2.20			0.000	0.000	0.000		1.25544	3.01012	2.35841		0.00
A154-GT-05	2.20	10.86	1.15		0.696	0.297	0.158		1.25544	3.01012	2.35841		2.14
A154-GT-05	10.86	26.88	3.68		2.226	0.949	0.504		1.25544	3.01012	2.35841		6.84
A154-GT-05	26.88	50.90	2.58		1.561	0.666	0.353		1.25544	3.01012	2.35841		4.80
A154-GT-05	50.90	78.50	4.46		2.698	1.151	0.611		1.25544	3.01012	2.35841		8.29
A154-GT-05	78.50	123.30	1.96		1.186	0.506	0.269		1.25544	3.01012	2.35841		3.64
A154-GT-05	123.30	123.80	2.00		1.210	0.516	0.274		1.25544	3.01012	2.35841		3.72
A154-GT-05	123.80	171.45	1.36		0.822	0.351	0.186		1.25544	3.01012	2.35841		3.53

Figure A3.34: Methodology for converting an undifferentiated P10 (column D) into a P32 potential value for subsequent geostatistical modelling

both including and excluding the kimberlite fracture intensity data, with the quality of the variography improving when the kimberlite was excluded, although the variability also increased. Good structures were identified and subsequently modelled.

The actual and modelled variograms are shown in Figure A3.38 with the vertical (D-90 purple line) being the

best variogram and is modelled with the best fit. The variogram model fitted has three main structures.

- Nugget (0.5).
- Spherical with a sill of 2.2 and a horizontal range of 25 m (green), and a vertical range of 90 m (purple).
- Spherical with a sill of 1.5 and a horizontal range of 300 m (green), and a vertical range of 600 m (purple).

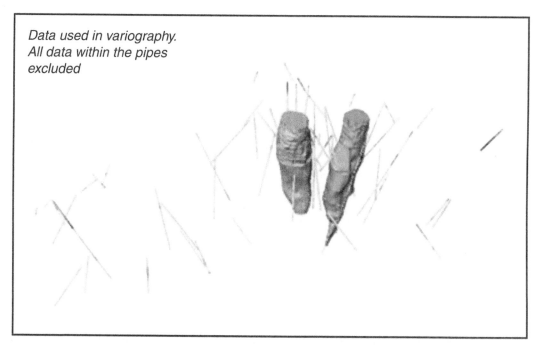

Data used in variography. All data within the pipes excluded

Figure A3.35: CE borehole used for geostatistical interpolation coloured by P32 potential. Hot colours represent high P32P and cool colours low P32P

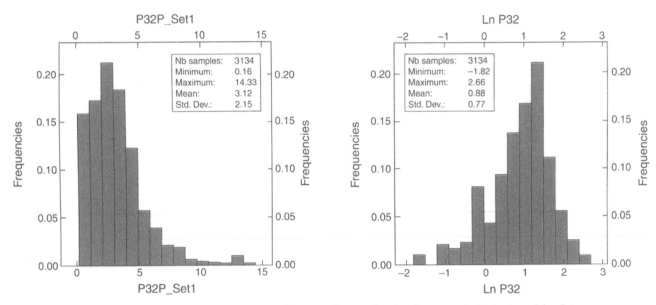

Figure A3.36: Histograms of P32 potential in base 10 and logspace showing the clear log-normal distribution of the data

This modelled variogram was then used as the primary input into the generation of the P32P block model using a turning bands simulation method. Simulation methods are preferred for this block model generation as the geological reality of the simulations is considered to be a better approximation of the distribution of fracture intensity within the rock mass.

To investigate the goodness of fit of the geostatistical analysis and modelling, a comparison between the simulated and actual P32P populations and also the

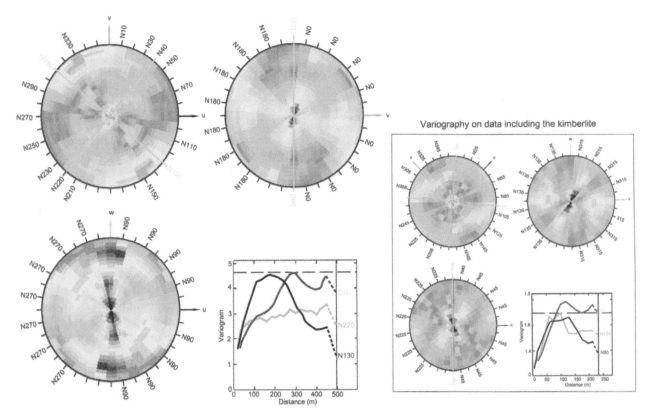

Figure A3.37: Anisotropic variography of the P32 potential property and modelled variograms for only the granitic rock mass (left); and the same analysis but incorporating data from the kimberlite pipes showing an increased variability but very consistent directional variography (right)

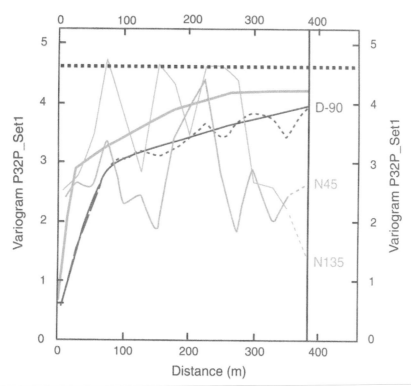

Figure A3.38: Actual and modelled variograms of the Diavik data

borehole sampled and raw P32P values was carried out and show strong agreement, see Figure A3.39.

Figure A3.40 and Figure A3.41 show images of the P32P fracture intensity block model.

3.2 DFN model building

3.2.1 Model volume and objects

The DFN model has been constructed within a sub set of the main model grid focused in the NW quadrant of the A154 pit, see Figure A3.42. The approximate elevation range of the DFN model is between 425 to 50 masl. Larger fractures will extend outside the main volume but these are subsequently clipped by the model shell.

3.2.2 Constraining fracture orientation

Fracture orientation has been constrained using a bootstrapping technique, using data available from the North-west wall bench mapping (Figure A3.20).

3.2.3 Constraining fracture size

Based upon the analysis of trace mapping and drift mapping form the A154 pit and A21 underground, fractures have been generated using a power law size distribution of radius with an exponent of D = 3.12, a minimum fracture size of 2.5 m and a maximum of 300 m.

3.2.4 Constraining fracture intensity

There are two main components to constraining fracture intensity.

- Honouring the intensity of fracturing at the borehole.
- Honouring the spatial variation of fracturing within the granitic rock mass (see Figure A3.43).

The borehole intensity is matched by FracMan allowing fractures to be generated until the average fracture frequency at the boreholes in the North-west wall has been honoured. For instance if there was 1 km of boreholes with an average fracture frequency of 2/m, FracMan would continue to generate fractures until 2000 fractures intersected the conditioning boreholes. The spatial variation in fracture intensity is controlled by the P32P block model property which provides a probability of occurrence of fractures. Thus where fracture intensity (P32P) is high, the probability of fractures being generated there will be high and consequently boreholes in that part of the model will have a higher fracture frequency.

3.2.5 Conditioning fracture transmissivity

Fracture transmissivity is primarily conditioned from the OXFILET analysis of hydraulic test data. However, there are two dependencies that need to be incorporated into the model to ensure reasonable reproduction of observed behaviour.

- Reproduction of the elevation sensitivity as shown in Figure A3.31.
- Correlation of transmissivity to size such that the largest fractures are also the most transmissive.

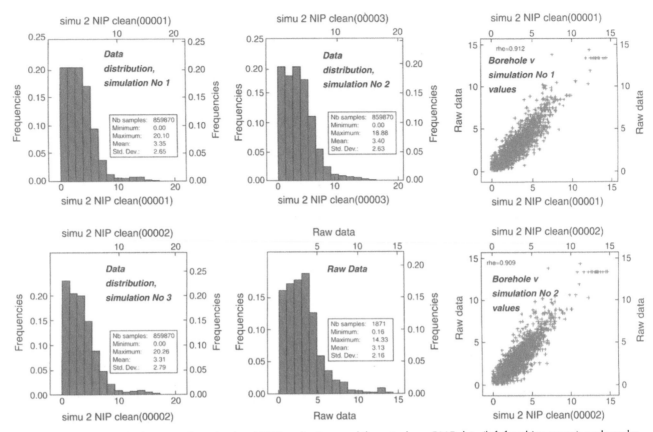

Figure A3.39: Comparison between three simulated P32P realisations and the actual raw P32P data (left four histograms); and graphs showing a comparison between the real (y axis) and simulated (x axis) borehole P32P properties for two realisations (right)

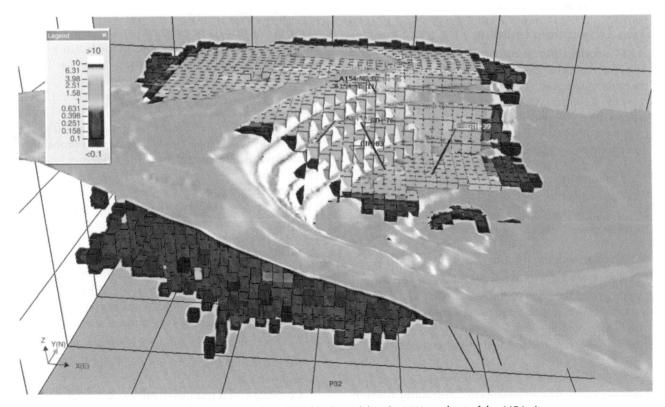

Figure A3.40: Image showing the general location of the P32P block model in the NW quadrant of the A154 pit

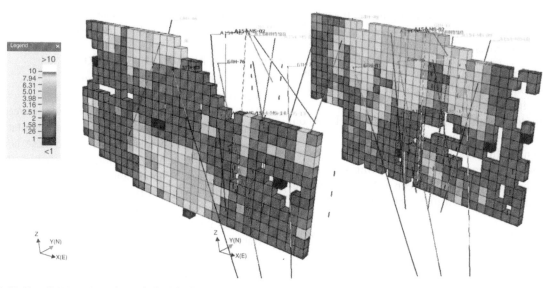

Figure A3.41: Two E–W sections through the block model showing the distribution of the P32P fracture intensity property

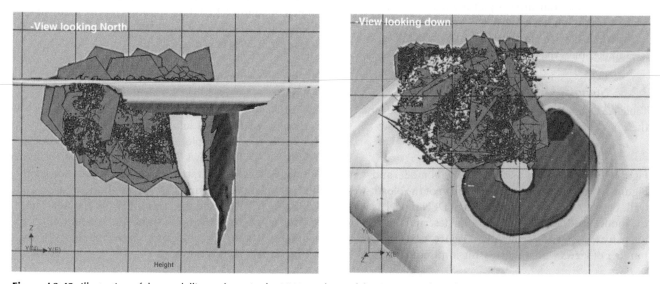

Figure A3.42: Illustration of the modelling volume in the NW quadrant of the A154 pit where the DFN model has been generated

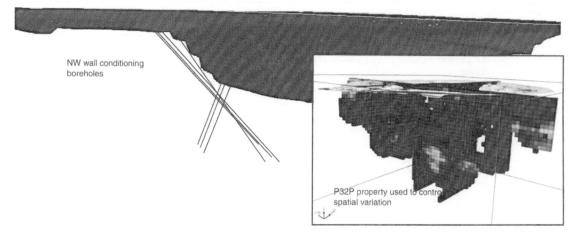

Figure A3.43: Figure showing the conditioning boreholes in the NW wall used for constraining the actual fracture intensity. The P32 Potential property (inset image) controls the spatial variation of fracture intensity, allowing more fractures to be generated where the property is high and fewer where it is low.

In order to achieve these two aims, the fracture transmissivity of the DFN model has been populated using the following approach.

- Using the relationship for transmissivity as a function of elevation, calculate the $\log_{10}$ of the transmissivity for each cell in the grid based on elevation.
- Map this $\log_{10}$ transmissivity to the fractures based on their location within the main model grid.
- Apply a noise function to the log10 transmissivity for each fracture independently by generating a value from a normal distribution with a mean of 1 and a standard deviation of 0.08 (this ensures a variably linear relationship).
- Calculate the actual transmissivity by adding the value mapped from the grid plus the noise term. A reduction factor of 80 (i.e. divide the transmissivity value by 80) is required to calibrate the simulated and actual transmissivity values.
- These actual transmissivity values were then scaled by a size function (Size/MeanSize) to ensure the larger fractures have the highest transmissivity, and the smallest ones the lowest transmissivity.

3.2.6 Summary of properties

Table A3.6 summarises the distribution, properties and their justifications used in the DFN model.

3.2.7 DFN model generation

Figure A3.44 shows two views of the DFN model and Figure 3.45 shows a simulated trace map of the North-west wall. Figure A3.46 shows a view of the DFN model as well the model upscaled into a continuum description of intensity (P32P) and permeability. Figure A3.47 shows a number of views with an increasing minimum fracture

size displayed and Figure A3.48 shows a view into the pit with just structures >100 m displayed.

3.3 DFN model validation

3.3.1 Orientation comparison

3.3.1.1 Introduction

A number of comparisons have been made between the simulated fracture orientations and actual orientations to test how well the DFN model has reproduced these data. These comparisons looked at (i) the total simulated fracture orientations vs the bootstrapping data set from North-west wall mapping, (ii) the orientation of simulated traces on the North-west wall vs the bootstrapping data set and (iii) all fractures intersecting boreholes in the model vs actual logged fractures.

3.3.1.2 Total fracture population

The overall fracture population was sub-sampled to 450 points to match the bootstrapping data. The stereonets shown in Figure A3.49 show the good reproduction of the strong sub-horizontal set, the ENE–WSW set, and with the weaker ESE–WNW set.

3.3.1.3 Trace map population

All fractures intersecting the North-west wall were identified and a trace map generated. From these intersecting fractures a sub-set of 450 fractures were randomly chosen and plotted on a stereoplot as shown in Figure A3.50. This shows the good agreement between the simulated and actual trace map derived orientations with both populations showing a strong sub-horizontal set and the ENE–WSW set and the weaker ESE–WNW set. The bootstrapped data were taken from North-west wall mapping data so this comparison shows the reproduction of the actual mapped results.

Table A3.9: Summary of DFN properties

Fracture property	Distribution	Definition	Justification
Fracture orientation distribution	Bootstrapped generation (i.e. pseudo-replicate sample)	Pit wall mapping data using Dispersion 80	The pit wall mapping data was chosen over the borehole data set as they show a lower degree of bias.
Fracture size distribution	Power law distribution	Exponent D = 3.12 minimum fracture radius generated = ~2.5 m, max 300 m	Determined from NW wall mapping data plus A21 underground.
Fracture intensity variation	Log normally distributed property	FracMan uses mean P10 intensity condition to control fracture generation	Fracture intensity conditioned to match mean fracture frequency P10 on all borehole intervals by sector, adjusted for minimum fracture size and T.
Transmissivity distribution	Log normally distributed	Correlated to size and elevation	Analysis of packer test data shows data are log normally distributed and reduce with elevation.

Visibility region set at 350 × 350 × 150 m

Fractures coloured by Size

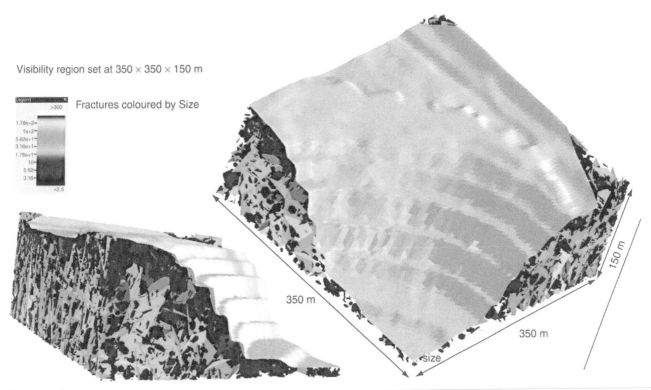

Figure A3.44: Cut away sections of the DFN model behind the pit shell with fractures coloured by fracture radius

3.3.1.4 Borehole intersection population

The last validation for the orientation data is from borehole intersections. Since the borehole logging data were not used to condition the DFN model, this gives an unbiased comparison between the model and actual data. Figure A3.51 shows the stereoplots for fractures intersecting borehole A154-MS-09 in the model (right) and logged (left). The modelled data capture all of the joint orientations well, although the intensity of simulated sub-vertical and sub-horizontal structures is respectively higher and lower than observed. Figure A3.52 shows the stereoplot for all fractures intersecting NW wall boreholes

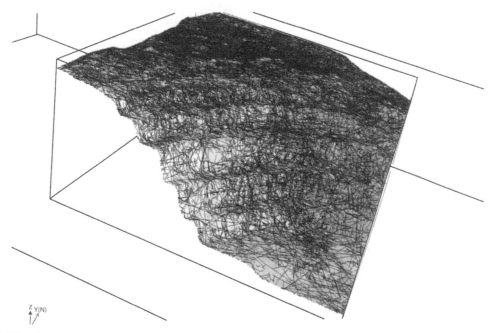

Figure A3.45: NW wall trace map

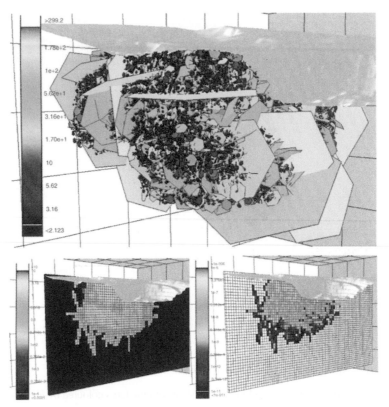

Figure A3.46: Different views of the DFN model from the south-west, looking north-east. DFN model with fractures coloured by radius (top); and east-west sections through grid properties of the model (bottom). Bottom left shows P32 (m2/m3) and bottom right shows equivalent permeability (m/s) broadly inversely related to elevation.

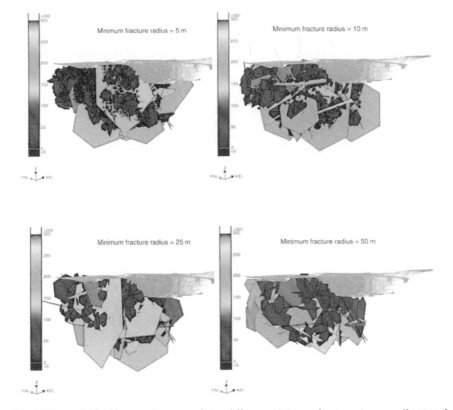

Figure A3.47: Views of the DFN model looking north-east applying different minimum fracture size cut offs. Note how as the minimum fracture size is increased, the network becomes sparser and connected at a larger scale.

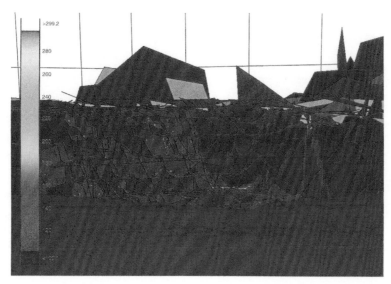

Figure A3.48: Simulated view into the pit of the north-west wall (looking towards the north-west) showing all the major structures >100 m

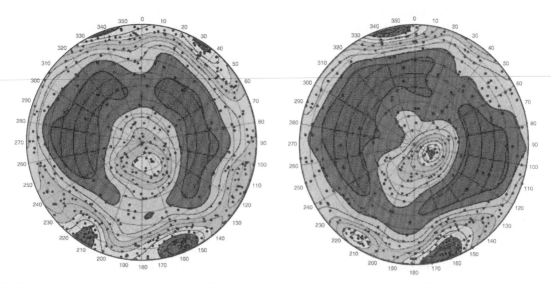

Figure 3.49: Stereoplot showing (a) sub-sample of total simulated population and (b) bootstrapping population

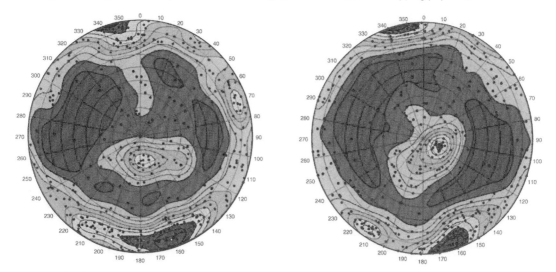

Figure A3.50: Stereoplot showing (a) sub-sample of fractures intersecting the pit wall and (b) bootstrapping population

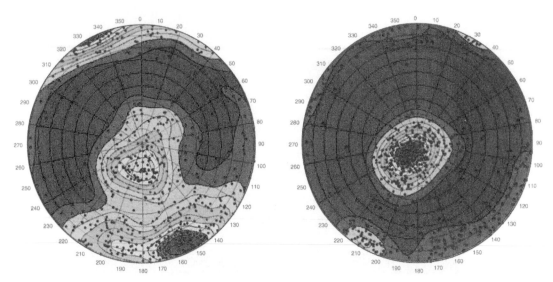

Figure 3.51: Stereoplot showing (a) A154-MS-09 simulated intersecting fractures and (b) A154-MS-09 logged fractures

and shows a similar trend of capturing all fracture orientations but with a stronger sub-vertical set and weaker sub-horizontal set.

3.3.2 Fracture size comparison

Fracture traces generated from intersection with the pit shell have been added to the actual mapping data from the A154 pit and also the A21 underground drift mapping. These data are plotted below in Figure A3.53 and show how the simulated mapping of the chosen power law fracture size has resulted in the accurate reproduction of the curve slope angle, and also shows the short and long length censoring effects common with all length data. The curves have not been normalised as the acquisition area was not recorded, which is why all of the curves are offset from each other.

To further explore whether the power law distribution of fracture size could reasonably reproduce key aspects of the

observed fracture trace distribution, simulated drift mapping was carried out by placing a 5 x 5 x 50 m long drift (box section tunnel) into the DFN model and deriving the trace map. When these data are plotted with the A21 underground drift mapping (normalised to match), it can be seen that now both the slope gradient and also the short and long length censoring effects also match (Figure A3.54).

Figure A3.54 shows that despite the actual underground data being best described with a log-normal distribution, the sampled power law distribution far better explains the observed data.

One potential issue with the size model used is the upper bound on fracture length. This has been fixed at a radius of 300 m but it is considered that this may result in too many longer structures being present. This can easily be fixed with the model by removing all very long fractures, but it raises the question of how best to define this limiting size when the measure of longer structures is

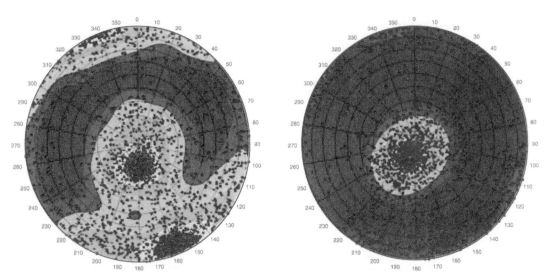

Figure 3.52: Stereoplot showing (a) simulated fracture log for all NW boreholes and (b) actual logged fractures for all NW boreholes

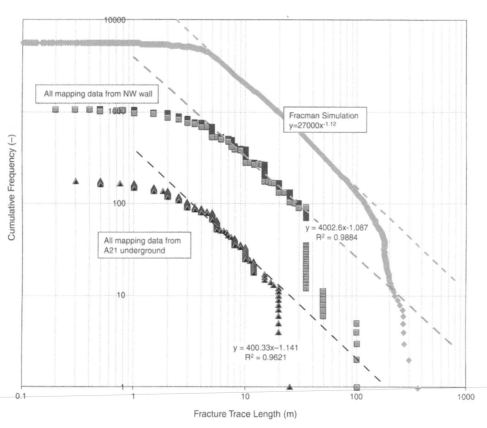

Figure A3.53: Fracture trace length distribution curves for logged data and simulated fractures. Curves are un-normalised by area hence their offset from each other.

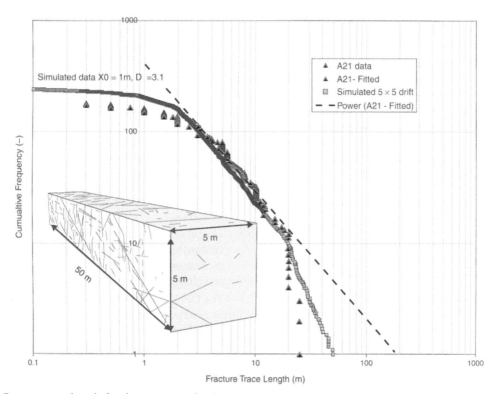

Figure A3.54: Fracture trace length distribution curves for the A21 underground drift mapping and simulated drift mapping with a 5 x 5 m drift. The simulated drift is inset in the bottom left hand corner.

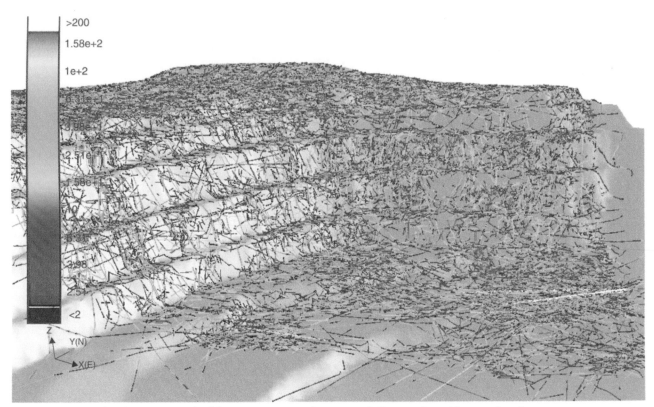

Figure A3.55: Simulated trace map with all fractures >150 m radius removed. Traces coloured by trace length.

difficult. An alternative trace map on the North-west wall has been generated with all fractures >150 m radius removed, Figure A3.55.

3.3.3 Fracture intensity comparison

The modelled fracture intensity has been tested by the 'drilling' into the model of all existing North-west wall boreholes and comparing the number of fractures observed in the simulation to the actual fracture count observed in the actual boreholes (adjusted for simulated fracture cut off sizes). Figure A3.56 shows the strong agreement between the simulated and actual fracture count for each P10 interval in the boreholes.

3.3.4 Fracture transmissivity comparison

Fracture transmissivity was tested by simulating hydraulic testing in the North-west wall boreholes. This has been undertaken using a geometric rather than a dynamic approach. This has been carried out as follows.

- Define for each borehole within the DFN model the top and bottom elevation of each hydraulic test interval.
- Sample the model to extract the transmissivity of all fractures intersecting the boreholes within a particular hydraulic test interval.
- Sum all fracture transmissivities within a simulated test interval to derive the total simulated test transmissivity.

The results of the simulated and actual hydraulic test transmissivities are shown below in Figure A3.57.

The majority of the data can be seen to be within approximately ± three-quarters of an order of magnitude of the 1:1 line, showing the strong agreement between the simulated and actual test results. A number of tests from higher elevations within the pit wall can be seen to have over predicted transmissivities as these particular tests do not fall on the transmissivity-elevation function shown in Figure A3.39.

It is unclear why these test results at higher elevations are anomalously low. A number of possible explanations have been suggested.

- Proximity to slope face resulting in some freeze-back influence.
- The presence of metasediment (biotite schist) associated within these higher elevation tests, resulting in the generation of excessive micaecous fines during drilling and clogging of the near-wellbore fractures.

The DFN model has been upscaled to equivalent directional permeability values to compare the permeability derived from the fracture model with the average hydraulic test results obtained for various elevation ranges. These results are shown in Figure A3.58.

The upscaling methodology uses an Oda tensor approach (Oda, 1984) and may over represent permeability where the network connectivity is low relative to the grid

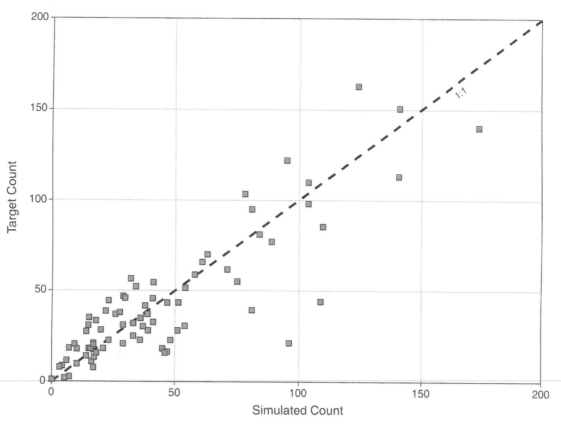

Figure A3.56: Graph of actual and simulated fracture count for each P10 intervals on the NW wall boreholes

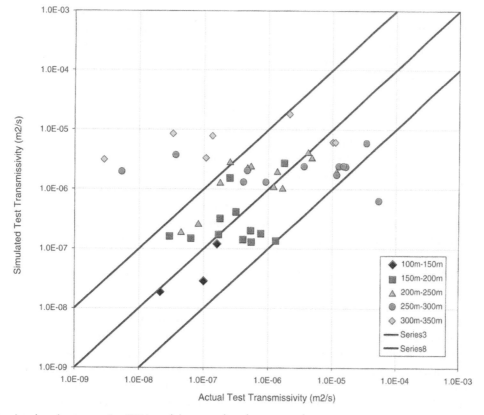

Figure A3.57: Simulated packer tests using DFN model vs actual packer test results

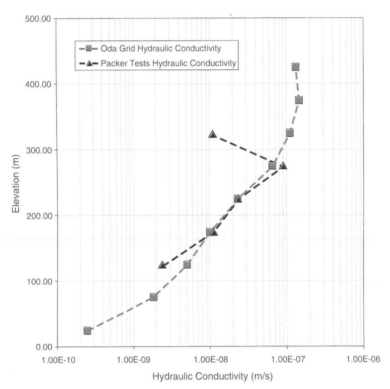

Figure A3.58: Upscaled permeability results lowered by a factor of 3 vs packer test hydraulic conductivity

size on which permeability is calculated. In the case of this model, a reduction factor of three in upscaled permeability was required to match the hydraulic test results.

3.3.5   Aperture comparison

Rather than directly using aperture data derived from logging, this has been derived indirectly from fracture transmissivity values for individual fractures using the formula:

$$\text{Aperture} = 0.5 \times T^{0.5} \qquad \text{(eqn A3.2)}$$

Using this approach, the DFN model has the sensible correlation of size-transmissivity–apertures, ensuring larger fractures are more likely to be wide and transmissive and smaller fractures to be narrow and poorly transmissive.

There are limited available aperture data from Diavik in general and specifically the North-west wall. However mechanical aperture data were recorded in three boreholes drilled through Dewey's Fault zone (DF16, 17 & 18). These data, as well as the simulated aperture distributions, are shown in Figure A3.59.

The distribution of simulated apertures (as calculated from transmissivity) and actual borehole data show that the data have the same general form (i.e. a log-normal distribution), but the calculated hydraulic apertures are approximately one order of magnitude lower than the mechanical apertures. This is in broad agreement with

observations on hydraulic vs mechanical apertures. As the permeability within Dewey's Fault zone is at least one order of magnitude greater than the country rock, this difference between simulated hydraulic and measured mechanical apertures is to be expected.

3.3.6   Fracture pathways

The DFN model has been interrogated to identify the most conductive pathways between a series of source-observation borehole pairs where actual interference responses have been detected, Figure A3.60. FracMan identifies the most conductive pathway by searching through the fracture network between a source and sink on a fracture by fracture basis to find the most conductive or geometrically shortest pathway. Whilst it has proven difficult to interpret the pathway results, it is apparent that the long and convoluted pathways between the source and sinks can help to explain the variable interference responses.

The basic hydraulic results, as well as DFN search results for all pathways from a signal at A154-MS-29, are shown in Table A3.10.

3.4   Summary

- A number of realisations of the Diavik North-west wall DFN model have been generated. These multiple realisations represent a reasonable description of the fracturing within the granitic rock mass and, by validation testing, have been shown to adequately

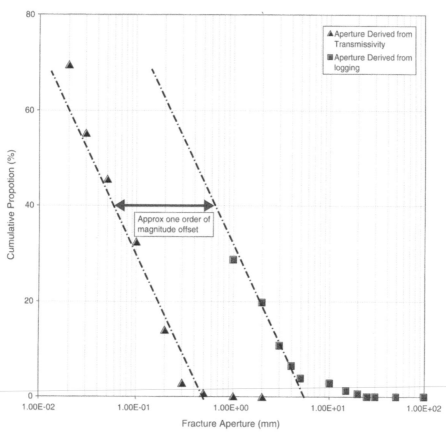

Figure A3.59: Comparison between logged aperture values from the Dewey's Fault area of the A154 pit and simulated apertures, calculated from transmissivity

Table A3.10: DFN simulation results and hydraulic test data for A154-MS-29 interference tests in the NW wall

Stress point	Observation point	DFN simulation			Hydraulic test data		
		Path length	No of fractures	Path conductance	Drawdown [m]	Norm drawdown [m] (d/r)*1E4	r
A154-MS-29	A154-MS-02A	3073.23	88.00	−522.33	0.03	1.54	195.20
A154-MS-29	A154-MS-02A	699.14	11.00	−232.23	0.03	1.54	195.20
A154-MS-29	A154-MS-02B	2686.36	39.00	−1164.30	0.04	2.80	142.83
A154-MS-29	A154-MS-02C	1007.65	10.00	−623.64	0.54	34.76	155.35
A154-MS-29	A154-MS-08A	1332.03	20.00	−1421.88	0.37	64.68	57.21
A154-MS-29	A154-MS-08B	1460.90	27.00	−1177.03			21.59
A154-MS-29	A154-MS-08C	3643.96	77.00	−220.71	9.10	3098.01	29.37
A154-MS-29	A154-MS-08D	4831.50	103.00	−220.71	2.60	388.74	66.88
A154-MS-29	A154-MS-09A	513.98	10.00	−700.20	0.31	62.95	49.24
A154-MS-29	A154-MS-09B	28.77	1.00	0.00	0.84	301.09	27.90
A154-MS-29	A154-MS-09C	2478.31	37.00	−1194.93	10.27	2031.01	50.57
A154-MS-29	A154-MS-09D	1914.92	23.00	−769.71	0.26	29.51	88.11
A154-MS-29	A154-MS-10B	65.87	1.00	0.00	0.57	86.04	66.24
A154-MS-29	A154-MS-10C	3396.12	82.00	−220.71	0.84	112.74	74.51
A154-MS-29	A154-MS-10D	2402.58	35.00	−1048.00	1.20	123.30	97.33

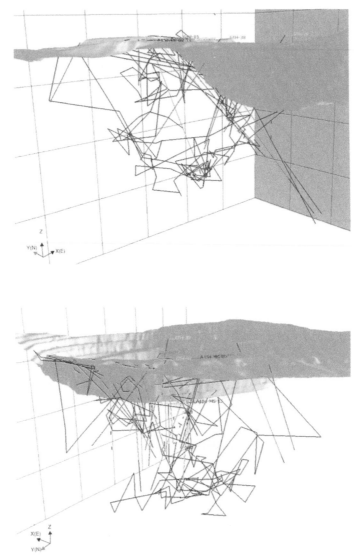

Figure A3.60: Mapped inter-borehole pathways. Upper model looking north-east and lower model looking south-east.

describe the distribution of fracture intensity, fracture size and fracture orientation within the rock mass.

■ There are two outstanding issues with the parameters used for the DFN model:

→ The orientation definition was derived from a bootstrapping technique using bench mapping data as the borehole derived data were highly biased, with an excessive frequency of low-angle structures. When the validation testing was undertaken, it was impossible to fully replicate the borehole orientation data using the bench mapping derived population although the general pattern of fracture sets was reproduced. There is a requirement for a work flow to be developed to demonstrate the appropriateness and accuracy of one orientation data set over another, to be utilised in order to ensure the reasonableness of a selected data set;

→ Whilst it is considered that the power law size distribution used in the DFN model is reasonable, there is limited constraint on the upper size limit. The obvious difficulties in measuring long structures within the pit when mapping is generally performed on a bench are well known. Thus the maximum fracture size does not correspond to the maximum identified trace length. A maximum fracture radius of 300 m was used that results in a theoretical maximum trace length of 600 m. This may be too long, although these larger fractures can be easily removed from the DFN model.

■ A key argument against the use of power law distributions being used to describe trace lengths is the reluctance to see a single physical process being responsible for fractures of all scales when it is known that multiple processes, potentially with multiple phases of stress are involved (e.g. explosive emplacement, transpressional fault movement, regional deformation, weathering). However, irrespective of the process argument, power law distributions are often acceptable for reproducing the main aspects of the observed size data.

- Each stochastic DFN model is equi-probable but different. Each model conforms to exactly the same general form, but is of different shape. This non-uniqueness becomes problematic when there are no deterministic major structures (as in the case of Diavik). If there are observed but not mapped long structures, the DFN model needs to be able to simulate long length scales in the order of 100s of metres. Thus the DFN model is generating a small number of very large structures that can act as potential failure planes or conductive pathways that will be in completely different locations and orientations between realisations. As a result, the discrete simulation of either flow or stability with differing DFN models could (and most likely will) result in very different outcomes. Using the upscaled properties (i.e. the conversion of discrete to equivalent properties, e.g. fracture permeability converted to equivalent permeability) will result in a more accurate description of the system with less extreme variation between realisations. If the longer structures are deterministically added, this produces a more realistic description of the fracture system.
- The effects of non-uniqueness can be reduced through better data acquisition and rock mass characterisation. Modern photogrammetry has provided a new way to do old work rather than providing a route to do new

analysis. If improvements could be made in the way that the results of photogrammetry are integrated into DFN models, making them more deterministic (i.e. placing structures where they really are, particularly large ones), the stochastic variability would diminish and the model behaviour would increasingly converge towards the real system.

- Few DFNs have been coupled to simulate hydrogeological and mechanical effects. If a DFN model is being constructed with a mechanical end use, it will be built with a smaller minimum size than if it is being constructed for a hydraulic application. The use of the *Slope Model* software requires a consideration of coupling, recognising that there is a minimum transmissivity and a minimum size that needs to be respected to ensure the model behaves correctly. Thus features <10 m might be hydrogeologically less significant, but mechanically are most likely important.

Fracture intensity is clearly related to rock mass strength, albeit perhaps in a less linear way than is often thought. The more fracture area in a given rock volume, the weaker the rock mass as a whole and the DFN property P32P (fracture area/volume) appears to be a good predictor of behaviour. However, a small number of key structures with low friction angles or stiffness can significantly weaken the rock more than a large number of smaller structures. This is similar to the situation involving flow, where one homogenous fracture with an aperture of 10 will have 100 times more flow capacity than 10 homogenous fractures each with an aperture of one, assuming the cubic law relationship.

Appendix 4

Case studies: Escondida East wall; Chuquicamata; Radomiro Tomic; Antamina West wall; Jwaneng; Cowal; Whaleback South wall; La Quinua (Yanacocha)

The purpose of the case studies is to provide supplemental information to support a methodology for developing a sector-scale conceptual hydrogeological model. The main research was carried out using data from the North-west wall of Diavik, as described in Section 3.3 and Appendix 3. The sites that are discussed in this section are intended to provide additional data for a range of rock types and a range of hydrogeological settings. Summaries of the case studies are presented in Section 3.5.

1 Escondida East wall

Manuel Rapiman, Rick Tunney, Rodrigo Gonzales, Geoff Beale, Rodrigo Quinteros and Daniela Alejandra Abarca Argomedo

Setting: Porphyry copper; situated adjacent to alluvial basin; with near-zero precipitation.

1.1 Background

Escondida is a large supergene-enriched porphyry copper deposit located in the Atacama Desert of northern Chile, some 160 km to the south-east of Antofagasta. The Escondida deposit falls within the 50 km wide Cordillera de Domeyko morpho-tectonic province of the northern Chilean Andes. The Escondida mine has been in operation since 1990. The average topographical elevation of the edge of the pit is about 3060 masl. Currently, the pit floor elevation is approximately 2550 masl.

The small and rare occurrences of precipitation do not contribute any significant recharge to the groundwater system. The majority of groundwater storage in the area occurs within tectonic depressions forming alluvial sedimentary basins ('salars'). As the pit has been progressively expanded and deepened by a series of pushbacks, so the East and South-east pit walls have been mined closer to saturated sediments that fill the Salar de Hamburgo basin.

1.2 Geology and hydrostratigraphy

The geology comprises a quartz-monzonite to granodiorite porphyry stock complex cutting Paleocene andesites with very little primary porosity. Groundwater flow within the bedrock is typically via secondary permeability associated with faults and fractures. The permeability of the ore body rocks is tightly controlled by the type of alteration and mineralisation, and by the distribution of fracturing in the major fault systems which are oriented N–S, NW–SE and NNE–SSW.

The primary ore body is overlain by a highly enriched leached cap of altered porphyry, which extends from the surface to depths of up to 300 m. Extensive drilling through the leached cap material has rarely encountered any discrete high-permeability fracture zones. These have become 'healed' as part of the alteration and weathering process. The softening of the rock in this zone has lead to some development of primary porosity, which has resulted in increased groundwater storage characteristics and more homogeneous hydraulic properties. The initial yield of wells and drain holes within the leached zone is typically 0.05 to 0.2 l/s, and the water is normally derived from multiple zones in the drill hole.

The deeper 'fresh' porphyry has a lower permeability rock mass and groundwater flow is predominantly controlled by the fracture systems. Discrete fracture zones of moderate to high permeability are regularly encountered, particularly within zones of potassic alteration. The yield of wells and drain holes within the fresh Escondida porphyry can vary between <0.01 and >2 l/s and the water is normally derived from one or two discrete fracture zones in the drill hole.

In the Hamburgo basin, 300 to 1500 m to the east of the open pit, variably cemented alluvial and colluvial deposits of gravel and sands form a groundwater storage reservoir that was originally up to about 300 m in thickness. The average drainable porosity value of the basin sediments is about 0.09, so they contain a considerable quantity of

Table A4.1: Hydraulic conductivity estimates of hydrostratigraphical units

Hydrogeological unit	Type of test	Hydraulic conductivity estimate (m/s)
Primary porphyry	Packer test	$2.3 \times 10^{-10} - 9.6 \times 10^{-8}$
	Falling head	$1.3 \times 10^{-9} - 6.1 \times 10^{-6}$
Highly enriched leached cap	Packer test	$1.3 \times 10^{-9} - 6.6 \times 10^{-8}$
Blast-damaged zone	Packer test	$1.4 \times 10^{-7} - 4.2 \times 10^{-6}$

groundwater in storage. In addition, a number of tailing dams are also located above the basin immediately to the east of the pit. These have recharged the basin system and have acted to 'top up' the groundwater system. Prior to implementing a dewatering system that pumps directly from the basin, groundwater levels in some areas of the basin had risen by more than 30 m from pre-mining conditions.

The rocks immediately underlying the basin sediments generally have very low permeability and storage, so their water-bearing potential is limited. However, layers of rhyolite porphyry rock occur between the East wall and the Hamburgo basin. These may create fractured and transmissive zones down to an elevation of about 2870 masl (190 m below ground). The rhyolites allow transmission of groundwater from the basin towards the open pit, but they do not occur directly within the current East wall.

Hydraulic testing of boreholes drilled for monitoring has enabled characterisation of hydraulic properties of the pit lithologies. Results of hydraulic testing for the main hydrostratigraphical units defined for pore pressure simulations of the pit walls (WMC, 2007) are shown in Table A4.1.

1.3 Dewatering and depressurisation

The hydrogeology of Escondida has been studied in detail over a period of about 20 years. The investigations have shown that the majority of the groundwater that is pumped from the pit for dewatering and slope depressurisation is ultimately derived from the Hamburgo basin. Because of this, a field of 30 to 50 interceptor wells pumping 300 to 600 l/s directly from the basin is operated to lower basin groundwater levels and reduce the amount of recharge flowing from the basin towards the pit. The goal of the interceptor wellfield is to reduce the amount of water entering the in-pit slope depressurisation systems and consequently make slope depressurisation more efficient.

In addition to the Hamburgo wellfield, Escondida employs the following dewatering and slope depressurisation measures.

- Bedrock interceptor wells around the North-east, East and South-east pit crest to remove residual groundwater flowing from the Hamburgo basin mostly in the shallow rhyolite porphyry layers.
- Horizontal drain hole arrays to reduce pore pressures in localised zones within the pit face. The mine typically maintains one horizontal drain drill operating full time.
- A small number of in-pit dewatering wells to locally reduce pore pressures in conjunction with the horizontal drain programs, where higher yielding fracture zones have been identified during test drilling.
- Drain hole fans installed from a drainage tunnel built behind the NE wall in 2001 and extended in 2008. The maximum yield from the tunnel drains was about 14 l/s during and immediately following initial construction. The current flow rate is about 7 l/s.
- Sumps in the pit floor, and also at some locations higher in the East wall.

Residual seepage into the pit mostly occurs in the NE, E, SE and South walls. However, there is also some seepage to the SW and W walls (the opposite side of the pit to the Hamburgo basin). The seepage is partly related to residual groundwater storage in the bedrock. However, groundwater recharge from the Hamburgo basin is also transmitted to the SW and W sectors along N–S trending structures behind the West wall.

Over the past six years, Escondida has significantly increased the number of horizontal drain holes drilled into the E and SE walls to help reduce pore pressures that are sustained by residual groundwater flow from the Hamburgo basin that is not captured by the interceptor system. Initial yields of individual drain holes may be >2 l/s; however, the individual drain yields decline from their initial values. The total groundwater flux intercepted by the East wall (drains plus seepage into the slope) is generally within the range 10–15 l/s.

Regular monitoring of the drain hole program has shown that:

- for each successive drain array there is a relatively rapid decline in total head and pore pressure as the drains remove water stored within the rock in their area of influence;
- as the groundwater heads fall locally, so the yield of the drains reduces from the initial values;
- as the drains' yields and total heads decline, so the system re-equilibrates to a new condition, until the next horizontal drain array is installed;
- the effect of the drains is superimposed on an overall slow decline in groundwater heads that is caused by ongoing seepage as new mining pushbacks are advanced downward in the slope.

1.4 Conceptual hydrogeological model

Prior to the operation of the pit, the natural groundwater level across the Escondida deposit was 2990 masl, with virtually no hydraulic gradient, which reflects the lack of any significant recharge. Heads along the East side of the mine were locally increased to about 3000 to 3020 masl because of the recharge caused by the tailing facilities. Since 2001, total heads in the pit area have been progressively lowered by the dewatering system. Current heads along the crest of the East wall typically range from 2920 to 3000 masl, with a strong downward head gradient. Current heads at the toe of the East wall are 2580 to 2620 masl, generally with an upward head gradient.

The East wall is dominated by a set of N–S (or NNE–SSW) trending faults and joints. These may be present as *en echelon* shear structures and fractures which open up through shearing movements within this zone. Monitoring of groundwater responses has confirmed there is a general anisotropy in permeability along the N–S trend. Faults of this trend are parallel to the slope and tend to act as 'dams' perpendicular to the overall hydraulic gradient from Hamburgo towards the pit crest, and to the toe of the slope. Prominent discrete NW–SE trending fault zones are also present in the slope. Three of these have been identified as prominent drainage conduits. They also define boundaries of hardness, mineralisation type, alteration and copper grade.

The location of recent piezometers and horizontal drain holes in the East sector of the pit is shown in Figure A4.1. Piezometer hydrographs for the East wall are shown in Figure A4.2. Figure A4.3 shows a comparison of conditions between the East wall (with high recharge) and the North-west wall (with lesser recharge).

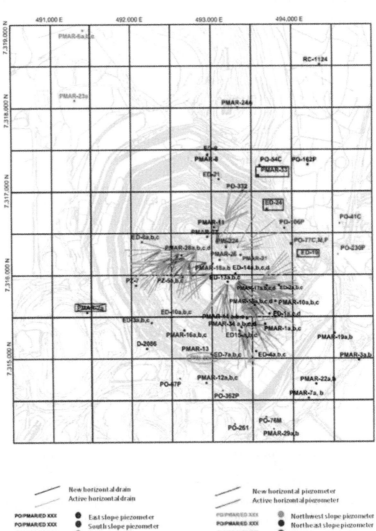

Figure A4.1: Locations of selected piezometers and horizontal drains

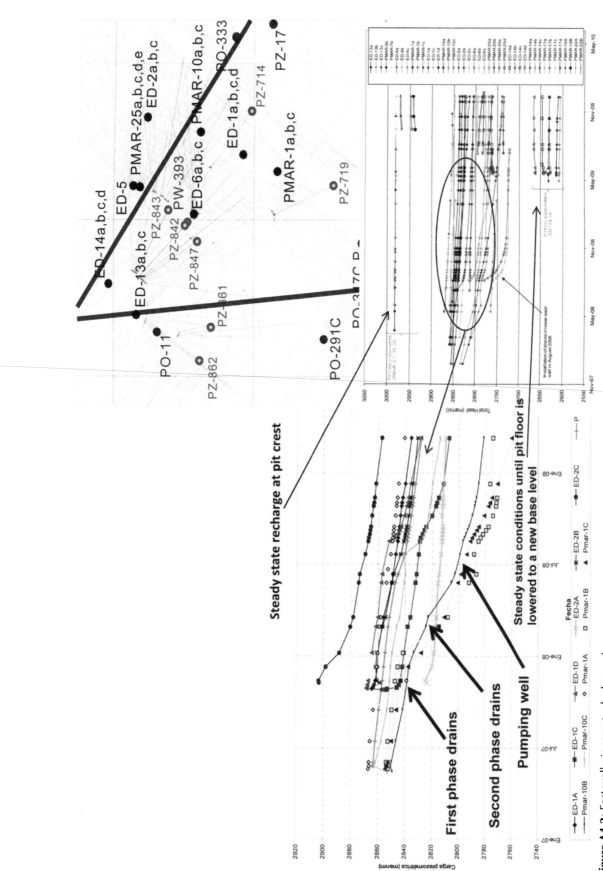

Figure A4.2: East wall piezometer hydrographs

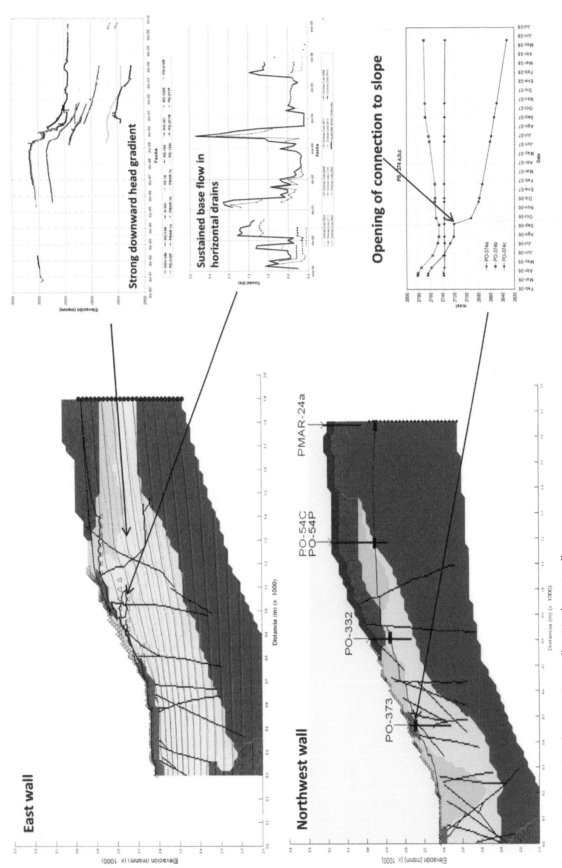

Figure A4.3: Comparison between East wall vs North-west wall

The East wall hydrographs show the following trends.

- Within the period shown, the 'pit floor' piezometers indicate minimal recent depressurisation. The piezometers indicate an upward hydraulic gradient. Because the piezometers are installed from the current pit floor out from the toe of the slope, they are below the level of drain drilling in the slope above.

- An upward hydraulic gradient also generally exists in the mid-slope piezometers, with head differences of around 20 to 60 m between the deep and shallow sensors. However, in some cases, the drain drilling program has reduced or reversed the upward head gradient.

- A strong response to drain installation, which causes around 50 m drop in head, is evident in piezometer ED-6 at mid-level within the wall. The upward hydraulic gradient between the upper two sensors in the hole is reduced, but the lower piezometer in the hole still records higher total head values.

- The upper-slope piezometers show very different pressure trends to the mid-slope piezometers, with strong downward vertical gradients indicated. Piezometers ED-4 and PMAR-1 show a head difference of 100 to 120 m between the shallow and deep sensors. In piezometer PMAR-10, the upper two sensors show a downward gradient, whilst the deeper sensor shows higher heads.

- Pit-crest piezometers show much higher heads than the mid-slope and upper-slope piezometers, with minimal long-term depressurisation. The piezometers respond more to the influence of the interceptor wells than the horizontal drain arrays. Observed hydraulic gradients at the pit crest are strongly downward.

Figure A4.4 is a schematic diagram illustrating the conceptual model of how depressurisation occurs in the Escondida East wall. In the upper-level 'pit crest' piezometers, only a limited amount of depressurisation has occurred because of the influx of recharge from the Hamburgo basin. These piezometers respond more to interceptor well pumping than the horizontal drains in the slope below. In addition, many of the upper-level sensors in each hole are installed within the more porous (higher storage) upper leached zone. Because fracturing in this zone is largely healed, direct rapid flow paths to the horizontal drain installations do not occur.

The downward hydraulic gradient shown by the mid-slope piezometers reflects a combination of (i) recharge moving down the slope at shallow levels (say the upper 50 to 100 m behind the slope) sustaining heads in the shallow installations, and (ii) the results of successive

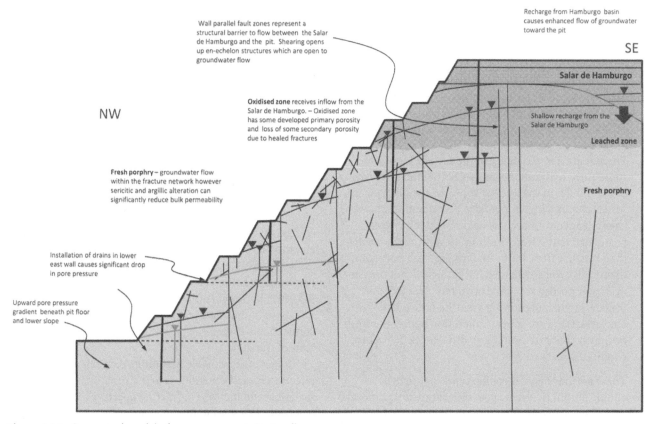

Figure A4.4: Conceptual model of pore pressures in East wall

horizontal drain arrays reducing heads in the deep installations. At some point about mid-slope, the vertical component of the hydraulic gradient changes from downward to upward.

The upward hydraulic gradient in the 'pit-floor' piezometer installations is mostly the result of the re-adjustment of heads following excavation of the lower slope. As the lower slope is progressively mined downward, it causes heads to drop within connected fractures at shallow depths behind the newly excavated slope. Deeper fractures are much less connected to the active flow system and therefore the deeper sensors take a much longer time to equilibrate.

1.5   Discussion

The over-riding factor controlling pore pressures in the East wall of the Escondida pit is recharge. However, Escondida has three important differences to the Diavik case study.

1. It is possible to greatly reduce recharge to the slopes by a combination of (i) lowering the general heads in the Hamburgo basin by pumping, and (ii) interception of residual recharge by interceptor wells installed in fractured rhyolite porphyry layers above the crest of the slope. The combination of these two factors has greatly reduced the recharge flux reaching the pit. Because of this, the yields than can be achieved with horizontal drains, in-pit pumping wells and underground drains can create significant localised depressurisation and 'make a difference' to the pit slope pore pressures.

2. In general, the yields achieved from horizontal drains at Escondida are higher than at Diavik. Short-term drain yields of >1 l/s can be achieved on a routine basis. Over 50 per cent of all drains in the fresh porphyry exhibit a yield that is high enough to achieve some measurable depressurisation (>0.2 l/s).

3. There is greater variability in the material types and consequent hydraulic behaviour of the fracture systems at Escondida. In the leached zone, the rock is typically softer, attains a higher porosity, prominent fractures have a tendency to become healed, and it has been postulated that porous-medium flow is more prevalent than fracture flow. In the underlying fresh porphyry, the rock is typically harder, has no significant primary drainable porosity, and the open fractures that are present produce significant yields to drains and wells. However, there are zones of alteration within the fresh porphyry that can create large differences in fracture conditions, as described below.

There are four main alteration types within the Escondida porphyry system: potassic; chloritic; sericitic; and argillic. There are large differences in fracture conditions and permeability between the alteration types. Overall, the horizontal drain yields are higher in zones of potassic alteration, and are generally lower in zones of sericitic and argillic alteration. However, in all the pit-wide piezometers that have been monitored over the past 15 or so years, there is no indication of any different long-term drainage response in the zones of sericitic or argillic alteration.

Throughout most pit slope sectors at Escondida there is a general downward total head gradient in the upper part of the slope and an upward gradient in the lower part of the slope. A similar pattern is observed at many large open pits. At Escondida, the downward head gradient in the upper slopes is strengthened because of (i) the residual recharge that is reaching the upper slopes, (ii) the presence of the poorly fractured and lower permeability leached (oxidised) porphyry in the upper slopes, which is seen to act as a 'header tank' in some locations, and (iii) the drainage of the underlying fresh porphyry that is being achieved by the horizontal drains and in-pit wells.

Field observations made in the NE wall drainage tunnel suggest that 'en masse' flow of groundwater is occurring through pervasive fracture systems that have developed within the Escondida porphyry rocks. Again, this is similar to that postulated for Diavik. (It should be noted that observations in the drainage tunnel are perhaps more applicable than those at the pit wall, which are influenced by the 'over-break' zone). There are no piezometers at Escondida that reflect 'D-type' fractures or an isolated rock mass. Where heads have been slow to respond to drainage, it is typically because (i) there are structures at adverse orientation to the slope creating 'dams' across the prevailing hydraulic gradient or causing 'compartmentalisation' of the groundwater system, and (ii) there is a high rate of recharge to that particular sector of the slope.

In addition to the piezometers that show a direct response due to horizontal drains, there are also piezometers that show a connection to the pit slope 'over-break' zone caused by blasting. Figure A4.3 includes a hydrograph of three piezometers installed in drill hole PO-191 on the North-west wall of the pit. The deeper piezometer in the set (PO-191P) shows a rapid response starting in October 2007. The sensor is installed at an elevation of 2705 masl (which is about the same elevation as the lower N12 wall at the time) and is about 165 m horizontally behind the slope. The drop in pressure in PO-191P is co-incident with a blast in the lower N12 wall. It is thought that the blasting created a drainage connection between the piezometer and the 'over-break' zone of the slope.

None of the available Escondida data sets indicate any effects that can be attributed to deformation of the slope materials and increase in fracture aperture. Any short-term depressurisation that may occur because of a change in fracture porosity due to unloading becomes quickly dampened by groundwater flow, and is not evident in any of the observed piezometer responses to date. Overall, it is thought the fracture permeability is too high for these effects to be evident.

2   Chuquicamata

Jorge Rodriguez, Eric Lopez, Luis Olivares, Paul Mansfield and Rodrigo Araya Gaete

Setting: Copper porphyry with distinct alteration types; near-zero precipitation.

2.1   Background

Chuquicamata is one of the two largest open pit copper mines in the world. The mine is located in the north of Chile, 215 km north-east of Antofagasta and near the city of Calama. The pit rim is at an elevation of about 3000 masl (Figure A4.5). The current pit floor is about 2100 masl.

Chuquicamata is also located in the Atacama Desert and receives negligible precipitation. However, unlike Escondida, the site is removed from the influence of any natural groundwater recharge. The only recharge that does occur is due to historical facility leakage, mostly from the old oxide plant above the South rim of the pit. This is estimated to be between 3 and 8 l/s, concentrated on the South wall. Pre-2002 hydrogeology programs focussed on the management of this recharge, and included:

- development of a drainage tunnel in the South wall (K1-S1). Drains drilled from the tunnel produced a combined yield of between 1 and 2 l/s;
- three dewatering wells south-west of the pit (PW-32, PW-33 and PW-34). The combined yield of these wells was about 7 l/s;

- six pit floor dewatering wells, with a combined flow rate of around 2 l/s.

Of these installations, only the K1-S1 tunnel is still in operation.

2.2   Geology and hydrostratigraphy

The ore body has evolved from the initial intrusions, through mineralisation (last major hydrothermal event at 31 Ma), to post-mineral brecciation and offset by the West fault system. The West fault system demonstrates post-mineral displacement that separates the largely mineralised Chuqui Porphyry to the east and the barren Fortuna complex to the west. The geotechnical units within the pit are largely determined by alteration type. They are summarised in Table A4.2 and shown in cross section in Figure A4.6, as follows.

- East wall. The upper wall is composed of the Granodiorite Elena (GES/GEN) complex. To the west of the granodiorite, the middle and lower wall moves progressively through north–south trending zones of metasediments (MET), chloritic alteration (PEC), potassic alteration (PEK), sericitic alteration (PES), with quartz-sericite (RQS) occurring in the lower wall and pit floor.
- West wall. The upper and middle wall is entirely composed of Granodiorite Fortuna rock (GDF). To the east of the West fault, the lower wall is comprised of shear zone material (ZCM, ZCI, BEF), with RQS at the lowest levels.

An intensive program of hydrogeological investigation, characterisation and monitoring has been carried out since 2003. Piezometers and drain hole installations are shown as Figure A4.5. The program was split into three stages, as follows.

Table A4.2: Geotechnical units at Chuquicamata

Geotechnical unit	Lithological unit	Alteration	Sector	Rock strength (according to IRS)
GDF	Granodiorite Fortuna	Weak supergene	West slope	Very strong
ZCM	Granodiorite Fortuna	Moderate shear attached to the West fault	West slope	Moderately strong
ZCI	Granodiorite Fortuna	Intense shear attached to the West fault	West slope	Weak
BEF	Undifferentiated	Tectonic breccia in West fault trace	East and West slopes	Strong and moderately strong
RQS	East porphyry	Quartz sericite	East slope	Strong and moderately strong
PES	East porphyry	Weak sericite	East slope	Strong and moderately strong
PEK	East porphyry	Potassic feldspar	East slope	Very strong
PEC	East porphyry	Chloritic	East slope	Strong and very strong
MET	Sedimentary and volcanic	Contact metamorphism	East slope	Strong and moderately strong
GES	Granodiorite Elena	Weak chloritic	East slope	Strong
LIX	All	Leached rock	East slope	Strong and moderately strong

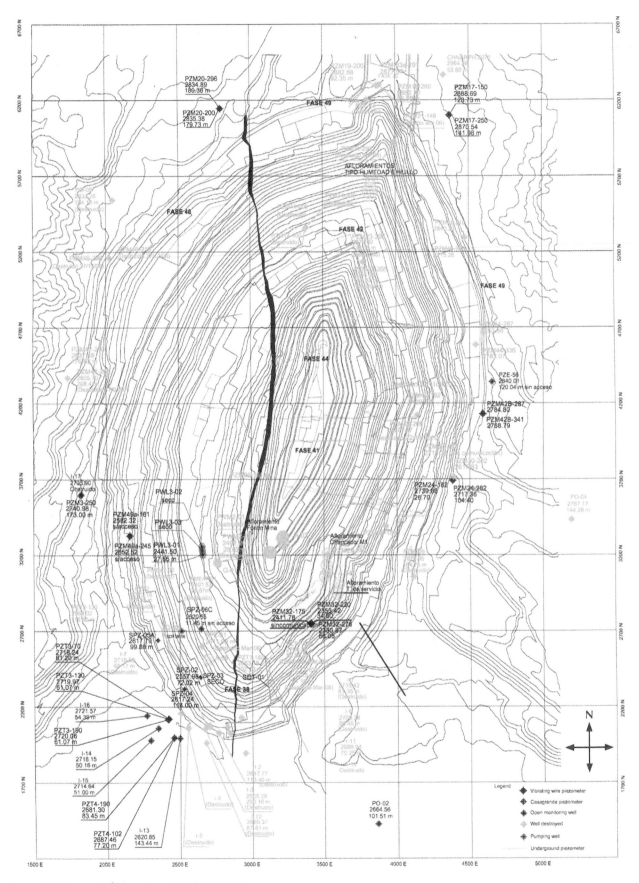

Figure A4.5: Layout of Chuquicamata piezometers

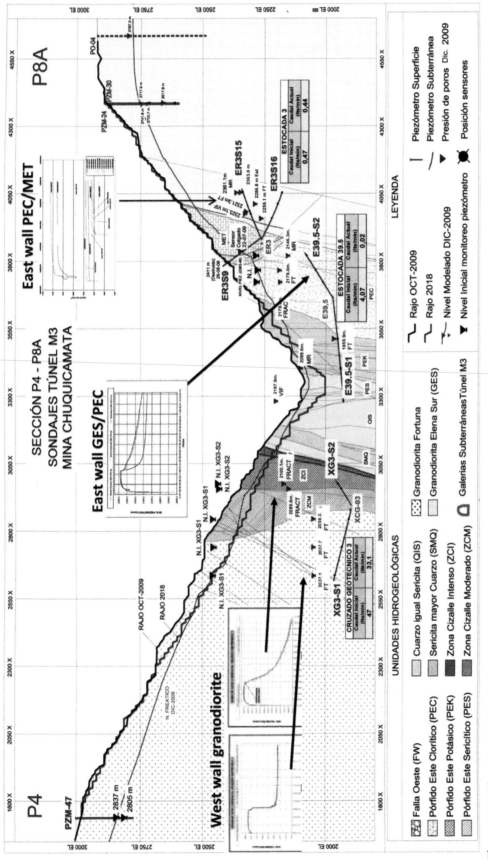

Figure A4.6: Chuquicamata Section P4–P8A

1. Stage 1 (through 2007) involved the installation of 29 hydrogeology test holes from the surface, including the installation of 42 VWP sensors. A total of 124 borehole packer tests was carried out as part of this work. In addition, 43 drain holes were installed from underground development behind the lower East wall. A hydrogeological conceptual model for the pit was developed.

2. Stage 2 (2008–2009) involved 28 new underground piezometers hole with 85 grouted-in VWP sensors, together with 68 addition drain holes.

3. Stage 3 involves additional underground piezometers and drain holes and was ongoing at the time of this case study.

The permeability data are summarised as follows.

- The overall calculated permeability values range from 2×10^{-12} to 1×10^{-5} m/s, with a geometric mean of 1×10^{-9} m/s.
- The geometric mean of the values of permeability in the East wall (Chuqui Porphyry) is 5×10^{-10} m/s (based on 208 tests).
- The geometric mean of the values of permeability in the West wall (Granodiorite Fortuna) below 100 m depth is 1×10^{-8} m/s (based on 52 tests).

In addition, with reference to specific units:

- many units such as PES, PEK and PEC show decreasing permeability within the first 100 to 150 m depth;
- PEC and PEK units show very low permeability, generally lower than 10^{-9} m/s, decreasing to below 10^{-10} m/s with depth (deeper than approx. 100–200 m);
- PEK is the highest quality rock from a geotechnical perspective with RQD of greater than 90 per cent, but the permeability of the unit is highly variable with values ranging from 10^{-12} to 10^{-8} m/s, suggesting important structural control;
- the GES unit, with permeability of around 1×10^{-9} m/s, does not show decreasing permeability with depth;
- the GDF (West wall) does show decreasing permeability with depth, but the permeability values are higher, with geometric means that range from 2×10^{-6} m/s within the upper 100 m and 1×10^{-8} m/s below 100 m.

2.3 Drain hole results

A total of 111 underground drain holes had been drilled up to the time of this case study (including underground mineral exploration holes that have been converted into drains). Of these, about 71 were drilled into the upper to middle units of the East wall, and 40 were drilled into the lower East wall and pit floor. In 2008/2009, three east–west cross cuts were developed beneath the pit floor at about the 1900 masl elevation. Drain holes were drilled from the end of the cross cuts to penetrate the West fault and intersect the GDF.

Overall, the drain holes results can be summarised as follows.

- Drainage flow rates in the RQS, PEC, PEK and PES (East wall) have been very low. Sixteen of the 28 original drains produced measurable flow, of which five have produced sustained flow.
- The combined flow rate of the 28 original drains was 0.0043 l/s (as on 18/06/09). Much of the flow originates from the central part of the wall (PEK).
- Drains drilled at lower elevations produced variable but somewhat higher yields, with an average flow of 0.115 l/s (as on 18/06/09). Much of the flow is derived from fracturing within the RQS, where residual recharge from the South wall is indicated to be moving northwards along N–S trending (West fault parallel) structures (mostly within the RQS).
- The drains drilled from the cross cuts to depressurise the lower West wall yielded individual short-term inflows of up to 0.55 l/s, and a peak combined flow of 1.72 l/s. This relates to the higher permeability of the GDF.

2.4 Conceptual hydrogeological model

2.4.1 General

The head distribution is unusual for an open pit mine because it shows a pervasive downward head gradient in all sectors of the pit, including the lower slopes and the pit floor. The exceptions to the pervasive downward gradient are (i) the southern part of the pit floor below the area of active recharge, and (ii) the lower part of the GF immediately upslope from the West fault. The following factors are considered important for controlling the total head distribution.

- The very low permeability and porosity of all units and the tendency for permeability and fracture porosity reduction with depth.
- Lithostatic unloading and deformation of the material remaining in the slopes due to progressive removal of material from the pit walls as successive pushbacks have been mined down. The influence of lithostatic unloading is seen in all units of the East wall because of the general low permeability, but is observed to be more of a factor for the MET, PEC, PEK and PES units, where the permeability is typically lower than 10^{-9} m/s.
- Localised head reductions due to drainhole inflows, most notably in the RQS, GDF and PEK units, where the drain holes typically show comparatively higher yields.
- Recharge from facilities at the south end of the pit which moves northward in the GDF behind the West fault, and in N–S trending structures in the RQS in the lower pit. The northward-moving groundwater is considered to be the reason for the upward total head gradient that occurs in the south part of the pit floor.

- The hydrogeological barrier created by the West fault and associated shear zone. In addition to recharge moving northward on the upslope side of the fault, groundwater moving slowly downslope in the GDF is forced to the surface by the fault zone. Seepage from above the fault zone is observed at a number of locations. The seepage discharges by evaporation (and formerly by pumping from PW-32, 33 and 34).

The available dataset indicates that deformation and increase in fracture aperture is the primary reason for the observed head distribution. As unloading of the slopes takes place, so fracture apertures increase and pore pressures drop. In most sectors of the pit, the lack of recharge and very low permeability and fracture porosity values means there is insufficient water available to allow any significant rebound in heads between active phases of mining.

2.4.2 Pervasive downward head gradient

The pervasive downward head gradient can exceed 0.6 m/m in certain parts of the pit (for example in VWP PZM-30). The downward gradient also occurs below the central and northern parts of the pit floor which, prior to installing cross cuts and tunnels for the deep sulphide project, was the lowest hydraulic point for over 40 km distant. Therefore, the downward gradient below the pit floor cannot be explained by any hydrological stresses.

The removal of a 900 m thickness of rock with an average bulk density of 2500 kg/m^3 would reduce the total stress on the underlying material by approximately 22 MPa. Simplistically, the underlying rock would expand in volume in proportion to its compressibility (reciprocal of bulk modulus of elasticity), α, as follows:

$$\alpha = \frac{-\Delta V_B / V_B}{\Delta \sigma_e} \qquad \text{(eqn A4.1)}$$

where V_B is bulk volume, ΔV_B is change in bulk volume, and $\Delta \sigma_e$ is the change in effective stress (total stress minus pore pressure). The minus sign is necessary because an increase in stress results in a decrease in volume and vice versa. In material of very low permeability (as determined by the test results), fluid migration will be so slow that the condition can be treated as effectively 'undrained' and the change in effective stress can be equated with the reduction in total stress caused by excavation of the pit.

Making this simplifying assumption and rearranging equation A4.1 gives:

$$\Delta V_B = -\alpha V_B \Delta \sigma \qquad \text{(eqn A4.2)}$$

which shows that a reduction in total stress will lead to a positive change (increase), ΔV_B, in bulk volume.

In groundwater studies, the rock matrix of the formation is usually considered to be incompressible, and all volume change is attributed to a change in fracture porosity. In the case of the rocks types at Chuquicamata,

the porosity is extremely low and compressibility values are also likely to be extremely low, so it is possible some of the volume change can be attributed to change in the crystal volume of the matrix. While this poses a problem in apportioning the volume change between solid and pore space, it is clear that part of the increase in volume will be a result of an increase in fracture porosity. This is further supported by the underground drain hole flows, some of which show an increase in flow that may be correlated with mining and unloading of the overlying benches. In an effectively 'undrained' condition, an increase in fracture porosity will result in an expansion of the water, and hence a reduction in pore pressure.

The compressibility of water, β, controls this change according to equation A4.3:

$$\beta = \frac{-\Delta V / V}{\Delta p} \qquad \text{(eqn A4.3)}$$

where V is volume of water (in this case essentially the original volume of pore space), ΔV is the change in that volume, and Δp is the change in pressure. Equation A4.3 can be re-arranged to give:

$$\Delta p = \frac{-\Delta V / V}{\beta} \qquad \text{(eqn A4.4)}$$

Equation A4.4 shows that, as expected, an increase in volume will lead to a decrease in pressure. As a result of the lack of recharge and low permeability, water is not available to flow into the expanded pore space. If a simple case of two layers is considered, both layers will have been subject to the same reduction in total stress from unloading. However, based on the deformation studies carried out for Chuquicamata, it is likely the lower layer will have lower porosity because the total increase in fracture aperture will likely have been less than for the upper layer.

Therefore, even if it also has lower compressibility and expands less, it is likely that the proportionate increase in pore water volume ($\Delta V / V$) will be greater in the lower layer than in the upper layer. Hence, the corresponding reduction in pore pressure (from Equation A4.4) will also be greater in the lower layer. In a more complex situation of several layers, or of more smoothly varying properties, the result would be a general downward gradient, as is currently observed. Because permeability is also seen to decrease with depth, the possibility of the flow of water occurring to dampen or equalise the pressure drop also decreases with depth.

2.4.3 East wall

Because the alteration units in the East wall are of such low permeability, groundwater flow to drains, and the resulting depressurisation, tends to be localised. Selected VWP hydrographs for the East wall are shown in Figures A4.7 and A4.8. In all piezometer sensors, an initial rapid pressure change is often evident following installation as

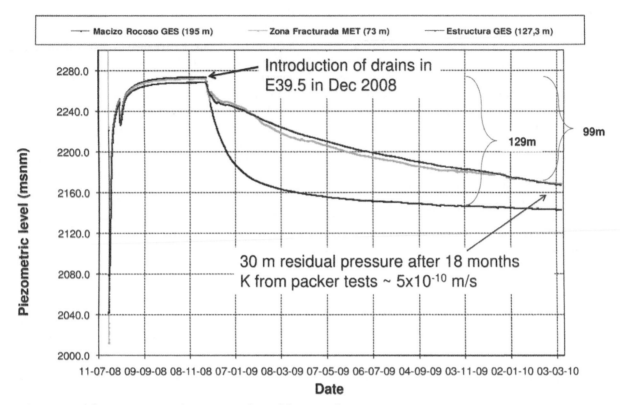

Figure A4.7: Lagged drainage in metasediment unit in the middle East wall

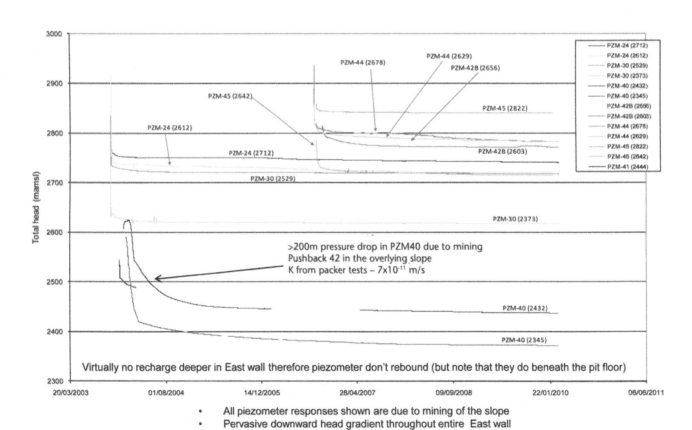

Figure A4.8: Unloading responses in middle East wall; sericitic alteration; chloritic alteration units

the grout cures and pore pressures in the formation around the drill hole equilibrate following drilling.

The following points can be drawn from interpretation of the hydrographs.

- All the piezometers show a downward total head hydraulic gradient (total heads decrease with depth).
- Some of the installed VWPs show specific depressurisation events from drain hole installation (Figure A4.7) although most of them do not.
- Many of the hydrographs show very slow 'ongoing' depressurisation.
- The two installed VWP sensors in PZM40 show considerable depressurisation (150 m) in comparison with the other hydrographs in Figure A4.8. The pressure drops have been correlated to the mining of pushbacks in the slope above the piezometer location.

In order to determine the pore pressure differences between the identified structures and the adjacent rock mass, five underground holes were selected for dual-level piezometer installations, with one grouted-in sensor installed within a known structural zone (VIF), and the second grouted-in sensor installed 20 m from the zone (measured along the trace of the drill hole). The piezometers were installed in PEC, PEK and PES units. For each installation, the two sensors showed a downward

head gradient, irrespective of their position relative to the structural zone. It was concluded that (i) there was no evidence that the structures were playing a role in the pore pressure distribution, and (ii) the site-wide downward head gradient was reflected by each piezometer pair.

2.4.4 Drainage of the granodiorite (GDF) in the West wall

Figure A4.9 shows hydrographs for underground VWP sensors installed in the GDF immediately upslope from the West fault (Figure A4.6). The piezometers show a response to the drains into the GDF drilled from the underground cross cuts. The response is rapid, with a total head drop of about 400 m from their initial condition in all piezometers. All piezometers respond in a similar manner, but XG352 shows a lag time of about 48 to 96 hours compared with the other installations, and a slightly flatter hydrograph.

The observed responses can be described as 'free drainage', in which pressures fall in response to a new drainage condition with no outside influence from recharge. The total flux producing the response was 0.55 l/s (peak short-term inflow of individual drains), with a peak combined total flow of 1.72 l/s. The results highlight the very sensitive nature of low-storage fracture networks to removal of water.

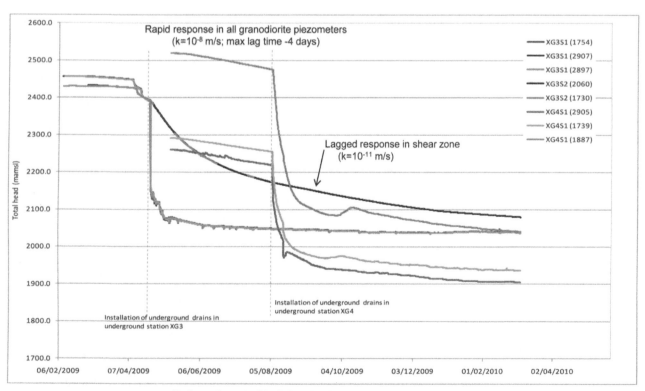

- Similar rock type to Diavik, but virtually no recharge
- Two orders of magnitude more depressurization; with 10% of flow (0.55 l/s)

Figure A4.9: Rapid drainage of granodiorite in the lower West wall

2.5 Discussion

Chuquicamata is unique amongst the case studies described for two reasons.

1. The low permeability, with only the West wall GDF exhibiting permeability values greater than 10^{-8} m/s.
2. The lack of any significant recharge, except for that produced by leaky facilities above the crest of the South wall.

The first factor means that the greatest influence on pore pressure is the result of unloading of the slopes by mining, and consequent deformation and expansion of fracture apertures in the slope materials. There is insufficient permeability for groundwater flow to equalise the pore pressure drops. The unloading response is strengthened by the inherently low fracture porosity of all formations, so a small increase in fracture aperture produces a relatively large decrease in pore pressure.

The second factor means that the depressurisation of the fracture systems is largely 'non-damped' (except for the RQS in the lowest part of the pit). There is insufficient water available to equalise the pressure changes that are created by the stress release and deformation of the materials.

In the more permeable West wall GDF, where free-drainage can occur in response to drain drilling, rapid pressure drops are observed across the entire volume of rock mass monitored by the piezometers. The available dataset indicates that the fracture systems are sufficiently permeable, and are sufficiently interconnected, for them to empty rapidly into the drain holes. The volume of water produced by the drain holes is less than 5000 m³. The entire monitored facture network can be estimated to encompass a volume of rock of at least 50 million m³. This suggests the drainable porosity is less than 10^{-4}.

3 Radomiro Tomic

Oscar Leiva, Claudio Rojas, Paul Mansfield, Jaime Leiva, Hernán Oyarzún and Geoff Beale

Setting: Copper porphyry, low permeability, with saturated alluvium in the upper West wall. Near-zero precipitation.

3.1 Background

The Radomiro Tomic (RT) pit (Figure A4.10) lies 3 to 4 km to the north of the Chquicamata pit and has broadly similar hydrogeology. Packer testing has indicated that all bedrock alteration units have a similar permeability range to that observed at Chuquicamata. The pit is in the early stages of operation (currently Year 7 of a >45 year mine plan), and therefore provides insight as to how initial unloading and depressurisation occurred at Chuquicamata. Important differences are that (i) the West fault is located to the west of the current crest of the West wall, and (ii) saturated low-permeability (10^{-7} to 10^{-8} m/s) cemented basin fill deposits occur to a depth of about 225 m in the West wall.

A total of 24 multi-level VWP installations has been completed to date around the RT pit. In addition, a total of six inclined VWP installations has been made in the East and West walls. All piezometers show a gradual reduction in total head with time. Several responses can be correlated with periods of active mining, particularly those to the east and south of the pit, away from the influence of the saturated alluvium. Three example hydrographs are shown in Figure A4.11, together with bench-levels for active mining in the vicinity of the piezometer.

The dataset indicates that:

- although the rate and magnitude of response is variable, all piezometers show a response to active mining which is attributable to stress release and unloading of the rock;
- in areas of where the most recent active mining has occurred (140 m vertical advance over a 12 month period), piezometers have shown a reduction in total head of between 14 and 37 m.

3.2 Response to mining pushbacks in the south-east area of the pit

The initial piezometric response to mining activity is demonstrated by PZMRT-04 (Figure A4.12 and Table A4.3). This piezometer was installed to the south of the pit in January 2006 and was mined out in July 2008 during the Phase 14 and 15 expansions. The hydrograph shows the initial rate of decline in piezometric level for the period up to July 2007 varies from 0.02 m/month for the shallow VW sensor (195 m depth) to 0.06 m/month for the deep VW sensor (323 m depth). From August 2007 to June 2008, the rate of decline increased to approximately 1 m over this period as a result of increased mining activity corresponding to the Phase 15 expansion.

Three VWPs (PZMRT-17, PZMRT-21 and PZMRT-24) are located on the south-east margins of the pit. The deep piezometers are installed at depths of 429 m, 332 m and 409 m, respectively (Table A4.3). The response of the deep VWP sensors can be broadly correlated with the Phase 15 and 16 pushbacks. In each piezometer, the deep sensor shows a stair-step response that can be related to the mining activity. For example, during the period August 2007 to June 2009, the Phase 16 pushback shows a deepening of 120 m in the South wall. During this period, the rate of decline in the deep PZMRT-21 sensor is approximately 10 m, corresponding to a rate of approximately 0.5 m/month. Over the corresponding period, the rate of decline for PZMRT-17 is 0.6 m/month and 1.0 m/month for PZMRT-24 (period July 2008 to June 2009).

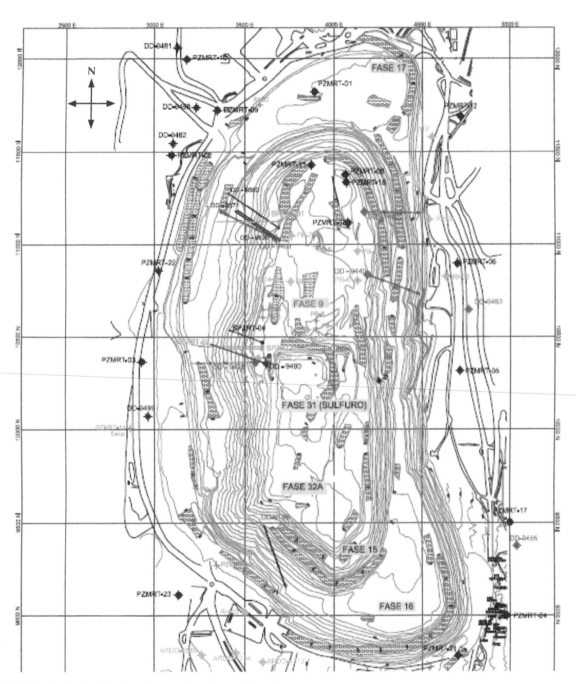

Figure A4.10: Radomiro Tomic mine pit expansion phases 2011–2012

During the period June 2010 to July 2011, the deep sensors show a stabilisation which corresponds to a period of reduced mining activity (30 m in Phase 16 and 45 m in Phase 15). During this period, there is no discernible decline in PZMRT-21 and the observed water level drops in PZMRT-17 and PZMRT-24 are 1.9 m and 0.7 m, respectively.

Moving along the East wall to the north, moving away from the mine expansion activity, the pore pressure response is much more subdued. For example, as shown by Figure A4.13, PZMRT-06 located on the crest of the

northern East wall shows a decline of around 0.2 m/month in both the shallow and deep VW sensors.

3.3 Piezometric responses in the West wall

The upper West wall of the RT pit includes gravels to a depth of 225 m, with saturation occurring in the lower part of the unit (Figure A4.14) and there has been continuous slope movement over the years. The observed piezometric response in the West wall is different to the rest of the RT pit and can be summarised as follows.

Table A4.3: Vibrating wire piezometer details

Piezometer	Sensor installation depth (m)	Period of record	Total head decline over period of record (m)	Monthly rate of decline (m/month)	Comments
PZMRT-04	195	Feb-06 to Jul-08	0.6	0.02	Piezometer destroyed by mine expansion in July 2008
	323		1.7	0.06	
PZMRT-06	241	Mar-06 to Aug-11	12.0	0.19	
	343		11.6	0.18	
PZMRT-17	289	Feb-07 to Aug-11	10.3	0.19	
	429		21.3	0.39	
PZMRT-21	225	Jul-07 to Aug-11	1.3	0.03	
	332		18.7	0.40	
PZMRT-24	409	Jul-08 to Aug-11	23.1	0.62	

- vertical recharge from the overlying partially saturated gravels acts to maintain elevated piezometric heads;
- continuous and sometimes accelerated slope displacements produce periodic 'instantaneous' rises in pore pressures in deep VW sensors;
- rising piezometric heads are evident in the area of the waste dumps and these are attributable to the waste dump construction (increasing total stress).

PZMRT-09 and PZMRT-10 are located in the Northwest margin of the RT pit. In both piezometers, the deeper VW sensor shows a higher piezometric head than the shallow VW sensor (upward gradient, Figure A4.15) which, given the location of the piezometers, is considered an anomalous condition. Possible explanations for the observed condition could potentially be (i) vertical recharge from the overlying partially saturated gravels, (ii)

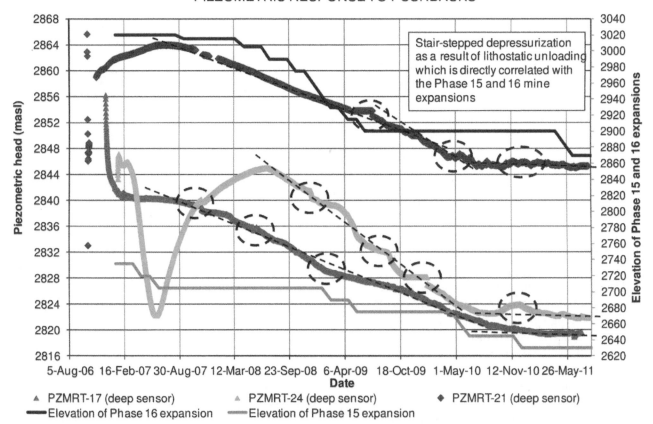

Figure A4.11: Piezometric response to mine pushbacks, South wall

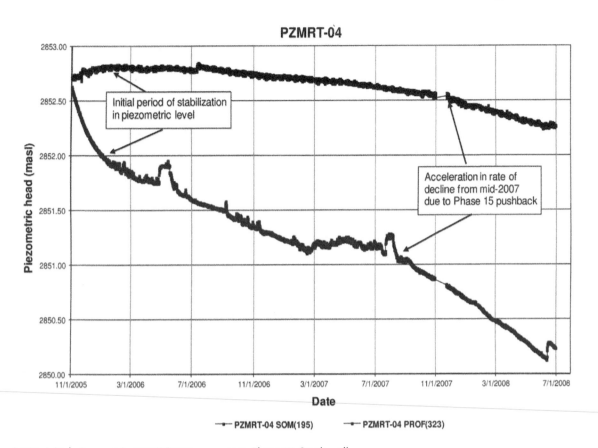

Figure A4.12: Initial piezometric (PZMRT-04) response to Phase 15, South wall

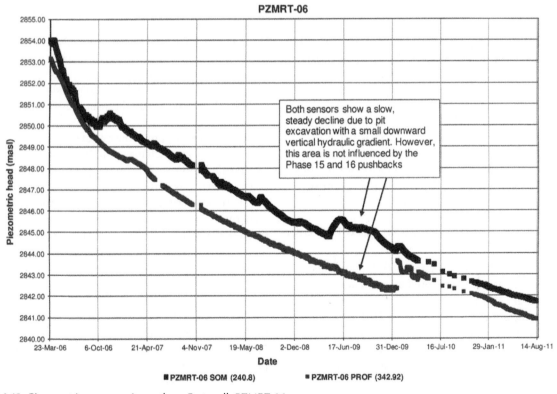

Figure A4.13: Piezometric response in northern East wall, PZMRT-06

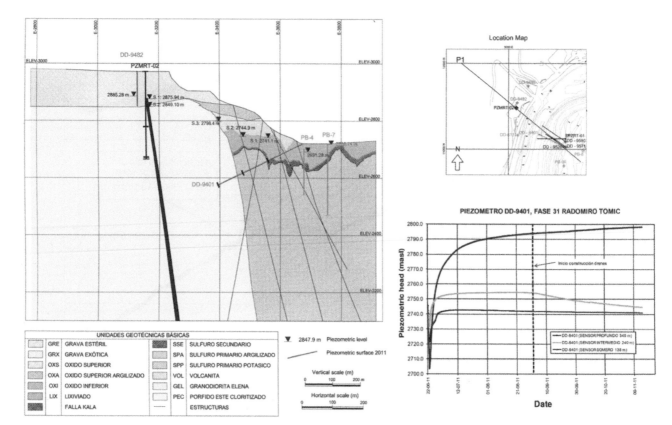

Figure A4.14: Hydrogeological condition, West wall

slope displacement, and (iii) lithostatic loading due to the continuous deposition of material in the area of the adjacent waste dumps.

Due to their location, it is considered that the elevated heads in some of the deeper VW sensors are a result of lithostatic loading effects due to the deposition of waste material on the adjacent dumps. This has the reverse effect to the lithostatic unloading observed in the area of the South wall. The additional load of the waste material produces additional stresses on the bedrock which cause fractures to close with a resultant rise in pore pressure. Similar to the unloading response, the effect is greater in the deeper VW sensors due to the lower fracture porosity and permeability with increasing depth.

3.4 Responses in the northern sector of the West wall

In the northern sector of the West wall, the continuous slope movement has been monitored using an array of prisms. There are also a number of vibrating wire piezometers, including PZMRT-02 and PZMRT-03, which were installed during the hydrogeological investigations carried out in 2005 and 2006. Figure A4.16 shows the locations of the prisms and vibrating wire piezometers.

Figure A4.17 shows the mid slope section of the wall directly downslope of piezometer PZMRT-02. One prism on

the figure shows a total slope displacement of up to 200 cm between January 2008 and September 2011. The other prisms show much lower rates of movement which do not exceed 40 cm for the same monitoring period. During this period, the deep sensor installed at a depth of 300 m shows two periods of 'instantaneous' increase in piezometric head of up to 30 m between March and June 2009 and around 20 m between November 2010 and February 2011. In both cases, the piezometric head increases observed in PZMRT-02 are preceded by an increase in the rate of decline in the piezometric head which may be indicative of increased slope movement. However, there is no apparent correlation with acceleration in any of the prisms in the vicinity.

These two periods indicate periodic groundwater recharge, most likely derived from the overlying alluvium through bedrock fractures, resulting in large short-term increases in pore pressure. The increases are followed by subsequent declines due to accelerated slope movement which results in the opening of fractures and the subsequent drop in pressure.

Figure A4.17 shows the piezometric head response in the shallow sensor installed at a depth of 163 m in PZMRT-03 together with slope displacement data from prisms located in the upper slope of the West wall in the vicinity of PZMRT-03. The prisms MRT-Z1A-2990-0321 and 0320 show the greatest movement of around 190 cm

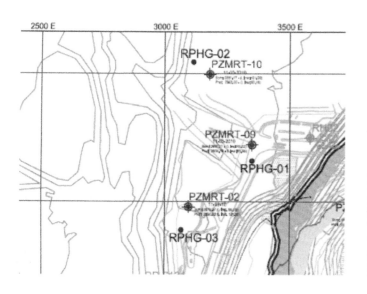

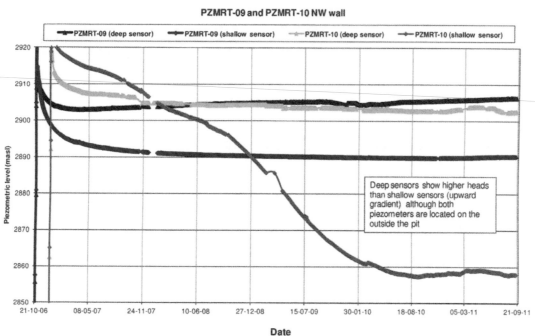

Figure A4.15: Piezometric response PZMRT-09 and PZMRT-10, NW sector

during the monitoring period from January 2008 to September 2011. During this period, the shallow sensor shows a general rise in the piezometric head of around 10 m. This may indicate constant recharge to fractures from the overlying partially saturated gravels due to the slope displacement.

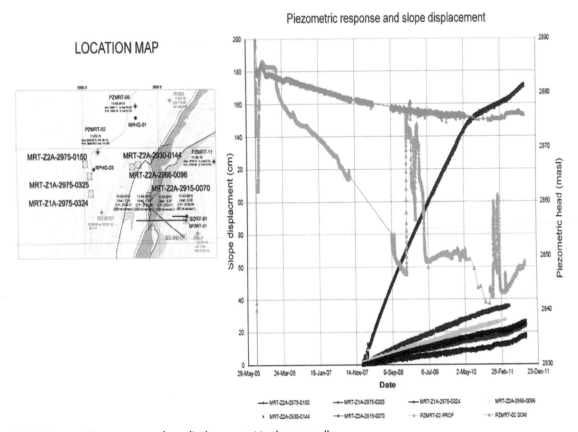

Figure A4.16: Piezometric response vs slope displacement, North-west wall

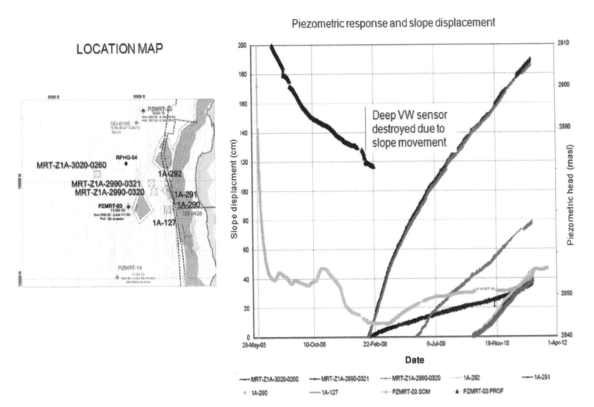

Figure A4.17: Piezometric response vs slope displacement, West wall

4   Antamina West wall

John Pottie, Alejandro Sanchez, Dave Gilbert and Geoff Beale

Setting: Quartz monzonite intrusive and skarn within moderate-permeability limestone country rock. Located in steep topography with about 1350 mm annual rainfall occurring within a six month wet season (between October and April).

4.1   Background

Minera Antamina is located in the Western Cordillera range of the Peruvian Andes, at an altitude of 4300 m, approximately 270 km north of Lima. The ore body is formed by a large multi-phased quartz-monzonite porphyry intrusive deposit surrounded by skarn alteration and is located at the bottom of a prominent NE–SW trending U-shaped glacial valley, surrounded by limestone ridges. The intrusive body has a SW–NE strike length of more than 2500 m, and a width of up to 1000 m.

4.2   Geology and hydrostratigraphy

The geology of the Antamina pit is shown in Figure A4.18. Geological characterisation within the open pit is an ongoing process for Antamina, with substantial effort put into characterisation of the geology within the open pit and surroundings

and surroundings, and updating of the geological block model. Antamina also puts high importance on the identification and determination of major and minor structures within the open pit and surroundings, and characterisation of the geological setting. Geological structure at Antamina is an important consideration for mineral evaluation, geotechnical stability, and hydrogeological evaluation

Major faulting and associated jointing was intruded by the porphyry, resulting in a large volume of intrusive rock in contact with the older Celendin Limestone host rock. The Celendin consists mainly of nodular grey marls and marly limestones, interbedded with well-stratified calcareous mudstones and siltstones and thinly bedded limestones. The contact zone is characterised by endoskarn, exoskarn and hornfels units, with marble occurring in parts of the North-west and North-east walls. The calcic skarn was mineralised by extensive hydrothermal solutions, forming at least six distinctive ore zones. Typically, the floor of the open pit consists of intrusive rock, the lower walls consist of skarn and transition material, and the upper walls consist of limestone, marble and hornfel. A typical hydrogeological cross section of the West wall is shown in Figure A4.19. A photograph of the West wall is shown in Figure A4.20.

The hydrostratigraphy of Antamina can be summarised as follows.

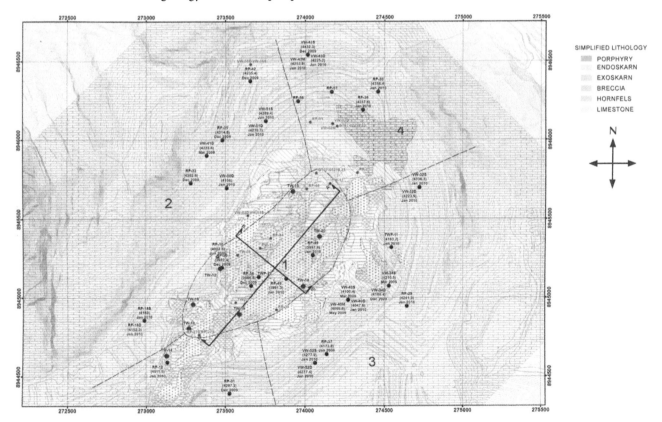

Figure A4.18: Antamina pit geology

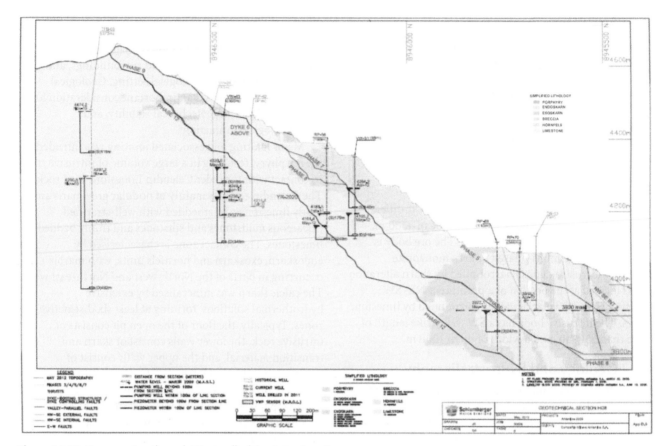

Figure A4.19: Cross-section through West wall of the Antamina pit

Figure A4.20: Photograph showing West wall of the Antamina pit

- Quartz-monzonite porphyry intrusive: the core intrusion represents the most permeable and drainable unit due to the presence of intense fracturing, particularly along recent fault zones and contacts with the adjoining endoskarn. The fracturing is very well connected, such that lateral and vertical hydraulic gradients are minimal, even when high pumping stress is applied. The drainable porosity of the unit is low.
- Endoskarn: found in direct contact with intrusive rock in the central pit area. The unit is fractured and the rock fabric has been highly altered and disaggregated to give a high permeability, and a higher drainable porosity than the intrusive. There is good fracture connection to the intrusive rock, such that the Endoskarn unit fully drains in response to pumping from the intrusive.
- Exoskarn: typically in contact with endoskarn. Comprises both hard and friable horizons. Fracturing is less intense and more localised than endoskarn. However, there is continuous fracturing between the exoskarn, endoskarn and intrusive, so that the exoskarn is also observed to fully drain in response to pumping from the intrusive, but often with a slightly lagged response.
- Hornfels: occurring as a discontinuous unit within the pit walls. The hornfels has lower permeability than the skarn units. It exhibits a response to pumping from the pit floor, but with a significant lag time and at a lower rate.
- Marble: a generally massive unit, with limited fracturing. Although not in direct communication with the skarn units, the marble does show a lagged drainage response to pumping from the pit floor.
- Celendin Formation: district scale groundwater flow does not typically occur in the Celendin Formation. The formation is well fractured around the area of the orebody, and may yield up to 10 l/s short-term inflows to horizontal drain holes. However, the fracturing is observed to be spatially variable and is also variable between beds. There are abundant small-scale bedding-parallel faults and minor thrusts due to the thinly bedded nature of the sequence, and the sequence includes abundant marly horizons. Locally, the Celendin does not show any karst development.

The fabric of all the lithologies in the pit area is of low matrix permeability. Groundwater movement is influenced by structural trends and, particularly, the degree of secondary fracturing and faulting of the individual lithological units. The structural control on groundwater movement can be summarised as follows.

- North-east–South-west trending (valley parallel) structures. These are observed to transmit groundwater along their strike for distances of up to 1000 m or more. They often form barriers to flow across strike. In the West wall, they may act as barriers to groundwater moving downslope towards the open pit floor.
- Bedding and bedding-controlled shear zones. Along the length of the West wall the bedding dips gently away from the slope. The bedding-parallel shears also dip away from the slope. Both features act to impede groundwater moving downslope towards the pit floor.
- North-west–South-east oriented joint sets. Structures of this orientation are observed to provide good drainage conduits for the East wall. They appear to be less of a control on the hydrogeology of the West wall. However, in some places, they may provide partial offsets to the valley-parallel structures.

4.3 Dewatering and depressurisation

Active dewatering at Antamina has been ongoing since 2004. An annual dewatering program is managed and implemented by the mine operations hydrogeological group. The program goals are to ensure mining conditions are dry and safe, to monitor pore pressures, and to ensure depressurisation of the pit walls. The program includes:

- installation of vertical dewatering wells with internal bore diameters of up to 18 inches (450 mm), and casing diameters of 12 inches (300 mm). Between two and three new production wells are installed each year to replace wells that are lost or to augment dewatering capacity;
- installation of standpipe piezometers within the drainable hydrogeological units (intrusive and skarns) to establish water levels within and around the pit floor. New piezometers are installed to replace damaged or lost piezometers, or to provide new monitoring points;
- installation of vibrating wire piezometers within the limestone, hornfels, or marble for measurements of pore pressures in the materials that form the pit slopes. Piezometers are normally targeted to intersect final pit walls, or are installed on intermediate pit walls that will not be affected by mining for several years;
- installation of horizontal drains to reduce pore pressures behind the pit walls to improve the overall wall stability.

At Antamina, the dewatering and piezometer installation program is strongly affected by the active mining operations and, in many cases, competes for available space. Dewatering is carried out using a series of in-pit vertical dewatering wells positioned principally within the intrusive unit close to the bottom of the pit. Initial well yields in the intrusive are up to 50 l/s, with

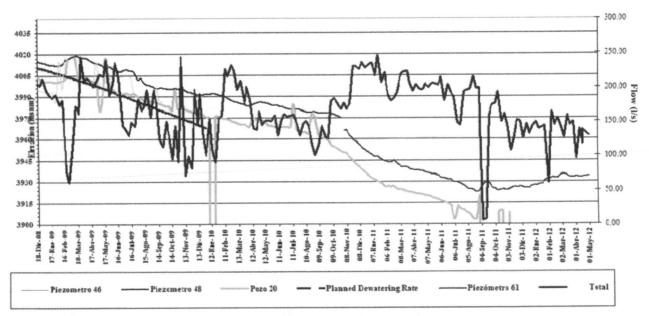

Figure A4.21: Response of intrusive piezometers to general dewatering

less than 10 m drawdown (giving a specific yield of more than 5 l/s per m). Two wells have also been completed within the endoskarn, which show a specific yield about 15 to 35 per cent of a typical intrusive dewatering well. Given that the intrusion is at the bottom of the open pit, and is the productive and drainable hydro-stratigraphic unit, dewatering wells are normally installed into the intrusion from lower mine catch benches. Since catch benches are constructed to catch overspill and falling rocks, well heads and conveyance pipelines require protection. Antamina has successfully employed well protection systems such as vertical steel A-frames or used haul truck tyres.

During the past two to three years, the total dewatering rate has been about 150–230 l/s, with typically four to seven wells operating at any one time. Antamina considers that the dewatering program is successful and predictable. During the period from the beginning of 2008 until June 2011, the rate of decline in the phreatic surface below the open pit floor averaged about 3 m/month or 125 m in total. Figure A4.21 shows the general dewatering rate from the pit floor and the typical response of pit floor piezometers installed in the intrusive and endoskarn rocks.

The goal of the dewatering program is to maintain the phreatic surface in the intrusive and skarn units at least 20 m below the lowest level of the operating pit floor. The monitoring data for the dewatering system indicate that,

even with the high applied pumping stress, no significant lateral or vertical hydraulic gradients have developed in either the intrusive or skarn units. Therefore, within and between these units, there is very good fracture interconnection.

As the phreatic surface in the intrusive and skarn units has been progressively lowered ahead of the pit floor, so drainage of the limestone in the pit walls has gradually occurred. This drainage is illustrated in Figure A4.19, where the phreatic surface is flat in the intrusive and skarn units and the contact between the exoskarn and the limestone forms a key hydrogeologic barrier. Drainage of the limestone rocks in the pit slope occurs to the limestone-skarn contact (behind the slope) rather than to the toe of the slope.

Gravity drainage of the limestones has occurred in all sectors, and at all levels. However, in some parts of the wall, the rate of depressurisation is less than required to achieve the desired target pore pressure targets. In these areas, horizontal drains are installed to increase the removal of water within the limestone. The drain installation program in the limestone has been mostly focussed in the northern sector of the pit. The target length of the drains is 300 m. Of the 32 horizontal drain holes installed into the limestone to date, all produced more than 0.1 l/s short-term yield, 18 produced more than 0.5 l/s, 11 produced more than 1 l/s, and four produced more than 5 l/s. Typically, individual drains produced 80% of their yield from one to three discrete fracture zones within the hole. The drains have created (i) a general increase in the overall rate of depressurisation, and (ii) a larger local-scale influence in their immediate area.

4.4 Conceptual hydrogeological model

Historical monitoring data indicate that the open pit area groundwater regime is isolated from the regional groundwater system. Currently, about half of all the water pumped from the dewatering system is derived from local recharge to the open pit floor during the rainy season. Seasonal recharge is also evident in the hydrographs of all pit slope (limestone) piezometers. The balance of the water is derived from local groundwater storage removal in the open pit area, and a small portion of groundwater flow from the south-west, along the axis of the Antamina valley.

Presently, there are about 30 piezometers that actively monitor pore pressures in the pit slopes. The hydrographs for piezometers installed in the West wall are displayed in Figure A4.22. The high seasonal precipitation at Antamina has a large impact on the hydrographs, in some cases with in excess of 50 m of seasonal variation in the piezometers installed at shallower depths. High seasonal variations in the hydrographs are often muted in the piezometers

installed at deeper levels (for example VW36D, VW51D). There is often a lag time of about one to three months between the time for recharge to influence the shallow installations and the time for recharge to affect the deeper piezometers.

The conceptual hydrogeological model for the Antamina West wall is illustrated by Figure A4.23. The phreatic surface throughout the West wall is relatively close to current topography because of the influence of seasonal recharge. There is a strong downward total head gradient throughout the slope. Deeper-level piezometer installations typically show total heads that are 100 to 200 m lower than the shallow-level installations. Thus, the materials in the West wall are saturated but primarily exhibit low pore pressure.

The downward total head gradient in the West wall is caused by a combination of (i) the seasonal recharge to the slope, and (ii) groundwater within the slope moving downward by gravity to its discharge point at the limestone/exoskarn contact. The downward gradient is strengthened by the three geological features that act as barriers to the flow of groundwater downslope, namely (i) the bedding that dips into the slope, (ii) the bedding-parallel shears, and (iii) the NE-SW (valley parallel) faults which are sub-parallel to the orientation of the West wall.

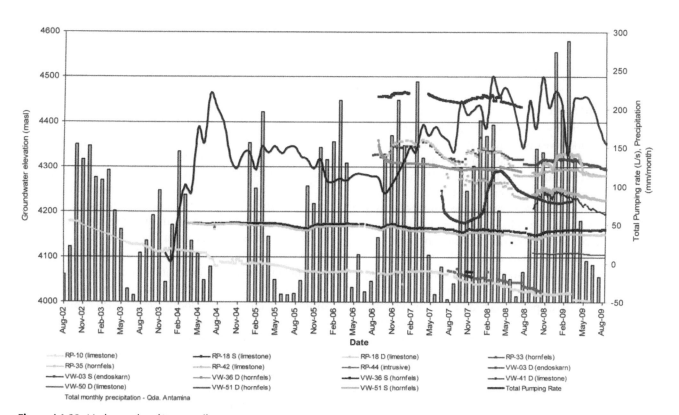

Figure A4.22: Hydrographs of West wall piezometers

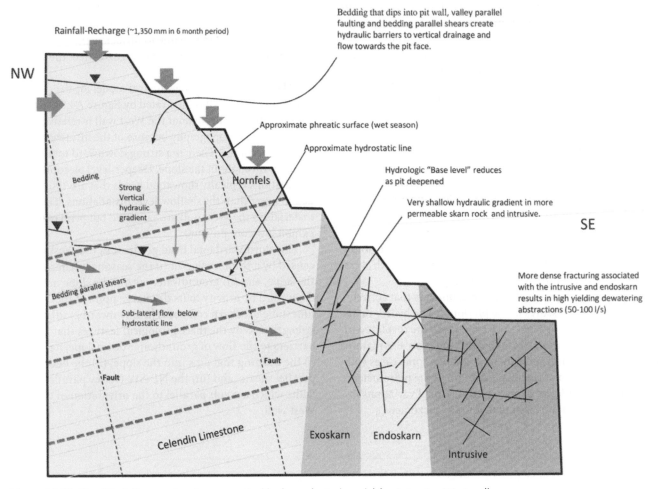

Figure A4.23: Simplified diagram showing conceptual hydrogeological model for Antamina West wall

4.5 Discussion

There is a sufficiently permeable and interconnected fracture network developed within the limestone wall rocks at Antamina to allow water to drain downslope to the limestone/exoskarn contact zone beneath the lower pit walls. This is observed in all sectors of the pit.

Elevated pore pressures occur, not because the limestone itself is poorly permeable or has insufficient local-scale fracture interconnection, but rather because of larger-scale structural features that act as 'dams' to the groundwater moving downslope. In addition, the available data suggest there is somewhat higher permeability and fracture interconnection developed along bedding. In the West wall, this is likely to be enhanced by the bedding parallel shears.

There is a network of larger-aperture ('A-type') fractures which provide short-term yields to horizontal drains of between 1 and 10 l/s. The frequency of these fractures is variable through the pit, depending on change in the lithology of the limestone itself, and on the intensity of larger structures. However, in the West wall, it appears these fractures may be intersected about every 50 to 100 m

of drain drilling (likely also depending on the orientation of the drain hole).

The structural 'barriers' alone are not sufficient to explain elevated pore pressures. The presence of recharge from precipitation acts to sustain the observed total heads in the pit walls. Most of the 'barriers' permit sufficient leakage across them that, without ongoing recharge, many of the slope sectors would naturally depressurise with time as a result of lowering the water level at the limestone/exoskarn contact.

The purpose of the horizontal drain program is to artificially pierce the structural barriers and induce drainage of groundwater locally impounded behind them. For Antamina, it is considered that 300 m is an adequate length for the horizontal drains to intersect a sufficient number of barriers and release 'impounded' groundwater.

There is no evidence that any limestone piezometers are installed in 'D-type' fractures with a large lag in drainage response. However, parts of the hornfels unit exhibit a generally slow drainage response because of the general lower permeability of the fracture systems within this unit.

5 Jwaneng Diamond Mine South-east wall

Mike Brook, Abel Tunono, Geoff Beale and Banda Maswabi

Setting: Kimberlite within a sedimentary sequence mostly consisting of strongly bedded shale; about 400 mm annual rainfall within a four to five month wet season.

5.1 Background

The Jwaneng diamond mine is located in southern Botswana, around 160 km south-west of the capital city, Gaborone, within a region that forms part of the semi-arid eastern Kalahari Desert. The mine is owned and operated by Debswana, a partnership between the Government of Botswana and De Beers. Production mining at Jwaneng is mostly within three diamond-bearing kimberlite pipes (centre, north, and south pipes) that converge near the surface. With an annual production of around 11 million carats of gem quality diamonds, it is the world's richest diamond mine.

5.2 Geology and hydrostratigraphy

The project area is masked by a 40 to 50 m thickness of unconsolidated Kalahari sand sequence, with calcrete and silcrete bands and layers. In the Jwaneng area, this is mostly unsaturated down to the underlying bedrock surface. The district geology comprises a Proterozoic depositional sequence of dolomites, shales and quartzites. In the vicinity of the mine, the strata dip 30° to 45° towards the north-west. The Jwaneng orebody itself intruded the host rocks along the predominant NE-SW trending orientation of a steep fault. The host rock is broken up by extensional shears, thrusts and steep faults. The geological strata of the Jwaneng pit comprise the following.

- **Kalahari Sands.** The Kalahari Sands are an aeolian surface deposit which covers the entire project area. The deposit occurs above the regional and pre-mining water table in the vicinity of the mine.
- **Shale sequence.** The shale sequence comprises (in downward sequence): the Laminated Shales (LS); the Carbonaceous Shales (CS); and the Quartzitic Shales (QS). The current slopes are predominantly cut through the shale sequence, with the LS exposed in the upper slopes (~100 m stratigraphic section), a relatively thick exposure of CS in the mid-slope, and QS exposed in the lower slopes. The QS is the dominant geologic unit exposed in the Jwaneng pit and has the greatest degree of observed fracturing and joint development.
- **Dolomites.** The dolomites occur below the shale sequence and are below the current pit floor in all

domains. They are not a factor in the current short-term depressurisation program. The dolomites are locally fractured and permeable, but do not appear to be connected to the regional dolomite groundwater flow system.
- **Kimberlites.** Prior to mining, the three pipes coalesced near the surface to form a tri-lobed linear body (1300 m by 500 m). The kimberlite is now exposed at the base of the Jwaneng pit as three separate pipes that exhibit higher porosity than surrounding country rocks.

Figure A4.24 shows a general view of the South-east wall. The shale sequence dips out of the slope so that bedding is an important control for the slope design.

Abundant bedding fractures are observed in the wall (Figure A4.25). However, inspection of core indicates relatively few bedding-controlled fractures, and those that are observed show little evidence of groundwater flow through them. A dominant feature of the core is the strong orthogonal joints, with two joint sets that are perpendicular to each other and to the bedding (Figure A4.26). Many of these fractures are observed to have strong oxidation on their fracture surfaces, indicative of groundwater flow. The orthogonal joint sets are best developed in the LS and QS, but are also observed in the CS.

'Shrinkage' fractures are evident to varying degrees at the contact between the kimberlite and the shale, and in the deeper contact between the kimberlite and the dolomite. Water is often found in the country rock and kimberlite contact zone (Figure 4.27). There are also zones of higher permeability within the kimberlites themselves.

The sequence is recharged from the surface at low rates. The recharge moves as diffuse unsaturated flow through the unsaturated Kalahari sand sequence and enters the underlying bedrock surface (LS). The natural recharge rate is very low, so most of the local recharge is derived from leaky site facilities, drainage diversion ditches, ponds and tailing dams to the south. The total recharge flux to the South-east wall has been estimated to be about 25–35 l/s.

Table A4.4 shows hydraulic conductivity estimates for the principal units, obtained through packer testing and pumping tests carried out between 2001 and 2007, predominantly from boreholes in the south-west and south parts of the pit. Based on the packer test results, the QS is the most permeable unit.

5.3 Dewatering program

Up until 2004, dewatering of the pit was achieved solely by pumping from sumps in the pit floor. In November 2004, dewatering well GH03 was brought on line. The well is located about 320 m away from the crest of the South-east wall (Figure A4.28).

Figure A4.24: General view of Jwaneng South-east wall (Source: Felix Ramsden)

Initial airlift testing of GH03 produced about 16 l/s with about 59 m of measured drawdown. The initial continuous pumping rate of the well was about 12 l/s, falling to a long-term rate of about 7.5 l/s as the amount of drawdown increased. The current pumping water level in the well is about 220 m. GH03 produces most of its water from two discrete fracture zones, one at a depth of 144 m below collar, the second at a depth of about 310 m below collar. The upper production zone is now dewatered (above the phreatic surface). Analysis of test pumping data from GH03 indicated transmissivity values of about 10 m²/day. Applying this to a fracture zone no more than 3 m wide would indicate a hydraulic conductivity of the producing fractures of about 10^{-5} m/s, which is two to three orders of magnitude higher than the values derived from the packer testing and the 'global' values derived from conventional pumping test analysis. The test program indicates the shales have an average drainable porosity value of about 0.001.

Table A4.4: Results of *in situ* hydraulic testing, 2001 to 2007

Geologic unit		Packer tests				Pumping tests	
		Permeability (m/s)				Transmissivity (m²/day)	Permeability (m/s)
		Min	Max	Geometric mean	Arithmetic mean		
Laminated shale		2×10^{-9}	7×10^{-8}	1×10^{-8}	2×10^{-8}	–	–
Quartzitic shale		2×10^{-9}	3×10^{-5}	3×10^{-8}	2×10^{-7}	0.3–15	3×10^{-7}
Dolomite		3×10^{-9}	1×10^{-7}	2×10^{-8}	3×10^{-8}	–	4×10^{-8}
Kimberlite	Overbreak	10^{-9}	1×10^{-5}	9×10^{-8}	1×10^{-6}	–	–
	Un-mined	10^{-11}	10^{-9}	–	–	–	–
	Breccia/Contact	9×10^{-9}	1×10^{-7}	5×10^{-8}	7×10^{-8}	–	–

The observed fracturing associated with the bedding is mostly the result of blast damage, so is less of an influence for depressurising the shales further behind the slope

An increase in aperture within the strongly developed sub-vertical joint sets is thought to be important for reducing pore pressure in the zone of deformation behind the slope

Minor seepage moving down the overbreak zone is brought to the face along blasting-induced bedding fractures

Figure A4.25: Pit face showing blast-induced bedding-parallel fractures and othogonal joint sets

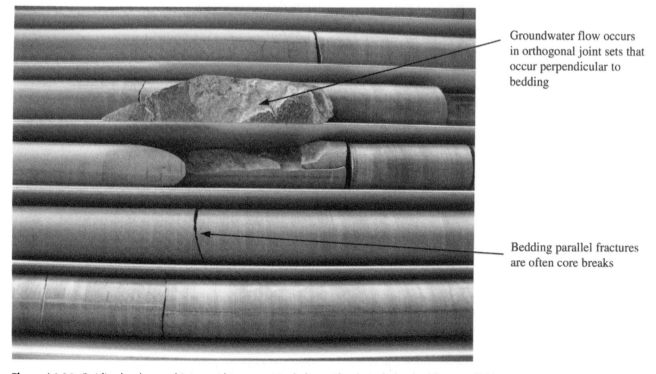

Groundwater flow occurs in orthogonal joint sets that occur perpendicular to bedding

Bedding parallel fractures are often core breaks

Figure A4.26: Oxidised orthogonal joints within quartzitic shales, with relatively few bedding-parallel fractures

Figure A4.27: Fracturing at the contact between the kimberlite and the shale country rock

Test drilling through the shales and dolomite sequence to date throughout the area of the South-east wall has failed to locate fracture zones that produce a higher production rate than the zones encountered in GH03. Therefore, it can reasonably be assumed that the fracture permeability in GH03 is representative of the most permeable fracture sets in the sequence.

5.4 Slope depressurisation

Groundwater levels around the pit are monitored by a network of standpipe and multi-level vibrating wire piezometers. Hydrographs for selected VWPs are shown in Figures A4.28 along with a general cross section through the South-east wall. Figure A4.29 shows the piezometers plotted onto a single section with a wide search distance.

GH03 has created a broad area of drawdown within the shales along the South-east wall of the pit. The drawdown is widespread both along the crest of the pit (along the principal structural strike) and also into the pit slopes (across the principal strike). There is little apparent influence of bounding structures or 'compartmentalisation' in the monitoring data, even though many prominent faults have been mapped and are evident in core logging.

The impact of pumping has had the effect of reducing seepage at the face. Historically, the most significant seepage face in the pit was in the South-east wall where a number of faults occur close together. Total seepage from the wall fell from about 3 to 4 l/s in July 2001 to less than 0.5 l/s in September 2005. Mapping in 2008 showed that seepage from the main shale sequence in the South-east wall was virtually non-existent. The main in-pit seepage now emanates from lower in the wall, along the shale-kimberlite contact zone.

There is a downward vertical head gradient evident in most VWP installations. Recharge at the bedrock contact moves downward through the shale sequence. Previously, the main mechanism for discharge was seepage faces in the shale sequence in the pit wall, and at the exposed shale-kimberlite contact zone in the lower pit. Since the start of pumping, a portion of the discharge now occurs to the deep fracture zone in GH03, rather than as seepage to the pit slopes.

The monitoring data indicate the entire hydrogeologic system in the shale sequence is relatively well connected. Many of the major fault zones do not appear to have any compartmentalising influence on the groundwater flow system. Although vertical gradients develop and are partially sustained by the recharge, the piezometer data indicates that the shale sequence is draining, as illustrated in Figure A4.29. Prior to installation of the VWPs, a horizontal drain hole program was planned in the lowest levels of the pit to dissipate pore pressures in the South-east wall. However, because of the observed inter-connected drainage, it has not been necessary, to date, to implement this program.

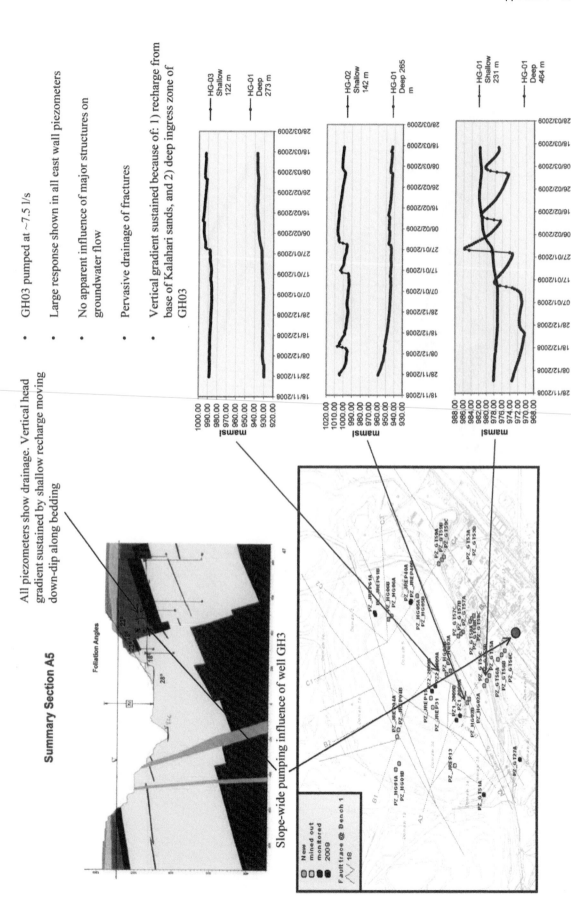

Figure A4.28: Hydrographs for VWPS

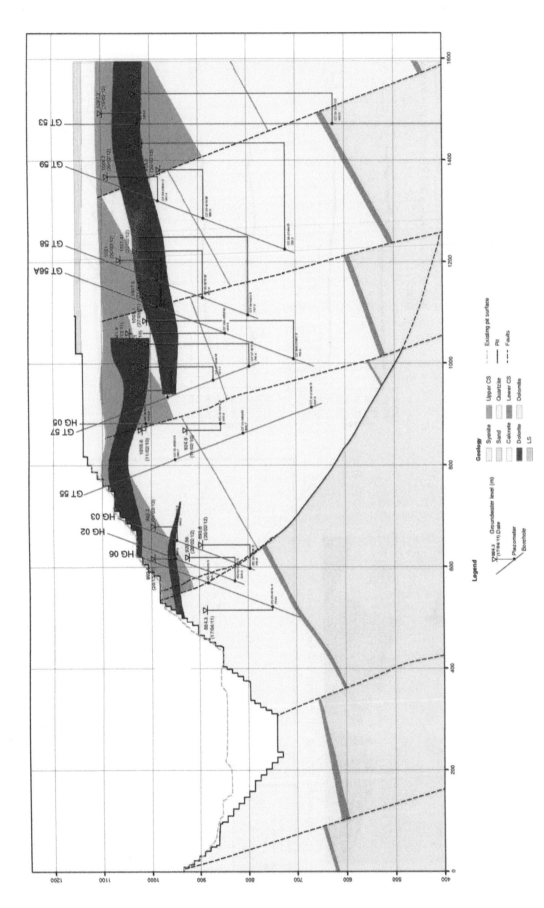

Figure A4.29: East wall piezometers plotted onto a single section line

Some minor residual 'perched' groundwater zones do exist in the shale sequence, but these are mostly related to isolated diorite sills that occur within the shales. There is one minor seepage face in the northern part of the East wall where dampness can be seen immediately above the diorite sill. A single VWP sensor installed in shale immediately above the diorite also shows a 'perched' groundwater condition.

5.5   Discussion

The entire shale sequence in the South-east wall is observed to possess extremely low, but generally well-connected storage (Figure A4.30). The small discharge flux (formerly to the seepage faces, and latterly to GH03) has induced a large depressurisation response. Over 120 m of depressurisation has been achieved to date along the entire wall (greater than 1000 m strike length) with a discharge flux of less than 10 to 15 l/s.

It is thought that the strongly developed orthogonal joint pattern creates a bedding-controlled Kh:Kv anisotropy, with the higher permeability parallel to the bedding. It is postulated that the reason for this is that the density and aperture of the jointing is controlled by variations in the brittleness of the rock in different shale beds. There are also shale beds where open jointing is observed in the core to be much less developed. Thus, the situation develops where local-scale permeability is dominated by joint sets that are perpendicular to bedding but, because the joints are inter-connected, the resultant groundwater flow is parallel to bedding.

In the South-east wall, the bedding dips towards the pit at typically between 30° to 45°. The bedding aligns the anisotropy favourably for drainage into the pit. However, although the bedding is clearly a control on groundwater flow, major influences of bedding are not generally evident in the piezometer results. This may be because of the vertical offsets to the shale sequence which create cross-bedding flow paths. Numerous old cored exploration drill holes are also likely to play a role in increasing the amount of cross-bedding connection.

Although considerable deformation of the shales is predicted by the geotechnical modelling results as mining progresses, and an increase in fracture aperture would consequently be expected, the phreatic surface and pressure profiles behind the wall are relatively flat (Figure A4.29), and there is no evidence that increasing fracture

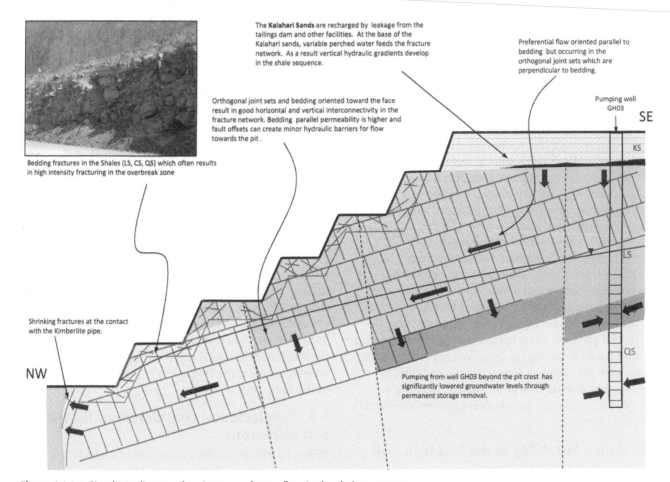

Figure A4.30: Simplistic diagram showing groundwater flow in the shale sequence

aperture is playing a role in groundwater flow. This may be partially because the initial fracture apertures are already wide enough to transmit the relatively small groundwater flux moving through the slope. It is also possible that most deformation occurs parallel to bedding, so the orthogonal joint sets are at an orientation which is least effected by deformation.

As for all other case studies, none of the piezometer installations at Jwaneng provide evidence that elevated pore pressures are being sustained because of non-connected fracture systems and/or because of low matrix permeability. If 'D-type' fractures and/or very low-permeability rock mass are present, they are in close enough proximity to higher order (more permeable) fractures to allow drainage to occur, at least at the scale that can be measured by the piezometers.

6 Cowal Gold Mine (CGM)

Bob Sharon, Andrew Bywater and Phil Greenhill

Setting: Volcanic rock with a well-developed saprolite weathering profile; about 500 mm annual rainfall generally on a constant monthly basis year around.

6.1 Introduction

The Cowal Gold mine is located in the western part of New South Wales. The operating experience at Cowal demonstrates that slope stability in the oxidised zone is sensitive to shear strength and groundwater conditions. To assist depressurisation of this upper saprolite zone, the operation has improved surface drainage around the pit perimeter, and installed vertical dewatering wells and horizontal drains. Development success is based on site experience, monitoring results from an extensive network of fully grouted vibrating wire piezometers (in vertical and horizontal holes) and slope stability analysis conducted during the previous several years. Key elements of these evaluations include interpretation of the mine area groundwater pressure distribution and development of a strategy for achieving slope depressurisation targets determined for the final mine design. Figure A4.31 shows a view of the South-east pit slope where horizontal drains are installed on most berms exposing the oxide and upper fresh rock zones, with drain collars connected to trunk lines to contain groundwater flow at the surface. Berms are out-sloped in the oxide zone to drain surface water and to minimise ponding.

6.2 Initial instability of the East high wall

Progressive instability of the East high wall developed in soft oxide materials starting as early as 2005. A review of

hydrogeology records indicates that a significant depressurisation event occurred in April 2007, interpreted to have resulted from ground movement. By June, clearly visible ground displacements were documented, as shown in Figure A4.32. This progressed to failure of the slope through transported sediments and the underlying saprolite zone by July 2007. Instability on the East high wall expanded southward in December of the same year. Seepage faces in the transported sediment zone were observed during this period.

Evaluation of as-built slopes and interpretation of back analysis results of the instability generated the following causal factors:

- low shear strength of the transported sediments and saprolite, including relict structural control in the saprolite zone;
- pore pressure conditions that exceeded original design tolerances;
- slopes exposing transported sediments and saprolite that were too steep for prevailing conditions that influence stability.

6.3 Geology and geotechnical domains

The geology of the oxide and fresh rock zones is divided into four geotechnical domains, as follows.

- *Transported Sediments (TS):* part of the soft oxide zone; 10 to 45 m thick in the mine area.
- *Saprolite (SOX):* oxidised residual soil with relict structure, defined as soft oxide; 10 to 55 m thick in the mine area.
- *Saprock (HOX):* oxidised fractured rock mass transitional between the saprolite and fresh rock, defined as hard oxide; 15 to 25 m thick in the mine area.
- *Fresh rock:* upper unaltered rock mass zone (primary rock) near the saprock contact is usually more fractured than at greater depth, with typical occurrence of oxidation of fracture surfaces.

The TS and SOX typically thicken eastward across the pit footprint. The South and East walls contain the thickest intervals of these weak units. In the current and final South and East high walls, the thickness of the TS varies from 35 m to 45 m; and SOX thickness varies from 35 m to 55 m. The combined thickness of these zones in the South and East walls is approximately 90 m. HOX zone thickness varies from 15 m in the East wall to 25 m in the South wall.

6.4 Geotechnical considerations for the soft oxide zone

General performance monitoring has identified the sensitivity of the weak, low-permeability transported sediments and saprolitic soils (TS and SOX) to pore

Figure A4.31: View of the south-east pit slope at CGM (2011)

Figure A4.32: Development of extension cracks and progressive failure conditions on the East wall (July 2007)

pressure. In the context of shear strength, permeability and deformation characteristics, these materials must be characterised as soils, and not as a rock mass.

6.5   Hydrogeological results

Data collected since 2008 have led to a generally good characterisation of the groundwater system. Drainage and depressurisation is variable, but is the most challenging in the low permeability TS. The collective HOX and upper primary zone is the most permeable and is targeted by the dewatering system to enhance under-drainage for the overlying less permeable SOX and TS.

As in many saprolite settings, it is possible to under-drain the upper slopes by pumping from the underlying fracture zones at the top of the unweathered (primary) rock. However, at Cowal, the obtainable yield from the upper fractured bedrock has generally been low. Nonetheless, experience indicates that depressurisation of the saprolite is achievable with vertical dewatering wells, despite the low production rates. Horizontal drains are also important for depressurising the ultimate slopes, and depressurisation of the TS may be enhanced with vertical drains.

At Cowal, the effective drainable porosity is attributed to partial alteration of intact basement rock and the presence of relict structural fabric. All dewatering wells are screened in the HOX and upper primary zone. Horizontal drains also target this zone. The dewatering network under-drains the SOX and to a lesser degree the TS. The TS responds slowly to under-drainage due to strong horizontal flow anisotropy of the sediments and the

prevailing clay mineralogy. Therefore, vertical drains may assist depressurisation of the TS in areas where under-drainage has been achieved.

Enhanced vertical flow anisotropy prevails in the HOX, upper primary zone and the lower SOX due to the predominantly steeply dipping structural fabric. Enhanced NNW flow anisotropy is attributed to a large-scale steeply dipping structural trend. However, targeting high-angle faults with dewatering wells is challenging; it is often more beneficial to target them with horizontal drains installed perpendicular to the structural trend. North-west oriented horizontal drains consistently produce more water than those drilled parallel to the structural trend. It is also recognised that major structure may also access groundwater storage adjacent to the mine that will continue to offset depressurisation achieved in the HOX and upper primary zones.

6.6   Prediction of pore pressures

It was determined that calibrated numerical models would not be realistic due to the lack of hydraulic control, data limitations and the many factors that influence the depressurisation response (e.g. mining rate, general mine seepage (sump pumping) and the active and passive components of the dewatering system). Prediction of future pore pressures was therefore carried out by taking the extensive piezometer database and by carrying out a forward extrapolation of the piezometer hydrographs on the five 2D geotechnical sections shown in Figure A4.33. Following interpretation of the pre-mining and current

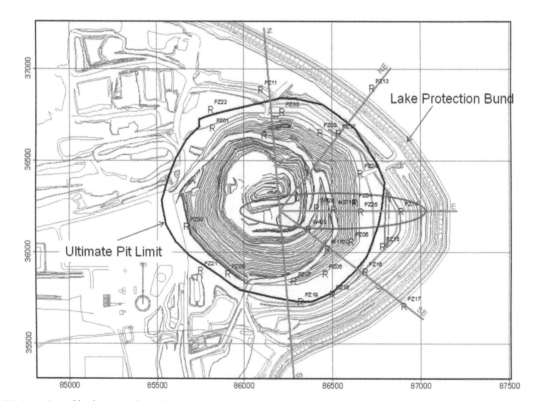

Figure A4.33: Location of hydrogeotechnical cross-sections

groundwater distribution, three predictive cases of future pressure distribution, based on pore pressure extrapolations, were developed.

- Case A: no additional drainage measures implemented.
- Case B: future drainage measures at the same level of intervention as in the past (depressurisation trends are maintained at a constant rate).
- Case C: doubling the dewatering effort going forward.

Resulting groundwater pressure distributions were input for geotechnical analysis using 2D pore pressure distributions developed empirically from the assessment of hydrographs (Figure A4.34). These indicated the development of enhanced permeability within the upper primary (green) and HOX (yellow) zones that are transitional to the overlying SOX (orange) and TS (brown) zones. The upper diagram shows the location of the current pit in purple. The ultimate design pit is shown in black in both diagrams. The upper figure shows locations of piezometers and the interpreted pressure distribution is unfiltered. It was determined that Case B interpretations satisfied minimum design criteria for the ultimate slopes at critical stages of mine development. The extrapolated pressurisation distribution for Case B dewatering in the lower diagram is filtered. The pressure distribution in both

diagrams indicates the relative permeability of the SOX relative to the TS.

In Figure A4.34, the upper diagram represents current (2010) pressure conditions, and the lower diagram represents an interpreted extrapolation of pressure distribution for the ultimate high wall for Case B.

6.7 Depressurisation response

Evaluation of depressurisation responses interpreted from hydrographs relative to the mine development rate indicates that a moderate level of depressurisation has been widely achieved in all sectors. A similar sustained dewatering effort was interpreted as being sufficient towards achieving targeted depressurisation levels for the ultimate slopes. Results are attributed to the effectiveness of horizontal drains and vertical dewatering wells and, to a lesser extent, from in-pit seepage. Figure A4.35 illustrates the typical response of vertically nested VWPs in a single borehole. The following is interpreted from these general observations.

- VWP grouting integrity appears to be good and data trends appear to be reliable.
- Except for the normal equilibration data spike immediately after installation, spikes in the graphs are interpreted to be telemetry system noise.

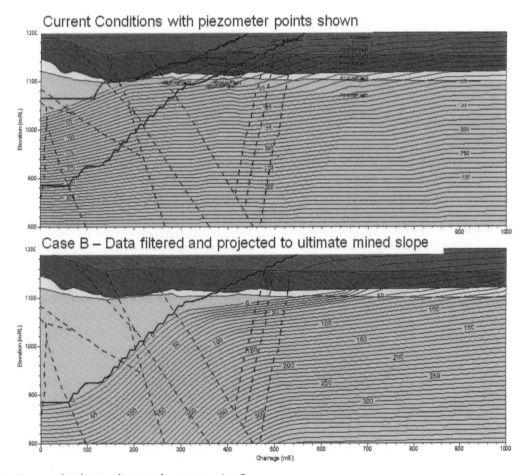

Figure A4.34: Pressure distribution diagrams for cross-section E

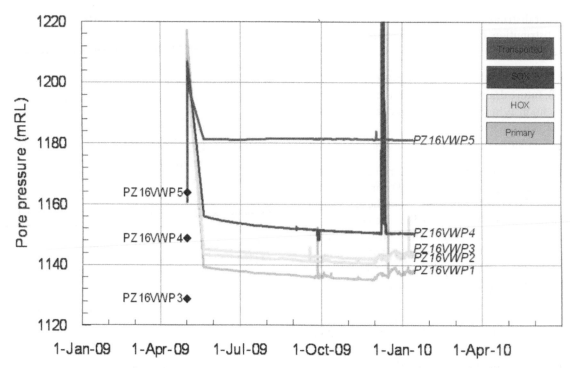

Figure A4.35: Typical response profiles of nested VWP hydrographs that indicate a downward head gradient through the oxide zone to the upper fresh rock zone

- A net downward head gradient is observed in nearly all VWP arrays.
- Similar pore pressure responses are evident for the HOX and primary rock zones, indicating a typical depressurisation rate of ~0.9 m/month.
- Depressurisation rate for the TS is typically the slowest at ~0.4 m/month.
- Pore pressure responses in the SOX zone is usually intermediate of the TS and HOX.

6.8 Dewatering well pumping and recovery test

A dewatering well pumping and recovery test was performed to further assess the hydrogeological system. The aim of the test was to apply a fixed hydraulic stress (constant head, or constant flow) and to monitor the resulting pressure drops. The test included:

- controlled pumping of existing dewatering wells for a six month period;
- shut-off the entire well dewatering system for a month with recovery monitoring;
- resume well pumping and measure local and general responses.

Continuous pumping occurred during the first half of 2010. The pump discharge rate during this period was less than 300 m³/day (3.5 l/s). All pumping wells were shut off on 2 June 2010; about half of them were turned back on

29 June 2010, and the remainder on 6 July 2010. The results of the test were as follows.

- a widespread response to the one month shut-down was measured;
- maximum recovery up to 10 m (equivalent pressure head increase) generally occurred within the horizon screened by dewatering wells in the HOX and upper primary zones;
- piezometer responses were particularly significant in the vicinity of major structures (e.g. Galway fault); the N–S trend of major fault structures produces a strong N–S groundwater flow anisotropy.

Hydraulic analysis results confirmed the bulk permeability of the upper primary and HOX zones, with a hydraulic conductivity of ~3 × 10⁻⁶ m/sec. Low material storage is inferred from the results with an estimated storativity value of ~5 × 10⁻⁴. The test results indicate the importance of dewatering wells as a component of the dewatering system for final mine development. However, additional wells will require strategic positioning which will be difficult to accommodate for final wall development, including the targeting of high-angle structures. Therefore, the strategy for achieving final slope depressurisation targets includes developing two access berms in the upper primary zone of the current cutback for drilling longer drains targeting the HOX and upper primary zones. Any shortfalls in achieving depressurisation targets would be mitigated by additional drainage measures or by allowing an increased time before mining of sensitive areas.

6.9 Horizontal drainage response example for ultimate slope depressurisation

Since 2009, most horizontal drains drilled at Cowal are in the 100 to 150 m depth range. In order to target the ultimate slopes for depressurisation, drain lengths drilled from the current pit must be increased to the 200 m plus range in some locations. Horizontal drains up to 400 m in length were drilled from the open pit eastward into the projected location of the final design high wall in May 2011 as part of the strategy to depressurise final design slopes in advance of mining. Figure A4.36 shows the locations of two horizontal drains relative to two nested vibrating wire piezometers (also shown in the figure). The two drains initially produced a flow rate of about 8 l/s and had a significant effect in lowering pressure heads in the two piezometers arrays. The drains were drilled through an interval of relatively permeable rock mass in the

vicinity of the saprock and fresh rock contact; they intersected northward trending high-angle permeable faults. Elevation heads decreased in the HOX and upper primary zones on the order of 35–50 m; and to a lesser degree in the SOX. As expected, elevation heads did not change significantly in the TS.

7 Whaleback South wall

Andrew Cottrell, Jed Youngs, Alice Dodman, Blair Douglas and Geoff Beale

Setting: Iron ore formation with low-permeability shales in North wall and complex stratigraphical sequence in South wall; about 310 mm annual rainfall.

7.1 Dewatering of the main ore body

Whaleback is one of the larger iron ore pits in Western Australia. The iron ore itself is easily dewatered by bores located at strategic areas inside the pit, with a total

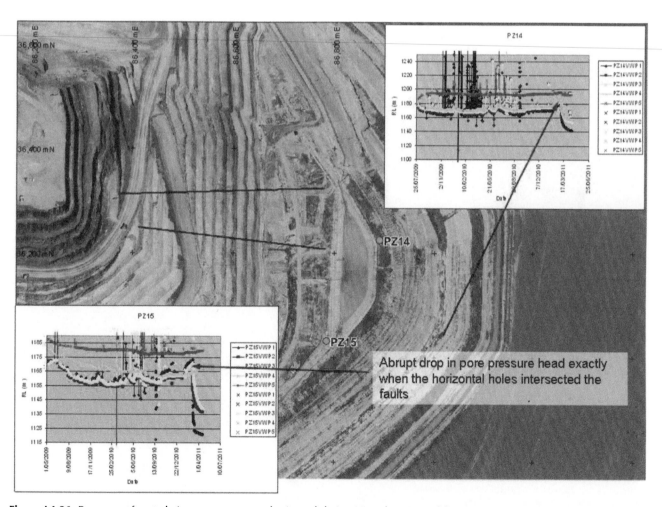

Figure A4.36: Response of nested piezometer arrays to horizontal drains. Note: locations of the piezometer collars are indicated by the green dots, relative to projected positions of two horizontal drain traces in red.

pumping rate varying between 30 and 50 l/s. Figure A4.37 shows the western part of the pit. The main issue for the dewatering bores is their longevity within an active mining area.

Abstraction targets for the dewatering bores within the main orebody include the permeable units of the Dales Gorge Member and Mt Sylvia Formation. The dewatering bores are operated on an intermittent basis due to incurred downtime for blasting and active mining. Recharge to the main ore zone is generally low and there is a relatively slow rebound in groundwater levels when the pumping rate declines. However, many of open standpipes and VWPs installed within the main iron ore sequence exhibit a rise in water level following rainfall events and in-pit runoff that accumulates on the pit floor.

The ore typically exhibits a dual porosity dewatering response, with most of the groundwater movement and ingress to the dewatering wells occurring along fracture zones and much of the storage occurring within the smaller fractures and porous rock mass. However, the margins of the pit to the north and south are excavated in more variable materials. Parts of the North wall include low-permeability shales. The South wall is comprised of a well-defined but complex stratigraphical sequence.

Initial monitoring of the dewatering system was carried out using a series of standpipe piezometers. As the pit was progressively expanded and the walls were pushed back into the more variable materials, a network of vertical and inclined VWPs was installed starting in 2009 to monitor pore pressures and delineate depressurisation responses to dewatering activities. Investigation work for the South wall initially focussed on the instrumentation of a type section to characterise the response of the various sedimentary units that occur behind the wall. Pore pressures are currently monitored at six hourly intervals. Figure A4.38 shows a view of the South wall. Figure A4.39 shows the cross section where the initial piezometers were focussed.

7.2 Conditions in the South wall

The sedimentary sequence behind the South wall exhibits folding and cross-faulting but the units are mostly

Figure A4.37: View of the Mt Whaleback pit, looking north-west

Figure A4.38: View of the South wall of the Mt Whaleback West pit (Source: Carmelo Bellia)

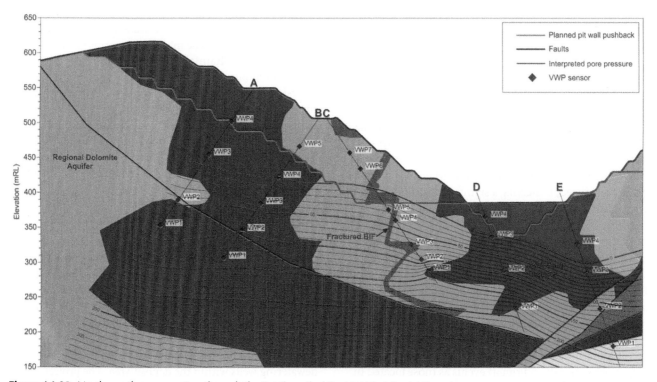

Figure A4.39: North–south cross-section through the South wall of the Mt Whaleback West pit

sub-parallel to the slope and steeply dipping. Most of the units are of low permeability and are not 'pumpable', and most of the piezometers in the South wall only show a limited response to the main in-pit dewatering program. Steep pore-pressure gradients are observed behind the planned South wall pushback, in contrast to the shallower pressure gradients that are observed within the main ore units.

However, the stratigraphical sequence within the South wall includes Bruno's Band which is the uppermost BIF unit of the Mt Sylvia Formation. Bruno's Band contains good fracturing and two active dewatering bores (WB40 and WB41) pump from this unit. The combined yield from these wells is about 20–30 l/s.

The current VWP installations for the South wall show that:

- a good pumping response is observed in Bruno's Band, but with a low specific drawdown, indicating that active recharge to the unit is occurring;
- there is good hydrologic connection along the strike of Bruno's Band, and therefore for a reasonable distance along the strike of the wall (behind the wall);
- the influence of pumping from Bruno's Band also extends across strike and into the inter-bedded shale units in the wall (Bee Gorge Member, Mt McRae Formation). The response within the adjacent units is lagged but a significant amount of drawdown occurs. The cause of the cross-connection is uncertain but may be related to cross-cutting structures;
- short-term recovery periods are observed in the hydrographs which can be correlated to periods of flow reduction in the Bruno's Band wells.

7.3   Interpretation

The main factor controlling the depressurisation of the South wall is the potential groundwater recharge moving along the strike of Bruno's Band from the west, from the direction of the Wittenoom Dolomite, which occurs to the west and south of the pit. Pore pressure data for the South wall exhibit a significant recovery rate when pumping rates are reduced, indicating a significant recharge component moving sub-parallel to the strike of the bedding.

Pumping of Bruno's Band also influences other units extending across the strike of the formations towards the south, where sensors show lower depressurisation rates. Recorded pore pressure decreases in the South wall are generally low, and a significant piezometric level recovery rate is observed when abstraction rate from the two active bores in Bruno's Band (WB40 and WB41) is reduced.

The South wall piezometers do not show the same response to rainfall and runoff events as those located in the central pit area.

8   La Quinua, Peru

Victor Perez, Hector Robles, Franz Soto, Geoff Beale, David Rios and Mike McGlone

Setting: Gold deposit hosted in debris flow material (layered gravels and ferricretes) underlain by strongly altered volcanic rock. About 1250 mm mean annual rainfall within a five to six month wet season.

8.1   General site setting

The Minera Yanacocha operations consist of multiple mining areas located within a district encompassing about 18 km east–west and about 14 km north–south. There are currently six pits being actively mined, of which La Quinua is the largest (Figure A4.40). The La Quinua ore body is hosted in a sequence of strongly cemented silty-sands and gravels which show strong geologic layering and exhibit a number of prominent ferricrete layers. It is thought that the deposit was laid down mostly as a debris flow eroded from the steep slopes above. The combined LSG (lower sequence gravel) sequence is typically up to 300 m thick in the central part of the mining area (Figure A4.41). The eastern boundary of the deposit area is controlled by the prominent north–south trending La Quinua fault, which dips steeply to the west and has played a major part in the design and performance of the La Quinua high wall.

Mean annual precipitation (about 1250 mm) mostly occurs as rain during the pronounced wet season between October and May. In addition to providing seasonal recharge to the site-wide groundwater system, intense rainfall events need to be managed in terms of their effect on slope stability. Rainfall events of greater than 20 mm/hr occur commonly during the wet season. Surface water diversions are required outside of the pits to prevent run-on to the slopes below. The weaker materials in the pit slopes themselves need protection to prevent erosion and back-cutting. In addition, the surface water needs to be shed from the slopes as quickly as possible to minimise the extent to which seasonal recharge sustains the pore water pressures in the underlying slope materials.

8.2   General mine dewatering

The pre-mining groundwater through-flow at the site was estimated to be about 150 l/s. Initial groundwater levels varied from about 3750 masl in the area of the East high wall to about 3500 masl in the lower parts of the site (Figure A4.42). The more permeable gravel layers typically show permeability values within the range 10^{-4} to 10^{-5} m/s, while the intervening layers show permeability values below 10^{-7} m/s. As such, it was possible to install production dewatering wells within the deposit area. The project was successfully dewatered using a field of up to 20 wells

Figure A4.40: La Quinua pit (looking west)

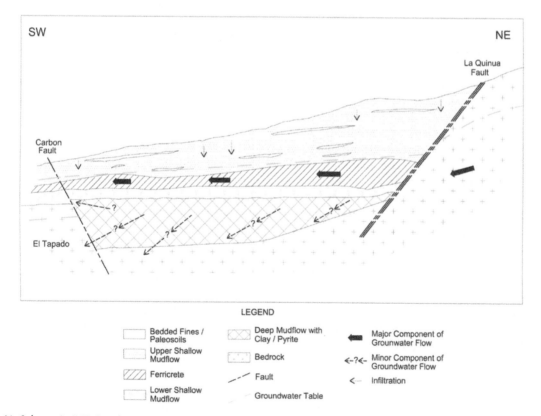

Figure A4.41: Schematic (NE–SW) long section through the La Quinua ore body

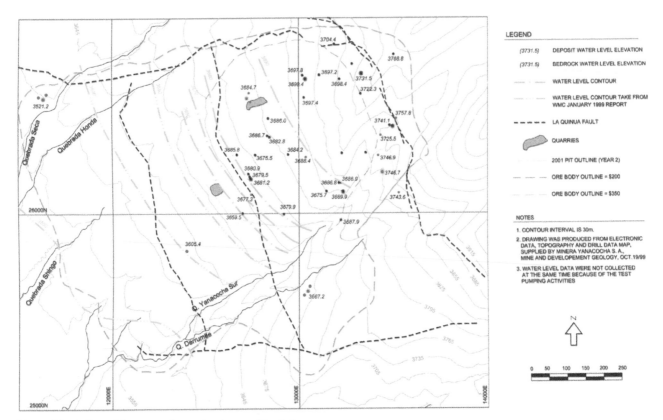

Figure A4.42: Pre-mining groundwater level map of La Quinua

producing a total of 250 to 300 l/s with a combination of interceptor wells in the bedrock upgradient of the La Quinua fault zone and in-pit dewatering wells within the Phase 1 gravel sequence itself. The large number of mineral exploration holes drilled through the deposit area helped cross-connect the permeable layers within the gravel sequence and create good vertical drainage of the orebody.

Although the dewatering system was installed according to design, it became increasingly difficult to create drawdown during the final phases of mining. Pilot hole drilling beneath the LSG orebody had revealed the presence of permeable silica alunite material below the base of the deposit area. It was therefore necessary to install a series of angled gravity drain holes along the toe of the East high wall to reduce the artesian bedrock pressures in the silica alunite behind the toe of the wall and also additional in-pit bedrock wells to depressurise the silica alunite below the base of the deposit. These wells were also artesian (flowing at the collar). With these additional measures, the original mine plan was successfully completed, including the mining of three additional benches below the base of the original pit.

8.3 Phase 2C area

The Phase 2C LSG ore body was hydraulically isolated from the main part of the La Quinua mining area (the Phase I area). A supplementary in-pit dewatering wellfield was therefore planned to lower the water levels in the 2C deposit area ahead of mining. A total of six in-pit dewatering wells was installed to lower the water levels in the Phase 2C area. The location of the wells is shown (in black) in Figure A4.43. The combined yield of the system typically ranged between about 50 and 120 l/s, as shown in Figure A4.44.

8.4 East high wall

The height of the 2C East high wall was progressively increased during the course of mining to a final height of about 450 m (Figure A4.45). The material exposed in the slope was predominantly cemented gravel material and ferricrete. During early 2010, cracking was observed around the crest of the slope, and acceleration was observed in prisms within the Zone 1 area. The initial movement in the prisms was up to about 7 mm/day and new prisms, inclinometers and VWPs were installed as part of a detailed investigation. The approximate location of the failure surface is shown on Figure A4.46. Behind the main part of the wall, the failure surface ran mostly along the contact zone between the bedrock and the overlying LSG material. Below the toe of the slope and the pit floor, the failure surface was about 70 m below bench level.

The investigation results identified strong artesian pressures in bedrock below the failure surface, as shown in

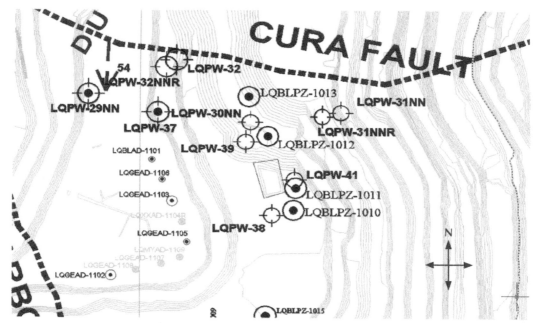

Figure A4.43: Map of Phase 2C showing dewatering wells (LQPW series) and piezometers

Figure A4.46. At the toe of the slope, the head in the material immediately below the level of the failure surface was up to 35 m above the bench level. Input of the observed water levels into a limit equilibrium analysis identified that the high water levels were creating a very low factor of safety for the slope (analysed at less than 1.1). Sensitivity analysis indicated the factor of safety could be significantly improved by lowering of the water pressures, particularly in the material immediately below the failure surface.

8.5   Remediation plan

The results of the geotechnical analyses were used to develop a remediation plan for the slope. The remediation plan had three basic components, as follows.

- Repair cracking in the surface water diversion channel around the crest of the pit in order to reduce the possibility of surface water directly entering the upper slope materials.

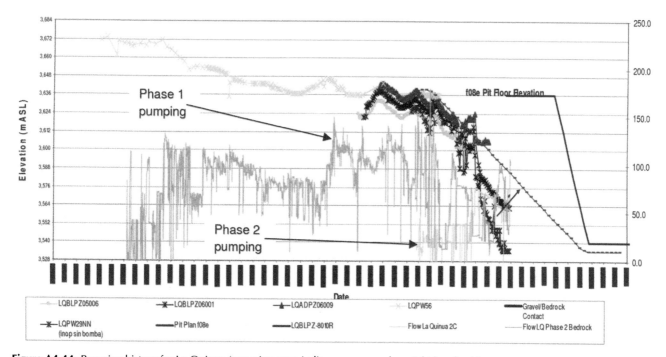

Figure A4.44: Pumping history for La Quinua (pumping rates in litres per second on right hand axis)

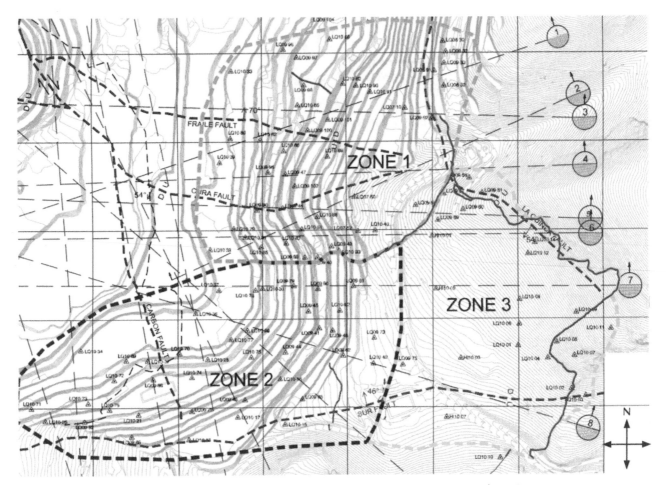

Figure A4.45: Map showing 2C remediation area (with prism locations) (Source: Gabriel Alfonso Becerra Abregu)

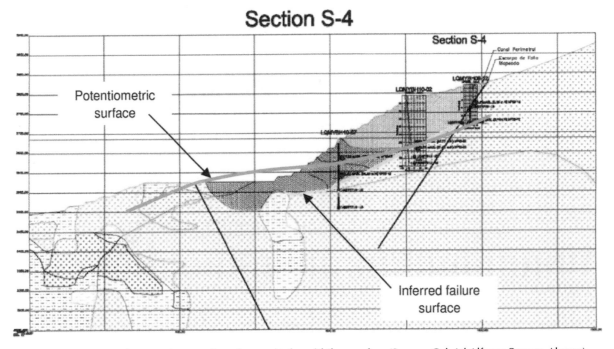

Figure A4.46: Cross section showing piezometric surface and inferred failure surface (Source: Gabriel Alfonso Becerra Abregu)

Figure A4.47: Photograph of the Phase 2C area looking south, showing the buttress and remediation wells on the 3588 L bench

- Construct a buttress immediately north of the Zone 1 area to help stabilise the movement in the short term.
- Carry out a pilot hole and production well drilling program to lower water pressures in the bedrock below the failure surface, and to gradually reduce water pressures in the overlying LSG material through the creation of a downward gradient in water pressure.

Repairs to the diversion channel around the crest of the pit were initially achieved by using a temporary dam in the upstream channel and placing bypass pipelines above the main area of cracking. The temporary measure provided immediate remediation for surface water infiltration while the movement and cracking was still occurring. More permanent repairs to the channel by infilling of the cracks and re-lining of the base of the channel were subsequently carried out following a reduction in the rate of movement and when the propagation of new cracks ceased.

Drilling of pilot holes identified the locations for five remediation wells to achieve the required slope depressurisation. All five remediation wells were installed on the 3588 L step-out bench which was 24 m

above the lowest Phase 2C pit floor level at the time. The wells were constructed by drilling 20-inch diameter holes and installing 14-inch diameter casing and screen. Their location is shown on Figure A4.47 (looking south) and Figure A4.48 (looking north).

The buttress was constructed immediately to the north of the Zone 1 area and the area of the production wells using 2.1 million tons of waste rock material. The toe was placed on the 3588 L step out bench, with some overspill occurring to the bench below. The maximum height of the buttress was 44 m. The goal was to first use the buttress to help provide initial stabilisation of the slope, then use depressurisation to provide long-term stabilisation in the Zone 1 and Zone 2 areas to the south of the buttress.

The remediation wells reached a maximum yield of about 100 l/s in Zone 1 and about 65 l/s in Zone 2. Figure A4.49 shows the pumping rate and the resulting water level response for Zone 2. A pressure reduction of up to 1500 Kpa (~150 m drawdown) was achieved within a period of two to three months in each zone and the pumping reversed the upward total head gradient in the lower part of the failure. Progressive reduction in the

Figure A4.48: Photograph of the Phase 2C area, looking north

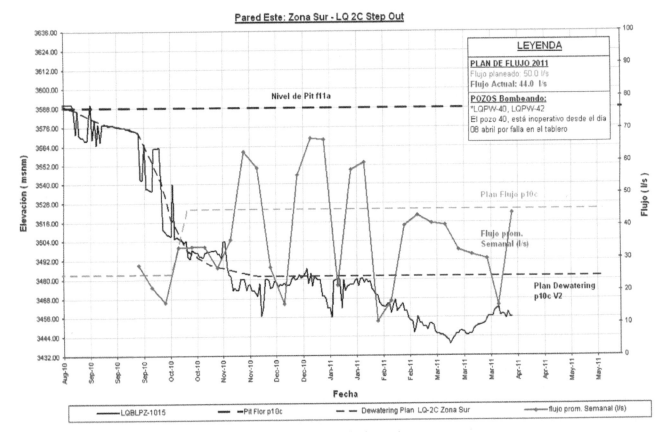

Figure A4.49: Hydrograph of pumping rate in Zone 2, plus piezometer hydrographs

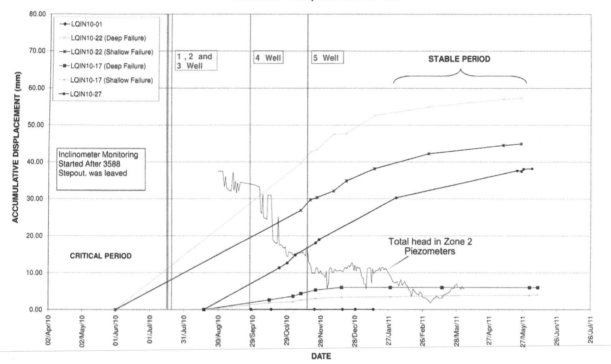

Figure A4.50: Plot of inclinometer data showing the reduction in the rate of movement (note: scale bar is tens of metres for the plot of total head)

inclinometer movements and slowing of the prisms was coincident with the increase in drawdown, as shown (for the inclinometers) in Figure A4.50. Following achievement of maximum of drawdown in each zone,

movement rates in the prisms generally slowed to rates of 2 mm/day or less. There was no further increase in movement and the remainder of the pit was successfully mined.

Appendix 5

Cases studies for numerical modelling

Loren Lorig and Jim Hazzard

Two case studies have been carried out to support the numerical modelling concepts described in Section 4 of this book. The first study was carried out using the dataset for the North-west wall of the Diavik A154 pit to provide numerical support for the use of equivalent porous medium (EPM) rather than fracture flow models. The second study uses data from a 140 m thick sequence of low permeability, layered Tertiary marls at the Cobre Las Cruces mine in Andalucia, Spain, to illustrate an example where hydromechanical coupled modelling is applicable and necessary. For the purposes of illustration, and to allow integration of geotechnical and hydrogeological properties, the case studies have been carried out using the groundwater flow packages included within the geotechnical codes *FLAC* and *UDEC*.

Case study 1: Numerical modelling of the North-west wall of the Diavik A154 pit

1.1 Analysis of the DFN

1.1.1 Introduction

The site-setting, geology and set-up of the discrete fracture network (DFN) for the North-west wall of Diavik is described in detail in Appendix 3. A DFN is a stochastic representation of a fracture network that may include larger features deterministically. The DFN definition provides statistical information about fracture spacing, length, orientation, spatial distribution and apertures. Based on this statistical information, different realisations of the fracture network can then be created. Values of fracture transmissivity can be estimated based on aperture values and field testing results. However, when interpreting the results, it should be noted that a large limitation for using DFN models in groundwater simulations is that the aperture of any given fracture may

have a wide variation and it is usually not possible to characterise the 'pinch points' (Section 1.1.2.5; Figure 1.16).

One realisation of the Diavik DFN was used in this case study. The location is shown in Figure A5.1. The goal is to help answer the questions:

1. can a discrete fracture network be represented by an EPM?
2. if an EPM can be used as an approximation for the DFN, what are the corresponding hydromechanical properties?

Distribution of fracture size and aperture are shown in Figure A5.2. This figure shows a power law distribution of fracture sizes and an approximately log-normal distribution of apertures, with apertures ranging over

Figure A5.1: DFN of North-west wall (shown in blue) and location of cross-section used in analysis (red line)

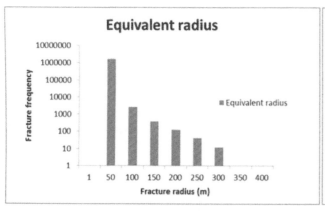
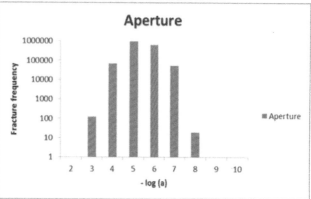

Figure A5.2: Fracture size distribution (left) and aperture distribution (right)

about 5 orders of magnitude. Figure A5.3 shows the spatial distribution of fractures in the 2D section of interest. From these figures, the following observations can be made.

1. The largest fracture size is in the range of the slope size; if the frequency of the large fractures is very low while their average aperture is significantly higher than the rest of the fractures, they need to be explicitly modelled.
2. The spatial distribution of fractures shows strong inhomogeneity on a macro scale. In other words, the density of fractures changes significantly within different locations in the model.

3. Aperture variations will likely result in an inhomogeneous hydraulic system, even for locations where fracture density seems to be uniform.

1.1.2 Applicability of an EPM approach to a DFN

To determine whether the fluid transport properties of a DFN can be approximated by an EPM, three dimensionless metrics have been used.

- The first metric, m_1, measures the difference between the steady-state pressure distribution in the EPM and the jointed rock medium.

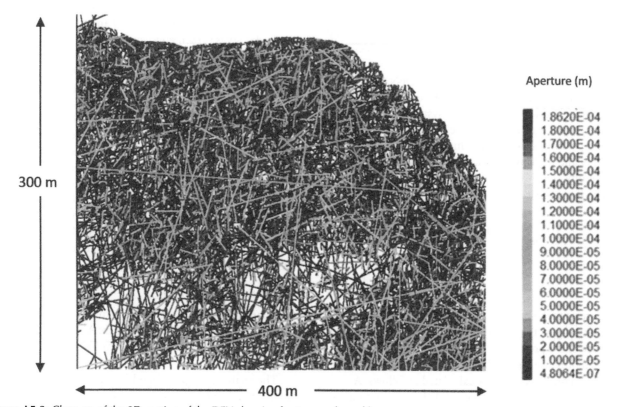

Figure A5.3: Close-up of the 2D section of the DFN showing fractures coloured by aperture

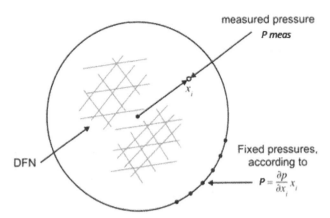

Figure A5.4: Sample DFN for evaluation of metric to test adequacy of continuum

- The second metric, m_2, measures the difference between the directional conductivity of the EPM and the DFN. In other words, it measures how different the conductivity tensor of the EPM is from hydraulic conductivities of a DFN.
- The third metric, m_3, measures the difference in evolution of pressures in an EPM and the DFN.

The study has focused on the steady-state behaviour of the two systems, i.e. the EPM and the DFN. While the use of a steady-state analysis is generally not appropriate for most active pit slopes, it represents a reasonable approximation for comparative purposes at Diavik because of the very close recharge boundary (Lac de Gras), the reasonably high conductivity of the fracture network, and the relatively long time between excavation of successive mine benches.

For a circular sample taken from the DFN, a uniform pressure gradient is imposed along the entire boundary of the sample (Figure A5.4). The flow system was simulated to steady-state, and a metric value was produced based on the comparison between the computed fluid pressure in the DFN and the fluid pressure field of the equivalent continuum.

The first metric, m_1, is calculated using

$$ m_1 = \frac{\left[\sum P_{meas} - P_{pres}^2\right]^{1/2}}{N\overline{P}} \quad \text{(eqn A5.1)} $$

where

$$ P_{pres} = p_0 + g_i x_i $$

represents the pressure of the continuum field at location x_i, g_i is the prescribed gradient, $\overline{P}$ is the average pressure of the pressure field and P_{meas} represents the sampled pressure within the DFN.

The second metric, m_2 is calculated as follows. A pressure gradient is imposed along different directions and the directional hydraulic conductivity in the direction of hydraulic gradient, K_I, is evaluated. K_I is defined as

$$ K_I = \frac{|q|\cos\theta}{|J|} \quad \text{(eqn A5.2)} $$

where $|J|$ is the magnitude of hydraulic gradient, and $|q|\cos\theta$ is the value of specific discharge in the direction of $\vec{J}$. In this procedure, K_I is calculated for different orientations, θ, of hydraulic gradient. For each orientation, a point, P, is plotted that represents a vector of magnitude and orientation of $1/\sqrt{K_I}$. If the conductivity matrix qualifies as a tensor, the set of points P will conform to an ellipse. In this case, the ellipse semi-axes are $1/\sqrt{K_1}$ and $1/\sqrt{K_2}$, where K_1 and K_2 are principal values of conductivity. Also, the principal directions of conductivity are estimated from the ellipse orientation.

The difference in the approximation of directional hydraulic conductivity with an ellipse and the directional hydraulic conductivity of the DFN is calculated using the following:

$$ m_2 = \frac{\sqrt{\sum_N (r_p - r_e)^2}}{\sum_N r_e} \quad \text{(eqn A5.3)} $$

where r_p is the modulus of P, and r_e is the modulus of the point on the ellipse that has the same radial coordinate angle as P.

1.1.3 Numerical evaluation of DFN

Knowing that the Diavik DFN is inhomogeneous due to spatial variation of fracture density and transmissivity, a methodology is presented to evaluate if an EPM can reasonably represent the Diavik DFN. First, analysis is performed to determine if regions of the DFN with a dimension less than the characteristic length of macro-scale inhomogeneities can be represented by a corresponding EPM. In this section, samples of the DFN of different sizes and at different locations are evaluated to determine EPM properties (if possible). Testing of different sizes represents an attempt to find the Representative Elementary Volume (REV) for which the fractured discontinuum can be represented by a continuum.

In Section 1.1.6 of this appendix, pressure distributions obtained from a slope-scale DFN discontinuum model are then compared to results from an EPM continuum model.

1.1.4 EPM approximation

Circular samples were extracted from the 2D DFN at 16 different locations as shown in Figure A5.5. The diameter of the samples, D, is taken to be 10 m, 20 m, 40 m, 60 m, 80 m and 100 m.

The analyses are performed for the steady-state flow condition using *UDEC*. It is assumed that the intact rock (i.e. *UDEC* blocks) are impermeable, and mechanical coupling is disregarded. In *UDEC*, solution time increases

Figure A5.5: Sample locations for DFN testing (elevations are compatible with mine elevations, while x coordinates are represented with UDEC's centre of coordinate system)

with increasing number of blocks. Thus, incorporation of all fractures in the samples slows down the solution. Therefore, samples are constructed from the DFN by imposing a minimum aperture threshold of 2×10^{-5} m. This was not thought to introduce significant error

because flow will be dominated by larger aperture fractures due to the fact that, in theory, fracture transmissivity is proportional to the aperture cubed (a^3).

Figure A5.6 shows the geometry of samples and the aperture distribution for the *UDEC* model with a

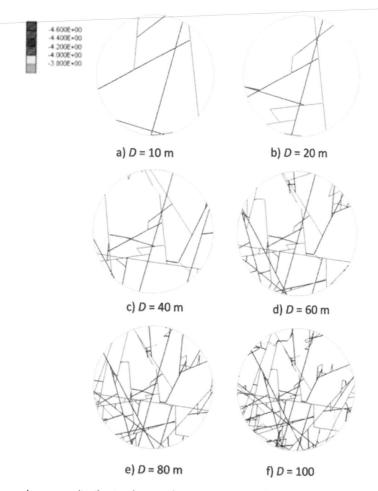

Figure A5.6: Block geometry and aperture distribution for samples at z = 150 m and x = –25 m. Fractures are coloured by –log(a) (see Figure A5.2)

Table A5.1: Results for samples at elevation (z) 150 m and x = –25 m

	D (m)	10	20	40	60	80	100
X = –25	m_1 (%):	0.03	0.03	0.11	0.05	0.16	0.08
	m_2 (%):	3.15	1.70	0.91	1.63	1.40	1.52
	Alpha (degrees):	158	3	40	51	52	46
	K_a (m/s):	4.5E-08	2.7E-08	1.5E-08	9.7E-09	7.3E-09	7.3E-09
	K_b (m/s):	1.8E-08	1.6E-08	7.7E-09	5.3E-09	4.5E-09	4.5E-09
	Scale_Ratio:	2.5	2.7	3.7	4.8	6.0	8.6
X = –100	m_1 (%):	0.1	0.03	0.17	0.11	0.10	0.27
	m_2 (%):	9.81	4.19	1.45	0.71	0.77	0.67
	Alpha (degrees):	86	152	162	149	140	129
	K_a (m/s):	4.5E-08	1.7E-07	1.1E-07	2.6E-08	2.1E-08	1.5E-08
	K_b (m/s):	7.8E-09	1.5E-08	3.1E-08	1.5E-08	1.2E-08	1.0E-08
	Scale_Ratio:	2.5	4.9	9.0	11.0	13.1	15.9
X = –175	m_1 (%):	0.01	0.02	0.18	0.09	0.05	0.06
	m_2 (%):	1.46	13.67	2.21	0.98	1.50	1.79
	Alpha (degrees):	18	75	8	17	165	164
	K_a (m/s):	1.5E-08	1.3E-08	2.2E-08	1.6E-08	1.9E-08	1.9E-08
	K_b (m/s):	6.9E-09	2.3E-09	4.5E-09	6.8E-09	6.2E-09	5.4E-09
	Scale_Ratio:	2.8	4.7	7.4	9.7	12.4	14.5
X = –250	m_1 (%):	no Ellipse	0.01	0.3	0.04	0.05	0.05
	m_2 (%):	no Ellipse	0.72	0.53	1.0	0.90	1.11
	Alpha (degrees):	no Ellipse	97	111	113	102	89
	K_a (m/s):	no Ellipse	2.9E-08	1.4E-08	1.1E-08	8.1E-09	5.6E-09
	K_b (m/s):	no Ellipse	1.1E-08	7.0E-09	4.2E-09	4.8E-09	5.6E-09
	Scale_Ratio:	no Ellipse	5.5	6.4	6.9	7.5	8.1

minimum aperture threshold of 2×10^{-5} m for a sample located at z = 150 m, and x = –25 m. Figure A5.7 shows the directional hydraulic conductivity values (i.e. $1/\sqrt{K_I}$) and the ellipse best fit to the discrete points representing $1/\sqrt{K_I}$ for various imposed gradient directions.

Table A5.1 shows values of metrics m_1 and m_2, maximum and minimum directional conductivity K_a and K_b, the anisotropy angle (alpha), and the scale ratio (defined as the ratio of average block size to sample diameter) for samples located at z = 150 m, and x = –25 m.

The metric m_1 measures the difference of the pressure field for the EPM and the DFN for the steady-state condition. In other words, m_1 measures directly the effects arising from inhomogeneity in hydraulic conductivity of the DFN compared to a homogeneous hydraulic conductivity field. Table A5.1 shows values of m_1 for all locations and sizes, except for one case in which the spacing was too sparse to extract EPM information. This is true for the other rows as well, suggesting that the effect of heterogeneity in hydraulic conductivity in the DFN is not considerable.

The metric m_2 measures the difference between the directional hydraulic conductivity of a DFN and the estimated hydraulic conductivity tensor of the EPM. Figure A5.8 shows values m_2 obtained for *UDEC* analyses of samples versus scale ratio (defined as the ratio between sample size and average block size). It must be noted that, for the DFN used in this study, determining scale ratio is very subjective. This is due to the fact that the average block size changes substantially depending on the imposed lower aperture threshold. Because most DFNs have non-uniform aperture distributions, it can be argued that the scale ratio must be determined based on the notion of dominant fractures. When fracture apertures have high variations, transmissivities show much higher variation due to the cube law relation. Thus, the transport properties are controlled by a network of connected fractures with higher aperture, which can be perceived as the dominant fractures. Therefore, it is thought that the scale ratios shown in Figure A5.8 are meaningful, even though smaller fractures have been filtered out.

The results show that in general, m_2 does not decrease with increasing sample size as might be expected. This is likely due to the fractal distribution of fracture sizes and the inhomogeneity in apertures and spacings. A true fractal distribution should be 'scale-free,' as long as a minimum

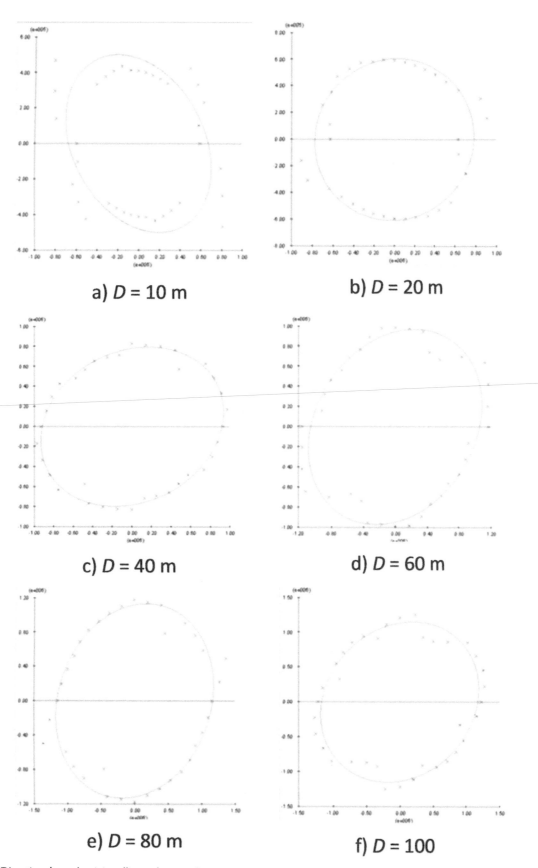

Figure A5.7: Directional conductivity ellipses for samples at z = 150 m and x = –25 m. Conductivities from the fracture model are shown as crosses. Circles represent the best-fit ellipse.

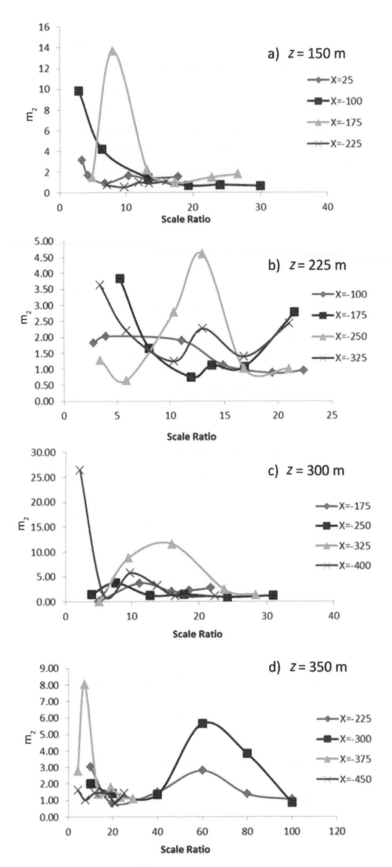

Figure A5.8: Metric m_2 vs scale ratio for different sample locations

sample size is surpassed (percolation threshold). If the smallest sample sizes are discounted, then all values of m_2 are less than 15 per cent.

1.1.5   Estimation of directional conductivity

Maximum and minimum components of the best-fit conductivity tensor are shown in Figures A5.9 and A5.10. These figures together indicate that, for each sample location, the variation in values of conductivity decreases for sample sizes more than 60 m (in general). In DFNs with variation of apertures, it is important to consider the convergence of values (or convergence properties of hydraulic conductivity), because it is possible that a small sample may give a low m_2 error (particularly if DFN is not filtered, or if a lower threshold for aperture is imposed). However, the hydraulic conductivities associated with the calculated m_2 are not representative of the true directional conductivities of the DFN in a larger scale because the sample does not contain any large fractures.

These results suggest that, for sample diameters more than 60 m or scale ratios more than five, a best-fit ellipse could be obtained. This agrees with analysis of Golder (2007) which shows representative hydraulic conductivities can be obtained for block sizes >50 m. However, due to the presence of macro-scale inhomogeneity, increasing sample diameters does not necessarily lead to decreasing values and convergence of the directional conductivity values. Thus, for selection of the best representative volume (REV) for DFNs, it is inferred that the following two criteria should be met.

1. Convergence in measured conductivity with increasing sample size.
2. Error threshold in value of proposed metrics.

The metric m_1 represents the error in the pressure field. For open pits, an error of less than five per cent is probably acceptable. The metric m_2 gives the error of a best-fit conductivity ellipse. The differences in the calculated conductivities both spatially and with different sample size are also important as described above. Given the large variation in conductivities measured in typical open pits, it seems reasonable that

variations in conductivity less than 1 order of magnitude are probably acceptable.

Figure A5.11 shows the range of calculated hydraulic conductivity values at each tested depth for the chosen REV samples. Conductivity values at each location are obtained from the sample with the minimum value of m_2 and a size greater than or equal to 60 m. This plot shows that the variation in conductivity is generally less than one order of magnitude. In this case, there is no obvious relationship between conductivity and depth. For comparison, values measured from packer tests are shown in Figure A5.12. Values obtained from the DFN are within the range of values measured in the field.

The anisotropy ratio and the anisotropy angle for the effective conductivities of the REV samples are presented in Table A5.2. The anisotropy ratio is defined as the ratio between maximum and minimum directional conductivity. These results suggest that in most cases, the anisotropy ratio remains below two. Considering the range of anisotropy ratio for rock masses, the range observed in this study is small. This implies that the DFN does not have a dominant anisotropic property. Also, there does not appear to be a dominant angle of anisotropy.

1.1.6   Comparative slope-scale analysis

In this section, a model representing the 2D section of interest (shown in Figure A5.1) is analysed. The fluid flow behaviour of the section is simulated using a fracture flow model and also an EPM model. Single phase fluid flow is simulated, and the pore pressure distribution is run to steady-state condition.

The geometry and boundary conditions for the *UDEC* DFN model are shown in Figure A5.13. It was assumed that, outside of the given DFN region, hydraulic conductivities are equal to the geometric mean value of hydraulic conductivities obtained from the testing of the previous sections.

The EPM model was created in *FLAC*. The model was constructed using the conductivities calculated for REV samples at each sample location as described in the previous section. A double linear interpolation was used to set the conductivities between measurement points.

Table A5.2: Anisotropy angle and anisotropy ratio for the REV samples at different locations.

z = 150 m	Alpha (degrees)	37.99	152.73	6.84	105.80
	K_a/K_b	1.38	1.63	2.36	1.82
z = 225 m	Alpha (degrees)	25.32	121.27	70.17	144.93
	K_a/K_b	1.30	1.36	1.40	1.77
z = 300 m	Alpha (degrees)	88.47	115.71	172.87	103.61
	K_a/K_b	3.05	1.45	1.81	1.98
z = 350 m	Alpha (degrees)	149.24	87.33	165.97	128.30
	K_a/K_b	3.76	1.67	1.95	1.40

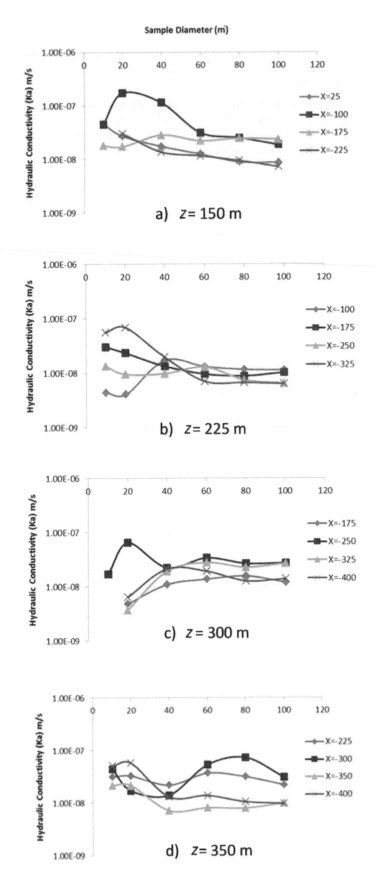

Figure A5.9: Maximum directional permeability values (K_a)

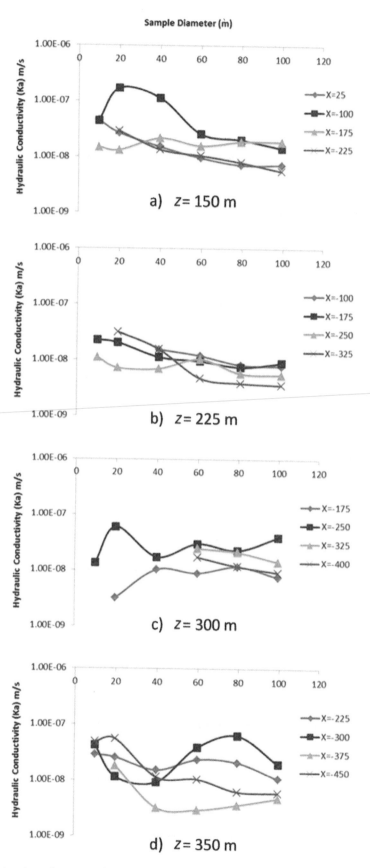

Figure A5.10: Minimum directional conductivity values (K_b)

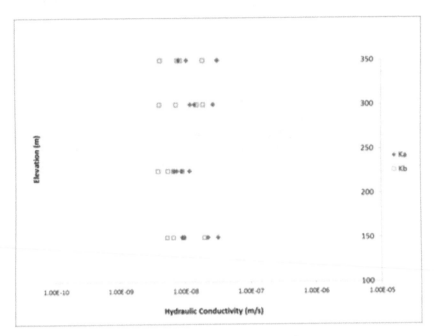

Figure A5.11: Calculated values of maximum and minimum components of the conductivity tensor

Outside of the DFN region, the geometric mean value was used. The model is shown in Figure A5.14.

Pressures from the DFN model superimposed on the EPM model are shown in Figure A5.15. This shows a generally good match except towards the bottom of the DFN. The bottom left shows pressures that are larger in the DFN model, and the bottom right shows pressures that are lower (see Figure A5.16). This is likely due to the very

low density of fractures at the bottom centre-left of the DFN causing a barrier to the recharge. This was not accounted for in the EPM model, because a sample could not be extracted from the region with a sufficient number of fractures for flow testing. The lack of any data control in this region should also be noted.

The DFN used in this study is only one possible representation of the fracture flow network (Section 1.1.1).

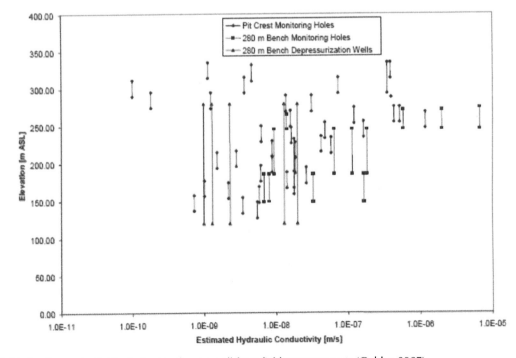

Figure A5.12: Hydraulic conductivities in the North-west wall from field measurements (Golder, 2007)

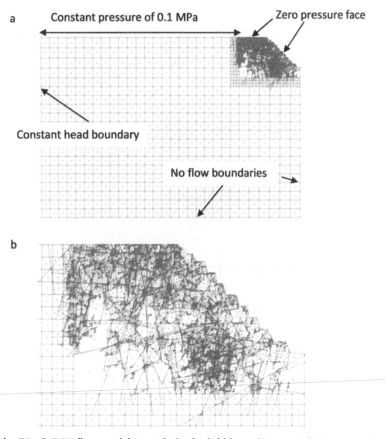

Figure A5.13: Geometry of the Diavik DFN flow model (a) with the far field boundaries and (b) close-up view of the slope region (a minimum aperture threshold of 2×10^{-5} m is imposed on the DFN)

Matching exactly the inhomogeneities in a particular DFN is probably not useful when there is no obvious trend or spatial correlation, as in this example. When this is the case, a homogeneous model may be sufficient. A second EPM model was constructed in which the conductivity was set to the geometric mean (m/s) throughout. The

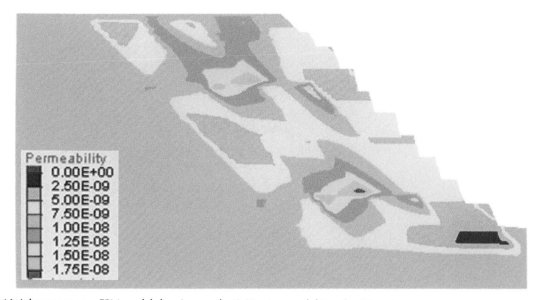

Figure A5.14: Inhomogeneous EPM model showing conductivities (permeability values) in m/s

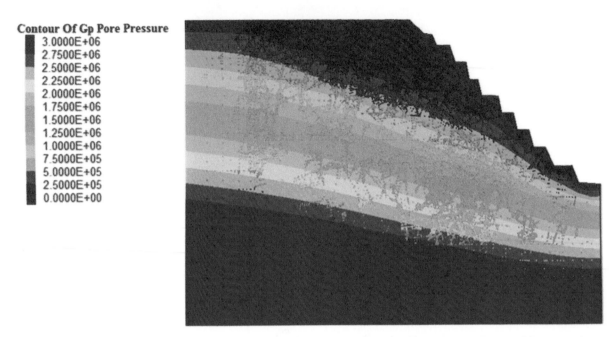

Figure A5.15: Pressures from the DFN model (dots) superimposed on pressures from the inhomogenous EPM model (contours) (pressures are in Pa)

pressures from this model compared with the fracture flow results are shown in Figure A5.17. The modelled pressures are very similar to those of the inhomogeneous model of Figure A5.15. Small differences in the contours can be observed at the toe of the slope where a region of low conductivity exists in the inhomogeneous model.

In general, the homogeneous EPM model gives a similar representation of the pressures in the slope compared to the DFN model. The EPM model is equally valid given the controlling data on which both models are based.

1.2 Use of the EPM Model

1.2.1 Geotechnical properties

The DFN has been used to derive equivalent continuum properties for the granite in the North-west wall. The granite is a strong, competent, moderate to well-fractured

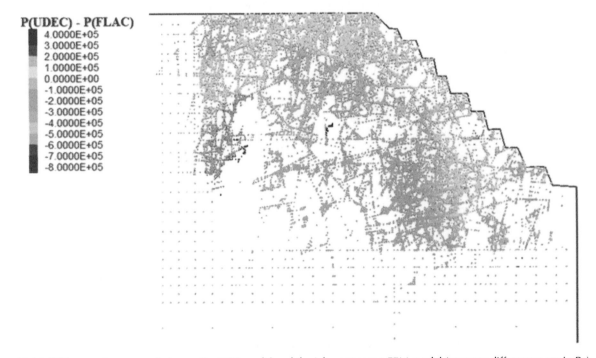

Figure A5.16: Difference in pressure between the DFN model and the inhomogenous EPM model (*pressure* differences are in Pa)

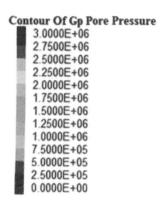

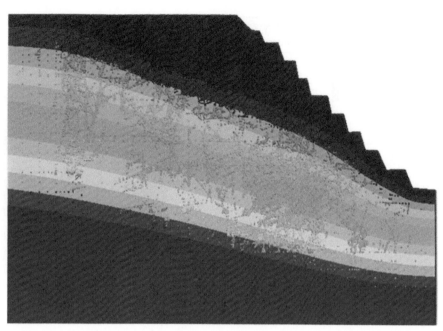

Figure A5.17: Pressures from the DFN model (dots) superimposed on pressures from the homogeneous EPM model (contours) (pressures are in Pa)

rock mass with about 15 per cent pegmatite and 5 per cent metasediments (biotite schist) mostly forming sub-horizontal rafts. Strength and stiffness results from laboratory testing on the wall rock are shown in Table A5.3. As noted in Section 3.3 of the main text, Dewey's Fault zone does not intersect the North-west wall and is therefore not part of the DFN.

Field estimates of rock strength place the granite in the 'strong rock' category according to ISRM strength rating system (Brown, 1981). This category suggests a UCS between 50 and 100 MPa, slightly lower than the value obtained from laboratory testing. Joints generally fall into three broad fracture sets: two steeply dipping sets (>80 degrees); and one flat-lying set (10 degrees to 20 degrees). Average joint length is about five metres and average spacing is on the order of one metre. Joints are generally clean and rough. All of the joint data are encompassed in the DFN.

Bieniawski's Rock Mass Rating (RMR) system of 1973, modified in 1976, was used to characterise the rock mass (Bieneawski, 1976). Block models were constructed to provide spatially variable RMR values throughout the A154 pit site. Average RMR_{76} is consistently in the 65 to 75

range throughout the mining depth with no clear trends laterally or vertically. The rock quality of the North-west wall is thought to trend towards the high end ($RMR_{76} \approx 75$).

For competent rock masses, it can be assumed that RMR_{76} is approximately equal to the Geological Strength Index (GSI) of Hoek (1994). Hoek and Diederichs (2005) provide the ratio of mass to intact Young's modulus for a rock mass based on Geological Strength Index (GSI):

$$\frac{E_{rm}}{E_i} = 0.02 + \frac{1 - D/2}{1 + e^{(60 + 15D - GSI/11)}} \quad \text{(eqn A5.4)}$$

Where D is the disturbance factor between 0 (undisturbed) and 1 (fully disturbed). If it is assumed that $D = 0$ for most of the slope, this yields $E_{rm} = 45$ GPa for the granite. This is larger than the value for E_{rm} used in other geotechnical models of the Diavik pit, in which $D = 1$ is assumed (e.g. Golder, 2006).

Similarly, Poisson's ratio for the rock mass is given in previous numerical models to be between 0.21 and 0.25. This is likely calculated from the relationship derived from data in Hoek and Brown (1997):

Table A5.3: Country rock properties from UCS testing (Golder, 2008)

Rock type	UCS			Avg. Young's modulus (GPa)	Avg. Poisson's ratio
	No. of tests	Average (MPa)	Std. Dev. (MPa)		
Pegmatite	13	124	37	65	0.31
Granite	74	136	31	55	0.34
Biotite Schist	6	81	48	45	0.24

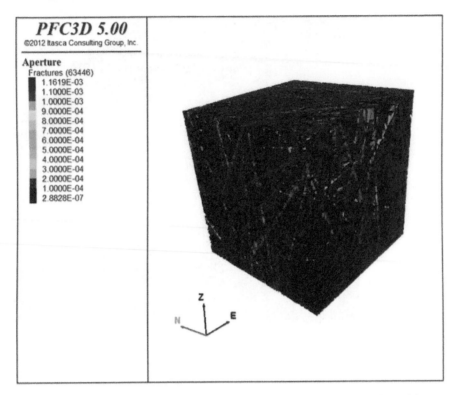

Figure A5.18: Volume of DFN used to calculate total fracture porosity (the cube is 80 m on each side, and fracture apertures are in metres)

$$v_{rm} = 0.32 - 0.0015 \cdot GSI \qquad \text{(eqn A5.5)}$$

It seems unlikely that the Poisson's ratio of a fractured rock mass would be less than that of intact rock, Therefore, the intact value for Poisson's ratio is adopted in this study.

1.2.2 Hydraulic properties

1.2.2.1 Hydraulic conductivity

Calculated hydraulic conductivity values for the North-west wall vary over several orders of magnitude, but the measurements generally show a log-normal distribution suggesting that the country rock can be considered statistically homogeneous over large volumes. For the purposes of this study, theoretical values of hydraulic conductivity at different locations in the slope were calculated by creating multiple fracture flow models from portions of the DFN and subjecting these volumes to 'flow tests' to obtain equivalent conductivity tensors and differences in assuming an EPM. The geometric mean of the conductivities calculated from the fracture flow models is 1×10^{-8} m/s. Therefore, for the EPM models described here, a homogeneous value of $K = 1 \times 10^{-8}$ m/s is used for the granite.

1.2.2.2 Porosity

An $80 \times 80 \times 80$ m volume of the DFN was chosen for calculating total fracture porosity. The centre of the volume corresponds to the centre of the DFN. The size of

the cube was chosen such that the edge of the cube does not intersect the slope face. The fractures are shown in Figure A5.18.

The DFN data include the theoretical transmissivity for each fracture. To calculate hydraulic apertures, a parallel plate model for flow was assumed and the aperture for each fracture was calculated by (e.g. Bear, 1972)

$$a = \sqrt[3]{\frac{12\mu T}{\rho_w g}} \qquad \text{(eqn A5.6)}$$

where T is the transmissivity in m²/s, μ is the dynamic viscosity of water and $\rho_w g$ is the unit weight of water. The porosity was then calculated from the volume of fractures per unit volume. This yields a value of $n = 0.0005$.

Golder (2010) note that the hydraulic aperture is generally about an order of magnitude less than the mechanical aperture. If the apertures used in the calculation of P33 are multiplied by 10, this yields a porosity of $n = 0.005$. It is assumed that the porosity of the matrix material (between fractures) is negligible.

1.2.2.3 Storage

For this study, a value of specific storage of 5×10^{-8} m⁻¹ has been used for most model runs based on the calculations of total fracture porosity calculated from the DFN and on literature values for granite with low total

Table A5.4: Hydraulic properties used the EPM models

Parameter	Value
Porosity, n	0.005
Biot coefficient, α_B	0.18
Conductivity, K	1×10^{-8} m/s
Storage, S	5.5×10^{-8} m^{-1}
Diffusivity, D	0.18 m^2/s

porosity. In addition, to evaluate the sensitivity of storage on the results, some of the model runs were carried out using values of 10^{-5} m^{-1}.

1.2.2.4   Hydraulic diffusivity

Values of hydraulic diffusivity of 0.18 m^2/s (using $K = 1 \times 10^{-8}$ m/s and $S = 5.5 \times 10^{-8}$ m^{-1}) and 0.001 m^2/s (using $K = 1 \times 10^{-8}$ m/s and $S = 10^{-5}$ m^{-1}) were used in the analysis.

1.2.2.5   Summary of hydraulic properties

Table A5.4 summarises the hydraulic properties used as input to the EPM models in the current study.

1.2.3   Slope geometry and excavation rate

A 2D numerical model was set up along the cross-section shown in Figure A5.19 and Figure A5.20. The cross-section passes through the centre of the pit and trends at an angle of –35 degrees from north. This section was chosen because of its proximity to piezometers that could be used in model calibration (A154-MS-10 which contains four piezometers, and A154-MS-02 which contains three piezometers).

Each elevation of the pit was excavated in a single phase and there were no pushbacks. Pit contours at six-month

intervals from 2006 to 2011 for the chosen cross-section are shown in Figure A5.21. The 2D excavation rate for the cross section was calculated to range between approximately 400–1300 m^2/month. The dimensionless excavation rate, R, was calculated by dividing the excavation rate by the diffusivity. For a storage value of 5.5×10^{-8} m^{-1}, $0.001 < R < 0.003$. For a storage value of 10^{-5} m/s, $0.3 < R < 0.5$. Both values are small.

1.2.4   Discussion of the numerical model

1.2.4.1   Model assumptions

For initial conditions, it was assumed that (i) the lake depth is 12 m on the outside of the water retaining dyke (Figure A5.20) with a leaky boundary condition at the bottom of the lake, (ii) there was no standing water in the pit – i.e. the pressure on the surface of the slope and the floor is 0, and (iii) the pressure on the ground surface on the pit side of the water retaining dyke is 0. Mean annual precipitation at Diavik is less than 300 mm and does not provide any significant recharge to the pit slopes (Section 3.3, main text).

1.2.4.2   Steady state vs transient

Figure A5.22 shows the Factor of Safety (FoS) calculated for the typical pit models using a transient flow analysis relative to a steady state analysis. For R values below about 0.1, the transient analysis and steady-state analysis produce essentially the same FoS. For R values greater than 0.1, the transient analysis produces lower FoS values. As the dimensionless excavation rate (R) increases, so the transient analyses produce lower FoS values because the water does not have time to drain, so pore pressures are higher and FoS values are lower.

The figure shows that for $R < 0.1$, a steady-state model may provide representative results. When compared to

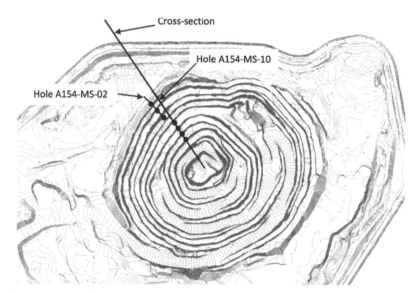

Figure A5.19: Location of cross-section and piezometers (shown as black circles along the borehole trace)

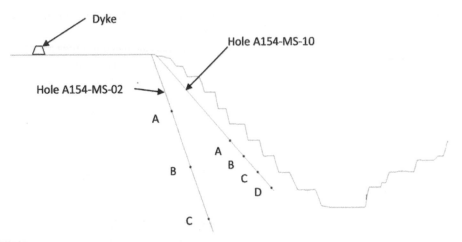

Figure A5.20: View of chosen cross section showing the piezometers

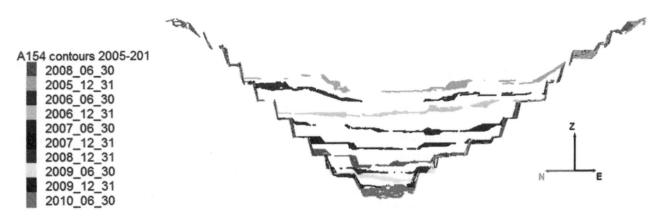

Figure A5.21: Cross-section of six-monthly pit contours

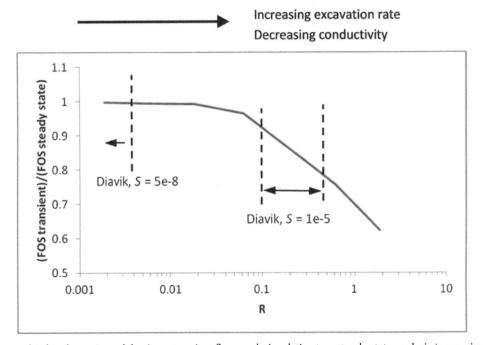

Figure A5.22: Factor of Safety for a pit model using a transient flow analysis relative to a steady state analysis (approximate values of R for Diavik models are shown)

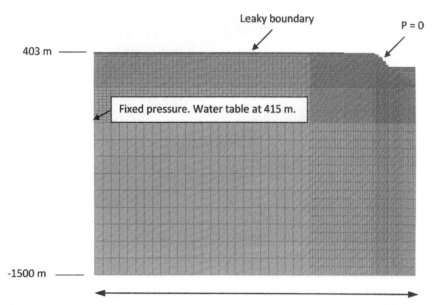

Figure A5.23: Numerical model for simulating flow in the North-west wall (slope geometry shown for January 2006

most pit slopes, the North-west wall of Diavik is unusual because of the proximity of the constant head boundary, the high diffusivity for the clean granite and the relatively low excavation rate. Nonetheless, a transient analysis was also performed for this study because it will always produce a lower FoS compared to a steady-state analysis and many of the calculated R values were greater than 0.1.

1.2.4.3 Model construction

The EPM model was constructed using *FLAC* and is shown in Figures A5.23 and A5.24. The rock excavated over different time periods is shown as different colours in

Figure A5.24. The excavation was simulated by removing rows of zones one at a time and running the model for the relevant time period divided by the number of rows in that block. The start of excavation from the surface was assumed to be January 2004. Therefore, the first block (not shown) is excavated over a period of two years.

The leaky boundary was modelled by assigning a pressure (in this case 12 m of water) and by assigning a leakage coefficient that is equal to the product of the thickness of the leaky layer and the hydraulic conductivity of the leaky layer. In all models the thickness of the lake-bed sediments is assumed to be 1 m. Different

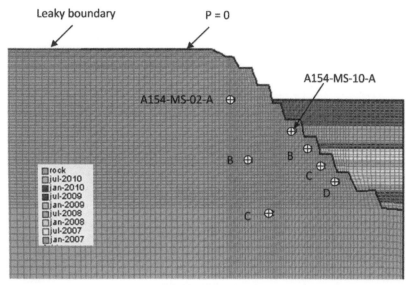

Figure A5.24: Close-up of the slope from Figure A5.23. Different coloured zones show the material excavated over 6-month intervals. Simulated piezometer locations are shown.

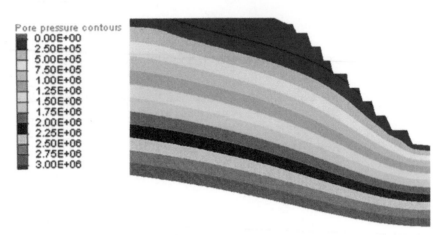

Figure A5.25: Modelled pressure contours (in Pa) in the Diavik North-west wall in July 2010

conductivities were tested ranging from 1×10^{-8} m/s (the same as the granite, i.e. a constant head boundary) to 1×10^{-11} m/s (essentially impermeable).

1.2.4.4 Model results

The pressures after the final excavation are shown in Figure A5.25. This figure shows the pressures for $S = 5.5 \times 10^{-8}$ m^{-1} and a constant head boundary representing the lake bottom (permeable lake sediments). Pressures recorded in piezometers compared to modelled pressures are shown in Figures A5.26 and A5.27. The modelled pressures agree reasonably well with the measured pressures for hole A154-MS02 (far behind the face). The modelled pressures for hole A154-MS10 (close to the face) are lower than the observed pressures which is likely due to a combination of seasonal freeze-back effects (impairing drainage) and blast-induced hydraulic connections of the near-slope

piezometers with the recharge sources (Section 3.3 of the main text). (Note that the piezometer data show a pressure drop in late 2009, which corresponds to the opening of the underground drainage galleries. These were not modelled in the current case study.)

The pressures in a steady-state model for each excavation stage are shown as block dots in Figures A5.26 and A5.27. The steady-state pressures are quite close to the transient pressures. This reinforces the discussion in Section 1.2.4.2 of this appendix which suggests the high diffusivity of the material, relative to the excavation rate, means that a steady-state solution is adequate for this particular slope based on the assumed conditions.

1.2.4.5 Effect of the leaky boundary

Changing the 'leakiness' of the boundary had little effect on pressure results. The pressures for the extreme case of

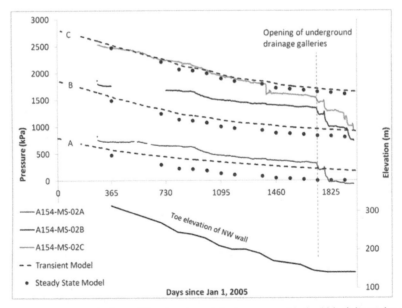

Figure A5.26: Pore pressures in piezometers (coloured lines) and modelled pressures (dashed black lines) for piezometer A154-MS02

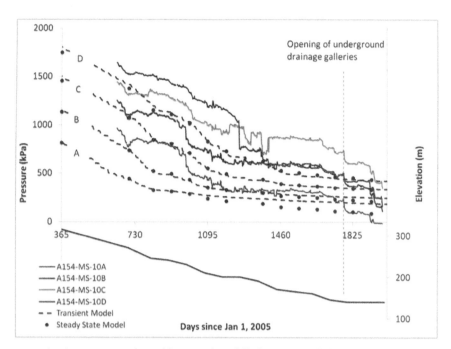

Figure A5.27: Pore pressures in piezometers (coloured lines) and modelled pressures (dashed black lines) for piezometer A154-MS10

an impermeable lake bed are shown in Figures A5.28 and A5.29. Although the shape of the water table is different from that of the model with a constant head boundary (Figure A5.25) as expected, the recorded pressures are only slightly lower. It should be noted, however, that the steady-state solution for the model with an impermeable boundary would be significantly different and would depend on the location of the constant head boundary. For an impermeable boundary, the transient solution no longer approximates the steady-state solution.

1.3 Conclusions

Overall, the results of the modelling case study for the Diavik North-west wall indicate the following.

▪ Similar pore pressure results can be achieved with an EPM model compared with a fracture flow model. The EPM model is therefore considered adequate for

simulating pore pressure at the scale and grid size used for the case study.

▪ The existence of a DFN model provides a means to define variations in the hydraulic conductivity field by analysing the fracture network. The DFN model also helps provide an understanding of the relative orientation and interaction of the various fracture sets. However, the use of a DFN model as the basis for groundwater flow modelling leads to a major source of uncertainty because of small-scale variations in the transmissivity of different fracture sets and within individual fractures.

▪ For the case of Diavik, the steady-state model produced generally similar pore pressure results to the transient model because of the proximity of the recharge boundary, the high diffusivity of the granite wall rock and the relatively slow excavation rate. Diavik is considered unusual in this regard and the

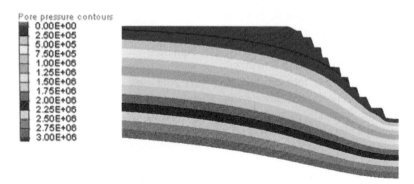

Figure A5.28: Pressures in the model with an impermeable lake bed

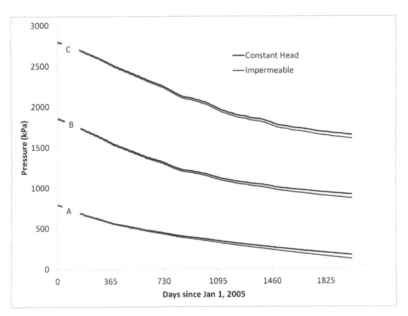

Figure A5.29: Pressures for simulated hole A154-MS02 in the model with a constant head boundary representing the lake bed, and the model with an impermeable lake bed

modification of the boundary condition (to a lower permeability or more distant boundary) invalidates the steady-state solution.

Case study 2: Numerical modelling of the marl sequence at the Cobre Las Cruces Mine, Andalucía, Spain

2.1 Background

2.1.1 General setting

Cobre Las Cruces is an open pit mine that extracts copper sulphides from a volcano-sedimentary Palaeozoic deposit that is overlain by 140 m of Tertiary soft marls (known as the Guadalquivir Blue Marls). The marl unit behaves geotechnically as a soil (over-consolidated clay), but bedding planes and other vertical joints play a major role. Between the marls and the ore there is a relatively thin permeable sandstone aquifer known as the 'Niebla-Posadas aquifer' that has a thickness ranging between 5 and 15 m within the area of the pit.

The pre-mining water table is located within 5 to 20 m of the ground surface (a value of 15 m has been assumed in the modelling for this case study). A peripheral dewatering system consisting of over 30 pumping wells surrounding the pit has been used to lower the piezometric surface within the sandstone and has been in operation since before the start of the pit excavation.

The marls also exhibit a weak strength and low deformational modulus, as well as pervasive discontinuities. For this reason, from the surface down to below the base of the marl sequence, a general slope design angle of 28 degrees was used. This average slope was phased in benches with a height of 10 m and a face angle of 60 degrees, except the first two benches in which a 45 degree bench face angle was adopted.

A contour map of the pit in June 2011 is shown in Figure A5.30. The maximum pit depth in the figure is 200 m, and the planned maximum depth is 250 m.

The marl sequence was chosen for the case study because of its low permeability and the coupled response which is observed in piezometer hydrographs. Initial testing of the marls indicated a permeability range of 10^{-8} to 10^{-10} m/s. Values of 10^{-8} m/s (lateral conductivity) and 10^{-9} m/s (vertical conductivity) were used for the case study. The study describes the decision making process for the type of modelling that is performed and provides example plots and results.

While winter rainfall values can be relatively high, the water is mostly lost to soil moisture storage and runoff, and the rainfall produces little infiltration into the marl sequence.

2.1.2 Geology and material properties

The following description of the geology is from Galera *et al.* (2009b).

a) Weathered marls: The following three sections were established during excavation.
   1. MET-1, highly weathered. Brownish coloured. The unit has 10 m thickness.

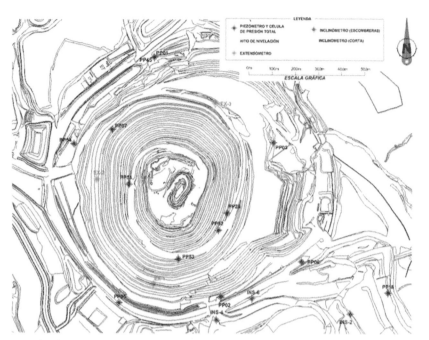

Figure A5.30: Contour map of Cobre Las Cruces pit in June 2011 (contour spacing is 10 m; piezometers are denoted by PP, extensometers by EX and inclinometers by INS; not all instrumentation is shown)

2. MET-2, down to 23 m depth. The marl is heavily weathered and presents vertical desiccation fractures spaced around one metre.
3. MET-3, down to 31 m depth. The marl is moderately weathered. The spacing of desiccation discontinuities is around 12 m. The strength parameters of this horizon are similar to MET-2.

b) Fresh marls: Below 31 m depth, there are no visible signs of weathering and the marls show a typical grey-blue colour. The following four levels can be distinguished.

1. Level-1, 'soft marl' from 31 to 80 m depth. There are no desiccation fractures, but several sub-vertical joint sets can be observed as well as horizontal bedding planes spaced 5 m apart.
2. Level-2, 'medium marl' from 80 to 110 m depth. The strength is a soft rock-type characteristic showing fragile failures.
3. Level-3, 'soft marl' from 110 to 115 m depth; presents the same characteristics as Level 1 with a thickness ranging from 5 to 10 m. Laterally, it disappears below Level-2. It shows strength and deformability properties similar to Level-2.
4. Level-4, from 115 to 130 m depth; 'strong marl' has a strength characteristic of a soft rock. There is an appreciable increase of the strength and stiffness of the marl.

c) Sandy marl; just above the sandstone of the 'Niebla-Posadas aquifer' there is a massive layer of sandy marls with an approximate thickness of 5 to 10 m.

d) Sandstones of the 'Niebla-Posadas aquifer', partially to well cemented with a thickness ranging from 5 to 15 m.

e) Paleozoic substrate, constituted by volcano-sedimentary rocks, and hosting the ore.

A general view of the pit is shown in Figure A5.31. Photographs of the bedded marls are shown in Figure A5.32. For the purposes of this case study, the geology is assumed to be horizontal (tabular) across the pit area.

The different material layers and the hydraulic properties are shown in Table A5.5. Typical properties for the rock/soil type were assumed where properties were not available. The conductivities were derived from 12 large scale pumping tests monitoring the water level in 55 piezometers, including vibrating wire piezometers within the marl sequence. An average horizontal and vertical value is shown for each layer. It is also likely that the conductivity within the marl sequence decreases with depth because fractures tend to close with increasing stress. However, in the absence of hard data, for the purposes of modelling, it is assumed that conductivity values are constant with depth.

Table A5.6 shows the calculated storage and horizontal and vertical diffusivity values. Most modelling programs calculate diffusivity automatically from input conductivity and storage, but the values of diffusivity for the different layers are shown because they are subsequently used to help determine what type of modelling approach is required.

Table A5.5: Properties for geotechnical units at Cobre Las Cruces. Grey table cells indicate values estimated from typical rock properties.

Layer[1]	Depth[1] (m)	Dry density[1] (kg/m^3)	Moisture content[1] (%)	Horizontal Conductivity[2] K_h (m/s)	Vertical Conductivity[2] K_v (m/s)
MET-1	0–10	1415	30.3		
MET-2	10–23	1459	28.5		
MET-3	23–31	1496	27.1		
Level-1	31–80	1528	25.5	1.15×10^{-8}	1.15×10^{-9}
Level-2	80–110	1585	24.2		
Level-3	110–115	1579	24.2		
Level-4	115–130	1620	23.1		
Sandy marls	130–140	1622	25.7		
Sandstones	140–150	1930	10 (porosity[2])	8.3×10^{-6}	8.3×10^{-7}
Paleozoic substrate	>150	2070	5 (porosity[2])	2.3×10^{-8}	2.3×10^{-9}

From (1) Galera *et al.* (2009b) and (2) Frasa (2001).

2.1.3 Geometry and excavation rate

The ultimate pit will measure approximately 1600 m long × 900 m wide × 250 m deep. To establish the appropriate modelling method, the dimensionless excavation rate was first calculated, as described in Section 4.2.5 of the main text. The mining rate was taken from contour plots of the pit at six-month intervals from January 2008 to June 2011. A cross-section of the pit was constructed at each time, and the excavation rate was determined by dividing the difference in excavated area by the time interval. The cross-section was taken at an azimuth of 127 degrees (Figure A5.33). Figure A5.34 shows the cross section at two different times.

From this analysis, mining rates were found to range from 500 to 1500 m^2/month. To derive the dimensionless mining rate (*R*), this was divided by the horizontal diffusivity. It was assumed that the larger horizontal diffusivity is the critical parameter, because flow towards the pit face in the more permeable horizontal direction will dominate flow-induced pressure changes. Using the median mining rate (1000 m^2/month) and the range of horizontal diffusivities in Table A5.6 yields an approximate range of $1 < R < 10$.

2.2 Model development

2.2.1 Selection of an EPM or fracture flow code

The use of an EPM code assumes that flow in the fractured rock mass can be represented by a continuum at the scale of interest, and that the pressures in the matrix and the fractures are in equilibrium. A rigorous way to assess the applicability of the EPM is to perform numerical tests on a DFN. However, for most mines,

Table A5.6: Calculated storage and diffusivity values. Young's moduli are from Galera *et al.* (2009a). Values of Poisson's ratio are assumed typical values for each rock type.

Layer	Young's modulus E (MPa)	Poisson's ratio, ν	Storage, S (m^{-1})	Horizontal Diffusivity, D_h (m^2/s)	Vertical Diffusivity, D_v (m^2/s)
MET-1	25		2.7E-04	4.3E-05	4.3E-06
MET-2	60		1.1E-04	1.0E-04	1.0E-05
MET-3	85		7.9E-05	1.5E-04	1.5E-05
Level-1	105		6.4E-05	1.8E-04	1.8E-05
Level-2	150	0.33	4.5E-05	2.5E-04	2.5E-05
Level-3	225		3.1E-05	3.8E-04	3.8E-05
Level-4	225		3.0E-05	3.8E-04	3.8E-05
Sandy marls	300		2.3E-05	5.0E-04	5.0E-05
Sandstones	100	0.3	7.3E-05	1.1E-01	1.1E-02
Paleozoic substrate	5000	0.25	1.7E-06	1.4E-02	1.4E-03

Figure A5.31: General view of the Cobre Las Cruces pit

including Cobre Las Cruces, a DFN is not available so rules of thumb are required to make the decision. Based on the small fracture spacing (5 m) relative to the slope height within the marls (~150 m), the EPM likely is applicable as an approximation of the fracture flow within the marl sequence, as discussed in Section 4.2.6 of the main text.

The choice as to how to treat pressures in the fractures and the matrix depends on the fracture spacing, the scale (i.e. slope height) and the relative hydraulic properties of the fractures and matrix material.One simple method of understanding pressures in the matrix and the fractures is to assume an equivalent continuum approximation of the fracture flow, and use a one-dimensional analytical

Figure A5.32: Bedded marls at Cobre Las Cruces (from Cooper *et al.*, 2011)

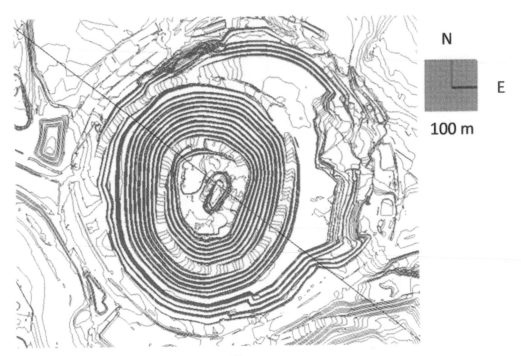

Figure A5.33: Location of the cross-section used in 2D modelling

solution to compute the change in head in the fracture network at some point behind the slope face:

$$\Delta h(x,t) = \Delta h_0 erfc\left(\frac{D_{eq}t}{x^2}\right)^{-0.5} \qquad \text{(eqn A5.6)}$$

where
$\Delta h(x,t)$ = change in head (drawdown) at point distance x from the slope face at time t
Δh_0 = the change in head at the slope face
D_{eq} = the equivalent diffusivity.

Using this equation, it can be shown that the pressure in the fractures will drop to 10% or less of its initial value when

$$\frac{D_{eq}t}{s^2} > 30$$

The drainage of a low-permeability matrix block surrounded by highly permeable fractures is discussed in Section 3.6.5.2. Assuming the block is a cube with dimension s, and that the pore pressures on the edges drop instantly to 0, the pressure at the centre of the block can be given by:

$$\frac{p}{p_0} = \left[2 \sum_{m=0}^{\infty} \frac{\sin a_m}{a_m} e^{-\frac{a_m^2 \hat{t}}{4}}\right]^3 \qquad \text{(eqn A5.7)}$$

where:

$$a_m = \frac{\pi}{2}(2m + 1)$$

$$\hat{t} = \frac{D_m t}{s^2}$$

D_m is the matrix diffusivity and p_0 is the initial pressure in the block.

From this equation, it can be shown that the matrix block remains at 90 per cent or greater of its initial pressure when

$$\frac{D_m t}{s^2}$$

is less than about 0.3.

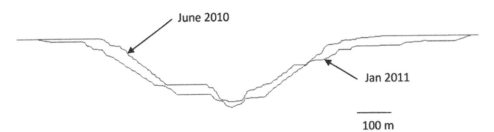

Figure A5.34: Cross section of the pit at two different times (the mining rate is calculated from the area between the curves divided by the time interval)

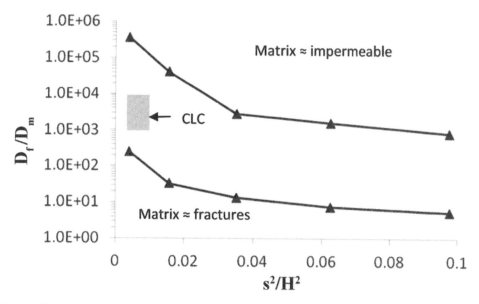

Figure A5.35: Curves delineating regions in which matrix can be considered impermeable (relative to the fractures) and regions in which the matrix and fractures can be considered essentially equivalent (from a numerical study by Lorig *et al.*, 2011). The curves shown are for dimensionless excavation rate $R = 1$. The likely range of properties for Cobre Las Cruces is shown as a grey box.

To find the conditions under which the fractures are mostly drained (i.e. pressure less than 10 per cent of the initial value) and the matrix is mostly undrained (i.e. pressure greater than 90 per cent of its initial value), a t equivalent for the two conditions can be determined (approximately) as follows:

$$\frac{D_{eq}}{D_m} > 100 \frac{x^2}{s^2} \qquad \text{(eqn A5.8)}$$

For the marls at Cobre Las Cruces, the fracture spacing (bedding) is on the order of $s \approx 5$ m and $D_{eq} \approx 10^{-4}$ m²/s. For a point x = 50 m behind the slope face (approximate location of a critical failure surface), then a matrix diffusivity of $D_m <10^{-8}$ m²/s would indicate a system in which there are different matrix and fracture pressures, suggesting the matrix blocks would exhibit an undrained response whereas the fractures would show pressure change with time. Using a typical diffusivity for stiff clay (2×10^{-8} m²/s – Taylor, 1948), it can be shown that an EPM model still may be suitable to approximate the fracture flow and pressures in fractures if the fracture spacing is small relative to the slope height, however the matrix pressure should be considered as well when assessing pore pressures to be used in slope stability calculations.

The simple method described above does not consider the excavation rate. To help address this, a typical large open pit excavation was modelled and high conductivity fractures were emplaced in the model at regular spacings. The water pressures in the fractures and in the matrix blocks were monitored for different diffusivity ratios,

different spacings and different dimensionless excavation rates. A set of charts was constructed to assist in determining how the pressures in matrix and fractures should be considered: (i) equal, in equilibrium; (ii) different, both changing during dissipation process (requiring dual porosity model); or (iii) different, but assuming that matrix pressure does not change during dissipation process.

Figure A5.35 shows curves derived from the study that delineate regions in which the matrix can be considered impermeable and when the matrix and fractures can be considered equivalent. These were derived by observing the pressure change in matrix blocks relative to pressure changes in fractures. If this ratio is small (<0.1), the matrix blocks can be considered essentially impermeable. This limit is shown by the upper curve in Figure A5.35. If the ratio of matrix pressure change to fracture pressure change is high (>0.9), the matrix and fractures can be considered essentially equivalent. This limit is shown by the bottom curve in Figure A5.35. The limit depends on the fracture spacing, s, the slope height, H, the ratio of fracture to matrix diffusivity, D_f/D_m and the dimensionless excavation rate, R. As the fracture spacing decreases or the slope height increases, the boundary curves move up, indicating that the slope is more likely to exhibit a continuum pore pressure. The curves also move up as the ratio of fracture to matrix permeability decreases. As the R value increases, the curves move down, indicating that faster excavation rates (or lower permeabilities) cause the matrix to appear more impermeable relative to the fractures. The curves shown in Figure A5.35 are for $R = 1$.

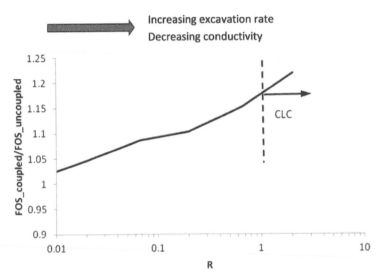

Figure A5.36: Factors of safety in models with two-way coupling (after nine months) relative to models with one-way coupling

Assuming the fracture spacing is 5 m, the slope height in the marl is 150 m and the D_f/D_m is between 10^3 and 10^4, the range of behaviours for the marl is plotted as a grey box in Figure A5.35. This figure shows that the marls lie in between the curves, which suggests that the matrix blocks should not necessarily be considered impermeable; and also that the pressures in the matrix and fractures are different, so a dual porosity response needs to be considered. Also, analyses in subsequent sections show that the response of this slope is practically undrained (i.e. no significant pressure change due to groundwater flow) and that an undrained response results in the same pressure changes in the matrix and fractures. Therefore, for the purpose of this case study, it was assumed that the matrix and fracture pressures are the same, in equilibrium.

2.2.2 Determination of the need to use a coupled approach

Once it was decided that an EPM approach is suitable, the next step was to determine what type of EPM model should be constructed. Section 4.2.8 of the main text provides guidelines for evaluating the conditions under which fully coupled modelling should be considered. To assist the analysis, models of a typical open pit were run with transient groundwater solutions and one-way coupling. Factors of Safety (FoS) were calculated using pore pressures to compute effective stresses. The same models then were run with solid-fluid coupling such that the stress relief due to excavations produced rapid drops in pore pressure. FoS values also were calculated for these models and were compared, as shown in Figure A5.36. For low dimensionless excavation rates, the uncoupled and coupled models produced essentially the same FoS values. However, as R increased, the coupled models produced higher FoS values, because when excavation rates are fast or conductivities are

low, then the pore pressures do not recover from the initial drops and the effective stress is greater.

At Cobre Las Cruces, the soft marls result in values of R towards the right of Figure A5.36. This indicates that an uncoupled model likely would underestimate the FoS by 15 to 20 per cent. Therefore, a coupled model is necessary in this case. Due to the very high values of R, it may be possible to perform an undrained-only analysis and assume no recovery of pore pressures at all (see Section 2.2.3 below).

2.2.3 Model set-up

A coupled 2D EPM model was constructed to represent the marl sequence along the cross-section at an azimuth of 127 degrees (Figure A5.33). This cross-section was chosen because it intersects two piezometers that will allow for calibration of the model to observed data (Figure A5.37). Figure A5.38 is a close-up of the pit and shows the location of the two piezometers (P07-75 and P07-95).

Material properties were assigned to the model as indicated in Tables A5.5 and A5.6. Boundary conditions are shown in Figure A5.37. A constant pressure boundary condition is assigned at a distance of 3 km from the centre of the pit. This location was chosen because water extracted by the dewatering system is re-injected at a distance of 2 to 3 km from the pit. However, the low conductivity of the marl means that the model results are insensitive to the location of the constant pressure boundary. The bottom boundary and the symmetry boundary are no-flow (perpendicular) boundaries.

The pressure at the surface of the excavation is assumed to be zero. Different software programs account for this in different ways. *FEFLOW* and *MODFLOW* use a time variant head condition to ensure that the hydraulic head on the boundary is equivalent to the elevation of the pit base.

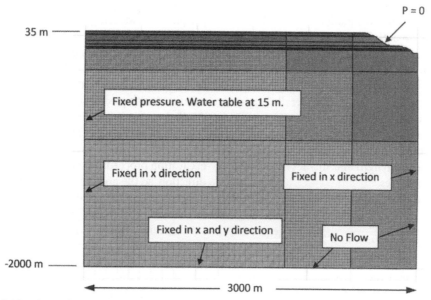

Figure A5.37: 2D model for shown for excavation state in January 2010

SEEP/W and *Slide* specify an 'unknown' condition that iteratively calculates the location of the seepage face. For this case study, *FLAC* was used with zero pressure assigned to the pit face and excavations were performed at one month intervals. Although it is possible that slight negative pore pressures may occur within any of the marl horizons that become de-saturated, this was not considered by the current study.

Figure A5.37 also shows the mechanical boundary conditions. Because a coupled model is being used, it is necessary to specify the mechanical boundary conditions, initial conditions and material properties. The initial vertical stresses are set equal to the stress due to the weight of the material. Horizontal stresses are calculated assuming they are equal to vertical stresses.

2.2.4 Approach for coupling

The marl sequence was modelled with a simplified coupling procedure. Simplified two-way coupling involves undrained analysis (e.g. change in pore pressure due to undrained deformation) followed by simulation of pore pressure dissipation. The assumption inherent in this type of analysis is that pressure changes due to pit slope unloading (i.e. excavation) occur more-or-less instantaneously, such that flow during excavation can be reasonably ignored.

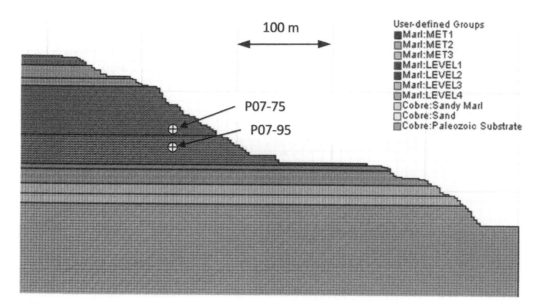

Figure A5.38: Close up of 2D model (January, 2010) showing piezometer locations

During the instantaneous mechanical response, the pore pressure drops according to equations 4.8 and 4.9 as described in Section 4.2.8 of the main text (see also Section 3.6.6.2). Biot's coefficient (α_B) ranges between 0 and 1 with the upper-bound value being assumed for incompressible grains or rock blocks. For low porosity rock, α_B generally varies between 0.03 and 0.5. In soil, α_B is generally assumed to equal 1. At Cobre Las Cruces, the marl exhibits a high total porosity ($\approx$ 30 per cent) and the rock/soil mass is fairly massive. A value of $\alpha_B = 1$ was adopted for this case study and laboratory strength and stiffness data were used in all rock mass analyses.

For $\alpha_B = 1$, the Skempton coefficient reduces to

$$B = \frac{K_w/n}{K_w/n + \kappa} \qquad \text{(eqn A5.9)}$$

For small K_w values, as in this example, $B \approx 1$, so essentially all of the stress change is translated into a drop in pore pressure according to equation 4.8 in Section 4.2.8 of the main text.

After the undrained pore pressure changes were calculated, a transient flow-only calculation was performed to simulate how pore pressures change according to the pressure differences, hydraulic diffusivities and time. The main advantage of this approach is reduced calculation time (compared to the two-way coupling) and representation of fluid pressure change due to instantaneous deformation (compared to one-way coupling).

2.2.5 Model results

The modelled pore pressures in January 2011 are shown in Figure A5.39. Pore pressures in the marl have dropped due to the stress relief caused by the removal of material. Because of the low diffusivity of the marl, the pore pressures do not recover over the modelled one month stress periods. The potentiometric surface of the underlying sandstone aquifer is shown as a horizontal band of low pore pressures about half way down the slope.

The rapid drainage of the sandstone is caused by the peripheral dewatering wells and by residual seepage to the pit face.

Pressures from the model are compared with actual marl piezometer results in Figure A5.40. The plot also shows the lowest depth of the pit floor. Note that the pit is excavated with pushbacks, so a constant pit depth in Figure A5.40 does not necessarily mean that excavation is not occurring. Each excavation in the 2D model is assumed to be a block of material that is infinitely long. In reality, the excavation proceeds in discrete blocks in the out-of-plane direction, so the pressure drops in the 2D model and observations in the piezometers will not align perfectly.

A sensitivity analysis was undertaken with conductivity values for the marls set 10 times lower than the initial values, as shown in Table A5.7. Figure A5.41 shows the model results. The pressure rebound between mining phases is less with the lower values of marl permeability and there is a better match between modelled and observed pressures.

Figure A5.42 shows the pore pressure histories for the original model when only the undrained response is considered. These pore pressures are very similar to those calculated when fluid flow is also considered (undrained-drained). The recovery of the fluid pressure after unloading is minimal due to the low conductivity of the marls. Depending on the purpose of the modelling, an undrained-only model (with no pore-pressure dissipation after unloading) may give satisfactory results. However, Figure A5.43 illustrates that, while pore pressures below the excavation are reduced due to stress and drop as expected, there is higher pressure at the toe of the slope due to a stress increase at this location. In reality, these pressure spikes will dissipate relatively quickly due to high gradients but, if no fluid flow calculation is performed, then these spikes remain and unrealistically low factors of safety values may be calculated.

Figure A5.44 also shows the pressure histories for a flow-only model (no undrained response). It is clear that

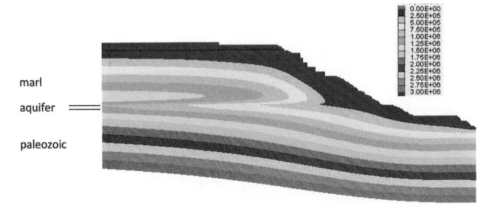

Figure A5.39: Modelled pressures (in Pa) in January 2011 (coupled undrained-drained model)

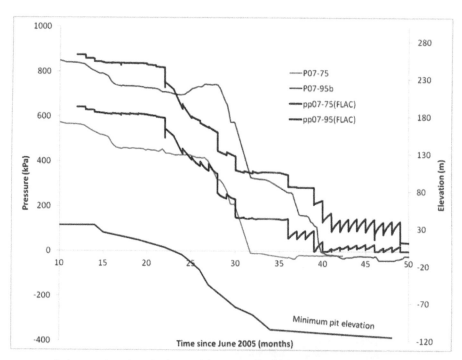

Figure A5.40: Pressures recorded at two locations in the model compared with pressures recorded with piezometers

Table A5.7: Conductivities used in sensitivity model

Layer	Depth (m)	Horizontal Conductivity, K_h (m/s)	Vertical Conductivity, K_v (m/s)
Marl	0–140	1.15×10^{-9}	1.15×10^{-10}
Sandstones	140–150	8.3×10^{-7}	8.3×10^{-8}
Paleozoic substrate	>150	2.3×10^{-8}	2.3×10^{-9}

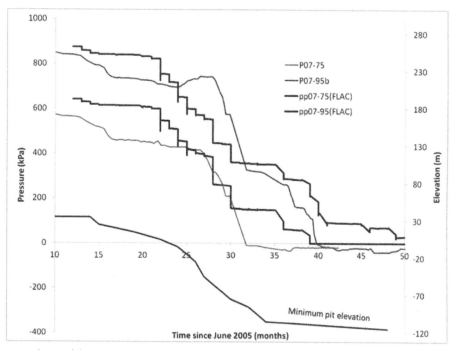

Figure A5.41: Pressures in the model using 10x lower conductivity in the marl

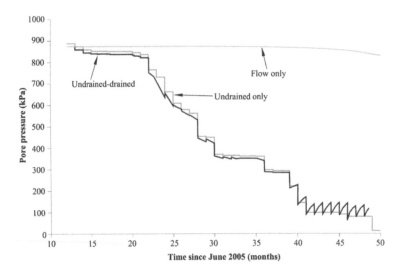

Figure A5.42: Modelled pore pressures at the location of piezometer P07-95 for different calculation methods

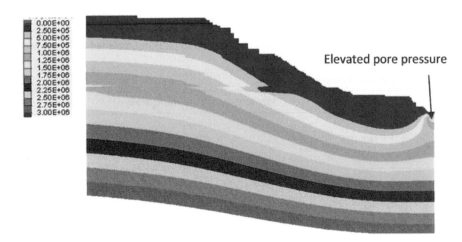

Figure A5.43: Pressures (in Pa) in the undrained-only model (January 2011)

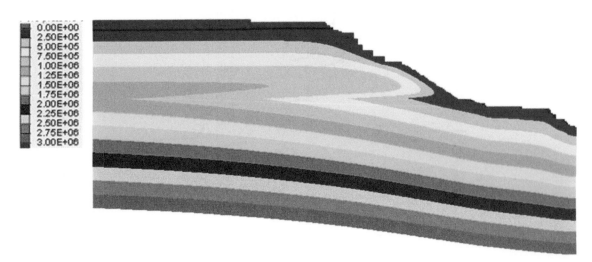

Figure A5.44: Pressures (in Pa) in the flow-only (drained) model (January 2011)

the pore pressures are significantly higher than in the coupled model. Figure A5.44 shows that there is no region of reduced pressure below the excavations in this case. Only the sandstone aquifer layer shows reduced pressures due to the dewatering wells.

2.3 Conclusions

The case study describes the modelling of a pit excavation through soft, low-permeability material. The low permeability and dual porosity storage result in a low hydraulic diffusivity relative to the excavation rate. It is shown that a coupled simulation is necessary to capture the water pressure behaviour. A fluid-only simulation overestimates the pore pressures.

The use of an EPM code was selected because of the hydraulic properties of the rock mass relative to the properties of the unfractured material (matrix blocks). Simulated pressures in the coupled model and in the undrained model are similar, so errors introduced by assuming an EPM are small.

For the marl sequence, the hydraulic conductivity is of secondary importance because it is low (10^{-8} to 10^{-10} m/s) and the observed pressure changes are largely due to mechanical (undrained) effects.

The results of the study support the general conclusion that coupled modelling should be considered when the permeability of the rock is less than 10^{-8} m/s and there is evidence of a coupled response from the field data.

Appendix 6

The lattice formulation and the *Slope Model* code

Loren Lorig and Bart Harthong

1   The lattice formulation

The lattice representation of a brittle solid is not new (e.g. Bolander & Sukumar, 2005), but its application to the failure of a jointed rock mass offers advantages over both continuum models and full discrete element (DEM) models in terms of both computational efficiency and flexibility. The explicit, time-marching formulation used in Itasca's lattice codes is documented by Cundall and Damjanac (2009). Newton's Second Law is used to compute the motion of nodes, with damping added for quasi-static solutions and a linear force-displacement relation used for each spring. The formulation includes optional shear springs in addition to springs in the normal direction, the latter supplying 'central forces' between nodes. The shear springs, and optional spin degrees of freedom, allow control of the Poisson's ratio of the bulk solid independently of the spring-connection topology. Fixed, free or stress boundary conditions are provided with

initialisation of spring forces to allow installation of a pre-existing stress state within the model domain.

A novel feature of the formulation is that joints (planar discontinuities) may be inserted anywhere in the lattice, at any angle. Figure A6.1 schematically depicts a joint plane cutting through an assembly of springs (black 'resistors') and lattice nodes (blue circles).

The constitutive equations for the cut springs respect the direction of the joint plane rather than the local spring directions and model slip and opening on the joint plane. For other springs, elastic normal and shear springs operate in the spring direction, with fracture occurring when loads in the springs reaches given limits (corresponding to the tensile strength of rock).

If the two nodes associated with a given spring are located on opposite sides of a joint trace, then the spring is designated as a 'joint spring.' The normal and shear directions of joint springs are set by the joint orientation

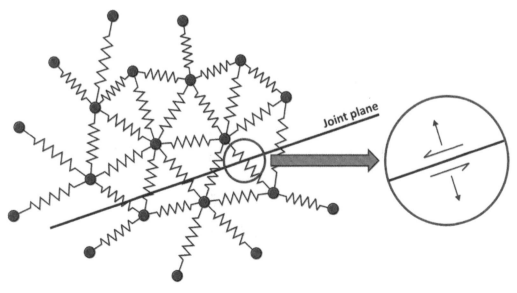

Figure A6.1: Schematic of a lattice array in which a through-going discontinuity cuts certain springs (shown as 'resistors')

rather than the node-to-node vector. Separation and slip on the joint plane are modelled as limits on spring forces, resolved in the joint directions.

If $F^n - pA < 0$ then $F^n = 0$, $F_i^s = 0$

else $F_i^s \leftarrow \dfrac{F_i^s}{|F_i^s|} \min(F^n - PA)\tan\phi\, \big|F_i^s\big|$ (eqn A6.1)

where ϕ is the friction angle, F^n is the normal force (compression positive), F_i^s is the shear force vector, p is the fluid pressure in the joint segment, and A is the apparent area of the segment.

In order to include fluid in the joints, each joint segment (associated with an intersected spring, as described above) is taken to be a fluid reservoir (or 'fluid element') that is regarded as a penny-shaped crack with an aperture that depends on its fluid content. Neighbouring fluid elements are connected by 'pipes' that are conceptually one-dimensional, but inherit the mean properties (such as aperture and extent) of the associated fluid elements. The flow resistances of the pipes are calibrated to recover the same resistance of the equivalent parallel plate of the same aperture and extent.

There is full coupling between mechanical and fluid elements using a formulation called Mechanical Incompressible Flow (MIF), which allows for an incompressible or stiff fluid without a penalty on timestep. The apparent stiffness (in terms of stress/displacement) of a small-aperture joint filled with fluid can be many orders of magnitude higher than that of the adjacent rock element, leading to very small timesteps in an explicit solution. The MIF scheme takes advantage of the fact that the fluid is in series with the stiffness of the rock element (see Figure A6.2, where the size of the depicted object corresponds to the lattice resolution).

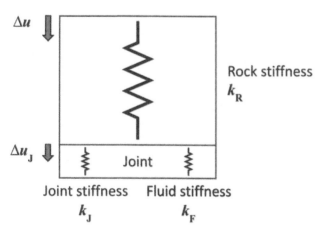

Figure A6.2: Schematic of a single cell in the lattice (of a dimension equal to the mean node spacing), with an embedded joint. Note that the fluid is in series with the lattice spring, thus avoiding the exposure of the full fluid stiffness to the lattice response, which results in a very small time step.

The incremental force response of the lattice spring, ΔF, is

$$\Delta F = \frac{\Delta u k_R(k_J + k_F)}{k_R + k_J + k_F} \to \Delta u k_R, \text{ as } k_F \to \infty \quad \text{(eqn A6.2)}$$

which is independent of the fluid stiffness in the limit of fluid incompressibility. The response to fluid flow (from surrounding pipes) into the element is similarly insensitive to fluid stiffness:

$$\Delta F = \frac{\Delta u_F k_R k_F}{k_R + k_J + k_F} \to \Delta u_F k_R, \text{ as } k_F \to \infty \quad \text{(eqn A6.3)}$$

where Δu_F is the 'fluid displacement' into the element ($\Delta u_F = QA\Delta t$, where Q is the volumetric flow rate and Δt the time step). The flow in pipes assumes, by default, laminar parallel-plate conditions:

$$q = \beta s^2(3 - 2s)\frac{a^3}{12\mu}\left[p^A - p^B + \rho_w g\left(z^A - z^B\right)\right] \quad \text{(eqn A6.4)}$$

Where
$a =$ is the hydraulic aperture
$\mu =$ the viscosity of the fluid
p^A and $p^B =$ fluid pressures at nodes 'A' and 'B', respectively
z^A and $z^B =$ elevations of nodes 'A' and 'B', respectively
$\rho_w =$ the fluid density
$S =$ saturation.

The dimensionless number β is a calibration parameter, a function of fluid resolution, used to match conductivity of a pipe network to the conductivity of a joint represented by parallel plates with aperture a.

2   Features of the lattice approach

The lattice approach differs in several important ways from more conventional numerical schemes, such as the finite-element or finite-difference methods. While the latter methods use polyhedral elements to fill a three-dimensional volume completely, the lattice method uses an assembly of zero-dimensional nodes that are placed at random (although uniformly in a statistical sense) within the solution domain. There is no attempt to tessellate space. After nodes have been placed, pairs of nodes are connected by springs if their spacing is less than a specified distance, which is some multiple of the (overall) mean node spacing.

The second difference between the lattice code and conventional element-based codes is that the macroscopic properties, such as elastic moduli and strength, are emergent features and not specified directly. Normally, a process of calibration is necessary to determine the micro-properties (spring stiffness and strength) that are necessary to recover the required macro-properties. However, in the implementation of the lattice scheme, calibration factors are built into the lattice codes, because the generated node arrays are derived by scaling from a master periodic cell which, in turn, is derived from a packed assembly of spheres created with *PFC3D* in periodic space. Thus, the calibration step is avoided in the implementation of the lattice scheme, while retaining the characteristic that macroscopic behaviour emerges from microscopic mechanisms and evolving micro-structure.

A third difference between the lattice codes and element-based codes relates to the handling of discontinuities. Placement of discontinuities (interface plane or joints) in the element-based codes often results in ill-conditioned meshes when joints intersect at small angles or form very small blocks. In contrast, any DFN (Discrete Fracture Network) may be overlaid on a given array of lattice nodes and springs. The lattice resolution determines whether small blocks or small intersection angles are resolved. If such objects are below the level of resolution, they simply do not exist. The onus is on the modeller to choose what object sizes are meaningful in any given system; any degree of fineness may be resolved at the expense of computer time and memory. Often, very fine resolution is unnecessary with regard to the response of a rock mass, but it is easy to check on the sensitivity by re-running a given model with different resolutions simply by changing one parameter.

Finally, a lattice code (in common with DEM codes) is able to accommodate new cracks in a natural way, in contrast to continuum codes, which either must modify element response in order to represent cracks in a smeared way or reconstitute the mesh locally to allow new discontinuities between elements. In the lattice code, a new micro-crack simply corresponds to a newly broken spring, with macro-cracks emerging automatically as micro-cracks coalesce or form contiguous chains. The fluid-flow network is extended automatically to encompass new crack pathways.

3 Example application

Fluid flow in *Slope Model* takes place in the pre-existing joints specified by a user at the beginning of a simulation, but also in the stress-induced cracks created during simulation. The pipe network, which is used to solve the fluid flow in joints, is updated automatically each time a new crack is formed. The fluid element is created at the location of the micro-crack: if it is within the crack radius from the existing fluid elements, it is connected into the pipe network. Thus, as fracturing grows throughout the model domain, the pipe network and potential for fracture flow extend simultaneously.

This example verifies the implementation of dynamic evolution of the pipe network with crack growth, and demonstrates potential effect of stress-induced cracks on flow field. The geometry of the model, which includes three pre-existing joints in a 500 m high slope, is shown in Figure A6.3. The model length of 50 m along the slope is selected to simplify the problem, making it

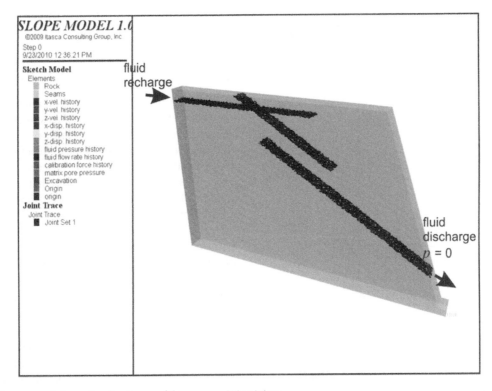

Figure A6.3: Geometry of slope showing traces of three pre-existing joints

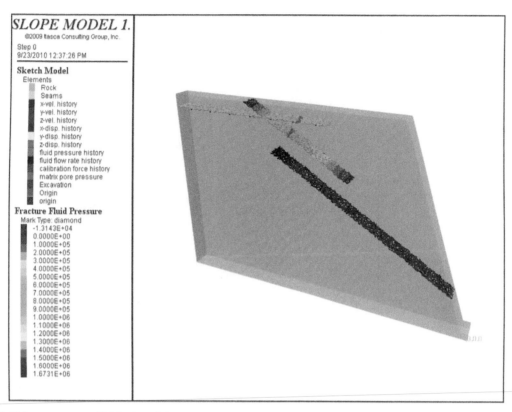

Figure A6.4: Initial pore pressure (Pa) in three joints

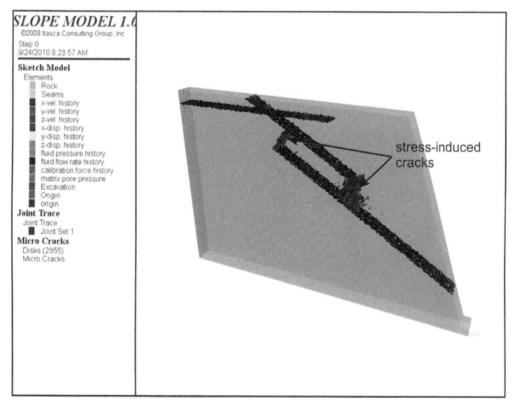

Figure A6.5: Micro-cracks induced after mechanical deformation

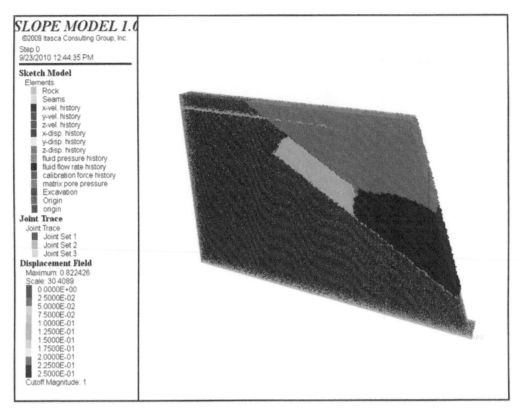

Figure A6.6: Failure mechanism caused by pre-existing joints and induced cracks

quasi-two-dimensional. The constant pressure of 0.5 MPa is applied on the left boundary. The free fluid discharge is specified on the slope face.

In the first simulation step, which involves flow calculation only, the initial pore-pressure field (shown in Figure A6.4) is determined. As expected, a hydrostatic state is achieved in the two fractures connected to the left boundary (i.e. the constant pressure boundary condition),

and the maximum pressure is at the lowest point of those two fractures. In the steady state, there is no fluid drainage on the slope face because three fractures, as specified, do not create a path from the left boundary to the slope face.

In the second stage of simulation, which involves simulation of the mechanical model only, the rock bridge between the upper and the lower joints breaks as a result of

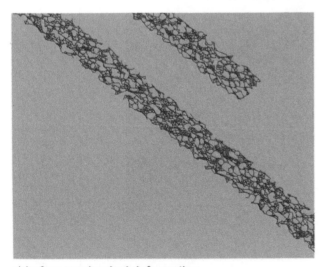

a) before mechanical deformation

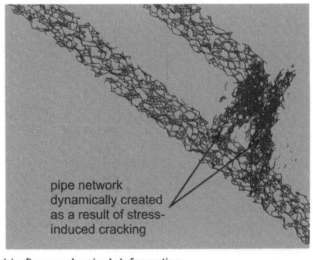

pipe network dynamically created as a result of stress-induced cracking

b) after mechanical deformation

Figure A6.7: Evolution of flow pipe network

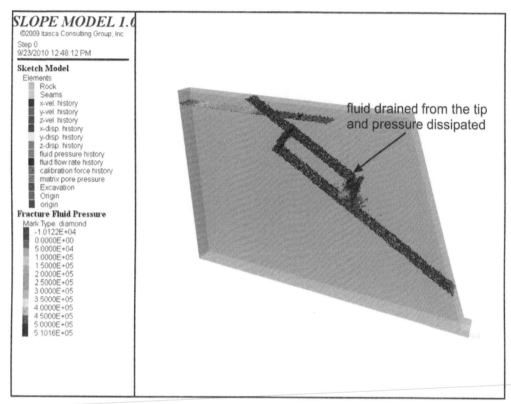

Figure A6.8: Final pore pressure (Pa) field in the joints

stress-induced cracks as shown in Figure A6.5. Consequently, a failure mechanism is formed resulting in instability of the large portion of the slope illustrated by the displacement field shown in Figure A6.6.

Another consequence of bridge failure is a change in the flow pipe-network. The newly created cracks are included in the network. The details of the pipe network, in the region where rock bridge fails before and after the failure, are shown Figure A6.7. After further simulation of the flow model, it can be seen that the failure of the rock bridge results in dissipation of hydrostatic pressure from the tip of the pre-existing joint connected to the left boundary (Figure A6.8).

4 Validation of *Slope Model* using experimental data from Coaraze test site

This section describes the main results of the validation process of *Slope Model* using an experimental data set from the Coaraze test site. The site and experimental results are described in Section 3.6.6.3.

4.1 Geometry

The geometry of the problem was built as a closed surface using the free software *Google Sketchup* (Figure A6.9).

These data were saved and converted into DXF format using the free software *Meshlab* and eventually read into *Slope Model*.

4.2 Boundary conditions

The mechanical boundary conditions applied are as follows.

- The topographic surface is free.
- All other surfaces are fixed for the sake of simplicity, as it is not yet possible to specify roller conditions in directions other than the axes. However, according to experimental results, the amplitude of the normal displacements of faults and bedding planes is very small compared with the distance between the measurement points and the boundaries (Figure A6.10). To a first approximation, it is expected that the mechanical boundary conditions should not affect the results.

When the water gate is closed, it is assumed that the reservoir below the level of the water gate is fully saturated (defined as altitude $z = 0$). However, the exact shape of the phreatic surface is unknown. Closing the water gate results in a pressure rise of up to 8 m (altitude $z = 8$ m). Above this level, water flows out from the open discontinuities such that the steady-state regime consists of a constant hydraulic head of about 8 m throughout the reservoir.

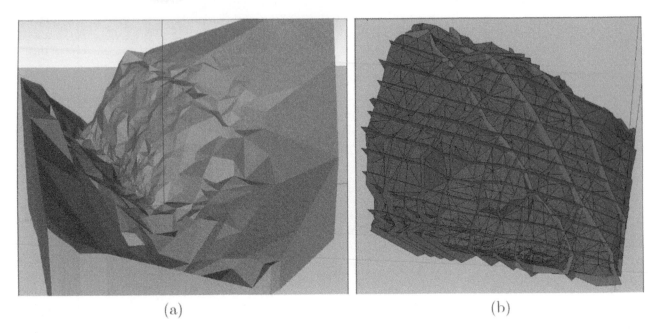

Figure A6.9: (a) View of the whole site; (b) view of the volume of interest, including discontinuities

Boundary conditions can be applied as a time-dependent increasing pressure from a horizontal surface at $z = 0$. This condition simulates the rise of the water from $z = 0$ to the maximum, $z = 8$ m. It is possible to apply any pressure-time evolution to the boundary. In particular, it is possible to apply the pressure evolution recorded during the experiments at one of the measurement points and study the response of other points, both in pressure and displacement. In this case,

both loading and unloading can be modelled using an imposed pressure-time evolution at the boundary.

4.3 Building the model

The geometry of the Coaraze site was imported into *Slope Model* through a previously created .dxf file (Figure A6.10b).

The discretisation is limited to a resolution of 750 mm to limit calculation time. This corresponds to a total of

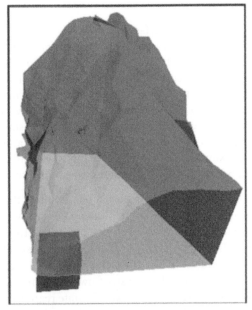

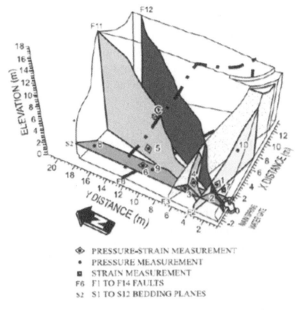

♦ PRESSURE-STRAIN MEASUREMENT
• PRESSURE MEASUREMENT
▪ STRAIN MEASUREMENT
F6 F1 TO F14 FAULTS
S2 S1 TO S12 BEDDING PLANES

Figure A6.10: Geometric model for boundary condition with a horizontal bottom surface with an applied a time-dependent rising pressure (left) and location of the experimental measurement points (right)

approximately 7000 nodes. As a result, some faults were removed because the distance between two neighbouring discontinuities must be several times larger than the resolution. For the same reason, only four bedding planes with a four metre spacing have been considered as a first approximation.

Changes in hydraulic apertures as a result of pressure changes were measured by Cappa *et al.* (2005) for faults F2, F11 and F12, and bedding planes S2 and S3. The measurements were performed at one to three different locations on the same faults and bedding planes. The results of the measurements indicate that the faults have similar hydraulic apertures ranging from 2.6×10^{-4} m to 5.3×10^{-4} m, and bedding planes have apertures ranging from 3.3×10^{-5} m to 9.6×10^{-5} m.

In the model, the hydraulic apertures of the faults and the bedding planes were set to $a_h^{\text{faults}} = 5 \times 10^{-4}$ and $a_h^{\text{BP}} = 5 \times 10^{-5}$ m, respectively.

The rock matrix is modelled as an isotropic linear elastic material, with an elastic modulus of 70 GPa, a Poisson's ratio of 0.29 and a density of 2400 kg m^{-3}. These values were taken from the previously cited papers by Cappa *et al.* (2005). They were obtained from laboratory measurements made on several intact samples taken at different locations on the site. The normal stiffnesses of some of the faults were determined by Cappa *et al.* (2005) through back analysis.

- Fault F2 where point 1 is located: 132 GPa m^{-1};
- Fault F12 where point 2 is located: 66 GPa m^{-1};
- Fault F11 where points 4 and 5 are located: 4.3 GPa m^{-1}.

When more than one value was given, the average was chosen, so that each fault had only one normal stiffness value. All other normal stiffnesses were set to 30 GPa m^{-1} for the faults and 400 GPa m^{-1} for the bedding planes. Shear stiffnesses were assumed to be 10 per cent of the normal stiffnesses in any case.

4.4 Simulation results

The exact experimental pressure-time curve measured at Pt. 1 was applied as a time-dependent boundary condition to the bottom surface of altitude $z = 0$. Consequently, all nodes located at an altitude between 0 and 0.75 m (0.75 m being the resolution) were applied to this pressure-time evolution. Pts 1 and 2 have z-coordinates of 0.264 m and 0.6253 m, respectively, which is less than 0.75 m, so that the imposed pressure-time evolution was applied on both of them. The results using this configuration are presented in Figure A6.11. As can be observed in Figure A6.11b, the discrepancy for Pt. 2 is quite acceptable; thus, this approach might be considered satisfactory. Note that Pt. 4 has a singular behaviour, but the overall numerical model results show fairly good agreement with the experiments (Figures A6.12 and A6.13). In particular, the time lag between the beginning of the loading phase and the time when the pressure starts to rise is reproduced at all points with a maximum discrepancy of one minute, and the final pressures match quite well.

4.5 Conclusions

The results illustrate the ability of the current version of *Slope Model* to simulate the hydromechanical response of a complex fracture network. However, although the Coaraze site is a well-constrained *in situ* experiment, it is much less constrained and less understood than

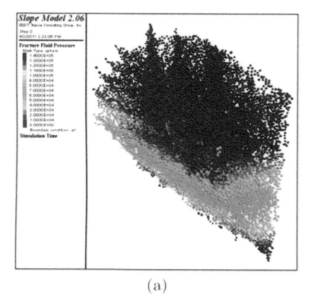

(a)

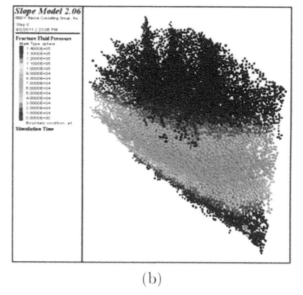

(b)

Figure A6.11: Fracture fluid pressure distribution in the model using [BC1] set for boundary conditions: (a) computed initial state; (b) filled reservoir after hydraulic loading phase

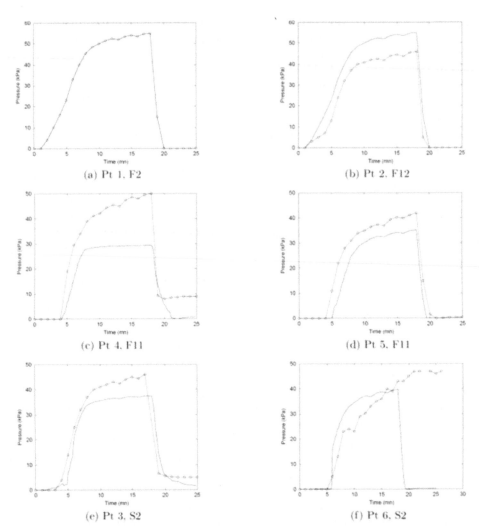

(a) Pt 1, F2

(b) Pt 2, F12

(c) Pt 4, F11

(d) Pt 5, F11

(e) Pt 3, S2

(f) Pt 6, S2

Figure A6.12: Comparison between the 5-bedding-plane model with the time-dependent boundary conditions [BC2] (black) and the experimental results (red), $a_h^{\text{faults}} = 5 \times 10^4$ m, $a_h^{\text{BP}} = 5 \times 10^5$ m

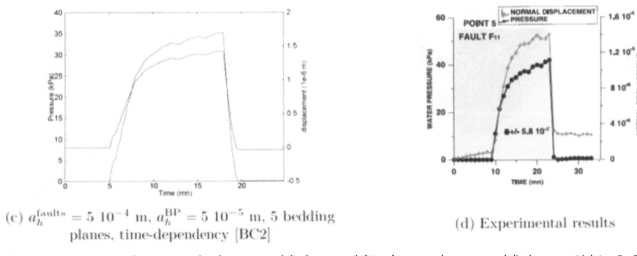

(c) $a_h^{\text{faults}} = 5 \ 10^{-4}$ m, $a_h^{\text{BP}} = 5 \ 10^{-5}$ m, 5 bedding planes, time-dependency [BC2]

(d) Experimental results

Figure A6.13: Comparison between simulated pressure and displacement (left) and measured pressure and displacement (right) at Pt. 5, Fault 11

laboratory-scale experiments. The mechanisms involve complex hydromechanical coupling and probably even more complex geometrical singularities. Because of this complexity, and also because of the size of the site, it is not possible to characterise such singularities. As a result, even a relatively simple model, such as the one set up here, requires an important set of parameters that is not defined completely. Only a few joint apertures could be measured locally and not all joints could be assigned an aperture.

Appendix 7

Lessons learnt and basic guidelines to monitoring for general dewatering

Gordon Maclear

1   Introduction

The intent of Appendix 7 is to provide a basic summary of the practical application of monitoring, together with lessons learnt, to achieve effective dewatering and general site water management. The document has been put together largely based on the experience of AngloGold Ashanti at their West Africa operations.

Various monitoring methods are discussed with practical guidelines provided, and common problems and faults related to monitoring are outlined based on field experience gained on mine site operations. In addition, a monitoring frequency is suggested to achieve the required results.

2   Water level monitoring

2.1   Reasons for monitoring

Water level monitoring in dedicated standpipes or piezometers (which may be equipped with pressure transducers) is a standard and essential tool for any mine site where the zone of drawdown around an actively dewatered pit needs to be known, as well as the extent of drawdown over time. In addition, monitoring and measurement of the dewatering-induced changes in piezometric pressure behind pit walls is critical for mine design to ensure safe working conditions. Monitoring also provides accurate data for assessing and implementing positive or negative intervention options where a high wall may be steepened or flattened respectively, which has direct impacts on the stripping ratio and thus mining costs of the operation.

A good spatial coverage of the monitoring points is required to ensure that the geometry of the piezometric surface can be accurately determined (Figure A7.1). In large deep open pits, a piezometer spacing of 100 to 150 m is ideal, however site-specific conditions will dictate the best spacing to achieve the desired results. Whilst in-pit piezometers are not ideal from a mining perspective (since they obstruct operations and require careful trimming with mining down around them), they are essential to determine the depth to water level below the active mining surface. This provides an indication of the 'mining buffer' or thickness of dry formation remaining below the pit bottom which should ideally be kept to a minimum of 10–20 m as a general rule, dependent on the bench sizes being mined at a particular operation.

If there are only a few monitoring points along design sections, the geometry of the piezometric surface cannot be accurately established (Figure A7.1) and any existing field conditions that could compromise the FoS of the high-wall may be masked as a result.

2.2   Water level measurement – practical guidelines

Water level monitoring is carried out by measuring the water level in a standpipe, piezometer or monitoring borehole using a dip meter, i.e. an electric water level meter or pressure transducer. In addition to monitoring of dedicated piezometers, water levels should be measured in as many supplementary points as possible to increase the monitoring density in and around an active pit. Disused or abandoned dewatering, water supply or environmental monitoring boreholes, and old diamond exploration boreholes, can be used as supplementary monitoring points, thereby establishing a spatially extensive site monitoring network.

With respect to field measurements, the following practical guidelines are relevant.

- Check that the dip meter is functioning and there are spare batteries before starting a monitoring round.
- Preferably always use the same dip meter, or ensure that the dip meter is correctly calibrated and functioning before starting monitoring.

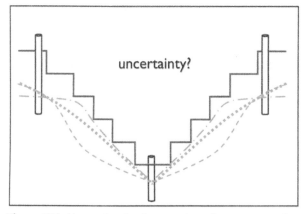 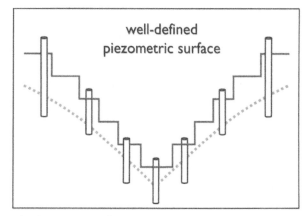

Figure A7.1: Uncertainty in piezometric surface geometry due to variations in piezometer spacing

- Whilst dip meters are robust, they can become damaged (poor contact) or an end section of the tape meter can break off and be lost down the hole (particularly if the borehole being measured is equipped with a pump). Corrections to the reading are then required to accommodate the lost tape section. This adjustment should be clearly marked on the meter and taken into account with all future readings.

- If the dip meter is equipped with a sensitivity adjustment (for variable salinity water), it should be calibrated to the site-specific water quality conditions.

- A reference mark should be made on each piezometer and this mark used as the measuring point for each reading.

- The water level reading at each point should be recorded in metres below measuring point and then reduced to total head (metres above sea level) by correcting for the elevation of the collar above ground level.

- Groundwater levels generally do not fluctuate dramatically and this should be taken into account when recording water levels and comparing them to previous records. In this regard, it is critical to have the water level readings from previous monitoring rounds for each piezometer in a field notebook that accompanies the technician during each monitoring round. By carrying out these cross-references, incorrect water level recordings (e.g. digit swopping, incorrect decimal place) can be identified and corrected at source, i.e. while the technician is still in the field.

- Ideally, the same technician should carry out each monitoring round. This will ensure continuity and the technician will become familiar with the water level history of each site. If a significant water level fluctuation from previous records is noted and confirmed as correct, then a relevant field observation should be made explaining the fluctuation. Within an active mining environment, these fluctuations are typically in response to increased or decreased pumping rates of a dewatering borehole in the vicinity of the piezometer, commissioning of a new dewatering borehole, or a change in the collar elevation due to mining down around the piezometer.

- Cross-checking results in the field can also identify problems with piezometer integrity, such as constant water level readings over an extended time period for a specific piezometer, compared with surrounding piezometers which are showing a response to dewatering. These constant readings typically indicate a blocked piezometer and actions can then be taken to remedy the situation.

The importance of intelligently assessing water level readings is illustrated in the following example of what can occur if all field data measurements are assumed to be correct. Two scenarios of pit piezometric surface geometries are shown in Figure A7.2. In Figure A7.2A, the water level from a blocked piezometer is included, compared with Figure A7.2B, where the blocked piezometer data is excluded. If the data from the blocked piezometer (shown in yellow) was assumed to be correct, a significantly higher than actual piezometric gradient behind the pit high wall would have been assumed.

In this example, if the incorrect gradient (i = 0.4 in Figure A7.2A) had been used in pore pressure and FoS calculations, instead of the actual gradient condition behind the pit high wall (i = 0.1 in Figure A7.2B), the pore water pressure – and thus probability of failure (PoF) – would have been over-estimated by a factor of four. Under practical operational conditions, and dependent on the geology (e.g. soft material versus hard rock), application of this incorrect assessment may have resulted in intervention measures being dictated such as a pushback to flatten the high wall angle to mitigate the apparent risk of wall failure.

2.3 Monitoring frequency

Monitoring frequency will depend on numerous factors such as the location and spacing of the piezometers within and around a pit, the scheduled mine-down rate of the pit, and whether the pit is being actively mined or not. Hydraulic factors including the formation storativity and transmissivity (thereby impacting on the water level drawdown or rate of dewatering), as well as the orientation of preferential flow zones (e.g. faulted or breccia zones) will also be a factor in determining the frequency of piezometer monitoring.

Typically, routine monthly monitoring of the entire piezometer network should be carried out around all pits. Even if mining activities have ceased, monitoring should continue to determine the water level rebound rate for mine closure purposes, for example, the rate of development of a pit lake.

In an actively mined pit, weekly or even daily water level monitoring should be carried out to determine the dewatering efficiency (rate of water level drawdown) as well as the amount of (dewatered) dry ground below the pit floor available to be mined.

Monitoring of other variables that have a direct impact on pit dewatering, such as dewatering borehole and sump pumping rates (discussed further in Section 2.5), should be carried out at the same frequency as the water level monitoring.

2.4 Location of piezometers

The location of water level monitoring sites is critical for ensuring the determination of both an accurate and representative piezometric surface across the entire footprint area of the pit. An open pit is a dynamic environment where mining occurs both vertically down and laterally outwards from the centre of the pit. This is particularly the case during times of increasing commodity prices, resulting in new design optimisations that typically comprise significant expansion pushbacks to open up blocks of ground previously considered to be marginal. These continual changes in pit-shell design, whilst obviously good for the life of mine, present challenges to the site hydrogeologist. As a result, the monitoring network design should remain flexible and consider future expansions to mining.

A complete and comprehensive monitoring network should comprise a network of ex-pit as well as in-pit piezometers. The ex-pit piezometers should be located in an array from as close to the pit crest as possible (taking into account the effect of possible pushbacks) to at least 100 to 200 m away from the pit crest, and potentially much further for a large dewatering system. The distance from the pit crest, as well as the number of piezometers located within that distance, should be based on the scale of the operation and the degree of accuracy required in determining the geometry of the piezometric surface away from the pit being dewatered (Figure A7.1). The

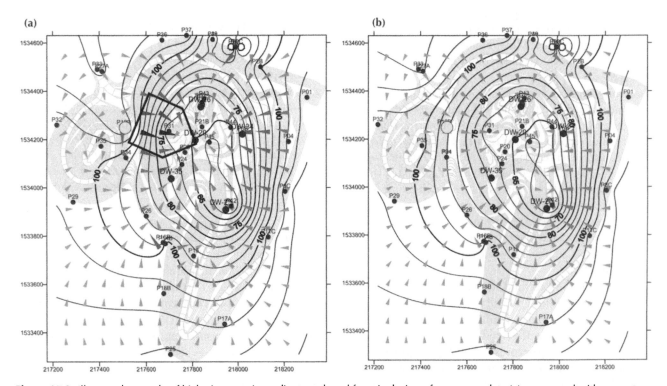

Figure A7.2: Illustrated example of high piezometric gradient produced from inclusion of erroneous data (a), compared with correct data (b)

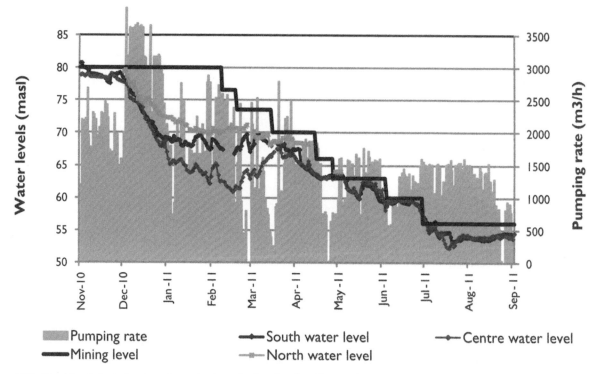

Figure A7.3: Combined chart showing all required monitoring data for effective dewatering

transmissivity of the formation will also have a determining effect on the piezometer location and spacing, with a highly transmissive formation having a laterally more extensive area of drawdown (and thus more spread out piezometer network) compared with a poorly transmissive formation.

With respect to in-pit piezometers, every effort must be made to ensure that they are located in areas that will remain accessible for the life of mine. Whilst it is understood that this is often an impossible task (regular pit shell design changes), piezometers located on large catch-benches, down ramps, along haulages or access roads, and in the vicinity of permanent pit infrastructure (e.g. stage-pumping system, generators) will have a better chance of long-term 'survival' than a piezometer network desk-designed on a theoretical grid of, say, 50 × 50 m.

For the reasons shown in Figure A7.1 it is critical to have intra high wall monitoring points to provide the necessary degree of accuracy for carrying out failure risk analyses. Typical suitable sites include accessible large catch benches or abandoned (cut-off) ramps or haulages.

2.5 Monitoring of pumping rates

In addition to water level monitoring, an equally effective and important variable requiring routine monitoring is the dewatering (pump-out) rate from each pit. This includes the combined pumping from dewatering boreholes, as well as sump pumping. Ideally, pumping rates should be measured by means of in-line flow meters on each pump (surface and submersible). Within all mining

environments, however, this level of data is often not available, in which case an estimate of pumping rate can be made by keeping a record of the pump motor run-time (hours) multiplied by the designed rating (delivery capacity in m^3/hr) of the pump.

If there is no active mining at a pit, and pumping is only being carried out e.g. to maintain the water level at a constant dewatered surface, then it is sufficient to monitor the pumping rate on a weekly or even monthly basis (provided there is either flow meter or hour meter data available). If, however, the pit is being actively mined then daily pumping rate monitoring is required. The benefit of this high frequency monitoring is two-fold: (i) to ensure that an efficient maximum dewatering pumping rate is being achieved according to a set target, and conversely to show areas that will require additional dewatering effort (e.g. installing additional dewatering boreholes), and (ii) to provide a management tool for the mining and technical team to highlight problem areas, such as pump break-downs and failures and servicing requirements for urgent action to re-establish dewatering efficiency to set targets.

The benefit of combining dewatering (pumping) rate monitoring together with water level monitoring is illustrated in Figure A7.3, where the impact of variations in pumping rate can be clearly seen on the water level and the ultimate long-term dewatering efficiency for a particular pit.

In addition to the variables requiring monitoring (as outlined above), it is also necessary to ensure that accurate

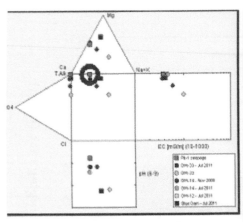
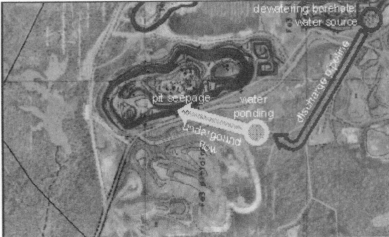

Figure A7.4: Hydrochemical fingerprinting to determine source of seepage water

long-term records are kept of site rainfall as part of the general water balance for a mine and for aquifer recharge rate determinations.

2.6 Monitoring for hydrochemical fingerprinting

Comparing hydrochemistry from various water points within an operating pit can assist in determining the source of 'problem' water, thereby contributing to the understanding of the hydrogeology and the dewatering system. A case in point from a West African open pit mine is discussed below.

A significant water inflow (~30 m³/hr) to an actively mined pit occurred from an exposed breccia zone in the high wall. After exploration drilling to target the assumed high-yielding structure proved unsuccessful (low yields only), the seepage water was analysed and compared to the existing hydrochemical database of the water points in the pit and surrounding areas. This identified the inflow to be from a dewatering borehole from an adjacent pit that was recharging the pit in question via a cross-cutting structure (Figure A7.4). It should be noted that this level of discrete sampling per borehole can only be done if taps or T-pieces are installed at the head of each dewatering borehole for purposes of routine sampling.

In this example, the benefit of monitoring the water quality from discrete sources is illustrated. If it were only possible to take samples of the combined discharge from various dewatering boreholes, this level of fingerprinting would not be possible. It also highlights the importance of general water management in operating pits. Due diligence in all aspects of monitoring related to water inflows or seepages will ensure both a high level of water awareness on a mine site as well as quick reaction to remedy any adverse situations.

3 Summary and conclusions

Monitoring is essential to both measure and manage the dewatering system. Planned, structured and relevant monitoring, as well as quick reaction to problems identified during the course of the monitoring, will ensure effective dewatering.

Important aspects for consideration in monitoring are summarised below.

- Design the monitoring system in accordance to the specific characteristics of the pit lithology and hydrogeology, also taking into account the requirements of the mining targets.
- Ensure that routine monitoring is not carried out 'blindly'. Proper care should be taken to ensure both the accuracy and validity of monitoring data by comparing with previous records.
- Ensure that the frequency of monitoring is in accordance with the mine plan – this will give the best results and data for effective management of the dewatering program.
- Monitor the discharge (dewatering pump-out rate) for all wells and sumps – this is essential to allow comprehensive management of the dewatering program. Combining the pumping rate data with the water level data provides a quick indication of the dewatering progress and efficiency.
- Take routine (quarterly) samples of discrete water points for hydrochemical analysis for purposes of source fingerprinting and to assist with the overall hydrogeological characterisation.

As a final consideration, due acknowledgement should be given to all the field technicians who carry out the task of routine monitoring. Without them there would be no data.

Symbols

The hydrogeological symbols used in this report follow, as far as practicable, the recommendations of the United States Geological Survey (Lohman *et al.*, 1972). Other symbols are those in general use. The Table below lists the symbols (in alphabetical order of first Roman followed by Greek characters), the dimensions and the name of the property represented, the preferred unit, the preferred unit reduced to SI base units, and some other commonly-used metric units.

Symbol	Dimensions	Property	Preferred unit	SI base unit	Other common units
a		hydraulic aperture, fracture aperture	m	[m]	(mm, μm)
b	[L]	formation thickness/fracture aperture	m	[m]	(mm, μm)
B	[–]	Skempton's coefficient	–		
c	[M/L.T^2]	cohesion	Pa	[kg/m.s^2]	(N/m^2; kPa, MPa, GPa, etc.)
d	[L]	distance/displacement	m	[m]	
D		Hoek-Brown disturbance factor			
D	[L^2/T]	hydraulic diffusivity	m^2/s	[m^2/s]	(m^2/day)
e	[–]	void ratio	–		
E	[M/L.T^2]	Young's modulus	Pa	[kg/m.s^2]	(N/m^2)
ε		strain			
F	[M.L/T^2]	force	N (newton)	[kg.m/s^2]	
F	[–]	Hvorslev's shape factor			
g	[L/T^2]	acceleration due to gravity	m/s^2	[m/s^2]	
h	[L]	hydraulic head	m	[m]	
k	[L^2]	intrinsic permeability	m^2	[m^2]	(μm^2, D (darcy), mD (millidarcy))
K	[L/T]	hydraulic conductivity ('permeability')	m/s	[m/s]	(m/day)
l	[L]	length/displacement	m	[m]	(m)
m	[M]	mass	kg	[kg]	(tonne)
n	[–]	porosity	–		
N	[–]	number of dimensions	–		
p	[M/L.T^2]	pressure	Pa	[kg/m.s^2]	(N/m^2; kPa, MPa, GPa, etc.)
P		probability	%		
q	[L/T]	specific discharge (rate of flow per unit area)	m/s	[m/s]	(m/day)
Q	[L^3/T]	volumetric rate of flow	m^3/s	[m^3/s]	(m^3/day)
r	[L]	radius (e.g. as of action of a well)	m	[m]	
s	[L]	drawdown	m	[m]	
S	[–]	storage coefficient	–		
S_s	[L^{-1}]	specific storage	m^{-1}	[m^{-1}]	
S_r	[–]	specific retention	–		
S_t	[M/L.T^2]	shear strength	Pa	[kg/m.s^2]	(N/m^2; kPa, MPa, GPa, etc.)
S_y	[–]	specific yield ('drainable porosity')	–		
t	[T]	time	second (s)	[s]	(minute, day, year)
T	[L^2/T]	transmissivity	m^2/s	[m^2/s]	(m^2/day)
u	[m/L.T^2]	pore pressure	Pa	[kg/m.s^2]	m head, kPa
v	[L/T]	speed (scalar), velocity (vector)	m/s	[m/s]	
W	[ML^2/T^2]	work or energy	J (joule)	[kg.m^2/s^2]	(N.m; kJ, MJ, etc.)

Symbol	Dimensions	Property	Preferred unit	SI base unit	Other common units
α	$[(L.T^2)/M]$	compressibility of rock mass	Pa^{-1}	$[m.s^2/kg]$	m^2/N
α_B	$[-]$	Biot's coefficient			
β	$[(L.T^2)/M]$	compressibility of water	Pa^{-1}	$[m.s^2/kg]$	m^2/N
β		slope angle	degree		
γ	$[M/L^2.T^2]$	specific weight (or weight density) (ρg)	N/m^3	$[kg/m^2.s^2]$	
κ	$[M/L.T^2]$	bulk modulus of drained rock mass	Pa	$[kg/m.s^2]$	$(N/m^2;$ kPa, MPa, GPa, etc.)
K_w	$[M/L.T^2]$	bulk modulus of water	Pa	$[kg/m.s^2]$	$(N/m^2;$ kPa, MPa, GPa, etc.)
μ	$[M/L.T]$	dynamic (absolute) viscosity	Pa.s	$[kg/m.s]$	
υ	$[L^2/T]$	kinematic viscosity	mm^2/s	$[m^2/s]$	
ν	$[-]$	Poisson's ratio	$-$		
ρ	$[M/L^3]$	mass density	kg/m^3	$[kg/m^3]$	(tonnes/m^3)
σ', σ_e	$[M/L.T^2]$	effective stress	Pa	$[kg/m.s^2]$	$(N/m^2;$ kPa, MPa, GPa, etc.)
σ, σ_T	$[M/L.T^2]$	total stress	Pa	$[kg/m.s^2]$	$(N/m^2;$ kPa, MPa, GPa, etc.)
τ	$[M/L.T^2]$	shear stress	Pa	$[kg/m.s^2]$	$(N/m^2;$ kPa, MPa, GPa, etc.)
ϕ		angle of internal friction	degree		
Φ	$[L^2/T^2]$	fluid (mass) potential (gh)	m^2/s^2	$[m^2/s^2]$	

Abbreviations

CAPEX = Capital Expenditure
DFN = Discrete Fracture Network
EPM = Equivalent Porous Medium
ET = Evapotranspiration
FoS = Factor of Safety
GSI = Geological Strength Index (Hoek-Brown)
MAE = Mean-Absolute Error
ME = Mean Error
NPV = Net Present Value
OPEX = Operating Expenditure
REV = Representative Elementary Volume (Bear, 1972)
RQD = Rock Quality Designation (Deere)
RMR = Rock Mass Rating (Bieniawski)
UCS = uniaxial compressive strength of intact rock
ZOR = Zone of Relaxation

Glossary

Notes:

cf (confer) = compare
q.v. (quod vide) = which see
Syn = synonym

Consequence = Outcome or impact of an event

Coupled flow = Process in which the fluid pressure and rock properties influence each other. For example, an increase (or decrease) in fluid pressure may cause an opening (or closing) of fractures and hence a change in rock permeability. A change in rock volume (for example, a volume expansion caused by unloading) may cause an expansion of fracture openings and an increase in pore volume, changing permeability and storage properties.

Darcy's law = An empirical law that describes the flow of fluid through a porous medium (*q.v.*). It states that the rate of flow of fluid is proportional to (i) the hydraulic gradient (*q.v.*) within the fluid, (ii) the cross-sectional area through which flow is occurring, and (iii) a property called the hydraulic conductivity (*q.v.*).

Diffusivity = *see* Hydraulic diffusivity

Discrete fracture network = DFN models capture the spatial geometry and connectivity of the fracture network within a fractured rock masses by explicitly representing key elements of the fracture system as discrete objects in space with appropriately defined geometries and properties. In groundwater studies, flow conditions in fractured rock can be determined by identifying and quantifying the flow through the individual fractures modelled. cf Equivalent porous medium

Double porosity = Said of a formation or geological unit in which fluid may occur and/or move in two types of pore space, typically intergranular or intercrystalline porosity and fracture porosity. The two types of porosity are interconnected and fluid is able to transfer between them in response to head differences.

Drainable porosity = *same as* Specific yield (*q.v.*)

Dual porosity = *same as* Double porosity

Effective stress = *see* Stress, effective

Epilimnion = The uppermost, warmest layer of water in a thermally stratified lake.

Equivalent porous medium = A concept or modelling approach in which a fractured rock is treated as behaving, in bulk, in the same way as a material in which the storage and flow occur in intergranular pore space with the same average properties. *cf* Discrete Fracture Network

Finite difference = A numerical method for approximating the solutions to differential equations using finite-difference equations to approximate derivatives.

Finite element = A numerical method for approximating the solutions to differential equations using numerical integration.

Fluid potential = The fluid potential at any point is the mechanical energy of a unit mass of the fluid at that point measured relative to an arbitrary state and datum level. The state is usually atmospheric pressure and a condition of zero velocity relative to the surface of the Earth, and the datum level is an arbitrarily chosen horizontal surface (for field conditions, usually sea level). The fluid potential Φ is proportional to the total head, h, the constant of proportionality being the acceleration due to gravity, g, i.e.

$$\Phi = gh$$

Groundwater = The water contained in the pore spaces of a rock or soil mass in a saturated state, i.e. below the water table.

Hazard = Source of potential harm; a potential occurrence or condition that could lead to injury, damage to the environment, delay or economic loss. *cf.* Risk

Hazard analysis = *see* Preliminary hazard analysis

Head, elevation = At any point is the portion of total head contributed by the elevation of the point in question above the datum level; it is equal to the elevation energy of the fluid per unit weight.

Head, pressure = At any point is the portion of total head of a liquid contributed by the fluid pressure at that point. It is the height of a column of the liquid that can be supported by the pressure at a specific point.

Head, static = The height above a reference or datum surface at which the water will stand in a piezometer. It represents the potential energy of the water per unit weight and is equal to the sum of the pressure head in the piezometer and the elevation above datum of the measuring point in the piezometer.

Head, total = The mechanical energy of the fluid per unit weight. It consists of three components: elevation head, pressure head and the component contributed by the kinetic energy of the water, which is termed velocity head, given by $v^2/2g$. The first two components when added together give the static head. Groundwater usually moves so slowly that its kinetic energy and therefore velocity head are negligible.

Holomictic lake = Lake with uniform temperature and density from top to bottom.

Homogeneous = Having the same properties at every point; of a porous medium, one in which any two volumes large enough to be a representative elementary volume will have the same properties, e.g. of porosity or permeability.

Hydraulic conductivity = The volume of groundwater (or other fluid), at the prevailing viscosity, that will move through rock or soil fully saturated with that fluid in unit time under a unit hydraulic gradient through a unit area measured at right angles to the direction of flow. It depends on the properties of the fluid saturating the pores as well as the properties of the medium and has dimensions of L/T. *See also* Permeability and Intrinsic permeability

Hydraulic diffusivity = The conductivity parameter divided by the storage parameter in a formation, e.g. hydraulic conductivity divided by specific storage or specific yield; transmissivity divided by storage coefficient. Hydraulic diffusivity is the major control on the speed with which a change in head propagates through a rock or soil. Dimensions are L^2/T.

Hydraulic gradient = The rate of change of static head with distance.

Hypolimnion = Colder and dense, bottom layer of water in a thermally stratified lake.

Interconnected porosity = *see* Porosity, interconnected

Intrinsic permeability = The property that controls the ability of a porous medium to transmit fluids under the influence of a hydraulic or potential energy gradient. Unlike hydraulic conductivity it is a property of the porous medium alone; it has dimensions of L^2 (area).

Isotropic = Said of a material whose properties (e.g. intrinsic permeability) do not vary with direction.

Likelihood = Probability or frequency of occurrence of an event, described in qualitative or quantitative terms.

Meromictic lake = Lake with layers that do not intermix.

Perched groundwater zones = Saturated parts of the rock or soil mass separated by unsaturated material from the main water table. Perched groundwater zones are normally caused by stratigraphic or structural horizons of low permeability that impede downward percolation of water. The vertical hydraulic gradient between the perched phreatic levels or water tables will be greater than unity.

Permeability = A term loosely used to mean intrinsic permeability (*q.v.*) (with dimensions of L^2) or hydraulic conductivity (with dimensions of L/T). In engineering practice, where there is usually little variation in the viscosity of the pore fluids, permeability is commonly used to mean hydraulic conductivity. It is used in this sense throughout this book.

Phreatic surface = Water table (*q.v.*)

Phreatic water = Groundwater (water that exists in a saturated state below the vadose zone, i.e. below the water table). The term 'phreatic surface' is used synonymously with water table throughout this book.

Pore pressure elevation = Static head (often above a particular failure surface or zone of weakness).

Porosity = The property of a rock or soil of possessing pores or voids; expressed quantitatively as the pore volume divided by the bulk volume.

Porosity, double = *see* Double porosity

Porosity, drainable = *same as* specific yield.

Porosity, interconnected = The fraction or percentage of the bulk volume of a rock or soil mass occupied by interconnected void space. Only part of the interconnected porosity will drain under gravity. Even after prolonged drainage, some water will remain adhering as films (pellicular water) to mineral grains or fracture surfaces, or completely filling the finer pores of the rock or soil. The fraction or percentage of the bulk volume that drains under gravity is termed the specific yield (S_y) (*q.v.*) and the fraction that remains water-filled is termed the specific retention (S_r). The sum of specific yield and specific retention equals total porosity (i.e. $S_y + S_r = n$).

Porosity, total = The fraction of the total volume of a rock or soil mass occupied by void space, usually expressed as a percentage. These voids may be intergranular spaces or fractures. Total porosity (n) may include pores such as fluid inclusions or isolated fractures that are not connected with other pores.

Porous medium = A material consisting of solid parts and holes. The term is also used loosely to describe a rock in which storage and flow take place within intergranular pore space rather than in fractures.

Potentiometric surface = Surface created by contouring static head values. It can apply to confined and unconfined aquifers. The term 'piezometric surface' is used synonymously with 'potentiometric surface' in this book. *See also* fluid potential

Pressure head = *see* Head, pressure

Representative elementary volume = The smallest volume of a porous or fractured medium that is large enough for the flow behaviour to be represented by a continuum approach such as Darcy's law rather than by a microscopic approach that deals with individual pore channels.

Risk = Chance of something happening that will have an impact on objectives.

Risk analysis = Systematic process to understand the nature of and to deduce the level of risk.

Risk assessment = Overall process of risk identification, risk analysis and risk evaluation.

Saturated zone = The zone below the water table, where the pore water is everywhere at a pressure greater than atmospheric. *cf* unsaturated zone

Specific discharge = The rate (volume/time) at which groundwater (or other fluid) flows through a unit area. It has dimensions of L/T but is not the true velocity of the fluid.

Specific storage = The volume of water released from storage by a unit volume of saturated aquifer for a unit decline in head. Specific storage, S_s, is related to porosity and compressibility by the equation

$$S_s = \rho g(\alpha + n\beta)$$

where ρ is the density of water, g is acceleration due to gravity, n is the porosity of the rock, α is the compressibility of the rock framework and β is the compressibility of water.

Specific yield = Fraction of the total volume of a rock or soil mass that can drain by gravity in response to lowering of the water table.

Static head = *see* Head, static

Storage coefficient = Volume of water that a formation releases from storage per unit surface area of the aquifer for unit decline in the component of head normal to that surface. In a confined aquifer, storage coefficient is the product of specific storage and aquifer thickness. In an unconfined aquifer it is the product of specific storage and saturated aquifer thickness plus the specific yield.

Stress = Force per unit area. Stress therefore has the same units as pressure, but is normally used for a body's internal response to the externally applied pressure.

Stress, effective = Difference between the total stress (the total pressure experienced as a result of the weight of the overlying material) and the pore water pressure. The effective stress is the pressure with which the grains or blocks of the formation are held in contact. If the pore pressure is reduced by lowering the potentiometric surface, the effective stress will increase. If the total stress is reduced (e.g. by removal of overlying material) the effective stress will reduce.

Total head = *see* Head, total

Total porosity = *see* Porosity, total

Transmissivity = Measure of the ability of a rock to transmit water; the rate at which water will move under a unit hydraulic gradient through a unit width of aquifer, measured at right angles to the direction of groundwater flow. Transmissivity is the product of hydraulic conductivity and saturated thickness.

Unsaturated zone = The zone between the ground surface and the water table, where the pore pressure is less than atmospheric. Some of the pores may, however, be completely filled with water. Also referred to as the vadose zone. *cf* saturated zone

Vadose zone = *see* unsaturated zone

Void ratio = The ratio of pore volume to grain volume. *See also* Porosity, which is the ratio of pore volume to bulk volume.

Water table = Surface in a rock or soil mass at which the pore pressure is exactly equal to atmospheric pressure. Frequently used interchangeably with the term 'phreatic surface'.

References

Allen DJ, Brewerton LJ, Coleby LM, Gibbs BR, Lewis MA, MacDonald AM, Wagstaff SJ & Williams AT (1997) *The Physical Properties of Major Aquifers in England and Wales*. British Geological Survey Technical Report, WD/97/34; Environment Agency R&D Publication 8.

Alvarez TA Jr., Cording EJ & Milhail RA (1995) Hydromechanical behavior of rock joints: a re-interpretation of published experiments. In: *Rock Mechanics* (35th U.S. Symposium, University of Nevada, Reno, June 1995). (Eds JJK Daemen & RA Schultz) pp. 665–671. Rotterdam, Balkema.

American Society for Testing and Materials (1991) Standard test method (Field Procedure) for instantaneous change in head (Slug Tests) for determining hydraulic properties of aquifers. *Groundwater and Vadose Zone investigations, ASTM D 4044-91*.

Anderson MP & Woessner WW (1992) *Applied Groundwater Modeling – Simulation of Flow and Advective Transport*. Academic Press, San Diego.

Appelo CAJ & Postma D (2005) *Geochemistry, Groundwater and Pollution*. 2nd edn. A.A Balkema Publishers, Amsterdam.

Australian Drilling Industry Training Committee Ltd (1997). *Drilling, the Manual of Methods, Applications, and Management*. ADITC, Sydney.

Azrag EA, Ugorets VI & Atkinson LC (1998) Use of a finite element code to model complex mine water problems. In: *Proceedings, Symposium on Mine Water and Environmental Impacts* (Johannesburg, September 1998), Vol. 1, pp. 31–41. International Mine Water Association, Johannesburg.

Back W, Rosenshein JS & Seaber PR (Eds) (1988) *Hydrogeology (The Geology of North America, Volume O-2)*. Geological Society of America, Boulder, Colorado.

Banks D & Robins NS (2002) *An Introduction to Groundwater in Crystalline Bedrock*. Norges geologiske undersøkelse (Geological Survey of Norway), Trondheim.

Banks D, Solbjørg ML & Rohr-Thorp E (1992) Permeability of fracture zones in a Precambrian granite. *Quarterly Journal of Engineering Geology* **25** (4), 377–388.

Barenblatt GI & Zheltov YP (1960) On the basic equations of fluid filtration in fissured rocks. *Dokl. AN SSSR* **132** (3).

Barker JA (1988) A generalized radial flow model for hydraulic tests in fractured rock. *Water Resources Research* **24**, 1796–1804.

Barker JA (1993) Modelling groundwater flow and transport in the Chalk. In: *The Hydrogeology of the Chalk of North-west Europe*. (Eds RA Downing M Price & GP Jones) pp. 59–66. Clarendon Press, Oxford.

Batchelor GK (1967) *An Introduction to Fluid Dynamics*. Cambridge University Press.

Beale G (2010) Hydrogeological model. In: *Guidelines for Open Pit Design*. (Eds J Read & P Stacey) pp. 141–199. CSIRO Publishing, Melbourne.

Beale G (2011) Creation of a Conceptual Model of Pore Pressures in a Fractured Rock Slope. Unpublished report to CSIRO Large Open Pit Project.

Bear J (1972) *Dynamics of fluids in porous media*. American Elsevier Inc., New York.

Bell FG (2000) *Engineering Properties of Soil and Rock*. 4th edn. Blackwell Science.

Berryman JG & Wang HF (1995) The elastic coefficients of double-porosity models for fluid transport in jointed rock. *Journal of Geophysical Research* **100** (B12), 24611–24627.

Bieniawski ZT (1976) Rock mass classification in rock engineering. In: *Exploration for Rock Engineering, Proceedings of the Symposium*. (Ed. ZT Bieniawski) **1**, 97–106. Cape Town, Balkema.

Bierschenk WH (1963) Determining well efficiency by multiple step-drawdown tests. *International Association for Scientific Hydrology, Publication* **64**, 493–507.

Biot M (1941) General theory of three-dimensional consolidation. *Journal of Applied Physics* **12**, 155–164.

Black JH & Kipp KL Jr. (1977) Observation well response time and its effect upon aquifer test results. *Journal of Hydrology* **34**, 297–306.

Black JH, Barker JA & Woodman ND (2007) An investigation of sparse channel networks: characteristic behaviours and their causes. ISSN 1402-3091 SKB Rapport R-07-35. Svensk Kärnbränslehantering AB (Swedish Nuclear Fuel and Waste Management Co.) (Available as a PDF from www.skb.se)

Bliss JC & Rushton KR (1984) The reliability of packer tests for estimating the hydraulic conductivity of aquifers. *Quarterly Journal of Engineering Geology* **17**, 81–91.

Bobeck P (2004) *The public fountains of the city of Dijon* (English translation of Darcy, 1856). Kendall/Hunt Publishing Company, Dubuque, Iowa.

Bolander JE & Sukumar N (2005) Irregular lattice model for quasistatic crack propagation. *Physical Review B* **71**, 094106-1–094106-12.

Botin JA (2009) *Sustainable Management of Mining Operations*. Society for Mining Metallurgy and Exploration.

Bouwer H & Rice RC (1976) A slug test for determining hydraulic conductivity of unconfined aquifers with completely or partially penetrating wells. *Water Resources Research* **12** (3), 423–428.

Braester C & Thunvik R (1984) Determination of formation permeability by double-packer tests. *Journal of Hydrology* **72**, 375–389.

Brahana JV, Thrailkill J, Freeman T & Ward WC (1988) Carbonate rocks. In: *Hydrogeology (The Geology of North America, Volume O-2)*. (Eds W Back, JS Rosenshein & PR Seaber) pp. 333–352. Geological Society of America, Boulder, Colorado.

Brassington R (1998) *Field Hydrogeology*. 2nd edn. *John Wiley & Sons*.

Brassington R & Walthall S (1985) Field techniques using borehole packers in hydrogeological investigations. *Quarterly Journal of Engineering Geology, London, 1985* **18**, 181–193.

Bredehoeft JD & Konikow LF (1993) Ground-water models: validate or invalidate. *Ground Water* **31** (2), 178–179.

British Standards Institution (1981) *Code of practice for site investigations. BS 5930*.

Brown ET (1981) *Suggested methods for rock characterization testing and monitoring*. International Society for Rock Mechanics.

Brown KE (1967) *Gas Lift Theory and Practice*. Petroleum Publishing Co, Tulsa, Oklahoma.

Cappa FY, Guglielmi P, Fénart V, Merrien-Soukatchoff & Thoraval A (2005) Hydromechanical interactions in a fractured carbonate reservoir inferred from hydraulic and mechanical measurements. *International Journal of Rock Mechanics and Mining Science and Geomechanics* **42**, 287–306.

Clemens A, Whal T, Bos M & Replogle J (2001) *Water Measurement with Flumes and Weirs*. ILRI.

Cooper HH & Jacob CE (1946) A generalized graphical method for evaluating formation constants and summarizing well field history. *American Geophysical Union Transactions* **27**, 526–534.

Cooper HH, Bredehoeft JD & Papadopulos IS (1967) Response of finite-diameter well to an instantaneous change of water. *Water Resources Research* **3**, 263–269.

Cooper S, Perez C, Vega L, Galera JM & Pozo V (2011) The role of bedding planes in Guadalquivir Blue Marls on the slope stability in Cobre Las Cruces open pit. In: *Proceedings, Slope Stability 2011: International Symposium on Rock Slope Stability in Open Pit Mining and Civil Engineering*, Vancouver, Canada.

Cundall PA (1983) *Numerical Modelling of Water Flow in Rock Masses*. Technical Report to Geognosis, United Kingdom DOE Report No. DOE/RW/83.059.

Cundall PA (2000) A discontinuous future for numerical modelling in geomechanics? *Geotechnical Engineering* **149** (1), 41–47.

Cundall PA & Damjanac B (2009) A comprehensive 3D model for rock slopes based on micromechanics. In:

Slope Stability 2009 (Universidad de Los Andes, Santiago, Chile, November 2009).

Dachler R (1936) *Grundwasserströmung*. Julius Springer, Vienna.

Darcy H (1856) *Les fontaines publiques de la ville de Dijon*. Victor Dalmont, Paris.

Davis SN & De Wiest RJM (1965) *Hydrogeology*. Wiley, New York.

Daw GP & Scott RA (1983) Ground water pressure relief wells in shaft sinking. *International Journal of Mining Engineering* **1**, 229–236.

Dawson EM, Roth WH & Drescher A (1999) Slope stability analysis by strength reduction. *Géotechnique* **49**, 835–840.

Deming D (2002) *Introduction to Hydrogeology*. McGraw-Hill, Boston.

Detournay E (1980) Hydraulic conductivity of closed rock fracture: an experimental and analytical study. In: *Underground Rock Engineering* (13th Canadian Rock Mechanics Symposium, Toronto, May 1980), pp. 168–173.

Detournay E & Cheng AH-D (1993) Fundamentals of poroelasticity. In: *Comprehensive Rock Engineering, Vol. 2*. (Ed. JA Hudson) pp. 113–171. Pergamon Press, Oxford.

DHI-WASY (2010) *FeFlow v 6.0*. Berlin, Germany.

Dingham L (2002) *Physical Hydrology*. Waveland Press, Illinois.

Domemico PA & Schwartz FW (1997) *Physical and Chemical Hydrogeology*. 2nd edn. Wiley, New York.

Doubek GR & Beale G (1992) *Investigation of Groundwater Characteristics using Dual Tube Reverse circulation Drilling*. Preprint No. 92-133. Society for Mining, Metallurgy, and Exploration, Inc., Littleton, Colorado.

Drever JI (1997) *The Geochemistry of Natural Waters*. 3rd edn. Prentice Hall, Englewood Cliffs, New Jersey.

Driscoll F (1986) *Groundwater and Wells*. 2nd edn. Johnson Filtration Systems Inc., St Paul, Minnesota.

Ekkerd JN, Ruest MR & Rankhododo NE (2011) DeBeers Venetia Mine Cut 4 Slope Optimization. In: *Proceedings, Slope Stability 2011, Vancouver, Canada*.

Ekstrom MP, Dahan CA, Chen MY, Lloyd PM & Rossi DJ (1986) Formation imaging with microelectrical scanning arrays. *Transaction of the SPWLA 27th Annual Logging Symposium, SPLWA, Houston, Texas, paper BB*.

Freeze RA & Cherry JA (1979) *Groundwater*. Prentice-Hall, Englewood Cliffs, New Jersey.

Faust CR & Mercer JW (1980) Ground-water modeling: numerical models. *Ground Water* **18** (4), 395–409.

FRASA Ingenieros consultores, S.L. (2001) Modelo hidrogeológico y gestión del sistema de drenaje – reinyección del acuífero. Universidad Autónoma de Barcelona, Spain.

Freyburg DL (1988) An exercise in ground-water model calibration and prediction. *Ground Water* **26** (3), 350–360.

Galera JM, Checa M, Perez C & Pozo V (2009) Enhanced characterization of a soft marl formation using in situ and lab tests, for the prestripping phase of Cobre Las Cruces open pit mine. In: *Slope Stability 2009 (Universidad de Los Andes, Santiago, Chile, November 2009)*.

Galera JM, Montero J, Perez C, Montego L & Varona P (2009) Coupled hydromechanical analysis of Cobre Las Cruces open pit. In: *Slope Stability 2009 (Universidad de Los Andes, Santiago, Chile, November 2009)*.

GEO-SLOPE International Ltd. (2007) *Seep/W 2007.* Calgary, Canada.

Gibbs, CW & Ingersoll Rand Company (1971) Compressed Air and Gas Data. Ingersoll Rand Company, Davidson, North Carolina.

Gibson RE (1963) An analysis of system flexibility and its effect on time-lag in pore water pressure measurements. *Géotechnique* **13**, 1–11.

Golder Associates (2006) Life of mine pit slope-underground interaction study A154 mine. Unpublished report submitted to Diavik Diamond Mines Inc., Project Number 05-1328-004/2200. Report 253.

Golder Associates (2007) Rock engineering, backfill and water management aspects of the Diavik underground mine feasibility study. Unpublished report submitted to Diavik Diamond Mines Inc., Project Number 05-1328-004, Report 393.

Golder Associates (2008) A154 & A418 geotechnical and hydrogeological data summary Diavik Diamond Mine. Unpublished report submitted to Diavik diamond Mines Inc., Project Number D01116, Report 612.

Golder Associates (2010) Construction & validation of a DFN flow model, A154 Pit, Diavik Mine. Unpublished report submitted to Stacey Mining Geotechnical Ltd., Project Number 10-1426-0006.

Johnson Screens (2008) *Groundwater and Wells.* 3rd edn. Johnson Screens, USA.

Guglielmi Y, Cappa F, Rutqvist J, Tsang CF & Thoraval A (2008) Mesoscale characterization of coupled hydromechanical behavior of a fractured-porous slope in response to free water-surface movement. *International Journal of Rock Mechanics and Mining Sciences* **45** (6), 862–878.

Haan C, Barfield B & Hayes J (1994) *Design Hydrology and Sedimentology for Small Catchments.* Academic Press, New York.

Hamm S-Y, Kim MS, Cheong J-Y & Kim J-Y (2007) Relationship between hydraulic conductivity and fracture properties estimated from packer tests and borehole data in a fractured granite. *Engineering Geology* **92**, 73–87.

Hantush MS (1964) Hydraulics of wells. In: *Advances in Hydroscience* **1**, 281–432. (Ed. VT Chow).

Hazzard J, Damjanac B, Detournay C & Lorig L (2011a) Developing rules of thumb for groundwater modelling in large open pit mine design. In: *2011 Pn-Am CGS Geotechnical Conference (Toronto, Canada).* Canadian Geotechnical Society.

Hazzard J, Damjanac B, Lorig L & Detournay C (2011b) Guidelines for groundwater modelling in open pit mine design. In: *Slope Stability 2011 (Vancouver, Canada).* (Eds E Eberhardt & D Stead), Canadian Rock Mechanics Association.

Heath RC (1983) *Basic ground-water hydrology.* US Geological Survey Water-Supply Paper 2220 (revised 2004).

Hem JD (1992) *Study and interpretation of the chemical characteristics of natural water.* US Geological Survey Water-Supply Paper 2254. US Government Printing Office, Washington, DC.

Hoek E (1994) Strength of rock and rock masses. *ISRM News Journal* **2**, 4–16.

Hoek E & Brown ET (1980) *Underground Excavations in Rock.* IMM, London.

Hoek E & Brown ET (1997) Practical estimates of rock strength. *International Journal of Rock Mechanics and Mining Sciences* **34**, 1165–1186.

Hoek E & Diederichs MS (2005) Empirical estimation of rock mass modulus. *International Journal of Rock Mechanics and Mining Sciences* **43**, 203–215.

Houlsby AC (1976) Routine interpretation of the Lugeon water-test. *Quarterly Journal of Engineering Geology, London* **9**, 303–314.

Hvorslev J (1951) *Time lag and soil permeability in groundwater observations.* US Waterways Experiment Station Bulletin 36, 1951.

Itasca Consulting Group, Inc. (2008) *PFC3D* (Particle Flow Code in 3 Dimensions), Version 4.0. Itasca, Minneapolis.

Itasca Consulting Group, Inc. (2010) *FLAC* (Fast Lagrangian Analysis of Continua), Version 7.0. Itasca, Minneapolis.

Itasca Consulting Group (2011) *FLAC* (Fast Lagrangian Analysis of Continua in Three Dimensions), v4.0. Itasca Minneapolis.

Itasca Consulting Group (2011) *UDEC* (Universal Distinct Element Code) Version 5.0. Itasca, Minneapolis.

Iwai K (1976) Fundamental studies of fluid flow through a single fracture. PhD Thesis, University of California, Berkeley.

Jacob CE (1940) On the flow of water in an elastic artesian aquifer. *Transactions American Geophysical Union* **22**, 574–586.

Jacob CE (1947) Drawdown test to determine effective radius of artesian well. *Transactions, American Society of Civil Engineers* **112**, 1047–1070.

Jacob CE & Lohman SW (1952) Nonsteady flow to a well of constant drawdown in an extensive aquifer. *Transactions American Geophysical Union* **33**: 559–569.

Jaeger JC, Cook NGW & Zimmerman RW (2007) *Fundamentals of Rock Mechanics*. 4th edn. Blackwell Publishing, Oxford.

Jeong W-C, Bruel D & Cho Y-S (2006) Numerical experiments of flow and transport in a variable-aperture fracture subject to effective normal stresses. *Journal of Hydraulic Research* **44**, (2), 259–268.

Journel AG & Huijbregts C (1978) *Mining Geostatistics*. Academic Press, New York.

Kazemi H, Seth MS & Thomas GW (1969) The interpretation of interference tests in naturally fractured reservoirs with uniform fracture distribution. *Journal of Petroleum Science & Engineering* **9**, 463–472.

Kozlovsky YeA (1987) *The Superdeep Well of the Kola Peninsula*. Springer-Verlag, Berlin.

Kruseman GP & de Ridder NA (1990) *Analysis and Evaluation of Pumping Test Data*. 2nd edn. Publication 47, Wageningen, The Netherlands, International Institute for Land Reclamation and Improvement.

Kutchling K, Chorley D & Zawadzki W (2000) Hydrogeological modeling of mining operations at the Diavik Diamond Project. Sixth International Symposium on Environmental Issues and Waste Management in Energy and Mineral Production.

Lancaster-Jones PF (1975) The interpretation of the Lugeon water-test. *Quarterly Journal of Engineering Geology* **8**, 151–154.

Lohman SW (1972) Ground-water hydraulics. US Geological Survey Professional Paper 708. US Government Printing Office, Washington, DC.

Lohman SW *et al.* (1972) Definitions of selected ground-water terms – revisions and conceptual refinements. US Geological Survey Water-Supply Paper 1988. US Government Printing Office, Washington, DC.

Long JCS (1983) Investigation of equivalent porous medium permeability in networks of discontinuous fractures. PhD thesis, University of California, Berkeley.

Long JCS & Witherspoon PA (1985) The relationship of the degree of interconnection to permeability in fracture networks. *Journal of Geophysical Research* **90**, 3087–3098.

Long JCS, Gilmour P & Witherspoon PA (1985) A model for study fluid flow in random three-dimensional networks of disc-shaped fractures. *Water Resources Research* **21**, 1105–1115.

Lorig LJ (2009) Challenges in current slope stability analysis methods. In: *Slope Stability 2009 (Universidad de Los Andes, Santiago, Chile, November 2009)*.

Lorig L, Damjanac B, Detournay C, Hazzard J, Riahi A & Cundall P (2011) Assessment of the Effect of Water Pressures on the Stability of Closely Jointed Rock.

Unpublished report 11-2462-24-49D to CSIRO LOP Project, Brisbane, Australia by Itasca Consulting Group, Minneapolis.

Lorig L & Varona P (2000) Practical slope stability analysis using finite difference codes. In: *Slope Stability in Surface Mining*. (Eds WA Hustrulid, MK McCarter & Van Zyl, Dirk JA) pp. 115–124. Society for Mining, Metallurgy, and Exploration, Inc., Littleton, Colorado.

Lugeon M (1933) *Barrage et Geologie*. Dunod, Paris.

Louis C (1974) Rock Hydraulics. Course presented at the Department of Mechanics of Solids, Udine, Rock Mechanics Symposium, pp. 299–387. (Ed. L Muller) Springer-Verlag, Berlin.

Maidment D (Ed.) (1992) *Handbook of Hydrology*. McGraw Hill, New York.

Matthews CS & Russell DG (1967) *Pressure buildup and flow tests in wells*. Storm Printing Corp., Dallas, Texas.

McDonald MG & Harbagh AW (1988) A modular three-dimensional finite difference groundwater flow model. Techniques of Water-Resources Investigations, 06-A1, U.S. Geol. Surv.

McKenna GT (1995) Grouted-in installation of piezometers in boreholes. *Canadian Geotechnical Journal* **32**, 355–363.

Mikkelsen PE & Slope Indicator (2000) Grouting-in Piezometers. Available from <http://www.slopeindicator.com/>.

Mikkelsen PE (2002) Cement-bentonite grout backfill for borehole instruments. *Geotechnical News* **20** (4; December), 38–42.

Moore DP & Imrie AS (1995) Stabilization of Dutchman's Ridge. *Proceedings of the Sixth International Symposium on Landslides, Christchurch, New Zealand, 10–14 February, 1992*. Vol. 3. Balkema.

Moye DG (1967) Diamond drilling for foundation exploration. *Civil Engineering Transactions*. The Institution of Engineers, Australia. pp. 95–100.

Muir-Wood AM & Casté G (1970) In situ investigations in soils and rocks. *Proceedings of the conference organised by the British Geotechnical Society in London, 13–15 May, 1969*. British Geotechnical Society, Institution of Civil Engineers, London. pp. 79–85.

Oda M (1984) Permeability tensor for discontinuous rock masses. *Geotechnique* **35**, 483–495.

Odling NWA, Elphick SC, Meredith P, Main I & Ngwenya BT (2007) Laboratory measurement of hydrodynamic saline dispersion within a micro-fracture network induced in granite. *Earth and Planetary Science Letters* **260**, 407–418.

Ophori DU, Brown A, Chan T, Davison CC, Gascoyne M, Scheier NW, Stanchell FW & Stevenson DR (1996) Revised model of regional groundwater flow.

Whiteshell Research Area, AECL Whiteshell Laboratories, Pinawa, Manitoba.

Oreskes N, Shrader-Frechette K & Belitz K (1994) Verification, validation, and confirmation of numerical models in the earth sciences. *Science* **263**, 641–646.

Pearson R & Money MS (1977) Improvements in the Lugeon or packer permeability test. *Quarterly Journal of Engineering Geology* **10**, 221–239.

Penman ADM (1961) A study of the response time of various types of piezometer. In: *Pore Pressure and Suction in Soils*. pp. 53–58. British Geotechnical Society, Butterworks, London.

Pierce M, Cundall P, Potyondy D & Mas Ivars D (2007) A synthetic rock mass model for jointed rock. In: *Rock Mechanics: Meeting Society's Challenges and Demands (1st Canada-U.S. Rock Mechanics Symposium, Vancouver, May 2007), Vol. 1: Fundamentals, New Technologies & New Ideas*. (Eds E Eberhardt *et al*.) pp. 341–349. Taylor & Francis Group, London.

Potyondy DO & Cundall PA (2004) A bonded-particle model for rock. *International Journal of Rock Mechanics & Mining Sciences* **41** (8), 1329–1364.

Price M (1996) *Introducing Groundwater*. 2nd edn. Chapman & Hall/CRC, Oxfordshire, UK.

Price M (2009) Barometric water-level fluctuations and their measurement using vented and non-vented pressure transducers. *Quarterly Journal of Engineering Geology and Hydrogeology* **42** (2), 255–260.

Price M & Williams AT (1993) Pumped double packer system for use in aquifer evaluation and groundwater sampling. *Proceedings of Institute of Civil Engineers, Water, Maritime and Energy Journal* **101**, 85–92.

Price M, Morris BL & Robertson A (1982) A study of intergranular and fissure permeability in Chalk and Permian aquifers, using double-packer injection testing. *Journal of Hydrology* **54**, 401–423.

Price M, Robertson AS & Foster SSD (1977) Chalk permeability, a study of vertical variation using water injection tests and borehole logging. *Water Services* **81**, 603–610.

Read J & Stacey P (Eds) (2009) *Guidelines for Open Pit Slope Design*. CSIRO Publishing, Melbourne.

Rocscience Inc. (2010) *Slide v 6.0*. Toronto, Canada.

Roscoe Moss Company (1990) *Handbook of Ground Water Development*. John Wiley & Sons, New Jersey.

Rorabaugh MI (1953) Graphical and theoretical analysis of step drawdown test of artesian well. *Proceedings American Society of Civil Engineers* **79** (Separate No. 362), 1–23.

Rutqvist J & Stephansson O (2003) The role of hydromechanical coupling in fractured rock engineering. *Hydrogeology Journal* **11**, 7–40.

Schlumberger (2011) Oilfield glossary. <http://www.glossary.oilfield.slb.com>.

Shepard D (1968) A two dimensional interpolation function for irregularly spaced data. *Proceedings of the 23rd National Conference ACM*. pp. 517–523.

Snow DT (1968) Fracture deformation and changes of permeability and storage upon changes of fluid pressure. In: *Geophysical and Geological Studies of the Relationships between the Denver Earthquakes and the Rocky Mountain Arsenal Well, (Quarterly of the Colorado School of Mines)* **63** (1), 201–224.

Singh RN, Ngah SA & Atkins AS (1985) Applicability of current groundwater theories for the prediction of water inflows to surface mining excavations. *International Mine Water Association Proceedings, Mine Water, Granada, Spain, 1985*, pp. 553–569.

Spitz K & Moreno J (1996) *A Practical Guide to Groundwater and Solute Transport Modeling*. John Wiley & Sons, New York.

Stacey TR, Xianbin Y, Armstrong R & Keyter GJ (2003) New slope stability considerations for deep open pit mines. *The Journal of the South African Institute of Mining and Metallurgy* July/August, 373–390.

Strack ODL (1989) *Groundwater Mechanics*. Prentice Hall, New Jersey.

Tate TK, Robertson AS & Gray DA (1970) The hydrogeological investigation of fissure-flow by borehole logging techniques. *Quarterly Journal of Engineering Geology* **2**, 195–215.

Taylor DW (1948) *Fundamentals of Soil Mechanics*. John Wiley & Sons, New Jersey.

Theis CV (1935) The relation between the lowering of the piezometric surface and the rate and duration of discharge of a well using ground water storage. *Transactions American Geophysical Union Part 2* **16**, 519–524.

Trainer FW (1988) Plutonic and metamorphic rocks. In: *Hydrogeology (The Geology of North America, Volume O-2)*. (Eds W Back, JS Rosenshein & PR Seaber) pp. 367–380. Geological Society of America, Boulder, Colorado.

U.S. Bureau of Reclamation (1951) *Permeability Tests Using Drill Holes and Wells*. Geology Report, vol. G-97. USBR, Denver, Colorado.

US Bureau of Reclamation (1977) Permeability tests in individual drill holes and wells. In: *Ground Water Manual*. Dept of the Interior, US Bureau of Reclamation, Washington, DC, pp. 317–342.

US Bureau of Reclamation (2001) *Water Measurement Manual*. <http://www.usbr.gov/pmts/hydraulics_lab/pubs/wmm/>.

Vaughan PR (1969) A note on sealing piezometers in boreholes. *Géotechnique* **19** (3), 405–413.

Ven Chow, Maidment D & Mays L (2013) *Applied Hydrology*. 2nd edn. McGraw Hill, New York.

Walsh JB (1981) Effect of pore pressure and confining pressure on fracture permeability. *International Journal* of Rock Mechanics and Mining Sciences, Geomechanics Abstracts **18**, 429–435.

Wang HF (2000) *Theory of Linear Poroelasticity with applications to Geomechanics and Hydrogeology*. Princeton University Press, New Jersey.

Wood WW & Fernandez LA (1988) Volcanic rocks. In: *Hydrogeology (The Geology of North America, Volume O-2)*. (Eds W Back, JS Rosenshein & PR Seaber) pp. 353–365. Geological Society of America, Boulder, Colorado.

Zangerl C, Eberhardt E & Loew S (2003) Ground settlements above tunnels in fractured crystalline rock: numerical analysis of coupled hydromechanical mechanisms. *Hydrogeology Journal* **11** (1), 162–173.

Ziegler T (1976) *Determination of Rock Mass Permeability*. Technical Report, vol. S-76-2. U. S. Army Engineers Waterways Experiments Station, Vicksburg, Mississippi.

Index

'A-B-C-D' concept, fracture flow   156–57, 168, 197, 202, 204, 318
active pore pressure control   15
active slope depressurisation program   15–16
advanced wireline geophysical logging   108–10, 113
air/foam drilling   399
air pressure, during drilling   86
airlift, and recovery tests   96, 121
airlift flow rate   86, 405, 406
airlift pumping   329, 401–403
airlift yield maps   131, 133
Andacollo pit (Chile)   296
angle drains   291, 365
angled drilling   401
anisotropy   26–28, 29, 200, 396–97, 415, 424, 543
Antamina West wall (Peru)   183–84, 507–12
aperture   482
aquifer tests   418–20
aquifers   392–93
auger drilling   398
axisymmetric models   233–34

base maps   131
baseline studies   80–81
Basin and Range province (USA)   63–64
Bingham Canyon Copper Mine (Utah, USA)   13–14
bit type   86
blast damaged zone   208–209
blast pattern design   340–41
blast-induced deformation, Diavik   168–69
blast-induced fracturing, Diavik   167–68
blasting   166–69, 208, 340–43
    Diavik   446, 447, 450–52, 454
    drainage pathways   337–38
borehole accelerometer arrays   106
borehole extensometers   106–107
boundary conditions, in numerical models   239–43, 573–74
brownfield sites   73, 79–80, 223, 225

cable tool drilling   398
catch benches   284, 349–50, 358
cementation   29, 46
Cerro Vanguardia Mine (Argentina)   342–43
channelling   32
choke points   32
Chuquicamata (Chile)   183, 189, 192, 199, 204, 493–500
    conceptual hydrogeological model   496–97, 499
    drain hole results   496

geology   493, 496
hydrogeology   183
hydrostratigraphy   493, 496
clay content   86
climatic data   48–49, 227
Coaraze (France)   171–72, 204, 573–75, 577
Cobre Las Cruces Mine (Spain)   556–67, 564, 567
collar casing   305, 309–10
compartmentalisation   36–37, 47, 147, 211
composite borehole logs   144–45
compressibility   23, 197–98, 202–204, 497
conceptual hydrogeological model   153–214
    Antamina   511
    Chuquicamata   496–97, 499
    Escondida   488, 491–92
    large-scale   155–56
    sector-scale   157–58
    slope-scale   185–92, 194–204, 207–11
Conceptual level, field investigations   74–75
conductivity
    directional   543
    hydraulic   24–25, 28, 32, 33, 89, 143, 198, 254, 267, 396, 410–12, 442, 550
    vertical   29
conductivity logs, and temperature   113, 115
confidence levels   74
confined storage   22–24
constant-head boundary   239
constant-head tests   91–92
constant-rate test   91, 125–26
Cooper-Jacob method   419–20
cost
    slope stability   8–10, 13–14
    water management   8–10, 13–14
cost–benefit analysis   8–10, 13–14
Cowal Gold Mine (Australia)   184, 290, 520–25
    depressurisation   523–25
    dewatering   524–25
    geology   520
    hydrogeology   184
    pore pressures   184, 522–23
cross-hole testing   119–22, 125–26

Darcy's law   24, 25, 32, 391, 392–97
data analysis, hydrogeological   145, 147, 149
data collection   68–71, 83, 129–52
data logging   372–73
data management systems   149–52
data presentation   130–31, 140, 149, 151

data validation   150–51, 150
data, one-off   130, 140–45
databases   149–52
datum level   392
deformation   168–69, 201–204, 207–209, 212, 213, 252, 377
depressurisation   *see* slope depressurisation
depressurisation holes   335–36
Design and Construction, field investigations   78
deviation logs   113
deviation surveys   307, 308
dewatering   51–54, 56
    Antamina   509–11
    Cowal Gold Mine   524–25
    Diavik   160
    early cost   10
    Escondida   487
    La Quinua   528, 530
    monitoring   578–82
    Whaleback   525–26
dewatering performance, well losses   421–23
dewatering rate   279–82, 291
dewatering targets   287, 289
dewatering tunnels   51, 333
dewatering wells   63, 291, 320, 325–27, 338–43
DFN   *see* Discrete Fracture Network (DFN) modelling
diamond (core) drilling   87–88
Diavik North-west wall   182, 158–79, 186–87, 192, 210, 290, 182, 436–85
    blast-induced deformation   168–69
    blast-induced fracturing   167–68
    blasting   166–69, 447, 450–51, 454
    climate   436
    depressurisation   161–62, 164, 437–39, 443–44, 446, 448, 449, 451–52
    dewatering   160, 445, 448
    DFN modelling   179–81, 456–85, 536–40, 543, 546–48
    EPM   536–40, 543, 546–56
    fracture flow   178
    freeze-back   169, 171–73, 444, 446, 447–48, 449, 451, 452
    hydraulic stress responses   172, 175, 178
    hydrogeological setting   436–37
    hydrograph analysis   440–56
    long-term drainage response   448–49
    numerical modelling   536–56
    piezometer installations   439–40
    pore pressure increase   450, 451
    pore pressure monitoring   160–62, 164
    pre-shear blasting   446–47
    pressure pulses   166–67, 168
    pressure responses   175
    pumping responses   454–56
    specific event analysis   443–56

    thawing   444, 446, 447–48, 449, 451, 452
    underground dewatering network   445, 448
    underground drainage galleries   175
    underground drainage system   450, 451, 452, 455–56
diffusivity, hydraulic   33–34, 180, 197–99, 201, 202
direct coupling   201
direct push method   398
directional conductivity   543
directional drilling   401
discharge areas   49
discharge pathways   211
Discrete Fracture Network (DFN) modelling   179–81, 214
    Diavik   179–81, 456–85, 536–40, 543, 546–48
dispersion   26
dissolved major ions   126
dissolved metals   126–27
distribution maps   375
diversion benches   349
downhole geophysical logging   107–10, 113, 115, 117–19
downhole video inspection   119
drain holes   51, 296, 364–65
drain trials   128–29
drainage, passive   295–96
drainable porosity   19–22
drainage pathway, creation   337–38
drainage trials, pilot   128–29
drainage tunnels   292, 333–37
drains
    angled   291, 312–16, 365
    horizontal   128, 183–84, 291, 300, 302–307, 309–12, 327–28, 358, 492, 522, 525
    vertical   128–29, 291, 312–16
drill holes, VWP installation   429–30
drilling methods   82–87, 398–401
drilling programs, hydrogeological   83–88
drill-stem injection tests   88–89, 408–10
dual porosity   22, 23, 46, 181
dual porosity materials   33
dual-tube reverse-circulation drilling   400–401

effective stress   8, 32, 41–43, 201–203, 497
elastic storage   22–24
elevation head   96, 392
environmental baseline studies   80–81
EPM   *see* equivalent porous medium
equipotentials   394
equivalent porous medium (EPM)
    Cobre Las Cruces Mine   556–67
    codes   230, 249–51
    Diavik   175, 178, 536–40, 543, 546–56
erosion, protection   360
Escondida East wall   182–83, 186, 486–93
    conceptual hydrogeological model   488, 491–92
    depressurisation   487

dewatering 487
geology 486–87
hydrostratigraphy 486–87
pore pressure 182–83
ex-pit wells 325–26
exploration drilling 82, 83, 84
external distant boundaries 240–42

faults 187–88
Feasibility level, field investigations 76–78
FEFLOW 232, 254, 258, 271 441, 562
field chemistry 126
field investigations
Conceptual level 74–75
confidence levels 74
Design and Construction level 78
design process integration 69–73
early stage 69
environmental baseline studies 80–81
Feasibility level 76–78
hydrogeology group 67–68
logistics 71–72
modelling 72
Operations level 78
planning 65–82
planning team 65, 67
Pre-Feasibility level 75–76
scale 68–69
water management 81–82
field programs, implementing 82–129
FLAC 272, 553, 563
flooded reverse-circulation drilling 399–400
flooding 48, 347, 349, 355
flow, preferential 30, 32–33, 36
three dimensional 394–97
flow paths 25, 35–36, 156, 158, 168, 169, 188, 192, 519
flow rate graphs 138–40, 374
flow reticulation 358
flow speed 25, 26, 178
fluid logs 113, 115, 117–19
fluid velocity logs 115, 116
fracture flow 32, 249
and 'A-B-C-D' concept 156–57, 168, 197, 202, 204, 318
Diavik 178
fracture flow conditions 21–22
fracture intensity 465–71, 480, 485
fracture network 156, 166–69, 172, 175, 178–79, 180–82, 186, 188, 194–97, 199–201, 208, 211–12
fracture orientation 160, 186, 457, 471, 474
fracture pathways 482
fracture radius distribution 458
fracture size 457–58, 471, 478, 480
fracture transmissivity 180–81, 462, 465, 471, 474, 480, 482

fractured rocks 30–32
fractured rocks, and production wells 317–20, 322–24
fracturing, degree of 86
freeze-back, Diavik 169, 171–73, 444, 446, 447–48, 449, 451, 452
Fullbore Formation Micro-Imager 109

geology 45–48, 156, 158, 186–89, 211–12, 227
Antamina 507, 509
Chuquicamata 493, 496
Cowal Gold Mine 520
Diavik 436
Escondida 486–87
geophysical logging 107–19
geostatistical modelling, of fracture intensity 468–71
geotechnical models 269–71
GIS systems 130, 140
Gold Quarry pit (Nevada, USA) 331, 333
greenfield sites 78–79, 223, 225
groundwater control programs
horizontal drain holes 296, 300, 302–307, 309–12
implementation 291–345
passive drainage 295–96
types 291–95
groundwater cut-off systems 292
groundwater flow model 215–77, 377–78
groundwater level data 228
groundwater models see numerical models
groundwater monitoring 96–107, 363–76
groundwater movement, speed 25–26
groundwater seepage, collection 346, 350
groundwater, flow paths 25, 35–36, 156, 158, 168, 169, 188, 192, 519
grout mixes 431
grouted-in vibrating wire piezometers 99–100, 155, 424–27, 429–31
grouting programs 294
guide-tremie pipe method 425–26

head 239, 391–92
head, lateral 35–36
head conditions 35–38
head contour map 147
head gradients, vertical 37–38
head loss 391, 415
heterogeneity 26, 27, 31
high-volume drain holes 51
hole diameter 305
horizontal drain holes 296, 300, 302–307, 309–12, 338, 364, 374
VWP installation 429–30
horizontal drain maps 133
horizontal drain trials 128
horizontal drilling 304, 401

horizontal piezometers 101–102
hydraulic conductivity 24–25, 28, 32, 33, 89, 143, 198, 254, 267, 396, 410–12, 442, 550
hydraulic diffusivity 33–34, 180, 197–99, 201, 202
hydraulic gradient 24, 27, 37–39, 47, 392
hydraulic parameters 141–42, 227–28
hydraulic stress responses, Diavik 172, 175, 178
hydraulics 156, 194–201, 212–13
hydraulic tests 83, 410–24
hydrochemical fingerprint 582
hydrogeological cross-sections 375
hydrogeological data analysis 145, 147, 149
hydrogeological drilling programs 83–88
hydrogeological models 45–49, 72, 273–74, 269–71
hydrogeology group, mine operations 67–68
hydrogeology, Chuquicamata 183
    Cowal Gold Mine 184
    Jwaneng 184
hydrograph analysis, Diavik 440–56
hydrographs 137–40
hydrological logging form 401, 402
hydrology 48–49, 156, 212, 227
'Hydrology group' 344–45
hydromechanical (HM) coupling 201–202, 204, 208
HydroPhysical logging 117–19
hydrostatic unloading 44
hydrostratigraphy, Antamina 507, 509
    Chuquicamata 493, 496
    Escondida 486–87

igneous rocks 46
inclined drill holes, VWP installation 429–30
inclinometers 102, 105–107
injection tests 88–89, 92, 121–22, 408–10
inner pit boundaries 242–43
in-pit dewatering installations, protection 338–43
in-pit seeps 365
in-pit wells 326–29, 331, 333, 338
interceptor wells 324–25
in-tunnel monitoring 336–37

joints 187–88
Jwaneng Diamond Mine (Botswana) 186, 188, 513–20
    dewatering 513–14, 515
    geology 513
    hydrogeology 184
    hydrostratigraphy 513
    slope depressurisation 516, 519

Kori Kollo pit (Bolivia) 60, 62

La Quinua (Peru) 528–35
lateral compartmentalisation 36–37
lateral head 35–36

lattice formulation 568–70
leakage 191
lithology 185–86
lithostatic unloading 44, 201–204, 212, 213

magnetic resonance logging 110, 113
maps
    head contour 147
    airlift yield 131, 133
    base 131
    distribution 375
    horizontal flow drain 133
    water chemistry 133
    water level 131, 374–75
measurements, while drilling 307
metamorphic rocks 46
micro-resistivity image logging 109–10
mine closure, considerations 58–64
mine design, level of effort 73–78
mine dewatering, general 51–54, 56–57, 279–82, 528
mine dewatering tunnels 323–37
mine planning data 225
mine water balance 57–58
mine water management, general planning 16–18
mineralisation 186
mine-scale models 221–23
model codes 271–72
models see numerical models
modelling limitations 217
MODFLOW 240, 245, 253, 254, 255, 271, 562
Mohr-Coulomb failure law 42
monitoring frequency 369
monitoring points, distribution 366–67
monitoring programs 366–67, 369–70
monitoring results, display 373–76
monitoring, short-term pore pressure changes 369–70
mud rotary drilling 399
multi-hole testing 119–22, 125–26
multiple-well testing 423–24

negative pore pressures 34, 274
no-flow boundary conditions 239
numerical models 215–77, 536–67
    active drainage measures 254
    boundary conditions 239–43
    budget considerations 223, 225
    calibration process 254–58
    Coaraze 570–75, 577
    Cobre Las Cruces 564, 567
    coupled modelling approach 252–54
    data requirements and sources 225–29
    development 229–62
    Diavik 536–56
    EPM codes 249–51

geometry   229, 231–34, 236, 238
interpreting results   258–59, 261
mesh or grid size   244–46
mine-scale   218–19
model domain   239–44
modelling workflow   225
pit slope (sector-scale)   219–20
planning   215–29
regional-scale   217–18
simulation types   246, 248–49
slope design process   220
Slope Model   277
steps   229–31
time step selection   251–52
timeframe   223, 225
transfer of outputs   272–75
transient pore pressures   275–77
use   262
validating results   261–62

observation wells   97, 137–38, 365
one-off data, analysis   140–45
Operations level, field investigations   78
orientation dependence   462
'over-break' zone   208–209
oxidation   187

packer tests   32, 92–96, 412–16, 418
packers   306
passive drainage   295–96
passive pore pressure control   14–15
penetration rate   85
performance assessment   376–79
permeability   24–34, 211
permeable fracture zones   212
phreatic surface   264, 265–69, 393
piezometers,   97–107, 130, 137–38, 365, 370, 393, 580–81
   in-tunnel   336–37
piezometer hydrographs   137, 140, 374
piezometer installations, Diavik   439–40
'piggy-backing', of data collection   83
pilot drainage trials   128–29
pilot holes   320–22, 405–406, 408
pit inflow prediction   279–80
pit slope boundary   243
pit slope depressurisation   *see* slope depressurisation
planar models   234, 236
planning team   65, 67
pore pressure   34–35, 41, 201, 208, 252–53
   Antamina   183–84
   controls   38–41
   Cowal Gold Mine   184, 522–23
   Diavik   160–62, 164, 450, 451
   and effective stress   42

Escondida   182–83
negative   34, 274
and shear strength   42–43
transient   209–11, 275–77
use in stability analyses   262, 264–77
Whaleback   184–85
pore pressure field   35, 249, 265
pore pressure gradients   35–36, 37–38, 233
pore pressure inputs   272
pore pressure models, work flow   271
pore pressure profiles   265–69
porosity   19–24, 550
   dual   22, 23, 46, 181
porous-medium conditions   19–21
potentiometric surface   23, 35, 36, 392, 393
power law scaling   461–62
precipitation   48, 183, 189, 192, 227, 239
prefabricated multi-level, VWP installation   430–31
Pre-Feasibility level, field investigations   75–76
preferential flow   30, 32–33, 36
   three-dimensional   394–97
pre-shear blasting, Diavik   446–47
pressure conditions   35–38
pressure head   392
pressure pulses, Diavik   166–67, 168
pressure responses, Diavik   175
pressure transducers   98–99, 370
production dewatering wells   51
production well constraints   317–20, 322–24
pumping rates, monitoring   581–82
pumping tests   82, 121–22, 125–26, 128, 418–24
pumping wells   291–92, 316–33
pumping, airlift   329, 401–403
pushback   13, 41, 302, 500–501

Radomiro Tomic   500–505
RC drilling   *see* reverse-circulation drilling
recharge   156, 158, 189–92, 211, 239
recharge areas   49
recharge reduction   283
recovery data   424
recovery period   126
regional-scale investigation   68
regional-scale models   220–21, 223
remediation plan, La Quinua   531, 533, 535
response time   33–34, 198–99, 202, 250
reverse-circulation drilling   84–87, 399–408
risk, water management   379–97
rivers   49, 156, 189, 191
rotary core drilling   398–99
runoff   345–50, 352–53

safety considerations, drilling   309–10
scale, of field investigation   68–69

sector-scale models   153–55, 222–23, 225
sedimentary materials, permeability   29
sedimentary rocks   46
sediments   45–46
seepage faces   52
SEEP/W   232, 271, 563
shear strength, and pore pressure   42–43
single hole testing   82, 88–96, 410–12
slice models   234, 236
slope depressurisation   52–54, 56–57, 192, 194, 197–201, 204, 207–208
    Antamina   509–11
    Cowal Gold Mine   523–25
    Diavik   161–62, 164, 437–39, 443–44, 446, 448, 449, 451–52
    Escondida   487
slope depressurisation systems   279–362
    drainage pathway creation   337–38
    drainage tunnels   323–37
    implementing   291–345
    integration with mine planning   283–86
    lead time   282–83
    ongoing maintenance   285–86
    organisational structure   343–45
    permanent access   285–86
    planning   279–91
    protection of in-pit dewatering installations   338–43
    pumping wells   316–33
    surface water control   345–60
    targets   286–87, 289–91
    tunnels   324–35
    work areas   284–85
slope failure   158, 209, 213, 254
Slope Model   277, 570–75, 577
slope performance   311, 376
slope-scale analysis   543, 546–47
slope-scale conceptual model   185–92, 194–204, 207–11
    'over-break' zone   208–209
    deformation   201–204, 207–209
    depressurisation   204, 207–208
    discharge   192, 194
    geology   185–89
    hydraulics   194–201
    recharge   189–92
    transient pore pressures   209–11
slope-scale investigation   68–69
slope stability   6, 14–16
    cost   8–10, 13–14
    and water pressure   41–44
slopes, erosion protection   360
slug tests   89–91, 410, 411
slurry walls   292–93
snow banks   191
soil-moisture deficit   189

sonic drilling   398
specific discharge   393–97
specific storage   23, 33, 143–44, 198–98
specific yield   144
spinner logging   115
springs   49
stability analysis   264
stability modelling, and use of pore pressure   262–77
staged pumping systems   365
standpipe piezometers   97–98, 100
static head   96, 113, 130, 131, 391, 392
steady-state conditions   35
steady-state calibration   255, 257
steady-state simulations   246, 248–49
step drawdown test   125, 423
steep-angled drains   312–16
stiff diagrams   136
storage coefficient   23, 34, 418, 421–22
storage properties   19–24, 143–44
stormwater management   346, 353–55
stream diversions   346–48
streams   49, 156, 191, 240, 346
stress periods   230, 251–52, 272–73
sub-horizontal layer   37, 47, 144
submergence   403–405
sumps   52, 355, 365
surface pumping   96
surface water
    drainage patterns   227
    features   48
    flow rate estimation   352–53
    in-pit stormwater management   353–55, 357–58
    recharge control   353
    sources   346
    system maintenance   355–57
surface water control   345–60
surface water monitoring   365–66

telemetry   150, 372–73
telescoping   243–44
temperature, and conductivity logs   113, 115
tests
    airlift   96, 121
    aquifer   418–20
    constant rate   125–26
    constant-head   91–92
    cross-hole   119–22, 125–26
    drill-stem injection   408–10
    hydraulic   410–24
    injection   88–89, 92, 121–22, 408–10
    multi-hole   119–22, 125–26
    multiple-well   423–24
    packer   32, 92–96, 412–16, 418
    pumping   121–22, 125–26, 418–24

single-hole variable-head   410–12
slug   89–91, 410, 411
step drawdown   125
variable-head   89–91
Westbay system   100, 122, 432–33, 435
Theis method   418–19
three-dimensional models   236, 238
Time Domain Reflectometry   102, 105–106
time steps   251–52, 254, 272–73
time, lead   282–84
time-series data   129–31
transient calibration   228–29, 257–58
transient pore pressure   209–11, 275–77
transient simulations   246, 248–49
transition materials   22
transmissivity   24, 28, 31, 33, 393
    fracture   180–81, 462, 465, 471, 474, 480, 482
trending heterogeneity   27, 31
tunnels
    dewatering   51, 333
    drainage   292, 323–37
    slope depressurisation   324–35
two-dimensional models   232–33

underground boreholes, VWP installation   430
underground drainage, Diavik   175, 445, 448, 450, 451, 452, 455–56
undrained simulations   230, 246, 248–49
unloading pressure   85–86

vacuum wells   315–16, 328
variable-head tests   89–91, 239
vertical compartmentalisation   37, 47, 137
vertical conductivity   29
vertical drain holes   365
vertical drain trials   128–29
vertical drains   128–29, 291, 312–16
vertical head gradients   37–38 68, 131

vertical hydraulic gradient   37–38
vertical variability, analysis   142–43
vibrating wire piezometers   98–100, 102, 105–107, 424–31
Victor Diamond Mine (Canada)   10
volumetric changes   43–44

water balance data   228
water chemistry   133, 136, 140
water control programs, goals   6, 8, 14–16
water level maps   131, 374–75
water level monitoring   370, 578–82
water level sounders   370
water management
    cost   8–10, 13–14
    field investigations   81–82
    in open pit mines   49–64
    site   279
water pressure, and slope stability   41–44
water quality monitoring   82, 126–28
water risk management   379–97
'Water Services Maintenance Group'   344–45
water transport, properties   24–34
weak materials, protection   350, 360
weathering   187
well losses   421–23
wells
    ex-pit   325–26
    in-pit   326–29, 331, 333, 338
    interceptor   324–25
    observation   365
    production dewatering   51
    pumping   291–92
Westbay multi-level system   100, 122, 432–33, 435
Whaleback South wall   184–85, 525–26, 528
wireline method, VWP installation   426–27, 429

Yanacocha Mine (Peru)   68, 192, 350

Printed and bound by CPI Group (UK) Ltd, Croydon, CR0 4YY

24/10/2024

01778285-0020